云南省林业和草原科学院建院60周年论文集

（上）

云南省林业和草原科学院 编

中国林业出版社

图书在版编目(CIP)数据

云南省林业和草原科学院建院60周年论文集.上/云南省林业和草原科学院主编.—北京:中国林业出版社,2019.11
ISBN 978-7-5219-0326-3

Ⅰ.①云… Ⅱ.①云… Ⅲ.①林业-文集②草原学-文集 Ⅳ.①S7-53②S812-53

中国版本图书馆CIP数据核字(2019)第238651号

出版	中国林业出版社(100009 北京西城区刘海胡同7号)
	http://www.forestry.gov.cn/lycb.html
	E-mail forestbook@163.com 电话 010-83143596
印刷	固安县京平诚乾印刷有限公司
版次	2019年11月第1版
印次	2019年11月第1次
开本	787mm×1092mm 1/16
印张	35
字数	852千字
定价	120.00元

《云南省林业和草原科学院建院 60 周年论文集》(上)
编 者 名 单

陈建洪　马建忠　李甜江　李云琴　刘　玲　胡光辉　罗　群
范　蓉　成伶翠　孙　睿

前 言

正值举国同庆伟大祖国 70 华诞之际,云南省林业和草原科学院迎来了建院 60 周年。60 年来,一批批科技工作者始终坚持以解决全省林业重大科技问题为己任,改革创新,潜心研究,戮力攻关,为我省林业生态建设与林产业发展提供了重要科技支撑。云岭大地的千山万水留下了他们的足迹,彩云之南的一抹抹绿色凝结着他们的汗水,云南省林业和草原科学院已成为在国内有一定影响力的林业科研机构。

科技论文是科技工作者展现科研成果的主要形式之一。60 年来,我院科技人员取得了丰硕的科研成果,在国内外各学术类刊物发表论文 5000 余篇。值此建院周年之际,结合我院学科特色,我们精心挑选了 60 篇优秀论文集册出版,涉及珍贵用材树种、生态环境修复、生物多样性保护、林木遗传育种、森林病虫害防治、经济林研究等 6 个版块内容。

本论文集的出版,得到了全院科技人员的大力支持,在此我们致以崇高的敬意,并对参与编撰本论文集的同志表示深深的感谢。

编者
2019 年 11 月

目 录

前言

第一篇 珍贵用材树种研究

云南松天然优良林分自由授粉混合种子子代测定 ················· 1
云南红豆杉的生物学与生态学特性 ················· 10
西南桦纯林与混交林生态系统 C 贮量的对比研究 ················· 19
12 个美国银合欢新品种在云南三地的引种初效 ················· 26
施肥对北美红杉幼林生物量、P 素含量及贮量的影响 ················· 37
思茅松林采伐迹地清理及其采伐剩余物的利用 ················· 43
松属 8 个树种的引种研究 ················· 48
基于数量化回归模型的秃杉优树选择 ················· 56
云南 2~7 年生直干桉人工林土壤物理性状研究 ················· 65
山桂花人工林林木材性与生长特性关系研究 ················· 73

第二篇 生态环境修复研究

封禁措施对云南金沙江流域主要林分的生态功能影响 ················· 79
香格里拉亚高山不同退化程度森林群落的空气温湿特征研究 ················· 86
滇西北纳帕海湿地景观格局变化及其对土壤碳库的影响 ················· 95
基于土地利用的局域社会生态系统动态平衡分析 ················· 105
云南高黎贡山中山湿性常绿阔叶林的群落特征 ················· 114
滇东南石漠化山地不同植被恢复模式下土壤地力变化和水土流失状况研究 ················· 126
岩溶山地不同植被恢复模式和恢复年限下土壤养分的变化 ················· 134
元谋干热河谷植物生态位特征研究 ················· 143
思茅松人工林不同空间配置模式下不同密度生长试验与分析 ················· 151
Similar Responses in Morphology, Growth, Biomass Allocation, and Photosynthesis
　in Invasive Wedelia Trilobata and Native congeners to CO_2 enrichment ················· 158

第三篇 生物多样性保护研究

从高山到河谷：德钦藏药植物资源的多样性及利用研究 ················· 174
野生中泰五味子果实的形态及营养成分分析 ················· 184
云南省藤黄属植物的地理分布及其区系特征 ················· 189

铁力木苗木分级研究 ……………………………………………………………… 201
印度尼西亚结香植物种质资源及沉香人工结香技术与借鉴 …………………… 207
云南珍稀濒危植物五裂黄连种群现状及生态习性初报 ………………………… 213
Phenotypic Plasticity of Lianas in Response to Altered Light Environment ……… 220
A New System for the Family Magnoliaceae ……………………………………… 236
In Vitro Germination and Low-temperature Seed Storage of *Cypripedium lentiginosum*
  P. J. Cribb & S. C. Chen, A Rare and Endangered Lady's Slipper Orchid ……… 257
Climate Change-induced Water Stress Suppresses the Regeneration of the Critically
  Endangered Forest Tree *Nyssa yunnanensis* ………………………………… 277
A Sophisticated Species Conservation Strategy for *Nyssa yunnanensis*, A Species with
  Extremely Small Populations in China ……………………………………… 293

第四篇　林木遗传育种研究

5 个乡土绿化阔叶树种不同育苗基质的当年育苗效应试验 …………………… 309
高松香思茅松无性系的选育 ……………………………………………………… 319
不同育苗基质对长蕊甜菜树苗木生长的影响 …………………………………… 324
辣木叶提取物的抗氧化活性研究 ………………………………………………… 333
七彩红竹二氢黄酮醇 4-还原酶基因 *IhDFR*1 的克隆及表达分析 ……………… 339
蓝桉 6×6 全双列交配生长性状的遗传效应分析 ………………………………… 348
Temporal Changes in Wetland Plant Communities with Decades of Cumulative Water
  Pollution in Two Plateau Lakes in China's Yunnan Province Based on Literature
  Review ……………………………………………………………………… 355
Somatic Embryogenesis in Mature Zygotic Embryos of *Picea likiangensis* (Franch.)
  Pritz ………………………………………………………………………… 367
Determination of Anthocyanins and Flavonols in *Paeonia delavayi* by High-Performance
  Liquid Chromatography with Diode Array and Mass Spectrometric Detection ……… 375
Identification of A Putative Polyketide Synthase Gene Involved in Usnic Acidbiosynthesis in
  the Lichen *Nephromopsis pallescens* ………………………………………… 385

第五篇　森林病虫害防治研究

不同日龄和吊飞过程中桔小实蝇成虫飞行肌能量代谢相关酶活性的变化 ……… 397
松实小卷蛾在云南生活史及其对思茅松的危害 ………………………………… 408
广南油茶虫害调查及糖醋液诱捕试验 …………………………………………… 413
针叶挥发物及营养物对楚雄腮扁叶蜂产卵选择影响的初步研究 ……………… 419
3 种菊科入侵植物叶片精油成分的 GC-MS 分析 ……………………………… 426
高原山区油茶茶苞病的发生与防治研究 ………………………………………… 432
蒜头果半寄生特性研究 …………………………………………………………… 439

Gravid Females of *Cephalcia chuxiongica*(Hymenoptera：Pamphiliidae)are Attracted to
Egg-carrying Needles of *Pinus yunnanensis* ··· 449
Attack Pattern and Reproductive Ecology of the Pine Shoot Beetle *Tomicus brevipilosus*
on Yunnan Pine（*Pinus yunnanensis*）in Southwestern China ································ 459

第六篇　经济林研究

板栗新品系的生物学特性 ·· 474
青刺果生物学特性观察及人工栽培技术 ·· 479
云南榛树资源及其开发利用 ··· 488
10个引种核桃品种嫁接成活率及苗期生长性状研究 ·· 493
珍稀濒危蒜头果资源保护与产业化发展瓶颈研究 ·· 505
云南早实早熟杂交核桃新品种——云新90303号的选育 ·· 514
美国山核桃在云南的引种表现及丰产栽培技术 ·· 517
腾冲红花油茶花器官的数量性状变异研究 ·· 521
油橄榄品种'皮瓜尔'的引种选育 ··· 528
Transcriptomics and Comparativeanalysis of Three Juglans Species,*J. regia*, *J. sigillata*
and *J. cathayensis* ·· 531

第一篇 珍贵用材树种研究

云南松天然优良林分自由授粉混合种子子代测定

陈强,常恩福,董福美,范国才,尹嘉庆

(云南省林业科学院,云南 昆明 650204)

摘要:1988 年在腾冲县古永林场采用完全随机区组设计开展试验。通过 7 年的观测,试验已显示出天然优良林分子代具有干型通直、木纹理扭转度小、生长量大的特点,其树高、胸径、材积和木纹理扭转度的实际增益分别达到 37.49%、65.60%、219.84% 和 269.57%,遗传增益分别为 20.14%、40.46%、107.11% 和 59.79%,目前已远远超过云南松天然优良林分子代遗传增益 3% 的攻关考核指标。通过试验反馈表明了所制定的"云南松天然优良林分选择方法及标准(滇西)"具有科学性和可靠性。

关键词:云南松;优良林分;子代测定;遗传增益

A study on the open pollination seeds progeny test of the original high-quality stands of *Pinus yunnanensis*

Chen Qiang, Chang En-fu, Dong Fu-mei, Fan Guo-chai, Yin Jia-qin

(Yunnan Academy of Forest Science, Kunming Yunnan 650204)

Abstract: The major aim of this test is that the actual increment and main genetic parameterof the original high-quality stands were evaluated by testing progeny of *Pinus yunnanensis*, so as to examine "The Selection Method and Criterion of District (Western Yunnan) of The Original High-quality Stands of *Pinus yunnanensis*". The authors adopted completely random district-group design to carry out the test in Guyong Forest Farm of Tengchong County in 1988. The results from the continous seven-year investigation show that the progeny of the original high-quality stands have the characteristics of straight tree trunk, minor twinst degree of grain on wood, high growth increment, the actual increment of the tree height, DBH, volume of timber and twinst degree of grain reached respectively 37.49%, 65.60%, 219.84% and 269.57%, their genetic gain reached respectively 20.14%, 40.46%, 107.11% and 59.79%, their genetic gain reached respectively 20.14%, 40.46%, 107.11% and 59.79%. The data mentioned above have farexceeded 3% of tackling-key-problem checkup index of progeny genetic gain of original high-quality stands of *Pinus yunnanensis*. The test feedback indicates that "The Selection Method and Criterion (western Yunnan) of the Original High-quality Stands of *Pinus yunnanensis*" is scientific and reliable.

Key words: *Pinus yunnanensis*; high-quality stands; progeny test; genetic gain

本项试验研究的目的是通过云南松(*Pinus yunnanensis*)林分子代测定试验,估算天然优良林分的实际增益及主要遗传参数,以此检验"云南松天然优良林分选择方法和分区(滇西)标准"。通过7年的观测,试验已显示出天然优良林分子代增益明显、干型通直、木纹理扭转度小、生长量已超过增益3%的考核目标。

1 试验方法及试验点

本试验以腾冲古永林场天然优良林分所设备固定样地为参试单元(A~J),商品种子作对照,进行林分子代测定。优良林分平均年龄为18.8年(仅部分林木开始结实),对优良林分所设置的9块固定样地,于每块样地内采集10株以上立木种子,经等量混合后作试料,分别以大理、保山两地各3批以上的商品种子等量混合后,作为2个对照(CK_1、CK_2),并以此估计两地的总体表型值。试验采用完全随机区组设计,10株单行小区,11次重复,11个处理,每个处理的株数为110株。定植株行距2m×2m,区组四周设保护行,试验面积1.67hm²。

试验地属北亚热带季风气候,具有空气湿度大、水热条件优厚等特点,年均气温11.9℃,年降水量1 843mm,年平均相对湿度大于80%。海拔2 100m,西南坡向,直线坡,坡度26°。土壤为花岗岩上发育的中、厚层山地黄壤,土质疏松肥沃,透水透气性好。原系云南松疏林地,皆伐后进行了火烧清林整地。于1988年按常规方式育苗造林,造林后加强管护工作。育苗前对参试种子进行了品质检验。

2 子代林生长观测及统计分析

苗木出圃前分别各处理测定出苗数和场圃发芽率,并抽取1/10的袋测定子叶数。1990年林木3年生时,机械抽取50%的植株作样本进行定株观测,开始测定树高和地径,以后每年年底进行一次总测,逐渐增测第1轮侧枝分枝数、冠幅、树干通直度和结实量等性状。1994年7年生时改测胸径,并在观测株中抽取60%的植株测定木纹理扭转度、第2轮侧枝10cm长度内的松针数,每株抽取30束松针测定每束松针的针叶数及针叶长[1~5]。

2.1 天然优良林分与对照子代生长状况比较

滇西地区云南松天然优良林分子代林长势良好,保存率达90%以上,优良林分子代具有生长量大、干型通直、木纹理扭转度小的特点,优良林分与对照子代生长状况见表1。

表1 子代林生长情况表
Table 1 Growth status of progeny forest

性状 Character	CK_1 大理 Dali	CK_2 保山 Baoshan	样地均值 Plot mean value
树高 Tree height(m)	2.01	2.27	2.94
胸径 DBH(cm)	2.72	3.51	5.17
单株材积 Individual volume(m³)	0.001 0	0.001 7	0.004 5
分枝数(枝) Number of branches	4.6	5.2	6.5
冠幅 Crown breadth(m)	1.62	1.66	2.06
松针数(束) Pine needle bunch number(bunches)	75.4	75.2	79.6
针叶数(针) Needle number(needles)	2.95	2.88	3.00
针叶长 Needle length(cm)	15.0	16.5	19.1

(续)

性状 Character	CK$_1$ 大理 Dali	CK$_2$ 保山 Baoshan	样地均值 Plot mean value
通直度 Straightness of stem	1.18	1.22	1.05
木纹理扭转度 Twist degree of timber grain	10.68	6.66	2.49
直干率 Straight stem percentage(%)	80.91	83.64	95.35
直纹率 Straight-grained percentage(%)	65.46	76.97	93.16
直干直纹率 Percentage of straight stem and grain(%)	51.52	68.49	89.01
单株结实量(个) Individual fruiting amount	1.8	3.4	0.5
结实株率 Percentage of fruiting plsnt(%)	29.55	50.91	17.69
千粒重 Weight per 1000 kernels(g)	14.350	13.765	13.893
发芽率 Germination percentage(%)	41.50	40.50	32.39
场圃发芽率 Nursery germination percentage(%)	62.58	74.30	48.26
子叶数(个) Number of cotyledon	6.5	6.4	6.5

由表1可见:①优良林分的子代,生长量明显高于对照。样地树高、胸径和单株材积平均值分别为:2.94m、5.17cm和0.0045m³,而对照的树高、胸径和单株材积平均值则分别为:2.14m、3.12cm和0.0014m³,优良林分分别超过对照37.38%、65.71%和221.43%。②优良林分的子代,具有干型通直、材质优良的特点。其直干率、直纹率和直干直纹率分别为:95.35%、93.16%和89.01%,而对照则分别为:82.28%、71.22%和60.01%,优良林分分别比对照高13.07%、21.94%和29.00%。③优良林分的子代,其分枝数、冠幅、松针数、针叶数和针叶长等性状观测值均大于对照。样地平均值分别为:6.5枝、2.06m、79.6束、3针和19.1cm,对照则分别为:4.9枝、1.64m、75.3束、2.92针和15.8cm。优良林分与对照相比较,分枝数多1～2枝、冠幅长25.61%、松针数多4束、针叶长20.89%、一般均为3针一束,而对照则有少数为2针一束。④优良林分子代的结实量和结实株率均比对照低。样地平均单株结实量为0.5个,结实株率为17.60%,比对照的2.6个和40.23%分别少2个和22.54%。⑤各供试样本种子千粒重较小,样地均值为13.893g,对照为14.058g,均低于云南松15～18g的水平。子叶数在4～9个之间,平均为6.5个,样地和对照的子时数和分布频率均相差不大。⑥样地的种子发芽率、场圃发芽率分别为32.39%和48.26%,低于商品种子的41.00%和68.44%,可能与样地采种母树年龄较小有关。

2.2 子代林表型相关分析

以各处理平均值作样本,计算林分各性状间的相关系数,表明:①云南松种子千粒重、苗木子叶数与林木的生长、材质及结实性状间相关不密切。②林木的营养生长愈旺盛,生长量和生物量产出也愈高,干型和材质也愈好,而结实量则愈低。分枝数、冠幅、松针数、针叶数和针叶长与树高、胸径和材积之间呈紧密的正相关,而与通直度、木纹理扭转度和结实量之间则呈紧密的负相关。

2.3 方差分析

根据调查所得数据,对各性状进行方差分析,并进行不同年龄间的差异比较(见表2)。

表 2 不同树龄各性状方差分析结果表

Table 2 The analytic results of the variance of the characters for different tree ages

性状	SU	DF	3年生 MS	3年生 $F_{检}$	4年生 MS	4年生 $F_{检}$	5年生 MS	5年生 $F_{检}$	6年生 MS	6年生 $F_{检}$	7年生 MS	7年生 $F_{检}$
树高	区组	10	88.372 0	2.144 1*	431.245 5	3.750 7**	0.221 9	8.459 0**	0.331 8	7.321 1**	0.683 5	6.907 7**
	处理	10	306.223 6	7.429 7**	907.790 2	7.895 4**	0.294 8	11.238 5**	0.720 6	15.897 3**	1.361 9	13.763 5**
	误差	100	41.216 0		114.977 0		0.026 2		0.045 3		0.098 9	
地(胸)径	区组	10	0.195 2	3.728 7**	0.858 6	5.977 7**	3.896 5	8.724 2**	6.014 4	11.824 6**	3.002 3	6.432 1**
	处理	10	0.259 4	4.955 1**	1.135 7	7.925 3**	3.009 4	6.738 0**	7.003 3	13.768 8**	8.728 3	18.699 2**
	误差	100	0.053 2		0.143 3		0.446 6		0.508 6		0.466 8	
分枝数	区组	10			2.192 9	2.615 0**	0.940 5	1.393 5	0.669 1	1.221 6	1.731 8	1.914 7*
	处理	10			5.139 5	6.128 7**	2.756 8	4.084 6**	3.037 9	5.546 2**	5.569 3	6.157 4**
	误差	100			0.838 6		0.674 9		0.547 7		0.904 5	
冠幅	区组	10					0.134 4	7.161 6**	0.196 4	8.032 8**	0.192 9	4.452 2**
	处理	10					0.088 7	4.729 5**	0.168 2	6.880 2**	0.369 8	8.535 1**
	误差	100					0.018 8		0.024 5		0.043 3	
通直度	区组	10									0.059 0	2.898 2**
	处理	10									0.040 8	2.005 0*
	误差	100									0.020 4	
结实量	区组	10							0.901 2	0.976 7	0.192 9	4.452 2**
	处理	10							9.357 6	10.141 0**	0.369 8	8.535 1**
	误差	100							0.922 7		0.043 3	

（续）

性状	SU	DF	3年生 MS	3年生 $F_{检}$	4年生 MS	4年生 $F_{检}$	5年生 MS	5年生 $F_{检}$	6年生 MS	6年生 $F_{检}$	7年生 MS	7年生 $F_{检}$
松针数	区组 处理 误差	10 10 100									178.719 9 70.997 3 63.435 2	2.817 4** 1.119 2
针叶数	区组 处理 误差	10 10 100									0.002 6 0.014 3 0.002 6	1.000 5 5.399 0**
针叶长	区组 处理 误差	10 10 100									7.607 6 23.862 0 2.077 5	3.661 9** 11.486 0**
木纹理扭转度	区组 处理 误差	10 10 100									40.668 6 81.453 9 19.695 6	2.064 9* 4.135 6**
材积	区组 处理 误差	10 10 100									1298.358 0 2162.429 0 188.813 2	6.876 4** 11.452 7**

由表2可见：子代林在树高、地(胸)径、材积、分枝数、冠幅、通直度、结实量、松针数、针叶数、针叶长和木纹理扭转度诸性状中，除松针数外，其余性状在处理间和重复间均存在着显著或极显著的差异。树高、地(胸)径、分枝数和冠幅在处理间和重复间的差异，一般随林龄的增加而加大。在方差分析基础上，为进一步分析优良林分各处理间及其与对照间的差异情况，分别对诸性状进行多重检验(LSR)及年度间的比较，结果表明：树高、胸径和材积在优良林分子代各处理与对照间存在着极显著的差异；而在优良林分内各处理间及两对照间均无显著差异。树高和胸(地)径两性状随林龄的增长，基因型效应(选择效果)愈加显现，但程度上不同。在木纹理扭转度上，优良林分各处理与对照的大理之间差异极显著；在树干通直度上，优良林分各处理与对照间的差异显著。冠幅和针叶长在优良林分各处理与对照间存在极显著的差异，而在优良林分各处理间及两对照间差异不显著；在针叶数上优良林分与保山的差异极显著；在分枝数上，优良林分与对照之间的差异，因林龄不同而有变化，规律性不强。

3 遗传参数的估算

3.1 遗传力

用方差分析法对各性状进行广义遗传力的估算，据公式：

$$H^2(\%) = (V_1 - V_2)/[V_1 + (r-1)V_2]$$

计算得到各性状7年生时的广义遗传力(见表3)。可见，树高、胸径、材积和针叶长的遗传力较高，而松针数的遗传力较低。

表3 各性状广义遗传力及排序

Table 3 Broad-sense heritability and sequence of every character

性状	树高	胸径	材积	分枝数	冠幅	结实量	松针数	针叶数	针叶长	木纹理扭转度
$H^2(\%)$	53.72	61.67	48.72	31.92	40.67	30.06	1.07	32.17	48.80	22.18
排序	2	1	4	7	5	8	10	6	3	9

3.2 遗传相关

为了解子代林个体各性状之间的相关程度，对诸性状进行方差、协方差分析，据遗传相关(Rg)、环境相关(Re)和表型相关(Rp)系数的分解计算公式：

$$Rg_{xy} = COVg_{xy}/(\delta^2 g_x \delta^2 g_y)^{1/2}$$
$$Re_{xy} = COVe_{xy}/(\delta^2 e_x \delta^2 e_y)^{1/2}$$
$$Rp_{xy} = COVp_{xy}/(\delta^2 p_x \delta^2 p_y)^{1/2}$$

计算得到各性状间的相关系数见表4。

表4 各性状间遗传相关系数

Table 4 Genetic correlation between the characters

性状	树高	X2	X3	X4	X5	X6	X7	X8	X9
胸径 X_2	1.017 2								

(续)

性状	树高	X2	X3	X4	X5	X6	X7	X8	X9
分枝数 X_3	1.005 5	0.975 2	0.971 3						
冠幅 X_4	1.040 9	1.013 0	0.971 3						
通直度 X_5	-1.273 5	-1.210 9	-1.310 1	-1.210 5					
木纹理扭转度 X_6	-1.432 4	-1.504 4	-1.479 3	-1.154 8	1.460 4				
结实量 X_7	-0.673 7	-0.646 6	-1.853 6	-1.201 4	1.758 0	0.623 5			
松针数 X_8	0.404 8	0.511 6	0.359 6	0.479 7	0.177 5	-0.569 3	-0.373 1		
针叶数 X_9	1.436 5	1.317 3	1.740 8	1.704 8	-2.035 2	-1.947 6	-2.162 5	-0.143 7	
针叶长 X_{10}	1.061 3	1.001 7	1.145 1	0.897 6	-1.154 9	-1.492 8	-0.244 5	0.400 2	1.179 1

由表可见:①各性状间的相关系数一般为 Rg>Rp>Re,除松针数外,各性状间的遗传相关都较为密切,说明诸性状之间的关系,较多地由于基因连锁或共效基因的作用,而相对受环境的影响较小。②生长量因子(树高、胸径)与营养生长因子(分枝数、冠幅、针叶数、针叶长)之间存在较紧密的正相关,而二者与质量性状(通直度、木纹理扭转度)和生殖性状(结实量)呈紧密的负相关,说明生长旺盛的植株其生产力和材质均较好,在选择上具有较高的一致性,但结实量是一个制约良种发展的因素。

为探讨树高、胸(地)径、分枝数和冠幅不同年龄间的关系,经方差、协方差分析后计算得到不同年龄间的遗传、环境和表型相关系数(表5)。由表可见:①性状不同年龄间相关系数中,以遗传相关(Rg)较为密切,说明其遗传稳定性较好。②4 个性状的表型相关均存在相邻两年之间的关系较密切,且年龄间隔愈大相关程度随之降低。树高、胸(地)径和冠幅的年度间的相关均较密切,而分枝数则相关不紧密。

表5 4个性状年度间遗传相关系数
Table 5 Genetic correlation between ages for four characters

性状	林龄 Tree-age	4 Rg	5 Rg	6 Rg	7 Rg
树高	3	0.972 4	0.974 7	1.044 3	1.141 2
	4		1.153 5	1.273 0	1.342 1
	5			1.023 0	1.053 8
	6				0.991 8
地(胸)径	3	0.996 1	1.091 0	1.240 3	1.205 4
	4		1.089 3	1.101 7	1.047 6
	5			0.998 3	0.992 1
	6				0.971 8
分枝数	4		0.783 3	0.740 1	0.624 2
	5			1.344 5	1.351 4
	6				1.256 1
冠幅	5			1.277 0	1.611 6
	6				1.010 5

3.3 实际增益和遗传增益的估算

因全部来自优良林分的子代在生长量上均大于对照,而木纹理扭转度均小于对照(计算增益时用倒数),据公式:

$$\Delta G_{\text{实}} = (Y_i - Y_{\text{对}})/Y_{\text{对}}$$

$$\text{选择差 SD} = Y_i - Y_{\text{对}}$$

$$\text{遗传增益 } \Delta G_{\text{遗}} = H^2 \times \text{SD}/Y_{\text{对}}$$

计算得到优良林分9个处理子代 A~J 树高、胸(地)径、材积和木纹理扭转度不同林龄的实际增益和7年生时的遗传增益(表6)。

表6 各性状的实际增益和遗传增益
Table 6 Actual average increment and genetic gain of every character

性状	增益	A	B	C	D	E	F	G	H	J	平均值 Mean
树高	$\Delta G_{\text{实}}$	41.12	28.50	42.06	28.97	37.38	42.52	37.85	31.78	47.20	37.49
	$\Delta G_{\text{遗}}$										20.14
胸径	$\Delta G_{\text{实}}$	77.24	57.05	72.76	47.44	60.58	70.51	68.27	61.54	75.00	65.60
	$\Delta G_{\text{遗}}$										40.46
材积	$\Delta G_{\text{实}}$	264.29	164.29	250.00	150.00	200.00	242.86	235.71	200.00	271.43	219.84
	$\Delta G_{\text{遗}}$										107.11
木纹理扭转度	$\Delta G_{\text{实}}$	218.75	219.93	132.44	351.56	215.27	325.00	327.09	431.90	204.21	269.57
	$\Delta G_{\text{遗}}$										59.79

4 结果与讨论

"滇西地区云南松天然优良林分选择的方法和标准"是可靠的。按该标准选择的优良林分,其子代与商品种子子代间在生长量和木纹理扭转度上存在显著差异,优良林分树高、胸径、材积和木纹理扭转度的遗传增益分别为:20.14%、40.46%、107.11%和59.79%,已超过原定遗传增益3%的考核目标;7年生时树高、胸径、材积和木纹理扭转度的实际增益分别为:37.49%、65.60%、219.84%和269.579%,高生长和径生长的实际增益出现随林龄增加而逐渐增大(高由30.45%增加为37.49%,径由19.11%增加为65.60%)的趋势。故云南松天然优良林分改建母树林所获得的良种具有极显著的增产效果。

在云南松各表型性状中:松针数遗传力较低[$H^2(\%)$ = 1.07%],分枝数在年度间的变化差异较大,胸径、材积和冠幅受郁闭度的影响较大,结实量、松针数、针叶数与生长量的表型相关不密切,均不宜作为优良林分的选择性状。树高是一个生长量性状,具较高的遗传力,与胸径和材积相关紧密,且受郁闭度影响小,是一个较为可靠的选择性状。木纹理扭曲是云南松木材的一大缺陷,对云南松木材的经济价值影响较大,具有一定的遗传稳定性[$H^2(\%)$ = 22.18%],所以木纹理扭转度也是一个重要的选择性状。针叶长和生长量相关密切,其遗传力也较高[$H^2(\%)$ = 48.80%],故可以作为间接选择性状考虑。所以云南松天然优良林分的选择采用树高和木纹理扭转度作为主要选择性状具有科学性和可靠性。

云南松种子千粒重和苗木子叶数与林木的生长及材质相关不密切。

云南松营养生长和生长量性状二者间呈紧密的正相关,而与质量性状和生殖性状间呈紧密的负相关,说明在生产和经济价值上的选择具有一致性,但结实状况对良种的生产是一个不利因素。

参考文献:

[1] 刘来福. 作物数量遗传[M]. 北京:农业出版社,1984.
[2] 顾万春. 林业试验统计[M]. 北京:农业出版社,1984.
[3] 汪企明. 马尾松不同种源苗期生长和生物量的初步研究[J]. 林业科技通讯,1988,2:3~5.
[4] 陈强. 滇西云南松天然优良林分子代测定试验初报[J]. 云南林业科技,1994,1:6~12.
[5] 顾万春. 森林遗传统计学[M]. 北京:科学出版社,1998.

(本文发表于《林业科学》,1998年)

云南红豆杉的生物学与生态学特性

王卫斌[1]，姜远标[2]，王达明[1]，周云[1]，景跃波[1]

(1. 云南省林业科学院，云南 昆明 650204；2. 思茅市林业科学研究所，云南 思茅 665000)

摘要：依据云南红豆杉的前期研究成果，从形态特征、物候学特性、种子特征、生长特性等方面论述了云南红豆杉的生物学特性，还从地理及垂直分布状况、适生环境条件、主要生态型等个体生态学特性方面阐明了云南红豆杉的生态学特性，为云南红豆杉研究工作的深入开展和其有效发展、利用提供理论基础。

关键词：云南红豆杉；生物学特性；生态学特性

Biological and Ecological Characteristics of *Taxus yunnanensis*

Wang Wei-bin[1], Jiang Iuan-biao[2], Wang Da-ming[1], Zhou Iun[1], Jin Yue-bo[1]

(1. Yunnan Academy of Forest Science, Kunming Yunnan 650204;
2. Forest Institute of Simao Profecture, Simao Yunnan 665000)

Abstract: Through reviewing the study results on *Taxus yunnanensis*, the biological characteristics of this species was described from the morphological characteristics, phenological characteristics, seed and growth characteristics. Besides, the ecological characteristics such as its distribution, suitable environmental condition and main eco-type were summarized.

Keyword: *Taxus yunnanensis*; biological characteristics; ecological characteristics

云南红豆杉(*Taxus yunnanensis*)别名土榧子、西南红豆杉，为第三纪孑遗植物，是我国红豆杉科(Taxaceae)红豆杉属(*Taxus*)植物中紫杉醇含量最高(树皮平均含量 0.01%～0.012%，枝叶平均含量 0.006%～0.008%)的树种[1]。由于其种群竞争力弱、天然更新缓慢和地理分布局限等原因，而在 20 世纪 80 年代云南红豆杉就被列入国家二级保护植物。20 世纪 90 年代以来，随着抗癌新药紫杉醇的开发利用，在巨额商业利益的驱使下，对云南红豆杉掠夺式的生产经营活动，加剧了其濒危程度，以至于 1999 年升格将云南红豆杉列为我国的一级保护植物。在保护的前提下，如何开发和利用好云南红豆杉资源，实现紫杉醇产业的可持续发展，已成为了全国关注的焦点。

我国于 20 世纪 80 年代初期对云南红豆杉的育苗和造林技术开始进行了试验研究，"九五"期间开展了其天然资源调查和药用人工原料林培育技术的试验示范。"十五"期间，重点针对人工药用原料林培育的需要，开展了云南红豆杉高紫杉醇优良单株选择及其采穗圃的营建工作，上述试验研究工作，为该树种的资源保护与开发利用提供了强有力的技术支撑。

本文依据云南红豆杉的前期研究成果，为配合国家林业局资助的六大林业重点工程关键技术应用研究与试验示范专项——"云南红豆杉高紫杉醇优良单株选择及其采穗圃的营建"研究工作的开展,对云南红豆杉的生物、生态学特性进行了初步总结。

1 云南红豆杉生物学特性

1.1 形态特征

云南红豆杉为常绿乔木或灌木,树高达 20m,胸径可达 1m,树冠倒卵形或广卵形,枝条密生。树皮红褐色或黄褐色,薄质,有浅裂沟,条状脱落。枝条水平展开,梢部下垂,柔软,小枝条不规则互生,1 年生枝平滑无毛,呈绿色,至 2 年生以上变为褐色或深褐色。叶质地薄,条状披针形,常呈镰状,边缘向下外卷,尖端渐尖或微急尖,叶柄短,顺小枝下延;长 1.5~4.7cm,宽 2~3cm;螺旋状排列,叶柄基部扭转,呈假二列羽状展开;叶面深绿色或绿色,有光泽,中脉稍隆起,中脉两侧外陷较深,叶背颜色较浅,呈淡灰绿色,中脉两侧各有一条淡黄色气孔带,中脉及气孔带上密生均匀微小角质的乳头状突起。雌雄异株,雄球花淡褐黄色,单生于叶腋,基部被数层鳞片所包,直立或小穗状,雄蕊 9~11 个,各有 5~8 个花药,呈盾状梅花型;雌花生于腋生的短枝上,其短枝被交互对生的数对鳞片叶所包,雌花上有 1 个卵形淡红色的胚珠,直生,有一层珠被,于胚珠下部遗存有假种皮的原始体。种子卵圆形至卵状广椭圆形,紫褐色,先端锐形或凸头,两侧微具钝脊,种脐椭圆形,着生于红色肉质杯状假种皮中,成熟种子先端露出,凹陷或凸出于假种皮 1~2mm;种子长 5~6mm,直径 3~4mm[2]。

1.2 物候学特性

云南红豆杉分布区较广,区内气候条件变幅较大,其物候节律随温度节律的变化而出现一定的差异[3]。云南红豆杉的物候年周期大致可分为生长期和休眠期,据对种植于昆明树木园植株的观测结果:云南红豆杉的休眠期为 11 月中旬到翌年 2 月中旬,营养生长期是从 2 月中旬到 11 月中旬,而繁殖生长期则从 2 月中旬开始延至 11 月下旬止,繁殖生长期比营养生长期长。

昆明树木园人工栽培的云南红豆杉叶芽和花芽的萌芽期均在 2 月中旬至下旬。从叶芽萌发到叶初展需要 30~35 天,展叶期为 4 月 5 日~20 日,展叶盛期为 4 月中旬。此时,新生叶的长度约为老叶的 1/2。花期为 3 月中旬到 4 月上旬,雌花与雄花的花期基本一致。种子成熟期 10 月下旬到 11 月下旬。

1.3 种子特征

1.3.1 种皮构造

云南红豆杉风干种子的种皮呈褐色,由外种皮、中种皮和内种皮 3 层构成。外种皮由一层厚壁细胞组成,表面角质化且具蜡质,呈鳞片状,凸凹不平,角质层细胞较小而紧实,细胞壁加厚;中种皮由 2~4 层木栓化的厚壁细胞组成,细胞排列极为紧密;内种皮膜质,由数层石细胞组成。

1.3.2 种胚发育状况

云南红豆杉种子千粒重 79~107g,种子内部几乎完全被胚乳所充满,胚乳浅黄色,油质。胚具有结构完整的胚根、胚轴和子叶,位于种子的中轴部,浅黄色,棒槌状;长约 1.6mm,仅为种子长度的 1/4~1/3,最初不具备萌发能力,需要后熟,只有当胚体充满胚腔,胚乳吸胀,种皮开裂,种子才能萌发[4~6]。

1.3.3 结实性状

人工种植的云南红豆杉10~12年生进入初果期,天然生长的云南红豆杉则需15年以上才开始结实,200年生以上仍能正常开花结实。云南红豆杉天然结实率很低,且年份间和单株间结实量差异显著。云南红豆杉属雌雄异株植物,天然分布的红豆杉雄株多、雌株少,雌株约是雄株的1/3,因此结实植株较少[7]。云南红豆杉单株结实量与光照有很大关系,一般位于林缘、林间空地或四旁植株结实较多,腾冲云华乡的一株生长于光照充足地的云南红豆杉,其结实量达7kg,而处于林下的云南红豆杉由于受上层荫蔽光照不足,结果稀少,其单株结实量为20~1 000g不等[8]。

1.4 生长特性

1.4.1 年生长节律

根据昆明和西畴2个试验点的观察(见表1),在昆明(25°01′N,102°41′E)人工引种栽培的云南红豆杉年生长期为5~11月,高生长主要集中于5~7月,占其年生长量的68.75%,全年有2次明显的抽梢,即春梢和夏梢,春梢约为夏梢长的1倍;地径生长主要集中于5~10月,占年生长量的95.83%。在西畴(23°27′N,104°40′E),由于地理位置偏南,年平均温(15.8℃)和年降水量(1 297mm)均比昆明点(年平均气温14.7℃、年降水量1 007mm)高,云南红豆杉的生长期为2~11月,比昆明点早且长。其高生长主要集中于2~7月,占年生长量的74.42%,秋梢长约0.11cm,占年生长量的25.58%;地径生长亦集中于2~10月,占年生长量的95.61%。2个点人工种植的云南红豆杉高生长和径生长的高峰期均为5~7月,占其全年生长量的30%~68.75%。

表1 昆明、西畴人工栽培的云南红豆杉年生长状况

Table 1 Annual growth of *Taxus yunnanensis* cultivated in Kunming city and Xichou county of Yunnan province

月份	昆明				西畴			
	树高		地径		树高		地径	
	生长量(m)	占年生长量(%)	生长量(cm)	占年生长量(%)	生长量(m)	占年生长量(%)	生长量(cm)	占年生长量(%)
2~4月	0.00	0.00	0.00	0.00	0.14	32.56	0.12	17.14
5~7月	0.11	68.75	0.12	50.00	0.18	41.86	0.30	42.86
8~10月	0.03	18.75	0.11	45.83	0.09	20.93	0.25	35.71
11月	0.02	12.50	0.01	4.17	0.02	4.65	0.03	4.29
12月至翌年1	0.00	0.00	0.00	0.00	0.00	0.00	0.00	0.00
年生长量	0.16	100	0.24	100	0.43	100	0.70	100

1.4.2 萌发特性

云南红豆杉具有较强的萌发能力。在云龙漕涧,其每个伐桩萌生植株1~40株不等。萌蘖更新的幼树多数生长较差,树干弯曲,呈灌木状。昆明树木园10年生的云南红豆杉幼树,距地面5~10cm处伐去主干后,当年有90%的伐桩萌生出新条,而每个伐桩的萌生条数

达 3~20 条不等,其 1 年生萌生条长 5~27cm[9]。

表 2 云南红豆杉样本单元
Table 2 Samples of *Taxus yunnanensis* inventory

序号	年龄(a)	树高(m)	胸径(cm)	序号	年龄(a)	树高(m)	胸径(cm)	序号	年龄(a)	树高(m)	胸径(cm)
1	60	8.1	18	41	50	6.8	18	81	250	13.1	40
2	65	8.1	20	42	50	6.9	20	82	165	13.1	40
3	60	7.4	16	43	45	6.0	14	83	80	8.7	26
4	60	6.8	18	44	50	7.1	18	84	200	12.5	44
5	70	7.8	18	45	40	5.2	12	85	210	13.3	44
6	70	7.8	22	46	80	8.8	20	86	210	10.7	38
7	45	6.7	16	47	50	6.5	16	87	105	9.5	24
8	70	8.0	20	48	90	7.5	18	88	90	10.3	28
9	75	7.8	20	49	70	6.5	18	89	90	11.6	26
10	100	7.1	30	50	60	7.9	20	90	100	7.9	22
11	100	11.6	26	51	70	7.7	24	91	150	14.0	34
12	140	12.6	32	52	70	7.0	18	92	230	14.0	44
13	50	6.1	16	53	60	7.0	18	93	85	6.5	16
14	60	7.3	20	54	50	7.1	18	94	110	8.8	26
15	50	7.5	8	55	40	6.3	14	95	125	11.4	32
16	65	6.3	16	56	50	7.4	16	96	140	13.3	36
17	25	4.7	8	57	70	7.3	22	97	250	14.5	56
18	50	5.5	12	58	40	7.0	14	98	115	10.4	30
19	50	7.0	16	59	60	6.6	16	99	150	8.2	28
20	30	5.4	8	60	60	7.3	18	100	95	8.8	26
21	50	6.8	18	61	50	7.0	16	101	130	10.9	36
22	60	8.3	18	62	50	7.3	20	102	125	10.6	32
23	35	6.4	12	63	50	7.5	18	103	95	8.9	24
24	40	6.9	14	64	65	7.0	18	104	110	11.2	30
25	70	7.3	18	65	60	7.3	18	105	90	8.3	22
26	80	12.0	20	66	40	6.7	14	106	200	11.6	30
27	60	7.0	18	67	80	8.3	36	107	150	11.2	28
28	50	7.5	16	68	60	8.0	18	108	160	12.6	36
29	75	8.3	22	69	50	7.5	14	109	80	9.8	28
30	50	7.5	18	70	50	6.4	14	110	200	12.7	34
31	60	7.8	18	71	55	6.4	18	111	180	12.7	34
32	60	6.9	18	72	50	6.2	14	112	60	6.8	18
33	60	7.7	18	73	30	6.0	14	113	230	15.8	64
34	70	7.3	18	74	25	4.3	8	114	55	5.7	14
35	50	6.2	16	75	120	13.3	46	115	300	15.0	56
36	45	6.2	14	76	180	11.4	32	116	200	12.4	38
37	50	7.4	16	77	100	8.4	24	117	300	15.8	48
38	50	6.9	16	78	160	13.8	38	118	100	10.8	32
39	60	7.2	18	79	135	10.7	30	119	250	13.6	48
40	40	6.5	14	80	95	9.5	24				

1.4.3 天然云南红豆杉的生长过程

根据云南省林业调查规划院1996年和1997年在云南省13个地(州)44个县开展的云南省天然云南红豆杉资源调查资料,从所调查的792个小班和6 107个样地中抽取了119个具有代表性,林分树龄为25~300年的样本单元,见表2。根据树高、胸径与树龄生长模型常用的直线式、幂函数式、指数式、对数式、双曲线式等经验公式进行回归拟合,在此基础上,采用实验形数公式 $V=0.41 \cdot G \cdot (H+3)$ 计算单株材积,并与树龄进行回归,选择相关系数值最大者作为回归模型。结果表明(表3):天然云南红豆杉树高、胸径、材积与树龄均成幂函数关系($y=A \cdot X^B$),其选择相关系数均在0.9以上。

表3 云南红豆杉树高、胸径、材积与树龄回归模型
Table 3 Regression models of tree height, DBH, volume and age of Taxus yunnanensis

项目	数学模型	相关系数	备注
树高	$y=1.099\ 37 \times X^{0.465\ 38}$	0.915 82	树龄≤300年
胸径	$y=1.021\ 16 \times X^{0.698\ 57}$	0.931 18	树龄≤300年
材积	$y=0.000\ 091 \times X^{1.728\ 52}$	0.999 96	树龄≤300年

依据表3数学模型,将树龄分别带入方程式进行计算云南红豆杉的树高、胸径、材积生长过程可知,天然云南红豆杉的生长极为缓慢,生长于林冠下30年生的植株树高生长量为5.4m,胸径仅为11.0cm,单株材积总生长量只有0.033m³。天然云南红豆杉林木的树高、胸径生长量在30年生以前相对较大,以后随树龄的增长逐渐减缓,160~200年达到自然成熟,生长停滞。年龄10~20年,树高生长量达每年12cm,20~30年每年10cm,30~100年每年5~7cm;胸径生长量10~20年每年0.32cm,20~30年每年0.27cm,30~100年每年0.18~0.24cm。

2 云南红豆杉的地理分布

2.1 水平分布

云南红豆杉天然分布于我国云南中部、西部、西北部以及西南部的新平、双柏、大姚、南华、景东、双江、耿马、临沧、云县、永德、凤庆、镇康、隆阳、腾冲、龙陵、弥渡、永平、云龙、鹤庆、剑川、漾濞、洱源、弥渡、祥云、宾川、南涧、玉龙、古城、永胜、宁蒗、香格里拉、德钦、维西、泸水、福贡、贡山、兰坪等37个县(市、区),四川西南部的木里、盐源、九龙、冕宁、西昌、德昌、普格一带以及西藏东南部的察隅、墨脱、波密、亚东等地。与我国西南部、西部接壤的缅甸、不丹、尼泊尔、印度等国亦有分布。在云南省,云南红豆杉的人工栽培已扩大到滇东南文山州的文山、丘北、马关、西畴,红河州的屏边、河口,楚雄市的武定、禄丰,西双版纳州的景洪、勐海,昆明市的安宁、富民、寻甸、石林、官渡、盘龙,保山市的施甸、昌宁,思茅市的翠云等19个县(市、区)。

其分布区的东界位于我国四川,大致为德昌、西昌、冕宁、九龙一线,在西昌以西;南界位于云南镇康、耿马、双江一线,为双江以北;西界在缅甸、不丹、尼泊尔境内;北界位于波密、德钦、九龙一线以南。

云南西部和西北部、四川西南部以及西藏东南部与缅甸的分布区连成一体,构成了云南红豆杉的中心分布区,其地理位置大致为23°12′~30°26′N,88°36′~102°16′E。

2.2 垂直分布

云南红豆杉属北温带区系,主要散生于海拔高度为2 000~3 200m的中山、亚高山针叶林、针阔混交林、湿性常绿阔叶林和季风常绿阔叶林内。天然分布的最低海拔高度为1 285m,位于云南腾冲县高黎贡山的西坡,其分布海拔高度最高达3 500m,位于丽江市高山乡大平坝。在云南,云南红豆杉主要集中分布于海拔高度2 500~3 200m的地带,四川和西藏分别集中于海拔高度2 500~3 200m和2 000~2 600m的地段。近年来,经人工栽培使其垂直分布不断下移。云南红豆杉的人工栽培最低点为西双版纳景洪市普文镇,海拔高度850m左右,且生长表现良好。

3 云南红豆杉个体生态学特性

3.1 适生环境条件

3.1.1 光照

云南红豆杉主要天然散生于中山、亚高山阴坡、半阴坡及沟谷的针叶林、针阔叶混交林及山地湿性常绿阔叶林的林冠下,形成该林分乔木第二、第三层,为荫蔽潮湿的生境。在林窗和上层林木稀疏(郁闭度0.3~0.6)有光照的地段,云南红豆杉呈集中分布,局部可形成以云南红豆杉为优势的林分。如在云南省永胜县大安乡海拔高度3 080~3 120m的梁子平台,生长有一片面积为4km²的以云南红豆杉为优势种的林分,林分郁闭度0.5~0.6,云南红豆杉种群密度在0.6以上。

试验与观察结果表明,云南红豆杉1~5年生的幼苗忌日晒,对直射光很敏感,在强光下易萎蔫或死亡,在散射光条件下生长良好,而绝对荫蔽生长也很差。幼苗期总的生长趋势是半荫蔽条件下比全光照的生长量大,成活率高,其随树龄增大所需光照逐渐增强。至幼树、成龄树则忌密闭,郁闭度以0.5~0.6时长势较好。

3.1.2 温度

云南红豆杉对温度的适应范围较宽,其天然中心分布区的年均温为5.3℃,而人工引种栽培的最南端已到景洪市普文镇。故其分布区(天然与人工种植区)的年均气温为4.7~20.1℃,极端最高气温38.3℃,极端最低气温-20℃,≥10℃的活动积温1 600~7 500℃。根据王达明所开展的云南红豆杉种植区划的研究结果,其最适温度范围为:年均气温10~18℃,≥10℃的活动积温1 600~6 000℃[12]。

3.1.3 湿度和降水

水分条件是限制云南红豆杉生长发育和分布的主导因子之一。水分不足使其生长缓慢,植株矮小,天然分布数量也少;水分充足而在土壤排水良好则生长较快,其植株可长成高大乔木,天然分布数量也多。

云南红豆杉为浅根性树种,主根浅,侧根发达,具有喜湿树种的特征,对水湿条件的要求较高。云南红豆杉分布区年平均相对湿度在70%以上,最高可达90%;年干燥度1.5以下,最低为0.6;年降水量一般在900mm以上,最高达1 600mm以上。其最适湿度和降水范围为:年干燥度1.0以下,年平均相对湿度大于80%,年降水量大于1 500mm。

表4 不同土壤类型云南红豆杉资源分布情况

Table 4 Distribution of *Taxus yunnanensis* resources on different types of soil

土类	面积(hm^2)		蓄积量(m^3)		株数(株)	
	数量	所占比例(%)	数量	所占比例(%)	数量	所占比例(%)
暗棕壤	4 879	2.2	43 869	6.2	127 380	3.6
黄壤	973	0.4	2 081	0.3	25 470	0.7
黄棕壤	59 929	27.4	262 254	37.1	762 910	21.8
棕壤	152 873	70.0	398 225	56.4	2 592 140	73.9
合计	218 654	100	706 429	100	3 507 900	100

3.1.4 土壤

云南红豆杉为浅根性树种,喜肥沃疏松、排水良好的土壤。在干旱瘠薄的土壤条件下生长较差,植株呈矮化状,生长量偏小。云南红豆杉天然分布区的土壤种类有暗棕壤、棕壤、黄棕壤和黄壤,以棕壤和黄棕壤为主(见表4)。若以红豆杉分布的数量计算,棕壤占73.9%,黄棕壤占21.8%,其余仅占5.3%。最适宜土壤是棕壤;其次是黄棕壤。分布区土壤呈酸性或微酸性,pH值4.5~6.5。

3.2 主要生态型

云南红豆杉虽属第三纪孑遗植物,但自然地理分布范围较广,且不连续,多为散生或群状散生,极少成纯林。其中心分布区垂直高差为2 215m,跨越了纬度7°14′、经度13°40′的4个气候带(亚热带、暖温带、温带、寒温带),4种土壤类型(暗棕壤、黄壤、黄棕壤、棕壤),12种森林类型(苍山冷杉林、怒江冷杉林、丽江云杉林、云南铁杉林、华山松林、云南红豆杉林、云南松林、黄背栎林、川滇高山栎林、槭树林、槲栎林、多变石栎林),生境条件复杂多样。云南红豆杉对不同特定生境的适应,形成了在形态结构、生理生态、遗传特性有显著差异的生态型。陈少瑜等对滇西北怒江、澜沧江和金沙江三江流域的3个天然云南红豆杉种群遗传多样性的研究,证实了云南红豆杉群体内的遗传多样性是比较高的。所研究的3个天然云南红豆杉种群平均基因多样性为0.325 3;云南红豆杉群体间存在一定的遗传分化,基因分化系数为0.146 6,平均遗传距离是0.118,遗传分化的大小与空间距离成正相关。依据其形态外貌差异,可将云南省天然分布的云南红豆杉划分为高大乔木型、小乔木型和灌木型3种生态型。

3.2.1 高大乔木型

主要分布于泸水、福贡、贡山、腾冲等县。海拔高度2 000~2 800m,为暖温性气候类型,降雨特别丰富(达1 400mm以上),湿度高,光照也较为充足,其云南红豆杉的高、径生长快,植株最高可达40m,胸径200cm,一般高10~14m,树冠呈塔形,分枝高,干形好。

3.2.2 小乔木型

主要分布于云龙、兰坪、丽江、永胜、维西等县。植株的海拔高度2 800~3 300m,为温凉性气候类型,降水相对较少,年均温较低。由于受上层荫蔽,光照不足,云南红豆杉高生长不旺盛,一般平均高7~8m。分枝低,常呈多干形,主干粗短,侧枝延伸很长而数量多,树冠呈球形或半球形。
3.2.3 灌木型

主要分布于宁蒗、中甸、德钦等县的高海拔地带或阳坡,海拔高度一般3 300m以上。气候特点是寒冷,湿度不高。该类型的植株高生长极差,树高3~4m,分枝低矮,侧枝也不发达,分枝细但不长,主干粗短,呈"小老头树"状。

4 讨论与建议

(1)云南红豆杉为雌雄异株植物,雌花和雄花的花期相遇,并可结实,不存在花期不遇导致不结果或结果量小的现象。天然生长的云南红豆杉结实率低、结实量少、雌株数量少(约是雄株的1/3),多为散状分布以及上层荫蔽光照不足等因素综合作用的结果。因此,对过度郁闭的云南红豆杉天然林分进行适当的透光抚育,伐除部分上层林木,以促进云南红豆杉的生长结实。

(2)云南红豆杉种皮角质化且坚硬致密,阻碍胚与外界的水汽交换。种子种胚体小,初时发育不全,需要生理后熟。云南红豆杉种子在天然条件下,需要两冬一夏才能萌发。大量种子在漫长的休眠期间丧失了生命力,种子深度休眠极大地限制了其实生繁殖,严重地威胁这一孑遗树种的种群生存和发展。因此,在进行云南红豆杉育苗时,种子必须经过催芽处理,完成其生理成熟阶段,使之尽快具有正常的发芽率。

(3)云南红豆杉具有较强的萌发能力,高生长和径生长的高峰期均为5~7月,可在这一阶段制定合理的抚育措施,促进其速生丰产。建议对其小枝叶采集时,要尽可能避开生长高峰期。

(4)云南红豆杉树高、胸径、材积生长量与树龄均成幂函数关系。天然的云南红豆杉树体生长极为缓慢,30年生树高为5.4m,胸径仅为11.0cm,单株材积只有0.033m^3,一旦破坏很难恢复。因此,必须加强保护现有天然红豆杉资源,通过大力发展人工原料林基地,实现资源的可持续利用。

(5)云南红豆杉能够耐受-20℃左右的低温和38.2℃的高温,对生境要求不太严格,我国大部分地区均能满足它生长发育所需条件,故通过引种、驯化可扩大云南红豆杉的分布范围。目前,其水平分布已南移至22°26′N(景洪市普文镇),垂直分布也随之下降到海拔高度850m的地带,且生长表现良好,进一步验证了云南红豆杉热带亚洲起源的论断。

(6)依据形态外貌差异,可将云南省天然分布的云南红豆杉划分为高大乔木型、小乔木型和灌木型3种生态型。

(7)在做云南红豆杉人工种植最适地选择时,应尽可能具备以下条件:①肥沃疏松、排水良好的酸性或微酸性土壤;②年均气温10~18℃,≥10℃的活动积温1 600~6 000℃;③年干燥度1.0以下,年平均相对湿度大于80%,年降水量大于1 500mm。

参考文献:

[1]陈振峰,张成文,寇玉锋,等.我国红豆杉资源及可持续利用对策[J].世界科学技术——中药现代化,2002,4(1):40~46.
[2]郑万钧.中国树木志[M].北京:中国林业出版社,1983.
[3]张茂钦,李达孝,左显东,等.云南红豆杉人工栽培及其生态生物学特性研究[J].林业科技通讯,1996

(3):8~12.
[4] 赵盛军.云南红豆杉种子休眠原因的初步研究[J].林业调查规划,1996(1):44~46.
[5] 程广有,唐晓杰,高红兵,等.东北红豆杉种子休眠机理与解除技术探讨[J].北京林业大学学报,2004,26(1):5~10.
[6] 朱念德,刘蔚秋,伍建军,等.影响南方红豆杉种子萌发因素的研究[J].中山大学学报(自然科学版),1999,38(2):75~79.
[7] 陈少瑜,吴丽圆,李江文.云南红豆杉天然种群遗传多样性研究[J].林业科学,2001,37(5):41~48.
[8] 王达明,李莲芳,周云,等.云南红豆杉人工药用原料林的经营技术[J].西部林业科学,2004,33(1):8~20.
[9] 王达明,李莲芳,周云.滇之云南红豆杉种植区划[J].西部林业科学,2004,33(4):1~6.

(本文发表于《西部林业科学》,2006年)

西南桦纯林与混交林生态系统 C 贮量的对比研究

杨德军,王卫斌,耿云芬,邱琼

(云南省林业科学院,云南 昆明 650204)

摘要:利用测定生物量的方法对 13 年生西南桦人工纯林、西南桦+肉桂混交林、西南桦次生林和热带次生林进行了 C 贮量的对比研究。结果发现 4 种林分的 C 密度分别为 148.42、140.33、108.25、129.38 t·hm^{-2};年固 C 量分别为 4.01、4.59、3.71、2.42 t·hm^{-2}·a^{-1};地上生物质 C 密度分别为 42.18、45.61、40.12、23.08 t·hm^{-2};地下生物质 C 密度分别为 9.73、14.06、8.10、8.41 t·hm^{-2};林分凋落物 C 密度分别为 6.03、8.81、3.03、2.61 t·hm^{-2};林地土壤 C 密度分别为 82.38、79.94、57.00、95.28t·hm^{-2}。结果表明西南桦是开展以固 C 为目标的生态造林项目的适合树种。

关键词:西南桦;人工林;混交林;西南桦次生林;热带次生林;C 密度;年固 C 量

Comparative analysis of ecosystem carbon stocks for pure and mixed stands of *Betula alnoides*

Yang De-jun[1], Wang Wei-bin[2], Geng Yun-feny[2], Qiu Qiong[1]

(1. Institute of Tropical Forestry Yunnan Academy of Forestry, Jinghong Yunnan 666102;
2. Yunnan Academy of Forestry, Kunming Yunnan 650204)

Abstract: Biomass measurement method was applied to conduct comparative study of carbon storage capacity between 13 years old Betula alnoides monoculture, mixed plantation with Cinnamomum cassia, B. alnoides secondary forest and tropical secondary forest. The results showed that the carbon density in monoculture, mixed plantation, B. alnoides secondary forest and tropical secondary forest were 148.42, 140.33, 108.25, and 129.38 t·hm^{-2} respectively. The annual rate of carbon sequestration of these 4 forests were 4.01, 4.59, 3.71, and 2.42 t·hm^{-2}·a^{-1} respectively. The carbon density of above ground part of these 4 stands were 42.18, 45.61, 40.12, and 23.08 t·hm^{-2} respectively. The root system carbon densities of these 4 stands were 9.73, 14.06, 8.10, and 8.41 t·hm^{-2} respectively. The litter carbon density of these 4 stands were 6.03, 8.81, 3.03, and 2.61 t·hm^{-2} respectively. The soil carbon density of 4 stands were 82.38, 79.94, 57.00, and 95.28 t·hm^{-2} respectively. This implies that B. alnoides is promising species for reforestatation taking the carbon sequestration as the management purpose.

Keyword: *Betula alnoides*; plantation; mixed plantation; B. alnoides secondary forest; tropical secondary forest; carbon density; annual rate of carbon sequestration

随着全球气候变暖速度的不断加快,对全球可持续的发展造成了越来越严重的影响,全球对以 CO_2 为主的温室气体的排放进行了越来越严格的限制,在这种背景下,CO_2 等温室气体的固定成为了一个可持续发展和生态环境建设研究中的热点。而通过人工造林对 CO_2 进行固定被认为是技术简单、回报率高的应对全球变暖的最佳选择之一。由一些发达国家或国际组织资助的旨在固定 CO_2 的造林项目已在一些国家相继开展[1]。在森林的 C 储量研究上,国外如 Lugo et al[2] 已做了大量工作,Lasco[3] 已对菲律宾的森林固 C 进行了详细的研究。李江等[4]也对云南热区几种人工幼林的固 C 作用进行了研究。但对同树种同林龄的西南桦(*Betula alnoides*)人工林(纯林与混交林)和次生林的固 C 作用对比研究还未见报道。西南桦属桦木科(Betulaceae)落叶大乔木。该树种干形通直,材质优良,是云南省制造胶合板的主要用材树种之一,也是优良的家具、装修用材,有着广阔的市场前景[1]。作为亚热带常绿阔叶林区次生林先锋树种,它不仅适应性强、耐贫瘠,还具有较强的天然更新能力。由于西南桦人工林生长速度超过天然林,每年树高生长 2m 以上,胸径生长 2cm 以上,近年来它作为短周期工业原料林树种,广泛用于云南热区山地造林,造林成本低,经济、生态效益表现良好[2~5]。对 13 年生的西南桦林(人工林与天然林)进行了 C 储量和年固 C 量的研究,以期对云南热区退化山地植被恢复的树种选择、混交林培育及次生林的经营等方面提供基础资料。

1 试验地概况

试验地位于普文试验林场内,该林场位于云南省西双版纳州景洪市北部,101°4′~101°6′E,22°24′~22°26′N,处于横断山帚状山系南缘,无量山南延末端;气候属热带北缘季风类型。1 年当中受潮湿的西南季风和干暖的西风南支急流交替控制,干湿季分明,11 月至翌年 4 月为干季,5~10 月份为雨季。年平均气温 20.1℃,≥10℃的日积温 7 459℃,持续 364.1d,最热月(7 月)平均气温 23.9℃,最冷月(1 月)平均气温 13.9℃,极端最高气温 38.3℃(1966 年 5 月,1969 年 5 月),极端最低气温 -0.7℃(1974 年 1 月)。年降水量 1 655.3mm,是西双版纳降水最多的地区。雨季降水量占全年的 86%,其月平均降水量可达 140~356mm,干季中,最少月降水量均在 20mm 以上。冬春多雾,雾日年均 145.5d,多雾不但可以缓解干旱,而且形成了山地逆温层。年平均相对湿度 83%,干燥度 0.71。在气候区划上,恰处于北热带北缘与南亚热带南缘的交界上。热带和南亚热带树种在此均能生长[5]。

西南桦人工纯林和混交林造林地主要是山地雨林迹地,海拔高度为 860~910m,长期以来受到乱砍滥伐及放牧的破坏,有用之材几乎采伐殆尽,林中牧道纵横;借助优越的水热条件,萌生树及次生林树种迅速更新,外貌虽茂密但林相已极度不整齐。林地与森林群落均已出现退化迹象,水土流失在局部地方开始发生。造林前对造林地的次生林实行皆伐,之后炼山,按田间试验要求选择试验地,各试验地块立地条件基本一致。试验林于 1992 年营造,西南桦纯林株行距 2m×3m,西南桦+肉桂(*Cinnamomum cassia*)混交林以间行形式混交,株行距为 2m×3m。均采用全垦穴状整地方式,规格 40cm×40cm×30cm,种植穴每穴施入复合肥 150g,以 6 月苗龄的袋苗定植[6]。

西南桦次生林与热带次生林均是普文试验林场内原有天然林经多次破坏所形成,其海拔为 900~950m。最后一次被完全破坏是在 1993 年由于修普文镇连接联合村的乡村公路而造成的,2 种林分的立地条件基本一致,但由于一个地段附近有西南桦的下种母树,现在形成了以西南桦为建群种的西南桦次生林;另一地段则形成了以黄牛木(*Cratoxylun cochinchinense*)、水锦树(*Wendlandia tinctoria*)、余甘子(*Phyllanthus emblica*)、红木荷(*Schima-*

wallichii)等树种为主的热带次生林。由于西南桦次生林所处地段当初修路时表土层受破坏比较严重,造成了西南桦次生林下土壤肥力与热带次生林的有较大差距。

2 研究方法

2.1 地上部分C密度的测定

乔木层每块标准地大小为20m×20m,设置在林分典型样地中。胸径5cm以上树木进行每木检尺,根据每木检尺的结果计算出每块样地的平均胸径和平均树高,然后在样地内选取平均木。2种人工林的乔木树种根据每木检尺结果均分为3个径级(西南桦为5~8cm、8~10cm、10cm以上;肉桂分为5~7cm、7~9cm、9cm以上)选取平均木,每径级选取1株;2种次生林根据每木检尺结果,按树种分别选取平均木,株数多的树种分为2个径级(5~8cm和8cm以上,株数少的树种就选1棵平均木)做解析木分析、生物量和生长量测定。调查时间为2006年11月至2007年1月。

灌木层和草本层的生物量测定同在乔木层样地中进行,灌木层用大小为5m×5m标准地收获,草本层用大小为1m×1m的5个标准地收获。

对各层的植物进行取样,实验室测定含水率,将鲜重换算为干重。根据有机质含C量为其干物质的50%将生物量换算为C密度[2,4,7]。

2.2 地下部分C密度的测定

乔木层以平均木作为根生物量测定的标准木。挖出全部根系并称取重量。取一些根系样品带回,放入烘箱中在85℃下烘干至恒重计算失水率,将鲜重换算成干重,从而推算整个乔木层地下部分生物量。

灌木层和草本层挖出样地中全部的根系称重,分别取一些根系样品带回,放入烘箱中在85℃下烘干至恒重计算失水率,将鲜重换算成干重,从而推算整个灌木层和草本层地下部分生物量。根据有机质含C量为其干物质的50%将生物量换算为C密度[2,4,7]。

2.3 枯枝落叶层C密度的测定

在乔木层20m×20m样地内机械设置5个1m×1m的小样方并收集所有的枯枝落叶,取回,放入烘箱中在75℃下烘干,称重。根据有机质含C量为其干物质的50%将生物量换算为C密度[2,4,7]。

2.4 土壤中C密度的测定

在每个20m×20m的乔木层样地中按"X"型选设5个样点,在各样点挖土壤剖面100cm深,每10cm取一个土样。土样在室内自然风干过2mm筛,测定其有机质含量;用环刀法每10cm取一土样测定其容重。各林分土壤中固定的C通过土壤有机质含量及土壤容重进行计算,应用的换算公式为[7],土壤C密度=土壤体积×土壤容重×土壤有机质含量÷1.724。

2.5 人工林分总C密度的测定

将地上部分、地下部分C密度,枯枝落叶及土壤中固定的C相加,即可获得单位面积人工林分的固 C 总量即 C 密度[4]。

2.6 人工林分年固 C 量的测定

因土壤监测结果显示土壤有机质含量年变化甚小[8]，不予考虑。用 13a 平均生物量变化来计算年固 C 量。

3 结果与分析

3.1 4 种林分乔木层标准木的选取

从表 1 可以看出，4 种林分的乔木层在密度、树高及胸径方面表现出较大差距，乔木株数最多的是热带次生林，乔木株数最少的是西南桦纯林，混交林与西南桦次生林介于两者之间。由于 2 种次生林具有较多的乔木种类，因此选取了较多的标准木。

表 1 各林分的乔木层特征
Table 1 Stand characteristics of the 4 forests

林分类型	标准木株数	种类数	密度(株·hm^{-2})	D(cm)	H(m)
西南桦+肉桂混交林 (Betulua alnoides+Cinnamomum cassia)	6	2	650	12.20	11.19
西南桦纯林 (Butula alnoides plantation)	3	1	350	11.89	12.37
西南桦次生林 (Betula alnoides secondary forest)	16	7	710	9.74	9.15
热带次生林 (Tropical secondary forest)	19	13	925	6.68	7.72

注：各林分样地乔木只统计胸径≥5cm 的。

3.2 地上部分生物质的 C 密度

从表 2 可以看出，4 种林分地上部分 C 密度最高的是西南桦+肉桂混交林，达 45.61t·hm^{-2}，其次是西南桦纯林(42.18t·hm^{-2})和西南桦次生林(40.12t·hm^{-2})，最低的是热带次生林(23.08t·hm^{-2})。乔木层的 C 密度占地上部分的比例最大，达到 95.03%～97.05%。灌木层和草本层的 C 密度占地上部分的比例较小，仅为 0.62%～3.45%。

表 2 林分地上部分生物质的 C 密度
Table 2 Carbon density of the aboveground live biomass in the forests

林分类型	乔木层			灌木层			草本层			总计		
	生物量(t·hm^{-2})	C 密度(t·hm^{-2})	百分比(%)	生物量(t·hm^{-2})	C 密度(t·hm^{-2})	百分比(%)	生物量(t·hm^{-2})	C 密度(t·hm^{-2})	百分比(%)	生物量(t·hm^{-2})	C 密度(t·hm^{-2})	百分比(%)
1	87.75	43.88	96.20	1.98	0.99	2.17	1.49	0.74	1.63	91.22	45.61	100
2	80.16	40.08	95.03	2.75	1.38	3.26	1.44	0.72	1.71	84.35	42.18	100
3	77.86	38.93	97.05	0.50	0.25	0.62	1.87	0.94	2.33	80.23	40.12	100
4	44.19	22.10	95.73	1.59	0.79	3.45	0.38	0.19	0.82	46.16	23.08	100

注：林分类型 1 为 13 年生西南桦+肉桂混交林；林分类型 2 为 13 年生西南桦纯林；林分类型 3 为 13 年生西南桦次生

林;林分类型 4 为 13 年生热带次生林。

3.3 地下部分(根系)生物质的 C 密度

从表 3 可以看出,4 种林分地下部分(根系)C 密度最高的是西南桦+肉桂混交林($14.06t \cdot hm^{-2}$),其次是西南桦纯林($9.74t \cdot hm^{-2}$)和热带次生林($8.41t \cdot hm^{-2}$),最小的是西南桦次生林($8.10t \cdot hm^{-2}$)。4 种林分地下部分 C 密度与地上部分 C 密度的比值为 0.20~0.36,比值最大的是热带次生林,这是因为热带次生林中多数小乔木或灌木,地上部分积累的生物量较其他林分相对较小,而地下部分由于林木株数较多造成了根系数量大,因此其比值较大。

表 3 各林分地下部分、凋落物及土壤有机 C 密度
Table 3 Carbon density of tree roots, litter and soil in the forests

林分类型	地下部分生物量($t \cdot hm^{-2}$)	地下部分 C 密度($t \cdot hm^{-2}$)	凋落物生物量($t \cdot hm^{-2}$)	凋落物 C 密度($t \cdot hm^{-2}$)	土壤有机质含量($t \cdot hm^{-2}$)	土壤有机 C 密度($t \cdot hm^{-2}$)
西南桦+肉桂混交林	28.11	14.06	17.61	8.81	137.82	79.94
西南桦纯林	19.48	9.74	12.06	6.03	142.02	82.38
西南桦次生林	16.20	8.10	6.05	3.03	98.27	57.00
热带次生林	16.81	8.41	5.22	2.61	164.27	95.28

3.4 凋落物层的 C 密度

从表 3 可以看出,凋落物层 C 密度最高的是西南桦+肉桂混交林,达 $8.81t \cdot hm^{-2}$,其次是西南桦纯林,为 $6.03t \cdot hm^{-2}$,西南桦林中最小的是西南桦次生林,为 $3.03t \cdot hm^{-2}$。凋落物层 C 密度最小的是热带次生林,为 $2.61t \cdot hm^{-2}$。大量的枯落物说明 2 种西南桦人工林都处于旺盛的发育期。西南桦次生林和热带次生林凋落物量与 2 种人工林相比,相对较少。

3.5 土壤中有机 C 密度

从表 3 可以看出,2 种西南桦人工林土壤的有机 C 密度差异不大($79.94t \cdot hm^{-2}$ 和 $82.38t \cdot hm^{-2}$),比热带次生林($95.28t \cdot hm^{-2}$)低,远高于当地西南桦次生林($57.00t \cdot hm^{-2}$)。西南桦次生林土壤 C 密度远低于其他 3 种林分类型,这与当初原生植被被破坏后,其表土层受到严重破坏,土壤有机质含量低的情况是相符的。

3.6 4 种林分的总 C 密度

从表 4 可以看出,4 种林分的总 C 密度为 108.25~$148.42t \cdot hm^{-2}$,土壤中的 C 密度占了较大比例,为 52.66%~73.64%。表明了该土壤是 4 种林分的主要 C 库。以吸收固定 CO_2 为主要目标的造林项目除了追求较高的生物量外,对林地土壤的管理也应引起足够的重视,应避免出现因大规模的水土流失而造成林分固 C 量的损失。3 种西南桦林生物质 C 密度为 48.22~$59.67t \cdot hm^{-2}$,占总 C 密度的 37.00%~44.55%,明显高于王效科等[9]对黑松、油松、马尾松、柳杉、杉木和水杉的幼中龄人工林的 C 密度总体估计值($<15t \cdot hm^{-2}$),表明西南桦林分有较高的固 C 潜力。4 种林分中 C 密度最高的是西南桦+肉桂混交林

(148.42t·hm^{-2}),其次是西南桦纯林(140.33t·hm^{-2})和热带次生林(129.38t·hm^{-2}),最低的是西南桦次生林(108.25t·hm^{-2})。

表4 4种林分的C密度及其组成与年固C量
Table 4 Ecosystem carbon density and sequestration in the 4 stands

林分类型	地上部分生物质C密度 (t·hm^{-2})	地下部分生物质C密度 (t·hm^{-2})	地上+地下部分生物质C密度		凋落物C密度		林地土壤C密度		总C密度 (t·hm^{-2})	年固C密度 (t·hm^{-2}·a^{-1})
			(t·hm^{-2})	%	(t·hm^{-2})	%	(t·hm^{-2})	%		
西南桦+肉桂混交林	45.61	14.06	59.67	40.20	8.81	5.94	79.94	53.86	148.42	4.59
西南桦纯林	42.18	9.74	51.92	37.00	6.03	4.30	82.38	58.70	140.33	3.99
西南桦次生林	40.12	8.10	48.22	44.55	3.03	2.80	57.00	52.66	108.25	3.71
热带次生林	23.08	8.41	31.49	24.34	2.61	2.02	95.28	73.64	129.38	2.42

3.7 4种林分的年固C量

从表4可以看出,4种林分年固C量最大的是西南桦+肉桂混交林,为4.59t·hm^{-2}·a^{-1},其次是西南桦纯林,为3.99t·hm^{-2}·a^{-1},西南桦次生林也达到3.71t·hm^{-2}·a^{-1},这与国际大气组织温室气体使用的热带人工林年固C量是一致的,在3.4~7.5t·hm^{-2}·a^{-1}之间[9]。同时也说明只要有合适的下种母树,在一定条件下次生林也能达到人工林的固C能力。也说明西南桦作为一种碳汇(carbon sink)树种的潜力是很大的,3种西南桦林每年吸收固定的C量都明显高于当地热带次生林(2.42t·hm^{-2}·a^{-1})和暖温带落叶阔叶林(2.19 t·hm^{-2}·a^{-1})。但是与典型的热带地区相比,云南热区的西南桦人工林的年固C量还是偏小的。如在菲律宾马占相思(*Mangium cassia*)纯林的年固C量测定值为20.8t·hm^{-2}·a^{-1},石梓(*Gamelina arborea*)林为9.8t·hm^{-2}·a^{-1}[10]。

4 结论与讨论

西南桦+肉桂混交林、西南桦纯林2种西南桦人工林分的C密度为140.33~148.42t·hm^{-2},明显高于当地同龄次生林的129.38t·hm^{-2}。3种西南桦林的年固C量在3.71~4.59 t·hm^{-2}·a^{-1},明显高于热带次生林的2.42t·hm^{-2}·a^{-1}。从林分现在的C密度和年固C量可以预测云南热区的西南桦人工林到中林阶段就可以形成相当规模的C库了。

无论是西南桦的人工纯林、混交林还是西南桦的次生林均表现出较高的固C能力,说明了西南桦这个树种具有较高的固C潜力,加之它又是一个乡土树种,因此可在云南热区大量应用于以吸固CO_2为目标的造林项目。同时西南桦为旱季落叶树种,枝条细而疏散,人工林下透光度较高,可形成具有繁茂下木层、灌草层的多层多种的结构,保持生物多样性。每年产生的大量枯枝落叶可减少地表径流和土壤冲刷,具有较强的保持水土的能力,西南桦也是一个作水源涵养林的好树种。

当然要全面评价西南桦人工林的价值与功能,还需要在多方面进行深入研究,如生物多样性保护、林分的稳定性、水土保持能力的比较、最佳经营措施的选择等。对13年生人工林生态系统的研究仅仅是一个阶段性的初步工作,以后随着时间的推移和西南桦人工林年龄的增长,还需要进一步的调查和研究,以期对西南桦人工林的发展给出一个更客观和科学的评价。

参考文献:

[1] Dixon R K. Forest sector carbon offset project:Near term opportunities to mitigate greenhouse gas mission [J]. Water,Air andSoilPolution,1993(70):561~577.

[2] Lugo A E,Brown S. Tropical forests as sinks of atmospheric carbon[J]. Forest Ecology and Management,1992(54):239~255.

[3] Lasco R D. Forests and land use change in the Philippines and climate change mitigation[J]. Mitigation and Adaptation Strate-gies for Global Change,2000(5):81~97.

[4] 李江,陈宏伟,冯弦. 云南热区几种阔叶人工林C储量的研究[J]. 广西植物,2003,23(4):294~298.

[5] 王达明. 西双版纳普文试验林场自然条件[C]∥云南省林业科学院. 热区造林树种研究论文集. 昆明:云南科学技术出版社,1996:7~11.

[6] 王达明. 西南桦造林技术研究[C]∥云南省林业科学院. 热区造林树种研究论文集. 昆明:云南科学技术出版社,1996:44~49.

[7] Lasco R D. Quantitative estimation of carbon storage and sequestration of tropical forest esosystem [J]. Professorial Chair Lec-ture UPLB College,Laguna,1999.

[8] 蒋云东,王达明,周云. 西双版纳几种人工林地力恢复趋势研究[J]. 云南林业科技,2002,21(3):50~54.

[9] 王效科,冯宗伟,欧阳志云. 中国森林生态系统的植物C贮量和碳密度的研究[J]. 应用生态学报,2001,12(1):13~16.

[10] 桑卫国,马克平,陈灵芝. 暖温带落叶阔叶林碳循环的初步测算[J]. 植物生态学报,2002,22(6):21~26.

(本文发表于《福建林学院学报》,2008年)

12个美国银合欢新品种在云南三地的引种初效

李江[1,2]，邱琼[1]，朱宏涛[3]，杨宴平[4]，刘海刚[5]，陈宏伟[1]，孟梦[1]，冯弦[1]，
刘永刚[1]，郭永清[1]

(1. 云南省林业科学院，云南 昆明 650204；2. 北京林业大学研究生院，北京 100083；
3. 昆明植物研究所，云南 昆明 650204；4. 保山市林业技术推广站，云南 保山 678000；
5. 云南省农业科学院热区生态农业研究所，云南 元谋 651300)

摘要：2007~2008年利用从美国引进的 KX2、KX3、KX4、K784、K636、K584、K608、K565、K29、K72、K156、K376 共12个银合欢新品种在云南的普文、鸡飞和开远三地开展了育苗和造林试验。育苗试验结果表明，采用热水浸种5min可以有效提高其种子的发芽率；采用两段式培育的袋苗比直播穴盘育的苗好，同龄出圃苗苗高前者明显高于后者，但地径差异不大。造林试验结果表明，在经历2008年低温寒害和2009~2010年的特大旱灾后，上述12个银合欢新品种在三地的造林成活率和保存率基本正常，林木生长良好，其中以开远试点表现最好。总的看来，引进的银合欢新品种具有较强的抗逆性，在云南推广种植的潜力很大。

关键词：美国银合欢；云南引种；育苗；造林

Introduction Experiments on New Varieties and Hybrids of Leucaena in Yunnan

Li Jiang[1,2], Qiu Qiong[1], Zhu Hong-tao[3], Yang Yan-ping[4], Chen Hory-wei[1], Meng Meng[1],
Feng Xuan[1], Liu Yong-gang[1], Guo Yong-qing[1]

(1. Yunnan Academy of Forestry, Kunming Yunnan 650204; 2. Graduate School, Beijing
Forestry University, Beijing 10083;
3. Kunming Institute of Botany, Kunming Yunnan 650204; 4. Baoshan Forestry Technology
Extension Station, Baoshan Yunnan 67800;
5. Institute of Tropical Eco-agricultural Science, Yuanmou Yunnan 651300)

Abstract: New varieties and hybrids of Leucaena introduced from Hawaii, including KX2、KX3、KX4、K784、K636、K584、K608、K565、K29、K72、K156 and K376 were experimental cultivated in Puwen, Jifei and Kaiyuan of Yunnan. Results from the experiment on seedling growing showed seed soaking with hot water before sowing increased germination rate remarkably; containerized seedlings undergone transplanting were better than those by direct sowing in nest containers; seedlings of same age grown in Puwen were better than those in Jifei in terms of height, but diameter at ground level did not vary significantly. The results of the planting experiments showed that the plantations in 3 sites survived 2008 cold weather damage and the 2009~2010 once in a century draught attack, still showed a satisfied growing performance, especially in Kaiyuan. In general, these introduced new varieties and hybrids have showed excellent adaptabilities to disadvan tageous conditions and have great potential for extensive planting in Yunnan.

Keyword: USA Leucaena; introduction to Yunnan; seedling growing; afforestation

银合欢属（*Leucaena*）为多年生乔木或灌木，多分布于美洲秘鲁到美国德克萨斯间南北约 7 000km 的广大地区。该属植物品种间从外表形态到基因组成都有较大的差异[1~2]。已发表的银合欢品种达 50 多个，但根据 Brew baker 等[3]的研究，可归并为 *L. collinsii*、*L. diversifolia*、*L. esculenta*、*L. greggii*、*L. lanceolat*、*L. leucocephala*、*L. macrophylla*、*L. pallida*、*L. pulverulenta*、*L. retusa*、*L. salvadorensis*、*L. shannonii* 和 *L. trichodes* 等 20 多个品种。其中 *L. leucocephala* 自 16 世纪由西班牙人带到世界各地，分布的面积最广。很久以来，银合欢属植物的茎、叶就被广泛用作饲料、绿肥，树体作杆材和薪材等用，是较为典型的多用途树种。总体来说，银合欢在中低海拔、季节性干旱、中性或微碱性土壤的热带和亚热带地区生长良好。银合欢具有良好的抗旱特性，能在年降水量仅 250mm 地区生长，但不耐水淹，在年降雨量 1 000~3 000mm，排水良好地区生长良好。自 20 世纪 80 年代以来，Brew Baker 等利用从世界各地收集的银合欢品种进行了大规模的杂交育种，育成了一批速生优质、用途广泛、抗虫、耐寒等抗逆性强的杂交品种。

云南元江于 1981 年从广东引种银合欢以来，现有成林面积约 540hm^2。银合欢已成为元江干热河谷四旁及荒山造林的先锋树种。另，元谋县于 20 世纪 90 年代引进 K8 银合欢。纵观云南干热河谷区，已引进的银合欢品种基本上为 *L. leucocephala*，该品种存在耐寒性差、虫害和病害严重的问题。此外有些银合欢呈灌丛状生长，结实量大，自我繁殖能力强，可能属于生物入侵风险大的基因型[4~6]。云南省林业科学院于 2007 年在"世界银合欢之父"Brew Baker 教授帮助下从美国夏威夷大学等机构引进了一批银合欢优良新品种，希望从这些品种中选择出耐寒、速生、抗虫害、抗酸性土壤且生物入侵风险小的新品种推广到云南干热河谷以外的地区。

1 引进银合欢品种及引种地概况

按照耐寒、速生、抗虫害、抗酸性土壤且生物入侵风险小的要求，共引进了 12 个银合欢新品种。其中 11 个品种引进种子，KX4 引进扦插苗。各个品种的特性参见表 1。

表 1　引进的银合欢新品种一览表
Table 1 List of introduced *Leucaena* varieties/hybrids

品种名称	用途	突出特性
L. leucocephala K636	饲料、用材	高产优质
L. leucocephala K29	饲料、用材	高产优质
L. leucocephala K565	饲料、用材	高产优质
L. leucocephala K584	饲料、用材	高产优质
L. leucocephala K608	饲料、用材	高产优质
L. leucocephala K72	饲料、用材	高产优质
L. diversifolia K156	饲料、用材	耐寒（海拔达 1 500m），抗木虱
L. diversifolia K784	饲料、用材	耐寒（海拔达 1 500m），抗木虱
L. pallida K376	多用途	重要的杂交母本
L. hybrid KX2	饲料、薪材	*L. leucocephala*×*L. pallida* 杂交四倍体，抗木虱，结实少，耐寒
L. hybrid KX3	优质用材	*L. leucocephala*×*L. diversifolia* 杂交四倍体，抗旱耐寒，结实少
L. hybrid KX4	用材、园林	*L. leucocephala* K636×*L. esculenta* K380 杂交三倍体，速生，不结实，抗虫耐寒

12个银合欢新品种引种的育苗及栽培试验分别在云南的普文、鸡飞、开远3个地点进行,其三地的基本概况如下。

(1)景洪市普文试验林场:普文试验林场(简称普文)地处101°6′E,22°25′N,海拔860m,属亚热带湿润季风气候类型。年平均气温20.1℃,≥10℃的积温7 459℃,持续日数364.1天,最热月(7月)均温23.9℃,最冷月(1月)均温13.9℃,极端最高气温38.3℃(1966年5月、1969年5月),极端最低温-0.7℃(1974年1月)。年降水量1 655mm,相对湿度83%。土壤以赤红壤为主,土层厚度在低山坡面达1m以上,只有在箐沟中较陡峭的局部坡面上,才出现0.5~0.8m的中厚度土壤。试验地土壤呈酸性,pH值4.5,有机质含量低,仅0.6g/kg,缺氮,尤其少磷,而钾较丰富。

(2)昌宁县鸡飞林场:昌宁县鸡飞林场(简称鸡飞)位于云南省保山地区的东南部,地理位置为99°30′E,24°57′N,试验地海拔1 450~1 550m,属南亚热带半湿润、半干燥立体气候。年平均气温12.9℃,年平均降水量1 059mm。土壤为红壤性冲积土,pH值6.0,土壤养分总状况是:有机质含量低,全氮含量少,全磷含量更少,土壤含速效氮中等,速效磷极缺乏,速效钾中等偏高。

(3)开远市果木林场:开远市果木林场(简称开远),位于云南省红河州中部,海拔1 150m,属南亚热带季风气候,年降水量750mm,雨季集中于5~10月,年均气温19.8℃,年日照2 200h,全年无霜期340天。土壤为紫色土,pH值7.1。

2 种子检验

(1)净度、千粒重和种子大小的检验:按国家《林木种子检验方法》对其种子进行检验。

(2)发芽试验:供试银合欢品种共6个(K584、K636、K608、K784、KX2、KX3)。用40%多菌灵可湿性粉剂1 000倍液浸种10min,然后分别用98%浓硫酸浸种5min,或用80℃热水浸种10min,并设置种子不作处理的对照。将处理后的种子置于培养皿中进行发芽试验,每处理50粒,3个重复。保持培养皿湿润,温度为20~28℃(室内),每2~3天观察1次,每5天记录统计1次各培养皿中的发芽种子数。种子的萌发以胚根出现为标准。

发芽率=30d内发芽数/各处理种子数×100%

3 育苗及造林方法

3.1 各引种地的育苗方法

3.1.1 普文试验林场育苗方法

育苗时间为2007年3月下旬,参试育苗品种有9个即K156、K584、K376、K784、KX2、KX3、K636、K29和K565。采用2种方式育苗。①两段式育苗:苗床底土翻松后上覆10cm腐殖土,播种前2天用1%高锰酸钾溶液和1‰的敌克松进行苗床消毒。种子播前用80℃热水恒温处理10min。每个品种为一播种小区。撒播密度约为1 000粒·m^{-2}。播种后用河沙覆盖,覆盖厚度为种子厚度的1~2倍,浇透水。待床苗展出2片子叶即移至营养袋继续培育。1袋1苗。营养袋规格为10cm×15cm,育苗基质为过筛、消毒的腐殖土。②穴盘直播育苗:基质为芬兰泥炭基质,用1%的高锰酸钾溶液消毒。种子经80℃热水处理10min直接点播在穴盘内,每穴2~3粒,播后浇透水。苗期进行常规的管理。

3.1.2 昌宁县鸡飞林场育苗方法

育苗时间为 2007 年 3 月下旬。育苗品种共 7 个即 K156、K584、K376、KX2、KX3、K636 和 K784。采用两段式育苗。苗床土壤翻松埋烧后打碎、整平。撒播前 1~2 天用 1‰的敌克松杀虫。种子用 80℃热水恒温处理 10min 后晾干,立即撒播,每个品种立一个播种小区。撒播密度约 1 000 粒·m^{-2}。播种后用火烧土覆盖,覆盖厚度为种子厚度的 1~2 倍,浇透水。苗床上面搭建 70cm 高的塑料拱棚。待床苗展出 2 片子叶即移至营养袋中(上袋)继续培育。1 袋 1 苗。营养袋规格为 10cm×15cm,袋内基质为火烧土。种子出芽时需在塑料拱棚上搭建 50%的遮阴网,上袋后也搭建 50%的遮阴网。

出苗调查方法为每小区随机抽取 0.2m×0.2m 的苗床,准确记录播种量和出苗量,出苗天数为播种日到第 1 株苗破土日之间的天数。齐苗天数是指从出苗之日到连续 3 天不再出苗之间的天数。生长量调查每品种采用"Z"型顺序抽样 50 株,重复 3 次,观测苗高与地径。

3.2 引种地的造林方法

3.2.1 普文试验林场

2007 年 8 月在普文试验林场育苗并分区种植 7 个银合欢引进新品种,分别为 KX2、KX3、KX4、K784、K636、K584、K608,种植面积 0.33hm^2。于 2008 年又种植了银合欢 K565、K29、K72、K156 共 4 个新品种和部分 KX2、KX3、KX4 和 K736 新品种,种植面积 0.33hm^2。造林方法采取穴状整地,种植塘规格 40cm×40cm×40cm,造林密度为 2m×2m。株施三元复合肥 200g 作底肥。每年雨季开始前和结束后对引种试验林进行两定砍除灌草抚育。年底分品种调查植苗造林成活率和苗木的生长情况(地径和树高)。每小区随机调查 100 株。

3.2.2 昌宁鸡飞国有林场

2008 年 7 月在昌宁鸡飞国有林场育苗并分区种植共 7 个银合欢引进新品种,分别为 KX2、K636、K156、K584、K784、KX3、K376,种植面积 0.67hm^2。采取穴状整地,种植塘规格为 40cm×40cm×40cm,造林密度 2m×2m。株施三元复合肥 200g 作底肥。每年雨季开始前和结束后对引种试验林进行两定砍除灌草抚育。年底分品种调查植苗造林成活率和苗木生长情况(地径和树高)。每小区随机调查 100 株。

3.2.3 开远市林业局苗圃

2007 年在开远市果木试验林场育苗种植了少量的 K156、K636、KX2 和 KX3 银合欢品种,面积 0.067hm^2。采取穴状整地,种植塘规格为 40cm×40cm×40cm,造林密度 2m×2m。株施三元复合肥 200g 作底肥。每年雨季开始前和结束后对引种试验林进行两定砍除灌草抚育。年底分品种调查植苗造林成活率和苗木生长情况(地径和树高)。每小区随机调查 100 株。

上述 3 地造林试验以来,除在普文点发现竹鼠啃食银合欢幼树外,各试验点都没有发生明显的病虫害。3 地的试验林都经历了 2008 年低温寒害和 2009~2010 年百年一遇的特大干旱而无大面积死亡,与此形成对比的是,相邻地区的西南桦(*Betula alnoidis*)、旱冬瓜(*Alnus nepalensis*)、云南松(*Pinus yunnanensis*)等的新造林地则出现大面积死亡。

利用 DPS 数据分析软件对参试的银合欢新品种苗木生长指标(苗高、地径)和幼树生长指标(树高、地径)进行均值统计和差异显著性分析。

4 结果与分析

4.1 种子品质检验及发芽试验结果

(1) 11个引进银合欢的新品种种子品质检验结果见表2。

表2　11个银合欢新品种的种子净度、千粒重和种子大小
Table 2 One thousand seed weight, seed purity and seed size of 11 *Leucaena* varieties/hybrids

品种号	净度(%)	千粒重(g)	种子大小		
			长(cm)	宽(cm)	厚(cm)
K156	91.94	18.42	0.85	0.57	0.16
K784	88.60	59.11	0.61	0.35	0.11
K584	89.37	58.73	0.86	0.52	0.19
K376	95.47	67.64	0.94	0.64	0.16
K29	90.10	52.91			
K565	74.90	51.68			
K608	88.61	59.11	0.88	0.55	0.20
K636	98.61	64.55	0.83	0.53	0.19
K72	82.41	47.30			
KX2	77.75	40.11	0.77	0.49	0.18
KX3	92.69	38.31	0.79	0.43	0.15

由表2可知，所引进的11个银合欢新品种的种子净度都比较高，为74.90%~98.61%，霉烂破碎种子较少，种子质量良好。从外观上看品种间种子大小和形状有差异但不明显，很难从外观上区别出不同品种的种子。种子的千粒重差别较大（18.42~67.64g），K376的千粒重接近K156的4倍。

(2) 经催芽处理的6个引进银合欢新品种的种子多在第2~3天开始发芽，发芽结束期约为17天；不经处理的种子发芽期相对较晚，6~7天开始发芽，发芽结束期为28天。说明用化学和物理方法对银合欢种子进行处理都能不同程度地加快种子的发芽。如表3所示，3种处理K608种子的发芽率都非常低（2%），表明该品种的种子已基本丧失了活力。除K608外，其他5个品种的种子经浓硫酸处理的发芽率在32%~55%之间，热水处理的发芽率在32%~78%之间，而种子不经任何处理（对照）的发芽率在22%~50%之间。表明银合欢新品种的种子经热水和浓硫酸处理后都能提高发芽率。相比较，热水处理比浓硫酸处理的效果要好。这说明破坏银合欢种子坚硬的外种皮有助提高种子的发芽率。热水处理银合欢种子比浓硫酸处理效果好的原因可能是部分种皮破损的种子仍具有生活力，但浓硫酸处理可能破坏了这些种皮破损种子内部结构，导致这一部分种子失去生活力。因此种皮破损多的种子宜用热水处理。

表3　6个引进银合欢新品种的种子经不同处理后的发芽率
Table 3 Seed germination rates of 6 *Leucaena* varieties/hybrids under different seed treation methods

品种号	浓硫酸处理的发芽率(%)	热水处理的发芽率(%)	不做处理（对照）的发芽率(%)
K584	52	68	45

(续)

品种号	浓硫酸处理的发芽率（%）	热水处理的发芽率（%）	不做处理（对照）的发芽率（%）
KX2	54	78	46
KX3	32	32	22
K636	48	66	42
K784	55	68	50
K608	2	2	2

4.2 育苗试验结果

4.2.1 普文和鸡飞两地育苗效果比较

普文和鸡飞采取两段式育苗方法，进行了 K156、K584、K376、KX2、KX3、K636 和 K784 共 7 个银合欢新品种的育苗效果试验，普文还增作了 K29 和 K565 两银合欢新品种的两段式育苗试验。其试验结果见表 4。普文共育的 7 个品种种子的出苗天数 2~11 天，比鸡飞的出苗天数 4~12 天少。其中 K156、K584、KX2 和 K636 4 个品种种子出苗天数普文比鸡飞少 4 天；K376、KX3 和 K784 3 个品种种子的出苗天数普文比鸡飞少 1~2 天；普文 7 个品种种子育苗的齐苗天数 5~20 天也比鸡飞齐苗天数 10~20 天少，2 个试验点最大相差 7 天（K584、K784）；普文出苗率（27%~70.1%）总体优于鸡飞（32.4%~40.5%），相差最大的品种为 K156，其种子的出苗率普文比鸡飞高出 28.7 个百分点。2 地银合欢新品种种子在出苗和齐苗天数上的差异可能是因为苗木培育期间普文的气温比鸡飞高导致的。

表 4 普文和鸡飞试验点参试银合欢品种种子两段式育苗的出苗情况
Table 4 Seedling growth performance of 9 Leucaena varieties/hybrids in Puwen

品种号	普文			鸡飞		
	出苗天数(d)	齐苗天数(d)	出苗率(%)	出苗天数(d)	齐苗天数(d)	出苗率(%)
K156	2	6	65.0	6	12	36.3
K584	2	7	49.6	6	14	32.4
K376	11	20	33.3	12	20	32.7
KX2	2	6	66.7	6	12	40.5
KX3	3	8	27.0	4	10	33.4
K636	2	7	65.0	6	12	36.6
K29	3	7	43.3	—	—	—
K565	3	6	70.1			
K784	3	5	46.3	4	13	37.2

表 5 普文试验点 2 种育苗方式 9 个银合欢新品种出圃苗木生长状况
Table 5 Seedling growth performance of 9 Leucaena varieties/hybrids in Jifei

品种号	腐殖土基质两段式营养袋育苗		泥炭基质穴盘直播育苗	
	苗高(cm)	地径(cm)	苗高(cm)	地径(cm)
K156	62.9a	0.37a	22.3a	0.35a

(续)

品种号	腐殖土基质两段式营养袋育苗		泥炭基质穴盘直播育苗	
	苗高(cm)	地径(cm)	苗高(cm)	地径(cm)
K584	34.5d	0.33b	16.6c	0.31b
K376	22.3d	0.22c	11.5e	0.23d
KX2	26.5d	0.20c	13.3d	0.25c
KX3	50.7b	0.30b	19.9b	0.25c
K636	60.2a	0.38a	14.9c	0.33a
K29	46.0c	0.33b	12.3d	0.30b
K565	45.7c	0.28bc	13.0d	0.28b
K784	47.4c	0.25c	25.8a	0.26b

注:同列数值后标注字母表示0.05水平差异性。

4.2.2 不同育苗方式的育苗效果比较

如表5所示,在普文试点,采用腐殖土基质的两段式育苗效果优于泥碳基质穴盘育苗,其出圃苗木苗高差异较大,前者比后者高出10.5~45.3cm;用芬兰基质进行穴盘直播育苗,3个月出圃时幼苗长势较差,苗高普遍在20cm以下,而采取两段式腐殖土基质育苗的苗木苗高绝大多数在30cm以上,最高者可达90cm。两段式育苗与直播育苗相比,可大大节约种子,苗木生长较为整齐,且袋苗有利于提高造林成活率。

由表5和表6可见,2试验点采取两段式育苗共育的7个银合欢新品种,3个月出圃袋苗的苗高普为22.3~62.9cm,普遍大于鸡飞袋苗的苗高21.8~43.8cm,其中以K156品种出圃袋苗的苗高差别最大,普文比鸡飞高36.3cm;地径相差不大,普文在0.22~0.38cm,鸡飞在0.24~0.40cm。

表6 鸡飞试验点7个银合欢新品种两段式育苗的出圃苗木生长状况
Table 6 Quality of lifted seedlings of 7 varieties in Jifei

品种	苗高(cm)	地径(cm)
K156	26.6c	0.33c
K584	41.7a	0.40a
K376	21.8e	0.24d
KX2	26.4cd	0.36b
KX3	24.1d	0.37b
K636	39.9b	0.43a
K784	43.8a	0.32c

注:同列数值后标注字母表示0.05水平差异性。

不同品种间出圃苗木的苗高在2个育苗试验点的差异显著(普文 $P=0.012$,鸡飞 $P=0.015$),在普文,K156和K636品种的出圃苗高较高,为62.9cm和60.2cm,在鸡飞,K784、K584和K636品种的出圃苗高较高,分别为43.8cm、41.7cm和39.9cm。不同品种间出圃苗木的地径在2个育苗试验点差异显著(普文 $P=0.022$,鸡飞 $P=0.015$)。在普文,K156和K636品种的地径较粗,为0.37cm和0.38cm,在鸡飞,K584和K636品种的地径较粗,为

0.40cm 和 0.43cm。不同品种间在苗高与地径上的差异可能是品种固有的差异,也可能是对引种地的适应性差异导致的。

4.3 造林试验结果

2008 年 11 个引进银合欢新品种植苗造林 6 个月后的测定结果（表7）显示,普文试验点所有品种当年的植苗造林成活率比较高,达 95%～100%,鸡飞试验点的植苗造林成活率情况总体良好但稍低于普文试验点,为 85.5%～100%。造林当年,普文试验点引种的银合欢新品种幼树树高为 58.8～163.7cm,不含 KX4 品种,明显高于鸡飞试验林（18.1～24.8cm）;普文试验林地径（0.60～1.78cm）明显高于鸡飞试验林（0.24～0.45cm）。在引进的银合欢新品种中 KX4 为银合欢三倍体,植株生长特别迅速。在普文植苗造林当年树高达 6.54m,地径达 4.55cm,胸径达 4.15cm。造林后 6 个月 2 引种地引种品种间的幼树生长已呈现显著的差异（P<0.05）。普文 KX4、K636 和 K784 3 个品种的幼树生长较快,鸡飞 KX3 和 K584 品种的幼树相对较快。造林初期普文银合欢幼树相较鸡飞生长较快,可能是苗期差异在定植后的持续表现。

表7 2008 年普文和鸡飞 2 地定植 6 个月的各银合欢品种造林成活率及幼树的树高和地径生长量
Table 7 Survival rate, height and diameter of the plantations of tested varieties established in 2008

品种号	普文试验点			鸡飞试验点		
	成活率（%）	树高（cm）	地径（cm）	成活率（%）	树高（cm）	地径（cm）
K156	100	114.5b	1.10bc	91.3	20.3bc	0.33c
K584	100	69.7d	0.79d	100	24.7a	0.48a
K736	95	58.8e	0.61d	88.2	21.2bc	0.24d
KX2	100	125.7b	1.23b	85.7	18.1d	0.37b
KX3	96.6	102.1b	1.09bc	95.5	24.8a	0.38b
K636	100	110.1b	1.78a	85.5	17.3d	0.45a
K29	100	62.3de	0.60e	—	—	—
K565	100	74.5c	0.72c	—	—	—
K784	100	163.7a	1.46a	95.8	23.0b	0.33c
K608	100	68.5de	1.06c	—	—	—
KX4	100	654	4.55	—	—	—

注：同列数值后标注字母表示 0.05 水平差异性。

根据 2009 年 12 月的观测结果（表8、表9,表10 和表11）,普文、鸡飞、开远引种的银合欢新品种试验林保存基本良好。普文 2007 年试验林的保存率为 45%～100%,2008 年试验林的保存率为 80.52%～100%;鸡飞 2008 年试验林保存率为 78.65%～95%;开远 2008 年试验林保存率为 85%～92.44%。在普文 2007 营造的银合欢新品试验林中,林木生长表现较好的品种为:KX4（树高 12.55m,地径 10.25cm）,K784（树高 2.29m,地径 1.78cm）和 KX2（树高 2.31m,地径 2.06cm）;在普文 2008 营造的银合欢新品种试验林中,KX4（树高 9.45m,地径 8.15cm）、K156（树高 2.29m,地径 2.02cm）、K565（树高 2.05m,地径 1.65cm）

和 KX3（树高 1.95m，地径 1.59cm）等品种表现较好。在鸡飞 2008 营造的银合欢新品种试验林中，林木生长表现较好的品种为：K584（树高 1.82m，地径 2.54cm）和 KX3（树高 1.64m，地径 1.84cm）。总的看来，定植 1.5 年后，鸡飞试验点与普文试验点引种的相同品种植株生长差距（表 7）已不明显，且在地径生长和林木的长势上呈现出超过普文的趋势。初步看来，土壤类型及其 pH 值的差异是其林木生长产生变化的主要原因。鸡飞试验点红壤性冲积土（pH 值 6.0）较普文试验点的赤红壤（pH 值 4.5）适合银合欢新品种林木的生长，因此在热量和降水显著低于普文的情况下，引种的银合欢新品种生长良好。开远 2007 年试验林（树高 2.88~3.65m，地径 2.28~3.22cm）生长在总体上明显高于普文的同龄试验林。

表 8 普文 2007 年种植的银合欢新品种林林木的保存及生长结实情况

Table 8 Survival rate, growth and seeding traits of the plantation planted in 2007 in Puwen

品种号	保存率(%)	树高(m)	地径(cm)	结实情况
KX2	85.00	2.31a	2.06a	少量
K784	88.45	2.29a	1.78a	少量
KX3	75.44	1.97b	1.79a	少量
K608	92.00	1.80b	2.11a	无
K636	82.13	1.80b	1.72a	无
K584	45.00	1.08c	1.07b	无
KX4	100.00	12.55	10.25	无

注：测定时间为 2009 年 12 月，表 9、表 10、表 11 同；同列数值后字母表示 0.05 水平差异性，KX4 因株数少且明显高于其他没做差异检验。表 9 同。

2009 年鸡飞试验林中的 K584 和 KX3 品种，普文银合欢新品种试验林中的 K784、K156、KX2 和 KX3 和开远银合欢新品种试验林中的 KX2 和 K156 品种的植株已少量结实，预计结实量将在最近几年内有较大增加，可以满足推广种植的需要。

表 9 普文 2008 年种植的银合欢新品种林林木的保存及生长结实情况

Table 9 Survival rate growth and seeding traits of the plantation planted in 2008 in Puwen

品种号	保存率(%)	树高(m)	地径(cm)	结实情况
K156	95.44	2.29a	2.02a	少量
K565	83.56	2.05b	1.65b	无
K29	81.49	1.94b	1.21d	无
KX3	90.12	1.95bc	1.59bc	少量
KX2	82.45	1.55cd	1.26cd	少量
K72	90.50	1.30de	1.34bcd	无
K636	80.52	1.06e	1.04d	无
KX4	100.00	9.45	8.15	无

表 10 鸡飞 2008 年种植的银合欢新品种林林木的保存及生长结实情况
Table 10 Survival rate growth and seeding traits of the plantation planted in 2008 in Jifei

品种号	保存率(%)	树高(m)	地径(cm)	结实情况
K584	95.00	1.822a	2.54a	少量
KX3	92.53	1.639a	1.84ab	少量
K784	88.54	1.600a	1.33b	无
K156	88.00	1.460ab	1.55ab	无
K636	78.65	1.335ab	1.32b	无
KX2	82.69	1.037bc	0.91b	无
K376	82.97	0.734c	1.09b	无

注:同列数值后标注字母表示 0.05 水平差异性。

表 11 开远 2007 年种植的银合欢新品种林林木的保存及生长结实情况
Table 11 Survival rate growth and seeding traits of the plantation planted in 2007 in Kaiyuan

品种号	保存率(%)	树高(m)	地径(cm)	结实情况
KX2	85.00	3.25b	2.86b	少量
KX3	92.44	3.65a	3.22a	无
K636	90.00	2.95c	2.95ab	无
K156	91.45	2.88c	2.28c	少量

5 结语

(1)普文、鸡飞和开远三地的银合欢新品种引种试验的初步结果显示,从夏威夷大学引进的 12 个银合欢品种在云南的亚热带和热带北缘地区都能利用其种子成功育苗。采用热水浸种处理可有效提高其种子发芽率,用两段式营养袋育苗的效果比穴盘直播育苗好。

(2)在 3 个试验点引种的银合欢新品种试验林中,相同品种开远点的表现比鸡飞和普文好,鸡飞试验林也呈现出超越普文的势头,证实所引种的银合欢新品种是喜中性和微碱性土壤,土壤的 pH 值应该是选择银合欢新品种推广种植地的重要指标。

(3)海拔 1 500m 的云南鸡飞国有林场引种的 7 个银合欢新品种生长均为良好,多数品种的林木在 2008 年低温寒害中表现出良好的抗寒性,特别是 L. diversifolia K156 和 L. diversifolia K784 两品种有望引种到海拔更高的山区。

(4)所引进的银合欢新品种中具三倍体的 KX4 生长量特别突出,但需要扦插才能繁殖,可以在立地条件较好的地方栽培。KX3、K584、K156 和 K784 品种均表现较好,适合按用材林的培育方向培育。KX2、K636 等品种则可进行密植丰产饲料栽培。

(5)引种试验林遭遇了 2009~2010 年的特大干旱,3 个引种试验点的银合欢品种的林木生长均受到一定影响,但生长基本正常,死亡很少,而与相邻地区西南桦、旱冬瓜、云南松等新造林地的林木却大面积死亡形成强烈的对比,表明引种的银合欢新品种有极强的抗旱能力。

(6)普文、鸡飞和开远 3 个引种试验点引种的银合欢新品种试验林都没有出现明显的病虫害，表明引进的12个银合欢新品种具有较强的抗病虫害能力。

（7）在云南省3地进行的美国银合欢新品种引种试验结果表明，12个引进品种都可以在云南的热带和亚热带地区正常生长，尤其在碱性和微酸性土壤上生长良好。但因本次引种试验开展的时间较短、地点不多，还需作长期及更多地方的引种试验，才能明确这12个银合欢新品种在云南的推广种植适宜地区。

参考文献：

[1] Sorensson, T. S. and Brew baker, J. L. Inter specific compatibility among 15 *Leucaena* specific- (Leguminosae: Mi-mosoideae) via artificial hybridizations[J]. American Journal of Botany, 1994, 81 (2): 240~247.

[2] Shi, X. B. and Brew baker, J. L. Vegetative propagation of *Leucaena* hybrids by cuttings[J]. Agroforestry Systems, 2006, 66: 77~83.

[3] Brew baker, J. L. and Sorensson C. T. New tree cropsfrom interspecific *Leucaena* hybrids. In Janick, J. and Simon, J. E. "Advances in newcrops"[M]. Portland: Timber Press, 1990.

[4] 刘海刚, 李江, 段日汤, 等. 银合欢扦插繁殖研究[J]. 山东林业科技, 2009, 5: 63~65.

[5] 刘海刚, 李江, 李桐森. 杂交银合欢的繁殖技术[J]. 安徽林业科技, 2008, 3: 17~18.

[6] 云南省林业科学院. 热区造林树种研究论文集[C]. 昆明: 云南科学技术出版社, 1996.

（本文发表于《西部林业科学》，2010年）

施肥对北美红杉幼林生物量、P素含量及贮量的影响

彭明俊[1],左显东[1],汪政初[2],陶国祥[2]

(1. 云南省林业科学院,云南 昆明 650204;2. 屏边县林科所,云南 屏边 661200)

摘要:2000~2002年在云南省红河州屏边县对北美红杉幼林进行了不同肥料品种(尿素、过磷酸钙、氯化钾)的施肥试验,共进行了8个处理,3次重复的随机区组设计。结果表明:施肥使北美红杉幼林生物量增加,其中处理7(P_2K_2)能明显增加北美红杉幼林各器官及林木总的生物量;施肥能促进根和叶的生长,增加根系对P的吸收,提高北美红杉幼林叶和根部P的含量和贮量;叶和根部成为北美红杉P素的主要贮存库。

关键词:北美红杉;幼林;施肥;生物量;P贮量

Effects of Fertilizations on Biomass, P Content and Storage in Young plantation of *Sequoia sempervirens*

Peng Ming-jun[1], Zuo Xian-dong[1], Wang Zheng-chu[2], Tao Guo-xiang[2]

(1. Yunnan Academy of Forestry, Kunming Yunnan 650204; 2. Pingbian Graduate school of Forestry, Pingbian Yunnan 661200)

Abstract: During 2000~2002, the fertilizations tests, including three kinds of fertilizations (urea, super phosphate, potassium chloride) in Pingbian, Yunnan Province were conducted. The tests include eight treatments and three replicates. The results revealed that fertilizations could increase the biomass of different components of the tree, especially the proportioned application of super phosphate and potassium chloride (the seventh treatment) could increase the amount of leaf and root as well as the total biomass of the three remarkably. Fertilizations could also increase the P content and storage in leaf and root. And leaf and root is the major P pool of *Sequoia sempervirens*.

Key words: *Sequoia sempervirens*, young plantation, fertilizations, biomass, P storage

北美红杉(*Sequoia sempervirens* Endll.)简称红杉。原产北美洲美国的太平洋沿岸,即从俄勒冈州到加利福尼亚州海拔24~945m的地带,是世界著名的速生用材树种。国外对北美红杉的研究较早,法国、美国、新西兰等对北美红杉的选育种进行了深入研究,法国对北美红杉在全分布区内的成活率和高生长进行了观测,也作了一些生理研究、生态因子观测[1~3]。我国对北美红杉的研究较晚,主要工作集中在引种、育苗及生态等方面[4~8]。

近年来,由于不合理的林业经营措施,引起林地地力衰退,林木生产质量下降,这已为许多研究者所指出[9,10]。为了维护和提高土壤肥力,促进林木生长,达到速生、丰产、优质的目的,林木施肥是一项极为重要的措施[9]。目前我国经济发展迅速,但森林资源贫乏,在一些地区对一些树种已进入到全株利用的阶段。研究不同施肥措施对林木生物量及养分元素的积累与分配的影响,对于指导北美红杉幼林林木施肥,提高投入养分的循环利用效率具有重要意义。

1 试验地概况

试验地位于云南省红河州屏边县,22°59′N,103°41′E,海拔1 330m,年平均气温16.4℃,绝对最高气温31.1℃,绝对最低气温-1.7℃,年均降水量1 653.4mm,年均蒸发量1 239.6mm,年平均相对湿度84%,1月份平均相对湿度89%,8月份平均相对湿度87%,霜日4d,雾日年平均109.4d,全年日照时数1 556.4h,属南亚热带季风气候。供试土壤系千枚状板岩发育的黄壤,土壤pH值5.2~5.6,有机质含量2.5%,速效N、P、K分别为43.3~83.7、0.22~2.72、21.40~115.24mg/kg。土壤P水平低。

2 研究方法

2.1 试验设计

选择土壤肥力中等立地,但土壤水分条件,土层厚度,土层通透性较正常,坡面比较完整,布置了施肥试验。试验共设3次重复,8个处理(见表1),按随机区组排列。每小区36株树为量测株,按4列9行排列,各处理间设2列保护行,供试区外围设2~3行作为保护行,郁闭前在每年4月底、9月底抚育除草2次,松土深度10cm,并及时防治病虫害。

表1 北美红杉施肥试验各处理施肥时间和配比量(kg·m^{-2})

Table 1 Times and assorted proportion of different fertilization treatments

处理	施肥量(N—P$_2$O$_5$—K$_2$O)		
	2000年	2001年	合计施肥量
1(P$_1$)	0—22.2—0	0—22.2—0	0—44.4—0
2(P$_2$)	0—44.4—0	0—44.4—0	0—88.8—0
3(P$_3$)	0—88.8—0	0—88.8—0	0—177.6—0
4(NP$_1$)	44.4—22.2—0	44.4—22.2—0	88.8—44.4—0
5(NP$_2$)	44.4—44.4—0	44.4—44.4—0	88.8—88.8—0
6(P$_1$K$_2$)	0—22.2—44.4	0—22.2—44.4	0—44.4—88.8
7(P$_2$K$_1$)	0—44.4—22.2	0—44.4—22.2	0—88.8—44.4
8(CK)	0—0—0	0—0—0	0—22.2—0

注:N肥—尿素,有效成分按46%N计;P肥—过磷酸钙,有效成分按14%P$_2$O$_5$计;K肥—氯化钾,有效成分按56%K$_2$O计。

2.2 施肥方法

在2000年3月对供试区整地,按60cm×60cm×60cm规格挖好树穴,在2000年6月造林时每穴中施入所需肥料,然后回表土混匀,进行造林,株行距3m×3m,1 110株·hm^{-2}。造林后第2年(2001年)进行抚育除草,然后在每木上坡、树冠投影外侧,挖一弧形沟,沟长60cm,宽15cm,深20cm,肥料施于沟中,随后立即混匀覆土(施肥量及时间见表1)。

2.3 生长调查及标准木生物量测定

本次试验所用材料是1999年3月从中国林木种子公司购买,编号为98C147的种子。于1999年3月进行种子育苗。育苗时种子通过检测其结果为:净度92.00%,含水率16.50%,千粒重5.87g,生活率87.00%。

进行施肥试验时对各小区(处理)量测株定株进行编号,施肥前(2000年6月)调查各小区本底苗高和地径,以后每年1、4、7、10月调查各小区苗高、地径,2003年8月根据各处理的平均苗高和平均地径分别选定3株标准木,齐地表伐倒,用分层切割法区分叶、枝、干等部分,测定各器官的鲜重,根据单株立木的营养面积,用壕沟法挖掘根系,测量其鲜重,同时采集各器官样品测定其干重及其P含量[11]。

2.4 样品分析方法

采集的植物样品经烘干,磨碎,过筛后,用浓H_2SO_4-$HClO_4$消化制备待测样品,用双酸消化—钼锑抗比色法测定P含量。

3 结果和分析

3.1 不同施肥对北美红杉幼林生物量的影响

从表2可以看出,各施肥处理的北美红杉幼林生物量与CK相比都有不同幅度的增加。如施肥后第三年沟施过磷酸钙22.2kg·hm^{-2}(处理1)较CK增加了29.83%,沟施过磷酸钙44.4kg·hm^{-2}(处理2)较CK增加了47.81%,沟施过磷酸钙88.8kg·hm^{-2}(处理3)较CK增加了30.18%,沟施尿素44.4kg·hm^{-2}和过磷酸钙22.2kg·hm^{-2}(处理4)较CK增加了53.44%,沟施尿素44.4kg·hm^{-2}和过磷酸钙44.4kg·hm^{-2}(处理5)较CK增加了58.10%,沟施过磷酸钙22.2kg·hm^{-2}和氯化钾44.4 kg·hm^{-2}(处理6)较CK增加了54.13%,沟施过磷酸钙44.4kg·hm^{-2}和氯化钾22.2kg·hm^{-2}(处理7)较CK增加了85.21%。生物量方差分析结果表明:施肥后对北美红杉幼林生物总量、苗干、根呈极显著差异。经多重比较表明(表3):处理7对北美红杉幼林的根、干、枝、叶、生物总量分别与其他处理相比呈显著或极显著水平。由此可见,处理7(P、K配合施肥)能有效促进北美红杉幼林各器官及林木总的生物量的增加。

表2 施肥第三年各处理标准木生物量(g)

Table 2 Biomass of standard tree in the third year of different fertilization treatment

处理	干	枝	叶	根	总量	%
1	13.89	14.89	12.37	14.12	56.27	129.83
2	13.42	13.38	19.37	15.89	64.06	147.81
3	12.19	9.38	11.83	20.02	56.42	130.18
4	14.30	10.24	24.96	13.00	66.50	153.44
5	15.59	13.94	15.53	18.46	68.52	158.10
6	15.96	17.35	12.85	14.64	66.80	154.13
7	19.70	13.77	27.16	12.64	80.27	185.21
8	11.27	8.16	6.28	9.63	43.34	100

表3 施肥第三年不同处理对北美红杉幼林生物量的多重比较
Table 3 Multiple comparisons of biomass in the third year of different fertilization treatment

干	枝	叶	根	总量	$F_{0.05}$	$F_{0.01}$
[7-8]=8.43**	[6-8]=9.19*	[7-8]=20.88**	[3-8]=10.39*	[7-8]=36.93**		
[7-3]=7.51**	[6-3]=7.97*	[7-3]=15.33*		[7-1]=24.00**		
[7-2]=6.28**	[6-4]=7.11*	[7-1]=14.79*		[7-3]=23.85**		
[7-1]=5.81**		[7-6]=14.31*		[7-2]=16.21**		
[7-4]=5.40**		[4-8]=18.68**		[7-4]=13.77**		
[7-5]=4.11**				[7-6]=13.47**		
[7-6]=3.74**				[7-5]=11.75*		
[6-8]=4.69**				[5-8]=25.18**		
[6-3]=3.77**				[5-1]=12.25*		
[6-2]=2.54**				[5-3]=12.10*		
[6-1]=2.07*				[6-8]=23.40**		
[5-8]=4.32**				[6-1]=10.53*		
[5-3]=3.40**				[6-3]=10.38*		
[5-2]=2.17*				[4-8]=23.16**		
[4-8]=3.03**				[4-1]=10.23*		
[4-3]=2.11*				[4-3]=10.08*		
[1-8]=2.62**				[2-8]=20.72**		
[2-8]=2.15*				[2-1]=7.79*		
				[2-3]=7.64*		
				[3-8]=13.08**		
				[1-8]=12.93**		

注:*表示差异显著;**表示差异极显著;[]中数字为各处理号。

3.2 不同施肥北美红杉幼林体内P含量和贮量的变化

由表4可见,北美红杉幼林各器官P的含量以叶、枝较高,根、干较低。在树冠下部(0~30cm),呈现出高低顺序为:叶、枝、干(处理3例外);在树冠上部(≥30cm)高低顺序为:叶、枝、干(处理2、3例外);相同器官,树冠上部高于下部。与对照(CK)相比,施肥后干和根等部位P的含量有所增加,而枝、叶等却大幅下降。说明施肥能增加根系对P素的吸收,提高根和干中P的含量。这一结论与俞元春及Adams[12~14]等的结论不一致。Adams认为,施肥后林木的生物量增加,但P的含量并不发生"稀释效应"。这种现象是否与北美红杉的P素营养生理有关,有待于进一步的研究。施肥处理北美红杉幼木体内各器官P的贮量,按不同区分段计算一般高于对照(处理8),但也有例外(见表4),如处理1、2、3、4、7(≥30cm)干的贮量低于对照;处理2、3、7(≥30cm)枝的贮量低于对照;处理6(0~30cm)干的贮量低于对照;处理4(0~30cm)枝的贮量低于对照;处理5(0~30cm)叶的贮量低于对照。但如按整株树计算,施肥各处理各器官P的贮量均高于对照(表5)。在各施肥处理中北美红杉体内P的贮量以处理4最高。这表明N、P肥配合施用能增加北美红杉对P的有效吸收,提高体内P的贮量。在林木体内的P贮量中,有约60%贮存在北美红杉的根、叶中(表5)。这说明:根、叶是P的主要贮存库,其次是干。表5表明:单施P肥的各处理叶的P贮量高于其他各部分,这说明单施P肥能促进北美红杉对P的吸收,且叶成为P的主要贮存库。但N、P配合施肥或P、K配合施肥的结果表明,有38%~54%的P贮存在根中,根又成为了P的主要贮存库。这一结果表明,随着施肥种类的不同而引起了北美红杉体内P的贮存发生了转移,这种现象是否与北美红杉N、P、K的营养生理有关,目前还不清楚,有待于进一步研究。

表4 施肥第三年各处理北美红杉各部位P含量(g·株$^{-1}$)

Table 4　P content of different parts of *Sequoia sempervirens* (g·tree^{-1})

高度(cm)	地表以下		0~30cm			≥30cm		
器官	主根	侧根	干	枝	叶	干	枝	叶
处理1	0.285 1	0.378 6	0.329 4			0.310 6	0.573 5	0.761 8
处理2	0.425 6	0.479 6	0.383 0	0.525 8	0.620 0	0.524 2	0.622 2	0.618 2
处理3	0.499 0	0.514 7	0.427 6	0.764 2	0.698 4	0.779 8	1.076 4	0.762 8
处理4	0.459 0	0.502 1	0.373 6	0.573 0		0.479 3	0.706 3	0.767 0
处理5	0.251 6	0.427 3	0.366 6	0.656 1	0.702 0	0.572 3	0.627 9	0.914 2
处理6	0.324 6	0.526 4	0.231 0	0.831 8		0.346 6	0.756 8	1.011 9
处理7	0.293 8	0.312 6	0.307 3	0.426 9	0.687 9	0.284 4	0.529 0	0.748 1
处理8	0.358 7	0.448 1	0.347 5	0.918 6	1.036 6	0.498 7	0.943 8	1.100 8

表5 施肥第三年各处理北美红杉各部位P贮量(g·株$^{-1}$)

Table 5　P storage of different parts of *Sequoia sempervirens* (g·tree^{-1})

	器官	叶	枝	根	干
处理1	贮量	0.009 4	0.007 6	0.004 6	0.004 5
	所占比例(%)	36.143 9	29.277 6	17.477 3	17.101 3
处理2	贮量	0.012 0	0.007 6	0.007 1	0.005 5
	所占比例(%)	22.147 9	16.997 4	37.249 7	23.605 0
处理3	贮量	0.008 7	0.008 3	0.010 1	0.005 8
	所占比例(%)	30.834 2	17.557 0	26.420 4	25.188 5
处理4	贮量	0.017 5	0.006 9	0.006 2	0.005 6
	所占比例(%)	17.132 7	15.595 9	48.225 7	19.045 8
处理5	贮量	0.014 0	0.007 4	0.005 6	0.006 8
	所占比例(%)	16.629 0	20.139 0	41.404 4	21.827 6
处理6	贮量	0.013 0	0.010 6	0.005 6	0.004 2
	所占比例(%)	16.747 1	12.627 8	38.902 0	31.723 1
处理7	贮量	0.019 1	0.006 3	0.003 9	0.006 0
	所占比例(%)	10.921 8	16.981 4	54.251 3	17.845 5
处理8	贮量	0.006 8	0.006 6	0.003 9	0.004 4
	所占比例(%)	17.903 5	20.332 6	31.394 6	30.369 3

4 结论

(1)施肥能促进北美红杉幼林各营养器官生物量的增长。其中以 P_2K_2 配合施用对北美红杉幼林各器官及林木部的生物量的影响最明显。施肥后第三年各处理北美红杉生物量差异显著,这将会给以后的幼林生长带来明显的促进作用。

(2)施肥能提高北美红杉幼林叶和根部 P 的含量的贮量。施肥各处理北美红杉各器官 P 的贮量一般都高于对照。在各施肥处理中,北美红杉体内的 P 贮量以处理 4 最高,即 N、P 肥配合施用能提高北美红杉幼林对 P 的吸收,增加体内 P 的贮量,特别是叶和根部成为北美红杉 P 素的主要贮存库。

参考文献:

[1] Hauxwell D H, S P B ulkin, C A Hanson. The influence of elevatation, vegetation type, and distance from the coast on soil temperature in Humboldt County, California[J]. Agronomy Abstracts, 1981:198.

[2] HELLMERS, H. Effects of soil and air temperatures on growth of redwood seeding[J]. Botanical Gazette, 1963 (124):172~177.

[3] ANEKONDA T S, R C CRIDDLE, W J LIBBY. Calorimetric evidence for site-adapted biosynthetic metabolism in coast redwood(Sequoia sempervirens)[J]. Canadian Journal of Forest Research, 1994(24):380~389.

[4] 王志清. 杉、红杉、日本扁柏引种试验初报[J]. 浙江林业科技, 1993, 13(1):44~46.

[5] 左显东,祁荣频,王懿祥. 等北美红杉在我国的引种及其生态适应性[J]. 云南林业科技, 2000(4):36~39.

[6] 张茂钦,刘世玉,李秀珍. 红杉引种及无性繁殖初步试验[J]. 云南林业科技, 1979(4):40~43.

[7] 邢章美. 北美红杉生物学特性的观察[J]. 浙江林业科技, 1990, 10(4):43~47.

[8] 诸葛强. N、P、K 对若干种木本植物离体培养繁殖的影响[J]. 林业科学研究, 1990, (1):41~46.

[9] 盛炜彤. 我国人工林的地力衰退及防治对策. 见:盛炜彤主编. 人工林地衰退研究[M]. 北京:中国科学技术出版社, 1992, 15~19.

[10] 李贻铨. 林木施肥是短期轮伐期工业用材林的基础技术措施. 见:盛炜彤主编. 人工林地衰退研究[M]. 北京:中国科学技术出版社, 1992, 43~45.

[11] 叶镜中,姜志林. 苏南丘陵杉木人工林的生物量结构[J]. 生态学报, 1983, 3(1):6~14.

[12] Adams M B, Campbell R G, Allen H L, et al. Root and foliar nutrient concentrations in loblolly pine: effects of season, site, and fertilization. For. Sci. , 1987, 33(4):984~996.

[13] 宇万太,陈欣,张璐,等. 不同施肥杨树主要营养元素内外循环比较研究[J]. 应用生态学报, 1995, 6(4):341~348.

[14] 俞元春,陈金林,曾曙才,等. 施肥对杉木幼林生物量和 P 素含量及贮量的影响[J]. 林业科学研究, 1996, 9(专刊):75~80.

(本文发表于《南京林业大学学报(自然科学版)》,2008 年)

思茅松林采伐迹地清理及其采伐剩余物的利用

孟梦,胡光辉,韩明跃,闫争亮,刘云彩,槐可跃,冯志伟

(云南省林业科学院,云南 昆明 650204)

摘要:对思茅松林区采伐剩余物的分类,相应的清理方法和存在问题进行了研究。提出在充分利用其采伐剩余物资源的前提下,对遗留于迹地上的其他剩余物应倡导自然腐烂法清理,减少烧除法清理。针对思茅松林采伐剩余物的利用途径,分析了思茅松林采伐剩余物的利用现状及存在的问题,从转变观念;加大政策上的引导和扶持入手;以及提高其采伐剩余物加工产品的技术含量;加强期采伐剩余物的管理等4方面提出了思茅松林采伐剩余物利用的改善措施。

关键词:思茅松;采伐迹地;采伐剩余物;清理;利用

思茅松(*Pinus kesiya* var. *langbianensis*)是云南省重要的速生用材及采脂树种。主要集中分布于云南省的普洱市,思茅松林是该市面积最大的森林。中华人民共和国成立以来,普洱林区作为全国重要的木材供给地之一,经对思茅松林的长期超量采伐,使可利用的思茅松天然林数量骤减,所存的思茅松天然林质量也差。另一方面,近年来大规模营造的思茅松人工林其后续资源培育时间还较短,10多年种植的林分基本未到主伐年龄,由此可见,现阶段思茅松林可利用的资源量较少。随着天然林保护工程的实施,思茅松作为云南热区当家的针叶速生用材树种,其在云南林业及社会经济中的重要作用将越来越突出。在上述背景情况下,对思茅松林采伐迹地进行科学清理,提高其采伐剩余物的利用率,缓解木材资源的短缺,将具有重大的现实意义。

1 思茅松林采伐迹地的清理

在思茅松林采伐后随即开展的是人工造林活动,因此对思茅松林采伐迹地的清理也就是造林地的清理。其清理工作主要是将采伐迹地的采伐剩余物枝丫、梢头、站杆、倒木等清除掉。采伐迹地清理工作的主要目的在于为迹地的森林更新创造条件。清理采伐迹地,有利于后续人工林的更新作业,有利于迹地卫生条件的改善,能有效阻止病虫害的传播漫延以及火灾的发生等。

1.1 可利用剩余物的清理

在清理过程中,对于有利用价值的思茅松林采伐剩余物一定要挑选出来加以利用,以提高其森林资源的利用率。如直径在2cm以上的枝丫、梢头等思茅松林采伐剩余物,粉碎后可生产细木工板;直径4cm以上的枝丫、倒木、梢头等,可加工成木片,用于制浆、造纸的原料,也可作为生产人造板的原料;对思茅松林采伐剩余物中长≥0.5m、直径≥6cm的短粗木、砸伤木、造材截头、站杆、伐区丢弃材等可进行小木加工,用来生产包装箱板、毛地板块、家具用原料块、短装修原条、原块等。针对以上不同规格的思茅松林采伐剩余物,可根据不

同的利用途径分别进行归堆,然后装车运出,至加工厂进行加工利用。

对于坡度较大的思茅松林采伐迹地中,还可利用枝丫等采伐剩余物沿等高线堆砌成简易的挡水坝,能有效降低采伐迹地即新造林地的水土流失。

1.2 其他采伐剩余物的清理

将可利用的采伐剩余物清理运出后,对思茅松林采伐迹地上的其他剩余采伐剩余物的清理方法目前主要有以下2种:

(1)烧除法。就是将此类思茅松林的采伐剩余物归成堆,或打散锯小后呈片状散铺于迹地地表,然后点火焚烧。该方法为思茅松林采伐迹地现阶段清理采伐剩余物主要使用方法。这种清理法能较彻底地改善造林地环境状况,清除非目的造林树种,还能增加林地土壤的速效养分,消灭附着在采伐剩余物或林地上的有害真菌、细菌和害虫(卵、蛹、幼虫和成虫),对新造的林木生长有利,同时省工省时。但由于思茅松林区基本都处于云贵高原的横断山地,山多坡陡,采伐迹地经火烧清理后易引起水土流失,同时亦造成林地生物多样性降低,因此此法在使用的时间、方式等方面都要根据采伐迹地的条件作适当安排,采取严密规范措施。

(2)自然腐烂法。思茅松林主要生于云南热带北缘及南亚热带地区,水、热条件较好,林内木腐菌种类丰富,可以利用这一森林生态系统中的微生物种群对其采伐剩余物进行腐解,以达到营养物质快速还地,维持地力的目的,以此保护迹地的生物多样性和生态环境,促进林地的可持续利用。

在思茅松林采伐迹地的清理过程中,将其采伐剩余物截成小段,归成长、宽、高规格约为 $0.5m \times 0.5m \times 0.2m$ 的小堆,堆放于今后所要开挖的两植苗穴间,任其自然腐烂。对于采伐剩余物多的采伐迹地,可把采伐剩余物按宽 $0.5 \sim 1.0m$、高 $0.5m$ 左右,沿等高线堆成带状进行带腐,其带间距可根据采伐剩余物的多少而定,一般 $5 \sim 10m$。带腐法因将采伐剩余物沿等高线堆放,能有效缓解水土流失。在有条件的采伐迹地,可利用机械将采伐剩余物打成碎片后铺散在迹地上,以加速其的腐烂;还可人为向采伐剩余物上喷洒经过筛选培养过的木腐菌菌剂,促进其分解。

对思茅松林采伐迹地采伐剩余物的清理方法应根据采伐迹地的具体情况区别对待。上述2种清理方法可根据采伐迹地实际情况单独使用或交叉结合应用。基于从采伐迹地可持续利用、生物多样性保护及森林环境的安全性出发,对于现阶段广泛使用的烧除法,应减少其使用的面积。特别是对坡度在15°以上的采伐迹地,经烧除法清理后迹地的地表植被亦被烧除,加之思茅松林分布区雨季的降水量较大且较集中,思茅松林采伐迹地在采伐当年乃至第二年的雨季期间,会造成林地(迹地)大面积的水土流失。因此,更宜使用自然腐烂法进行其他采伐剩余物的清理。

而对于坡度较大或较为贫瘠的采伐迹地,应尽可能地保留采伐剩余物,待其分解后可补充林地的营养元素,维持林地的生产力。对思茅松成过熟林采伐迹地的采伐剩余物,可因其带有较多的病虫害,因此在将有利用价值的采伐剩余物运出,而对其他的采伐剩余物则采用烧除法清理掉。

2 采伐剩余物的利用途径

通过对思茅松林采伐迹地留下的枝丫、梢头、倒木、伐根等采伐剩余物开发利用,可以生产出用于经济建设和人们生活的木质产品,同时对生态环境和社会也非常有益,其主要表现在:①提高森林资源的利用率,充分节约木材;②增加企业经济效益;③创造巨大的社会效益;④改善林地状况,促进森林更新。

在现阶段,思茅松林采伐剩余物的利用途径主要有以下几个方面:

(1)削片。随着木材生产工艺的改进,对直径4cm以上的枝丫、倒木、梢头、伐根等思茅松林采伐剩余物都可利用而加工成片。这大大提高了其森林资源的利用率,同时也改善了采伐迹地卫生状况。其采伐剩余物加工后的削片,可作为制浆、造纸的原料,也可作为生产人造板的原料,因而近年来普洱市建立了许多利用思茅松林采伐剩余物的削片加工厂。

(2)生产细木工板。用思茅松林采伐剩余物的枝丫生产细木工板(芯板)又是另一个重要的利用途径。一般直径在2cm以上的枝丫均可利用,且枝丫不用削皮,这样对其采伐剩余物的利用更为充分。枝丫粉碎后的原料价格每吨200~250元,仅此一项每年普洱林区可创造上亿元的财富。

(3)小木加工。对思茅松林采伐剩余物中的长≥0.5m、直径≥6cm的短粗木,如砸伤木、造材截头、站杆、伐区丢弃材等都可进行小木加工,可用来生产包装箱板、毛地板块、家具用原料块、短装修原条、原块等。

(4)种植茯苓。茯苓(*Poria cocos*)为寄生在松树根上的菌类植物,其菌体形状甘薯,外皮黑褐色,里面白色或粉红色,为优良的药材。近年来,普洱市茯苓种植规模不断扩大,专门有个人或企业在思茅松林的采伐迹地上承包种植茯苓。此举一方面利用了松树伐根,创造经济价值,以每棵思茅松树伐桩收入 6~13 元,一个农户种植茯苓 13.33hm², 以 300 个伐桩·hm^{-2}计,每年可为农户增加 2.4万~5.2万元的经济收入。同时种植茯苓还加速了伐根的腐烂,可改善迹地的卫生状况。

(5)烧柴。普洱思茅松林采伐区周边的村民可捡拾少量的思茅松林采伐剩余物用作烧柴,以解决民众的部分燃料需求。

3 采伐剩余物利用的现状及存在问题

近年来,普洱市紧紧把握国家实施西部大开发的战略机遇、在林纸一体化和云南省建设绿色经济强省的战略方针引领下,走"林纸为龙头,林板、林化为两翼,森林资源培育为基础,林下资源开发为补充",生态建设与产业发展相互协调、相互促进的可持续发展路子。通过进一步调整和优化产业结构,扶持了一批重点企业,使得国有、集体、民营各种林业产业群体不断发展壮大。至2006年年底,全市共有448个林业企业,其中锯材业211个,胶合板业14个,刨花板业4个,中纤板业11个,地板条业8个。"十五"期间,普洱市累计生产商品木材360.1万m³、锯材78.9万m³、人造板156万m³、火柴15万件、纸浆39.1万t,实现林业总产值121.5亿元,林产工业产值62.5亿元,林业上缴的税利9.2亿元,分别占全市工农业总产值、工业总产值和地方财政收入的三分之一。

在这一大的发展背景下,普洱市思茅松林伐区采伐迹地采伐剩余物的生产和加工利用范围也正逐步扩大,有很大的发展前景。但就现在而言,加工利用还存在许多问题。

(1)缺少相关扶持政策。目前,在国家和地方层面上对于森林采伐剩余物利用的相关政策法规较少,缺乏相关的技术规程和具体的操作指南。这就使得在思茅松林采伐剩余物利用上存在注重经济利益,而在其环境和社会效应上考虑较少。使得在有利可图的情况下就加大其的采集利用力度,而无利可图时就随意丢弃,造成思茅松林采伐剩余物的极大浪费,还对环境造成负面影响,如采伐迹地病虫害的滋生、漫延、火灾的发生等。此外,对思茅松林采伐剩余物的资源量、计算方法和具体利用情况也没有规范的统计和方法。这无疑会对思茅松林采伐剩余物的有效利用产生一定的影响。

(2)搜集困难,利用率低。思茅松林区多为山区,其的许多采伐迹地山高坡陡,道路状况较差,且枝丫等剩余物的粗长不一,形态各异,散布在采伐迹地上,为收集运输工作带来很大困难。其加工利用的生产成本提高,使得许多企业放弃使用,因而导致思茅松林采伐剩余物资源利用率低下。

(3)技术落后,深加工程度低。目前思茅松林采伐剩余物的加工利用仍处于初级原材料生产阶段,技术含量低,大多加工企业都只能进行较简单的削片、初条和毛木块的生产,产品附加值较低,且产品单一、生产规模较小,缺乏竞争力。

4 改善采伐剩余物利用的措施

针对思茅松林采伐剩余物利用现状和存在的诸多问题,必须从多方入手,逐步改变现有情况,挖掘其采伐剩余物的利用潜力,缓解木材的供需矛盾,以繁荣地方经济。

(1)转变观念。目前,人们普遍认为森林采伐剩余物是林木砍伐后没有更大用途的物质,往往把森林采伐剩余物的利用同用作烧柴联系起来,即在思想上还没有认识到森林采伐剩余物的利用同木材利用具有同等重要的意义。因此,要强化思茅松林采伐剩余物的利用,首先应加大相关宣传力度,使民众明确有效利用森林采伐剩余物的意义和重要性;其次是树立对森林采伐剩余物进行深加工和综合利用的观念,在思想上为科学合理地利用思茅松林采伐剩余物奠定基础。

(2)加大政策上的引导和扶持。加强对思茅松林采伐剩余物加工利用的引导和扶持,通过在计划、资金、项目安排等方面加大政策的引导和扶持力度,为思茅松林采伐剩余物的更有效地利用创造宽松环境。例如,增加银行对思茅松林采伐剩余物利用项目的贷款,并适当降低贷款利息;制定相关的优惠政策,吸引民间资本投资于思茅松林采伐剩余物的技术开发,以鼓励人们对思茅松林采伐剩余物进行更广泛的加工和综合利用。

(3)提高产品的技术含量。加大科技投入,提高产品的科技含量是提高思茅松林采伐剩余物利用的一项关键措施。随着木材采伐量的即伐区的减少,森林的各种采伐剩余物数量自然也会减少,所以要想提高思茅松林采伐剩余物的利用率,主要取决于其采伐剩余物加工生产的技术含量的高低。可以通过积极开展相关研究项目以求获取高附加值的加工利用技术,应尽量限制或禁止原木或锯材等初级产品的生产。同时对木材工业实行结构调整,致力于发展装饰型材、家具、细木工部件等高附加值产品,开发能生产思茅松林采伐剩余物这些高附加值产品的加工技术,从有限的资源中获得更多的利润。通过对思茅松林采伐剩余物再加工技术的提高,使其采伐剩余物的利用达到集约化和规模化。

(4)加强对采伐剩余物的管理。首先要建立和健全思茅松林采伐剩余物数量及其利用的统计程序和方法。这是一项非常重要的基础工作。普洱市林业局应尽快摸清思茅松伐区采伐剩余物的数量和利用情况,各县林业局应成立专门的管理部门,负责思茅松采伐剩余物的搜集和管理。管理部门要制定合理的思茅松林采伐剩余物枝丫等合理收购价格,以调动生产者的积极性;在条件允许的情况下,在伐区就近设立加工场地,将思茅松采伐剩余物在伐区就近加工成片,以降低其加工成本。此外,对人员进行技术培训;招商引入高效率的思茅松林采伐剩余物加工企业,以提高思茅松林采伐剩余物的综合利用效率。

参考文献:

[1] 云南省林业科学研究所. 云南主要树种造林技术[M]. 昆明:云南人民出版社,1985.13~16.
[2] 王达明,李莲芳. 思茅松速生丰产林培育的关键技术[J]. 云南林业科技,1999(4):6~8.
[3] 孟梦,陈宏伟,王达明. 优良的纸浆材树种——思茅松[J]. 西南造纸,2004(3):13~15.
[4] 姜远标,王发忠,赵文书,等. 思茅松种源/家系造林试验初报[J]. 西部林业科学,2007,36(2):110~113.
[5] 胡光辉,杨斌,姜远标,等. 不同地理种源思茅松对松实小卷蛾抗性调查[J]. 云南林业科技,2003(4):67~71.
[6] 胡光辉,雷玮,槐可跃,等. 松实小卷蛾在云南生活史及其对思茅松的危害[J]. 中国森林病虫,2005,24(2):13~15.
[7] 付国君,杨贵海,王文凯,等. 试论伐区清理方法和剩余物利用途径[J]. 防护林科技,2003(1):47~48.
[8] 夏景涛,姚贵宝,庞传洪. 伐区剩余物的生产与综合利用[J]. 森林工程,2001(5):18~19.
[9] 邓泽,唐学明,张雪梅. 思茅松伐桩不同接种位置对茯苓产量的影响[J]. 西部林业科学,2008(增刊):75~77.
[10] 冯志伟,胡光辉,韩明跃,等. 思茅松林采伐迹地采伐剩余物的自然及接种白腐菌腐解效果对比试验[J]. 部林业科学,2009,38(2):67~70.
[11] 万志芳,王飞,李明. 林区森林采伐剩余物利用状况分析[J]. 中国林业经济,2007(7):17~19.

(本文发表于《西部林业科学》,2009 年)

松属8个树种的引种研究

郑畹[1], 舒筱武[1], 蔡雨新[1], 李思广[1], 黄春良[2]

(1. 云南省林业科学院, 云南 昆明 650204;
2. 建水县林业技术推广所, 云南 建水 654300)

摘要: 为丰富云南造林树种资源, 提高单位面积的木材产量和质量; 改变长期以来树种单一带来的病虫害严重、产量和质量下降的状况, 在1981年云南松地理种源试验中, 引入了湿地松、火炬松、海岸松、马尾松等8个松类树种一并参试。经过20年来试验示范、跟踪观测、分析比较, 评选出适应性广、抗性强、生长量高的马尾松树种。其蓄积量超过海岸松、湿地松、火炬松的3.5~11.2倍, 超过云南松对照种源的1.5倍, 收到较好的引种效果。

关键词: 松属8个树种; 引种; 适应性; 生长量

Study on the Introduce of Eight *Pinus* species

Zheng Wan[1], Shu Xiao-wu[1], Cai Yu-xin[1], Li Si-guang[1], Huang Chun-liang[2]

(1. Yunnan Academy of Forestry, Kunming Yunnan 650204; 2. Forestry Extending Station of Jianshui County, Jianshui Yunnan 654300)

Abstract: In order to enrich forestation species of Yunnan, raise quantity and quality of the timber and release diseases and insects damage owing to single species, the test on introducing 8 Pinus species, including *P. elliottii*, *P. taeda*, *P. pinaster*, *P. densata*, *P. griffithii*, *P. sylvestris*, *P. massoniana* and *P. kesiya* var. *langbianensis*, was conducted together with the provenance test of *Pinus yunnanensis* in 1981. Through demonstration, observation, comparison and analysis of more than 20 years, *P. massoniana* was selected as the species of wide adaptability, strong stress resistance and fast growth. Its timber volume is 3.5~11.2 times that of *P. elliottii*, *P. taeda*, *P. pinaster*, and 1.5 times that of the check provenance of *Pinus yunnanensis*.

Key words: 8 *Pinus* species; introduce; adaptability; fast growth

引种是良种选育的一个重要环节, 是实现林木速生、丰产、优质的一个重要手段[1]。本研究于1981年在开展云南松(*Pinus yunnanensis*)地理种源试验过程中, 为丰富松类造林树种资源, 提高林地生产力, 培育速生、丰产、优质林分; 故从外地引进了湿地松(*P. elliottii*)、火炬松(*P. taeda*)、海岸松(*P. pinaster*)、高山松(*P. densata*)、乔松(*P. griffithii*)、樟子松(*P. sylvestris*)、马尾松(*P. massoniana*)、思茅松(*P. kesiya* var. *langbianensis*)等树种。除樟子松苗期不适被淘汰外, 其他7个树种与云南松各种源一并参试, 进行栽培对比试验。经过20年来的观测、分析比较, 马尾松以旺盛的生长优势, 在树高、直径、材积等主要经济性状上明显超过其他引进树种和云南松各种源。相继在生产上建立的示范区和大面积的推广造林, 在经济、生态、社会等方面都取得了较好的效益。实践证实, 马尾松具有广泛的适应性, 在新的引种区能顺利地生长发育, 以旺盛的生长优势位居其他树种之上。这对丰富云南的松类造林树种,充实和完善现有的树种布局,不断提高林业经营水平,增加单位面积产量,扩大森林产品,改善生态环境都具有重要的经济价值。

1 引种和示范点的自然概况

试验点设在滇中地区的昆明北郊大马山试验林场,示范点设在蒙自、建水等地,位于北纬 23°23′~25°04′,东经 102°04′~103°23′(表1),海拔高度 1 600~2 000m,年降水量 783.5~1 011.8mm,干湿季明显,降水量主要集中在 5~10 月,年相对湿度 72%~76%,冬季较为干旱,霜期 10 月至翌年 3 月,无霜期 285.3~359.3 天,全年日照 2 224.8~2 481.2h,平均气温 14.7~19.7℃,1 月均温 7.7℃,≥10℃ 积温 4 479.7~6 271.1℃,年极端最低气温 -2.4~-5.4℃,极端最高气温 31.5~38.2℃。昆明试验点有的年份有降雪,多在 12 月底至翌年 1~2 月。3 月初往往有较大的降温天气,被称为倒春寒,对正在萌动的植物有致命威胁。而其他各引种示范点极少降雪。

昆明引种点原为乔灌混生的次生林地,乔木有云南松、华山松(Pinus armandii)、圆柏(Cupressus duclouxiana)、旱冬瓜(Alnusne palensis)、麻栎(Quercus acutissima)、小铁仔(Myrsine africona);禾本科草类有:扭黄茅(Heteropagon contartus)、四脉金茅(Eulalia guadrinervis)、菜蕨(Pteriduma guilinum)。土壤为发育在石灰岩上的酸性红壤,pH 值 5~6.5,较干燥贫瘠。引种试验示范点与原产地纬度相近,昆明试验点与福建德化的纬度均在 25°~26°,示范推广点的蒙自、建水、开远、个旧与原产地的广西百色、田阳的纬度更加接近,都属亚热带气候类型。不同之处在于引种试验、示范、推广点的海拔偏高,经度偏西。蒙自、建水等引种示范点为荒山荒地和云南松、华山松的采伐迹地。

表1 试验示范及推广点的自然概况表
Table 1 Natural conditions of demonstration and extension sites

试验及示范点	纬度	经度	高温(℃)	低温(℃)	海拔高度(m)	年平均气温(℃)	≥10℃积温(℃)	年降水量(mm)	年相对湿度(%)	日照时数(h)	有霜期日数(d)
昆明	25°01′	102°41′	31.5	-5.4	1 970	14.7	4 479.7	1 011.8	73	2 481.2	79.7
蒙自	23°23′	103°23′	36.0	-4.4	1 600~2 000	18.6	6 271.1	827.1	72	2 224.8	5.7
建水	23°42′	102°52′	35.1	-2.9	1 600	18.4	6 249.8	828.3	72	2 276.2	8.6
开远	23°44′	103°15′	38.2	-2.5	1 500	19.7	6 864.0	813.9	71	2 195.2	5.7
个旧	23°29′	103°10′	30.3	-4.7	1 650	15.8	4 851.9	1 096	78	1 944.9	7.2
禄丰	25°04′	102°04′	36.1	-4.5	1 870	16.7	5 214.2	783.5	74	2 166.4	55.4

2 材料与方法

2.1 试验材料

8 个参试树种的原产地是:马尾松种源引自福建的德化和邵武,湿地松引自广东,火炬松引自美国,海岸松引自阿尔巴尼亚,樟子松引自内蒙古,乔松来自西藏,高山松引自丽江,思茅松引自澜沧、景谷、思茅等 9 个产地,以及云南松分布区的 41 个种源。

2.2 试验方法

纳入本引种试验的松属 8 个树种和云南松分布区的 41 个种源及 9 个思茅松种源,共计

57个处理。其田间设计采用的是BIB平衡格子设计[2]，即$V=57$、$K=8$、$b=57$、$r=8$、$\lambda=1$的设计安排试验。林地的规划严格按设计要求，同一区组内的造林定时、定人按统一的技术要求实施，使每个参试的种和种源在外界条件基本一致和管理措施相同的条件下进行栽培对比试验，使环境造成的误差减少到最低限度，以获取较为可靠的遗传增益。

引种试验林地的株行距采用树种及种源间距为2.0m，小区内为1.5m，每小区6株，单行排列。示范林和大面积推广林株行距采用1.5m×1.5m或2.0m×2.0m。

整地方式采用40cm×40cm×30cm的块状。整地时间在造林前2~3个月进行，土壤翻挖后经过一段时间的风化曝晒，杀死土壤中的部分害虫，土壤也较疏松，容易保水，有利幼苗的生长。造林后的林地管理、病虫害防治等项措施均一致。

2.3 抚育管理

6年后，林分进入郁闭状态，森林环境初步形成，此时人为地对林分内的个体作适当的调整，进行抚育间伐时每塘保留1株，间伐对象坚持去劣留优、砍小留大、去弱存强、砍密留稀、砍弯留直的原则。在引种试验林内，为保证样本数的统计量，使每一个种或种源均达到大样本的要求，小区内的每塘均保留1株。在示范林和大面积推广林的株行距多采用1.5m×1.5m，单位面积上的株数随林龄的递增，个体间竞争加剧，根据上述间伐原则开展间伐抚育管理，扩大保留木的营养空间，使其更好的生长。

于成林期对林木进行修枝，修除树干下部生长弱的枝条，保障主干的粗生长，形成圆满的主干。修枝用锋利的刀具，采取不留长桩的修枝方法，使修枝切口与树干齐平，保持切口平滑，可加速伤口的愈合，实现无节良材的培育。修枝以始终保持树冠部分有较大的营养面积，利于地上部分生长为原则，以此调整冠长与树高形成适当比例。在10年生以前冠长为树高的2/3，10~15年间冠长为树高的1/2~1/3。修枝时间在每年的10~11月。第一次修枝和间伐同步进行，以后视需要单独开展。

2.4 观测内容与方法

在造林后的头3年，每年年终作一次生长量观测，以后隔2年又连续观测3年，视需要和根据生长变化情况进行。每次观测均对参试的每一个种和种源的适应性、抗逆性及树高、胸径（地径）的生长量作全面观测。所获材料作系统全面的统计分析和反复的比较，从比较中找差别，从差别中评选出适应广、抗性强、生长旺盛，遗传增益高的种和种源，为生产应用提供依据。

3 结果与分析

3.1 树种间的生长量比较

3.1.1 树高生长

在参试的树种中，马尾松高生长，从苗期开始就显现出优势，居2~3位，海岸松在1~2年生时树高生长表现最好，到第3年生以后落后于2个马尾松种源，2个马尾松种源由原来的2~3位上升到1~2位（表2），这一位秩20年来都比较稳定。对历年树高的变量分析，由表3可以看出，树种间高生长的差异是极显著的。20年生时2个马尾松种源的平均高是

表 2 参试树种各林龄的生长量比较
Table 2 Growth comparison of various aged trees of the species tested

树种	1a生 苗高(m)	2a生 树高(m)	3a生 树高(m)	6a生 树高(m)	6a生 胸径(cm)	6a生 蓄积量(m³·hm⁻²)	7a生 树高(m)	7a生 胸径(cm)	7a生 蓄积量(m³·hm⁻²)	8a生 树高(m)	8a生 胸径(cm)	8a生 蓄积量(m³·hm⁻²)	11a生 树高(m)	11a生 胸径(cm)	11a生 蓄积量(m³·hm⁻²)	20a生 树高(m)	20a生 胸径(cm)	20a生 蓄积量(m³·hm⁻²)
马尾松(德化)	0.23	0.60	1.27	3.21	3.69	9.1320	3.52	4.63	15.095	4.32	5.37	22.797	6.07	10.19	101.71	10.80	15.70	382.74
马尾松(邵武)	0.28	0.66	1.17	3.05	3.95	10.041	3.38	4.64	14.835	4.25	5.60	24.555	5.94	10.10	98.492	10.43	15.60	380.04
湿地松	0.22	0.49	0.86	1.89	2.67	3.7650	2.19	2.93	4.8120	2.36	3.13	5.6715	3.78	5.27	20.336	6.30	7.30	54.345
火炬松	0.22	0.39	0.79	1.80	1.71	1.5165	1.95	2.13	2.4255	2.28	2.71	4.1880	3.30	4.79	15.611	4.50	6.00	31.305
海岸松	0.35	0.69	0.94	1.57	2.29	2.5890	1.91	2.80	4.1580	2.29	3.32	6.2970	3.12	4.81	16.418	6.50	8.60	84.0
乔松	0.09	0.20	0.35	0.94	1.81		1.22	1.47		1.53	2.17	1.0620	2.50	2.25	3.0075	4.50	4.90	20.310
高山松	0.07	0.23	0.49	1.32	2.06		1.52	2.07	2.0910	1.85	2.35	2.8920	2.69	3.34	7.1775	4.50	4.90	20.985
云南松(ck)	0.09	0.35	0.83	1.93	3.32	5.9040	2.43	3.96	9.1965	3.27	4.60	12.044	4.56	6.60	35.567	8.50	10.90	155.82

表 3 各年度树种间高、径生长变量分析结果
Table 3 Analysis on variable of height and diameter growth for the species of years

年度	变异来源	树高 平方和	树高 自由度	树高 均方	树高 F值	胸径 平方和	胸径 自由度	胸径 均方	胸径 F值	F临界值 $F_{0.05}$	F临界值 $F_{0.01}$
1984	种间	4.640	5	0.928	21.737	103.230	5	2.065	15.445	2.603	3.86
	区组内	0.496	5	0.099	2.325	1.835	5	0.367	2.745		
	误差	1.067	25	0.043		3.342	25	0.134			
	总变异	6.204	35			15.499	35				

（续）

年度	变异来源	树高				胸径				F 临界值	
		平方和	自由度	均方	F 值	平方和	自由度	均方	F 值	$F_{0.05}$	$F_{0.01}$
1988	种间	37.873	7	5.410	31.428	67.177	7	9.597	14.141	2.203	3.03
	区组内	3.724	7	0.532	3.090	8.333	7	1.190	1.754		
	误差	8.435	49	0.172		33.255	49	0.679			
	总变异	50.023	63			108.766	63				
1989	种间	47.018	7	6.717	29.352	91.744	7	13.106		2.285	3.20
	区组内	2.808	5	0.562	2.455	4.137	5	0.827			
	误差	8.009	35	0.229		25.271	35	0.722			
	总变异	57.835	47			121.152	47				
1992	种间	93.072	6	15.512 0	44.584	360.687	6	60.115	23.740		3.35
	区组内	1.314 4	6	0.219 0	0.629	18.306	6	3.051	1.205		
	误差	12.525	36	0.347 9		91.161	36	2.532			
	总变异	106.912	48			270.155	48				
2000	种间	403.129	9	44.792	27.987	1 043.459	9	115.940	26.341		3.759
	区组内	5.750	6	0.958 4	0.599	30.252	6	5.042	1.145		4.449
	误差	86.426	54	1.600		237.679	54	4.401			
	总变异	495.305	69			1 311.391	69				
2000	种源间	1 121.722	56	20.030 75	13.427 65	2 438.83	56	43.550 54	9.221 32		1.789
	重复间	12.012 83	6	2.002 138	1.342 137	39.677 45	6	6.612 908	1.400 207		3.847
	误差	501.229 3	336	1.491 754		1 586.864	336	4.722 808			
	总变异	1 634.964	398			4 065.371	398				

10.62m。超过高山松、乔松(4.5m)的1.4倍;比火炬松、湿地松、海岸松树种的平均高4.5~6.5m提高135.9%~68.5%;比对照云南松(8.45m)提高24.9%。马尾松的高生长至今仍居于7个树种之首,而海岸松、湿地松高生长仅为云南松对照种源的76.5%~74.1%,高山松、乔松、火炬松生长已进入衰退状态。树种间高生长的t检验结果表明(表4),马尾松旺盛的生长优势,位居参试树种之首。参试树种思茅松经过历次低温大部分植株死亡,保留下来的少量植株虽有一定的生长量,但作为试种点的滇中地区是不宜引种的,故在本文中不予论述。

表4 参试树种间20年生时树高均数差异t检验
Table 4 T-test on difference of average height of tested species at 20 years old

树种名	平均树高(m)	各树种间差异比较					
马尾松(德化)	10.8						
马尾松(邵武)	10.4	0.4					
思茅松	10.0	0.8	0.4				
云南松(ck)	8.5	2.3**	1.9*	1.5*			
海岸松	6.5	4.3**	3.9**	3.5**	2.0*		
湿地松	6.3	4.2**	4.1**	3.7**	2.2*	0.2	
火炬松	4.5	6.3**	5.9**	5.5**	4.0**	2.0*	1.8*
乔松	4.5	6.3**	5.9**	5.5**	4.0**	2.0*	1.8*
高山松	4.5	6.3**	5.9**	5.5**	4.0**	2.0*	1.8*

注:$t_{0.05}(N=9)S_D=1.5$;$t_{0.01}(N=9)S_D=2.2$。

3.1.2 直径生长

直径生长与高生长密切相关,3年生以后树高生长迅速加快,直径生长也随之跟上。所获材料证实,20年生时马尾松在德化和邵武两个种源的差异很小,平均直径为15.65cm(表5),平均年生长量为0.78cm,种源内个体间生长差异很大,个体间直径的变幅在10.4~24.0cm,马尾松的直径生长量在参试树种间生长最快,且差异显著,由表3知树种间的差异显著。马尾松直径生长比高山松、海岸松等树种的平均直径(4.9~8.6cm)提高2.2~0.8倍,比对照的云南松(10.9cm)提高43.6%。均数间t检验的比较结果马尾松极显著高于其他树种(表5)。

表5 参试树种间20年生平均胸径差异的t检验
Table 5 T-test on difference of average DBH of tested species at 20 years old

树种名	平均胸径(cm)	各树种间差异比较					
马尾松(德化)	15.7						
马尾松(邵武)	15.6	0.1					
思茅松	12.6	3.1*	3.0*				
云南松(ck)	10.9	4.8**	4.7**	1.7			
海岸松	8.6	7.1**	7.0**	4.0**	2.3		
湿地松	7.3	8.4**	8.3**	5.3**	3.6**	1.3	
火炬松	6.0	9.7**	9.6**	6.6**	4.9**	2.6*	1.3
乔松	4.9	10.8**	10.7**	7.7**	6.0**	3.7**	2.4
高山松	4.9	10.8**	10.7**	7.7**	6.0**	3.7**	2.4

注:$t_{0.05}(N=9)S_D=2.5$;$t_{0.01}(N=9)S_D=3.6$。

3.1.3 蓄积生长

从单位面积上的蓄积量,更加体现出树种间的生长优劣。在每公顷蓄积量计算中统一采用公式 $V=0.000036\times d^2\times(H+3)$ 计算,以每公顷蓄积作比较。每公顷造林 3 330 株,保存率为 91.0%,故以每公顷保留 3000 株计算。6 年生以后参试的各树种的树高已达 1.5m 以上。历年的分析材料表明,树种间蓄积的差异很大。对 20 年生时的变量分析材料得到 $F=15.663 > F_{0.001}=3.760$,树种间的蓄积量差异在 0.001 的水平上显著。为生产上的应用提供了极为可靠的依据。17 年来马尾松的蓄积生长量位秩一直排于参试树种之首,2 个马尾松种源 20 年生时的平均蓄积生长量是 $381.39 m^3 \cdot hm^{-2}$。在蓄积均数间多重比较检验结果表明,马尾松蓄积生长量极显著优于其他树种,超过海岸松、湿地松、火炬松($84.0\sim31.305 m^3 \cdot hm^{-2}$)$3.5\sim11.2$ 倍;超过乔松、高山松($20.648 m^3 \cdot hm^{-2}$)17.5 倍,超过云南松对照种源($155.82 m^3 \cdot hm^{-2}$)1.5 倍。

3.2 不同树种的抗病虫性能

在昆明测试点上,参试的各树种都不同程度的出现过病虫危害,云南松的各种源时有松梢螟(*Dioryctrias plendidella*)、云南松毛虫(*D. latipennis*)、松卷叶蛾(*Petrova cristata*)危害;松赤枯病(*Pestalotia funerea*)、松落针病(*Lophodermium pinastri*)始终威胁着火炬松、湿地松、海岸松树种的生长发育;而在同一环境条件下的马尾松未受病虫感染。1988 年课题组在蒙自芷村林场蚌德林区被小蠹虫危害后的采伐迹地上利用马尾松和云南松较优种源进行更新,14 年来马尾松树种和多数云南松种源生长旺盛,特别是个别种源如双江种源已发现有小蠹虫危害的情况下,可靠近它的马尾松则未遭侵害,因此马尾松的抗虫能力较强。

为深入了解引入的几个松类树种对纵坑切梢小蠹虫(*Tomius piniperda*)的抵抗程度。将昆明引种的马尾松、湿地松、火炬松、海岸松等树种及云南松部分种源一并纳入室内接种小蠹虫试验作比较。经过 3 个月的观测结果表明,所有参试树种中,云南松仍然是纵坑切稍小蠹虫的主要侵害对象,在置放的 3 个笼内,每个笼内的云南松都遭侵入,其侵入孔 $5.5\sim7.7$ 个,平均 6.5 个。在参试树种中,侵入孔数较多的另一树种是华山松,在 2 个笼内发现侵入孔 $1\sim3$ 个,平均 1.3 个。马尾松和火炬松只在 1 个笼内发现有 2 个侵入孔,其他 2 个笼内没有侵入,平均侵入孔是 0.7 个。其中海岸松和湿地松,对纵坑切稍小蠹虫有较强抗性,在 3 个笼内均没有小蠹虫的侵入孔,由此可以看出在室内环境基本一致的条件下,马尾松、火炬松、海岸松较少被纵坑切稍小蠹虫攻击。反映出它们对小蠹虫有一定的抗性,特别作为适应性广、生长快、增益大的马尾松树种更加反映出其引种优势。

4 结论与建议

(1)经过 20 年的跟踪观测和对材料完整系统的分析,发掘出了增益较大的马尾松树种。栽培实践证实,马尾松具有广泛的适应性、较强的抗逆性及速生性,在同等条件下,它的蓄积生长量超过湿地松、火炬松、海岸松、乔松、高山松的 3.5 倍以上,超过云南松对照种源的 1.5 倍,可见引种带来的良好效益。

(2)云南是一个多山的省份,山地占全省面积的 94%。长期以来一直以种植云南松、华山松、思茅松为主,由于管理和采伐利用上的不合理,负向选择造成新造林分生长量和质量每况愈下,特别是大面积的云南松人工纯林发生小蠹虫、云南松毛虫危害,严重制约着云南林业的发展。马尾松的引种成功,对于丰富引种地区树种资源的多样性和遗传物质的多样性,提高林业经营水平,完善造林树种结构的合理搭配,充分利用地力,提高单位面积木材产量和质量,增强森林对病虫害的抵抗能力,实现山区生产的良性循环具有十分重要的价值。

(3)根据省外对马尾松地理变异的研究及开展的种源区划,结合云南引种的试验示范林的生长情况,造林用种应做到适地适种源,云南省海拔在1 900m以下的广大山地,以引用广西宁明、田阳等种源为主,并适当采用四川南溪、蒲江、涪陵种源。为保证干形的通直度,纹理、材质的优良和生长的速生性,尽量采用上述种源区经过疏伐后建立起来的母树林或采种基地上的种子,把种源、林分、家系三部分的遗传增益联合并用,使引种收益更大。

(4)湿地松和火炬松的蓄积生长量仅为云南松对照种源的34.9%～20.1%,因此外来树种如果不经过试验就在生产中大量推广,势必要冒很大的风险,甚至花了大量的人力物力和时间,结果还不如当地的乡土树种。

参考文献:

[1]南京林产工业学院. 树木遗传育种学[M]. 北京:科学出版社,1982.
[2]中国科学院数学研究所概率统计室. 常用数理统计表[M]. 北京:科学出版社,1979.
[3]云南省林业厅. 云南主要林木种质资源[M]. 昆明:云南科学技术出版社,1996.
[4]杨文云,李昆. 云南森林树种种质资源保存策略初探[J]. 林业科学研究,2002,15(6):706～711.
[5]王明怀,陈建新,殷祚云,等. 五种木麻黄的种源引种初报[J]. 林业科学研究,2002,15(6):751～755.
[6]陈建新,王明怀,殷祚云,等. 广东省屠杉引种栽培效果及栽培区划分研究[J]. 林业科学研究,2002,15(4):399～405.
[7]秦国峰. 马尾松地理起源及进化繁衍规律的探讨[J]. 林业科学研究,2002,15(4):406～412.

(本文发表于《西部林业科学》,2004年)

基于数量化回归模型的秃杉优树选择

王庆华[1],陈强[1],刘永刚[1],苏俊武[1],沈立新[2],刘云彩[1],
毕波[1],许彦红[3],周筑[1],段成波[4],杨锐铣[5],赵永红[1],孙志刚[1],孙宏[1]

(1. 云南省林业科学院,云南 昆明 650201;2. 西南林业大学亚热森林组织昆明培训中心,
云南 昆明 650224;3. 西南林业大学林业调查规划设计研究院,云南 昆明 650224;
4. 腾冲市林业局,云南 腾冲 679100;5. 大理农林职业技术学院,云南 大理 671003)

摘要:对秃杉分布区13个县(市、区)的纯林、混交林、散生木、孤立木进行调查,选出236株作为初选优树,并观测这些候选优树的立地因子、生长性状等指标。经相关分析,选择单株材积作为优树复选的主要指标。采用数量化回归的方法,建立经度、纬度、海拔、坡度、黑土层厚度、树龄6个数量因子和坡向、坡位、坡形、基岩、土壤类型、起源、立木类型7个定性因子与秃杉单株材积的回归方程,其复相关系数为0.799。秃杉单株材积实测值与理论值的差值(I_i)代表基因型值,其频率分布成正态分布。以I_i与差值平均值I±标准差δ相比较作为划分优树等级的依据。秃杉以候选优树70%的入选率统计,差值$I_i \geq I+0.3\delta$,即$I_i \geq 1.6446$为Ⅰ级优树;$I+0.3\delta>I_i>I-0.3\delta$,即$1.6446>I_i>-1.6446$为Ⅱ级优树;$I_i \leq I-0.3\delta$,即$I_i \leq -1.6446$为Ⅲ级优树(一般林木),Ⅲ级优树淘汰不选。用此标准对秃杉236株野外初选优树进行复选,Ⅰ级优树54株,占候选优树的22.88%;Ⅱ级优树114株,占候选优树的48.31%。Ⅰ、Ⅱ级优树预估遗传增益达20.82%。

关键词:秃杉;优树;选择指标;多元数量化模型;材积

Selection of Superior Trees of Taiwania flousiana Based on the Multiple Entry Quantity Model

Wang Qing-hua[1], Chen Qiang[1], Liu Yong-gang[1], Su Jun-wu[1], Shen Lixin[2], Liu Yun-cai[1], BI Bo[1], Xu Yan-hong[3], Zhou Zhu[1], Duan Cheng-bo[4], Yang Rui-xian[5], Zhao Yong-hong[1], Sun Zhi-gang[1], Sun Hong[1]

(1. Yunnan Academy of Forestry, Kunming Yunnan 650201; 2. APFNet - Kunming Training Center, Southwest Forestry University, Kunming Yunnan 650224; 3. Research Institute of Forestry Inventory and Planning, Southwest Forestry University, Kunming Yunnan 650224; 4. Forestry Bureau of TengchongCounty, Tengchong Yunnan 679100; 5. Dali Vocational and Technical College of Agriculture and Forestry, Dali Yunnan 671003)

Abstract: The site factors and growth traits of 236 candidate *Taiwania flousiana* trees of pure forest, mingledforest, scattered wood and isolated wood in 13 counties were measured. The correlation analysis of candidate trees showed that volume could be used as the main indexes for selection of superior trees of *T. flousiana*. The regression equation with individual volume was established by using 6 quantitative factors including longitude, latitude, altitude, slope, the depth of black soil layer, ages of trees and 7 qualitative factors including slopeaspect, slopeposition, slopeshape, bedrock, soiltype, origin, tree type. Its multiple correlation coefficient was 0. 799 . There was a difference between theoretical and realistic volume , the frequency of difference was in normal distribtuion. The superior tree selection was based on the comparison of I_i and the mean value of difference I ± standard deviation δ. The superior tree selection standards with 70% selective ratio were as follows: Class Ⅰ superior tree, $I_i ≥ I+0.3δ$, that is $I_i ≥ 1.6446$. Class Ⅱ superior tree, $I+0.3δ > I_i > I-0.3δ$, that is $1.6446 > I_i > -1.6446$. Class Ⅲ superior tree(normal tree) , $I_i ≤ I-0.3δ$, that is $I_i ≤ -1.6446$. 54 strains of class Ⅰ superior trees and 114 strains of class Ⅱ superior trees were selected by the standards . They accounted for 22.88% and 48.31% in all candidate superior trees respectively. The genetic gain of the class Ⅰ and class Ⅱ superior trees reached 20.82%.

Key words: *Taiwania flousiana*; superior tree; selective index; Multiple Entry Quantity Model; standing volume

秃杉(*Taiwania flousiana*)寿命长,树干圆满通直、树皮薄、出材率高、材质好[1,2],是培育大径材、特大径材的优良用材树种。秃杉的野生资源主要集中在23°53′~30°20′N,98°50′~108°55′E,海拔600~3 000m。云南西部、西北部的澜沧江、怒江流域为集中分布区;湖北利川石门乡中心村、星斗山的花板溪,贵州雷公山东南斜坡的沟谷两侧,四川武陵山地西部的酉阳县等区域[3~6]呈间断性零星分布。引种和人工造林试验表明,秃杉在河南、安徽、江苏、浙江、江西、湖南、湖北、福建、广西、广东、四川、贵州、云南等省份天然分布区以外地区亦生长良好,能耐-13℃的低温,可在我国22°15′~33°48′N,98°15′~122°06E′,海拔3.8~3 400m的范围内选择水湿条件好、透水性能强、土层深厚的阴坡、半阴坡进行大面积人工造林[7~12]。用RAPD技术、ISSR标记分析秃杉不同天然群体的遗传多样结果表明,秃杉种源间和种源内、居群内遗传多样性丰富[13~15],具有较大的遗传改良潜力。

1990年采用"优势系数选择法"在云南省的昌宁、龙陵、腾冲3个县秃杉人工林内进行优树选择,共选出了40株优树,通过嫁接的方式建立了第一个面积约20hm²秃杉初级种子园,现已挂果[16]。2011年起为了收集、保存秃杉优良的种质资源和扩大优树数量,云南省林业科学院在云南、贵州、湖北、重庆等省市的秃杉分布区13个县(市、区)开展了优树的选择工作。由于秃杉的结实单株以纯林、混交林、散生木、孤立木等多种形式存在,树龄差异大,分布区域广、立地条件复杂多样,无法采用优势木对比法、小标准地法、绝对生长量法、标准差法等优树选择方法,在野外只能根据秃杉单株表型特征进行优树初选。为了有效剔除环境因子的影响,减少表型选择的误差,使复选指标能客观反映优树的基因型表现,选出基因型优良的优树,提高遗传增益,本研究采用数量化回归模型建立秃杉单株材积与诸环境因子的回归方程,用候选优树单株材积的实测值与理论值的差值做为优树复选指标,以差值与差值平均值相比较作为划分优树等级的依据,探讨秃杉优树选择的新方法。

1 研究方法

1.1 选优区域

在收集秃杉相关资料,摸清其资源分布状况的基础上,确定野外调查范围为云南西部、西南部和西北部的保山、德宏、临沧、怒江、迪庆等地的人工林和天然林区;贵州东南部的雷公山自然保护区;湖北西南部利川星斗山自然保护区和沙溪石门村及重庆酉阳毛坝。

1.2 优树候选标准

选择标准:结实良好;干形圆满通直,树干通直度为Ⅰ级(主干无弯);生长健壮,无病虫害;候选优树间距≥50m;天然起源的秃杉应具较强的生长势,树冠呈尖塔形、圆锥形,树皮产生的新裂纹明显;人工起源的秃杉树高和胸径应明显大于周围立木等标准选定候选优树。

1.3 选优方法

(1)踏查:深入选优区域,了解树种分布状况与特点,确定具体的选优线路。

(2)调查标记:按候选优树选择标准确定候选优树,填写候选优树基本情况记录表,描绘优树位置示意图,拍摄照片。在树干 1.5m 处用红漆涂环,写明调查序号。调查内容包括树高(精确到0.1m)、胸径(精确到0.1cm)、冠幅(精确到0.1m)、枝下高(精确到0.1m)、树龄、起源、立木类型、通直度、冠型、健康状况、结实状况及经度、纬度、海拔、坡向、坡度、坡位、坡形、基岩、土壤类型、黑土层厚度等因子。其中树高用 VERTEX Ⅳ 超声波测高仪测量;胸径用太平洋牌钢围尺测量;冠幅用博世 BOSCH DLE40/4000 激光测距仪测量;树龄用瑞典 Haglof 树木生长锥在树干60cm处钻取木心测定;经度、纬度、海拔用 Garmin60csxGPS 测定;坡向、坡度用地质罗盘仪 DQY-1 测定;挖取宽0.8m,深1m的土壤剖面确定基岩、土壤类型、黑土层厚度。

1.4 数据处理

1.4.1 候选优树实测材积计算

以调查所得候选优树胸径、树高值用公式(1)计算候选优树实测材积[17]。

$$v_i = 0.000058777 \times d_i^{1.9699831} \times h_i^{0.89646157} \quad (1)$$

式中:v_i是材积实测值(m^3);d_i是胸径(m);h_i是树高(m)。

1.4.2 候选优树理论材积计算

数量化理论是多元统计学的一个分支,始于20世纪50年代,该理论根据研究问题目的的不同分为4种,分别称为数量化理论Ⅰ,Ⅱ,Ⅲ,Ⅳ。数量化理论Ⅰ主要研究含有定性变量的建模问题。

设自变量有 h 个定量因子,它们在第 i 个样本中的数据为 $x_{i(u)}(u=1,2,\cdots,h;i=1,2,\cdots n)$,有 m 个定性因子(项目),其中第 j 项目有 r_j 个类目,它们在第 i 个样品中的反应是 $\delta_{i(j,k)}(j=1,2,\cdots,m;i=1,2,\cdots,n)$,因变量为 y,它在第 i 个样本中的数据为 $y_i(i=1,2,\cdots,n)$,则有如下的线性模型[18]:

$$y_i = \sum_{u=1}^{h} b_u X_{i(u)} + \sum_{j=1}^{m}\sum_{k=1}^{r_j} \delta_{i(j,k)} b_{jk} + \varepsilon_i \quad (2)$$

式中:$b_u(u=1,2,\cdots,h)$,$b_{jk}(j=1,2,\cdots,m;k=1,2,\cdots,r_j)$是未知常数,$\varepsilon_i(i=1,2,\cdots,n)$是随机误差。其中$\delta_{i(j,k)}$是第$i$个样本在第$j$个项目上的反应值,该值按如下法则确定:

$$\delta_{i(j,k)} = \begin{pmatrix} 1 & 当第\,i\,个样本属于第\,j\,个项目的第\,k\,个类目 \\ 0 & 其他 \end{pmatrix}$$

候选优树调查因子包含定量因子和定性因子，因此应用多元数量化回归模型Ⅰ进行分析，并据此建立候选优树理论材积回归方程，计算候选优树理论材积。

2 结果与分析

2.1 优树初选结果

在秃杉分布区共选出了 236 株候选优树，分布在 13 个县(市、区)，地理位置在 98°11′01″~109°15′17″E，23°43′39″~30°03′37″N 之间，海拔 708~2 313m。树龄 25~500a，平均树龄为 84a；树高 15.0~48.0m，平均树高为 30.8m；胸径 46.0~211.0cm，平均胸径为 83.9cm；单株实测材积 1.8~74.24m³，平均单株实测材积($V_{均}$)为 9.123 7m³。各地候选优树基本情况见表1。

表1 各地候选优树基本情况
Table 1 Basic information of the candidate tree

地点	株数	树龄(a)	树高(m)	胸径(cm)	枝下高(m)	冠幅(m)	立木类型	起源
腾冲	92	29~80	22.0~43.0	58.0~106.0	1.7~16.0	7.0~15.0	纯林、混交林、散生木、孤立木	人工林
盈江	22	25~46	21.0~34.0	46.0~75.0	2.0~12.0	4.0~15.0	纯林、孤立木	人工林
梁河	18	38~43	23.0~32.0	53.0~75.0	1.3~11.0	6.0~16.0	散生木	人工林
龙陵	20	31~65	28.0~35.0	53.0~91.0	1.4~20.0	6.0~15.0	纯林、散生木	人工林
昌宁	7	50~63	22.0~35.0	69.0~92.0	2.7~12.0	9.0~12.0	纯林、混交林、孤立木	人工林
凤庆	5	93	15.0~26.0	55.0~83.0	2.0~3.3	9.0~12.0	散生木	人工林
临翔	9	90~150	19.5~31.0	53.0~191.0	1.4~6.3	8.0~24.0	散生木、孤立木	人工林
福贡	15	75~467	26.0~47.0	65.0~211.0	2.0~33.0	8.0~20.0	纯林、散生木、孤立木	天然林
泸水	6	33~150	19.0~38.0	55.0~90.0	2.0~22.0	3.0~13.0	纯林、散生木、孤立木	天然林、人工林
利川	6	150~191	32.0~44.0	74.0~120.0	9.0~16.0	10.0~18.0	纯林、散生木	天然林
雷山	21	102~500	30.0~48.0	72.0~150.0	5.4~14.0	7.0~17.0	混交林、孤立木	天然林
剑河	8	62~208	25.0~42.0	61.0~178.0	4.0~8.0	6.0~22.0	散生木、孤立木	天然林
台江	7	86~198	28.0~37.0	86.0~130.0	6.0~16.0	12.0~16.0	散生木、孤立木	天然林
合计	236							

2.2 优树复选指标性状确定

以 236 株候选优树为样本，对单株材积、树高、胸径、径高比、冠幅和枝下高等 6 个性状进行相关分析，结果见表2。由表2可知，除径高比与树高和枝下高间相关不密切外，其他性状间均存在极显著的正相关，其中胸径与树高、径高比、冠幅和枝下高等 4 个性状间呈显著正相关，树高与胸径、冠幅和枝下高间为显著正相关，说明这些性状间受连锁基因的作用存在着多重共线性[17]，表明这些性状在选择上具有较高的一致性。材积是树高与胸径的综合体现，其较高的生长量也是生产上所追求的目标，因此选择材积作为优树复选的主要性状。调查发现秃杉树干通直，病虫害很少，因此将这些性状作为候选优树的控制指标，而不作为优树复选的指标性状。

表 2　6 个性状间相关系数

Table 2　Correlation coefficient among 6 characteristics

	材积	树高	胸径	冠幅	枝下高
树高	0.640 1**	1			
胸径	0.932 7**	0.565 7**	1		
冠幅	0.626 6**	0.398 5**	0.666 7**	1	
枝下高	0.446 2**	0.421 8**	0.405 7**	0.299 7**	1
径高比	0.538 2**	−0.107 6	0.731 2**	0.463 2**	0.111 0

注：**表示差异极显著。

2.3　多元数量化回归模型 I 统计结果

应用多元数量化回归模型 I，以经度、纬度、海拔、坡度、黑土层厚度、树龄 6 个定量因子及坡向、坡位、坡形、基岩种类、土壤类型、起源、立木类型 7 个定性因子为自变量，单株材积为因变量进行多元数量化回归。其复相关系数为 0.799，方差为 568.050，剩余方差为 33.233，F 值为 17.093**。有关统计值见表 3。

表 3　材积多元数量化回归方程有关统计值

Table 3　Statistics values of multiple entry quantity regression model

项目	类　目	回归系数	偏相关系数
		$b_0 = 5.050$	
坡向	阴坡	$b_{1,1} = -0.062$	0.029 4
	半阴坡	$b_{1,2} = 0.361$	
	半阳坡	$b_{1,3} = 0$	
	阳坡	$b_{1,4} = 0.920$	
坡位	上	$b_{2,1} = -1.547$	0.061 4
	中	$b_{2,2} = -0.250$	
	下	$b_{2,3} = 0$	
坡形	直形坡	$b_{3,1} = 0$	0.075 1
	凸形坡	$b_{3,2} = -5.083$	
	凹形坡	$b_{3,3} = 0.489$	
基岩种类	石灰岩	$b_{4,1} = 0.597$	−0.012 7
	玄武岩	$b_{4,2} = 0.357$	
	页岩、泥质页岩	$b_{4,3} = -0.318$	
	砂岩、砂页岩、片麻岩、火山岩	$b_{4,4} = 0$	
土壤类型	黄壤	$b_{5,1} = 0.321$	0.016 6
	棕壤	$b_{5,2} = 0$	
	火山土	$b_{5,3} = 1.590$	
起源	人工林	$b_{6,1} = 0$	0.180 0
	天然林	$b_{6,2} = 6.437$	
立木类型	混交林	$b_{7,1} = -0.339$	−0.151 8
	纯林	$b_{7,2} = 0$	
	散生木	$b_{7,3} = -2.452$	

(续)

项目	类目	回归系数	偏相关系数
立木类型	孤立木	$b_{7,4}=-2.331$	
坡度		$b_8=-0.093$	-0.1241
黑土层厚度		$b_9=0.008$	0.0678
海拔		$b_{10}=0.002$	0.0507
经度		$b_{11}=0.313$	0.0704
纬度		$b_{12}=-1.485$	-0.1684
树龄		$b_{13}=0.087$	0.5781

将各统计值代入(2)式,得到候选优树的材积理论值回归方程,即材积理论值计算公式:

$$y_i = 5.050 + 0.313E - 1.485N + 0.002A - 0.093s + 0.008t + 0.087a + c \quad (3)$$

式中:y_i 是候选优树的材积理论值;E 是经度(°);N 是纬度(°);A 是海拔(m);s 是坡度(°);t 是黑土层厚度(cm);a 是树龄(a);c 是各定性因子得分之和,定性因子包括:坡向、坡位、坡形、基岩、土壤类型、起源、立木类型。

定性因子得分见表4。将各因子得分和观测值代入公式(3),计算得到各候选优树的材积理论值。

表4 定性因子得分表
Table 4 The qualitative factor score

因子	得分	
坡向	阴坡	$n_{1,1}=-0.062$
	半阴坡	$n_{1,2}=0.361$
	半阳坡	$n_{1,3}=0$
	阳坡	$n_{1,4}=0.920$
坡位	上	$n_{2,1}=-1.547$
	中	$n_{2,2}=-0.250$
	下	$n_{2,3}=0$
坡形	直形坡	$n_{3,1}=0$
	凸形坡	$n_{3,2}=-5.083$
	凹形坡	$n_{3,3}=0.489$
基岩	石灰岩	$n_{4,1}=0.597$
	玄武岩	$n_{4,2}=0.357$
	页岩、泥质页岩	$n_{4,3}=-0.318$
	砂岩、砂页岩、片麻岩、火山岩、花岗岩	$n_{4,4}=0$
土壤类型	黄壤	$n_{5,1}=0.321$
	棕壤	$n_{5,2}=0$
	火山土	$n_{5,3}=1.590$
起源	人工	$n_{6,1}=0$
	天然	$n_{6,2}=6.437$
立木类型	混交林林木	$n_{7,1}=-0.339$
	纯林林木	$n_{7,2}=0$
	散生木	$n_{7,3}=-2.452$
	孤立木	$n_{7,4}=-2.331$

2.4 优树复选结果

将236株秃杉候选优树的单株材积实测值(v_i)减去理论值(y_i)得到其差值(I_i)。根据差值的频率分布绘制出频率分布图(图1),其频率分布呈正态分布。

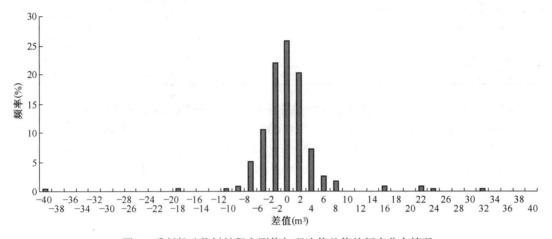

图1 秃杉候选优树材积实测值与理论值差值的频率分布情况

Figure 1 The frequency distribution of differences between theoretical and realistic volumes of candidate *T. flousiana*

秃杉单株材积实测值与理论值间的差值反映了秃杉在去除树龄、立木类型、起源及立地因子等因素影响后的生长表现,在很大程度上代表其基因型值,可据此差值进行优树复选,以减少表型选择的误差,故以 I_i、I、δ 相比较作为划分优树等级的依据[19]。

按照既能保证入选优树的质量,又能满足生产上有足够的秃杉优株可供利用来确定不同的选择强度,根据选择强度确定标准差 $\delta(\delta=5.481\ 9)$ 的倍数值。按入选率70%计算,Ⅰ级优树:$I_i \geq I+0.3\delta$,即 $I_i \geq 1.644\ 6$;Ⅱ级优树:$I+0.3\delta > I_i > I-0.3\delta$,即 $1.644\ 6 > I_i > -1.644\ 6$;Ⅲ级优树(一般林木,淘汰):$I_i \leq I-0.3\delta$,即 $I_i \leq -1.644\ 6$。

用以上选择标准对236株候选优树进行复选,共选优树168株,占候选优树的71.19%,平均树高31.8m,平均胸径85.8cm,平均单株材积9.984 7m³。其中Ⅰ级优树为54株,占候选优树的22.88%;Ⅱ级优树为114株,占候选优树的48.31%,Ⅲ级优树(一般林木)为68株,占候选优树的28.81%,Ⅲ级优树淘汰不选。Ⅰ级优树、Ⅱ级优树共168株,其 I_i 值均值 $I_{优}$ 为 1.899 9m³。

2.5 遗传增益预估

表型值是基因型值与环境型值之和,秃杉单株材积实测值为其表型值,多元数量化回归模型Ⅰ计算出的理论值为其环境型值,I_i 值是秃杉单株材积实测值与理论值间的差值,代表其基因型值,在不区分基因的加性和非加性效应时,其广义遗传力为1,因而据此对复选出的秃杉优树群体进行遗传增益预估。遗传增益(ΔG)等于168株复选优树 I_i 值均值($I_{优}$)除以236株候选优树平均单株实测材积($V_{均}$),即:$\Delta G = I_{优} / V_{均} \times 100\%$。代入 $I_{优}$ 值 1.899 9m³,$V_{均}$ 值 9.123 7m³,遗传增益(ΔG)预估值为20.82%。

3 结论与讨论

（1）在13个县（市、区）选出候选优树236株,复选出Ⅰ级优树为54株,Ⅱ级优树为114株,共168株,预估遗传增益达20.82%。为秃杉遗传改良储备了珍贵的种质材料。

（2）秃杉开花结实树龄在不同起源林木上差别较大,人工起源的林木开花结实较早,天然起源的林木开花结果较晚。对236株候选优树树龄与结实状况分析结果表明,秃杉优树的结实树龄≥25a。

（3）采用多元数量化回归模型Ⅰ对秃杉各立地因子、起源、立木类型和树龄与材积进行多元数量化回归,其复相关系数为0.799。偏相关系数大小可表征因子对秃杉生长的影响及大小,在诸因子中,对秃杉材积生长影响较大的主要是树龄、起源、立木类型、坡度和纬度等5个因子,而经度、海拔、坡向、坡位、坡形、基岩、黑土层厚度、土壤类型对秃杉材积生长的影响相对较小。

（4）采用多元数量化回归模型Ⅰ建立秃杉候选优树单株材积理论值回归方程,用候选优树材积实测值与理论值间的差值I_i做为优树复选指标,有效剔除了立地因子、起源、立木类型、树龄等因子的影响,减少表型选择的误差及误选率。这一方法有效解决了以纯林、混交林、散生木、孤立木多种形式存在,树龄差异大,起源多样,立地条件复杂的秃杉优树复选难题。

（5）针叶树种在人工林或同龄的天然纯林中选优常采用综合评分法、5株(3株或4株)优势木对比树法、小样地法,在天然林特别是异龄林或混交林中选优常采用回归法、绝对生长量法[17],在秃杉人工异龄林中选优采用"优势系数选择法"[16],这些方法所建立的优树复选指标都是基于表型选择的指标值。基于多元数量化回归模型Ⅰ的秃杉优树选择方法,其依据单株材积建立的优树复选指标值基本上代表了秃杉在材积生长量方面的基因型表现。从理论上说基于基因型的秃杉优树选择方法较基于表型选择的秃杉优树选择方法在优树的复选上更能真实地反映优树的遗传差异,准确性更高。基于数量化回归模型Ⅰ的优树选择方法是对行业标准中回归法的进一步细化和创新。

参考文献：

[1]罗良才,徐莲芳.秃杉木材物理力学性质的研究[J].云南林业科技,1982(1):24-35.
[2]云南省林业科学研究所.云南主要树种造林技术[M].昆明:云南人民出版社,1985,74~77.
[3]陶国祥.秃杉[M].昆明:云南科学技术出版社,2001.
[4]洪菊生,潘志刚,施行博,等.秃杉的引种与栽培研究[J].林业科技通讯,1997(1):5~12.
[5]湖北省林业厅.湖北林木种质资源[M].武汉:湖北科学技术出版社,1993.
[6]漆荣.秃杉地理种源变异的研究[D].武汉:华中农业大学,2005.
[7]李晓储,黄利斌,杨继明,等.秃杉种源苗期地理变异的研究[J].江苏林业科技,1993(2):1~8.
[8]张新华,鄢洪星,王律军,等.豫南引种秃杉生长规律研究[J].河南林业科技,2011(3):18~20.
[9]陈建新,王明怀,殷祚云,等.广东省秃杉引种栽培效果及栽培区划分研究[J].林业科学研究,2002,15(4):399~405.
[10]愈慈英.舟山海岛秃杉引种试验[J].浙江林学院学报,1994,11(1):26~32.
[11]李乃铨,杨启邦.秃杉引种试验[J].江苏林业科技,1996(1):24,49.
[12]刘亚俊.晚松、桤木、秃杉、福建柏引种试验简报[J].江西林业科技,1993(6):27.

[13] 宋丛文,张新叶,胡兴宜,等. 秃杉种源遗传多样性的 RAPD 分析[J]. 湖北林业科技,2004(4):1~4.
[14] 杨琴军,陈光富,刘秀群,等. 湖北星斗山台湾杉居群的遗传多样性研究[J]. 广西植物,2009,29(4):450~454,567.
[15] 陈光富,杨琴军,陈龙清. 台湾杉 DNA 提取及 RAPD 反应体系的建立[J]. 湖北农业科学,2008,47(10):1108~1110,1121.
[16] 白中萍. 秃杉优树选择及无性系种子园营建技术[J]. 林业调查规划,2004,29(2):61~63,70.
[17] 陈强,常恩福,董福美,等. 云南松天然优良林分自由授粉混合种子子代测定[J]. 林业科学,1998(5):40~46.
[18] 董文泉,周光亚,夏立显. 数量化理论及其应用[M]. 长春:吉林人民出版社,1979:1~48.
[19] 李淡清. 蓝桉、直干桉优树选择研究[J]. 林业科学,1990,26(2):167~174.
[20] LY/T 1344—1999. 主要针叶造林树种优树选择技术[S]. 1999.

(本文发表于《西南林业大学学报》,2017 年)

云南2~7年生直干桉人工林土壤物理性状研究

蒋云东[1]，周云[1]，阚忠明[1]，许林红[1]，石忠强[2]，王一新[2]，杨忠丽[2]，李思广[1]

(1. 云南省林业科学院，云南 昆明 650201，2. 富民县林业局，云南 富民 650400)

摘要：以云南省2~7年生直干桉人工林为研究对象，采用环刀法测量其林下土壤的物理性状及粘粒含量，为直干桉人工林可持续经营提供科学依据。结果表明：云南滇中、滇南一带的直干桉人工林土壤结构一般较好，土壤容重普遍在1.1~1.4g·cm^{-3}，表现出由上到下逐渐变大的规律和随树龄的增加有先降后升的趋势；土壤总孔隙度随树龄的增加表现出先升后降再上升的特点；毛管孔隙度平均为42.90%~53.19%，具有较好的持水性；土壤饱和持水量平均为44.24%~55.22%，而且随树龄的增加也有先升后降再上升的趋势；土壤黏粒含量普遍大于25%，质地多为重黏土，土壤容易板结，干旱时变得坚硬。

关键词：直干桉；土壤容重；孔隙度；黏土含量

Soil Physical Properties of *Eucalyptus maidenii* Plantations in Yunnan Province

Jiang Yun-dong[1], Zhou Yun[1], Kan Zhong-ming[1], Xu Lin-hong[1],
Wang Yi-xin[2], Yang Zhong-li[2], Li Si-guang[1]

(1. Yunnan Academy of Forestry, Kunming Yunnan 650201, P. R. China;
2. Fumin County Forestry Bureau, Fumin Yunnan 650400, P. R. China)

Abstract: In this paper, the soil physical properties of *Eucalyptus maidenii* 2-7 years old in Yunnan Province were studied. The results showed that the soil structure of *E. maidenii* plantations in central Yunnan and southern Yunnan was generally good, and the soil bulk density was generally 1.1-1.4g·cm^{-3}, a gradually increasing tendency from top to bottom and decreasing with age increasing at the beginning then increasing at later; The total soil porosity showed the tendency of increase-decrease-increase mode; The average capillary porosity was 42.90%-53.19%, with good water binding capacity; The average soil saturated water content was 44.24%-55.22%, and with the increase of tree age, showed a tendency of increase-decrease-increase mode; The soil clay content was generally more than 25%, the soil texture was mostly heavy clay, easy to harden, hard in drought.

Key words: *Eucalyptus maidenii*; soil bulk density; soil porosity; clay content

桉树(*Eucalyptus*)生长迅速、经营周期较短，并且单位面积内出材率高，因此，其经济价值显著[1]。滇中、滇南一带的桉树种植以直干桉(*Eucalyptus maidenii*)居多。直干桉利用率高，其树干可生产木材或造纸，梢头可做活性炭，树根可做根雕，树皮能制栲胶，枝叶可蒸烤桉油。在云南许多地方甚至将直干桉树下的枯枝落叶都收集，用于蒸油，因此直干桉经营过程对其林下土壤肥力的消耗很大[2,3]。土壤物理性质是评价土壤肥力和生态效益的重要指标，土壤物理性状的好坏在很大程度上受制于土壤有机质含量的多寡，由于直干桉几乎全树被利用，枝叶等有机物归还土壤较少，所以，种植直干桉等对土壤物理性质的影响程度有多大,值得认真研究。国内许多学者与此相关的报道有:谢直兴等桉树人工林现状及其可持续发展[4],徐柳斌等滇西山地桉树林土壤物理性质研究[5],黄影霞等马尾松($Pinus$ $massoniana$ Lamb.)改植桉树后对土壤物理性质的影响[6],李东海等桉树人工林林下植被、地面覆盖物与土壤物理性质的关系[7],吕祥涛的尾巨桉($E.$ $urophylla×E.$ $grandis$)人工林土壤物理性质变化的研究[8]等。但是,直接全面系统针对直干桉土壤物理性状的研究未见报道。本文以云南省昆明、曲靖、楚雄和红河等州(市)种植的2~7年生直干桉林为研究对象,用环刀法测量其土壤容重、孔隙度等物理性状[9],并采集土样,在实验室测量土壤黏粒含量(方法采用GB/T50123—1999)[10],为直干桉人工林可持续经营评价提供科学依据。

1 材料与方法

1.1 直干桉林地基本情况

云南直干桉林地土壤为红壤和紫色土,pH值4.5~6。其主要土壤养分含量和直干桉生长量见表1。

表1 直干桉调查点的树高、胸径和0~40cm土层的土壤养分含量
Table 1 Tree height, DBH of E. maidenii plantations and soil nutrient contents within the soil layer of 0~40cm

树龄(a)	树高(m)	胸径(cm)	pH值	有机质 (g·kg^{-1})	水解性N (mg·kg^{-1})	有效P (mg·kg^{-1})	速效K (mg·kg^{-1})	有效B (mg·kg^{-1})
2.0	5.30	5.38	5.05	15.80	36.15	12.25	83.27	0.21
3.0	7.78	6.09	5.33	9.82	26.51	12.76	89.11	0.13
4.0	9.99	7.88	4.92	11.01	21.69	14.89	74.19	0.24
5.0	10.60	8.65	5.62	18.42	43.68	7.59	70.80	0.16
6.0	12.13	10.38	6.46	17.98	48.35	19.86	92.08	0.18
7.0	12.30	10.39	5.36	17.27	37.20	14.48	54.81	0.17

云南省的直干桉种植密度为2 500~5 000株·hm^{-2},穴状整地40cm×40cm×40cm或撩壕整地60cm×40cm等,定值时间多在6~7月。一般底肥施农家肥5~10kg/株,年追施尿素200~300g·株$^{-1}$;或者底肥施复合肥600kg·hm^{-2}、钙镁磷肥750kg·hm^{-2},年追施尿素150kg·hm^{-2}等,每年抚育1~2次。直干桉林下主要植物有:棠梨树($Pyrus$ $calleryana$ Decne.)、厚皮香[$Ternstroemiagy$ $mnanthera$(Wight et Arn.)]、黄锁梅($Rubusellipticus$ Smith var. $obcordatus$ Focke)、三颗针($Berberis$ $pruinosa$ Franch.)、野坝子($Artemisia$ $lavandulaefolia$ DC.)、鬼针草($Bidens$ $pilosa$ Linn.)、紫茎泽兰($Eupatorium$ $adenophora$ Spreng.)、粘粘草($Cynoglossum$ $wallichii$ var. $glochidiatum$)、白蒿($Arteimisia$ $scoparia$ waldst. et Kit.)、犁头筋($Viola$ $japonica$ var. $stenopetala$ Franch. ex H. Boissieu.)、马蹄金($Dichondra$ $repens$ Forst.)、酢浆草($Oxalis$ $corniculata$ L.)、铁满蕨($Neottopteris$ $nidus$ J. Sm.)、狼毒($Stelleracha$ $maejasme$ Linn.)、翻白叶($Potentilla$ $griffithii$ Hook. f. var. $velutina$ Card.)、地桃花($Urenal$ $obata$ L.)、水锦树($Wendlandia$ $cavaleriei$ Lévl.)等。

1.2 直干桉林下土壤物理性质调查方法

2017年1~3月对滇中、滇南桉树种植区昆明、曲靖、楚雄和红河等州(市)的主要品

种—直干桉开展了标准地调查。共设置45块20m×20m的标准地,其中2年生、3年生和4年生的直干桉林各5块,5年生、6年生和7年生的直干桉林各10块。调查内容将主要包括:地理位置(经纬度、海拔)、地形因子(包括坡度、坡向、坡位)、林龄、品系、密度(株行距)、直干桉进行每木检尺(调查内容:树高、胸径、冠幅)。然后在标准地内依据土壤采样规则挖3个土壤剖面做土壤调查,采用环刀法测定0~20cm、20~40cm、40~60cm、60~100cm的土壤容重、总孔隙度、毛管孔隙度、饱和持水量等指标。并取0~40cm土层混合样和40~100cm土层混合样共计90个土样,送检测定黏土含量等。

黏土含量由云南省农业科学院(云南悦分环境检测有限公司)测定,测定方法采用GB/T50123—1999。

2 结果与分析

2.1 直干桉各土层土壤物理性质的测定结果

将云南2~7年生直干桉林下各土层土壤物理性质的测定结果列入表2。由表2可以看出:直干桉人工林的土壤容重大多在1.1~1.4g·cm^{-3},土壤结构较好。其土壤容重的平均取值区间为1.13~1.46g·cm^{-3},其中0~20cm土层的土壤容重1.13~1.37g·cm^{-3},平均1.25g·cm^{-3};20~40cm土层土壤容重的1.20~1.36g·cm^{-3},平均1.31g·cm^{-3};40~60cm土层土壤容重的1.20~1.43g·cm^{-3},平均1.33g·cm^{-3};60~100cm土层土壤容重的1.22~1.46g·cm^{-3},平均1.36g·cm^{-3}。呈现出由上到下土壤容重逐渐增加的规律,60cm以下土层土壤容重部分超过1.4g·cm^{-3},土壤结构变差。

表2 2~7年生直干桉林下4个土层的土壤物理性质
Tab. 2 Physical properties of soil from different profile layers of 2~7 years old *E. maidenii* plantations

测定土层(cm)	树龄(a)	土壤容重(g·cm^{-3})	总孔隙度(%)	毛管孔隙度(%)	饱和持水量(%)	取样深度(cm)	树龄(a)	黏粒(%)
0~20	2.0	1.37	48.37	42.90	44.24	0~40	2.0	48.26
	3.0	1.23	53.69	46.56	51.52		3.0	54.5
	4.0	1.13	57.32	45.99	48.85		4.0	50.02
	5.0	1.29	51.16	44.66	47.50		5.0	44.14
	6.0	1.29	51.33	47.08	49.07		6.0	44.56
	7.0	1.20	54.60	45.43	47.49		7.0	50.53
20~40	2.0	1.34	49.57	46.04	47.28			
	3.0	1.32	50.06	44.31	50.11			
	4.0	1.20	54.56	51.70	54.59			
	5.0	1.36	48.84	45.10	47.38			
	6.0	1.36	48.62	46.37	48.87			
	7.0	1.25	52.65	50.08	51.44			

(续)

测定土层(cm)	树龄(a)	土壤容重(g·cm^{-3})	总孔隙度(%)	毛管孔隙度(%)	饱和持水量(%)	取样深度(cm)	黏土含量	
							树龄(a)	黏粒(%)
40~60	2.0	1.43	45.93	43.80	45.27	40~100	2.0	54.32
	3.0	1.30	50.78	46.45	52.93			
	4.0	1.20	54.80	53.19	55.22		3.0	55.84
	5.0	1.43	45.99	42.98	45.20			
	6.0	1.35	48.88	47.20	50.38		4.0	51.72
	7.0	1.28	51.89	50.52	52.56			
60~100	2.0	1.46	45.01	43.22	44.82		5.0	49.3
	3.0	1.41	46.92	43.30	50.17			
	4.0	1.22	54.10	51.99	54.57		6.0	54.79
	5.0	1.42	46.30	43.74	45.50			
	6.0	1.36	48.57	47.28	50.42		7.0	59.6
	7.0	1.27	52.16	50.28	52.66			

直干桉人工林的土壤总孔隙度,一般为45%~60%,为黏质土。土壤总孔隙度的平均取值区间为45.01%~57.32%,其中0~20cm土层的土壤总孔隙度48.37%~57.32%,平均52.75%;20~40cm土层的土壤总孔隙度为48.62%~54.56%,平均50.72%;40~60cm土层的土壤总孔隙度层为45.93%~54.80%,平均51.89%;60~100cm土层的土壤总孔隙度为45.01%~54.10%,平均48.84%。直干桉林下呈现出由上层到下层其土壤总孔隙度逐渐减少的规律。

直干桉人工林土壤毛管孔隙度的平均取值区间为42.90%~53.19%,其中0~20cm土层的土壤毛管孔隙度为42.90%~47.08%,平均45.44%;20~40cm土层的土壤毛管孔隙度为44.31%~51.70%,平均47.27%;40~60cm土层的土壤毛管孔隙度为42.98%~50.52%,平均47.36%;60~100cm土层的土壤毛管孔隙度为43.22%~51.99%,平均46.64%。除0~20cm土层的土壤毛管孔隙度较低外,其他3个土层基本相近。

直干桉人工林土壤饱和持水量的平均取值区间为44.24%~55.22%,其中0~20cm土层的土壤饱和持水量为44.24%~49.07%,平均48.11%;20~40cm土层的土壤饱和持水量为47.28%~54.59%,平均49.95%;40~60cm土层的土壤饱和持水量为45.20%~55.22%,平均50.26%;60~100cm土层的土壤饱和持水量为44.82%~54.57%,平均49.69%。除0~20cm土层的土壤饱和持水量较低外,其他3个土层变化不大。

直干桉人工林土壤黏粒含量平均取值区间为44.14%~59.60%,其中0~40cm土层的土壤黏粒含量44.14%~54.5%,平均48.67%;40~100cm土层的土壤黏粒含量为49.3%~59.60%,平均54.26%。

2.2 不同树龄的土壤物理性状

由图1可见,直干桉林下各土层的土壤容重随树龄的增加均呈现先降后升的趋势,但土壤结构普遍较好,土壤容重一般在 1.1~1.4g·cm⁻³。2~7 年生直干桉林下 0~100cm 土层的平均土壤容重分别为:2 年生 1 411.68g·cm⁻³、3 年生 1 331.68g·cm⁻³、4 年生 1 191.68 g·cm⁻³、5 年生 1 391.68g·cm⁻³、6 年生 1 351.68g·cm⁻³ 和 7 年生 1.25g·cm⁻³,其中土壤容重最低值 0.82g·cm⁻³;最高值 1.68g·cm⁻³。

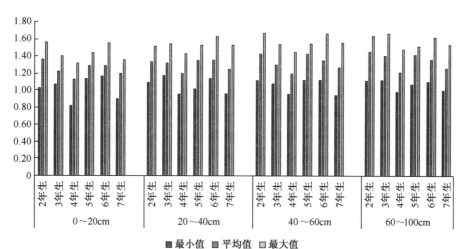

图 1 2~7 年生直干按各土层的土壤容重(g·cm⁻³)

Figure 1 Soil bulk densities of different soil layers of 2~7 years old I. *maidenii* plantation

由图 2 可见,直干桉各土层的土壤总孔隙度随树龄的增加均有先升后降再上升的趋势,土壤总孔隙度一般在 40%~60% 之间,具有较好的通透性。2~7 年生直干桉林下 0~100cm 土层的平均土壤总孔隙度分别为:2 年生 46.78%、3 年生 49.67%、4 年生 54.98%、5 年生 47.72%、6 年生 49.19% 和 7 年生 52.69%,其中土壤总孔隙度最高值 69.03%、最低值 36.79%。

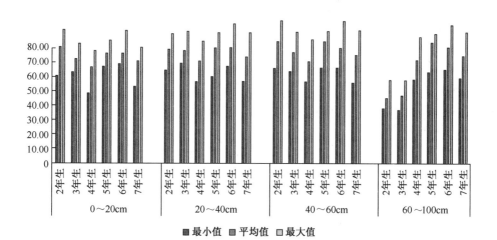

图 2 2~7 年生直干桉林下各土层的土壤总孔隙度(%)

Figure 2 Porosity of different soil layers of 2~7 years old *E. maidenii* plantations

注:图 2 中纵坐标为土壤总孔隙度(%);横坐标为树龄(年)。

2~7年生直干桉林下各土层的土壤毛管孔隙度随树龄的增加均有先升后降再上升的趋势,土壤毛管孔隙度一般在40%~60%之间,具有较好的通透性(见图3)。2~7年生直干桉林下0~100cm土层的平均土壤毛管孔隙度分别为:2年生43.84%、3年生44.79%、4年生50.97%、5年生44.04%、6年生47.04%和7年生49.32%,其中土壤毛管孔隙度最高值63.14%、最低值25.57%。毛管孔隙是土壤水分贮存和水分运动最强的地方,毛管孔隙的数量直接反映出土壤质地、结构等性状。云南2~7年生直干桉其林地多为黏质土,由图3可以看出,其土壤毛管孔隙度较高,说明土壤细孔隙较多,持水性较强。

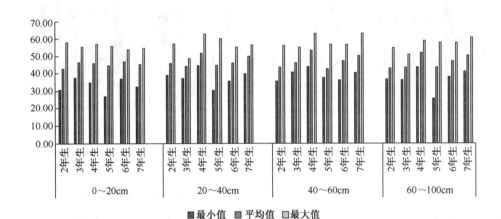

图3　2~7年生直干桉林下各土层的土壤毛管孔隙度(%)

Figure 3　Capillary porosity of different soil layers of 2~7 years old *E. maidenii* plantations

注:图3中纵坐标为土壤毛管孔隙度(%);横坐标为树龄(年)。

2~7年生直干桉林林下各土层的土壤饱和含水量随树龄的增加也有先升后降再上升的趋势,平均土壤饱和含水量一般在44.24%~55.22%,具有较好的持水性(见图4),但持水性过强的林地,会造成直干桉湿害。2~7年生直干桉林下0~100cm土层的平均土壤饱和含水量分别为:2年生45.29%、3年生50.98%、4年生53.56%、5年生46.21%、6年生49.83%和7年生51.36%,其中土壤饱和含水量最高值68.38%、最低值25.81%。

直干桉人工林2~7年生树龄的林下平均土壤黏粒含量0~40cm土层分别为:2年生48.26%、3年生54.5%、4年生50.02%、5年生44.14%、6年生44.56%和7年生50.53%,其中土壤黏粒含量最高值70.8%,最低值21.0%;40~100cm土层分别为:2年生54.32%、3年生55.84%、4年生51.72%、5年生49.3%、6年生54.79%和7年生59.6%,其中土壤黏粒含量最高值76.2%,最低值17.6%,黏粒含量普遍大于25%,质地多为黏土,土壤容易板结,干旱时坚硬。仅少数林地的黏粒含量17%~25%,为壤土。

3　结论

(1)云南2~7年生直干桉人工林林下的土壤容重大多在1.1~1.4g·cm^{-3},土壤结构普遍较好。直干桉林下呈现出由上到下土壤容重逐渐增加的规律,但60cm以下土层土壤容重部分超过1.4g·cm^{-3},土壤结构变差。此外,随树龄的增加土壤容重呈现先降后升的趋

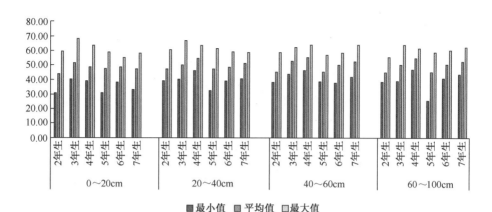

图 4 2~7 年生直干桉各土层的土壤饱和含水量(%)

Figure 4 Saturated soil water contents of different soil layers of 2~7 years old *E. maidenii* plantations

注:图 4 中纵坐标为土壤饱和含水量(%);横坐标为树龄(年)。

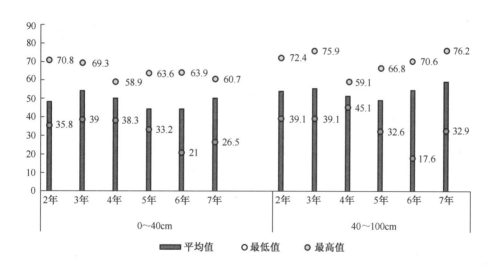

图 5 2~7 年生直干桉各土层的土壤黏粒含量(%)

Figure 5 Clay particle contents of different soil layers of 2~7 years old *E. maidenii* plantations

注:图 5 中纵坐标为土壤黏粒含量(%);横坐标为树龄(年)。

势,2~7 年生直干桉林下 0~100cm 土层的平均土壤容重分别为:2 年生 1 411.68g·cm^{-3}、3 年生 1 331.68g·cm^{-3}、4 年生 1 191.68g·cm^{-3}、5 年生 1 391.68g·cm^{-3}、6 年生 1 351.68g·cm^{-3}和 7 年生 1.25g·cm^{-3}。

(2)云南 2~7 年生直干桉人工林的土壤总孔隙度,一般在 45%~60%,为黏质土。呈现出由上到下土壤总孔隙度逐渐减少的规律,而且,随树龄的增加土壤总孔隙度有先升后降再上升的趋势,2~7 年生直干桉林下 0~100cm 土层的平均土壤总孔隙度分别为:2 年生 46.78%、3 年生 49.67%、4 年生 54.98%、5 年生 47.72%、6 年生 49.19%和 7 年生 52.69%。

(3) 2~7年生直干桉林地土壤毛管孔隙度的平均取值区间为42.90%~53.19%,具有较好的持水性性,林下各土层的土壤毛管孔隙度随树龄的增加均有先升后降再上升的趋势。2~7年生0~100cm土层的平均土壤毛管孔隙度分别为:2年生43.84%、3年生44.79%、4年生50.97%、5年生44.04%、6年生47.04%和7年生49.32%。

(4) 2~7年生直干桉林林地土壤饱和持水量的平均取值区间为44.24%~55.22%,除0~20cm土层较低外,其他3层变化不大。各土层的土壤饱和含水量随树龄的增加也有先升后降再上升的趋势,土壤饱和含水量一般在44.24%~55.22%%之间,具有较好的保水性,2~7年生直干桉林下0~100cm土层的平均土壤饱和含水量分别为:2年生45.29%、3年生50.98%、4年生53.56%、5年生46.21%、6年生49.83%和7年生51.36%。

(5) 直干桉人工林的土壤黏粒含量平均取值区间为44.14%~59.60%。2~7年生直干桉林下的平均土壤黏粒含量0~40cm土层分别为:2年生48.26%、3年生54.5%、4年生50.02%、5年生44.14%、6年生44.56%和7年生50.53%;40~100cm土层分别为:2年生54.32%、2年生55.84%、2年生51.72%、2年生49.3%、2年生54.79%和2年生59.6%。2~7年生直干桉人工林林下土壤黏粒含量普遍大于25%,质地多为重黏土,土壤容易板结,干旱时变得坚硬。

参考文献:

[1] 石忠强,蒋云东,周志忠,等. 云南桉树研究现状和存在的问题[J]. 西部林业科学,2015,44(1):152~156.

[2] 张荣贵,蒋云东. 桉树与环境[J]. 云南林业科技,1998(1):52~56,73.

[3] 杨民胜,陈少雄. 桉树生态问题的来源与对策[J]. 桉树科技,2002(2):9~16.

[4] 谢直兴,严代碧. 桉树人工林现状及其可持续发展[J]. 四川林业科技,2006(1):75~81.

[5] 徐柳斌,陆梅,向仕敏. 滇西山地桉树林土壤物理性质研究[J]. 山东林业科技,2007,173(6):41~43.

[6] 黄影霞,张凤梅,黄承标. 马尾松改植桉树后对土壤物理性质的影响[J]. 现代农业科技,2013(17):236~237,240.

[7] 李东海,杨小波,邓运武,等. 桉树人工林林下植被、地面覆盖物与土壤物理性质的关系[J]. 生态学杂志,2006,25(6):607~611.

[8] 吕祥涛. 尾巨桉人工林土壤物理性质变化的研究[J]. 内蒙古林业调查设计,2011,34(4):121~122.

[9] 中国科学院南京土壤研究所. 土壤理化分析[M]. 上海:上海科学技术出版社,1981:511~514.

[10] 中华人民共和国国家标准编写组,GB/T50123—1999 土工试验方法标准[S]. 北京:中国计划出版社,1999.

(本文发表于《西部林业科学》,2018年)

山桂花人工林林木材性与生长特性关系研究

冯 弦,陈宏伟,刘永刚,李莲芳,周云,孟梦,李江

(云南省林业科学院,云南 昆明 650204)

摘要:对山桂花人工林林木木材基本密度和纤维长度径向变异的研究表明,山桂花人工林林木的木材材积加权平均基本密度为 0.482 6g·cm^{-3},低于山桂花天然林林木的木材基本密度(0.530 0g·cm^{-3});山桂花人工林林木木材纤维长度为 1 182.4~1 596.4μm,材积加权平均值为 1 375.45 μm,低于山桂花天然林林木的木材纤维长度(1 200~2 250μm)。通过研究还得出了基本密度(BD)、纤维长度(FL)与生长轮(年龄 CA)和生长轮宽度(RW)。数学模型为 FL= 1269.8641+ 27.0207CA- 5.4443RW;BD= 0.4782- 0.0024CA+ 0.0029 RW,以此分析了木材性质与树木生长特性之间的关系,为速生用材林的培育提供依据。

关键词:山桂花;人工林;木材材性;生长特性

A Study of the Relationship between Wood Characteristics and Growth of *Paramichelia baillonii* Plantation

Feng Xian, Chen Hong-wei, Liu Yong-gang, Li Lian-fang, Meng Meng, Li Jiang

(Yunnan Academy of Forestry, Kunming Yunnan 650204, China)

Abstract: The radial variation of wood density and fiber length of *Paramichelia baillonii* was studied. The resultsshowed that theweighted average of wood basic density of Paramichelia baillonii in plantation was 0.482g·cm^{-3}, whichis lower than that of natural forest (0.530g·cm^{-3}). The weighted average of wood volume was 1 375.45μm. Theweighted average of f iber length of was 1 182.4 ~ 1 596.4μm, which was lower than that of natural forest (1 200~2 250μm). The mathematic model of basic density, fiber length and cambium age and ring width was deduced as FL = 1269.8641 + 27.0207CA- 5.4443RW;BD= 0.4782- 0.0024CA+ 0.0029 RW. The analysis on the relationshipbetween wood characterist ics and growth, provided the cultivation and of rapid growing timber using plantat ion with ascientific basis.

Key words: *Paramichelia baillonii*; plantation; wood characteristics; growth characteristics

山桂花(*Paramichelia baillonii*)也称拟含笑、合果木(《中国树木志》)、大果白兰、埋洪(傣),为木兰科拟含笑属的一种高大常绿乔木,是云南南部主要的优良速生用材树种之一。山桂花生长迅速,树干通直圆满,其木材花纹美观,是胶合板、木地板、刨切单板和家具的优良用材。西双版纳州普文试验林场 20 世纪 80 年代营造的山桂花人工试验林,现已基本成材。过去对该树种的材性研究仅限于天然林,随着人工林的大量营造对山桂花人工林材性的研究尤显重要。作为工业用材林要求生长迅速、培育周期短,而树木的生长速度对木材性质有一定的影响。通过对山桂花人工林植株不同部位木材基本密度、纤维长度变异的分析,

得出山桂花人工林木材性质与树木生长特性间变异规律的数学模型,为速生用材林的培育提供依据。

1 材料与方法

1.1 试验材料

试验所用的山桂花人工林林木采自云南省西双版纳普文试验林场1987年定植的山桂花纯林,初植密度2m×3m。按照国家标准《木材物理力学试材采集方法》GB1927—91的规定进行取样。标准地为14龄山桂花人工纯林,坡向西向,平均树高15.87m,平均胸径13.94cm;在同一标准地中取不同径级的10株山桂花作样木,在树干1.3m处截取5cm厚的圆盘作试材。样木情况见表1。

表1 山桂花样木的试材情况
Table 1 Test wood of *Paramichelia baillonii* plantation

样木号	胸径(cm)	树高(m)	枝下高(m)	冠幅(m × m)
A1	14.0	15.1	9.8	2.5×3.2
A2	14.7	15.0	10.0	3.0×3.8
A3	9.8	14.9	9.0	2.8×3.5
A4	13.2	15.7	9.9	3.0×3.5
A5	19.0	19	12.4	3.5×4.0
A6	13.5	15.9	9.3	2.5×3.5
A7	12.0	15.1	9.1	3.5×3.0
A8	21.5	20.4	11.8	4.8×4.8
A9	11.8	16.2	9.8	2.3×2.8
A1	8.0	11.6	8.4	3.4×2.0

1.2 取样及试验方法

分别从圆盘的髓心朝树皮南北向的每一生长轮上取试样,每一试样包含一个年轮(早材和晚材),共取试样480件。木材基本密度试验采用饱和含水率法[1]进行;纤维长度采用HNO_3、$KClO_3$离析法,在显微投影仪上测定,每个生长轮随机测量50个纤维长度。采用EXCEL、SPSS1010、朗奎健BASIC程序集[2]等进行数据处理和分析。

表2 山桂花人工林样木木材基本密度径向变异
Table 2 Radial variation of basic density of *Paramichelia baillonii* plantation

年轮	1	2	3	4	5	6	7	8	9	10	11	12	13
X	0.4125	0.4454	0.4574	0.4804	0.4801	0.4768	0.5005	0.5076	0.5138	0.4883	0.4856	0.4885	0.4653
Dmax	0.0645	0.1146	0.0869	0.0387	0.0564	0.0569	0.0823	0.0885	0.0677	0.0999	0.0888	0.0873	0.04
C.V.	7.2	10.6	7.8	3.4	4.3	4.7	7	7.8	5.7	7.4	7.2	7.9	3.5

注:X-均值(g/cm^3),D_{max}-极差(g/cm^3),C.V.变异系数(%)。下同。

2 结果与分析

2.1 木材性质的径向变异

2.1.1 基本密度

山桂花人工林样木木材基本密度的径向变异状况如表 2,其方差分析结果见表 3。表 3 方差分析结果表明不同年龄木材基本密度存在着显著的差异($F = 4.62 > F_{0.05} = 1.84$)。山桂花人工林林木木材材积加权平均基本密度为 $0.482 \text{g} \cdot \text{cm}^{-3}$,低于山桂花天然林林木的木材基本密度($0.530 \text{g} \cdot \text{cm}^{-3}$)[3]。

表 3 山桂花人工林样木基本密度方差分析结果

Table 3 Variance analysis of basic density of *Paramichelia baillonii* plantation

差异源	SS	df	MS	F	Fcrit
不同年龄	0.04	12	0.003	4.62	1.84
不同单株	0.1	9	0.114	15.57	1.96
误差	0.08	108	0.007		
总计	0.22	129			

表 4 山桂花人工林样木木材纤维长度径向变异

Table 4 Radial variation of fibre length of *Paramichelia baillonii* plantation

年轮	1	2	3	4	5	6	7	8	9	10	11	12	13
X	1 182	1 301	1 299	1 319	1 375	1 448	1 479	1 478	1 509	1 511	1 524	1 596	1 561
Dmax	84.2	216.6	275.7	207.9	384.8	123.4	228.4	262.9	228.9	112.3	186.5	113.1	131.4
C.V.	5.04	8.63	8.78	7.65	13.12	3.9	7.03	8.06	6.5	3.32	5.35	3.17	3.78

2.1.2 纤维长度

样木的木材纤维长度的径向变异见表 4。经方差分析,不同年龄纤维长度间有极显著的差异($F = 7.40 > F_{0.05} = 1.84$)(表 5)。山桂花人工林林木的纤维长度为 $1\,182.4 \sim 1\,596.4 \mu m$,材积加权平均值为 $1\,375.45 \mu m$,低于山桂花天然林林木的木材纤维长度($1\,200 \sim 2\,250 \mu m$)[3]。

表 5 山桂花人工林样木纤维长度方差分析

Table 5 Variance analysis of fibre length of *Paramichelia baillonii* plantation

差异源	SS	df	MS	F	Fcrit
不同年龄	2 535 526	12	211 293.8	7.40	1.84
不同单株	390 098.7	9	43 344.3	1.52	1.97
误差	3 084 134	108	2 555 679		
总计	6 009 758	129			

2.2 生长轮宽度的径向变异

山桂花人工林林木生长轮宽度的径向变异是从髓心向外呈逐渐减小的趋势。第一年生长较为迅速,到第 3、4 轮(年)增加到最大为 9mm,第 5 轮(年)后逐渐减小,趋于稳定。变化趋势与树木径生长规律相符。生长轮宽度与生长年轮之间回归方程为 $y = 10.157 e^{-0.1243x}$,回归相关系数 $R^2 = 0.8627$,相关显著(图 1)。山桂花天然林林木的生长轮宽度在 20 年时达到最大为 4.2mm,以后逐渐减小[4](图 2)。人工林林木的生长轮宽度明显大于天然林 (2.7~4.2mm)。

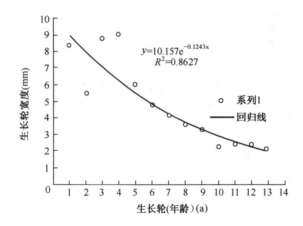

图 1　山桂花人工林林木木材生长轮宽度与年龄的回归曲线

Figure 1　Regression curve of ring width with cambium age of *Paramichelia baillonii* plantation

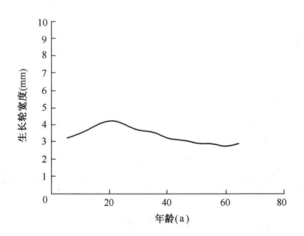

图 2　山桂花人工林林木木材生长轮宽度径向变异曲线

Figure 2　Radial variation curve of ring width with cambiumage of *Paramichelia baillonii* plantation

2.3　木材性质与生长轮(年龄)和生长速率的关系

生长轮(年龄)和生长程度是影响木材材质的两大因子,而基本密度和纤维长度是木材性质的重要指标[1,5],它们直接或间接影响木材的使用价值和加工利用性能。通过对纤维长度、基本密度、年轮宽度等的变化,建立山桂花人工林林木木材纤维长度、基本密度随生长轮(年龄)和生长速度变化的数学模型[6],以便研究材性与树木生长速度的关系。

采用多对多线性回归[2,7]对山桂花人工林林木的木材基本密度和纤维长度与生长轮

(年龄)和生长速率的关系进行分析,建立如下回归方程:
$$y_1 = b_0 + b_{11}x_1 + \cdots\cdots b_{1p}x_p$$
$$\cdots\cdots$$
$$\cdots\cdots$$

观测 n 个样本,令
$$X = \begin{bmatrix} x11\cdots x1p \\ \vdots \\ xn1\cdots xnp \end{bmatrix} \quad Y = X = \begin{bmatrix} x11\cdots x1p \\ \vdots \\ xn1\cdots xnp \end{bmatrix} \quad B = \begin{bmatrix} b_{10}\cdots b_{q0} \\ \vdots \\ b_{1p}\cdots b_{qp} \end{bmatrix}$$

正规方程为:$X'XB = X'Y$
$$B = (X'X)^{-1}X'Y$$

剩余协方差矩阵:$Q = Y'Y - B'X'Y$

表6 山桂花人工林林木木材纤维长度、基本密度与生长轮(年龄)、生长轮宽度的相关系数

Table 6 Relativecoefficient of fiber length and basic density withcambium age and ring width in *Paramichelia baillonii* plantation

	项目	纤维长度	基本密度
生长轮 (年龄)	相关系数	0.960 0**	-0.603 2*
	偏相关系数	0.811 949 8**	-0.193 410 4
生长轮 宽度	相关系数	-0.882 1**	0.596 8*
	偏相关系数	-0.188 648 1	0.161 036 4
	复相关系数 R	0.961 437 4	0.616 684 5

注:** 在 0.01 水平显著相关,* 在 0.05 水平显著相关。

结果表明,山桂花人工林林木木材纤维长度、基本密度与生长轮(年龄)、生长轮宽度简单相关系数较为显著($r = 0.596\ 8 \sim 0.960\ 0$),见表6。但由于是两个变量,故只有偏相关系数才能反映其本质联系,生长轮年龄与纤维长度相关显著;$rFL,CA = 0.811\ 949\ 8$,通过回归方程得出回归系数 $b_{10} = 1\ 269.864\ 1$,$b_{11} = 27.020\ 7$,$b_{12} = -5.444\ 3$;$b_{20} = 0.478\ 2$,$b_{21} = -0.002\ 4$,$b_{22} = 0.002\ 9$,木材材性与生长轮(年龄)和生长速度相关模型为:

$$FL = 1\ 269.864\ 1 + 27.020\ 7CA - 5.444\ 3RW \tag{1}$$
$$BD = 0.478\ 2 - 0.002\ 4CA + 0.002\ 9RW \tag{2}$$

其中:FL—纤维长度(um);BD—基本密度;CA—生长轮(年龄);RW—年轮宽度。

3 结论与讨论

山桂花人工林林木木材基本密度靠近髓心部较小($0.412\ 5\text{g}\cdot\text{cm}^{-3}$),随后逐渐增大,第9年达到最大($0.513\ 8\text{g}\cdot\text{cm}^{-3}$),并缓慢减小。由于试材年龄有限,只有13年之内的数据,按Pashin等的木材密度径向变异分类标准,应属于(Ⅰ)类,即自髓心至树皮,密度值呈直线或曲线增加,或在成熟材区域趋向平缓及至老树则有所下降;不同年龄的木材基本密度差异显著。山桂花人工林林木木材材积加权平均基本密度为 $0.482\ 6\text{g}\cdot\text{cm}^{-3}$,低于山桂花天然林林木的木材基本密度($0.530\ 0\text{g}\cdot\text{cm}^{-3}$)。

山桂花人工林林木木材的纤维长度在髓心附近最短为 1 182μm，并由髓心向外呈递增趋势，不同年龄的木材纤维长度差异极显著。管胞长度由髓心向外呈递增趋势，1~6 轮之间增加迅速，之后趋于稳定，到 13 轮达到最大值，明生长年轮与材性有一定的关系。在髓心附近管胞最短，这是由于树木生长初期，形成层原始母细胞垂周分裂速度快，但尚未成熟，表示树木处于幼龄期；随着树龄增大，形成层原始母细胞垂周分裂速度缓慢，子细胞的生长使管胞变长，当形成层原始母细胞成熟时，管胞长度趋于稳定，树木进入成熟期，管胞长度随树龄的变化曲线基本趋于平缓和稳定。山桂花人工林林木木材纤维长度为 118 214~159 614μm，材积加权平均值为 1 375 145μm，低于山桂花天然林林木的木材纤维长度（1 200~2 250μm）。

山桂花人工林林木的生长轮宽度有从髓心向外逐渐减小的趋势，第 1 年生长轮宽度较大（8.35mm），到第 2 年生长轮宽度明显减小（5.45mm），3、4 年生长轮宽度达到最大（9mm），5 年后趋于平稳并逐渐减小；这与山桂花天然林林木的生长轮宽度的径向变异有一定差异，山桂花天然林林木的生长轮宽度由髓心向外增大，到第 20 年达到最大（4.2mm）后逐渐减小。可以认为山桂花人工林生长速度明显大于天然林，人工林第 1 年在苗圃中育苗，水肥条件较好，生长迅速；而在天然林分中，种子自然萌芽，生长较为缓慢。人工林在第 2 年经移苗造林后形成蹲苗期，生长速度降低。

山桂花人工林林木的木材纤维长度（FL）、基本密度（BD）与生长轮年龄（CA）、生长轮宽度（RW）的相关模型为：$FL = 1 269.864\ 1 + 27.020\ 7CA - 5.444\ 3RW$；$BD = 0.478\ 2 - 0.002\ 4\ CA + 0.002\ 9\ RW$。山桂花人工林林木木材纤维长度、基本密度与生长轮（年龄）、生长轮宽度简单相关系数较为显著。

参考文献：

[1] 成俊卿. 木材学[M]. 北京：中国林业出版社，1985.

[2] 郎奎健. IBM PC 系列程系集：数理统计、调查规划、经营管理[M]. 北京：中国林业出版社，1987.

[3] 罗良才. 云南经济木材志[M]. 昆明：云南人民出版社，1989.

[4] 云南省林科所优良速生珍贵树种调查组. 云南优良速生珍贵树种专刊[J]. 云南林业科技通讯，1977（4）：57.

[5] 郭明辉. 天然林杉木材质变异规律的研究[J]. 世界林业研究，1995(8)：426~432.

[6] 鲍甫成. 人工林杨树材性与生长轮年龄和生长速度关系的模型[J]. 林业科学，1999，35(1)：77~82.

[7] 中国科学院数学研究所数理统计组. 回归分析法[M]. 北京：科学出版社，1975.

[8] 柴修武. 柠檬桉树幼龄材与成熟材构造的差异[J]. 世界林业研究，1995(8)：57~63.

（本文发表于《云南林业科技》，2003 年）

第二篇 生态环境修复研究

封禁措施对云南金沙江流域主要林分的生态功能影响

李贵祥,孟广涛,方向京,柴勇,和丽萍,张正海

(云南省林业科学院,云南昆明,650204)

摘要:通过封禁措施对林分的生态功能影响分析,认为:封禁后的云南松林和常绿阔叶林,在胸径或树高上都表现出明显的优势,且乔木上层盖度较大而均匀,优势种优势度明显,各优势种种群的年龄组成亦处于稳定状态,乔木树种多数更新良好,在垂直分布上呈连续分布,使得整个群落在结构上处于稳定的状态;实施封禁措施的林分物种丰富度和多样性指数都明显高于未封禁的林分,且封禁时间越长,多样性指数越高。封禁后物种多样性指数总体上表现为常绿阔叶林>针阔混交林>针叶林>灌丛;实施封禁措施,土壤得到改良,通透性能较好,封禁后的腐殖质层厚度、土壤总孔隙度、毛管孔隙度比未封禁的高。在有机质、有效氮、有效磷方面,封禁的比未封禁的高,封禁的阔叶林比封禁的针叶林高。体现了封禁后群落土壤具有较高的肥力,有利于群落生物生产力的提高,使群落生物生产力始终维持在较高水平,封禁后土壤疏松,持水能力增强,枯落物层多,对封禁前后的群落土壤最大持水量、枯落物最大持水量进行分析,得出封禁后林分的土壤最大持水量、枯落物最大持水量都高于未封禁的林分。

关键词:封禁措施;云南金沙江流域;生态功能

Effects of closing measures on ecosystem function of main forest in Jinshajiang Watershed of Yunnan province

Li Gui-xiang, Meng Guang-tao, Fang Xiang-jiang, Chai Yong,
He Li-ping, Zhang Zheng-hai

(Yunnan Academy of Forestry, Kunming Yunnan 650204)

Abstract: Effects of closing measures on ecosystem function of forest were analyzed. The results showed that the stability of community of Pinus yunnanensis forests and evergreen broad-leaved forest was improved byclosing measures, which mainly included increasing of breast diameter and height, improving of conspicuousnessand regeneration of most species of tree layer and continuity of its vertical distribution. Comparing with forest that of not bing closed, the abundance and biodiversity of that of being closed were higher. With closing measures, the abundance and biodiversity increased rapidly. The order of the biodiversity was evergreen broad-leaved forest>mixedconiferbroad-leavedfores>coniferousfores>shrub. Soil quality was also improved by closing measures. The thickness of soil humus, soil total porosity, soil capillary porosity, soilorgainic matter, available nitrogen and available phosphorus were higher than that of not being closed, whichindicated closing measures werecontribute to improving soil fertility, productivity of community. The results also showed that soil maximumwaterholdingcapacity, littermaximum waterholdingcapacity were higher than that of not being closed.

Key words:closing measures;Jinshajiang watershed of Yunnan;ecosystem function

封禁是封山育林的主要类型和主要技术手段。其实质就是通过封禁来减少人类对林地的干扰和破坏,给物种创造一个休养生息的机会,让其沿着自然演替的方向发展,并在发展中不断地改善自己的生存条件,从而使种群不断地繁衍和扩大[1]。采用封禁措施把天然(次生)植被生长状况较好、地块比较偏远、人和牲畜活动难以到达的地块[2],以及水土流失极其严重的地块。对一定规模的荒山、迹地、残林、灌丛等宜林地划界封禁,施以一定的技术措施,为一定的目的培育和管理森林[3]。利用林木天然更新能力、植物群落自然演替规律,使疏林、灌木林、散生木林、荒山等林业用地自然成林[4]。通过封禁,能使林地上的乔、灌、草等同时增长,形成种类多、层次复杂的群落结构[1,5]。

云南金沙江流域的植被大多受到一定程度的破坏,在群落结构、森林组成等方面发生了一定程度的退化,对这些在自然条件下或虽受到轻微干扰但经自然恢复仍能稳定发展的森林实行封禁,最大限度地减少人为破坏,让它在自然条件下沿着群落本身演替的规律进行发展,一定程度的退化,对这些在自然条件下或虽受到轻微干扰但经自然恢复仍能稳定发展的森林是保护与恢复这些植被最为直接而有效的途径,更有利于物种的保存和维持整个群落的稳定性,最大限度的发挥森林群落的功能。

1 研究区概况

研究区——永仁县,位于滇中高原北缘,与川西深切高中山一江(金沙江)之隔。地理位置为25°52′~26°32′N,101°19′~101°52′E。地势西北高,东南低,以中低山丘陵为主,最低海拔926m,最高海拔2 885m。金沙江流经永仁境段,长157km,县内主要河流有万马河、永定河、白马河、羊蹄河、江底河、永兴河,均为金沙江的一、二级支流。土壤主要为棕壤、黄红壤、中性紫色土、红壤、酸性紫色土。腐殖质层厚2~30cm,pH5.7~6.7,有机质含量0.349%~3.467%。气候为北亚热带西南季风气候,冬暖夏凉,干湿季分明,因相对高差较大而垂直气候明显,年均温11.6~19.5℃,年降水量900.0~1 295.3mm,90%左右集中在6~10月的雨季。

研究地植被主要有暖温性针叶林类型、温凉湿润常绿阔叶林、暖温性针阔混交林、干热河谷稀树草坡等。除干热河谷区外,以暖温性云南松(*Pinus yunnanensis*)针叶林类型为主。

2 研究方法

2.1 时空替代法

为扩展研究的时间尺度,研究采用"时空替代法",即以空间代时间对封禁30年、封禁35年、封禁50年和未采取封禁措施的不同林分类型,进行调查和对比分析[6]。把研究区设置的4块20m×20m的样地,对乔木进行每木调查。样地中每10m×10m样方中再设置2m×2m、1m×1m小样方各4块,分别调查灌木及草本植物。

2.2　群落多样性的测度

选用丰富度指数(S)、物种多样性指数和均匀度指数进行测定分析[7]。

2.3　土壤的物理及化学性能测定

在标准样地内随机设置3个测样点,采用国家标准方法,分析测定土壤理化性质[8]。

2.4　持水能力的测定[9]

2.4.1　枯枝落叶层含水量测定

在标准地内随机设置1m×1m的小样方,把小样方内的枯枝落叶全部收集,进行烘干称重,测得枯枝落叶层的干重和含水量,然后把枯枝落叶浸入水中2h以上,取出后称重,测得最大含水量。

2.4.2　土壤最大持水量测定

用环刀法测定最大持水量。

3　结果与分析

3.1　封禁措施对群落结构的影响

不论是云南松原始林还是半湿润的常绿阔叶林,在实施封禁措施后,群落结构均维持在较稳定的状态,且封禁时间越长,群落结构稳定性越明显。

3.1.1　封禁对云南松针叶林的结构影响

封禁后的云南松林,郁闭度达到0.65,密度667株·hm^{-2},树高31.29m,胸径达到40.55cm,见表1。垂直结构可分为乔木层、灌木层和草本层,乔木层以云南松为主,有时伴生光叶石栎(Lithocarpus mairei)、黄毛青冈(Cyclobalanopsis delavayi)、高山栲(Castanopsis delavayi)、锥连栎(Quercus franchetii)等而形成乔木亚层,这些混交树种林冠相互镶嵌,组成较稳定的复合结构。灌木层则以上层乔木的幼树(苗)居多,特别是云南松更新状况良好,各径阶均有一定比例存在,这也说明了封禁后林分结构较为稳定。未封禁的云南松林,由于受到人为干扰较为严重,其林下枯枝落叶层被用作肥料或燃料,其生长较为缓慢,平均树高为11.45m,平均胸径为11.19cm;林木株数为3 067株·hm^{-2},但很少有更新的云南松幼苗,在垂直结构上,基本没有灌木层或者灌木非常少。上述分析说明了未封禁的云南松林不如封禁后的云南松林稳定。

表1　实施封禁措施的群落结构特征表

群落	封禁年限(年)	地点	海拔(m)	土壤类型	树高(m)	胸径(cm)	密度(株·hm^{-2})	乔木层盖度(%)	灌木层盖度(%)	草本层盖度(%)	郁闭度
高山栲、锥连栎林	50	方山小黑箐	2 310	黄红壤	8.44	9.57	3 167	80	30	2	0.95
滇石栎、厚皮香林	35	方山静德寺	2 190	黄红壤	7.10	5.90	5 700	85	60	5	0.90

(续)

群落	封禁年限（年）	地点	海拔(m)	土壤类型	树高（m）	胸径（cm）	密度（株·hm^{-2}）	乔木层盖度(%)	灌木层盖度(%)	草本层盖度(%)	郁闭度
黄毛青冈、云南松林	35	白马吃水箐	2315	黄红壤	17.58	20.40	1 125	85	30	40	0.90
锥连栎、云南松林	未封禁	幸福水库	1860	黄红壤	3.30	4.65	2 567	60	40	70	0.65
云南松林	30	白马吃水箐	2365	黄红壤	31.29	40.55	667	65	15	25	0.65
云南松林	未封禁	森林经营所	1940	黄红壤	11.45	11.19	3 067	60	1	50	0.60
锥连栎灌丛	未封禁	永定林场	2020	黄红壤	2.20	2.50	—	—	50	90	0.15

3.1.2 封禁对阔叶林的结构影响

封禁后的常绿阔叶林，以封禁50年的林分来看，见表1，郁闭度达到0.95，密度3 167株·hm^{-2}，树高8.44m，胸径9.57cm，垂直结构乔木层可分为两个亚层，乔木上层高7~12m，以常绿的壳斗科植物为主，如高山栲、元江栲（*Castanopsis orthacantha*）、滇青冈（*Cyclobalanopsis glaucoides*）、光叶石栎等。乔木下层高3~7m，常见的有厚皮香（*Ternstroemia gymnanthera*）、云南木犀榄（*Olea yunnanensis*）、银木荷（*Schima argentea*）等。相比而言，未经封禁或封禁时间较短的林分则胸径和树高都较小，结构简单。可见，封禁后的林分无论在胸径或树高上都表现出明显的优势，且乔木上层盖度较大而均匀，优势种优势度明显，各优势种种群的年龄组成亦处于稳定状态，乔木树种多数更新良好，在垂直分布上呈连续分布，使得整个群落在结构上处于稳定的状态。未封区林树冠较小，层次简单，林下植物较少。

3.2 封禁后群落生物多样性变化状况

采取封禁措施，使林分在无干扰状态下自然演替，最终它将向着地带性植被常绿阔叶林的方向发育，而群落生物多样性也随着林分的发育而不断变化。表2是封禁条件下不同演替阶段的林分生物多样性调查结果。

表2 实施封禁措施的群落物种多样性

群落名称	封禁年限（年）	层次	物种丰富度 S	Simpson 指数	Shannon-Winener 指数	均匀度指数 Jsw	均匀度指数 Jsi
高山栲、锥连栎林	50	T	14	0.861 8	2.177 8	0.826 2	0.928 1
		S	12	0.870 4	2.219 6	0.893 2	0.949 5
		H	9	0.838 1	1.960 6	0.892 3	0.942 9
滇石栎、厚皮香林	35	T	9	0.788 9	1.731 0	0.787 8	0.887 5
		S	11	0.861 2	2.143 7	0.894 0	0.947 4
		H	6	0.724	1.510 7	0.843 1	0.868 8
黄毛青冈云南松林	35	T	6	0.776 4	1.631 8	0.910 7	0.919 7
		S	6	0.743 8	1.550 6	0.865 5	0.892 6
		H	12	0.788 2	1.950 0	0.784 7	0.859 9
锥连栎云南松林	未封禁	T	4	0.641 0	1.151 2	0.830 4	0.854 7
		S	8	0.809 7	1.874 0	0.901 6	0.925 5
		H	10	0.828 2	1.978 0	0.859 1	0.920 2

(续)

群落名称	封禁年限（年）	层次	物种丰富度 S	Simpson 指数	Shannon-Winener 指数	均匀度指数 Jsw	均匀度指数 Jsi
云南松林	30	T	4	0.296 3	0.633 7	0.457 1	0.395 1
		H	23	0.911 2	2.809 1	0.895 9	0.952 6
云南松林	未封禁	T	6	0.153 1	0.384 1	0.214 4	0.183 7
		H	21	0.901 4	2.564 5	0.871 0	0.951 5
锥连栎灌丛	未封禁	S	4	0.513 1	0.940 9	0.678 7	0.684 1
		H	8	0.801 8	1.789 8	0.860 7	0.916 3

从表中可看出，物种多样性指数总体上表现为常绿阔叶林>针阔混交林>针叶林>灌丛，实施封禁措施的林分物种丰富度和多样性指数都明显高于未封禁的林分，且封禁时间越长，多样性指数越高。从林分的各层次看，常绿阔叶林、云南松林封禁后乔、灌、草三层次的物种多样性都在增加，而针阔混交林封禁后乔木层的物种多样性指数在增加，而灌木层和草本层的物种多样性指数在降低。这是因为未封禁的林分由于光照条件较好，林下发育了较多的物种，物种多样性指数较高；封禁后林冠逐渐郁闭，林下光照条件变弱抑制了林下植物的生长，导致物种多样性降低；但随着林下阴湿环境的形成，耐阴性物种将逐渐侵入，林下灌草层的物种多样性又将逐渐升高。

3.3 封禁对群落土壤的理化性质影响

实施封禁后的林分在土壤物理性状方面明显比未封禁的林分好，见表3，主要表现为土壤腐殖层厚、土壤容重轻、总孔隙和非毛管孔隙度较大，这些在物理性状方面的特点都更有利于树木的生长。腐殖质层厚度、土壤总孔隙度及毛管孔隙度均是封禁50年的高山栲、锥连栎林最高，依次为高山栲、锥连栎林（封禁50年）>云南松林（封禁30年）>滇石栎、厚皮香林（封禁35年）>黄毛青冈云南松林（封禁35年）>云南松林（未封禁）>锥连栎灌丛（未封禁）>锥连栎云南松林（未封禁），而容重恰好相反，说明封禁后土壤得到改良，通透性能较好，其腐殖质层厚度、总孔隙度、毛管孔隙度比未封禁的高。在土壤化学性质方面，实施封禁后的林分土壤有机质及N、P、K含量都较高，体现了较高的土壤肥力。通过对有机质、有效氮、有效磷的比较，则是封禁的比未封禁的高，封禁的阔叶林比封禁的针叶林高。

表3 实施封禁措施的群落土壤理化性质

群落类型	封禁年限（年）	腐殖质层（cm）	容重（g·cm^{-3}）	总孔隙（%）	毛管孔隙（%）	非毛管孔隙（%）	pH 值	有机质（%）	水解氮（mg·kg^{-1}）	有效磷（mg·kg^{-1}）	速效钾（mg·kg^{-1}）
高山栲、锥连栎林	50	8	1.20	57.22	41.62	15.60	6.24	2.62	49.71	1.32	71.87
滇石栎、厚皮香林	35	7	1.26	53.68	40.26	13.42	5.87	2.66	90.73	0.94	69.10
黄毛青冈、云南松林	35	4	1.25	50.18	38.94	11.24	5.51	1.36	45.83	0.95	85.72

(续)

群落类型	封禁年限(年)	腐殖质层(cm)	容重($g \cdot cm^{-3}$)	总孔隙(%)	毛管孔隙(%)	非毛管孔隙(%)	pH值	有机质(%)	水解氮($mg \cdot kg^{-1}$)	有效磷($mg \cdot kg^{-1}$)	速效钾($mg \cdot kg^{-1}$)
锥连栎、云南松林	未封禁	2	1.45	45.72	33.04	12.68	5.69	0.44	47.91	0.85	45.45
云南松林	30	6	1.25	54.96	41.25	13.71	6.47	0.83	37.16	0.70	81.99
云南松林	未封禁	2	1.36	44.10	34.58	9.52	6.45	0.37	36.96	0.75	37.65
锥连栎灌丛	未封禁	1	1.40	40.24	33.35	6.89	5.26	1.16	39.43	0.76	39.43

3.4 封禁对群落生物生产力的影响

封禁后林分的生物生产力比未封禁的林分高出许多。如封禁50年、35年常绿阔叶林蓄积量分别达到220.1$m^3 \cdot hm^{-2}$、204.9$m^3 \cdot hm^{-2}$,而未封禁的锥连栎经人为反复砍伐后已呈现灌丛状,树高2.2m,胸径2.5cm,蓄积量为3.19$m^3 \cdot hm^{-2}$,封禁50年和封禁35年后阔叶林的蓄积量分别是未封禁的69倍和64倍。封禁后的云南松林蓄积量为850.2$m^3 \cdot hm^{-2}$,封禁后的针阔混交林蓄积量为334.5$m^3 \cdot hm^{-2}$,均明显高于未封禁或封禁时间较短的林分。由此可见,实施封禁措施,有利于群落生物生产力的提高,使群落生物生产力始终维持在较高水平。

3.5 封禁对群落的土壤及枯落物持水能力分析

森林的水源涵养作用,主要依赖于地被物和土壤。通过地被物吸收降水、削弱雨滴的冲击能力和减缓地表径流,可以起到水土保持的作用。而良好的土壤结构,孔隙度大,土壤的透水和持水能力也强,因而可以在水土保持中起到极大作用。通过对封禁前后的群落土壤最大持水量、枯落物最大持水量进行分析,50cm土壤深处最大持水量依次是云南松林(封禁30年)2 980t $\cdot hm^{-2}$,高山栲、锥连栎林(封禁50年)2 861t $\cdot hm^{-2}$,滇石栎、厚皮香林(封禁35年)2 684t $\cdot hm^{-2}$,黄毛青冈云南松林(封禁35年)2 509t $\cdot hm^{-2}$,锥连栎云南松林(未封禁)2 286t $\cdot hm^{-2}$,云南松林(未封禁)2 205t $\cdot hm^{-2}$,锥连栎灌丛(未封禁)2 012t $\cdot hm^{-2}$。枯落物最大持水量则依次为高山栲、锥连栎林(封禁50年)11.28t $\cdot hm^{-2}$,滇石栎、厚皮香林(封禁35年)8.55t $\cdot hm^{-2}$,云南松林(封禁30年)7.44t $\cdot hm^{-2}$,黄毛青冈云南松林(封禁35年)6.71t $\cdot hm^{-2}$,锥连栎云南松林(未封禁)3.17t $\cdot hm^2$,云南松林(未封禁)2.4t $\cdot hm^{-2}$,锥连栎灌丛(未封禁)1.78t $\cdot hm^2$。从上述顺序可看出,封禁后的天然林分的土壤最大持水量、枯落物最大持水量都高于未封禁的林分。

4 结 论

(1)封禁后的云南松林和常绿阔叶林,无论在胸径或树高上都表现出明显的优势,且乔木上层盖度较大而均匀,优势种优势度明显,各优势种种群的年龄组成亦处于稳定状态,乔木树种多数更新良好,在垂直分布上呈连续分布,使得整个群落在结构上处于稳定的状态。未封禁的林分,由于受到人为干扰,其生长较为缓慢,层次简单,林下植物较少。

（2）实施封禁措施的林分物种丰富度和多样性指数都明显高于未封禁的林分，且封禁时间越长，多样性指数越高。封禁后物种多样性指数总体上表现为常绿阔叶林>针阔混交林>针叶林>灌丛。

（3）通过封禁对土壤理化性质影响分析，说明封禁后土壤得到改良，通透性能较好，封禁后的腐殖质层厚度、土壤总孔隙度、毛管孔隙度比未封禁的高。通过对有机质、有效氮、有效磷的比较，则是封禁的比未封禁的高，封禁的阔叶林比封禁的针叶林高。体现了封禁后群落土壤具有较高的肥力。有利于群落生物生产力的提高，使群落生物生产力始终维持在较高水平。

（4）通过对封禁前后的群落土壤最大持水量、枯落物最大持水量进行分析，认为封禁后林分的土壤最大持水量、枯落物最大持水量都高于未封禁的林分。

参考文献：

[1] 刘志良．白龙江林区封山育林生态效益初探[J]．甘肃林业科技，2005，30(1)：62~63．

[2] 官凤英，孟宪宇，梁永伟，等．封山育林类型划分标准的探讨——密云县古北口镇潮关西沟示范区封山育林类型划分[J]．林业资源管理．2003(2)：8~12．

[3] 李铁华，项文化，徐国祯，等．封山育林对林木生长的影响及其生态效益分析[J]．中南林学院学报．2005，25(5)：28~32．

[4] 王永安．封山育林的生态经济作用[J]．世界林业研究．2000，13(3)：19~25．

[5] 杨梅，林思祖，曹子林，等．中国热带、亚热带地区封山育林研究进展[J]．北华大学学报(自然科学版)．2003，4(4)：342~346．

[6] 方代有，朱东伟，杨富权，等．封禁措施对窿缘桉水土保持林群落结构的影响[J]．中国水土保持科学，2005，3(4)：119~123．

[7] 钱迎倩，马克平．生物多样性研究的原理和方法[M]．北京：中国科学技术出版社，1994．

[8] 孟广涛，郎南军，方向京，等．云南金沙江流域山地圣诞树人工林水土保持效益[J]．水土保持学报，2000，(4)：60~63．

[9] 林业部科技司．森林生态系统定位研究方法[M]．北京：中国科学技术出版社，1994．

（本文发表于《水土保持学报》，2007年）

香格里拉亚高山不同退化程度森林群落的空气温湿特征研究

张劲峰,李勇鹏,马赛宇,耿云芬,景跃波

(云南省林业科学院,云南 昆明 650201)

摘要:选取滇西北香格里拉亚高山的极度退化、重度退化、中度退化和轻度退化等4种退化程度的森林群落,以未退化的原生森林群落为参照,采用典型样地法设置观测样地,测定其林内不同季节(月)及一日各时点的空气温度和湿度因子,以研究本区不同退化程度森林群落的空气温湿特征。研究结果表明:林内空气温度在同一季节和同一时点的变化均表现为随森林群落退化程度的增加而显著上升,而林内空气湿度则表现为随森林群落退化程度的增加而显著下降的特点。同时,空气温度和湿度两因子不论是日变化还是季节变化,其变化幅度基本上呈现极度退化森林群落>重度退化森林群落>中度退化森林群落>轻度退化森林群落>未退化森林群落的规律。即随着退化程度的降低,原森林群落结构的变化越小、越稳定,森林群落调节空气温、湿度的作用越显著。明显地反映出该地区退化程度越低、结构复杂和林冠层较高的森林群落,其降低辐射、降温、保湿的作用越显著。

关键词:香格里拉;亚高山;退化程度;森林群落;空气温湿特征

Research on Characteristics of Air Temperature and moisture in Degraded Forest Communities in Subalpine of Shangri-la

Zhang Jin-feng, Li Yong-jie, Ma Sai-yu, Geng Yun-feng, Jing Yue-bo

(Yunnan Academy of Forestry, Kunming Yunnan 650201, P. R. China)

Abstract: Four degraded forest communities of extreme degradation, heavy degradation, medium degradation, light degradation in sub-alpine of Shangri-la were selected as study targets, while primary forest (none degradation) was chosen as the control. Sample method was adopted to carry out comparative study on seasonal and daily changing characteristics of air temperature and moisture in the five degraded forests. The study results showed that with the alleviation of forest degradation, the air temperature within the forest community decreased significantly at the same season and same time during a day, whereas, the air moisture showed a significant increasing trends. In addition, the seasonal and daily deviation of both air temperature and moisture in five degraded forest communities shows trends of extreme degradation > heavy degradation > medium degradation > light degradation > non degradation, which imply along with mitigation of the degradation, environmental factors in community are getting stable, their changing deviation are smaller, and vegetation have stronger function in adjusting microclimate condition. We conclude that the lighter of forest degradation is the more complicity of community and higher of canopy layer would be, which would result in low exposure, low air temperature and high air moisture in the forest community.

Keyword: Shangri-la; subalpine; degradation; forest community; air temperature and moisture

不同的森林是不同环境因子作用的产物。森林也在不断地影响和改变着环境，形成特有的不同于森林外环境的森林小环境。所形成的森林小环境能综合反映森林群落的质量，而森林小环境中的空气温、湿状况则是评价森林群落退化程度的一个重要指标。森林群落的退化，改变了森林的结构和功能，也改变了森林空气的温湿特征（蒋有绪，1996；刘世荣，1996；吴兴宏，1985；郝佳波，2007）。通过对香格里拉亚高山不同退化程度森林群落林内空气温度和湿度特征的研究可以揭示和评价该地区不同退化程度的森林群落与环境的相互关系。

1 研究方法

本项研究根据退化森林群落的外貌，参照包维楷（1995）和李博（1997）对森林生态系统退化程度的划分方式，将香格里拉亚高山的退化森林群落划分为极度、重度、中度和轻度4个退化等级，并以未被破坏的当地神山——五凤山原生森林群落为未退化森林群落参照系，来对比分析该地区不同退化程度森林群落的空气温湿特征值。各退化类型森林群落样地特征见表1。其林内的空气温、湿度数据收集分别于2007~2008年的1月冬季、4月春季、6月夏季、8月秋季和11月秋末冬初的上旬进行。在每个退化程度的森林群落样地内布设5个样点，共25个点，各样点的海拔、坡向和坡位尽可能一致。在观测样点的1.5 m处固定放置温湿度计。用通风干湿表和自记温湿度计测定林内的空气温度和湿度。每次观测选择晴天，连续3天进行，每天从9:00~17:00每隔2个小时（9:00、11:00、13:00、15:00和17:00）观测记录温度、湿度各1次。

表1 香格里拉亚高山不同退化程度森林群落样地的基本特征

退化程度	经纬度	海拔(m)	坡度	基本特征
极度退化Ⅰ	99°41′06″ 27°50′59″	3 349	7°~12°	长时间过度放牧，缺少木本植物，禾草(Kobresia spp.)、狼毒(Euphorbia spp.)和草莓(Fragaria spp.)等阳性草本植物为优势
重度退化Ⅱ	99°43′19″ 27°48′17″	3 399	17°~21°	草本植物出现一些中性植物的科属，如桔梗科(Campanulaceae)，败酱科(Valerianaceae)。灌木层植物有所增加，但仍属于有毒、耐动物啃食的物种，如橙黄瑞香(Daphna aurantica)、华西蔷薇(Rosa moyesii)等
中度退化Ⅲ	99°43′08″ 27°47′49″	3 422	17°~23°	群落乔木层植物主要为大果红杉(Larix potaninii)、白桦(Betula platyphylla)等；灌木层以杨柳科(Saliaceae)及杜鹃花科(Ericaceae)的植物为主；草本主要以豆科(Orchidaceae)及鸢尾科(Iridaceae)植物种
轻度退化Ⅳ	99°43′14″ 27°47′49″	3 345	16°~19°	群落上层以大果红杉为主；灌木层主要以杜鹃花科、蔷薇科(Rosaceae)植物为主；草本以报春花科(Primulaceae)、毛茛科(Ranunculaceae)及兰科(Orchidaceae)中的阴性植物为主
未退化Ⅴ	99°43′18″ 27°47′50″	3 460	23°~25°	上层树种为大果红杉及川滇高山栎(Quercus aquifolioides)；灌木层植物主要为大白花杜鹃(Rhododendron decorum)、米饭花(Lyonia ovalifolia)等。上层植被郁闭度在85%以上，林下环境潮湿，枯枝落叶层较厚

2 结果与分析

2.1 不同退化程度森林群落的林内空气温度变化特征

2.1.1 空气温度的日变化特征

计算香格里拉亚高山不同退化程度森林群落的林内空气温度年平均日变化值,并对其同一时点的年平均气温值进行差异显著度分析,结果见表2。表2数据表明,香格里拉亚高山不同退化程度森林群落的林内空气温度日变化具有明显的规律。

表2 香格里拉亚高山不同退化程度森林群落的林内空气温度年平均日变化状态分析

时间		9:00	11:00	13:00	15:00	17:00
退化程度	极度退化Ⅰ	6.40Aa	12.84Aa	21.60Aa	20.64Aa	14.12Aa
	重度退化Ⅱ	5.84ABa	11.40Aa	20.08Aa	17.20Bb	13.56ABa
	中度退化Ⅲ	5.92ABa	9.30Bb	15.78Bb	16.44Bb	13.76ABa
	轻度退化Ⅳ	5.02Ba	7.88Cb	14.20Bb	14.84Cbc	12.90Ba
	未退化Ⅴ	5.90ABa	8.80BCb	12.00Cb	13.78Cc	13.42ABa
平均		5.82	10.04	16.73	16.58	13.49
标准差		0.50	2.03	4.02	2.63	0.56
变异系数%		8.60	20.21	24.03	15.86	4.15

注:大写字母代表0.05水平上存在显著差异;小写字母代表0.01水平上存在显著差异;相同字母表示差异不显著。表3~表5同。

首先,到达峰值的时间不同。随着日出后气温的逐渐上升,极度退化和重度退化森林群落的林内空气温度一般在正午时分最先出现峰值,随着森林群落退化程度减弱的状态,即中度退化、轻度退化和未退化的顺序其林内空气温度到达峰值的时间渐慢,而推迟至15:00。表明结构复杂未退化的森林群落和轻度退化的森林群落对林内气温日变化的缓冲作用较强,其林内空气温度的日变化比较平缓。相反,退化程度越高,结构越简单的森林群落,林内气温的日波动快,变化剧烈。

其次,不同退化程度的森林群落同一时刻林内空气温度之间的差异值随大气温度的升高而增加,表现为早晚气温差异小,午间气温差异大的特点。这是由于退化程度较高的森林群落林内气温在太阳辐射作用下的变化较退化程度低的森林群落更为剧烈之故。

第三,各个季节林内空气温度的日最高值均出现在极度退化的森林群落内,重度退化森林群落和中度退化森林群落林内的日最高温度次之,轻度退化和未退化森林群落林内气温的最高值最低。这是由于正午太阳直射,缺乏良好植被覆盖的极度退化森林群落(荒草地)的空气温度迅速上升,而在未退化的原生森林群落内光线被层层遮挡反射,林内空气温度只能通过气体循环传递,故而上升缓慢。

表2的方差分析结果表明:上午9:00,除极度退化森林群落林内的空气温度在0.05水平上与轻度退化森林群落存在显著差异外,与其他退化程度的森林群落间的差异不显著。而从11:00~15:00,极度退化和重度退化森林群落林内的气温在0.05和0.01水平上同中度、轻度退化及未退化森林群落林内的气温均有显著差异。至17:00,各退化程度森林群落

林内的气温又趋于接近。从各时间点气温的变异系数也可以看出,不同退化程度森林群落林内的日气温存在较大差异的时间是 11:00~15:00,其中差异最大的时间是正午 13:00,随后为 11:00 和 17:00,即太阳辐射越强,退化森林群落林内的气温差异也越显著,森林群落对气温日变化的调节作用越明显。

香格里拉亚高山不同退化程度森林群落林内各季节空气温度日较差即一天中气温最高值与最低值之差见图 1。

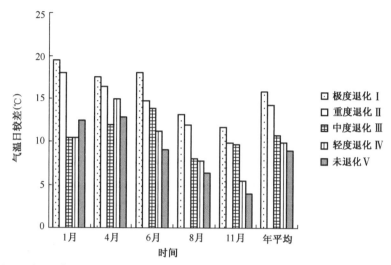

图 1 香格里拉亚高山不同退化程度森林群落各季节林内空气温度的日较差状况

Figure 1 Daily deviation of air temperature of different degraded forests in different season in Shangri-la subalpine

由图 1 可看出,各个季节林内空气温度的日较差在不同退化程度的森林群落间存在明显差异,其总趋势为:极度退化森林群落>重度退化森林群落>中度退化森林群落>轻度退化森林群落>未退化森林群落。年平均气温日较差在极度退化的森林群落内达 16℃,重度退化的森林群落内为 14.3℃,中度退化的森林群落内为 10.8℃,轻度退化的森林群落内为 10℃,未退化的森林群落内仅 9℃。造成这种显著差异的原因是退化程度高的森林群落,因缺少上层植被的庇护,白天,对太阳辐射的削弱作用小,到达地面的太阳辐射量大,升温快;晚上大气逆辐射强,降温快,所以气温日较差较大。相反,退化程度低的森林群落,白天,浓密的植被对太阳辐射的削弱作用强;晚上大气逆辐射弱,所以林内的气温日较差较小。

2.1.2 空气温度的季节变化特征

以 1 月、4 月、6 月、8 月和 11 月林内的平均气温,表示香格里拉亚高山不同退化程度森林群落的不同季节林内空气温度的变化状况。分析其季节性的林内空气温度的变化特征,并对同一季节不同退化程度森林群落林内的气温差异做显著性方差分析,结果见表 3。

表 3 香格里拉亚高山不同退化程度森林群落林内空气温度季节变化状态分析

	时间	1月	4月	6月	8月	11月
退化程度	极度退化 I	1.40Aa	14.50Aa	25.20Aa	24.16Aa	10.34Aa
	重度退化 II	0.60Bb	13.40Aa	23.68ABa	21.80ABa	8.60ABab

(续)

	时间	1月	4月	6月	8月	11月
退化程度	中度退化Ⅲ	-2.14Dc	14.74Aa	21.26Bab	21.00ABa	6.34BCbc
	轻度退化Ⅳ	-1.60Cc	15.02Aa	16.44Cb	18.88Ba	5.80CDc
	未退化Ⅴ	-1.60Cc	14.00Aa	16.92Cb	19.72Ba	4.86Dc
平均		-0.67	14.33	20.70	21.11	7.19
标准差		1.56	0.64	3.93	2.04	2.24
变异系数(%)		105.98	4.47	18.99	9.66	31.16

分析结果显示,香格里拉亚高山各退化程度森林群落的林内气温不仅在不同季节间存在显著差异,且在相同季节内亦表现出差异性。4月份春季,各退化程度森林群落间的林内平均气温差异最小,变异系数仅4.47%,最大差值仅为1.34℃。1月和11月份冬季和秋末冬初,极度退化和重度退化森林群落林内的平均气温同轻度和未退化森林群落相比其变异系数最大,分别达到105.98%和31.16%,林内气温差异极其显著,是各个季节中不同退化程度森林群落林内气温差异最大的月份。这是由于秋季高原气候转凉,太阳辐射量直接决定了森林群落内气温的高低,缺少乔木覆盖的极度退化和重度退化的森林群落林地,能直接受到阳光的照射,其气温上升快,显著高于上层植被覆盖率高而林下日照辐射量较小的中度退化、轻度退化和未退化的森林群落。香格里拉亚高山不同退化程度森林群落林内空气温度季节变化的总趋势是极度退化和重度退化森林群落在各个季节(月份)的平均气温显著高于轻度退化和未退化的森林群落,再次证明退化程度低的森林群落由于具有相对浓密的林冠层,可有效地阻挡和吸收太阳辐射,降低风速,削弱空气的热量交换,对林内气温的升降具有明显的缓解作用。

2.2 不同退化程度森林群落的林内空气相对湿度变化特征

2.2.1 空气相对湿度的日变化特征

计算香格里拉亚高山不同退化程度森林群落的林内空气湿度年平均日变化值,并对其在同一时点的年平均空气湿度值进行差异显著度分析,结果见表4。

表4 香格里拉亚高山不同退化程度森林群落的林内空气湿度的年平均日变化状况分析　%

	时间	9:00	11:00	13:00	15:00	17:00	变异系数
退化程度	极度退化Ⅰ	48.60Bb	34.20Cc	22.96Cc	22.56Cb	22.56Cb	32.79
	重度退化Ⅱ	49.00Bb	36.58Cc	22.38Cc	24.24Cb	36.20Bb	32.07
	中度退化Ⅲ	53.40AB	46.44Ba	34.10Bb	34.72Ba	38.78Bab	19.98
	轻度退化Ⅳ	59.10Aa	47.28Ba	39.14Bab	36.86ABa	44.22Aa	19.26
	未退化Ⅴ	58.82Aa	52.98Aa	44.96Aa	39.96ABa	44.56Aa	15.63
平均		53.78	43.50	33.08	31.30	39.63	
标准差		5.09	7.86	9.48	8.27	4.62	
变异系数		9.46	18.07	28.66	26.43	11.66	

分析结果看出，香格里拉亚高山不同退化程度森林群落林内空气相对湿度的日变化趋势为早晚高、午间低。清晨（9:00）林内空气湿度出现一个较高值，随后逐渐下降，到午间（13:00~15:00）即气温最高、太阳辐射较强时，其林内空气湿度达到最低值。之后，随着气温的降低和太阳辐射强度的减弱，林内空气湿度又逐渐回升。这种变化规律同林内气温的日变化状况正好相反。

退化程度较低的轻度退化和未退化森林群落林内具有较高的相对湿度，而退化程度较高的重度退化和极度退化森林群落林内空气相对湿度较低。明显地反映出在香格里拉亚高山森林群落中生态系统的退化程度越低、结构越复杂和林冠层较高的群落，其保湿与增湿作用显著。

从表4还可以看出，香格里拉亚高山不同退化程度森林群落林内的相对湿度具有不同的日变化规律。早晨随日出退化程度较高的森林群落，其林内相对湿度急速下降，一般在13:00林内相对湿度降至最低点，然后回升。相比之下，退化程度较轻的森林群落林内相对湿度的日变化比较缓慢，明显滞后，林内日相对湿度的最低点一般出现在15时，且林内各个时点的相对湿度均显著大于退化程度高的森林群落，反映出森林群落结构的复杂性及覆盖度与空气湿度的正相关性。

在香格里拉亚高山不同退化程度的森林群落中，林内空气相对湿度在同一时刻的表现为从9:00~17:00，极度退化与重度退化的森林群落间不存在显著差异，而与轻度退化和未退化森林群落在0.05和0.01水平上都有显著差异。从11:00~15:00，重度退化和极度退化森林群落与中度退化森林群落在0.05和0.01水平上均有差异；在17:00，不同退化程度的森林群落林内空气湿度的差异状况与上午9:00相似。

在相同时点上，5种退化程度森林群落间林内空气相对湿度的变异系数分别为9:00 9.46%、11:00 18.07%、13:00 28.66%、15:00 26.43%、17:00 11.66%，表明不同退化程度的森林群落在同一时刻林内空气湿度的差异程度表现为早晚小、午间大的特点。

各退化程度森林群落林内的空气相对湿度日变化变异系数分别为极度退化森林群落32.79%、重度退化森林群落32.07%、中度退化森林群落19.98%、轻度退化森林群落19.26%和未退化森林群落15.63%。表明极度退化和重度退化的森林群落林内空气湿度的日变化幅度比退化程度低的森林群落大而剧烈。

2.2.2 空气湿度的季节变化特征

以1月、4月、6月、8月和11月林内的平均空气相对湿度，表示香格里拉亚高山不同退化程度森林群落林内不同季节的空气相对湿度变化状况，分析其林内空气相对湿度的季节性变化特征（图2），并对同一季节不同退化程度森林群落林内的空气相对湿度差异做显著性方差分析，结果见表5。

从图2可看出，香格里拉亚高山不同退化程度森林群落林内的空气相对湿度在各个季节均表现为：未退化森林群落>轻度退化森林群落>中度退化森林群落>重度退化森林群落>极度退化森林群落的规律。在轻度退化和未退化的森林群落中，由于树木林下蒸发和林冠蒸腾作用的改变，从而提高了林内空气的湿度，形成了湿润的森林小环境。

表5的数据显示，在香格里拉亚高山的不同退化程度的森林群落中，林内空气相对湿度季节变化的差异性各不相同，但也显示出一定的规律性，并可分为两组，极度退化和重度退化森林群落为一组，他们之间的各季节林内空气相对湿度的差异不显著；轻度退化和

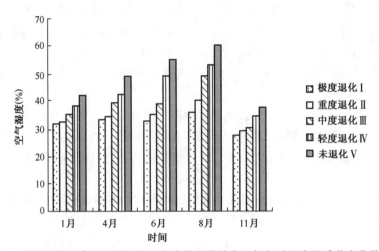

图 2 香格里拉亚高山不同退化程度森林群落林内空气相对湿度的季节变化状况

Figure 2 Seasonal variation of air moisture in different degraded forests in Shangri-la subalpine

未退化森林群落为另一组,其差异也不显著,但两个类型之间各季节林内的空气相对湿度存在显著差异。在 4~8 月的春夏两季,不同退化程度森林群落间林内空气相对湿度的差异加大且更复杂,而极度退化和重度退化森林群落间无显著差异。

表 5 香格里拉亚高山不同退化程度森林群落各季节林内空气相对湿度比较　%

	时间	1月	4月	6月	8月	11月	平均值	变异系数
退化程度	极度退化 I	31.67Cb	33.07Cb	32.70Cb	36.27Cc	27.60Cb	32.262	9.67
	重度退化 II	32.60Cb	34.65Cb	35.50BCb	40.53Cc	29.43BCab	34.542	11.81
	中度退化 III	35.53BCab	39.47BCb	38.93Bb	49.07Bb	30.67BCab	38.734	17.43
	轻度退化 IV	38.17ABab	39.47BCb	38.93Bb	53.60Bab	34.87ABab	43.576	17.67
	未退化 V	41.87Aa	49.40Aa	55.30Aa	60.40Aa	37.82Aa	48.958	18.98
平均		35.968	39.752	42.30	47.974	32.078		
均方差		4.17	6.51	9.55	9.74	4.18		
变异系数		11.59	16.38	22.58	20.30	13.03		

香格里拉亚高山的各退化程度森林群落林内空气相对湿度的年变幅变异系数值随退化程度的增加而降低,其季节间的变幅表现为未退化森林群落>轻度退化森林群落>中度退化森林群落>重度退化森林群落>极度退化森林群落的规律。

表 5 还显示出,香格里拉亚高山的各退化森林群落林内空气相对湿度的季节变化趋势表现为 8 月>6 月>4 月>1 月>11 月,这与当地的干湿季相吻合,表明空气湿度与降雨量密切相关。6~10 月为雨季,空气湿度较大,导致了 6 月和 8 月各退化森林群落林内的相对湿度高于其他季节。处于滇西北的香格里拉亚高山地区属云南省少雨区,全年空气较为干燥,其各季节平均相对湿度分别为冬季 1 月 35.97%,春季 4 月 39.75%,夏季 6 月 42.30%,秋季 8 月 47.97%,秋末冬初 11 月 32.08%。在最湿润的秋季 8 月未退化森林群落林内的相对湿度也仅 60.4%,而在林内最低的相对湿度出现在 11 月的极度退化森林群

落中仅 27.6%。

3 结论与讨论

3.1 空气温度的变化特征

香格里拉亚高山不同退化程度的森林群落林内空气温度变异状态的研究结果表明,森林群落对林内空气温度的变化具有明显的缓冲作用。群落结构越复杂,其林冠层越浓密,截留太阳辐射,降低风速,削弱空气热量交换,对气温变化起缓冲作用的能力越强,使林内空气温度的变化越平缓。而森林群落退化的程度越高,覆被率越低,结构越简单,环境的热容量越小,升温和降温的速度越快,林内气温的波动越快,变化越剧烈。因此,随着森林群落退化程度的上升,一天内林内空气温度到达峰值的时间渐快;且同一时刻不同退化程度的森林群落林内空气温度间的差异,随大气温度的升高而增加,表现为早晚气温差异小、正午大。

在不同季节,香格里拉亚高山不同退化程度的森林群落间林内气温差异最小的季节在春夏季;最大的季节在秋冬季。这是由于秋冬季高原大气温度最低,太阳的辐射量直接决定了森林群落林内气温的高低,缺少乔木层覆盖的极度退化和重度退化的林地,直接受到阳光的照射,气温上升快,其林内气温显著高于上层植物覆盖率高而林下辐射量较小的中度退化、轻度退化和未退化的森林群落。

总之,退化程度越低的森林群落由于林冠的阻挡,使到达林内的太阳辐射能比退化程度高的森林群落少,加之其林内的空气湿度大,为提高林内温度需要的热量也多,故林内温度在日照强时也低于退化程度高的森林群落。同时,由于进入林内的乱流涡旋体受到林木枝叶的阻截和摩擦,被分解为更小的形状不一、大小不等的乱流涡旋体,使乱流交换强度趋小,在加上林冠层阻挡地面的长波辐射到达外层空间,缓和地面夜间所释放的热量,因而其林内气温的变化缓和(彭少麟,2003)。使得香格里拉亚高山不同退化程度森林群落林内气温变化幅度呈现极度退化森林群落>重度退化森林群落>中度退化森林群落>轻度退化森林群落>未退化森林群落的规律。随着退化程度的降低,森林群落内的空气温度变幅越小、越稳定,群落调节气温的作用越显著。

3.2 空气相对湿度的变化特征

此外,香格里拉亚高山不同退化程度森林群落的林内空气相对湿度变异状态的研究结果则表明:各退化程度森林群落林内空气相对湿度的日变化趋势为早晚大、中午小,与气温日变化正好相反。退化程度较高的森林群落林内空气湿度的日变化幅度比退化程度低的森林群落大而剧烈;其日变化出现最低值的时间为退化程度较高的森林群落早于退化程度较低的森林群落,即完好的原生森林群落对空气湿度的变化起重要的缓冲作用。

香格里拉亚高山不同退化程度森林群落同一时刻林内空气湿度的差异程度表现出早晚小,午间大的特点。香格里拉亚高山不同退化程度森林群落林内空气相对湿度在各个季节均表现为未退化森林群落>轻度退化森林群落>中度退化森林群落>重度退化森林群落>极度退化森林群落的规律。明显地反映出退化程度越低、结构越复杂和林冠层较高的森林群落,其保湿与增湿作用越显著。

香格里拉亚高山各退化程度森林群落林内空气相对湿度的季节变化幅度也表现出未退化森林群落>轻度退化森林群落>中度退化森林群落>重度退化森林群落>极度退化森林群落的规律,同热带和亚热带地区,未退化森林群落林内的相对湿度的季节变化比退化森林群落更和缓的结论相反(彭少麟,2003;温远光,1998;郑科,2005)。这可能是由于香格里拉亚高山处滇西北亚高山地区属云南省的少雨区之故,其年平均空气相对湿度仅30%~50%,显著低于热带、亚热带地区。加之本地区干湿分明的气候特点,退化程度较高的森林群落,即使在雨季,局部降雨后空气中所含的水分也会因缺乏植被的保护作用而很快丧失,导致其干湿季空气湿度差异较小;而在退化程度较低的森林群落林内降雨后增加的湿度,能更有效地被保持住,因而同干季能产生更大的差异。

香格里拉亚高山退化程度高的森林群落林内显示出高温低湿,且温、湿度的年变化和日变化幅度较大的特点;随着退化程度的降低,森林群落林内的温度逐渐下降,湿度有所上升,其变幅进一步缓和;到未退化的原生森林群落则显示温度低,湿度大,变化幅度小,环境状况最稳定的特点,而最有利于该植被的充分发展。

参考文献:

[1] 蒋有绪. 中国森林生态系统结构与功能规律研究[M]. 北京:中国林业出版社,1996.
[2] 刘世荣. 中国森林生态系统水文生态功能规律[M]. 北京:中国林业出版社,1996.
[3] 吴兴宏. 热带亚热带森林生态系统研究[M]. 海口:海南人民出版社,1985.
[4] 郝佳波,司马永康,徐亮. 云南拟单性木兰的地理分布与气候因子的关系[J]. 西部林业科学,2007, 36(2):105~109.
[5] 包维楷,陈庆恒,刘照光. 山地退化生态系统中生物多样性恢复与重建研究[A]//中国科学院生物多样性委员会,林业部野生动物和森林植物保护司. 生物多样性研究进展[M]. 北京:中国科学技术出版社,1994.
[6] 李博. 中国北方草地退化及其防治对策[J]. 中国农业科学,1997(30):1~9.
[7] 斯珀尔 S H,巴恩斯 B V. 森林生态学[M]. 北京:中国林业出版社,1982.
[8] 彭少麟. 热带亚热带恢复生态学研究与实践[M]. 北京:科学出版社,2003.
[9] 温远光,梁宏温,和太平,等. 大明山退化生态系统群落的温湿特征[J]. 广西农业大学学报,1998, 17(2):204~210.
[10] 郑科,郎南军,郭玉红,等. 元谋干热河谷降雨、温度、蒸发量的监测分析[J]. 西部林业科学,2005, 34(3):57~62.

(本文发表于《西部林业科学》,2012年1期)

滇西北纳帕海湿地景观格局变化及其对土壤碳库的影响

李宁云[1,2]，袁华[3]，田昆[2]，彭涛[3]

(1. 云南省林业科学院，云南 昆明，650204；2. 国家高原湿地研究中心，云南 昆明，650224；3. 西南林业大学，云南 昆明，650224)

摘要：采用3S技术和In-situ原状土就地取样技术，对滇西北纳帕海湿地26年来的景观格局变化及其驱动下的湿地土壤碳库变化研究表明：纳帕海景观格局变化显著，与1974年相比，景观破碎化程度增强、斑块形状趋于复杂、呈离散分布，湿地景观类型总面积比例呈略有增加(1994年)至大幅度减小(2000年)的变化，非湿地景观类型总面积比例则呈略有减小(1994年)至大幅增加(2000年)的变化并取代湿地景观成为基质景观。响应景观类型面积变化，土壤碳储量由1974年的 33.46×10^4 t 增至1994年的 36.91×10^4 t，2000年降至 32.92×10^4 t；随景观类型的转化，1974～1994年土壤碳库积累量为 6.08×10^4 t，释放量为 2.63×10^4 t，1994～2000年积累量为 2.01×10^4 t，但碳释放量为 5.99×10^4 t，是前20年的2.28倍。纳帕海湿地景观格局和土壤碳库的变化是自然和人为因素共同作用的结果，在地质、水文和气候等自然因素提供的变化背景上，排水、垦殖、过度放牧、无序旅游、汇水区植被破坏等强烈的人为活动干扰加剧了变化。

关键词：纳帕海湿地；景观格局变化；土壤碳库；汇源变化

Landscape pattern change and its influence on soil carbon pool in Napahai wetland of Northwestern Yunnan

Li Ning-yun[1,2], Yuan Hua[3], Tian Kun[2], Peng Tao[3]

(1. Yunnan Academy of Forestry, Kunming Yunnan 650204, China; 2. National Plateau Wetland Research Center, Kunming Yunnan 650224, China; 3. Southwest Forestry University, Kunming Yunnan 650224, China)

Abstract: The wetland landscape dynamics and its environmental effects have been considered as a research hotspot. But researches about soil carbon pool changes driven by landscape pattern changes were rarely seen. To address this issue, a study was carried out in the Napahai wetland, a sensitive region of global changes in northwestern Yunnan, adopting the In-situ intact soil sampling methodology and supported by the "3S" tools. Results showed that the landscape altered significantlywithin 26 years. Compared with 1974a, the fragmentation of Napahai landscapes was increasing, the landscape shapes became more complicated, dispersed, and dominated by large patches, the matrix of the Napahai landscapes has been converted from wetland types to the ever-increasing human land use types. At the land level, the patch number and landscape shape index increased 42% and 12.19% by 1994a, as well as 40% and 1.02% by 2000a. The aggregation index decreased 0.56% (by 1994a) and 0.52% (by 2000a). Landscape diversity and landscape dominance firstly increased 0.844% and 0.847% by 1994a and then decreased 3.130% and 3.134% by 2000a. At the class level, the changes of wetland types trended to complexity, the area percentage of water body, marsh and swampy meadow increased form 70.29% (1974a) to 72.20% (1994a) then decreased to 48.79% (2000a), however, that of meadow and farmland decreased a bit in 1994a then largely increased in 2000a. The landscape pattern changes of the wetland had impacts on the wetland's soil carbon storage fluctuation. Responding to the landscape area changes, soil carbon storage increased from $33.46×10^4$ T (1974a) to $36.91×10^4$ T (1994a) and then decreased to $32.92×10^4$ T (2000a). With the landscape type transformations, the soil carbon sequestration and emission form 1974a to 1994a reached respectively $6.08×10^4$ T and $2.63×10^4$ T. The soil carbon sequestration form 1994a to 2000a reached $2.01×10^4$ T, nonetheless, the emission increased sharply, which is up to 2.28 times of that of 1974–1994a. The landscape pattern changes, the soil carbon storage and its "source/sink" shifts have been imposed by both natural factors and human impacts. Under the context of geology, hydrology and climate changes, human disturbances such as wetland drainage, reclamation, overgrazing, and vegetation destruction of catchments have further intensified the changes of wetland.

Key Words: Napahaiwetland; Landscape pattern changes; Soil carbon pool; Sequestration and emission changes

土壤有机碳是全球气候变化的敏感指标[1]。湿地生态系统较低的分解特性,使其拥有陆地上各种生态系统中最高的单位面积碳储量,成为重要的"碳汇"[2]。滇西北位于青藏高原东南缘,属于全球变化敏感区[3]。该区湖泊湿地众多,且湖盆沼泽、湖滨沼泽发育较为普遍,积累了大量有机物质,有些甚至堆积了极其深厚的泥炭层[4],对于全球 CO_2 平衡,降低温室效应都具有重要的意义。

湿地景观格局变化会对区域乃至全球气候产生深刻影响[5],国内外的研究主要关注湿地景观格局变化对湿地环境[6~8]、水禽生境[9~11]及生物多样性[12~14]的影响,但对湿地土壤碳库影响的研究还未见报道。本文以滇西北纳帕海为研究对象,对湿地景观格局变化及其驱动下的土壤碳库变化进行研究,揭示湿地景观格局变化的生态环境效应,以期为高原湿地固碳能力评价和湿地管理提供科学依据。

1 研究区概况

纳帕海湿地(99°37′~99°43′E,27°49′~27°55′N)地处云南西北部,海拔 3 260m,面积 2 552hm²,发育在石灰岩母质上,第三纪陷落成湖,受喀斯特作用的强烈影响,湖盆底部被蚀穿形成落水洞,湖水潜流 10km 后汇入金沙江;区域气候属寒温带高原季风气候区西部型季风气候,由于地理位置偏北且海拔较高,冬季又受青藏高原寒流影响,年均温较低为 5.4℃;水量补给主要依靠降雨、地表径流、冰雪融水和多条河流的注入以及湖盆两侧金沙江—中甸大断裂上涌的泉水;年均降雨量 612.8mm,在降雨集中的 6~9 月(占年降水量的 76%)湖水上涨,水生、沼生植物大量繁殖;受西南季风影响,9 月后湖水退落,随后大量植物死亡,由于气温较低、水循环不畅,导致植物残体堆积,不能彻底分解,有机质大量积累,发生泥炭化及潜育化,形成沼泽土类型的湿地土壤[15](表1)。

表1 纳帕海湿地土壤理化特征
Table 1 Soil physicochemical propertiesof Napahai wetland

层次 soil horizon	质地 Texture	pH	有机质 OM (g·kg⁻¹)	全氮 TN (g·kg⁻¹)	全磷 TP (g·kg⁻¹)	全钾 TK (g·kg⁻¹)	碱解氮 AN (mg·kg⁻¹)	速效磷 AP (mg·kg⁻¹)	速效钾 AK (mg·kg⁻¹)
T	中壤	4.90	288.0	11.8	0.7	1.4	530.00	7.00	65.00
TG	重壤	5.30	107.0	6.4	0.5	8.6	472.00	5.00	68.00
G	重壤	5.00	47.0	2.2	0.2	2.3	559.00	21.00	148.00

注:T——泥炭层;G——潜育层;TG——泥炭层向潜育层的过渡层。

2 研究方法

2.1 景观数据来源及格局变化

2.1.1 景观数据来源与处理

选用1974年MSS、1994年LandsatTM与2000年LandsatETM影像为遥感数据源,采用RGB543彩色合成,经几何校正后,将三期图像统一到北京54坐标系统下。根据纳帕海湿地演替规律[16],结合遥感影像颜色、形状、质地、结构及其与周边环境的关系特征和GPS(TRIMBLE GEOCE,精度0.5 m)野外实地调查结果,建立景观类型解译标志(表2)。

表2 景观类型解译标志
Table 2 Visual interpretation key system of the landscape types

景观类型 Landscape types	影像特征 The character of satellite imagery
水体 Water body	蓝色至深蓝色,纹理平滑,呈大片分布
沼泽 Marsh	深绿至浅蓝色,不光滑
沼泽化草甸 Swampy meadow	淡黄色至浅蓝色,影像结构复杂
草甸 Meadow	土黄色,一般位于湿地边缘或地势相对较高处
耕地 Farmland	形状规整,紫褐色,影像结构单一,位于村子周围

在ArcGIS9.0中进行目视解译,将纳帕海景观划分为水体、沼泽、沼泽化草甸、草甸和耕地,并生成3个时期景观类型图(图1)。通过ENVI4.3,三期影像分类精度分别为86.77%、92.36%和85.74%。

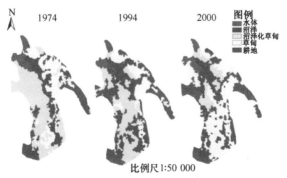

图1 纳帕海湿地景观分类图(1974年、1994年、2000年)
Figure 1 Landscape classification map of Napahai wetland (1974a,1994a,2000a)

2.1.2 景观格局变化

利用空间分析模块生成景观类型转移矩阵,参照相关研究成果[17~18],选取面积指数(CA)、斑块数(NP)、聚集度指数(AI)、形状指数(LSI)、Shannon 多样性指数(SHDI)及 Shannon 均匀度指数(SHEI)从景观水平、景观类型两种尺度上进行景观格局变化分析。利用 Fragstats3.3 进行各项景观指数的计算,计算方法及其生态意义见文献[17~18]。

2.2 湿地土壤碳库变化

2.2.1 土壤碳储量

依据典型性和代表性原则,对沼泽、沼泽化草甸、草甸和耕地 4 种景观类型分别设置 4~5 个定位点,应用 In-situ 原状土就地取样技术[19],根据湿地水文特征,分别于 2002 年 11 月,2003 年 3 月、6 月、8 月、10 月,2004 年 1 月进行土样采集,利用重铬酸钾法[20]测定有机碳含量,取各景观类型 0~20cm 土样均值计算土壤碳储量(表3);水体沉积有机碳数据来源于殷勇等的研究,以平均值 $6.5\mathrm{g\cdot kg^{-1}}$ 计算碳储量[21]。

表3 2002~2004年不同景观类型土壤有机碳含量/(平均值±标准差,g·kg⁻¹)
Table 3 Soil organi carbon content in different landscape types form 2002 to 2004a/(Mean±S.E,g·kg⁻¹)

土层深度 Soil depth(cm)	沼泽 Marsh	沼泽化草甸 Swampy meadow	草甸 Meadow	耕地 Farmland
0~20	59.39±12.94	40.89±18.66	26.91±14.55	29.75±4.07

土壤容重根据自然土壤(水体、沼泽、沼泽化草甸、草甸),耕作土壤(耕地)的属性,分别采用相应土壤有机碳和容重关系进行计算[22]。

2.2.2 土壤碳汇源变化

以景观类型变化造成的土壤有机碳排放通量来反映湿地土壤碳的汇源变化[23],计算公式如下:

$$Ec = \rho(Pc_0 - Pc_1)$$

式中:Ec——景观类型演变引起的土壤碳储量变化(t),Ec 为正值表示土壤碳的释放,Ec 为负值表示土壤碳的积累;

ρ——景观类型面积转化量(hm^2);

Pc_0,Pc_1——分别代表原有景观类型与转化后景观类型 0~20cm 土层深度单位面积土壤有机碳含量($kg\cdot m^{-2}$)。

3 结果与分析

3.1 景观格局变化特征

3.1.1 景观类型面积

26年来,纳帕海景观面积发生了较大变化(表4),1974年湿地景观类型面积比例为70.29%,至1994年,为72.20%,但2000年却降至48.79%,表明景观基质由湿地类型转变为非湿地类型。从景观类型来看,水体和沼泽化草甸面积呈减小趋势,水体2000年虽略有回升,但萎缩仍十分明显;沼泽和耕地面积总体呈增加的趋势,表明逆向生态演替和人类活动干扰加剧;草甸面积在1994年减少,2000年却大幅增加,这可能与水文条件等因素有关,但也体现出湿地强烈的陆地化进程。

表4 景观类型面积比例变化

Table 4 Landscape type area percentage changes

时期 Year	景观面积比例(%) Landscape type area percentage(%)				
	水体 Water body	沼泽 Marsh	沼泽化草甸 Swampy meadow	草甸 Meadow	耕地 Farmland
1974年	6.35	24.78	39.17	24.47	5.23
1994年	4.04	36.84	31.32	17.84	9.96
2000年	4.99	33.59	10.21	41.30	9.91

3.1.2 景观空间格局

与1974年相比,纳帕海景观斑块数增加了42%(1994年)和40%(2000年),水体的萎缩和破碎,导致湿地水量减少,加剧湿地逆向生态演替,沼泽、草甸和耕地斑块数增加,沼泽化草甸斑块数降低(图2a,图2b);景观形状指数增加了12.19%(1994年)和1.02%(2000年),景观斑块形状趋于复杂,但至2000年,沼泽化草甸斑块形状却变得相对规则和简单,草甸景观则一直向简单与规则发展(图2c,图2d);景观聚集度指数分别减小了0.56%(1994年)和0.52%(2000年),景观斑块呈现出离散度较高的空间分布格局,但沼泽和草甸景观却有连接成片的趋势。

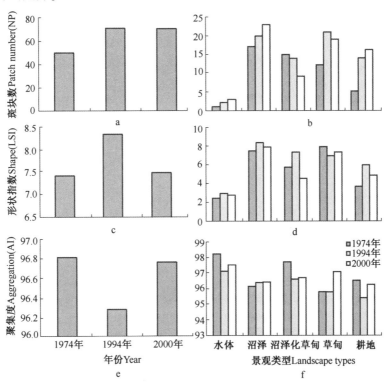

图2 景观空间格局变化特征

Figure 2 Landscape pattern changes

a、c、e分别为景观水平斑块数、形状指数、聚集度;b、d、f分别为类型水平斑块数、形状指数、聚集度

3.1.3 景观多样性

景观多样性指数和均匀度指数的变化趋势具有较好的一致性(图3)。与1974年相比,1994年两者均呈增加的趋势,分别增加了0.844%和0.847%,表明景观多样性增加,各景观类型所占面积比例差异相对较小,趋向相对均匀分布;但2000年,两者均呈减小的趋势,比1974年分别减少了3.130%和3.134%,表明景观多样性降低,各类景观类型所占面积比例差异相对较大,呈不均匀分布,体现出单类景观占主导地位的趋势。

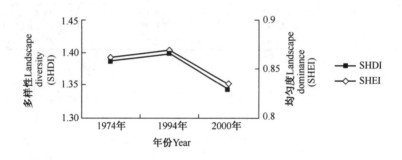

图3 景观多样性变化

Figure 3 Landscape diversity changes

3.2 土壤碳库的响应特征

3.2.1 土壤碳储量

响应景观面积的变化,土壤碳储量由1974年的33.46×10^4t增至1994年的36.91×10^4t,但2000年,却降至32.92×10^4t(图4),表明逆向生态演替导致的沼泽化进程有利于土壤碳的积累,但随湿地环境的丧失,不利于碳的沉积,碳储量随之减小。

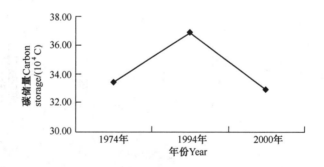

图4 土壤碳储量变化

Figure 4 Soil carbon storage changes

3.2.2 土壤碳的积累与释放

在景观类型的演变中,随沉积环境的改变,土壤碳发生积累与释放(表5,表6)。1974~1994年,景观类型转变中共积累了6.08×10^4tC,释放了2.63×10^4tC;1994~2000年,共积累了2.01×10^4tC,释放了5.99×10^4tC;1994~2000年释放量是前20年的2.28倍,表明土壤碳沉积的湿地环境发生了巨大变化,导致碳的大量释放。

表5 1974~1994年景观格局变化的土壤C积累与释放/(C:10⁴t)

Table 5 Sequestration and emission of soil organic C driven by landscape pattern changes from 1974 to 1994a/(C:10⁴t)

1974年景观类型 Landscape type of 1974a	1994年景观类型 Landscape type of 1994a				
	水体 Water body	沼泽 Marsh	沼泽化草甸 Swampy meadow	草甸 Meadow	耕地 Farmland
水体 Water body	0	-0.573	-0.302	-0.037	-0.026
沼泽 Marsh	0.056	0.000	0.814	0.825	0.117
沼泽化草甸 Swampy meadow	0.036	-2.511	0	0.430	0.340
草甸 Meadow	0	-1.749	-0.803	0	-0.046
耕地 Farmland	0	-0.019	-0.016	0.016	0.000

表6 1994~2000年纳帕海湿地景观格局变化的土壤C积累与释放/(C:10⁴t)

Table 6 Sequestration and emission of soil organic C driven by landscape pattern changes from 1994 to 2000a/(C:10⁴t)

1994年景观类型 Landscape type of 1994a	2000年景观类型 Landscape type of 2000a				
	水体 Water body	沼泽 Marsh	沼泽化草甸 Swampy meadow	草甸 Meadow	耕地 Farmland
水体 Water body	0	-0.026	-0.043	-0.068	0
沼泽 Marsh	0.568	0	0.407	2.055	0.072
沼泽化草甸 Swampy meadow	0.065	-0.964	0	2.656	0.087
草甸 Meadow	0.009	-0.408	-0.379	0	-0.032
耕地 Farmland	0.025	-0.087	0	0.049	0

4 讨 论

4.1 自然因素对湿地景观格局及土壤碳库的影响

纳帕海是在特定地质地理条件下水源补给和水源储存处于极限平衡状态的自然历史产物,其所处的横断山区,新构造运动呈上升趋势,流域内河流不断向上源侵蚀,导致侵蚀基准面下降,有效水源补给不断减少;加之湖盆发育在石灰岩母质上,一直受喀斯特作用的强烈影响,两者的作用使得湖盆的正常蓄水受到威胁。与20世纪80年代以来全球性气候变暖的趋势一致,纳帕海地区的气候也呈现出转暖的迹象。与70年代、60年代和50年代相比[24],纳帕海区域的气温80年代分别上升了0.4℃、0.6℃和0.2℃,1998年以来的年平均气温超过准30年(1961~1990年)气候距平均值的5.6℃(图5);降水呈现出减少的趋势,2003年降至多年来的最低值,年降雨仅486.8mm,比50年的降雨平均值低22.16%,从温度与降水的综合趋势来看,纳帕海区域气候呈现出暖干的趋势;响应气候变化,作为纳帕海水量补给的雪山和冰川会发生一定程度的消融,虽带来一定的水量,但却使高原的表面反射率降低,进一步加速了高原的变暖趋势[25]。湿地水量的极大减少,加剧了湿地逆向生态演替,非湿地景观类型逐渐演变为优势景观,湿地碳沉积环境随之改变甚至丧失,最终导致碳储量减小、排放量增加。

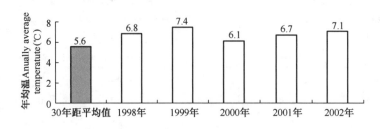

图 5　纳帕海区域气温变化
Figure 5　Temperature variations in Napahairegion

纳帕海湿地所处的中甸地区由于喜马拉雅构造运动的影响,断裂带构造密布,地震较为频繁,历史上四次强烈地震有两次震中均在纳帕海。地质构造活跃导致湿地周围山体及地层结构破碎,地表物质处于不断的侵蚀、搬运和堆积过程中,为湖盆的淤积提供了丰富的物质,半封闭的湿地环境使纳帕海在流域景观中成为淤积物质的"汇"[26],加速了湿地景观陆地化进程,削弱了其碳汇功能。

4.2　人为干扰对湿地景观格局及土壤碳库的影响

相对低平的湖盆环境,在促进社区经济发展的同时,传统资源利用方式对湿地环境产生了极大影响。20世纪60年代以前,纳帕海保持着较为原始的状态,但自20世纪70年代初起,为获得耕地与牧场,开挖了大量排水沟渠并扩大了出水口,直至近年,降低水位的活动仍没有停止。2000年相对于1994年为特丰水年[27],年均温也相对较低(图5),但湿地景观类型面积却大幅锐减,非湿地景观成为景观的基质,表明排水等降低水位的人为活动干扰是导致湖泊萎缩、湿地生态系统退化演替加剧的重要原因,并驱动湿地景观格局发生变化,导致湿地碳储量减少、释放量增加。

经排水或开垦草甸获得耕地,改变了景观类型和景观格局特征,减弱或终止了湿地土壤的泥炭化、潜育化过程,并随耕作活动的进行,改善了土壤通气性,使有机质分解加快,碳排放增加;以放牧为主要利用方式的沼泽化草甸和草甸类型,由于超出理论载畜量(2 600羊单位)132.5%[16]的放牧,减少了有机物质的归还量,影响碳的积累,牲畜的践踏,尤其是过度放养的家猪对土层的翻拱,加速了土壤有机物质的矿化分解;无序的骑马观光旅游,马匹的重度践踏将使土壤有机质减少68.83%[28],加剧了碳排放;人口快速增长以及传统生活方式对木材的需求,致使汇水面山植被覆盖率从80%锐减至20%~30%[15],森林大多退化演替为灌丛,削弱了森林生态系统的水土保持和水源涵养功能,泥沙沉积加剧了纳帕海的陆地化进程,湿地景观类型发生改变,湿地碳沉积环境改变,甚至丧失。

5　结　论

(1)纳帕海景观格局26年来变化显著。景观水平尺度上,景观破碎化程度增强、斑块形状趋于复杂、呈离散分布,并体现出单类景观占主导地位的趋势;景观类型尺度上,各景观类型格局变化较为复杂,水体、沼泽和沼泽化草甸3类湿地景观总面积比例呈略有增加(1994年)至大幅度减小(2000年)的变化,草甸、耕地2类非湿地景观总面积比例则呈略有减小(1994年)至大幅增加(2000年)的变化并成为基质景观。

(2)湿地景观格局变化驱动了土壤有机碳的"汇""源"变化。非湿地景观向湿地景观的转变,增加了土壤有机碳的积累,体现出碳"汇"效应;反之,土壤有机碳积累量减少,释放量增加,碳"源"效应增强。

(3)纳帕海湿地景观格局及土壤碳库变化是在自然因素的作用下,叠加人为干扰造成的。水文、气候和地质等自然因素提供了变化背景,排水、垦殖、过度放牧、无序旅游、汇水区植被破坏等强烈的人为活动干扰则加剧了湿地景观格局及土壤碳库的变化。

参考文献:

[1] 肖辉林. 气候变化与土壤有机质的关系[J]. 土壤与环境,1999,8(4):304.

[2] 刘子刚,张坤民. 黑龙江省三江平原湿地土壤碳储量变化[J]. 清华大学学报:自然科学版,2005,45(6):788~791.

[3] 王建林,欧阳华,王忠红,等. 高寒草原生态系统表层土壤活性有机碳分布特征及其影响因素——以贡嘎南山—拉轨岗日山为例[J]. 生态学报,2009,29(7):3501~3508.

[4] 孙广友. 横断山滇西北地区沼泽成因、分布及主要类型的初步探讨[C]// 中国科学院长春地理研究所. 中国沼泽研究[M]. 北京:科学出版社,1988:275~283.

[5] 白军红,欧阳华,杨志锋,等. 湿地景观格局变化研究进展[J]. 地理科学进展,2005,24(4):36~45.

[6] Frost P C, Larson J H, Johnston C A, Young K C, Maurice P A, Lamberti G A, Bridgham S D. Landscape predictors of stream dissolved organic matter concentration and physicochemistry in a Lake Superior river watershed. Aquatic Sciences-Research Across Boundaries, 2006, 68(1): 40~51.

[7] Mita D, DeKeyser E, Kirby D, Easson G. Developing a wetland condition prediction model using landscape structure variability. Wetlands, 2007, 27(4): 1124~1133.

[8] 刘明,王克林. 洞庭湖流域中上游地区景观格局变化的水文响应[J]. 生态学报,2008,28(12):5970~5979.

[9] Naugle E D, Johnson R R, Estey E M, Higgins F K. A landscape approach to conserving wetland bird habitat in the Prairie Pothole region of Eastern South Dakota. Wetlands, 2000, 20(4): 588~604.

[10] Taft O W, Haig S M. Importance of wetland landscape structure to shorebirds wintering in an agricultural valley. Landscape Ecology, 2006, 21(2): 169~184.

[11] 刘红玉,李兆富,李晓民. 湿地景观破碎化对东方白鹳栖息地的影响——以三江平原东北部区域为例[J]. 自然资源学报,2007,22(5):817~823.

[12] Lopez R D, Davis C B, Fennessy M S. Ecological relationships between landscape change and plant guilds in depressional wetlands. Landscape Ecology, 2002, 17(1): 43~56.

[13] Vermaat J E, Goosen H, Omtzigt N. Do biodiversity patterns in Dutch wetland complexes relate to variation in urbanization, intensity of agricultural land use or fragmentation? Biodiversity and Conservation, 2007, 16: 3585~3595.

[14] 刘红玉,吕宪国,张世奎. 三江平原流域湿地景观多样性及其50年变化研究[J]. 生态学报,2004,24(7):1472~1479.

[15] 西南林学院. 云南碧塔海自然保护区综合科学考察报告[D]. 昆明:西南林学院,2002.

[16] 田昆,陆梅,常凤来,等. 云南纳帕海岩溶湿地生态环境变化及驱动机制[J]. 湖泊科学,2004,16(1):35~42.

[17] 白军红,欧阳华,崔保山,等.近40年来若尔盖高原高寒湿地景观格局变化[J].生态学报,2008,28(5):2245~2252.
[18] 鲁韦坤,杨树华.滇池流域景观格局变化研究[J].云南大学学报:自然科学版,2006,28(S1):201~208.
[19] 田昆,陈宝昆,贝荣塔,等.In-situ方法在研究退化土壤氮库时空变化中的应用[J].生态学报,2003,23(9):272~278.
[20] 刘光崧.土壤理化分析与剖面描述[M].北京:中国标准出版社,1996:31~37.
[21] 殷勇,方念乔,胡超涌,等.云南中甸纳帕海古环境演化的有机碳同位素记录[J].湖泊科学,2001,13(4):289~295.
[22] Song G H, Li L Q, Pan G X, Zhang Q. Topsoil organic carbon storage of China and its loss by cultivation. Biogeochemislry, 2004, 74(1): 47~62.
[23] 王根绪,卢玲,程国栋.干旱内陆流域景观格局变化下的景观土壤有机碳与氮源汇变化[J].第四纪研究,2003,23(3):270~279.
[24] 田昆.云南纳帕海高原湿地土壤退化过程及驱动机制[D].长春:中国科学院东北地理与农业生态研究所,2004.
[25] WWF.长江源区气候变化及其生态水文影响[M].北京:气象出版社,2008:1~12.
[26] 陈利顶,傅伯杰,徐建英,等.基于"源-汇"生态过程的景观格局识别方法——景观空间负荷对比指数[J].生态学报,2003,23(11):2406~2413.
[27] 李杰,胡金明,董云霞,等.1994~2006年滇西北纳帕海流域及其湿地景观变化研究[J].山地学报,2010,28(2):247~256.
[28] 罗姗,张昆,彭涛,等.旅游活动对高原湿地纳帕海土壤理化性质的影响研究[J].安徽农业科学,2008,36(6):2391~2393.

(本文发表于《生态学报》,2011年)

基于土地利用的局域社会生态系统动态平衡分析

陈剑[1], 华梅[1], 杨文忠[1], 李灿雯[1], 赵莉莉[2], 蒋宏[1]

(1. 云南省林业科学院 国家林业局云南珍稀濒特森林植物保护和繁育重点实验室, 云南 昆明 650201;
2. 昆明理工大学 环境科学与工程学院, 云南 昆明 650504)

摘要: 社会生态系统的发展演化是社会子系统和自然子系统相互作用的结果。社会生态系统(SES)中社会和自然子系统的相互作用处于动态平衡之中, 山地 SES 两类子系统的动态平衡关系, 主要取决于自然植被的演替和人类活动的干扰。本研究在村域尺度上, 运用 3S 技术, 结合社会经济入户调查, 构建了基于土地利用的社会生态系统动态平衡指数, 分析村域社会生态系统(V-SES)在从裸地(或人工建筑)到顶级群落演替过程中所处的位置, 揭示 V-SES 两类子系统相互作用关系。对高黎贡山东坡不同海拔三个自然村旱龙、芒岗、芒晃的 V-SES 动态平衡分析, 结果显示: V-SES 动态平衡指数排序为旱龙(-0.50)>芒岗(-0.71)>芒晃(-0.75); 其中人口压力较小、土地生产力较高的旱龙, 处于最佳的 V-SES 动态平衡之中; 芒岗的土地生产力较高但人口压力最大, 芒晃的人口压力小但土地生产力最低, 二者均处于 V-SES 动态平衡较差的状态。研究结果为 V-SES 动态平衡分析提供了一种新的方法, 且对开展更大尺度上的 SES 研究具有参考价值。

关键词: 局域社会生态系统, 系统动态, 动态平衡, 土地利用

Dynamic Equilibrium of Local Social-Ecological Systems Based on Land Use

Chen Jian[1], Hua Mei[1], Yang Wen-zhong[1], Li Can-wen[1], Zhao Li-li[2], Jiang Hong[1]

(1. Key laboratory of Rare and Endangered Forest Plant of Yunnan Academy of Forestry, Kunming Yunnan 650201, P. R. China;
2. Environmental Science and Engineering School of Kunming University of Science and Technology, Kunming Yunnan 650504, P. R. China)

Abstract: Evolution of Social-ecological systems is a result of interactions between social sub-system and natural sub-system. Interactions between two sub-systems of SES are at a dynamic equilibrium. The dynamic equilibrium of mountainous SES results from the natural vegetation successions and human disturbances on lands. This means that every patch of land in SES marks its position in a series from barren field or the artificial to natural climax vegetation. In other words, the dynamic equilibrium can be revealed from land uses. Using 3S techniques and data gained from household survey and land use investigation, we built a dynamic equilibrium index (DEI) based on information extracted from land uses. The DEI can be applied to analyze the dynamic equilibrium situations of village SES (V-SES). We took 3 villages on eastern slope of the Gaoligong Mountains as cases to test the method. Results show that DEI of 3 V-SESs ranks Hanlong village (-0.50) > Manggang village (-0.71) > Manghuang village (-0.75). Hanlong village with lower population pressure and higher land productivity has the best dynamic equilibrium. Manggang village has the highest population pressure and higher land productivity, and Manghuang village has the lower population but lower land productivity. Both of them are in worse dynamic equilibrium situation. This study provides a new method to analyze dynamic equilibriumof V-SES. The results can be a reference for SES study at a larger scale.

Keywords: local social-ecological system; system dynamics; dynamic equilibrium; land use

社会生态系统(Social-Ecological Systems, SES), 是由具社会属性的人和具生态学意义的自然环境要素联结而成的系统[1,2]。SES 是一个多层次嵌套系统, 涵盖从小型社区与其自然环境构成的局域系统, 到全球人与自然系统等多个圈层。村域社会生态系统(Village SES, V-SES)是该嵌套系统的最低层次局域系统, 是全球人与自然系统的基本组成单位。以山地为主的西部省区地处我国地势较高的第一、二阶梯[3], 在国家可持续发展战略中发挥着生态安全屏障和"水塔"的功能和作用[4], 研究认识山地 V-SES 社会子系统与自然子系统的动态平衡关系, 是实现自然资源合理利用、环境经济和谐共生及山区可持续发展的基本前提, 对科学推进我国西部大开发战略的实施具有重要现实意义。

人与自然的相互作用关系是 SES 研究的基本内容。21 世纪初, 不断完善的复杂适应系统理论和人类可持续发展的迫切需要, 共同推动着 SES 研究快速发展[5~8]。SES 研究一方面在于探索其结构、功能特征和演化规律, 另一方面致力于协调人与自然关系的管理实践; 但研究认识社会子系统和自然子系统之间的关系、促进 SES 可持续性, 是 SES 研究的核心内容; 且 SES 可持续性可等价为自然资源对人类的满足程度和人类需求对环境压力之间的动态平衡。

SES 研究涉及生态、人口、社会、经济等各方面的指标, 跨学科研究是攻克 SES 难题的基本方法[9]。Liu J. G. 等在生态学领域总结 10 篇关于生态、人口、社会、经济综合模型论文的基础上, 提出跨学科研究是认识 SES 的基本方法[10,11]; Ostrom E. 通过总结近年的 SES 实证研究, 构建了包含资源系统、资源单元、管理系统和使用者的 SES 分析框架[12,13]。尽管 Liu J. G. 等指出在全球各地开展实证研究的重要性, 但当前的 SES 研究以全球尺度和海岸、海湾区域尺度的 SES 研究居多[14~16]。村域尺度研究相对较少。事实上, 村域水平的 SES 研究可根据具体情况对大量的研究参量进行简化, 达到对村域社会生态系统进行分析评价的目的, 并对更大尺度上的社会生态系统研究具有参考价值。以农业为主的 V-SES, 村民的各种创收和消费活动都建立在对土地资源的利用上, V-SES 平衡动态可通过系统内土地利用表现出来, 本文通过定量和定性相结合的方法, 分析研究这一类 V-SES 的动态平衡问题。

1 研究区概况

百花岭村地处全球生物多样性热点地区之一的高黎贡山东坡, 毗邻高黎贡山国家级自然保护区, 地理坐标位于 25°18′33″~25°15′40″N, 98°47′27″~98°50′17″E, 行政上隶属于保山市隆阳区芒宽乡, 海拔为 700~2 000m, 气候主要属山地中、北亚热带气候, 全年盛行西偏南风, 干湿季显著, 四季不分明, 全年日照时数 2 000~2 100h, ≥10℃积温 4 200~6 000℃, 持续日数 260~350 天, 年平均气温 13~18℃, 年降水量 1 100~1 700mm。海拔 1 100~2 000m

自然植被为季风常绿阔叶林,1 100m 以下为河谷稀树灌丛,除自然植被类型及分布格局的差异外,各自然村的小气候、经济作物、土地利用、村民收入来源等各方面,都随海拔变化存在显著差异。本研究选取百花岭上、中、下三个海拔地段的3个自然村为研究对象,分别是旱龙、芒岗和芒晃,均为汉族、傈僳族、白族、彝族、傣族等多民族混居村落。

2 研究方法

2.1 土地利用/覆被数据

参照 SPOT-5 卫星高清影像进行现地 GPS 采点和土地利用斑块勾绘,对土地类型、植物群落结构、物种构成、植被演替阶段等详情进行记录;并下载当地 SRTM 影像进行海拔地形分析(http://srtm.csi.cgiar.org/SELECTION/inputCoord.asp)。

2.2 村民经济收入数据

根据各自然村人口数量,按人口比例的25%通过半结构入户访谈方式获取,并考虑到访谈对象覆盖不同收入水平。

2.3 数据处理

根据 SPOT-5 卫星影像和勾绘的各类斑块进行土地利用状况,运用 Arcgis 9.3 进行统计分析和可视化分析;通过 Global Mapper 13 对 SRTM 影像进行高程分析和等高线制作;运用 SPSS 19 与 Excel 2003 完成数据分析、统计和制图。

2.4 动态平衡分析

社会生态系统的可持续性,取决于人类需求对自然环境的压力和自然环境对人类需求的支撑之间的平衡状态。对以农业为主的局域社会生态系统而言,这种平衡状态与土地利用状况密切相关,因此可通过对土地利用的量化分析揭示人与自然的动态平衡关系。其他影响 V-SES 动态平衡的因素,如政策法规、民族习俗、文化程度、生活水平等实际上已经体现在土地利用状况之中,可不直接对其进行量化分析,但可作为辅助量化分析的验证手段。基于这一理念构建 V-SES 动态平衡指数,并通过 Arcgis 平台完成环境压力梯度的可视化分析。

2.4.1 V-SES 动态平衡指数

V-SES 动态平衡指数基于当前 V-SES 中各种土地类型斑块处于"裸地"到"当地顶级群落"演替系列中的位置,以此来计算整个 V-SES 是处于平衡状态或是存在明显的偏自然或偏人工干扰演替的动态趋势,这种趋势以 V-SES 动态平衡指数进行表征,其生态意义为V-SES 在从裸地/人工建筑到当地顶级群落演替系列中所处的位置。

$$DEI = \sum_{i=1}^{n}\left[(E_i - 1) \times \frac{S_i}{S}\right] \quad (1)$$

式中 DEI 为 V-SES 动态平衡指数($-1 \leq DEI \leq 0$),$DEI = 0$ 表示生态系统为天然顶级群落,$DEI = -1$ 表示生态系统中无植被,全为裸地或人工建筑;E_i 为当地第 i 个土地类型演替进程,S_i 为当地第 i 个土地类型的面积;S 为研究区域总面积。结合社会经济因素,在[0,1]区间内对处于不同演替进程的土地类型进行赋值(表1)。

表1 土地类型演替进程赋值条件

Table 1 Conditions for valuation of different land succession

		赋值条件	赋值指标
土地类型演替进度（E_i）	1	顶级群落植被：群落结构稳定，物质能量输入输出达到平衡	0.7~1
	2	演替中期植被：群落能自行更新，先锋物种比例下降，以地带性物种为优势种	0.4~0.6
	3	演替初期植被：植被以草本或先锋树种幼苗为主	0.1~0.3
	6	长期强烈干扰不能正常演替或在人工干扰下稳定在某一演替阶段的植被（如四旁林、农地、经济作物）	0.1~0.4
	4	裸地	0
	5	人工建筑	0

注：系统中存在一个以上气候顶级群落，则以生物量最大的顶级群落为准。旱龙和芒岗的顶级群落为亚热带季风常绿阔叶林，芒晃的顶级群落是干热河谷稀树灌丛；从土地生产力的角度，稀树灌丛的生物量远不及亚热带季风常绿阔叶林，为比较分析自然资源对社会发展的支撑作用，把芒晃的顶级群落视作从裸地到季风常绿阔叶林演替的一个阶段。

2.4.2 V-SES 动态平衡可视化分析

对每个土地斑块依表1赋值标准对其斑块质心赋值，赋值标准为表1对整个自然村进行插值计算，得到各自然村各地段的 V-SES 动态平衡可视化分布。

可视化分析用到局部插值相对误差较小的 IDW（反距离权重）IDW 插值计算方法为[17]：

$$Z_0 = \left[\sum_{i=1}^{n} \frac{z_i}{d_i^k}\right] \bigg/ \left[\sum_{i=1}^{n} \frac{1}{d_i^k}\right]$$

式中 Z_0 为点0的估计值，Z_i 为控制点，d_i 为控制点到点0的距离，n 为控制点数目，k 为指定的幂，插值点距离控制点距离越远，所受其影响越低。

3 结果与分析

3.1 三个自然村的土地利用情况

高黎贡山各种生境条件随海拔有明显的区别，本研究依海拔梯度选择的三个自然村（旱龙 1 200~2 000m，芒岗 1 120~1 600m，芒晃 700~1 500m），由于气候土壤条件差异，导致自然植被分布、土地利用类型和利用格局都存在明显差别（图1）。

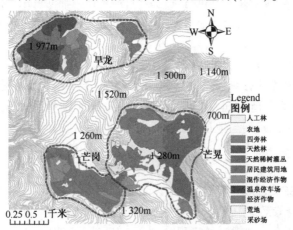

图1 三个 V-SES 的土地利用分布格局

Figure 1 Land use pattern in 3 village SES

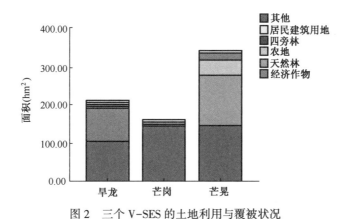

图 2　三个 V-SES 的土地利用与覆被状况

Figure 2　Land use and land cover in 3 village SES

旱龙村土地面积 211.28hm², 接近高黎贡山国家级自然保护区边界, 在土地利用上最显著的特征是保存有干扰较小的天然季风常绿阔叶林, 面积 88.94hm², 接近其总面积的一半; 经济作物以核桃(较高海拔地段)、咖啡、板栗(海拔较低地段)为主, 核桃种植面积共 50.27hm², 其中 14.96hm² 与板栗混作; 板栗纯林 3.05hm²; 咖啡种植面积 44.65hm²; 另有少量甘蔗、草果、桑树等经济植物; 用于粮食作物种植的农地少, 只有 2.32hm², 但未成林核桃林内常有玉米混作; 四旁树种有核桃、竹类、芭蕉、构树等; 居民建筑用地 5.67hm² (图 2)。

芒岗村土地面积 161.14hm², 土地人工利用充分, 几无自然植被; 经济作物以咖啡为主, 共 101.25hm², 其中 18.72hm² 混作核桃; 核桃纯林 12.9hm²; 另有少量板栗、荔枝、橘子等经济植物; 用于粮食作物种植的农地仅有 0.9hm²; 四旁林以竹类和杉木为主, 共 5.65hm²; 居民建筑用地 6.15hm²。

芒晃村土地面积 343.42hm², 也存在大片未经人工利用的自然植被, 即干热河谷稀树灌丛, 共 132.59hm²; 经济作物以咖啡为主, 咖啡纯林共 132.76hm²; 橘子、荔枝、龙眼、杧果等各种经济作物以不同比例相互混作模式共 14.39hm²; 农地 39.62hm², 三个自然村中面积最大; 四旁林以杉木和竹类为主, 共 18.69hm²; 居民建筑用地 3.63hm²。

3.2　基于土地利用的 V-SES 动态平衡

简单的土地利用类型分类, 并不能说明土地利用的动态趋势, 主要是基于以下几个原因:①同样的土地利用类型, 如成熟核桃林与初植核桃幼树林、粗放管理的核桃林与人工干扰强度大的核桃林, 可能存在演替阶段的巨大差别(体现在生物量,群落结构、物种多样性等方面);②不同土地利用类型斑块的边界, 可能没有明确的边界;③不同斑块边界存在自然植被演替与人工干扰的相互影响。因此在分析 V-SES 动态平衡时, 本研究以不同土地利用状态下各个土地类型斑块的演替阶段作为分析基础, 通过(1)式计算出各自然村的 V-SES 动态平衡指数, 在对各斑块演替阶段赋值的基础上(表1), 通过(2)式对各自然村土地进行演替阶段分 10 个梯度进行插值计算, 分析潜在的社会发展压力或自然资源支撑能力在土地上的梯度分布。

各斑块在从裸地/人工建筑到地带性顶级群落不同演替阶段中所处的位置, 以 0.1 为区间划分不同演替阶段的土地斑块,构成土地斑块演替系列(图3),分布梯度位置越高(深绿色),说明该局部地段人为干扰小,反之分布梯度位置越低(深红色)说明局部地段人为干扰大。各类斑块在V-SES中所占的比例,反映了该村社会经济发展对自然资源的压力或自然环境对社会经济的支撑能力。图3直观显示出三个V-SES中旱龙村有较全面的梯度分布和较多的高梯度斑块,说明旱龙的社会发展对自然资源的压力较小,或自然环境的支撑能力较强;而芒岗和芒晃的梯度构成和分布,表明这两个V-SES的土地大多处于较低的梯度位置,社会发展对自然资源的压力大,或自然环境的支撑能力不足。

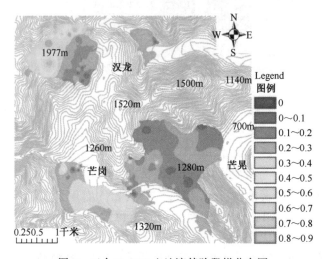

图3 三个V-SES土地演替阶段梯分布图

Figure 3 Gradientdistribution of land succession in 3 village SES

三个自然村V-SES动态平衡指数DEI的顺序为旱龙(-0.50)>芒岗(-0.71)>芒晃(-0.75)(表2),根据(1)式定义可知,在-1~0范围内,旱龙处于比芒岗与芒晃更居中的平衡位置,芒岗与芒晃的DEI值接近,处于人为干扰较严重的状态。

表2 三个V-SES的动态平衡状况

Table 2 Dynamic equilibrium of 3 village SES

村名	不同土地演替梯度的面积分布(hm²)										动态平衡指数DEI
	0	0~0.1	0.1~0.2	0.2~0.3	0.3~0.4	0.4~0.5	0.5~0.6	0.6~0.7	0.7~0.8	0.8~0.9	
旱龙	0.01	3.59	38.5	78.15	46.86	23.42	11.23	5.39	2.72	1.41	-0.50
芒岗	0.01	0.7	3.03	51.71	105.18	0.51					-0.71
芒晃		15.25	168.34	156.96	2.87						-0.75

3.3 V-SES动态平衡成因

V-SES动态平衡分析,可根据各斑块演替状态判断其在从人工系统(如裸地或人工建筑)到天然顶级群落演替过程中所处的位置。但仅根据V-SES动态平衡指数DEI值,难于判断该动态的成因,如对一个DEI值低的村域社会生态系统而言,就难以判断该系统是因为社会发展对自然资源的压力大(如人口过多、经济收入要求高),还是自然环境支撑能力不足(如生境贫瘠,生产力弱),这时需要结合人口、经济收入、土地生产力等进行分析(表3)。

表3 三个 V-SES 人口、经济收入和土地利用状况
Table 3 Population, income, and land use of 3 village SES

村名	人口	土地面积 (hm²)	人均土地面积 (hm²)	人均已开发利用土地* (hm²)	土地开发利用率 (%)	人均年收入 (万元)	已开发利用土地的年平均产值 (万元·hm⁻²)	年平均土地面积产值 (万元·hm⁻²)
汉龙	206	211.28	1.03	0.57	55.80	1.10	1.92	1.07
芒岗	420	161.14	0.38	0.38	98.03	1.45	3.83	3.77
芒晃	131	343.42	2.62	1.60	60.96	1.48	0.93	0.56

*已开发利用土地含经济作物种植地、四旁林、农地、居民建筑用地。

通过对人口、经济收入和土地利用状况等因素的综合分析,认为影响村域社会生态系统动态平衡的因素主要如下几个方面。

(1)经济收入趋同性。三个自然村的生境、人口、土地面积、土地的人工利用率、平均土地经济产值等都存在显著差异,但人均年收入差异较小,这有可能是经济收入的"趋同性"[18],即特定区域内的人群在价值观上追求同等经济收入的观念。在满足基本生存条件的基础上,"趋同性"驱使区内人群通过各种途径达到经济收入上的平等,从而造成对生态环境的不同依赖和影响。为达到人均年收入1.10万~1.48万元这个水平,三个自然村人均对土地开发的面积相差最大约为4倍;但是,由于各个自然村受天然资源的限制,不可能无限开发,以农业为主且物质能量流动较少的村域社会生态系统,最终的经济收入会受到环境资源的限制达到极限,如芒岗对土地的开发利用率已达98.03%,与芒岗和芒晃相比,经济增收的空间已经非常有限。

(2)社会发展对自然环境的压力。由于土地划分有着复杂的历史背景原因,各自然村之间、各农户之间或各村民之间,都不可能获得面积、生产力、自然资源完全一致的土地。芒岗和芒晃两个V-SES的人均土地面积相差约达7倍,造成其社会发展对自然资源压力的显著差异。芒岗人口最多而土地最少,对土地的开发利用率高达98.03%,形成以果树种植园为主的集约化经营产业,已开发土地的年均产值分别约为旱龙和芒晃的2倍和4倍,基本已无天然林分布。而人均土地面积较多的旱龙和芒晃,天然林保存率为44.2%和39.04%。

(3)环境资源对社会发展的支撑能力。本研究中环境资源对社会发展的支撑能力,主要体现在自然生态系统生产力上,芒晃的自然生态系统受水热条件限制,生产力较旱龙与芒岗的差,体现为天然植被的地带性顶级群落为干热河谷稀树灌丛,旱龙与芒岗的地带性顶级群落为亚热带季风常绿阔叶林;在年平均土地面积产值上,芒晃为0.56万元·hm⁻²,旱龙为1.07万元·hm⁻²,芒岗为3.77万元·hm⁻²,已开发利用土地的年平均产值也反映了与此类似的结果,即土地生产力越差,环境资源对社会发展的支撑能力越低。芒晃村由于土地生产力较低,人均开发利用土地面积是芒岗的4倍,才达到与芒岗接近的人均年收入。旱龙在三个自然村中土地利用开发率最低,原因是本村良好的天然林为大量鸟类提供了栖息地,催生了涵盖导鸟、摄影、农家旅店等的生态观鸟旅游业,使村民获得了比直接进行土地开发更高的收入。加之前来观鸟的游客多具保护意识,与村民形成了良性的互动,促使村民逐步减少

了核桃种植面积,提高天然林保护和恢复的意识。

综上可知,V-SES 动态平衡指数 DEI 值的高低,受到社会发展对自然资源压力和自然环境支撑能力的共同影响,即在同样面积的土地范围内,人口越少,土地生产力越高,则 DEI 值越高;反之,人口越多,土地生产力越低,则 DEI 值越低。

4 结论与讨论

4.1 结论

(1)作为村域社会生态系统,对土地资源的直接开发利用是当地村民基本的生存和发展方式,土地利用状况(包括利用类型和演替阶段)可直接反映 V-SES 动态平衡,该动态平衡的状态主要取决于社会发展对自然资源的压力和自然环境的支撑能力,通过开展理论探索和案例分析,本研究成功构建了 V-SES 动态平衡分析方法。

(2)通过对研究区三个自然村的 V-SES 动态平衡指数分析测算,DEI 值排序为旱龙(-0.50)>芒岗(-0.71)>芒晃(-0.75),社会发展和人口压力较小、土地生产力较高的旱龙处于最佳的 V-SES 动态平衡状态,人口压力最大、土地生产力较高的芒岗和人口压力较小、土地生产力最低的芒晃,均处于较差的 V-SES 动态平衡状态,且前者稍优于后者。

(3)通过对旱龙、芒岗和芒晃三个自然村的分析检验,V-SES 动态平衡分析方法,能准确反映村域社会生态系统自然子系统和社会子系统之间的动态关系,并可获得量化和可视化结果。

4.2 讨论

(1)V-SES 动态平衡与可持续发展。V-SES 动态平衡分析的目的是为可持续发展提供科学参考,在村域水平上,可持续发展需要重视提高当地村民生活水平、防止生物多样性丧失、防止水土流失、防止土地侵蚀和肥力退化、增加植被碳汇功能这样的基本关注点。百花岭三个自然村各自典型的动态平衡特征对村域可持续发展这一类关注点具有非常重要的借鉴意义,即达到可持续发展,必须综合考虑社会发展对自然资源的压力和自然环境的支撑能力,如在社会发展压力大、环境资源贫瘠的地区,土地开发利用与生物多样性保护之间是相冲突的,需要在宏观规划上制定替代经济产业,转移社会发展对土地的压力。

(2)V-SES 动态平衡指数的适用范围。V-SES 动态平衡分析,主要适用于经济收入主要依赖土地资源(农业)的村域社会生态系统,这样的系统相对封闭,系统内自然植被演替与人为干扰的消长关系是系统动态平衡的决定因素,故根据土地利用方式及演替阶段即可较便捷地对系统动态平衡进行分析。对开放性高的社会生态系统而言,存在大量的物质能量流动,系统内社会经济发展可能不仅依靠土地资源,如很多城郊村域系统,土地利用率高,自然植被少,失去了自然植被演替与人类干扰相消长的动态平衡基础。因此,在存在明显的跨系统物质能量流动时,如水污染、空气污染、人口流动等,必须扩大至更广范围或更高等级的社会生态系统,直到自然演替与人类干扰存在消长动态关系,方可开展社会生态系统动态平衡分析。

参考文献:

[1] Berkes F, C Folke. Linking social and ecological systems: management practices and social mechanisms for

building resilience[M]. Cambridge University Press, Cambridge, UK. 1998:1~26.
[2] Walker B, Gunderson L, Kinzig A, et al. A handful of heuristics and some propositions for understanding resilience in social-ecological systems[J]. Ecology and Society. 2006, 11(1):13.
[3] 蒋捷, 杨昕. 基于DEM中国地势三大阶梯定量划分[J]. 地理信息世界. 2009, 7(1):8~13.
[4] 陈国阶. 对中国山区发展战略的若干思考[J]. 中国科学院院刊. 2007, 22(2):126~131.
[5] Kates R W, Clark W C, Corell R, et al. Sustainability science[J]. Science. 2001, 292(5517):641~642.
[6] 苗东升. 系统科学大学讲稿[M]. 北京:中国人民大学出版社, 2007. 11:381~397.
[7] 张知彬, 王祖望, 李典谟. 生态复杂性研究——综述与展望[J]. 生态学报, 1998, 18(4):433~441.
[8] Clark, W. C. Sustainability science: A room of its own [J]. Proceedings of National Academy of Sciences of USA, 2007, 104(6):1737~1738.
[9] Liu J G. Special Issue:Integration of ecology with human demography, behavior, and socioeconomics [J]. Ecological Modeling, 2001, 140(1-2):1~8.
[10] Liu J. Integrating ecology with human demography, behavior, and socioeconomics:Needs and approaches [J]. Ecological Modelling. 2001, 140(1-2):1~8.
[11] Alberti M, Asbjornsen H, Baker L A, et al. Research on Coupled Human and Natural Systems (CHANS):Approach, Challenges, and Strategies[J]. Bulletin of the Ecological Society of America. 2011, 92:218~228.
[12] Ostrom E. A general framework for analyzing sustainability of social-ecological systems[J]. Science. 2009, 325(5939):419~422.
[13] Ostrom E. A diagnostic approach for going beyond panaceas[J]. Proceedings of National Academy of Sciences of USA. 2007, 104(39):15181~15187.
[14] Latif M, Keenlyside N S. El Niño/Southern Oscillation response to global warming[J]. Proceedings of National Academy of Sciences of USA, 2009, 106(49):20578~20583.
[15] Leslie H M, Schlüter M, Cudney-Bueno R, Levin S A. Modeling responses of coupled social-ecological systems of the Gulf of California to anthropogenic and natural perturbations [J]. Ecological Research, 2009, 24(3):505~519.
[16] Radeloff V C, Stewart S I, Hawbaker T J, Gimmi U, Pidgeon A M, Flather C H, Hammer R B, Helmers D. P. Housing growth in and near United States protected areas limits their conservation value [J]. Proceedings of National Academy of Sciences of USA, 2010, 107(1):940~945.
[17] 邓晓斌. 基于ArcGIS两种空间插值方法的比较[J]. 地理空间信息, 2008, 6(6):85~87.
[18] 洪会明, 王家顺. 倒U假说与收入趋同[J]. 经济论坛, 2006, 20(1):49~50.

(本文发表于《山地学报》,2014年)

云南高黎贡山中山湿性常绿阔叶林的群落特征

孟广涛[1],柴勇[1,3],袁春明[1],艾怀森[2],李贵祥[1],王骞[1],李品荣[1],蔺汝涛[2]

(1. 云南省林业科学院,云南 昆明 650201;2. 高黎贡山国家级自然保护区保山管理局,云南 保山 678000;3. 中国科学院西双版纳热带植物园,云南 昆明 650223)

摘要:2009~2010年在高黎贡山国家级自然保护区南段生物走廊带建立1块4hm^{2}的中山湿性常绿阔叶林动态监测样地,分析其群落特征。结果表明:样地内胸径≥1.0cm的木本植物有10 546株,分属于35科64属95种,在同森林类型样地中物种丰富度较高,但稀有种所占比例较小;样地中含种数和个体数较多的科为樟科、山茶科、五加科、壳斗科和山矾科,5个科包括了44.21%的种和56.70%的个体数;重要值较大的植物物种有多花山矾、肖樱叶枎、长蕊木兰、龙陵新木姜子、硬壳柯和薄片青冈,但优势均不明显,为多优种的群落类型;植物区系以热带区系成分明显占优势,具有热带起源的背景;样地群落最小面积为1.32hm^2,能够包括样地内80%以上的树种;样地内整体径级结构及优势种径级结构均表现为基部宽、顶部窄的金字塔结构,幼树储备充足,能够保证群落的稳定发展;存活曲线大都表现为Deevey-Ⅲ型,小径级个体死亡率高,大径级个体死亡率低而稳定。

关键词:群落特征;地理成分;动态样地;常绿阔叶林;高黎贡山

Community Characteristics of Mid-Montane Humid Ever-Green Broad-Leaved Forest in the Gaoligong Mountains, Yunnan

Meng Guang-tao[1], Chai Yong[1,3], Yuan Chun-ming[1], Ai Huai-sen[2],
Li Gui-xiang[1], Wang Qian[1], Li Ping-rong[1], Lin Ru-tao[2]

(1. Yunnan Academy of Forestry, Kunming Yunnan 650201; 2. Administrative Bureau of Gaoligongshan Mountains National Nature Reserve Baoshan, Baoshan Yunnan 678000; 3. Xishuangbanna Tropical Botanical Garden, Chinese Academy of Sciences, Kunming Yunnan 650223)

Abstract: A 4hm^2 permanent dynamic plot of mid-montane humid evergreen broad-leaved forest located in ecological corridor in southern region of Gaoligong Mountains National Nature Reserve had been established during 2009~2010, and the community characteristics of it had been studied. The results showed that there were 10 546 free-standing individuals with $DBH \geq 1.0$cm in the 4 hm^2 dynamic plot, belonging to 95 species, 64 genera and 35 families. The species richness was higher and the proportion of rare species was lower in the plot, compared with other plots of the same kind forest. Lauraceae, Theaceae, Araliaceae, Fagaceae, and Symplocaceae were dominant families in the plot and they comprised 44.21% of all species and 56.70% of all individuals. *Symplocasramosissima*, *Euryapseudocerasifera*, *Gordonialongicarpa*, *Neolitsealunglingensis*, *Lithocarpushancei*, and *Cyclobalanopsislamellosa* had lager importance value in the plot, but no one had obvious advantage, namely the plot did not contain an obviously dominant species. Floristic characteristics of the community indicated that the tropical elements were much more than temperate elements, implying that this area could have an origin of tropics. The minimal area of the community is 1.32 hm², that could comprise more than 80% of all species. The structure of *DBH* size class of all species and some dominant species in the plot showed a typical pyramid structure with a wide bottom and a thin top, implying a good regeneration in the community. The survival curve of most of dominant species trended to be of the Deevey-Ⅲ type, with a higher mortality rate at small DHB size class and a low and stable mortality rate at large *DBH* size class.

Key words: community characteristics; geographical distribution; dynamic plot; evergreen broad-leaved forest; Gaoligong Mountains

森林植被与其生存环境相互制约、相互依存,一直处于不断的发展变化之中。对森林植被的动态进行监测,通过考察植被的组成、结构、功能与其生境关系的时空变化过程,揭示出它们之间相互作用的规律,并在此基础上对森林植被及其生境的演变方向和机理进行科学研究并作出综合评价,有利于森林植被的有效保护和管理(李玉媛,2003)。20世纪80年代初,Stephen Hubbell 和 Robin Foster 等在巴拿马 BCI(Barro Colorado Island)建立了第一块 50hm² 的森林大样地(Condit,1995);随后美国 Smithsonian 研究院热带森林科学研究中心(Center for Tropical Forest Science,CTFS)在全球热带地区的15个国家建立了20多个森林生物多样性监测大样地,形成了热带森林生物多样性监测网络(http://www.ctfs.si.edu)。中国科学院生物多样性委员会也于2004年开始实施了生物多样性监测网络建设,分别在长白山(张健等,2007;郝占庆等,2008)、古田山(祝燕等,2008)、鼎湖山(叶万辉等,2008)、天童山(杨庆松等,2011)、哀牢山(巩合德等,2011)、东灵山(刘海丰等,2011)和西双版纳(兰国玉等,2008)等地建立了多个动态监测样地,对我国的温带针阔混交林、暖温带落叶阔叶次生林、中亚热带常绿阔叶林、南亚热带常绿阔叶林和热带雨林的森林生态系统的物种组成、分布格局等进行研究。尽管如此,由于中国地跨热带到寒温带,是世界上生物多样性最丰富的国家之一,现有的监测样地网络仍没能全面地反映出全国生态系统和生物多样性的现状。

高黎贡山位于云南西部,是中国生物多样性最丰富的地区之一。特殊的地形、地貌和气候使得它成为了野生动植物南北过渡的走廊和第三、四纪冰期中野生动植物的"避难所",形成了"动植物种属复杂、新老兼备、南北过渡、东西交汇"的格局,获得了"世界物种基因库""世界自然博物馆"等美誉。中国科学院生物多样性委员会将其列为"具有国际意义的陆生生物多样性关键地区"和"重要模式标本产地"(熊清华等,2006)。在过去的几十年中,已有众多植物学家、生态学家在高黎贡山生物多样性调查与研究方面做了大量工作(石天才等,1994;尹五元,1994;薛纪如,1995;徐志辉,1998;李恒等,2000;刘经伦,2003;朱振华等,2003;王志恒等,2004;柴勇等,2007;李嵘等,2007;2008;欧光龙等,2008;汪建云等,2008a;2008b;徐成东等,2008)。由于受研究尺度和方法的限制,以往的研究大多在相对较小的样地上进行,一些须在较大尺度下研究的生态现象较难涉及。因此,在高黎贡山建立较大的固定监测样地,对其主要植被类型进行长期动态监测和深入研究是十分必要的。2009~2010年,云南省林业科学院在高黎贡山国家级自然保护区赧亢保护站建立了1块4hm²的森林动态监测样地。本研究以此样地的第一次本底调查数据为基础,分析群落的物种组成、区系特征、径级结构等群落特征,为今后开展物种空间格局、物种共存机制、森林植被更新及群落演替动态等方面的研究提供基础资料。

1 研究区概况

高黎贡山国家级自然保护区位于中国云南省西部(98°08′~98°50′E,24°56′~28°22′N),分为南、中、北互不相连的三段,是云南省面积最大的自然保护区。高黎贡山北依青藏高原,南向中印半岛俯冲,形成北高南低的地势地貌,境内最高海拔5 128m,最低海拔720 m。属亚热带高原季风气候,全年盛行西南风,四季不分明,干湿季显著,气温日较差大、年较差小。年均气温14~17℃,最冷月气温在7℃以上,年降雨量1 000mm以上。11月下旬至翌年4月为干季,日照充足,降水少,空气相对湿度50%~60%。5~10月为雨季,降水量占全年的87%。境内高差悬殊,立体气候明显,从而也发育了较完整的自然植被垂直带,如南段的基带为热带性的半常绿季雨林或季风常绿阔叶林(1 700 m以下),向上依次出现中山湿性常绿阔叶林(1 800~2 600m)、温凉性针叶林或山顶苔藓矮林(2 700~3 100m)、寒温性针叶林(3 100~3 400m)和高山灌丛与草甸(3 300m以上)等,东西坡差异较小。

监测样地位于高黎贡山国家级自然保护区南段生物走廊带内(98°45′53.1″~98°46′1.3″E,24°50′9.8″~24°50′17.3″N),样地内地形较复杂,最高海拔2 229.65m,最低海拔2 135m,相对高差为94.65 m。样地土壤以黄棕壤为主,森林植被为典型的中山湿性常绿阔叶林。乔木层主要树种有长果大头茶(*Gordonia longicarpa*)、多花山矾(*Symplocos ramosissima*)、肖樱叶柃(*Eurya pseudocerasifera*)、硬壳柯(*Lithocarpus hancei*)、龙陵新木姜子(*Neolitsea lunglingensis*)和薄片青冈(*Cyclobalanopsis lamellosa*)等。灌木层中竹子层片明显,局部地段层盖度可达90%以上,种类以带鞘箭竹(*Fargesia contracta*)和滇川方竹(*Chimonobambusa ningnanica*)较常见,其他的灌木种类还有聚果九节(*Psychotria morindoides*)、长小叶十大功劳(*Mahonia lomariifolia*)、梗花粗叶木(*Lasianthus biermannii*)、长梗常春木(*Merrilliopanax listeri*)及一些上层乔木树种的幼树(苗)等。草本层不发达,除一些低凹沟谷处有宽叶楼梯草(*Elatostema platyphyllum*)成片分布外,其他种类都在林中呈零星分布。林中不乏藤本植物及附生植物,常见的有上树蜈蚣(*Rhaphidophora lancifolia*)、多种树萝卜(*Agapetes* spp.)和多种兰科植物(Orchidacee)附生或悬挂于树枝上。

2 研究方法

在研究区域中山湿性常绿阔叶林分布相对较连片集中的地段建立1块4hm²(边长为200m的正方形)的永久固定样地,建设技术参照巴拿马BCI样地建设规范(兰国玉等,2008;Condit,1998;Cao,et al.,2008)。

用全站仪将整个样地划分为100个20m×20m的样方,每个样方再划分为16个5m×5m的小样方。以20m×20m样方为基本样方,对每个基本样方内胸径≥1.0cm的所有木本植株个体进行编号,并记录其名称、胸径、树高。整个样地以东西方向为X轴,南北方向为Y轴,测量每株树木在5m×5m小样方的坐标值以进行定位。样地勘测及植物调查工作于2009年10月至2010年11月进行。

乔木重要值为相对多度、相对显著度和相对频度3者之和(宋永昌,2001)。本研究中相对多度仅计算独立个体的数量,相对显著度计算包括分枝的断面积,计算相对频度的总样方数为100个(20m×20m)。

从样地西南角原点开始分别选择400,800,1 200,1 600,2 000,……、40 000m²为取样面

积,统计各面积内的树种数量,绘制种-面积曲线,并分别用对数模型和幂函数进行模型拟合(宋永昌,2001)。

由于测定树木的年龄比较困难,故用树木的径级结构来代替年龄结构。样地中各优势树种径级结构划分以5cm为径阶,共划分为21个径级,它们分别为:1cm≤DBH<5cm、5cm≤DBH<10cm、10cm≤DBH<15cm、……、95cm≤DBH<100cm和DBH≥100cm。分别统计这些树种在各径级的存活个体数,并换算成标准化存活个体数(一般标准化为1000),以径级为横坐标,以各径级标准化存活个体数为纵坐标,绘制成各树种的存活曲线。各径级存活个体数标准化公式为:$l_x = a_x/a_0 \times 1000$,$l_x$为X径级的标准化个体数,$a_x$为X径级的存活个体数,$a_0$为开始时的存活个体数(覃林,2009)。

3 结果与分析

3.1 物种组成

样地中共记录到DBH≥1.0cm的木本植株个体10 546株(包括分枝和萌枝共计14 498株),平均每公顷2 637株,包括1种蕨类植物(中华桫椤 Alsophila costularis)和94种种子植物,隶属于35科64属(表1)。其中常绿树种85种,占总种数的89.47%;落叶树种10种,占总种数的10.53%。从科的组成看,含种数最多的是樟科(Lauraceae)(14种),其次为山茶科(Theaceae)(10种)、五加科(Araliaceae)(8种)、壳斗科(Fagaceae)(6种)和冬青科(Aquifoliaceae)(6种),含4种的科有木兰科(Magnoliaceae)、蔷薇科(Rosaceae)、山矾科(Symplocaceae)和茜草科(Rubiaceae),含3种的科有杜鹃花科(Ericaceae)和芸香科(Rutaceae),另含2种的有5科,含1种的有19科。含个体数最多的科为山茶科(1 716株),其次为樟科(1 613株)、山矾科(1 169株)、壳斗科(1 097株)、蔷薇科(676株)和紫金牛科(Myrsinaceae)(578株);含100~500株的科有14科,如槭树科(Aceraceae)、山龙眼科(Proteaceae)、木兰科和五加科等,含100株以下的有15科,如八角科(Illiciaceae)和杜鹃花科等。从属的组成看,含种数最多的是冬青属(Ilex)(6种),其次为山矾属(Symplocos)(4种)、柏那参属(Brassaiopsis)(3种)、鹅掌柴属(Schefflera)(3种)、木姜子属(Litsea)(3种)、新木姜子属(Neolitsea)(3种)、山茶属(Camellia)(3种)和石栎属(Lithocarpus)(3种)。另含2种的有11属,含1种的有45属。从科属的组成上可以看出,该群落是比较典型的常绿阔叶林,樟科、山茶科、壳斗科、木兰科、山矾科和五加科的植物在其中明显占有优势。

表1 高黎贡山常绿阔叶林动态监测样地个体数量≥300的优势科
Table 1 Dominant families with number of individuals ≥300 in the evergreen broad-leaved forest dynamic plot in the Gaoligong Mountains

科名 Name of families	属数 Number of genera	种数 Number of species	个体数量 Number of individuals
山茶科 Theaceae	6	10	1 716
樟科 Lauraceae	7	14	1 613
山矾科 Symplocaceae	1	4	1 169
壳斗科 Fagaceae	3	6	1 097
蔷薇科 Rosaceae	3	4	676

(续)

科名 Name of families	属数 Number of genera	种数 Number of species	个体数量 Number of individuals
紫金牛科 Myrsinaceae	2	2	578
槭树科 Aceraceae	1	2	470
五加科 Araliaceae	4	8	385
山龙眼科 Proteaceae	1	1	364
木兰科 Magnoliaceae	3	4	315
其他25科 Other 25 species	33	40	2 163
合计 Total	64	95	10 546

3.2 科属区系特征

根据吴征镒(1991;2003)、吴征镒等(2003)和李锡文(1996)对中国种子植物科、属分布区类型的划分,样地中35个科可划分为7个类型2个变型,64个属可划分为10个类型7个变型(表2)。其中,世界分布科有2科,分别为蔷薇科和远志科(Polygalaceae),世界分布属有1属(远志属 Polygala);热带分布科有25科(占75.76%),热带分布属有46属(占73.02%);温带分布科8科(占24.24%),温带分布属有17属(占26.98%);没有出现中国特有分布的科和属。科属区系组成情况反映出样地群落具有明显的热带起源背景。

表2 高黎贡山常绿阔叶林动态监测样地木本植物地理成分
Table 2 Distribution patterns of species in the evergreen broad-leaved forest dynamic plot in the Gaoligong Mountains

分布区类型 Areal type	科数 Number of families	属数 Number of genera
1 世界分布 Cosmopolitan	2	1
2 泛热带分布 Pantropic	20	14
2-2 热带亚洲、非洲和中、南美间断分布 Trop. Asia, Africa & C. to S. Amer. disjuncted		2
3 热带亚洲和热带美洲间断分布 Trop. Asia & Trop. Amer. disjuncted	3	5
4 旧世界热带分布 Old World Tropics		1
4-1 热带亚洲、非洲(或东非、马达加斯加)和大洋洲间断分布 Trop. Asia., Africa (or E. Afr., Madagascar) & Australasia disjuncted	1	1
5 热带亚洲至热带大洋洲分布 Tropical. Asia & Trop. Australasia		3
6 热带亚洲到热带非洲分布 Trop. Asia to Trop. Africa		2
6-2 热带亚洲和东非或马达加斯加间断分布 Trop. Asia & E. Afrca or Madagascar disjuncted		1
7 热带亚洲(印度—马来西亚)分布 Trop. Asia (Indo-Malaysia)	1	12
7-1 爪哇、喜马拉雅间断或星散分布到华南、西南 Java, Himalaya to S., SW. China disjuncted or diffused		2
7-4 越南(或中南半岛)至华南(或西南)分布 Vietnam (or Indo-Chinese Peninsula) to S. China (or SW. China)		3

分布区类型 Areal type	科数 Number of families	属数 Number of genera
热带小计 Subtotal of tropical	25	46
8 北温带分布 North Temperate	4	3
8-4 北温带和南温带（全温带）间断分布 N. Temp. & S. Temp. disjuncted (Pan-temperate)	1	1
9 东亚和北美洲际间断分布 E. Asia & N. Amer. disjuncted	2	8
14 东亚分布 E. Asia	1	4
14-1 中国-喜马拉雅分布 Sino-Himalaya (SH)		1
温带小计 Subtotal of temperate	8	17
合计 Total	35	64

3.3 重要值组成

样地中重要值≥1.0的物种共有38种，占总种数的40%，但它们的个体数和胸高断面积分别占总个体数和总胸高断面积的86.68%和90.75%（表3）。从重要值的分布来看，所有树种的重要值均在5%以下，没有特别优势的树种。其中，重要值最大的是多花山矾（4.15%），其他较大的还有肖樱叶柃（3.95%）、长果大头茶（3.69%）、龙陵新木姜子（3.53%）、硬壳柯（3.45%）、薄片青冈（3.42%）、毛柄槭（Acer pubipetiolatum）（3.41%）、尖叶桂樱（Laurocerasus undulata）（3.40%）、多沟杜英（Elaeocarpus lacunosus）（3.21%）、星毛鹅掌柴（Schefflera minutistellata）（3.11%）和针齿铁仔（Myrsine semiserrata）（3.09%）等，它们之间的重要值差异也不明显。多花山矾因其个体数最多而在群落中占有较重要的地位，其相对多度对其重要值的贡献率达63.82%。但多花山矾的植株个体较小（平均胸径仅4.49cm），大部分处于高度10m以下的层次，因而它对群落的作用和影响也主要发生在乔木下层。群落中因个体数量相对较多而重要值较大的树种还有针齿铁仔、龙陵新木姜子、硬壳柯和毛柄槭，它们的相对多度对重要值的贡献率分别达到58.83%、46.39%、43.43%和41.99%。这些树种虽少见大树（平均胸径均不超过10cm），但却是群落中最常见的种类，对构建林下环境起着至关重要的作用。肖樱叶柃、尖叶桂樱的个体数量有所减少，但出现了一定数量的较大胸径个体，且出现频率也较高，因而在群落中也占有较重要的地位。相较而言，长果大头茶、多沟杜英、薄片青冈和星毛鹅掌柴的个体数量明显更少，但植株个体较高大（分别有43.09%、33.85%、24.88%和27.80%的个体胸径超过20cm），它们的重要值主要由相对显著度贡献，贡献率分别达到68.99%、55.03%、53.78%和43.83%。这些树种都能到达乔木上层，对群落环境的影响和作用最大，是群落中最重要的优势树种。

其他树种的重要值均在3%以下，其中2%~3%的有8种，1%~2%的有19种，<1%的有57种。它们中以个体形态较占优势（相对显著度占重要值的比例50%以上）的树种有南亚含笑（Michelia doltsopa）、红锥（Castanopsis hystrix）、滇琼楠（Beilschmiedia yunnanensis）和红花木莲（Manglietia insignis）等，以个体数量较占优势的树种有怒江山茶（Camellia saluenensis）和疏花卫矛（Euonymus laxiflorus）等，以个体分布（相对频度）较占优势的树种有平顶桂花（Osmanthus corymbosus）和红梗润楠（Machilus rufipes）等。

样地中的稀有种[平均每公顷的个体数少于 1 的树种(Hubbell, et al., 1986)]有 19 种,占总种数的 20%,其中个体数仅有 1 株的有 11 种,如印度木荷(Schima khasiana)、香面叶(Lindera caudata)、多脉冬青(Ilex polyneura)、西域青荚叶(Stachyurus himalaicus)、粉苹婆(Sterculia euosma)、长叶水麻(Debregeasia longifolia)、云南木姜子(Litsea yunnanensis)和尖叶野漆(Toxicodendron succedaneum)等。

表3 高黎贡山常绿阔叶林动态监测样地重要值≥1 的优势物种组成
Table 3 Species composition with importance value ≥1 in the evergreen broad-leaved forest dynamic plot in the Gaoligong Mountains

种名 Species	个体数 Number of stems	平均胸径 MeanDBH(cm)	胸高断面积 Basal area(m^2)	频度 Frequency	重要值 Importance value
多花山矾 Symplocos ramosissima	837	4.49	2.671 2	82	4.15
肖樱叶柃 Eurya pseudocerasifera	454	10.08	7.857 1	82	3.95
长果大头茶 Gordonialongicarpa	181	21.69	13.053 7	48	3.69
龙陵新木姜子 Neolitsea lunglingensis	518	6.88	4.559 5	84	3.53
硬壳柯 Lithocarpus hancei	474	6.93	5.109 9	80	3.45
薄片青冈 Cyclobalanopsis lamellosa	213	16.57	9.425 1	76	3.42
毛柄槭 Acer pubipetiolatum	453	9.38	5.733 6	72	3.41
尖叶桂樱 Laurocerasus undulata	337	11.03	6.538 3	89	3.40
多沟杜英 Elaeocarpus lacunosus	192	16.47	9.046 7	70	3.21
星毛鹅掌柴 Schefflera minutistellata	277	14.03	6.986 3	73	3.11
针齿铁仔 Myrsine semiserrata	575	4.03	1.748 2	78	3.09
红锥 Castanopsis hystrix	206	16.81	9.145 8	42	2.94
怒江山茶 Camellia saluenensis	460	5.69	2.546 7	78	2.88
瑞丽山龙眼 Helicia shweliensis	364	5.76	2.470 1	74	2.52
金平木姜子 Litseachinpingensis	246	9.67	4.356 5	70	2.46
少花桂 Cinnamomum pauciflorum	279	8.71	3.666 3	72	2.46
南亚含笑 Michelia doltsopa	79	29.55	8.454 4	42	2.40
双齿山茉莉 Huodendron biaristatum	152	13.91	5.410 4	57	2.22
红花木莲 Manglietia insignis	96	19.37	6.284 3	54	2.17
森林榕 Ficus neriifolia	200	8.55	2.290 4	76	1.99
腾越枇杷 Eriobotrya tengyuehensis	226	6.74	2.337 3	59	1.87
滇琼楠 Beilschmiedia yunnanensis	89	19.44	5.573 7	38	1.82
长蕊木兰 Alcimandra cathcartii	119	15.06	4.401 7	48	1.81
马蹄荷 Symingtonia populnea	178	11.09	4.279 5	28	1.73
星毛柯 Lithocarpus petelotii	136	10.33	3.219 8	53	1.69
黑皮插柚紫 Linociera ramiflora	196	6.98	1.824 0	55	1.63
大理茶 Camellia taliensis	207	4.34	0.565 1	63	1.52
黄牛奶树 Symplocos laurina	211	3.87	0.453 9	59	1.46

(续)

种名 Species	个体数 Number of stems	平均胸径 MeanDBH(cm)	胸高断面积 Basal area(m²)	频度 Frequency	重要值 Importance value
瑞丽茜树 Aidia shweliensis	142	8.28	1.349 1	55	1.37
团花新木姜子 Neolitsea homilantha	140	7.73	1.457 7	48	1.30
波叶新木姜子 Neolitsea undulatifolia	113	9.41	1.696 7	48	1.26
疏花卫矛 Euonymus laxiflorus	263	2.22	0.222 6	28	1.21
蜂房叶山胡椒 Lindera foveolata	83	14.41	2.135 1	39	1.14
红梗润楠 Machilus rufipes	105	6.00	0.572 9	57	1.12
微脉冬青 Ilex venulosa	92	10.66	2.477 0	29	1.12
泥柯 Lithocarpus fenestratus	66	15.57	2.365 8	35	1.09
平顶桂花 Osmanthus corymbosus	77	10.63	1.454 9	45	1.06
油葫芦 Pyrularia edulis	105	8.57	1.305 2	39	1.05
其他 57 种 Other 57 species	1 405	−	15.803 2	568	14.30
合计 Total	10 546	8.68	170.849 8	2 793	100

3.4 种—面积曲线

取样面积为 400m² 时树种有 15 种,取样面积为 1 600m² 时,树种增加到 47 种,接近总种数的 50%。取样面积为 1 000m² 时,树种增加到 71 种,占总种数的 74.74%,取样面积为 13 200m² 时,树种增加到 76 种,占总种数的 80.00%。取样面积再增加时,树种数量增加趋于平缓(图 1)。可见,要覆盖样地 80% 的树种,需要的取样面积大致为 13 200m²。对数模型的种—面积曲线拟合方程为 $S=16.143\ln(A)-76.744$ ($R^2=0.984$, $P<0.001$);幂函数模型的种—面积曲线拟合方程为 $S=9.291A^{0.221}$ ($R^2=0.984$, $P<0.001$)。其中,S 为树种数,A 为取样面积(m²)。检验结果显示 2 种模型的拟合效果均较好。

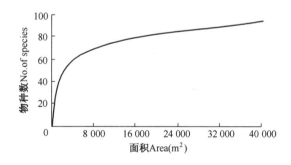

图 1 高黎贡山常绿阔叶林动态监测样地种—面积曲线图

Figure 1 Species-area curve in the evergreen broad-leaved forest dynamic plot in the Gaoligong Mountains

3.5 优势树种的径级结构与存活曲线

样地内 DBH≥1.0cm 的个体有 10 546 株,其中 DBH≥10cm 的个体有 2 784 株,DBH≥20cm 的个体有 1 207 株,DBH≥50cm 的个体有 146 株。样地内最大胸径 104cm(树种为长蕊木兰(Alcimandra cathcartii),平均胸径 8.68cm。选取了样地中重要值≥2.0 的 19 个树种进行种群结构及存活曲线的分析,其中有 5 个树种的胸径可达 80cm 以上(图 2A),有 7 个树种的胸径可达 50cm 以上,但不超过 80cm(图 2B)。有 7 个树种的胸径不超过 50cm(图 2C)。从图 2 可以看出,选取的 19 个树种及全部树种的个体数在各径级的分布呈现大致相

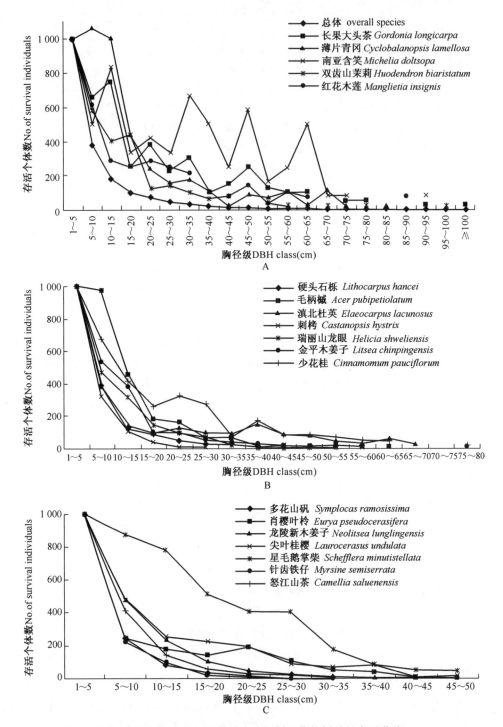

图 2　高黎贡山常绿阔叶林动态监测样地优势树种的存活曲线

Figure 2　Survival curve of species with importance value ≥2 in the evergreen broad-leaved forest dynamic plot in the Gaoligong Mountains

同的趋势,表现为小径级个体数量较多,中等径级及大径级的个体数量逐渐减少,呈基部宽、顶部窄的金字塔结构。这样的径级结构储备了足够的幼树资源,有利于群落的稳定发展。其中,几乎全部树种胸径在20cm以下的个体所占比例都超过50%,甚至有10个树种超过80%。相较而言,个体较大树种的幼树储备较少,如长果大头茶、薄片青冈和南亚含笑胸径在5cm以下的个体数分别仅占22.10%、21.60%和15.19%。

存活曲线是通过各径级存活个体数量的变化来反映种群动态的发展趋势。Deevey把存活曲线大致分成3种类型(吴承祯等,2000):Deevey-Ⅰ型,曲线呈凸型,表示种群的大多数个体均能实现其平均的生理寿命,在达到平均寿命时几乎同时死亡;Deevey-Ⅱ型,曲线呈对角线型,表示各龄级具有相同的死亡率;Deevey-Ⅲ型,曲线呈凹型,表示幼苗的死亡率高,以后的死亡率低而稳定。图2中,全部树种及红花木莲、双齿山茉莉(*Huodendron biaristatum*)表现为Deevey-Ⅲ型,它们胸径20cm以下特别是5cm以下的个体死亡率很高,超过20cm后死亡率则渐渐降低,至40cm以后已基本趋于平稳。长果大头茶、南亚含笑、薄片青冈这3个树种也表现为Deevey-Ⅲ型,但死亡率波动较大。薄片青冈在5~15cm阶段出现负死亡情况,个体数量反而上升,但在15~20cm阶段死亡率突然增加,个体数量又迅速降低,至30cm时累积死亡率已高达85%。长果大头茶及南亚含笑的存活曲线波动更加明显,在10~15cm,30~35cm,45~50cm和55~65cm等阶段都出现个体数量上升的情况,但每一次个体数量增加都伴随一次大量死亡,死亡率高峰分别发生在15~20cm,35~45cm,50~55cm和65~70cm阶段。存活曲线的连续波动是否因为这2个树种的生命周期存在大小年还有待深入研究。星毛鹅掌柴的存活曲线几乎成一条直线,接近Deevey-Ⅱ型,说明该树种在各阶段的死亡率大致相同,在群落中属于较稳定的种群。其他树种的存活曲线都表现为Deevey-Ⅲ型,死亡率的最高峰都发生在5~10cm阶段,以后死亡率逐渐下降,至20cm以后死亡率趋于平稳,但累积死亡率都已达到80%以上。其中少花桂(*Cinnamomum pauciflorum*)和肖樱叶枹在20~25cm阶段发生了负死亡现象。

4 结论与讨论

高黎贡山中山湿性常绿阔叶林4hm²动态监测样地内胸径≥1.0cm的木本植株有10 546株(包括分枝和萌枝共计14 498株),分属于35科64属95种,其植株密度和物种丰富度均高于纬度稍偏南的哀牢山样地,但稀有种所占比例较小。样地内胸高断面积总和达到170.849 8m²,平均为42.71m²·hm⁻²。样地中含种数和个体数较多的科为樟科、山茶科、五加科、壳斗科和山矾科。优势种为多花山矾、肖樱叶枹、长蕊木兰、龙陵新木姜子、硬壳柯和薄片青冈,但优势均不明显,为多优种的群落类型。在属级水平上,热带区系成分占总属数的73.02%,温带区系成分占26.98%,样地热带性质明显。样地的最小表现面积为1.32hm²,能够包括样地内80%以上的树种。样地内整体径级结构及优势种径级结构均表现为基部宽、顶部窄的金字塔结构,幼树储备充足能够保证群落的稳定发展。存活曲线大都表现为Deevey-Ⅲ型,小径级个体死亡率高,大径级个体死亡率低而稳定。

高黎贡山样地中热带成分科和属所占比例分别高达75.76%和73.02%,热带起源性质十分明显。在地质地理因素方面,在第三纪造山运动以前和初期,高黎贡山地区位于现代位置的450km以南(23°~24°N),即北回归线以南,为典型的热带植被和热带植物区系。在二叠纪至第三纪早始新世或稍晚,掸马板块(滇西—掸邦—马来亚板块为冈瓦纳古陆的边缘部分,简称掸马板块,高黎贡山为其中部板块的一个板片)发生北移和右旋,促使高黎贡山热带植物区系向温带植物区系的蜕变和演化(李恒等,2000)。而在其南端,仍可能保留较多的热带植物区系的成分。从所处地理位置来看,高黎贡山样地所处区域的植被在云南植被区划(吴征镒等,1987)中正好处于高原亚热带南部季风常绿阔叶林地带与高原亚热带北部常绿阔叶林地带的过渡地带,南北走向的山体及河谷为南北植物的交流提供了通道,因而它既具有典型的亚热带林所特有的以樟科、山茶科、壳斗科、木兰科为主体的物种组成特征,又具有南部季风常绿阔叶林所具有的以热带成分为主的区系性质特点。从科属的地理成分来看,高黎贡山样地除拥有上述四大主体科外,还拥有较多热带分布的科如山矾科、卫矛科(Celastraceae)、山龙眼科、芸香科以及亚热带常见的科如五加科、杜英科(Elaeocarpaceae)、冬青科、野茉莉科(Styracaceae)、水东哥科(Saurauiaceae)、省沽油科(Staphyleaceae)、虎皮楠科(Daphniphyllaceae)和紫金牛科等。而且这些科中的许多属也都具有热带起源的性质,如黄肉楠属(*Actinodaphne*)、润楠属(*Machilus*)、樟属(*Cinnamomum*)、山胡椒属(*Lindera*)、木姜子属、新木姜子属、柃木属(*Eurya*)、山茶属、杨桐属(*Adinandra*)、青冈属(*Cyclobalanopsis*)、长蕊木兰属(*Alcimandra*)、含笑属(*Michelia*)、杜英属(*Elaeocarpus*)、猴欢喜属(*Sloanea*)、山茉莉属(*Huodendron*)等。众多热带起源的科、属在这里都得到了较充分的发展。由此可见,本区域植被起源于热带的性质十分明显,并有其地质地理演化的历史根源。

参考文献:

[1] 柴勇,孟广涛,武力,2007. 高黎贡山自然保护区国家重点保护植物的组成特征及其资源保护[J]. 西部林业科学,36(4):57~63.

[2] 郝占庆,李步杭,张健,等,2008. 长白山阔叶红松林样地(CBS):群落组成与结构[J]. 植物生态学报,32(2):238~250.

[3] 巩合德,杨国平,鲁志云,等,2011. 哀牢山常绿阔叶林树种多样性及空间分布格局[J]. 生物多样性,19(2):143~150.

[4] 兰国玉,胡跃华,曹敏,等,2008. 西双版纳热带森林动态监测样地——树种组成与空间分布格局[J]. 植物生态学报,32(2):287~298.

[5] 李恒,郭辉军,刀志灵,2000. 高黎贡山植物[M]. 北京:科学出版社.

[6] 李嵘,刀志灵,纪运恒,等,2007. 高黎贡山北段种子植物区系研究[J]. 云南植物研究,29(6):601~615.

[7] 李嵘,纪运恒,刀志灵,等,2008. 高黎贡山北段东西坡种子植物区系的比较研究[J]. 云南植物研究,30(2):129~138.

[8] 李锡文,1996. 中国种子植物区系统计分析[J]. 云南植物研究,18(4):363~384.

[9] 李玉媛,2003. 莱阳河自然保护区定位监测[M]. 昆明:云南大学出版社.

[10] 刘海丰,李亮,桑卫国,2011. 东灵山暖温带落叶阔叶次生林动态监测样地:物种组成与群落结构[J]. 生物多样性,19(2):232~242.

[11] 刘经伦,2003. 高黎贡山植物资源开发[J]. 保山师专学报,22(5):6~10.

[12] 欧光龙,彭明春,和兆荣,等,2008. 高黎贡山北段植物群落 TWINSPAN 数量分类研究[J]. 云南植物研究,30(6):679~687.

[13] 覃林,2009. 统计生态学[M]. 北京:中国林业出版社.

[14] 石天才,熊汝泰,1994. 高黎贡山自然保护区药用植物考察[J]. 西南林学院学报,14(1):13~21.

[15] 宋永昌,2001. 植被生态学[M]. 上海:华东师范大学出版社.

[16] 汪建云,母其爱,蔺如涛,2008a. 高黎贡山自然保护区南段生物走廊带维管植物名录(1)[J]. 云南师范大学学报,28(3):57~63.

[17] 汪建云,母其爱,蔺如涛,2008b. 高黎贡山自然保护区南段生物走廊带维管植物名录(2)[J]. 云南师范大学学报,28(4):52~59.

[18] 王志恒,陈安平,朴世龙,等. 高黎贡山种子植物物种丰富度沿海拔梯度的变化[J]. 生物多样性,2004,12(1):82~88.

[19] 吴承祯,洪伟,谢金寿,等,2000. 珍稀濒危植物长苞铁杉种群生命表分析[J]. 应用生态学报,11(3):333~336.

[20] 吴征镒,1991. 中国种子植物属的分布区类型[J]. 云南植物研究(增刊Ⅳ):1~139.

[21] 吴征镒,2003.《世界种子植物科的分布区类型系统》的修订[J]. 云南植物研究,25(5):535~538.

[22] 吴征镒,周浙昆,李德铢,等,2003. 世界种子植物科的分布区类型系统[J]. 云南植物研究,25(3):245~257.

[23] 吴征镒,朱彦丞,1987. 云南植被[M]. 北京:科学出版社.

[24] 熊清华,艾怀森,2006. 高黎贡山自然与生物多样性研究[C]//高黎贡山研究文丛:第一卷[M]. 北京:科学出版社.

[25] 徐成东,冯建孟,王襄平,等,2008. 云南高黎贡山北段植物物种多样性的垂直分布格局[J]. 生态学杂志,27(3):323~327.

[26] 徐志辉,1998. 怒江自然保护区[M]. 昆明:云南美术出版社.

[27] 薛纪如,1995. 高黎贡山国家自然保护区[M]. 北京:中国林业出版社.

[28] 杨庆松,马遵平,谢玉彬,等,2011. 浙江天童20 hm^{2}常绿阔叶林动态监测样地的群落特征[J]. 生物多样性,19(2):215~223.

[29] 叶万辉,曹洪麟,黄忠良,等,2008. 鼎湖山南亚热带常绿阔叶林20公顷样地群落特征研究[J]. 植物生态学报,32(2):274~286.

[30] 尹五元,1994. 高黎贡山自然保护区珍稀保护植物[J]. 西南林学院学报,14(1):6~12.

[31] 张健,郝占庆,宋波,等,2007. 长白山阔叶红松林红松与紫椴的空间分布格局及其关联性[J]. 应用生态学报,18(8):1681~1687.

[32] 祝燕,赵谷风,张俪文,等,2008. 古田山中亚热带常绿阔叶林动态监测样地——群落组成与结构. 植物生态学报,32(2):262~273.

[33] 朱振华,毋其爱,杨礼攀,2003. 高黎贡山自然保护区野生动植物资源现状及保护[J]. 林业科技,28(6):63~65.

[34] Cao M, Zhu H, Wang H, et al, 2008. Xishuangbabna tropical seasonal rainforest dymamics plot: Tree distribution maps, disameter tables and species documentation. Kunming: Yunnan Science and Technology Press.

[35] Condit R, 1995. Research in large, long-term tropical plot. Trends in Ecology and Evolution, 10: 18~22.

[36] Condit R, 1998. Tropical forest census plots: methods and results from Barro Colorado Island, Panama and a comparison with other plots. Berlin: Springer.

[37] Hubbell S P, Foster R B, 1986. Commonness and rarity in aneotropical forest: implications for tropical tree conservation//Soulé ME. Conservation biology: the science of scarcity and diversity. Sunderland, UK:. SinauerPress, 205~231.

(本文发表于《林业科学》,2013年)

滇东南石漠化山地不同植被恢复模式下土壤地力变化和水土流失状况研究

李品荣,孟广涛,方向京

(云南省林业科学院,云南 昆明 650204)

摘要:通过对滇东南石漠化山地不同植被恢复模式下土壤的理化性质进行定点观测和地力变化及水土流失状况的分析,结果表明:封山育林地土壤的肥力较佳,其次是耕地土壤,这是由于封山育林地土壤多年来进行封山育林,耕地土壤进行了平衡施肥;三年后人工林地土壤的理化性质有很大改善,土壤肥力有很大提高,这是退耕还林的结果;从水土流失状况来看,耕地是产流产沙、固体和液体养分流失最严重的类型,人工林地水土流失最低,说明退耕还林和封山育林有利于石漠化山地土壤肥力的改善和水土保持功能的提高。

关键词:石漠化山地;植被恢复模式;土壤理化性质;地力;水土流失

Study on Land Capacity Change and Soil Erasionon Different Vegetation Recover Patterns in Rock Desertification Mountains of Southeastern Yunnan

Li Pin-rong, Mong Guan-tao, Fang Xiang-jing

(Yunnan Academy of Forestry, Kunming Yunnan 650204, China)

Abstract: A study on the physicochemical property and soil erasionof different vegetation patterns in Xichong county of southeastern Yunnan was conducted by ocean weather station observation. The results showed that the land capacity of closing hillsides soil is the highest andthe land capacity of cultivation land is higher, which resulted by closing hillsides to facilitate afforestation for many years, and balance fertilizing of cultivation land. Asreturning land for farming to forestry, the physicochemical property and land capacity of soil has been greatly improved after three years. Cultivation land is the most severe type from water and soil erasionanalysis, the planted forest lands are the smallest. Closing hillsides to facilitate afforestation and returning land for farming to forestry are good approaches of land capacity improvement and soil and water conservation in karst mountains desertification soil of southeastern Yunnan.

Key words: rock desertifitationmountains; vegetation recover pattern; soil physicochemical property; land capacity; water and soil erasion

岩溶石山地区地质环境脆弱,在不合理的人类活动和自然因素作用下,一些地区植被退化乃至消亡,导致水土严重流失,最终形成了连片分布的裸露石漠,我们将这一过程及其结果称之为石漠化;石漠化类型以碳酸盐岩石漠化为主,主要受控于纯碳酸盐岩,夹层或互层型则相对较少[1]。云南全省石漠化土地面积占国土面积的 2.2%,近 25 年来石漠化土地面积增长速度约 147.5km^2·a^{-1};滇东南石漠化土地面积 3 440.0km^2,占国土面积的 5.41%;

文山州石漠化土地面积 2 906.3km²,占国土面积的 8.31%,在面积和比例上居全省各地州市第一[2]。本文通过对滇东南文山州西畴县石漠化山地的川滇桤木(*Alnus ferdinandi-coburgii*)林地、墨西哥柏(*Cupressus lusitanica*)林地、花椒(*Zanthoxylum bungeanum*)地、封山育林地和耕地土壤的理化性状和水土流失状况进行定点观测,在 2002 年和 2005 年观测的基础上,探讨滇东南石漠化山地不同植被恢复模式下土壤的地力变化和水土流失情况,并作对比和分析,然后对不同植被恢复模式下土壤的地力和水土流失状况作出综合评价,对石漠化山地植被恢复和水土保持具一定指导意义。

1 研究方法

1.1 自然概况

试验区位于滇东南西畴县东北部的法斗乡,104°46′12″~104°47′36″E,23°25′26″~23°26′46″N,海拔 1 160~1 691m,平均温度 14.8~17.6℃,最热月平均温 19.6~22.3℃,极端最低温-4.3℃,无霜期 350~360d,年降雨量 1 075.7~1 615.3mm,日照时数 1 500~1 600h,空气相对湿度 82%。该区虽然雨量充沛,但降水的季节分布不均匀,80%以上的降水集中在雨季(6~9 月),且降水多由岩隙渗入地下,区域内人畜饮水和农业用水仍十分困难,但气候特点却有利于林木的生长发育;该区属滇东南岩溶山原区,为深切割的中山山原地貌[3],区内地质构造复杂,寒武系、奥陶系、石炭系、泥盆系、二叠系、三叠系的地层均有分布。该区属亚热带湿润季风气候区域,具有"冬无严寒,夏无酷暑,干湿分明,冬春早,夏季涝,全年多雾"的气候特征;土壤为石灰岩、页岩发育形成的荒地土壤和石灰土,地带性植被为以壳斗科、樟科、木兰科等为主的亚热带季风常绿阔叶林。

川滇桤木林是 2002 年在退耕还林地上营造的纯林,2005 年已郁闭,平均树高 5.19m,平均胸径 6.51cm,郁闭度达 0.85,林下植物以荩草(*Arthraxon hispidus*)、臭蒿(*Artemisia hedinii*)、小叶蕨(*Lemmaphyllum microphyllum*)、贯众(*Cyrtomium fortunei*)、糯米团(*Gonostegia hirta*)和车前草(*Plantago asiatica*)等植物为主;墨西哥柏林也是 2002 年在退耕还林地上营造的纯林,2005 年已郁闭,平均树高 3.42m,平均胸径 5.30cm,郁闭度达 0.65,林下植物以荩草、臭蒿(*Artemisia hedinii*)、小叶蕨、贯众、糯米团和车马蹄金(*Dichondra repens*)等植物为主;花椒地也是 2002 年开始退耕还林,但生长缓慢,平均树高 1.13m,平均地径 1.38cm,地表植物主要以臭蒿、白茅(*Imperata cylindrica*)、车前草和马蹄金等耐旱植物为主,总盖度约 75%;封山育林地是对 20 世纪 90 年代的次生灌丛开始封禁,植被以剥皮鼠李(*Ziziphus xiangchengensis*)、滇丁香(*Luculia intermedia*)、华西小石积(*Osteomeles schwerinae*)、马桑(*Coriaria nepalensis*)和平枝栒子(*Cotoneaster horizontalis*)等萌生的常绿灌木为主,具旱生性、岩生性、喜钙性等特点,总盖度达 80%,灌丛平均高 2.35m,地表有大量阴生植物如苔藓(*Mnium spinosum*);耕地种的农作物主要是玉米(*Zea mays*),地表植物种类主要有鬼针草(*Bidens bipinnata*)、矮脚苦蒿(*Conyza blinii*)、荩草(*Artheaxon hispidus*)等。

1.2 研究方法

根据不同植被恢复模式,对川滇桤木林地、墨西哥柏林地、花椒地、封山育林地和耕地进行定点取样,共挖掘 13 个土壤剖面,并作野外现场观察记载,每个土壤类型 2~3 个剖面,共采集了43个样品;每种恢复模式内设置一个径流场,共5个径流场,径流场面积为5m×10m,每天定时观测径流量和泥沙量,每月分析径流中的固体和液体养分;室内分析采用中国科学院南京土壤研究所分析方法[4]和中国分析标准方法[5]对不同土类的物理化学性质进行研究。

2 结果与分析

2.1 土壤物理性状及其变化

土壤物理性状及其变化见表1,从表中看出:

(1)一般来说,容重小,土壤疏松,有利于拦渗蓄水,减缓径流冲刷,容重大则相反[6,7]。从表1看出:不同植被恢复模式下的土壤容重有较大差异,其排列顺序为:墨西哥柏林>川滇桤木林>花椒林>封山育林>耕地,说明墨西哥柏林和川滇桤木林林地的土壤黏性较重,这一点从土壤结构分析结果可得到较好说明。土壤孔隙度及大于0.25mm的土壤颗粒含量均为川滇桤木林<墨西哥柏林<耕地<花椒林<封山育林。川滇桤木林和墨西哥柏林林地土壤颜色从暗黄棕到棕红色,土壤紧实、结构为小块状及核状,质地轻黏到黏;耕地土壤暗棕到棕色,结构多为粒状,较疏松,质地轻黏;封山育林地土壤为暗棕、棕黑色,多团粒结构,疏松,质地轻壤到中壤;花椒地土壤土层深厚,颜色淡黄到黄棕,容重中等,但质地黏重,土壤紧实,结构不良,小于0.25mm土壤黏粒含量高达73.4%,总孔隙度虽大,但以毛管孔隙为主,通气性较差。几种不同植被恢复模式下土壤物理性状比较,封山育林地土壤物理性状最好,具有良好的拦渗蓄水和减缓径流冲刷能力,这是多年来一直封山育林的结果;其次是耕地,最差是花椒地。

表1 不同植被恢复模式下土壤物理性状比较

植被恢复模式		川滇桤木林		墨西哥柏林		花椒林		封山育林		耕地	
		2002年	2005年	2002年	2005年	2002年	2005年	2002年	2005年	2002年	2005年
颜色		淡黄棕、暗黄棕		淡黄棕、暗黄棕		黄棕		棕黑、暗棕		棕、暗棕	
耕层厚(cm)		8.1~15.5		8.3~14.8		8.5~15.6		9.3~18.5		10.9~19.7	
紧实度		上松下紧		上松下紧		紧实		疏松		稍松	
质地		黏壤		黏壤		黏壤		轻壤		轻黏至中壤	
结构		块状、核状		块状、核状		块状、粒状		团粒		小块、粒状	
容重(g·cm⁻³)		1.21	1.15	1.21	1.18	1.15	1.09	1.06	1.02	1.13	1.12
比重(g·cm⁻³)		2.64	2.69	2.64	2.65	2.75	2.75	2.49	2.52	2.69	2.65
总孔隙度(%)		54.2	57.3	55.4	56.1	59.8	58.3	61.6	60.5	58.0	57.7
毛管孔隙(%)		36.2	38.1	36.5	38.3	40.5	40.1	44.1	43.8	42.7	42.2
水稳性团聚体(%)	>5mm	17.67	17.58	15.24	16.14	3.99	4.27	9.96	9.47	9.19	11.76
	5~2mm	11.83	16.70	12.31	13.53	4.84	4.77	16.05	14.47	15.04	14.76
	2~1mm	7.86	10.67	8.72	13.17	3.39	3.34	16.51	12.62	10.40	14.01
	1~0.5mm	12.68	12.40	12.63	12.15	6.08	6.70	18.2	20.71	20.06	16.13
	0.5~0.25mm	10.39	6.37	13.81	8.26	8.29	9.97	8.93	11.64	11.60	9.11
	<0.25mm	39.57	36.27	37.29	36.75	73.41	71.93	30.38	31.08	33.71	34.23
	>0.25mm	60.43	63.73	62.71	63.25	26.59	28.07	69.62	68.92	66.29	65.77

(续)

植被恢复模式		川滇桤木林		墨西哥柏林		花椒林		封山育林		耕地	
		2002年	2005年	2002年	2005年	2002年	2005年	2002年	2005年	2002年	2005年
干筛团聚体(%)	>5mm	40.87	42.15	41.34	43.07	41.02	39.85	26.92	28.52	29.14	27.55
	5~2mm	28.42	27.43	27.84	26.53	31.21	30.54	29.14	27.88	29.02	28.54
	2~1mm	9.38	10.5	9.68	10.24	8.86	11.71	14.61	13.91	12.62	14.57
	1~0.5mm	11.1	9.7	12.15	10.72	7.84	9.93	16.01	15.89	14.91	15.26
	0.5~0.25mm	4.45	5.01	4.46	5.31	6.92	4.38	6.89	8.44	7.42	7.74
	<0.25mm	5.78	5.21	4.53	4.13	4.15	5.36	6.43	5.36	6.89	6.34
	>0.25mm	94.22	94.79	95.47	95.87	95.85	96.41	93.57	94.64	93.11	93.66
破坏率(%)		35.86	32.77	35.26	33.77	72.26	70.88	25.60	27.18	28.80	29.78

(2)三年后川滇桤木林和墨西哥柏林林地的土壤物理性状变化较大,其他土壤类型物理性状变化不大。这主要是因为人工林地的林木已经郁闭,植物的根系不断增加,有利于改善土壤物理结构,增强土壤抗蚀性。川滇桤木林林地样点三年前后数据汇总对比,土壤容重从 1.21g·cm^{-3}降到 1.15g·cm^{-3},而孔隙度从 54.2%增加到 57.3%,>0.25mm 水稳性团聚体从 60.4%变为 63.7%,表明大颗粒在水力作用下消散力减弱,大团聚体含量增加,大孔隙数量增多,土壤通透性增强;同时稳定性增加,抗蚀性也随之增强,总体物理性状有较大改善。所以从某种程度来说,减少人为活动是改善石灰土物理性质,减少水土流失,减缓石灰土退化的一项切实可行的措施,也是石漠化治理简单易行的有效举措。

(3)不同植被恢复模式下土壤之间风干团聚体的差异不大,而水稳性团聚体有较大变化,因此各级水稳性团聚体的比例应该能够较好反应土壤团聚体的质量,湿筛后团聚体结构破坏率说明这些土壤间团聚体有较大差别,这是不同土类抗蚀性和储水性有较大差别的原因。花椒地土壤团聚体的破坏率高达 70%以上,抗蚀性和储水性最差;封山育林地土壤最低,说明封山育林有利于土壤改良。三年后,川滇桤木林和墨西哥柏林林地土壤团聚体破坏率减少,土壤结构有所改善。

2.2 土壤化学性状及其变化

从表2中看出:

(1)土壤 pH 值为弱酸至中性,三年后 pH 值变化不大;土壤有机质含量较高,封山育林地土壤有机质最高,达 13.6%,三年后川滇桤木林地土壤有机质增加了 47.12%,墨西哥柏林林地土壤有机质增加了 37.8%,另三种变化不大。

(2)土壤速效氮、全氮含量丰富,速效磷、全磷缺乏,速效钾、全钾含量中等偏低,各项指标均以封山育林地>耕地>川滇桤木林>花椒林>墨西哥柏林地,但全钾含量差别不明显,这主要与处于同一地区,矿物源相同有关;人工林地土壤磷、钾缺乏较严重,其他三类含量也不高,所以试验区均应重视磷钾的施用。

(3)三年后人工林地土壤各养分均增加,川滇桤木林地土壤有机质、全氮及速效磷含量变化较明显,分别比三年前增加 47.12%、56.67%及 27.56%,其次是墨西哥柏林地,分别比三年前增加 37.79%、70.83%及 19.23%,对于这三项成分而言,这样的增幅是很可观的,其他成分变化不明显,这与前述土壤物理性状的变化原因一致,再一次验证植被恢复及生长对防止土壤退化的重要性；耕地土壤的速效磷、速效钾增加较大，分别提高13.14%和51.39%，主要原因是近两年农业生产中强调平衡配方施肥；其次是封山育林地土壤，速效磷、速效钾分别提高54.19%和14.30%，主要原因是多年来一直进行封山育林。

（4）阳离子代换量中等偏低，最高是封山育林地土壤，达23.46cmol·100g^{-1}，人工林地土壤最低，但墨西哥柏林地和花椒林地增加最高，三年后阳离子代换量分别增加了29.58%和11.38%，说明造林后土壤利用率提高。

表2　不同植被恢复模式下土壤常规养分比较

时间	植被恢复模式	有机质（%）	pH	速效氮（mg·kg^{-1}）	速效磷（mg·kg^{-1}）	速效钾（mg·kg^{-1}）	全氮（%）	全磷（%）	全钾（%）	阳离子（cmol·100g^{-1}）
2002年	川滇桤木林	2.78	6.93	150.51	2.54	65.21	0.18	0.033	1.277	8.86
	墨西哥柏林	2.62	6.80	101.85	1.82	73.55	0.168	0.034	1.201	8.96
	花椒林	3.22	7.25	130.87	2.34	82.80	0.181	0.029	1.183	8.96
	封山育林	13.60	7.32	569.48	4.30	78.22	0.644	0.219	1.198	23.46
	耕地	4.23	6.84	182.23	8.75	98.28	0.272	0.127	1.249	13.12
2005年	川滇桤木林	4.09	7.03	161.42	3.24	67.01	0.282	0.033	1.286	8.83
	墨西哥柏林	3.61	6.73	120.52	2.17	68.85	0.287	0.034	1.197	11.61
	花椒林	3.41	7.19	146.76	2.75	84.98	0.251	0.049	1.122	9.98
	封山育林	12.97	7.26	570.21	6.63	89.41	0.618	0.219	1.244	25.16
	耕地	4.40	6.88	180.13	9.90	148.79	0.294	0.127	1.302	13.07

2.3　土壤中微量元素及其变化

微量元素在植物体内含量虽少，但它们与大量元素一样同等重要，具有不可替代性。土壤中微量元素及其变化见表3，从表中看出：几种模式土壤钙含量丰富，镁含量中等，微量元素硼、钼极缺乏，硫、氯、铜、锌含量偏低，镁、钼含量极低，其他微量元素含量中等；三年后几种恢复模式下土壤中的部分微量元素略有上升，这与不同林木选择性吸收和长期封山育林有关。

表3　不同植被恢复模式下土壤中微量元素比较　　（mg·kg^{-1}）

时间	土壤类型	有效锌	有效硫	有效钙	有效镁	有效硼	氯离子	有效铜	有效钼
2002年	川滇桤木林	3.10	11.51	2083.60	54.68	0.13	10.26	0.47	0.09
	墨西哥柏林	2.41	13.66	2287.6	67.52	0.16	6.13	0.549	0.09
	花椒林	2.63	12.24	2049.80	64.58	0.12	7.81	0.61	0.13
	封山育林	6.09	11.07	6013.10	53.61	0.22	7.53	1.05	0.12
	耕地	2.12	12.02	2648.08	88.22	0.26	10.12	1.35	0.15
2005年	川滇桤木林	2.62	13.47	2148.60	57.36	0.15	9.63	0.48	0.09
	墨西哥柏林	2.71	11.21	1598.4	56.33	0.16	8.24	0.41	0.07
	花椒林	2.18	14.31	1982.70	51.44	0.11	7.21	0.40	0.09
	封山育林	7.15	11.05	7984.10	31.46	0.22	4.71	0.69	0.09
	耕地	2.28	12.20	2389.57	93.11	0.22	9.61	1.30	0.15

2.4 地力综合评价

在土壤肥力研究中,单项肥力指标往往不能全面地反映出土壤的肥力水平,灰色系统理论采用关联分析的方法对土壤养分、pH 值、孔隙度等作系统分析,能较好地综合反映土壤的肥力状况[8]。各恢复模式下土壤的灰关联度分析结果见表4。

表4 土壤肥力灰关联度(r_i)

植被恢复模式	川滇桤木林	墨西哥柏林	花椒林	封山育林	耕地
2002 年	0.561 36	0.548 41	0.591 08	0.935 46	0.745 74
2005 年	0.581 63	0.561 99	0.588 20	0.935 51	0.749 76
增加(%)	3.61	2.48	-0.49	0.01	0.54

从表4中看出:不同恢复模式下土壤的土壤肥力依次是封山育林地>耕地>花椒地>川滇桤木林>墨西哥柏林,说明次生灌丛经若干年封育后,土壤肥力保持较高的水平,封山育林有利于土壤肥力的提高;三年后除花椒地外土壤地力均上升,地力上升最高是川滇桤木林(3.61%),其次是墨西哥柏林(2.48%),说明营造人工林在短时间内有利于土壤地力的迅速提高;耕地地力提高是由于年年进行施肥,保证农作物的生长;封山育林地力提高仅0.01%,说明封山育林提高土壤肥力是一个长期而缓慢的过程;花椒地地力有所下降,说明人工林郁闭前,由于林木生长需要养分,短时间内地力下降,但从长期来看,郁闭后地力也会逐步提高。

2.5 水土流失状况

不同植被恢复模式下的产流产沙量和土壤侵蚀模数见表5,从表中看出:产流量的顺序是耕地>花椒地>墨西哥柏林地>川滇桤木林地>封山育林地,产沙量的顺序是耕地>墨西哥柏林地>封山育林地>花椒地>川滇桤木林地,耕地是产流产沙和土壤侵蚀最严重的类型,川滇桤木林地水土流失最低,说明人工造林有利于水土保持,林地郁闭度与土壤的水土流失有直接关系,郁闭度越高,土壤水土流失越小。

表5 不同植被恢复模式下水土流失比较

植被恢复模式	2004 年					2005 年				
	川滇桤木林	墨西哥柏林	花椒林	封山育林	耕地	川滇桤木林	墨西哥柏林	花椒林	封山育林	耕地
产流量 (t·hm^{-2})	141.80	146.20	209.81	140.27	214.26	98.03	111.42	149.66	96.50	165.17
产沙量 (t·hm^{-2})	0.013	0.50	0.45	0.47	2.07	0.009	0.38	0.32	0.32	1.60

不同植被恢复模式下的养分流失见表6,从表中看出:固体养分中的有机质、全氮、全磷、速效氮、速效磷和速效钾的流失顺序是耕地>封山育林>墨西哥柏林>花椒地>川滇桤木林,耕地的液体养分流失最高,说明耕地是固体和液体养分流失最严重的类型,退耕还林和封山育林措施均可减少水土流失。

表6 不同植被恢复模式下养分流失比较

项目	时间	固体养分流失量（kg·hm^{-2}）							液体养分流失量（kg·hm^{-2}）		
		有机质	全氮	全磷	全钾	速效氮	速效磷	速效钾	氮	磷	钾
川滇桤木林	2004年	0.503	0.031	0.011	0.187	0.003	0.0002	0.004	7.779	0.479	2.257
	2005年	0.348	0.022	0.007	0.129	0.002	0.0001	0.002	5.378	0.331	1.560
	平均	0.426	0.027	0.009	0.158	0.003	0.0002	0.003	6.579	0.405	1.909
墨西哥柏林	2004年	21.600	1.250	0.415	7.560	0.113	0.008	0.121	7.573	0.200	1.717
	2005年	16.416	0.950	0.315	5.746	0.086	0.006	0.092	5.772	0.153	1.309
	平均	19.008	1.100	0.365	6.653	0.100	0.007	0.107	6.673	0.177	1.513
花椒地	2004年	16.920	1.170	0.387	7.704	0.110	0.008	0.118	11.162	0.470	1.549
	2005年	13.184	0.832	0.275	5.478	0.079	0.005	0.084	7.962	0.335	1.105
	平均	15.052	1.001	0.331	6.591	0.095	0.007	0.101	9.562	0.403	1.327
封山育林	2004年	39.668	1.631	0.479	3.859	0.125	0.022	0.196	8.287	0.505	1.224
	2005年	27.008	1.110	0.326	2.627	0.085	0.015	0.133	5.701	0.347	0.843
	平均	33.338	1.371	0.403	3.243	0.105	0.019	0.165	6.994	0.426	1.034
耕地	2004年	85.284	6.024	1.697	34.797	0.458	0.028	0.562	12.269	0.636	3.067
	2005年	65.920	4.656	1.312	26.896	0.354	0.021	0.434	9.458	0.491	2.364
	平均	67.794	4.665	0.682	13.902	0.377	0.023	0.707	9.738	0.505	2.434

3 结 语

（1）石漠化山地土壤养分不平衡现象较突出，有机质及氮含量丰富，磷缺乏，钾偏低，所以既要重视大量元素磷、钾的施用，又要重视微量元素硼、钼施用；三年后土壤养分不平衡现象有所改善。

（2）通过人工造林，三年后物理性状有较大改善，表现为土壤容重降低，水稳性大团聚体增加，稳定性增强，大孔隙数量增多，土壤通透性增强，同时抗蚀性增强。

（3）从地力综合评价来看，不同恢复模式下土壤的土壤肥力依次是封山育林地＞耕地＞花椒地＞川滇桤木林＞墨西哥柏林，说明次生灌丛经若干年封育后，土壤肥力保持较高的水平，封山育林有利于土壤肥力的提高；三年后除花椒地外土壤地力均上升，地力上升最高是川滇桤木林（3.61%），其次是墨西哥柏林（2.48%），说明营造人工林在短时间内有利于土壤地力的迅速提高；耕地地力提高是由于年年进行施肥，保证农作物的生长；封山育林地力提高仅0.01%，说明封山育林提高土壤肥力是一个长期而缓慢的过程；花椒地地力有所下降，说明人工林郁闭前，由于林木生长需要养分，短时间内地力下降，但从长期来看，郁闭后地力也会逐步提高。

（4）从水土流失状况来看，耕地是产流产沙、固体和液体养分流失最严重的类型，说明退耕还林和封山育林是改善土壤物理性状，减少水土流失，减缓石灰土退化的一项切实可行的措施，也是石漠化治理简单易行的有效举措。

参考文献:

[1] 王世杰. 喀斯特石漠化概念演绎及其科学内涵的探讨[J]. 中国岩溶, 2002, 21(2): 101~104.
[2] 谭继中, 张兵. 云南省土地石漠化特征初步研究[J]. 地质灾害与环境保护, 2003, 14(1): 32~37.
[3] 杨一光. 云南省综合自然区划[M]. 北京: 高等教育出版社, 1991: 197~207.
[4] 中国科学院南京土壤所. 土壤理化性质分析[M]. 上海: 上海科学技术出版社, 1987.
[5] 刘光崧. 土壤理化分析与剖面描述[M]. 北京: 中国标准出版社, 1996: 1~41.
[6] 徐岚. 利用马尔可夫过程预测东陵区土地利用格局的变化[J]. 应用生态学报, 1993, 4(3): 272~278.
[7] 刘海燕. GIS在景观生态学研究中的应用[J]. 地理学报, 1995, 50(增刊): 105~111.
[8] 蒋云东, 何蓉, 陈娟. 灰色关联分析在杉木人工林土壤肥力研究中的应用[J]. 云南林业科技, 1998, 2: 34~38.

(本文发表于《水土保持学报》, 2009年)

岩溶山地不同植被恢复模式和恢复年限下土壤养分的变化

常恩福[1,3]，李娅[1]，李品荣[1,3]，侬时增[2]，刘永国[2]，王竣[2]

(1. 云南省林业科学院，云南 昆明 650201；2. 文山州林业科学研究所，云南 文山 666300；3. 云南建水荒漠生态系统国家定位观测研究站，云南 建水 654399)

摘要：2002 年、2005 年、2009 年、2016 年 8 月，分别对西畴县法斗乡岩溶山地 11 块植被恢复模式固定样地上、中、下部位的土样进行 4 次调查测定，分析不同树(草)种 11 种植被恢复模式和恢复年限下林地土壤养分变化及其恢复状况。结果表明：人工植被恢复 14 年后，恢复年限及土壤有机质含量、速效 N 含量、速效 P 含量、速效 K 含量、土壤 pH 之间存在紧密的相关关系；土壤有机质含量在植被恢复后增幅及恢复效果明显；土壤速效 N 含量也可得到明显的恢复；土壤速效 P 含量呈"缓增—急增—缓降"的变化规律，土壤速效 P 含量未能得到明显恢复和改善；土壤速效 K 含量呈现出"缓降—急降—缓增"的变化规律，其恢复效果并不明显；土壤 pH 值在 7 年后呈明显的下降趋势，14 年后林地土壤 pH 值得到有效降低并趋于稳定。限制性养分速效 P 及速效 K 含量的恢复明显滞后于有机质及速效 N 的恢复，说明限制性养分要得到明显的恢复需要更长的时间。基于植被恢复进程中土壤养分变化及恢复状况，建议选择模式 7(香木莲+墨西哥柏木+清香木)等 6 种模式为云南岩溶山地人工植被恢复的优化模式。

关键词：岩溶山地；植被恢复；恢复年限；土壤养分

Soil nutrients of karst mountainous areas in Yunnan province along a 14 years chronosequence under different vegetation restoration models

Chang En-fu[1,3], Li Ya[1], Li Pin-rong[1,3], Nong Shi-zeng[2], Liu Yong-guo[2], Wang Jun[2]

(1. Yunnan Academy of Forestry, Kunming Yunnan 650201, P. R. China; 2. Forest Institute of Wenshan Prefecture, Wenshan Yunnan 666300, P. R. China; 3. Jianshui Station for Desert Ecological System Observation and Research, Jianshui Yunnan 654399)

Abstract：In 2002 2005 2009 and August 2016 soil samples from the upper middle and lower parts of 11Fixed vegetation restoration sites in Fadou were determined by 4 surveys. The changes of 11 kinds of vegetation of different species of tree (grass) and the changs of soil nuterients and restoration of forestl and under restoration years were analyzed. The results showed that along the14 years succession, there was significant correlation between restoration time and soil nutrient parameters such as organic matter, available N, available P, available K and pH. Since the third year after vegetation restoration treatment, thecontent of soil organic matter showed the tendency ofcontinuously increasing, the restoration effect wassignificant. The increase extend of N was not significant, however the content was not low, implying the significant recovery. The content of available P showed the tendency of mild increase, sharp increase and mild decrease along the time, the content increased significantly, however the soil still in the status of phosphorus deficiency, meaning the soil available P was not significantly improved. The content of available K showed the tendency of mild decrease, sharp decrease and mild increase along the time, the content was similar to the status before restoration treatment, implying the effect of revegetation on K was not significant. As far as the soil pH was concerned, since the seventh year, the pH showed the significant tendency of decrease, 14 years after restoration treatment, the pH was significantly decrease and gradually to a stable status. There were extremely significant differences of soil organic matter, available N, available P and available K among 11 restoration models, the accumulation of available P and available K significantly slower than available N, indicated that the recovery of the limiting nutrients P and K need more time. Based on the change and restoration of soil nutrients in the course of vegetation restoration 6 species of fragrant wood such as Mangliatiaaromatica+ Cupressus lusitanica and Pistaciaweinmannifolia were optimized for artificial vegetation restoration in Yunnan karst mountainous areas.

Key words: karst mountainous area; vegetation restoration; chronosequence; soil nutrient

滇东南岩溶山原地貌区是云南岩溶地貌最典型、分布最集中的区域，岩溶面积达36 697.49km², 占土地总面积的51.78%, 属典型的岩溶生态脆弱环境区。在岩溶石山地区，土地退化日益严重，石漠化快速扩张，制约着这一地区社会经济的持续发展[1]。在岩溶生态系统中，土壤既是物理基础，又是生态系统物质和能量流通的媒介和动力，是生态系统中生命支持系统的根本依托。岩溶生态系统的退化与恢复的评价不但应该着重考察土壤的自然存在(是否剥蚀及其程度)，而且应该考察土壤作为生命支持系统的功能是否存在和变化[2]。因此，土壤作为植被恢复的基础及影响岩溶生态系统退化和石漠化发展的主要因素，近年来受到了众多研究者的关注。土壤退化是形成石漠化的本质，主要表现为理化性质的劣化[3]。针对喀斯特石漠化地区的土壤性质已有大量研究，主要围绕石漠化形成过程中土壤性质变化，不同程度石漠化土壤理化性质及不同退耕模式对土壤肥力的影响[4~8]。司彬等[9]则针对石漠化后喀斯特植被自然恢复演替过程中土壤性质的变化进行了研究，揭示了黔中喀斯特植被自然恢复过程中土壤特性的变化规律。云南省针对岩溶地区生态恢复的研究主要集中在造林物种及造林模式筛选、石漠化综合治理技术及植被恢复技术等方面，而对植被恢复进程中土壤肥力的研究也仅限于植被恢复的初期。本研究在李品荣等[10]及张云等[11]对2002年设置的11种造林模式下林地土壤肥力早期变化进行研究的基础上，通过对固定样地土壤养分的连续测定及分析，以探寻并揭示滇东南岩溶山地不同人工造林模式植被恢复进程中土壤养分变化的特征及规律，从而为云南岩溶地区造林物种及造林模式的选择、生态修复提供科学依据。

1 研究区概况

研究区为云南省文山壮族苗族自治州西畴县东北部的法斗乡，海拔1 350~1 673m, 地形为典型的滇东南岩溶山地。该区地处亚热带湿润季风气候区，年均温14.8~17.6℃, 最热月均温19.6~22.3℃, 极端最低温-4.3℃, 年降雨量1 075.7~1 615.3mm, 空气相对湿度82%, 年日照时数1 500~1 600h。无霜期350~360d。虽然雨量充沛，但降水分布不均匀，80%以上的降水集中在6~9月，且降水多由岩隙渗入地下，区域内人畜饮水和农业用水十分困难。研究区土壤为石灰岩发育形成的石灰土和黄壤，地带性植被以壳斗科(Fagaceae)、樟科(Lauraceae)、木兰科(Magnoliaceae)等科的树种为主组成的亚热带季风常绿阔叶林。

其耕地的农作物以玉米(Zea mays)为主[11,12]。

2 材料与方法

2.1 植物配置模式

本研究所采用的树(草)种、配置模式和造林方法详见张云等[11]及陈强等[12]的研究。11种人工植被恢复模式见表1。

表1 植被恢复模式
Table 1 Vegetation restoration models

编号	配置树(草)种	配置比例	配置类型	配置方式
1	墨西哥柏(Cupressus lusitanica)+苦刺(Solanum deflexicarpum)+紫花苜蓿(Medicago sativa)	1:1:1	乔灌草	带状
2	川滇桤木(Alnus creraastogyne cv.Yanshan)+红三叶(Trifolium pratense)	—	林草	全面
3	墨西哥柏	—	纯林	
4	川滇桤木+清香木(Pistacia weinmannifolia)+苦刺	2:1:1	乔灌	带状
5	川滇桤木	—	纯林	
6	花椒(Zanthoxylum bungeanum)+大百脉根(Lotus corniculatus)	—	林草	全面
7	香木莲(Manglietia aromatica)+墨西哥柏+清香木	2:1:1	乔灌	随机
8	云南拟单性木兰(Parakmeria yunnanensis)+黄柏(Phellodendron amurense)	1:1	林药	带状
9	墨西哥柏+金银花(Lonicera japonica)	1:1	林药	带状
10	任豆(Zenia insignis)+黑荆(Acacia mearnsii)+白三叶(T. repens)	1:1:1	林草	全面
11	川滇桤木+栾树(Koelreuteria paniculata)	1:1	阔阔	随机

2.2 土壤样品采集及测定

2002年、2005年、2009年和2016年8月,分别在11块植被恢复模式固定样地的上、中、下部位设置6个样点,每个样点挖1个30cm深的土壤剖面,取10cm土层的土样250g左右,将6个样点的土样(共计1.5kg左右)放入袋中,然后混合均匀带回实验室测定。土壤样品由文山州土壤肥料测试中心进行测定,土壤化学指标含量测定方法如下:速效N采用碱解扩散法;土壤pH、有机质含量、速效P含量和速效K含量分别按NY/T1377.8—2007、NY/T1121.6—2006、GB1229—1990及NY/T889—2004标准进行测定。

2.3 数据处理与分析

应用Excel、DPS7.05软件进行统计分析与试验数据处理。

3 结果与分析

3.1 植被恢复年限与土壤化学指标含量的相关性分析

对恢复年限及所测定的5个土壤化学指标含量进行相关分析,其结果见表2。恢复年限与土壤pH值呈极显著（$P<0.01$）负相关,与土壤速效P含量、有机质含量分别呈显著（$P<0.05$）和极显著（$P<0.01$）正相关;土壤pH值与有机质含量呈显著（$P<0.05$）负相关,土壤有机质含量与速效N、速效P、速效K含量均呈极显著（$P<0.01$）正相关,速效N含量与速效P、速效K含量分别呈显著（$P<0.05$）和极显著（$P<0.01$）正相关。其结果与张云等[11]对植被恢复后0~7年土壤养分含量间的相关性分析结果一致。相关性分析的结果表明在无人为抚育施肥,人工植被处于自然生长的状况下,林地土壤有机质含量是影响土壤肥力的主要因素,会影响土壤的其他养分含量。

表2 恢复年限及土壤养分含量指标间的相关性
Table 2 Correlation between restoration time and soil nutrients

相关性	恢复年限	pH值	有机质含量	速效N含量	速效P含量	速效K含量
恢复年限	1.0000					
pH值	-0.6224**	1.0000				
有机质含量	0.6161**	-0.3763*	1.0000			
速效N含量	0.1605	-0.1013	0.6154**	1.0000		
速效P含量	0.3309*	0.2419	0.3965**	0.4098**	1.0000	
速效K含量	-0.1783	0.0843	0.3834**	0.3276*	-0.0542	1.0000

＊＊表示极显著相关,＊表示显著相关。

3.2 土壤有机质含量的变化特征

土壤有机质是植物矿质营养和有机营养的源泉,其含量是土壤肥力高低的重要指标之一,对土壤肥力具有重要影响[13,14]。11种植被恢复模式下土壤有机质含量的差异显著性分析结果（表3）表明,模式7与6的有机质含量差异极显著（$P<0.01$）,模式7与9、模式6与3、4的有机质含量差异显著（$P<0.05$）。

不同恢复年限下土壤有机质含量的差异显著性分析结果（表4）表明,在不同的恢复年限下,林地土壤有机质含量在植被恢复后0~3年无显著差异;3年后其增速加快,7年时较恢复前有明显的增加且差异显著（$P<0.05$）,但与恢复3年时相比无显著差异;14年时土壤有机质含量与恢复前及恢复3年、7年时相比,差异均极显著（$P<0.01$）,较恢复前增加了52%,达5.2291%,处于丰富状态。这表明,人工植被随着恢复时间推移,林地生物量不断增加,地表的枯落物不断累积提高,且随着其分解,根系分泌物和微生物的增加,丰富了有机物来源,从而使林地的有机质含量得到有效补充和提高。这与向志勇等[15]、樊文华等[16]、王友生等[17]在不同区域、不同立地条件下研究得出的土壤有机质含量变化规律一致。

表3 11种植被恢复模式和不同恢复年限下土壤有机质含量差异显著性
Table 3 Difference significance of soil organic matter of 11 revegetation models

模式	有机质含量均值(%)	差异显著性水平	
7	5.2150	a	A
4	5.0125	ab	AB
3	4.9700	ab	AB

模式	有机质含量均值(%)	差异显著性水平	
1	4.747 5	abc	AB
5	4.512 5	abc	AB
11	3.942 5	abc	AB
2	3.910 0	abc	AB
8	3.855 0	abc	AB
10	3.647 5	abc	AB
9	3.510 0	bc	AB
6	3.227 5	c	B

注:同一列不同小写字母表示差异显著,同一列不同大写字母表示差异极显著。

表4 不同恢复年限下土壤有机质含量差异显著性
Table 4 Difference significance of soil organic matter of various restoration periods

恢复年限	有机质含量均值(%)	差异显著性水平	
14	5.229 1	a	A
7	4.410 0	b	B
3	3.854 5	bc	BC
0	3.433 6	c	C

注:同一列不同小写字母表示差异显著,同一列不同大写字母表示差异极显著。

3.3 土壤速效N、速效P及速效K含量的变化特征

N、P、K是土壤的3种大量元素,是植物生长所必须的3大营养元素,速效N、速效P及速效K则是最易被吸收利用的有效养分。11种植被恢复模式和不同恢复年限下土壤速效N、速效P及速效K含量差异显著性分析结果分别见表5、表6。

表5 11种植被恢复模式土壤速效N、速效P及速效K含量差异显著性
Table 5 Difference significance of soil available N, P and K of 11 revegetation models

模式	速效N含量均值 (mg·kg^{-1})	差异显著性水平		模式	速效P含量均值 (mg·kg^{-1})	差异显著性水平		模式	速效K含量均值 (mg·kg^{-1})	差异显著性水平	
7	224.100 0	a	A	7	8.092 5	a	A	7	101.680 0	a	A
4	218.405 0	a	A	1	5.375 0	b	AB	3	77.700 0	b	AB
1	179.902 5	ab	A	2	4.780 0	b	AB	4	73.475 0	b	AB
3	170.852 5	ab	A	10	4.612 5	b	AB	1	72.005 0	b	AB
10	166.657 5	ab	A	6	4.522 5	b	AB	11	71.222 5	b	AB
11	163.397 5	ab	A	8	4.407 5	b	AB	5	70.332 5	b	AB
2	160.260 0	ab	A	4	3.885 0	b	B	8	66.210 0	b	B
5	156.310 0	ab	A	5	3.635 0	b	B	9	66.175 0	b	B
8	150.152 5	ab	A	11	3.260 0	b	B	2	61.765 0	b	B
6	136.657 5	b	A	3	3.107 5	b	B	6	60.995 0	b	B
9	124.842 5	b	A	9	2.997 5	b	B	10	53.630 0	b	B

注:同一列不同小写字母表示差异显著,同一列不同大写字母表示差异极显著。

表6 不同恢复年限下土壤速效N、速效P及速效K含量差异显著性
Table 6 Difference significance of soil available N, P and K of various restoration periods

恢复年限	速效N含量均值(mg·kg^{-1})	差异显著性水平		恢复年限	速效P含量均值(mg·kg^{-1})	差异显著性水平		恢复年限	速效K含量均值(mg·kg^{-1})	差异显著性水平	
14	176.727 3	a	A	7	6.636 4	a	A	0	77.495 5	a	A
7	173.090 9	a	A	14	4.727 3	b	B	3	76.347 3	a	A
3	164.165 5	a	A	3	3.269 1	c	B	14	71.363 6	a	AB
0	159.302 7	a	A	0	3.067 3	c	B	7	57.509 1	b	B

注：同一列不同小写字母表示差异显著，同一列不同大写字母表示差异极显著。

3.3.1 土壤速效N含量的变化特征

11种植被恢复模式下土壤速效N的差异显著性分析结果(表5)表明，模式7的速效N含量分别与模式6、9的差异显著($P<0.05$)，模式4的速效N含量分别与模式6、9的差异显著($P<0.05$)，其余模式之间差异均不显著。这主要与树(草)种配置及恢复前土壤自身速效N的含量差异有关。

不同恢复年限下土壤速效N含量的差异显著性分析结果(表6)表明，在不同恢复年限下，林地土壤速效N含量呈缓慢增加之势且无显著差异，恢复14年后土壤速效N平均含量由159.320 7mg·kg^{-1}提高至176.723 7mg·kg^{-1}，增加了11%。说明植被的恢复使林地土壤速效N含量得到增加，而在无人为外来N源补充的条件下，还能满足林木生长吸收消耗的需求并维持在丰富的状态，表明林地土壤速效N也可得到有效的恢复。这与任京辰等[2]所研究的岩溶土壤经过3~5年以上的人工或自然植被的恢复，土壤速效N含量也可以得到较快的恢复的结论一致。

3.3.2 土壤速效P含量的变化特征

11种植被恢复模式下土壤速效P的差异显著性分析结果(表5)表明，模式7的速效P含量分别与模式3、4、5、9、11的差异极显著($P<0.01$)，模式7的速效P含量分别与模式1、2、6、8、10的差异显著($P<0.05$)。

不同恢复年限下土壤速效P含量的差异显著性分析结果(表6)表明，在植被恢复后3~7年，其土壤速效P的含量快速增加，至第7年达到峰值，相较于恢复前增加了116%，速效P的平均含量达6.636 4mg·kg^{-1}。这可能是由于人工植被在3~7年处于生长旺盛期，林木和土壤微生物为适应这种低P生境，通过形成菌根和分泌根系分泌物等方式提高对土壤P素的利用率，从而表现为土壤P素有效性的提高[7]；7年后逐步下降，14年时土壤速效P平均含量为4.727 3mg·kg^{-1}，分别与恢复前、恢复后3年相比差异显著($P<0.05$)，但与恢复前一样仍处于缺P状态。这说明植被恢复后土壤速效P含量虽有一定的增加，但并没有得到明显提高，土壤缺P的状况并未得到明显改善。这与任京辰等[2]对岩溶土壤研究的结论一致，其原因是岩溶土壤中P素的存在状况具有一定的特殊性，石灰岩发育的土壤缺P，P有效性较低[2]，植物和微生物的生长和活动也受到P素限制[18]。土壤速效P含量变化的总体趋势呈"缓增(0~3年)—急增(3~7年)—缓降(7~14年)"的变化规律。

3.3.3 土壤速效K含量的变化特征

11种植被恢复模式下土壤速效K的差异显著性分析结果表明，模式7的速效K含量与模式 2、6、8、9、10 的差异极显著($P<0.01$),模式 7 与模式 1、3、4、5、11 的速效 K 含量差异显著($P<0.05$)。

不同恢复年限下土壤速效 K 含量的差异显著性分析结果(表6)表明,在植被恢复后的 3~7 年,其土壤速效 K 含量显著下降,至第 7 年达到最低值并较恢复前差异极显著($P<0.01$),土壤速效 K 平均含量较恢复前下降了 26%,其含量仅为 57.509 1mg/kg,这主要是林木生长吸收土壤中大量的速效 K 所致。7 年后逐步增加,第 14 年土壤速效 K 的含量较第 7 年增加了 19%,含量为 71.363 6mg/kg 则是因有机质逐渐积累,使得土壤有机质含量大幅提高,从而使岩石中结晶 K 释放多,提高了速效 K 含量[14],但其含量仍与恢复前相当,处于稍缺状态,这也说明植被的恢复对土壤速效 K 的恢复效果并不明显。而魏媛等[13]的研究也表明不同树种配置模式对喀斯特山地土壤 K 含量的改善效果较小。主要原因是 K 素受成土母质制约较大,迁移能力不强,K 素是主要养分限制因子[19~21]。土壤速效 K 含量变化的总体趋势呈"缓降(0~3 年)—急降(3~7 年)—缓增(7~14 年)"的变化趋势。

3.4 土壤 pH 的变化特征

土壤 pH 值直接影响着土壤酶活性和土壤微生物数量等,从而影响土壤腐殖质的形成与分解,以及养分矿质化和土壤养分的积累,进而影响土壤肥力[22]。从表7~表8可以看出,11 种植被恢复模式下林地土壤 pH 值在 0~7 年无明显变化,7 年后呈明显的下降趋势,第 14 年的土壤 pH 值分别与恢复前及恢复后 3 年、7 年相比差异极显著($P<0.01$)。这表明,人工植被随着恢复时间的推移,能有效降低土壤 pH 值。而司彬等[9]、崔晓晓等[14]、昊海勇等[23]的研究结果也表明,随着喀斯特植被自然恢复进程的发展,土壤 pH 值逐渐降低。究其原因,主要是地表枯落物不断累积及分解,导致土壤有机质含量提高并释放大量酸性物质进入土壤。

表7 11 种植被恢复模式和不同恢复年限下土壤 pH 差异显著性
Table 7 Difference significance of soil pH of 11 revegetation models under various restoration periods

模式	土壤 pH 均值	差异显著性水平	
7	7.005 0	a	A
6	6.912 5	a	A
2	6.875 0	a	A
1	6.842 5	a	AB
8	6.730 0	ab	AB
5	6.715 0	ab	AB
9	6.672 5	ab	AB
10	6.627 5	ab	AB
4	6.572 5	ab	AB
3	6.540 0	ab	AB
11	6.195 0	b	B

注:同一列不同小写字母表示差异显著,同一列不同大写字母表示差异极显著。

表8 不同恢复年限下土壤有机质含量差异显著性

Table 8 Difference significance of soil pH of various restoration periods

恢复年限	土壤pH均值	差异显著性水平	
7	6.9409	a	A
0	6.8727	a	A
3	6.8273	a	A
14	6.1545	b	B

注:同一列不同小写字母表示差异显著,同一列不同大写字母表示差异极显著。

4 结论和讨论

(1)土壤有机质是影响土壤有效养分的主要因素。会影响土壤的其他养分含量及土壤的pH值植被恢复时间长短。对土壤pH、有机质及速效P含量的变化具有明显影响人工植被的恢复可使岩溶地区有机质含量得到有效增加及提高。

(2)11种植被恢复模式下的岩溶区林地14年后土壤有效养分变化的研究结果表明:土壤速效N的含量得到明显的恢复;土壤速效P的含量呈"缓增—急增—缓降"的变化规律,岩溶山地人工植被未能使土壤速效P的含量得到明显恢复,土壤缺P的状况并未得到明显改善;土壤速效K含量呈现出"缓降—急降—缓增"的变化规律,植被的恢复对土壤速效K的恢复效果并不明显。因此,在人工植被恢复的进程中选择适当的时机人为施用适量P肥和K肥以加快土壤有效养分库N、P、K的恢复。

(3)岩溶山地的人工植被随着恢复时间的推移,能有效降低土壤pH值。11种植被恢复模式间的土壤有机质、速效N、速效P及速效K的含量存在一定差异,主要与树(草)种配置及恢复前土壤自身的含量差异有关;而土壤限制性养分速效P及速效K的恢复明显滞后于有机质及速效N的恢复,说明岩溶山地土壤限制性养分要得到明显的恢复需要更长的时间。基于植被恢复进程中土壤养分变化及恢复状况,建议选择模式7(香木莲+墨西哥柏+清香木)、3(墨西哥柏)、4(川滇桤木+清香木+苦刺)、1(墨西哥柏+苦刺+紫花苜蓿)、5(川滇桤木)、11(川滇桤木+栾树)6种植被恢复模式作为云南岩溶山地人工植被恢复的优化模式。鉴于岩溶地区生态环境地质分异、植被退化和石漠化严重等问题,着眼于生态系统功能提升,应深入开展植被退化与恢复生态学机制、耐旱耐瘠薄乡土植物材料的筛选、人工林草植物群落构建技术以及仿自然植物群落优化配置等技术等方面的研究,以期为云南岩溶地区生态恢复提供理论和技术支撑。

参考文献:

[1] 中国科学院学部. 关于推进西南岩溶地区石漠化综合治理的若干建议[J]. 地球科学进展,2003,18(4):489~492.

[2] 任京辰,张平究,潘根兴,等. 岩溶土壤的生态地球化学特征及其指示意义:以贵州贞丰—关岭岩溶石山地区为例[J]. 地球科学进展,2006,21(5):504~512.

[3] 龙健,江新荣,邓启琼,等. 贵州喀斯特地区土壤石漠化的本质特征研究[J]. 土壤学报,2005,42(3):419~427.

[4] 涂成龙,林昌虎,何腾兵,等. 黔中石漠化地区生态恢复过程中土壤养分变异特征[J]. 水土保通报,

2004,24(6):22~25,89.
[5] 罗海波,宋光煜,何腾兵,等.贵州喀斯特山区石漠化治理过程中土壤质量特性研究[J].水土保持学报,2004,18(6):112~115.
[6] 刘方,王世杰,刘元生,等.喀斯特石漠化过程土壤质量变化及生态环境影响评价[J].生态学报,2005,25(3):639~644.
[7] 朱海燕,刘忠德,钟章成.喀斯特退化生态系统不同恢复阶段土壤质量研究[J].林业科学研究,2006,19(2):248~252.
[8] 龙健,李娟,黄昌勇.我国西南地区的喀斯特环境与土壤退化及其恢复[J].水土保持学报,2002,16(5):5~8.
[9] 司彬,姚小华,任华东,等.黔中喀斯特植被恢复演替过程中土壤理化性质研究[J].江西农业大学学报,2008,30(6):1122~1125.
[10] 李品荣,陈强,常恩福,等.滇东南石漠化山地不同退耕还林模式土壤地力变化初探[J].水土保持研究,2008,15(1):65~68,71.
[11] 张云,周跃华,常恩福.滇东南岩溶山地不同人工林配置模式对其林地土壤肥力的影响[J].西部林业科学,2010(1):62~68.
[12] 陈强,李品荣,常恩福,等.滇东南岩溶山区树种配置的初步研究[J].云南林业科技,2002(4):1~10.
[13] 魏媛,吴长勇,孙云,等.不同树种配置模式对喀斯特山地土壤理化性质的影响[J].贵州农业科学,2014,42(9):81~85.
[14] 崔晓晓,王圳,王纪杰,等.喀斯特峡谷区植被恢复过程中土壤性质变化[J].福建林学院学报,2011,31(2):165~170.
[15] 向志勇,邓湘雯,田大伦,等.五种植被恢复模式对邵阳县石漠化土壤理化性质的影响[J].中南林业科技大学学报,2010,30(2):23~28.
[16] 樊文华,李慧峰,白中科.黄土区大型露天煤矿不同复垦模式和年限下土壤肥力的变化:以平朔安太堡露天煤矿为例[J].山西农业大学学报(自然科学版),2006,26(4):313~316.
[17] 王友生,吴鹏飞,侯晓龙,等.稀土矿废弃地不同植被恢复模式对土壤肥力的影响[J].生态环境学报,2015,24(11):1831~1836.
[18] 胡忠良,潘根兴,李恋卿,等.贵州喀斯特山区不同植被下土壤C、N、P含量和空间异质性[J].2009,29(8):4187~4195.
[19] 鲁剑巍,陈防,刘冬碧,等.成土母质及土壤质地对油菜施钾效果的影响[J].湖北农业科学,2001(6):42~44.
[20] 谭宏伟,周柳强,谢如林,等.广西亚热带岩溶地区石灰性土壤钾素特征研究[J].中国生态农业学报,2006,14(3):58~60.
[21] 刘淑娟,张伟,王克林,等.桂西北喀斯特峰丛洼地表层土壤养分时空分异特征[J].生态学报,2011,31(11):3036~3043.
[22] 高国雄,李得庆,贾俊姝,等.退耕还林不同配置模式对土壤养分的影响[J].干旱区资源与环境,2007,21(5):104~107.
[23] 吴海勇,彭晚霞,宋同清,等.桂西北喀斯特人为干扰区植被自然恢复与土壤养分变化[J].水土保持学报,2008,22(4):143~147.

(本文发表于《西南林业大学学报》,2018年)

元谋干热河谷植物生态位特征研究

郭永清[1],郎南军[1],江期川[1],杨旭[1],谷晓萍[2]

(1. 云南省林业科学院,云南 昆明 650204;2. 沈阳农业大学,辽宁 沈阳 110161)

摘要:本文采用样方法对云南元谋干热河谷区20个物种的30块样地进行调查研究,并计算了物种的生态位宽度、生态位相似性比例和生态位重叠值。结果表明,扭黄茅是干热河谷区的主要优势种,其重要值和生态位宽度最大,分别为0.81和0.938。扭黄茅与其他物种的生态位相似性比例相对较低,生态位重叠值也较小,相反,生态位宽度较小的物种之间生态位重叠值和生态位相似性比例较高。在植被演替过程中,生态位较小的物种容易被替代。

关键词:干热河谷;生态位;生态位重叠;演替

Study on Niche Characteristics of Plant in the Dry and Hot Valley of Yuanmou County

Guo Yong-qing[1], Lang Nan-jun[1], Jiang Qi-chuan[1], Yang Xu[1], Gu Xiao-ping[2]

(1. Yunnan Academy of Forestry, Kunming Yunnan 650204, China; 2. Shenyang Agricultural University, Shenyang Liaoning 110161, China)

Abstract: In this paper the sampling plots were taken for the investigation on 20 plant species in 30 plots of the dry and hot valley in Yuanmou County of Yunnan Province and the calculation for the niche breadth, percentage similarity and niche overlap value of the species have been done. The result shows that *Heteropogon contortus* is the dominant species in the dry and hot valley and its importance value and niche breadth is the biggest one (0.81and 0.938 respectively). Based on the comparison to the other species, the percentage similarity of *Heteropogon contortus* is lower and the niche overlap value is smaller. However, among the species the species with smaller niche breadth bigger niche overlap value and percentage similarity. During the succession, the species that has small niche will be easilyreplaced by the others.

Key words: dry and hot valley;niche;niche overlap;population

Joseph Grinnell 1917年首先应用生态位(niche)一词来描述对栖息地再划分的空间单位,他强调的是生态位的空间概念[1],生态位后来被定义为"指种群在时间、空间的位置以及种群在群落中的地位和功能作用"[2]。生态位理论经过发展后在种间关系、群落结构、种的多样性及种群进化等方面的研究获得了广泛的应用,成为解释自然群落中种间的共存与竞争机制的基本理论之一,生态位理论研究的基本内容之一是生态位宽度与生态位重叠的测度[3]。近年来,国内外许多学者对生态位理论、生态位的计测进行了大量的研究,对不同类型生态系统植物生态位特征进行了研究,致力于解释生物与环境之间的相互关系[4~7]。元谋干热河谷位于金沙江上游,由于受到自然、人为等因素的干扰,出现了严重水土流失和土地退化等现象[8]。因此,国内学者对元谋干热河谷区从土壤水分[9,10]、土壤组成[11]、植被组成[12]、恢复技术和方法[13~16]等方面进行了大量的研究,对元谋干热河谷植被恢复起到了积极的作用,但对植物生态位特征的研究较少。研究该地区的植物生态位特征能够揭示植物与环境之间的关系,为生态恢复提供理论依据,同时也可以全面了解生态恢复演替过程中植物的个体适应性。

1 材料与方法

1.1 研究区自然概况

研究区位于云南省金沙江流域干热河谷核心区元谋县苴林大黑山,金沙江一级支流龙川江下段,其地理位置为25°31′~26°07′N,101°36′~102°07′E,平均海拔在1 500m以下,年平均降水量为614.0mm,年蒸发量却达到3 847.6mm,为降水量的6倍多,且年降水量差异较大,最大年份达906.7mm,最小年份仅有287.4mm,相差约3倍[17]。年降水量分布也不均匀,雨季(6~10月)平均降水量为583.8mm,占全年降水量的92%,旱季(11月至翌年5月)降水仅50.2mm,仅占8%。自然植被属于典型稀树灌草丛[10],有扭黄茅(*Heteropogon contortus*)、滇橄榄(*Phyllanthus emblica*)、芸香草(*Cymbopogon distans*)等一年生、旱生灌木或草本植物,此外还有部分加勒比松(*Pinus caribaea*)、山合欢(*Albizzia kalkora*)、桉树(*Eucalyptus* spp.)等人工乔木树种,其中一年生植物占33.3%[18]。研究区从1998年开展生态恢复研究,目前植被盖度达到了85%以上,土壤以燥红壤为主,其次是褐红壤。

1.2 研究方法

试验于2007年7~8月份开展,此时是干热河谷的雨季,植被生长旺盛,从研究区沟底按照"V"形设置两条样带,在两条样带上根据植物种类在最具有代表性的植被群落中共设置30个样方,每个样方面积为4m×5m,然后在每个样方中再设置3个面积为1m×1m的小样方,共计90个小样方。调查小样方内植被种类、盖度、高度、冠幅、频度等。

1.3 计测方法

1.3.1 重要值的计算

重要值采用公式IV(%)=(相对盖度+相对多度)/2计算[5]。由于所调查样方中未出现高度大于2m的植株,所有植被重要值均按此公式计算。其中,

相对多度=某物种出现的样方数/全部种出现的样方数之和×100%

相对盖度=某物种的底面积/全部种的底面积之和×100%

1.3.2 生态位宽度

生态位宽度采用Levins生态位宽度计测公式:

$$\text{Levins 生态位宽度 } B_i = 1/\sum_{j=1}^{r} P_{ij}^2$$

式中,B_i为种i的生态位宽度,P_{ij}为种i对第j个资源的利用占它对全部资源利用的频度,即$P_{ij} = \frac{n_{ij}}{N_i}$,而$N_i = \sum_{j=1}^{r} n_{ij}$,$n_{ij}$为种$i$在资源$j$上的优势度(即样方中物种的重要值),$r$为样方数。

Hurlbert[19]修正后的 Levins 生态位宽度：

$$B_a = \frac{B_i - 1}{r - 1}$$

式中，$B_i = 1/\sum_{j=1}^{r} p_{ij}^2$，$B_a$ 为生态位宽度，其值域为 [0,1]。

1.3.3 生态位重叠

生态位重叠是指一定资源序列上，两个物种利用同等级资源而相互重叠的情况，采用 Pianka 的生态位重叠指数公式计测。

$$O_{ik} = \sum_{j=1}^{r} P_{ij} P_{kj} / \sqrt{(\sum_{j=1}^{r} P_{ij})^2 (\sum_{j=1}^{r} P_{kj})^2}$$

式中，O_{ik} 为物种 i 和物种 k 的生态位重叠值，P_{ij} 和 P_{kj} 为种 i 和种 k 在资源梯度级 j 的数量特征，本文中为种 i 和种 k 在样方 j 的物种重要值，r 为样方数，其值域为 [0,1]。

1.3.4 生态位相似性比例

$$C_{ik} = 1 - \frac{1}{2} \sum_{j=1}^{r} |P_{ij} - P_{kj}|$$

式中，C_{ik} 表示物种 i 和 k 的相似程度，具有 $C_{ik} = C_{ki}$，值域 [0,1]，P_{ij} 和 P_{kj} 分别为物种 i 和 k 在资源 j 上的重要值百分率。

2 结果与分析

2.1 物种重要值

元谋属于典型的干热河谷区，植被以稀树灌草丛为主，对元谋干热河谷 30 个样方调查结果表明，30 个样方中出现植物种共计 20 种，其中出现频率达到 30% 以上的种仅有 3 个，其中以扭黄茅出现频率最高，达到 96.7%。从表 1 可以看出，扭黄茅的重要值最大，平均 0.81，明显高于其他物种，是干热河谷区的主要优势种。按照重要值大小排序，最大的 5 个物种分别为扭黄茅、大叶千斤拔、蔓草虫豆、车桑子和龙须草（表 1）。而其他物种重要值介于 0.02~0.1 之间，远远低于扭黄茅，属于典型的伴生种。

2.2 物种生态位宽度

从表 2 看出，元谋干热河谷 20 种主要物种的生态位宽度最大的 5 个物种为扭黄茅、野古草、大叶千斤拔、车桑子和蔓草虫豆，生态位宽度分别为 0.938、0.405、0.310、0.308 和 0.275，其中生态位宽度最大的物种为扭黄茅，达到 0.938，说明其在干热河谷的分布比较广，对环境资源的利用率比较高，是干热河谷的优势种。在植被演替过程中，扭黄茅与其他物种相比具有明显的竞争优势，能充分利用环境资源，表明其在生态适应方面更趋向于泛化，而小蓬、茜草和滇橄榄 3 个物种生态位宽度为 0，属于典型的伴生植物，在干热河谷的分布面积很小，对环境资源具有较高的选择性，植被演替过程中容易被其他物种所取代。

表 1 元谋干热河谷主要物种重要值
Table 1 The important value of main species in dry and hot valley in Yuanmou county

物　种	重要值
s1 车桑子 *Dodonaea viscosa*	0.18
s2 朴叶扁担杆 *Grewia celtidifolia*	0.07
s3 大叶千斤拔 *Flemingia macrophylla*	0.20
s4 银合欢 *Leucaena glauca*	0.06
s5 川楝 *Melia toosendan*	0.05
s6 扭黄茅 *Heteropogon contortus*	0.81
s7 蔓草虫豆 *Atylosia scarabaeoides*	0.19
s8 芸香草 *Cymbopogon distans*	0.10
s9 白花叶 *Porana henryi*	0.03
s10 龙须草 *Eulaliopsis binata*	0.11
s11 小蓬 *Conyza canadensis*	0.02
s12 茜草 *Rubia cordifolia*	0.02
s13 野古草 *Arundinella grandiflora*	0.06
s14 胡枝子 *Lespedeza bicolor*	0.07
s15 南莎草 *Cyperus niveus*	0.07
s16 毛叶黄杞 *Engelhardtia colebrookiana*	0.03
s17 九死还魂草 *Selaginella pulvinata*	0.03
s18 青紫藤 *Cissus javana*	0.05
s19 滇橄榄 *Phyllanthus emblica*	0.02
s20 桔草 *Cymbopogon goeringii*	0.05

表 2 元谋干热河谷主要物种生态位宽度
Table 2 The niche breadth of main species in dry and hot valley in Yuanmou county

物种	生态位宽度	物种	生态位宽度	物种	生态位宽度	物种	生态位宽度
s1	0.308	s6	0.938	s11	0	s16	0.034
s2	0.032	s7	0.275	s12	0	s17	0.034
s3	0.310	s8	0.136	s13	0.405	s18	0.069
s4	0.034	s9	0.034	s14	0.069	s19	0
s5	0.069	s10	0.165	s15	0.103	s20	0.069

注:代号表示的物种与表 1 相同。

2.3　物种生态位重叠分析

从表 3 看出,在元谋干热河谷区植被的生态位重叠程度不高,出现生态位重叠的种对有 78 对,占全部种对的 41.1%,生态位重叠值最高的物种是桔草—野古草,达到 0.334,其次是南莎草—胡枝子,达到 0.25。生态位重叠值超过 0.1 的种对有 22 对,占全部种对的 11.6%。扭黄茅是元谋干热河谷的优势种,生态位平均宽度最大,但与其他物种的生态位重叠值很小,说明生态位重叠值大小与生态位宽度大小并不成正比关系。相反,生态位宽度较小的物种之间有较大的生态位重叠,说明它们之间的生态适应性和生物学特性比较相近。银合欢、白花叶、小蓬、茜草和滇橄榄这4个物种的生态位与其他物种基本不出现重叠,说明这些物种在干热河谷出现的频率低,分布范围很小,是典型的伴生物种。

2.4 物种生态位相似性比例

物种生态位相似性比例小主要取决于种间对资源利用的相似性程度和物种生物学特性,从表4看出,元谋干热河谷主要物种的生态位相似性比例很高,相似性比例大于0.9的种对有173对,占总种对的91.1%,表明干热河谷大部分物种的生态特征都基本相似。而扭黄茅的生态位与其他物种的生态位相似性比例相对较低,只有5个物种与其生态位相似性比例大于0.9,这也说明了扭黄茅作为干热河谷的主要优势种,其生态位比较宽,生态适应性比其他的物种较高,分布范围广,而生态位小的物种之间生态位相似性比例较高。

3 结论与讨论

生态位宽度是指物种资源开发利用的程度,在现有的资源谱中,仅能利用其中一小部分的生物称为狭生态位种,而能利用较大部分的,则称为广生态位种。物种的生态位宽度越大,表明该物种利用资源的能力越强,其特化程度就越小,对生态因子的适应幅度就越大,在群落内的分布范围也就越大[20]。生态位宽度的大小取决于物种对环境的生理适应性、生境特征、物种间的竞争关系等[21,22]。本研究区中,植被中扭黄茅的生态位宽度和重要值分别为达到0.938和0.81,是20个物种中最大的,这与金沙江干热河谷植被的实际分布是一致的,说明扭黄茅数量多,分布广,生态幅较大,对资源环境的利用也较为充分,所以它们的重要值和生态位宽度最高。相反,有许多物种的重要值介于0.02~0.2之间,它们的生态位宽度则相对较小,经测算,小蓬、茜草和滇橄榄3个物种的生态位宽度为0,说明这些物种仅利用资源序列中的一个等级,物种的生态位就最小,分布范围也最窄。

生态位相似性比例是指两个物种在资源序列中利用资源的相似程度,元谋干热河谷20个物种的生态位相似性比例较高,大部分物种间生态位相似性比例都在0.9以上,充分说明了大多数物种之间对资源的利用较为相似。但是,生态位宽度较大的扭黄茅与其他物种的相似性不显著,而生态位宽度较小的物种之间的生态位相似性却很高,由于物种间的生物学特性不同,则对资源的要求也不相似,从而导致生态位宽度较大的物种间可能会出现较小的生态位相似性;同时,生态位宽度较小的物种,对环境资源的利用极其相似,也可能会出现较高的生态位相似性。

生态位重叠是指在一个资源序列上,两物种利用相同等级资源而相互重叠的状况,也用来表示两个或多个种群的生态位相似性、或两个或多个种群对同一类环境资源共同利用或联合利用的程度[23]。当两个或多个物种共同占有同一资源时就会出现生态位重叠现象,一般来说,生态位比较宽的两个物种间会出现较大的生态位重叠,但是并不意味着生态位宽度小的物种间就不会出现生态位重叠现象,在元谋干热河谷区扭黄茅属于优势种,生态位平均宽度最大,但与其他物种的生态位重叠值很小,相反,生态位宽度较小的物种之间有较大的生态位重叠,这与物种的生物学特性有很大的关系。

表 3 元谋干热河谷主要物种生态位重叠
Table 3 The niche overlap of main species in dry and hot valley in Yuanmou county

种类	s1	s2	s3	s4	s5	s6	s7	s8	s9	s10	s11	s12	s13	s14	s15	s16	s17	s18	s19	s20
s1	1	0.044	0.042	0.049	0.104	0.034	0.021	0.044	0	0	0	0	0.029	0.030	0.024	0.051	0.047	0	0	0.030
s2		1	0.080	0	0	0.026	0	0	0	0.095	0	0	0.039	0	0	0	0	0	0	0
s3			1	0	0	0.035	0.032	0.041	0.047	0	0	0	0.034	0.049	0.049	0.052	0	0.032	0.097	0.034
s4				1	0.165	0.039	0.055	0	0	0	0	0	0	0	0	0	0	0	0	0
s5					1	0.037	0	0	0	0	0	0	0.107	0.084	0.084	0	0.167	0	0	0.107
s6						1	0.034	0.030	0.039	0.022	0.039	0.040	0.034	0.036	0.036	0.032	0.033	0.029	0.027	0.034
s7							1	0	0.061	0.049	0.114	0	0	0	0	0	0	0	0.106	0
s8								1	0	0	0	0	0	0.048	0.048	0.110	0	0	0	0
s9									1	0	0	0	0	0	0	0	0	0.078	0	0
s10										1	0.152	0	0	0	0	0	0.118	0	0	0
s11											1	0	0	0	0	0	0	0	0	0
s12												1	0	0	0	0	0	0	0	0
s13													1	0.170	0.170	0.161	0.114	0	0	0.334
s14														1	0.250	0.123	0.126	0.082	0	0.170
s15															1	0.124	0.126	0.082	0	0.170
s16																1	0	0	0	0
s17																	1	0.166	0	0.161
s18																		1	0	0.115
s19																			1	0
s20																				1

表 4 元谋干热河谷主要物种生态位相似性

Table 4 The niche percentage similarity of main species in dry and hot valley in Yuanmou county

种类	s1	s2	s3	s4	s5	s6	s7	s8	s9	s10	s11	s12	s13	s14	s15	s16	s17	s18	s19	s20
s1	1	0.974	0.971	0.976	0.980	0.906	0.959	0.968	0.968	0.956	0.970	0.970	0.970	0.968	0.968	0.973	0.973	0.964	0.970	0.970
s2		1	0.971	0.987	0.986	0.886	0.965	0.974	0.984	0.978	0.986	0.986	0.981	0.979	0.979	0.984	0.984	0.981	0.986	0.981
s3			1	0.969	0.968	0.908	0.962	0.966	0.970	0.954	0.968	0.968	0.967	0.971	0.971	0.971	0.965	0.968	0.973	0.967
s4				1	0.994	0.887	0.973	0.976	0.987	0.975	0.989	0.989	0.983	0.982	0.982	0.987	0.987	0.984	0.989	0.983
s5					1	0.886	0.968	0.977	0.987	0.976	0.990	0.990	0.989	0.987	0.987	0.987	0.992	0.9845	0.990	0.989
s6						1	0.898	0.893	0.884	0.884	0.881	0.881	0.887	0.889	0.889	0.884	0.884	0.887	0.882	0.887
s7							1	0.961	0.976	0.969	0.978	0.973	0.968	0.966	0.966	0.971	0.971	0.968	0.978	0.968
s8								1	0.980	0.968	0.982	0.982	0.977	0.980	0.980	0.985	0.980	0.977	0.982	0.977
s9									1	0.978	0.992	0.992	0.987	0.985	0.985	0.990	0.990	0.987	0.992	0.987
s10										1	0.986	0.981	0.975	0.973	0.973	0.978	0.983	0.981	0.980	0.975
s11											1	0.995	0.989	0.987	0.9874	0.992	0.992	0.990	0.995	0.989
s12												1	0.989	0.987	0.987	0.992	0.992	0.990	0.995	0.989
s13													1	0.992	0.992	0.987	0.992	0.989	0.989	1
s14														1	1	0.990	0.990	0.987	0.987	0.992
s15															1	0.990	0.990	0.9872	0.987	0.992
s16																1	0.990	0.987	0.992	0.987
s17																	1	0.992	0.992	0.992
s18																		1	0.989	0.989
s19																			1	0.989
s20																				1

参考文献：

[1] 戈峰. 现代生态学(第一版)[M]. 北京:科学出版社,2002:266~267.

[2] Elton C S. Animal Ecology [M]. London : Sidgewick and J ackson,1927:63~68.

[3] 祖元刚. 非线性生态学模型[M]. 北京:科学出版社,2004:234~238.

[4] 陈子萱,田福平,牛俊义,等. 玛曲高寒沙化草地植物生态位特征研究[J]. 草地学报,2007,15(6):525~530.

[5] 程小琴,韩海荣,魏阿沙,等. 山西省庞泉沟自然保护区森林群落主要物种生态位特征[J]. 北京林业大学学报,2007,29(2):283~287.

[6] 杨自辉,方峨天,刘虎俊,等. 民勤绿洲边缘地下水位变化对植物种群生态位的影响[J]. 生态学报,2007,27(11):4901~4906.

[7] Sanna L L, Maria P, Helena K. Niche breadth and niche overlap in three epixylic hepatics in a boreal old-growth forest, southern Finland[J]. Journal of Bryology,2005,27(2):119~127.

[8] 张建平,王道杰,王玉宽,等. 元谋干热河谷区生态环境变迁探讨[J]. 地理科学,2000,20(2):148~152.

[9] 黄成敏,何毓蓉. 云南省元谋干热河谷土壤水分的动态变化[J]. 山地研究,1997,15(4):234~238.

[10] 李昆,陈玉德. 元谋干热河谷人工林地的水分输入与土壤水分研究[J]. 林业科学研究,1995,8(6):651~657.

[11] 张信宝,陈玉德. 云南元谋干热河谷区不同岩土类型荒山植被恢复研究[J]. 应用与环境生物学报,1997,3(1):13~18.

[12] 金振洲. 滇川干热河谷种子植物区系研究[J]. 广西植物,1999,19(1):1~14.

[13] 杨忠,张信宝,王道杰,等. 金沙江干热河谷植被恢复技术[J]. 山地学报,17(2):152~156.

[14] 高洁,张尚云,傅美芬. 干热河谷主要造林树种旱性结构的初步研究[J]. 西南林学院学报,1997,17(2):59~63.

[15] 余丽云. 元谋干热河谷植被恢复造林树种选择研究[J]. 西南林学院学报,1997,17(2):49~54.

[16] 周蛟,张明友. 元谋抗旱耐热造林树种的定量选择研究[J]. 云南林业科技,1998,(3):32~36.

[17] 郑科,郎南军,郭玉红,等. 元谋雨季中期加勒比松林林分水文气象研究[J]. 中南林学院学报,2005,25(3):34~38.

[18] 金振洲,欧晓昆. 元江、怒江、金沙江、澜沧江干热河谷植被[M]. 昆明:云南大学出版社,云南科技出版社,2000:141~146.

[19] Hurlbet S H. The measurement of niche overlap and some relatives[J]. Ecology,1978,9(1):67~77.

[20] 张桂萍,张峰,茹文明. 山西绵山植被优势种群生态位研究[J]. 植物研究,2006,26(2):176~181.

[21] 赵永华,雷瑞德,何兴元,等. 秦岭锐齿栎林种群生态位特征研究[J]. 应用生态学报,2004,15(6):913~918.

[22] 闫淑君,洪伟,吴承祯,等. 万木林中亚热带常绿阔叶林林隙主要树种的高度生态位[J]. 应用与环境生物学报,2002,8(6):578~582.

[23] 余树全,李翠环. 千岛湖水源涵养林优势树种生态位研究[J]. 北京林业大学学报,2003,25(2):18~23.

(本文发表于《西北林学院学报》,2009年)

思茅松人工林不同空间配置模式下不同密度生长试验与分析

徐玉梅[1], 杨德军[1], 史富强[2], 许林红[1], 刘永刚[1], 陈伟[1], 陈绍安[1]

(1. 云南省林业科学院,云南 昆明 650201; 2. 普洱市林业科学研究所,云南 普洱 665000)

摘要:以 12 年生思茅松人工林为研究对象,采用标准地每木检尺方法,研究 3 种不同空间配置模式(组合式配置、正方形配置、长方形配置)下不同造林密度对其保存率、树高、胸径、冠幅、单株材积和活立木蓄积量的影响。结果表明:①组合式配置不同密度的树高、胸径、冠幅和单株材积,各处理间均存在极显著差异;区组间胸径和单株材积的分析说明,组合式配置不同密度对其胸径和单株材积有较大的影响,不同立地间对其也有一定的影响,但没有不同密度对其影响明显。组合式配置的保存率、树高、胸径、冠幅、单株材积均与密度成反比,活立木蓄积量与密度不成比例。②正方形配置和长方形配置不同密度的树高、胸径、冠幅和单株材积在各处理间均存在极显著差异,区组间差异不显著。不同密度对正方形配置和长方形配置的树高、胸径和单株材积有明显影响,正方形配置的保存率、树高和胸径均与密度成反比,活立木蓄积量与密度成正比,冠幅和单株材积与密度不成比例。③组合式配置培育大径材最适宜的密度是 2m×2m×5m,培育原料林最适宜的密度是 1m×2m×4m;正方形配置培育大径材最适宜的密度是 2m×2m,培育原料林最适宜的密度是 1m×1m;长方形配置培育大径材最适宜的密度是 2m×3m,培育原料林最适宜的密度是 1m×2m。

关键词:思茅松;组合式配置;正方形配置;长方形配置;密度效应

Study on planting density of *Pinus kesiya* var. *langbianensis* plantation under different spacial allocation modes

Xu Yu-mei[1], Yang De-jun[1], Shi Fu-qiang[2], Xu Lin-hong[1], Liu Yong-gang[1], Chen Wei[1], Chen Shao-an[1]

(1. Yunnan Academy of Forestry, Kunming Yunnan 650201, China; 2. Forest Institute of Pu'er City, Pu'er Yunnan 665000, China)

Abstract: Taking 12 years old *Pinus kesiya* var. *langbianensis* plantation at Qingshuihe of Pu'er city as the study materials, choosing three spacial allocation modes namely assembled allocation, square allocation and rectangle allocation, through field individual measurement, the effects planting density on reserving rate, tree height, DBH, crown width, individual volume and standing stock were studied and analyzed. The results are as follows: 1) Extremely significant differences on reserving rate, tree height, DBH, crown width and individual volume exist amongdifferent planting density treatments under assembled allocation mode, among blocks, the differences of DBH and individual volume were significant at 0.05 level, which indicated that under different assembled allocation, planting density significantly affected the DBH and individual volume, site condition was also an relevant factor, however the impact was not that significant as the planting density. The reserving rate, tree height, DBH, crown with, individual volume were all inversely proportional to planting density, however the standing stock was out of proportion to the planting density. 2)As far as the square allocation and the rectangle allocation were concerned, among different treatments, the tree height, DBH, crown width and individual volume of different planting densities were significantly different. Under square allocation, the reserving rate, tree height and DBH was reversely proportion to the planting density, the standing stock was proportion to the planting density, however the crown width and individual volume were out of proportion to the planting density. 3)Based on the study results, it was concludethat under assembled allocation mode, the recommended planting densities for big size timber cultivation and forest industrial raw material were 2m×2m×5m and 1m×2m×4m respectively. Refer the square allocation mode, the proper planting densities for big size timber cultivation and industrial raw material were 2m×2m and 1m×1m respectively, and for rectangle allocation mode, he proper planting densities for big size timber cultivation and industrial raw material were 2m×3m and 1m×2m respectively.

Key words: *Pinus kesiya* var. *langbianensis*; assembled allocation; square allocation; rectangle allocation; planting density

思茅松(*Pinus kesiya* var. *langbianensis*)为常绿乔木,强阳性树种,是云南省特有的乡土速生用材和产脂树种,可用于纸浆材、坑木、枕木和采脂[1~2],是材、脂兼用树种,具有速生、优质、高产脂和生态适应性强等特点[3]。思茅松自然分布于云南南部、西南部和东南部海拔600~1 600m 的热带及亚热带地区,以普洱的思茅、景东、宁洱、墨江、镇源、景谷以及临沧市东北部海拔700~1 200m 为分布中心[4~6],是云南省思茅林区重要林产工业主要原料林基地。但由于思茅松人工林培育的时间不长,还缺乏人工林培育方面的成熟技术,特别需人工林不同空间配置下不同密度控制方面的成果来支撑其人工林的发展,该研究通过对普洱市思茅区清水河 12 年生 3 种不同空间配置模式下不同密度的保存率、树高、胸径、冠幅、单株材积和活立木蓄积量进行统计分析,分别得出 12 年生林分组合式配置、正方形配置和长方形配置模式下的最佳种植密度和不同空间配置模式下密度对保存率、树高、胸径、冠幅、单株材积和活立蓄积量的影响,以期为思茅松人工原料林的发展提供技术支持。

1 材料与方法

1.1 概况

研究区地处横断山帚状山系南缘,无量山南延末端(100°57′E,22°41′N),海拔 1 236m,在气候区划上属于亚热带季风气候。受湿润的西南季风和干暖的西南风支急流交替控制,半年为雨季,半年为干季,年均气温为 17.7℃,最热月(7 月)均温 21.7℃,最冷月(1 月)均温 11.4℃。年降雨量为 1 547.63mm,雨水主要集中在 5~10 月,降雨量为 1 351.8mm,占全年降雨量的 87.3%。地貌以低山为主,主要土壤类型为赤红壤,其土层厚度在低山坡面达 1m 以上,只有在箐沟较陡峭的局部坡面上,才出现厚 0.5~0.8m 的中层土壤。土壤呈酸性,pH 为 4.3~6.3,有机质含量较低,仅 0.6~2.7g·kg^{-1},氮缺乏,尤其少磷,而钾较丰富,原生植被为季风常绿阔叶林。

1.2 实验设计

研究对象于2000年7月根据试验设计要求,选Ⅰ、Ⅱ级苗进行整地造林(穴规格:60cm×40cm×40cm),按常规方法进行抚育管理,现长势良好。目前,林分乔灌木主要有思茅松、水锦树(*Wend landiapaniculata*)、红木荷(*Schima wallichii*)、红椎(*Castonopsis hystrix*)、云南黄杞(*Engelhardita spicata*)等。实验林分中设3种不同配置方式,在不同空间配置及密度林中选择具有代表性的12年生思茅松林分为研究对象,对不同密度林分的生长量进行比较分析、方差分析和多重比较分析,以找出各组合式配置的最佳林分密度。

(1)组合式配置:适宜的宽窄行距可充分发挥林木的边缘效应,兼具提高单位面积木材产量和维护林地地力的双重功效,是实现思茅松人工林可持续经营的较为理想的人工林培育技术措施。对组合式配置6种不同密度,即2m×2m×5m(A1)、1m×2m×10m(A2)、1m×1m×11m(A3)、1m×2m×4m(A4)、1m×1m×5m(A5)、1m×1m×3m(A6)。

(2)正方形和长方形配置:正方形配置其株行距相等,苗木之间的距离均匀,利于树冠均匀地生长发育,正方形配置的设3种不同密度,即2m×2m(B1);1.5m×1.5m(B2);1m×1m(B3)。

(3)长方形配置:长方形配置有利于行间进行抚育和间作,设2种不同密度,即2m×3m(C1);1m×2m(C2)。

1.3 生长量调查及数据分析

在选定的林分内设置标准地,标准地大小20m×20m,相同林分各设置3块标准地,在标准地内进行每木检尺,调查树高,胸径、冠幅、保存率和林下植被等。采用Excel进行数据统计,用DPS7.05分析软件进行数据分析处理。

2 结果与分析

2.1 3种配置方式不同密度林分生长量的方差分析

12年生思茅松在3种配置方式下不同密度的生长量方差分析结果见表1。

表1 思茅松组合式配置、正方形配置和长方形配置不同密度的生长量方差分析(F值)
Table 1 Variance analysis on growth increment of different planting densities within different spacial allocation modes

配置方式	树高 height		胸径 DBH		冠幅 crown		单株材积 volume		$F_{0.05}$	$F_{0.01}$
	区组间	处理间	区组间	处理间	区组间	处理间	区组间	处理间		
组合式	1.226	17.912**	2.290*	66.171**	1.703	47.709**	2.749*	75.076**	2.22	3.04
正方形	1.226	17.912**	1.322	53.212**	1.146	53.703**	1.270	747.105**	3.03	4.68
长方形	1.350	55.106**	1.304	140.765**	0.928	131.848**	1.3360	133.595**	3.90	6.81

从表1可看出:组合式配置的树高、胸径、冠幅和单株材积各处理间均存在极显著差异,区组间胸径和单株材积在0.05水平差异显著,在0.01水平差异不显著。处理间差异远远大于区组间差异,说明组合式配置不同密度对其胸径和单株材积有较大的影响,不同立地间

对其也有一定的影响,但其影响没有不同密度明显。正方形配置的树高、胸径、冠幅和单株材积各处理间均存在极显著差异,区组间差异不显著。表明正方形配置的思茅松,不同密度对其树高、胸径和单株材积有明显影响。长方形配置的树高、胸径、冠幅和单株材积各处理间均存在极显著差异,区组间差异不显著。表明长方形配置的思茅松,不同密度对其树高、胸径和单株材积有明显影响。

综上,不同配置方式的思茅松人工林,其树高、胸径和单株材积均受到不同密度的明显影响,不同立地条件也有一定影响,但没有不同密度间影响显著。

2.2 3种配置方式不同密度林分生长量的多重比较

12年生思茅松在3种配置方式下不同密度的多重比较结果见表2。

表2 不同配置方式下不同密度林分生长量多重比较分析
Table 2 Multiple comparison on growth increment of different planting densities under assembled allocation

处理号	密度(株·hm^{-2})	树高(m)	胸径(cm)	冠幅(cm)	材积(m^3)
A1	1 429	12.065aA	14.616aA	3.466aA	0.1107aA
A2	1 667	11.594aA	12.204bB	2.939bB	0.0760bB
A3	1 667	11.449bB	11.894bB	2.951bB	0.0724bB
A4	3 334	11.195bB	11.821bB	2.653cC	0.0695bB
A5	3 334	11.268bB	10.996cC	2.632cC	0.0611cC
A6	5 000	10.683cC	10.346cC	2.411dD	0.0529dD
B1	2 500	10.577aA	11.575aA	2.667aA	0.0624Aa
B2	4 445	10.035bB	10.148bB	2.317bB	0.0466bB
B3	5 000	10.017bB	10.108cC	2.342cC	0.0477cC
C1	1 667	10.943aA	13.330aA	3.241aA	0.0842aA
C2	5 000	10.017bB	10.108bB	2.342bB	0.0477bB

从表2看出:组合式配置6种不同密度的保存率、树高、胸径、冠幅和单株材积长势最好的均是A1处理,长势最差的均是A6处理。保存率、树高、胸径、冠幅和单株材积的生长均与密度成反比。正方形配置的保存率、树高和胸径最好的处理均是B1,最差的均是B3,保存率、树高和胸径均与密度成反比,冠幅和单株材积与密度不成比例,冠幅和单株材积与密度不成比例,有可能是种植时间短,密度效应还没有完全体现出来,有待于进一步的观测研究。长方形配置的树高、胸径、冠幅和单株材积长势最好的均是C1处理,最差的均是C2处理。

组合式配置保存率、树高、胸径、冠幅和单株材积均随密度的增加而降低,正方形配置保存率、树高和胸径均随着密度的增大而递减,这种变化规律在林分发展过程中的不同阶段有明显差异。在幼林郁闭前,平均胸径、平均冠幅和平均材积随密度的变化规律不明显。林分进入郁闭后,林木之间开始发生激烈的竞争。随着林龄的增加,平均胸径随着密度的增加而减小的现象愈加明显。

2.3 3种配置方式下不同密度林分保存率和活立木蓄积量的差异

3种配置方式下不同密度林分保存率和活立木蓄积量的差异见图1。

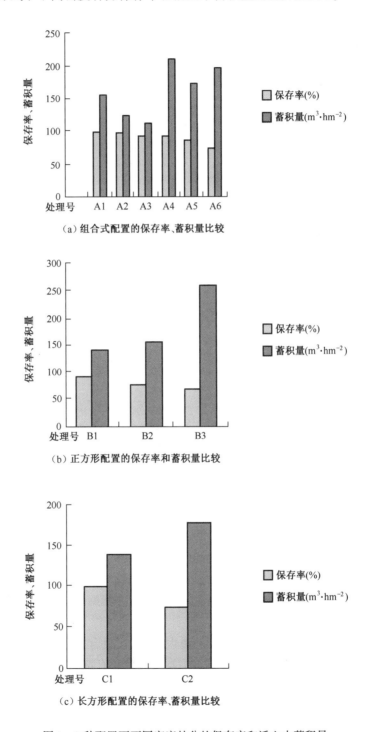

图1 3种配置下不同密度林分的保存率和活立木蓄积量

Figure1 Reserving rate and volume of different planting densities undera ssembled allocation

从图1中可看出：组合式配置下，林分的保存率随着密度的增加而变小，保存率与密度成反比，活立木蓄积量与密度不成比例，活立木蓄积量最好的是A6处理，最差的是A3处理，选择A6处理的优点是林分郁闭快，可减少抚育管理费用，单位面积内蓄积量大，缺点是造林投资大，保存率低，单株蓄积量小，不利于培养大径材。保存率最好的处理是A1处理，最差的是A6处理[图1(a)]。活立木蓄积量与密度不成比例，有可能是造林时间短，密度效应还没有完全体现出来。还有待于进一步的观测研究。

正方形配置下，活立木蓄积量最好的是B3处理，最差的是B1处理。思茅松活立木蓄积与密度成正比；保存率与密度成反比[图1(b)]。

长方形配置下，保存率最好的处理是C1处理，最差的处理是C2处理；活立木蓄积量最好的处理是C2处理，最差的处理是C1处理[图1(c)]。因此，种植时需根据培育目的，选择不同的种植密度。

3 讨 论

（1）组合式配置、正方形配置和长方形配置的不同密度树高处理间均存在极显著差异，密度均与树高生长均成反比。密度对树高的影响比较复杂，结论也不一[7~10]。刘杏娥等认为初植密度对小黑杨树高影响不大[11]。郑海水[12]等的研究表明密度对西南桦树高生长有影响但不显著。

（2）思茅松不同密度对胸径具有显著影响，组合式配置、正方形配置的胸径与密度均成反比。揭建林[10]等报道，胸径随着密度的增加而减小，王凌晖[13]等报道，随着密度的增大，胸径表现出与密度间的负相关，费本华等报道随着密度的减小，胸径明显增大[14]。本研究结果与黄旺志，王凌晖等人的研究结果一致。

（3）思茅松不同密度对冠幅具有显著影响，组合式配置的冠幅与密度成反比，正方形配置的冠幅与密度不成比例。黄旺志[15]等报道，冠幅随密度的增加而减小，本研究结果的组合式配置与黄旺志等人的研究结果一致，正方形配置与黄旺志的研究结果不一致，有可能是正方形配置的密度效应还没有体现出来，有待于进一步的观测研究。

（4）思茅松不同密度对单株材积具有显著影响，组合式配置的单株材积与密度成反比，正方形配置的单株材积与密度不成比例。幼林时期造林密度对单株材积的作用还不太明显，但随着林分年龄的增加，林木之间生存竞争日趋激烈[16]，密度对单株材积生长的影响逐渐增加。温佐吾[17]的试验结果表明：单株材积随密度的增大而减小，不同处理间的差异从6~8年起达到显著或极显著差异。本文的组合式配置的研究结论与温佐吾的研究结果相符，正方形配置的研究结论与其不相符，有可能是正方形配置的密度效应还没有体现出来，随着树龄的增加需进一步观测研究。

（5）思茅松的活立木蓄积量随着密度的增大而增大，组合式配置的活立木蓄积量与密度不成比例，这只是造林前期的现象，对造林后期的现象还有待于观测研究。正方形配置的活立木蓄积量与密度成反比。但密度大的保存率低，种植成本高。密度小的虽蓄积量小，但材质好，保存率高，种植成本低。种植时需根据木材需要选择不同的种植密度，如果种做纸浆材，可选密度大的，如果种做大径材需选密度小的，本研究结果只能代表12年生思茅松的活立木蓄积量，对于思茅松幼树和大径材的活立木蓄积量还有待于进一步的调查研究。

参考文献：

[1] 云南省林业科学研究所. 云南主要树种造林技术[M]. 昆明：云南人民出版社, 1985: 13~16.
[2] 王达明, 李莲芳. 思茅松速生丰产林培育的关键技术[J]. 云南林业科技, 1999, 28(4): 6.
[3] 李思广, 付玉嫔, 张快富, 等. 高产脂思茅松半同胞子代测定[J]. 浙江林学院学报, 2008, 25(2): 158~162.
[4] 徐永椿, 毛品一, 伍聚奎, 等. 云南树木图志: 上册[M]. 昆明：云南科学技术出版社, 1988: 90.
[5] 吴中伦. 中国森林: 第二卷. 针叶林[M]. 北京：中国林业出版社, 1999: 983.
[6] 吴兆录. 思茅松研究现状的探讨[J]. 林业科学, 1994, 30(2): 151~157.
[7] 沈国舫, 罗菊春, 翟明普, 等. 森林培育学[M]. 北京：中国林业出版社, 2001.
[8] 林开敏, 俞新妥, 邱尔发, 等. 杉木造林密度生长效应规律的研究[J]. 福建林学院学报, 1996, 16(1): 53~56.
[9] 黄宝灵, 蒙钰钗, 张连芬. 不同造林密度对尾叶桉生长及产量的影响[J]. 广西科学, 1997, 4(3): 202~207.
[10] 揭建林, 骆昱春, 龙蔚, 等. 南岭山地杉木人工林密度试验研究[J]. 江西农业大学学报, 2007, 29(2): 203~208.
[11] 刘杏娥, 王小青, 江泽慧, 等. 初植密度对小黑杨人工林生长和材质的影响以及材质评价模型的建立[J]. 北京林业大学学报, 2007, 29(6): 161~166.
[12] 郑海水, 黎明, 汪炳根, 等. 西南桦造林密度与林木生长的关系[J]. 林业科学研究, 2003, 16(1): 81~86.
[13] 王凌晖, 吴国欣, 施福军, 等. 不同造林密度对杂交相思生长的影响[J]. 南京林业大学学报(自然科学版), 2009, 33(2): 134~136.
[14] 费本华, 王小青, 刘杏娥, 等. 栽植密度对小黑杨人工幼林材和成熟材生长量的影响[J]. 南京林业大学学报(自然科学版), 2007, 31(5): 44~48.
[15] 黄旺志, 赵剑平, 王昌薇, 等. 不同造林密度对杉木生长的影响[J]. 林业科技, 2005, 32(2): 14~16.
[16] 郭小军. 巨桉纸浆原料林造林密度效应研究[D]. 成都：四川农业大学, 2005.
[17] 温佐吾, 谢双喜, 周运超. 造林密度对马尾松林分生长、木材造纸特性及经济效益的影响[J]. 林业科学, 2000, 36(1): 36~43.

(本文发表于《南京林业大学学报》, 2015年)

Similar Responses in Morphology, Growth, Biomass Allocation, and Photosynthesis in Invasive Wedelia Trilobata and Native congeners to CO_2 enrichment

HE Li-ping[1], KONG Ji-jun[1], LI Gui-xiang[1], MENG Guang-tao[1], CHEN Ke[2]*

([1] Yunnan Academy of Forestry, Kunming 650201, China; [2] School of Life Science and Engineering, Southwest University of Science and Technology, Mianyang 621010, China)

Abstract: Both global change and biological invasions threaten biodiversity worldwide. However, their interactions and related mechanisms are still not well elucidated. To elucidate potential traits contributing to invasiveness and whether ongoing increase in CO_2 aggravates invasions, noxious invasive *Wedelia trilobata*, native *W. urticifolia* and *W. chinensis* were compared under ambient and doubled atmospheric CO_2 concentrations in terms of growth, biomass allocation, morphology, and physiology. The invader had consistently higher leaf mass fraction (LMF) and specific leaf area (SLA) than the natives, contributing to a higher leaf area ratio, and therefore to faster growth and invasiveness. The higher LMF of the invader was due to lower root mass fraction (RMF) and higher fine root percent. Meanwhile, the invader allocated a higher fraction of leaf nitrogen (N) to photosynthetic apparatus, which was associated with its higher photosynthetic rate, and resource use efficiency. All these traits collectively contributed to its invasiveness. CO_2 enrichment increased growth of all studied species by increasing actual photosynthesis, although it decreased photosynthetic capacities due to decreased leaf and photosynthetic N contents. Responses of the invasive and native plants to elevated CO_2 were not significantly different, indicating that the ongoing increase in CO_2 may not aggravate biological invasions, inconsistent with the prevailing results in references. Therefore, more comparative studies of related invasive and native plants are needed to elucidate whether CO_2 enrichment facilitates invasions.

Keywords: CO_2 enrichment; Growth; Invasiveness; Morphology; Nitrogen and biomass allocation

Abbreviations AC, ambient atmospheric CO_2 concentration; A_{growth}, actual photosynthetic rate measured at growth CO_2 concentration; EC, doubled atmospheric CO_2 concentration; FRP, fine root percent; G_s, stomatal conductance; LA : RM, the ratio of leaf area to root mass; LA : FRM, the ratio of leaf area to fine root mass; LAR, leaf area ratio; LMF, leaf mass fraction; $N_{bioenerg}$, N_{carbox}, N_{LHC}, and $N_{photosynth}$, nitrogen contents in bioenergetics, carboxylation, light-harvesting components, and all components of the photosynthetic apparatus, respectively; $N_{bioenerg}/N_L$, N_{carbox}/N_L, N_{LHC}/N_L, and $N_{photosynth}/N_L$, the fractions of leaf nitrogen allocated to bioenergetics, carboxylation, light-harvesting components, and all components of the photosynthetic apparatus, respectively; N_L, total leaf nitrogen content; P_{max}, light-saturated photosynthetic rate; PNUE, photosynthetic nitrogen-use efficiency; RGR, relative growth rate; RMF, root mass fraction; SLA, specific leaf area; SMF, support mass fraction; WUE, water-use efficiency.

Introduction

Global mean CO_2 concentration has increased from 290 to 375 $\mu mol \cdot mol^{-1}$ during the last

100 years and is conservatively projected to be doubled by the end of 21st century, strongly dependent on future scenarios of anthropogenic emissions (Nagel et al., 2005). The ongoing increase in atmospheric CO_2 may cause changes in species composition of ecosystems, either by altering global climate (Chapin et al., 1995) or more directly, by favoring certain photosynthetic pathways (Arp et al., 1993) or changing competition dynamics within ecosystems (Owensby et al., 1999). Invasive species, that may exploit the new environmental conditions caused by global change such as CO_2 enrichment, may gain footholds in previously inhospitable ecosystems, changing species composition, and biological invasions have become a serious environmental and socioeconomic problem and hot topic of ecological research worldwide (Dukes and Mooney, 1999). However, the interactions between biological invasions and global change (Bond and Midgley, 2000; Rogers et al., 2008) and the mechanisms underlying invasiveness are still not well elucidated (Daehler, 2003; Feng et al., 2009). Identifying the factors that contribute to success of invasive alien plants is important for predicting and controlling potentially invasive plants.

It has been found that some successful invasive plants have higher light-saturated photosynthetic rate (P_{max}), specific leaf area (SLA), and leaf area ratio (LAR) than native plants (Nagel and Griffin 2004; Zou et al., 2007). Pattison et al., (1998) and Zheng et al., (2009) found that P_{max} is positively correlated with relative growth rate (RGR) in some invasive plants. Higher RGR may confer competitive advantages on invasive species, facilitating invasions (Zheng et al., 2009). LAR, the product of SLA and leaf mass fraction (LMF), is the most important determinant of RGR especially at low irradiance (Feng et al., 2009). High SLA may also contribute to invasiveness of alien plants by decreasing leaf construction cost and increasing nitrogen (N) allocation to photosynthesis and photosynthetic N-use efficiency (PNUE) (Feng et al., 2009). Higher LMF and lower root mass fraction (RMF) were indeed found in some successful invasive plants in comparison with native plants (Wilsey and Polley, 2006). This pattern of biomass allocation may promote irradiance capture but impair water and nutrient absorptions, suggesting that biological invasions are environment-dependent (Zheng et al., 2009). It is well known that increased availabilities of resources such as irradiance, nutrients and water often facilitate alien plant invasions (Daehler, 2003; Zheng et al., 2009).

CO_2 is necessary for photosynthesis, and increased atmospheric CO_2 supply generally increases photosynthetic rate and plant growth (Long et al., 2004). However, the effects of elevated CO_2 are significantly different among plant species and functional groups (Ainsworth et al., 2007). Many studies found that growth of invasive plants is more strongly stimulated by elevated CO_2 than growth of native plants (Raizada et al., 2009; Song et al., 2009). For example, doubled atmospheric CO_2 concentration increases biomass accumulation by 56% in invasive Rhododendron ponticum versus 12% in understorey native plants (Hättenschwiler and Körner, 2003). Furthermore, the intrinsically broader environmental tolerance, higher growth rate and phenotypic plasticity, characteristic of many invasive plant species (Jia et al., 2016) may enable them to respond more positively to environmental changes that result in increased resource availability (elevated levels of water supply, atmospheric CO_2 concentrations and N deposition) than native plants adapted to low resource conditions (Nackley et al., 2017; Zhang et al., 2017). The different responses of C_3 and C_4 plants to elevated CO_2 have been suggested as a potential explanation for invasions of native C_4 grasslands by woody C_3 plants in North America (Bond and Midgley, 2000). However, the mechanisms by which these C_3 plants spread at the expense of existing native C_4 plants are poorly understood, and relatively few studies have compared the differences in response to elevated CO_2 between invasive and native plants, especially the differences between phylogenetically related invasive and native plants.

Wedelia trilobata (L.) Hitchc. [syn. *Sphagneticola trilobata* (L.) Pruski] (creeping oxeye), native to the tropics of South America (Qi et al., 2014), is a perennial evergreen creeping clonal herb. It has been listed as one of the 100 world's worst invasive alien species (IUCN 2001). This noxious weed was introduced to South China on a large scale as a common groundcover plant in the 1970s, but it rapidly spread to the field (Li and Xie, 2003). Fast dispersal through vegetative propagation (clonal growth) is one of the pivotal factors for the successful invasion of *W. trilobata*. Once established in plantations, *W. trilobata* can overgrow into a dense groundcover and prevent the regeneration of other species, including some native congeners, which are typically used as important traditional Chinese medicines (Song et al., 2010). In our study, *W. trilobata* was compared with two sympatric native congeners, *W. urticifolia* DC. and *W. chinense* L. under ambient and doubled atmospheric CO_2 concentrations. The main aims of this study were to explore: (1) the traits contributing to invasiveness of the invader; (2) how the studied plants acclimate to CO_2 enrichment in terms of growth, biomass allocation, morphology, and photosynthesis; (3) whether CO_2 enrichment aggravates invasion of the invader and related mechanisms.

Materials and methods

Plant materials and treatments

Seeds of each studied species were collected from a minimum of 15 individuals distributing around Kunming (25°06′ N, 102°50′ E, 2 200 m a.s.l.), Yunnan Province, southwest China and mixed. The seeds were germinated on a seedbed in a greenhouse in March 2013 with average air temperature of 25℃ and relative humidity of 42% during the experimental period. In May 2013, when the seedlings were approximately 10cm tall, similar-sized individuals were singly transplanted into 5 L pottery pots filled with 4 kg homogenized forest topsoil. After one month growth at an open site, 40 similar-sized seedlings per species were selected and randomly divided into two groups. Each group was moved into closed-topchambers (E-sheng Tech. Co., Beijing, China) located outdoors at Ailaoshan Station for Subtropical Forest Ecosystem Studies (24°32′ N, 101°01′ E, 2 490 m a.s.l.), Jingdong County, southwest China. Detailed information on the chambers can be found in our previous study (Meng et al., 2013). Seedlings of each species in each chamber were randomly divided into five groups, four seedlings per group. One group of each species was put together and the 12 seedlings of the three studied species were randomly arranged and watered when necessary. No fertilizer was added during the experiment.

One chamber was supplied with compressed CO_2 gas to obtain a doubled atmospheric CO_2 concentration treatment (EC), and another chamber was used as control (AC, $320 \mu mol \cdot mol^{-1}$ CO_2). CO_2 concentration in EC chamber was controlled automatically with a computer-controlled CO_2 supply system. CO_2 concentration and temperature in each chamber were recorded at a 15 s interval. Hourly mean CO_2 concentrations were 280~340 and 590~670 $\mu mol \cdot mol^{-1}$ in AC and EC chambers, respectively. There was no significant difference in temperature between chambers. Three months after CO_2 treatments, measurements were taken on five individuals per species per treatment.

Photosynthesis measurements

Under saturating photosynthetic photon flux density (PPFD, $2000 \mu mol \cdot m^{-2} \cdot s^{-1}$), photosynthesis was measured on the youngest fully expanded leaf of each sample plant using a Li-6400 Portable Photosynthesis System (Li-Cor, Lincoln, NE). Relative humidity of the air in the leaf chamber was controlled at ≈ 70% and leaf temperature at 25 ℃. Actual photosynthetic rates (A_{growth}) under growth ambient atmospheric CO_2 concentrations were measured at 320 and $640 \mu mol \cdot mol^{-1}$ CO_2 in the reference chamber for plants grown under AC and EC, respectively. For determining photosynthetic responses to intercellular CO_2 concentration, gas exchanges were measured at 380, 300, 260, 220, 180, 140, 110, 80, and $50 \mu mol \cdot mol^{-1}$ CO_2 in the reference chamber. P_{max} and stomatal conductance (G_s) were the values measured at $380 \mu mol \cdot mol^{-1}$ CO_2 and saturating PPFD. Afterwards, light- and CO_2-saturated photosynthetic rate was measured after 500 s under saturating PPFD and $1500 \mu mol \cdot mol^{-1}$ CO_2. Before measurement, each sample leaf was illuminated with saturating PPFD provided by the LED light source of the equipment for 10~30 min to achieve full photosynthetic induction. No photoinhibition occurred during the measurements.

Two 10mm diameter leaf discs were taken from each sample leaf, oven-dried at 60 ℃ for 48 h. SLA was calculated as the ratio of leaf area to mass. Leaf N content (N_L) was determined with a Vario MAX CN Element Analyzer (Elementar Analysensysteme GmbH, Hanau, Germany). Leaf chlorophyll content was measured following the method of Lichtenthaler and Wellburn (1983). Water-use efficiency and PNUE were calculated as the ratios of P_{max} to G_s and N_L, respectively.

Calculations of P_n-C_i curve-related variables

The P_n-C_i curve was fitted with a linear equation ($P_n = kC_i + i$) within 50~200 $\mu mol \cdot mol^{-1}$ C_i. Maximum carboxylation rate (V_{cmax}), dark respiration rate (R_d) and maximum electron transport rate (J_{max}) were calculated according to Feng et al., (2009) and Zheng et al., (2009) as follows:

$$V_{cmax} = k[C_i + K_c(1 + O/K_o)]^2 / [\Gamma^* + K_c(1 + O/K_o)] \tag{1}$$

$$R_d = V_{cmax}(C_i - \Gamma^*) / [C_i + K_c(1 + O/K_c)] - (kC_i + i) \qquad (2)$$

$$J_{max} = [4(P'_{max} + R_d)(C_i + 2\Gamma^*)] / (C_i - \Gamma^*) \qquad (3)$$

Where K_c and K_o were the Michaelis–Menten constants of Rubisco for carboxylation and oxidation, respectively; Γ^* was CO_2 compensation point; O was the intercellular oxygen concentration, close to 210mmol · mol^{-1}.

The fractions of total leaf N allocated to carboxylation (P_C, g · g^{-1}) and bioenergetics (P_B, g · g^{-1}) of the photosynthetic apparatus were calculated as:

$$P_C = V_{cmax} / (6.25 \times V_{cr} \times N_A) \qquad (4)$$

$$P_B = J_{max} / (8.06 \times J_{mc} \times N_A) \qquad (5)$$

$$P_L = C_C / (N_M \times C_B) \qquad (6)$$

Where V_{cr} and J_{mc} were 20.78μmol CO_2 · g^{-1} Rubisco S^{-1} and 155.65μmol electrons · μmol^{-1} cyt f · s^{-1}, respectively. C_B was 2.15mmol · g^{-1}. 6.25 (g Rubisco g^{-1} nitrogen in Rubisco) was the conversion coefficient between nitrogen content and protein content in Rubisco, and 8.06 (μmol cyt f · g^{-1} nitrogen in bioenergetics) was the conversion coefficient between cyt f and nitrogen in bioenergetics. Nitrogen contents in carboxylation (N_C) and bioenergetics (N_B) were calculated as the products of N_A and P_C, P_B, respectively.

Growth measurements

Five seedlings per species per treatment were harvested after measurements of height and ramet (> 5cm branches originating from root collar) number. All samples were separated into leaves, support organs (including stems, branches and petioles), and fine (diameter < 1mm) and coarse roots. Then total leaf area was determined using Li-3000C leaf area meter (Li-Cor, Lincoln, NE). Finally, all the organs were oven-dried at 60°C for 48 h, and weighed. Support mass fraction (SMF, support organ mass/total mass), LMF (leaf mass/total mass), RMF (root mass/total mass), fine root percent (FRP, fine root mass/total root mass × 100), leaf area to root mass ratio (LA : RM, total leaf area/total root mass), leaf area to fine root mass ratio (LA : FRM, total leaf area/fine root mass), LAR (total leaf area/total mass), and RGR (mass increase per unit mass per unit time) were calculated according to Poorter and Remkes (1990).

Statistical analyses

Effects of species, treatment and their interactions on variables measured in this study were tested using two-way ANOVA. Differences among species grown at both CO_2 treatments were tested using one-way ANOVA. Difference between CO_2 treatments in correlation between each pair of variables was tested using one-way ANCOVA. Treatment (AC vs. EC) was used as a fixed factor; variables indicated by y- and x-axes in each figure as dependent variable and covariate, respectively. If the difference was significant, we then tested for significances of the correlations (Pearson correlation, two-tailed) for CO_2 treatments separately; otherwise, we pooled data from both treatments to test for significance of the correlation. All analyses were carried out using SPSS

13.0 (SPSS Inc., Chicago, Il). Principal component analysis (PCA) of eco-physiological traits was used to identify the most discriminatory effects of elevated temperature and drought. PCA analyses were performed using Canoco 5.0 (Microcomputer Power, USA).

Results

Morphology, growth, and biomass allocation

Invasive *W. trilobata* was significantly higher in biomass, RGR, total leaf area, LAR, SLA, LMF, FRP, and LA:RM than native *W. urticifolia* and *W. chinense* (Table 1). The invader was also higher in height than the natives, although the difference was not statistically significant. In contrast, the invader was lower in RMF and LA:FRM (significant only for *W. urticifolia* under EC) than the natives. The invader showed 4.00 ± 0.32 and 4.60 ± 0.68 ramets under AC and EC, respectively, while the natives had no ramets (data not shown).

Table 1 Differences in morphology, growth, and biomass allocation among invasive *Wedelia trilobata*, and native *W. urticifolia* and *W. chinense* grown under ambient (AC) and doubled (EC) atmospheric CO_2 concentrations

Variables	*W. trilobata*		*W. urticifolia*		*W. chinense*		F-values	
	AC	EC	AC	EC	AC	EC	Species	Treatment
Height (cm)	43.8±2.20ab	47.8±2.65a	41.6±1.94ab	46.2±2.62ab	39.6±2.21b	43.4±2.21ab	1.730	4.775*
Total biomass (g)	13.7±0.53b	17.6±0.76a	5.4±0.46d	6.8±0.38c	4.2±0.28d	5.2±0.18d	317.537***	29.924***
Relative growth rate ($mg \cdot g^{-1} \cdot d^{-1}$)	34.7±0.43b	37.5±0.49a	24.3±0.98d	26.9±0.65c	21.6±0.74e	23.9±0.39d	236.185***	24.354***
Total leaf area (m^2)	1.25±0.11b	1.49±0.14a	0.23±0.02c	0.27±0.02c	0.19±0.01c	0.22±0.01c	164.704***	2.920
Leaf area ratio ($cm^2 \cdot g^{-1}$)	91.2±6.53a	84.0±5.72a	44.3±4.64b	40.1±2.53b	44.9±4.03b	41.4±1.93b	65.621***	1.814
Specific leaf area ($cm^2 \cdot g^{-1}$)	215±16.3a	191±7.8b	162±15.2cd	149±16.4d	179±27.1bc	163±13.2cd	19.212***	8.114**
Leaf mass fraction	39.6±0.67a	38.9±0.67a	27.8±0.81b	28.1±1.27b	26.3±0.63b	25.8±1.72b	93.767***	0.097
Support mass fraction	34.1±0.23b	31.6±0.42b	35.1±1.59ab	33.5±1.63ab	36.9±1.07a	35.4±1.84ab	3.365	3.230
Root mass fraction	26.3±0.61b	29.4±0.54b	37.1±1.64a	38.3±1.57a	36.7±1.17a	38.8±0.60a	51.375***	5.540*
Fine root percent (%)	17.5±1.04b	21.1±0.45a	6.9±0.53c	6.9±0.41c	7.9±0.63c	7.8±0.16c	263.360***	5.624*
Leaf area root mass ratio ($cm^2 \cdot g^{-1}$)	346±22.1a	285±17.9b	121±14.4c	104±3.9c	122±8.0c	107±4.9c	145.979***	7.565*
Leaf area fine root mass ratio ($cm^2 \cdot g^{-1}$)	530±54.4ab	400±33.4b	649±65.8a	584±37.9a	572±52.6a	526±19.5ab	5.328*	4.450*

Interactions between species and CO_2 treatment were not significant for all variables except total biomass ($F = 5.896**$) and fine root percent ($F = 6.262**$). Means ± SE ($n = 5$). Different letters indicate significant differences among species and treatments ($P < 0.05$). *: $P < 0.05$; **: $P < 0.01$; ***: $P < 0.001$.

Although CO_2 enrichment significantly increased RGR in all studied species, there is no influences on other morphological and growth traits including height, TLA, LAR, SLA, LMF, SMF, RMF, FRP, LA:RM and LA:FRM of *W. urticifolia* and *W. Chinese* except for the increased biomass of *W. urticifolia* (Table 1). However, In *W. trilobata*, elevated CO_2 significantly decreased SLA, LA:RM and increased biomass, RGR, TLA, FRP and LA:RM.

The correlations between LAR and LMF, SLA, and RGR were significant (Figure 1). RGR increased significantly with increasing LMF, SLA, FRP, and total leaf area (Figure 2a–d). At given values of LMF, SLA, and LAR, plants grown under EC had higher RGR than plants grown under AC.

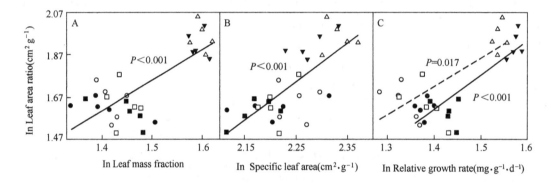

Figure 1　Correlations between leaf area ratio (LAR) and leaf mass fraction (LMF, a), specific leaf area (SLA, b), and relative growth rate (RGR, c) in invasive *Wedelia trilobata* (triangles), and native *W. urticifolia* (squares) and *W. chinense* (circles) grown under ambient (open symbols) and doubled (closed symbols) atmospheric CO_2 concentrations. Data were transformed into natural logarithms. Lines fitted for ambient (dashed line) and doubled (solid line) atmospheric CO_2 concentration treatments were given, respectively, if the difference between treatments was significant according to the result of ANCOVA. Otherwise, only one line fitted for pooled data was given.

Photosynthesis

Invasive *W. trilobata* was significantly higher in A_{growth}, G_s, P_{max}, PNUE, WUE (not significant for *W. urticifolia*), $N_{photosynth}/N_L$, N_{carbox}/N_L, $N_{bioenerg}/N_L$, $N_{photosynth}$, N_{carbox}, and $N_{bioenerg}$ than native *W. urticifolia* and *W. chinense* (Table 2). N_L and N_{LHC}/N_L were not significantly different between the invasive and native species. The invader was higher in N_{LHC} than the natives under AC but not under EC.

CO_2 enrichment significantly increased A_{growth} in *W. trilobata* and *W. urticifolia* (Table 2). In *W. trilobata*, CO_2 enrichment significantly increased WUE. In contrast, CO_2 enrichment significantly decreased G_s, N_L, $N_{photosynth}$, N_{carbox} (not significant for *W. urticifolia*), and N_{LHC}. P_{max} in *W. urticifolia* and $N_{bioenerg}$ in *W. chinense* were significantly decreased by CO_2

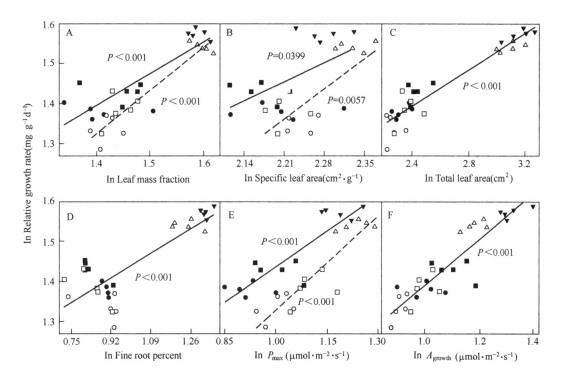

Figure 2   Relative growth rate (RGR) as a function of leaf mass fraction (LMF, a), specific leaf area (SLA, b), total leaf area (c), fine root percent (FRP, d), light-saturated photosynthetic rate (P_{max}, e), and photosynthetic rate measured at growth ambient CO_2 concentration (A_{growth}, f) in three species and two CO_2 concentrations treatments. Species and treatment codes, as well as statistical analyses were shown as in Figure 1.

enrichment. In contrast, CO_2 enrichment did not influence P_{max} and $N_{bioenerg}$ in invasive *W. trilobata*. The effects of CO_2 enrichment on PNUE, $N_{photosynth}/N_L$, N_{carbox}/N_L, $N_{bioenerg}N_L$, and N_{LHC}/N_L were not significant.

Table 2   Differences in physiological traits among invasive *Wedelia trilobata* and native *W. urticifolia*, and *W. chinense* grown under ambient (AC) and doubled (EC) atmospheric CO_2 concentrations

Variables	W. trilobata		W. urticifolia		W. chinense		F-values	
	AC	EC	AC	EC	AC	EC	Species	Treatment
A_{growth}	15.3±0.60b	21.0±1.06a	10.0±0.44d	12.9±0.88c	8.2±0.21d	10.0±0.66d	90.707***	37.910***
P_{max}	17.3±0.82a	15.6±0.83a	12.8±0.75b	10.3±0.66c	10.1±0.46c	8.4±0.45c	58.350***	12.428**
G_s	0.53±0.02a	0.40±0.01bc	0.49±0.03b	0.37±0.02d	0.43±0.02c	0.33±0.02d	50.790***	107.31***
PNUE	14.6±1.32a	16.4±2.09a	10.9±1.27bc	11.7±2.38b	8.7±0.26d	9.2±0.55cd	48.177***	2.405
WUE	32.5±3.92b	39.0±4.55a	29.6±1.84bc	32.2±4.06b	26.1±1.99c	27.2±2.02c	20.487***	8.479**
N_L	1.20±0.09a	0.96±0.07bc	1.18±0.06a	0.92±0.06c	1.16±0.06ab	0.93±0.06c	0.146	17.461***
$N_{photosynth}/N_L$	0.51±0.035a	0.49±0.017a	0.37±0.018b	0.37±0.018b	0.35±0.014b	0.34±0.013b	44.076***	0.813
N_{carbox}/N_L	0.33±0.018a	0.31±0.017a	0.23±0.031b	0.23±0.038b	0.21±0.006b	0.19±0.004b	44.051***	0.959
$N_{bioenerg}/N_L$	0.07±0.003a	0.07±0.004a	0.05±0.005bc	0.05±0.003b	0.04±0.003c	0.04±0.003bc	56.652***	3.121

(续)

Variables	W. trilobata		W. urticifolia		W. chinense		F-values	
	AC	EC	AC	EC	AC	EC	Species	Treatment
N_{LHC}/N_L	0.11±0.005a	0.10±0.005a	0.09±0.004a	0.091±0.007a	0.10±0.009a	0.10±0.011a	1.915	1.039
$N_{photosynth}$	0.62±0.12a	0.45±0.05b	0.43±0.04b	0.32±0.02cd	0.40±0.02bc	0.31±0.01d	30.692***	29.601***
N_{carbox}	0.39±0.05a	0.30±0.06b	0.26±0.04bc	0.21±0.04cd	0.24±0.02c	0.18±0.01d	27.347***	18.97***
$N_{bioenerg}$	0.08±0.01a	0.07±0.01a	0.05±0.01b	0.05±0.01b	0.05±0.01b	0.04±0.01c	54.666***	11.659**
N_{LHC}	0.13±0.02a	0.09±0.01cd	0.11±0.01bc	0.08±0.01d	0.12±0.02b	0.09±0.01cd	4.627*	38.491***

Interactions between species and CO_2 treatment were not significant for all variables except A_{growth} ($F = 4.093*$). Statistical analyses and replications were shown as in Table 1.

RGR increased with increasing P_{max} and A_{growth} (Figure 2 e, f). At a given value of P_{max}, plants grown under EC had higher RGR than plants grown under AC. The correlations between $N_{photosynth}/N_L$ and SLA, P_{max}, A_{growth}, and RGR were significant (Figure 3). At a given value of $N_{photosynth}/N_L$, plants grown under EC had higher A_{growth} and RGR but lower P_{max} than plants grown under AC.

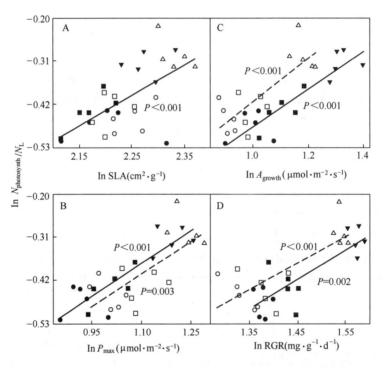

Figure 3 Correlations between fractions of leaf nitrogen in photosynthetic apparatus ($N_{photosynth}/N_L$) and specific leaf area (SLA, a), light-saturated photosynthetic rate (P_{max}, b), photosynthetic rate measured at growth ambient CO_2 concentration (A_{growth}, c), and relative growth rate (RGR, d) in three species and two CO_2 concentrations treatments. Species and treatment codes, as well as statistical analyses were shown as in Figure 1.

According to the PCA, native and invasive species were separated along the first axis of the PCA, which were strongly correlated with biomass accumulation, leaf area, fine root ratio, leaf area ratio, and accounted for 69.65% of the observed variance; meanwhile, CO_2 treatment showed modest differentiation (Figure 4).

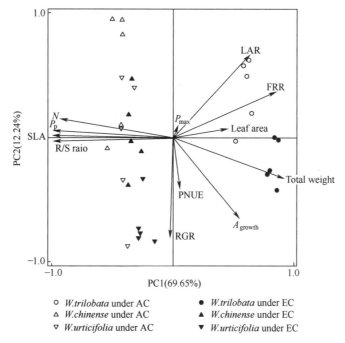

Figure 4 Principal component analysis (PCA) based on eco-physiological traits in invasive *Wedelia trilobata* (circles), and native *W. urticifolia* (triangles) and *W. Chinese* (reverse triangles) grown under ambient (open symbols) and doubled (closed symbols) at atmospheric CO_2 concentrations.

Discussion

Traits contributing to invasiveness

Higher RGR and biomass accumulation of *W. trilobata* in comparison with native congeners may contribute to invasiveness. High RGR can facilitate capture of available resources (Grotkopp and Rejmánek, 2007), which is important for alien plant invasions (Davis et al., 2000). Higher RGR of the invader contributed to higher total leaf area and ramet number, and therefore to invasiveness (Table 1). Both higher LMF and SLA of *W. trilobata* contributed to higher LAR, one the determinants of RGR (Poorter and Remkes, 1990; Zheng et al., 2009), and therefore to higher RGR (Fig. 1, 4). Positive correlations between RGR and LMF, SLA, and total leaf area were indeed found (Fig. 2 a–c). SLA is an important determinant of RGR (Poorter and Remkes, 1990); Daehler (2003) found through reviewing published references that invasive plants have significantly higher SLA and total leaf area than co-occurring natives. Positive correlation between RGR and LMF was also found by Poorter and Remkes (1990).

The higher LMF of *W. trilobata* was due to lower RMF as SMF was not significantly different between the invader and natives (Table 1). Lower RMF of the invader contributed to higher RGR not only by increasing LMF, but also by decreasing root respiratory carbon loss (D'Antonio et al., 2001; Feng et al., 2009). The invader supported more leaves with fewer roots, as indicated by higher LA : RM (Table 1), which did not influence growth, N_L and photosynthesis, indicating that roots of the invader were more efficient in physiological functions than those of natives. Higher FRP of the invader may explain efficient root functions; LA : FRM was even lower in the invader than in the natives (Table 1). Decreasing root diameter can increase the ratio of surface area to mass, promoting water and nutrient absorptions (Akinnifesi et al., 1998; Bauhus and Messier, 1999). The significantly positive correlation between RGR and FRP confirmed the role of fine roots in invasion success of the invader (Figure 2 d).

The higher SLA of *W. trilobata* contributed to higher RGR not only by increasing LAR, but also by increasing P_{max} which is positively correlated with net assimilation rate, one of the determinants of RGR (Feng et al., 2009). Generally, SLA is negatively correlated with cell wall mass (Onoda et al., 2004). Feng et al., (2009) found that 3.5%-9.3% of leaf N is allocated to cell walls in *Eupatorium adenophorum* which was mediated by SLA, and the proportion of leaf N in cell walls decreases with increasing SLA, leaving more N available for allocation to photosynthesis. Higher SLA of the invader indeed contributed to higher $N_{photosynth}/N_L$, and therefore to higher RGR through higher P_{max} and A_{growth} (Figure 2 e, f; 3). The higher stomatal conductance may also contribute to higher photosynthesis in the invader, while similar N_L of the invader and natives may not (Table 2). The invader had both higher PNUE and WUE, breaking the trade off between them (Feng et al., 2009), which may confer competitive advantages on the invader especially under barren environments. It is a potential novel mechanism underlying alien plant invasions that invasive plants allocate higher fractions of leaf N to photosynthesis than native plants and native conspecifics (Feng et al., 2009).

Effects of CO_2 enrichment on invasiveness

Growth of *W. trilobata* and natives was significantly stimulated by EC treatment (Table 1), consistent with results of many other studies (Ainsworth and Long, 2005; Hättenschwiler and Körner, 2003; Raizada et al., 2009; Smith et al., 2000). LAR and SLA could not be used to explain the increased growth, which showed decrease trends under EC. The increased growth could be attributed to increased A_{growth}, which was caused by increased C_i (Table 2). The increased C_i was mainly caused by the elevated atmospheric CO_2 concentrations. CO_2 enrichment decreased N_L but did not significantly affect N allocation to photosynthesis, leading to decreased N contents in photosynthesis, and therefore to decreased photosynthetic capacity, i.e. P_{max} (Table 2). Reduced stomatal conductance may also contribute to the decreased P_{max} under EC. It has been found that P_{max} is significantly correlated with N content in photosynthesis and stomatal conductance (Feng et al., 2009). Down-regulation of photosynthetic capacity is common under prolonged elevated CO_2 concentration (Ainsworth and Long, 2005; Medlyn et al., 1999), which could be explained by the decreased foliar N concentrations.

Elevated CO_2 tended to increase RMF and decrease SMF, resulting in a reallocation of biomass from support organs to roots (Table 1). The increased allocation to roots under elevated CO_2 may be driven by an increased need for belowground resources such as N to meet the increased demand associated with faster growth and additional carbon sequestration (Chapin et al., 1995), which are highly dependent on availability and cycling of N (Norby et al., 2010). However, the potential increase in N uptake may only support the increased root production and may not help improve N nutrition at the whole plant level (Johnson et al., 2004). This was confirmed by the decreased N_L (Table 2), which may be due to the dilution effect caused by faster growth. Walch-Liu et al., (2001) found that CO_2 enrichment leads to a preferential N partitioning into roots over shoots in tobacco, reducing leaf Rubisco concentration. Moreover, Mc Guire et al., (1995) observed a decrease of 21% in leaf and 9% in root N concentrations under CO_2 enrichment, which was confirmed by our results (Table 1).

The responses of *W. trilobata* and natives to elevated CO_2 were not significantly different, as judged by non-significant interactions between species and CO_2 treatment for RGR and many other variables (Table 1). For example, EC increased RGR in *W. trilobata*, *W. urticifolia*, and *W. chinense* by 8%, 11%, and 11%, respectively. Similar results were also found by Dukes (2002) under competitive conditions but not under noncompetitive conditions. The results suggest that CO_2 enrichment may not exaggerate *W. trilobata* invasion in the future with elevated CO_2. Our results are not consistent with those of many other studies, which found that CO_2 enrichment increases growth more strongly in invasive plants than in natives (Baruch and Jackson, 2005; Hättenschwiler and Körner, 2003; Raizada et al., 2009; Smith et al., 2000; Song et al., 2009). However, almost all these studies compared phylogenetically unrelated invasive and native species. It has been recognized that responses to CO_2 enrichment are species specific (Ainsworth and Long, 2005). The responses are also significantly different between invasive plants (Rogers et al., 2008) and between natives (Ainsworth and Long, 2005). Phylogenetically related plants may share more common traits and more overlapping resource requirements than unrelated plants (Goldberg, 1987). Comparisons between related invasive and native plants may shed more light on invasiveness of alien plants (Feng et al., 2009), and some recent comparative studies indeed control phylogeny (Grotkopp and Rejmánek, 2007; Penuelas et al., 2010).

Recently, Liu et al., (2017) summarized that invasive plants showed a slightly stronger positive response to increased N deposition and precipitation than native plants, but these differences were not statistically significant ($P=0.051$ for N deposition; $P=0.679$ for increased precipitation) through meta-analysis with 74 alien and 117 native species. Furthermore, Liu and Van Kleunen (2017) found that alien plant species produced more biomass only when nutrients were supplied as a single pulse in the middle of growth period instead of supplied at a constant rate, whereas the reverse was true for the native species. The findings was also supported by Godoy et al., (2011), who compared 20 invasive alien and 20 widespread native congeners in Spain across nutrient gradients, and found that both groups responded to environmental variation with similar levels of plasticity.

As for CO_2 enrichment, Hager et al., (2016) found that differences in trait means between invasive and noninvasive species tended to be similar across CO_2 levels, which was well in agreement with our results, as CO_2 enrichment showed modest differentiation (Figure 4). The lack of response to CO_2 may be due to indirect effects of CO_2 on N, for elevated CO_2 can commonly reduce N availability, thus indirectly limit CO_2 effects on invasion (Luo et al., 2004). For example, CO_2 reduced resin-available soil N by 47%, and Bromus tectorum N concentration by 30% (Blumenthal et al., 2016). In our study, significant decreases in foliar N concentrations were observed in both native and invasive species (Table 2). Thus, these studies collectively provide evidence, albeit circumstantial, that CO_2-induced reductions in N can limit CO_2 effects on invasion, and unlikely result in dramatically changes in competitive hierarchy. A more complete model of invasive species responses to CO_2 enrichment will require knowledge of how ecophysiological responses are likely to be mediated by factors such as light, nutrients, competition, and herbivory.

In conclusion, a suite of traits such as consistently higher LMF, SLA, LAR, total leaf area, FRP, N_{carbox}/N_L, $N_{bioenerg}/N_L$, $N_{photosynth}/N_L$, N_{carbox}, $N_{bioenerg}$, $N_{photosynth}$, P_{max}, A_{growth}, and PNUE, and lower RMF contributed to higher RGR and biomass accumulation in W. trilobata in comparison with native W. urticifolia and W. chinense, and therefore to invasiveness. CO_2 enrichment increased growth of all studied plants by increasing actual photosynthesis, which was due to increased CO_2 supply rather than increased photosynthetic capacity. The stimulation effect of elevated CO_2 was similar for the invader and natives, indicating that the ongoing increase in CO_2 may not enhance invasion of the invader. Our results were not consistent with the prevailing results that CO_2 enrichment stimulates growth of invasive plants more strongly than growth of natives. The difference may be associated with the fact that most studies in references compared phylogenetically unrelated invasive and native plants. Therefore, more comparative studies of related invasive and native plants are needed to elucidate whether CO_2 enrichment aggravates invasion success of alien plants. On the other hand, many other factors including light, nutrients, competition, and herbivory should be taken into consideration for a more complete understanding on the comparative responses of invasive and native species to CO_2 enrichment.

Acknowledgement

This study was financially supported by Natural Science Foundation of China (41361076) and the Science and Technology Foundation of Sichuan Province, China (2015JY0015). We are grateful to the anonymous reviewers and the journal editors for their comments that have helped to improve this manuscript.

References

Ainsworth EA, Long SP, 2005. What have we learned from 15 years of free-air CO_2 enrichment (FACE)? A meta-analytic review of the responses of photosynthesis, canopy properties and plant production to rising CO_2. New Phytol 165: 351~371.

Ainsworth EA, Rogers A, Leakey ADB, Heady LE, Gibon Y, Stitt M, Schurr U, 2007. Does elevated atmospheric CO_2 alter diurnal C uptake and the balance of C and N metabolites in growing and fully expanded soybean leaves? J Exp Bot 58: 579~591.

Akinnifesi FK, Kang BT, Ladipo DO, 1998. Structural root form and fine root distribution of some woody species evaluated for agroforestry systems. Agroforest Syst 42: 121~138.

Arp WJ, Drake BG, Pockman WT, Curtis PS, Whigham DF, 1993. Interactions between C_3 and C_4 salt-marsh plant-species during 4 years of exposure to elevated atmospheric CO_2. Vegetatio 104: 133~143.

Baruch Z, Jackson RB, 2005. Responses of tropical native and invader C_4 grasses to water stress, clipping and increased atmospheric CO_2 concentration. Oecologia 145: 522~532.

Bauhus J, Messier C, 1999. Soil exploitation strategies of fine roots in different tree species of the southern boreal forest of eastern Canada. Can J Forest Res 29: 260~273.

Blumenthal DM, Kray JA, Ortmans W, Ziska LH, Pendall E, 2016. Cheatgrass is favored by warming but not CO_2 enrichment in a semi-arid grassland. Global Change Biol 22: 3026~3038.

Bond WJ, Midgley GF, 2000. A proposed CO_2-controlled mechanism of woody plant invasion in grasslands and savannas. Global Change Biol 6: 865~869.

Chapin FS III, Shaver GR, Giblin AE, Nadelhoffer KJ, Laundre JA, 1995. Responses of arctic tundra to experimental and observed changes in climate. Ecology 76: 694~711.

D′Antonio CM, Hughes RF, Vitousek PM, 2001. Factors influencing dynamics of two invasive C_4 grasses in seasonally dry Hawaiian woodlands. Ecology 82: 89~104.

Daehler CC, 2003. Performance comparisons of co-occurring native and alien invasive plants: Implications for conservation and restoration. Annu Rev Ecol Evol Syst 34: 183~211.

Davis MA, Grime JP, Thompson K, 2000. Fluctuating resources in plant communities: a general theory of invasibility. J Ecol 88: 528~534.

Dukes JS, 2002. Species composition and diversity affect grassland susceptibility and response to invasion. Ecol Appl12: 602~617.

Dukes JS, Mooney HA, 1999. Does global change increase the success of biological invaders? Trends Ecol Evol 14: 135~139.

Feng YL, Lei YB, Wang RF, Callaway RM, Valiente-Banuet A, Inderjit, Li YP, Zheng YL, 2009. Evolutionary tradeoffs for nitrogen allocation to photosynthesis versus cell walls in an invasive plant. Proc Natl Acad Sci USA 106: 1853~1856.

Godoy O, Valladares F, Castro-Diez P, 2011. Multispecies comparison reveals that invasive and native plants differ in their traits but not in their plasticity. Funct Ecol 25: 1248~1259.

Goldberg DE, 1987. Neighborhood competition in an old-field plant community. Ecology 68: 1211~1223.

Grotkopp E, Rejmánek M, 2007. High seedling relative growth rate and specific leaf area are traits of invasive species: phylogenetically independent contrasts of woody angiosperms. Am J Bot 94: 526~532.

Hager HA, Ryan GD, Kovacs HM, Newman JA, 2016. Effects of elevated CO_2 on photosynthetic traits of native and invasive C_3 and C_4 grasses. BMC Ecol 16:28.

Hättenschwiler S, Körner C, 2003. Does elevated CO_2 facilitate naturalization of the non-indigenous Prunus laurocerasus in Swiss temperate forests? Funct Ecol 17: 778~785.

IUCN, 2001. IUCN red list categories and criteria. Version 3.1. IUCN, Species Survival Commission, Gland, Switzerland and Cambridge, United Kingdom.

Jia JJ, Dai ZC, Li F, Liu YJ, 2016. How will global environmental changes affect the growth of alien plants? Front Plant Sci 7: 1623.

Johnson D, Cheng W, Joslin J, Norby R, Edwards N, Todd D, 2004. Effects of elevated CO_2 on nutrient cycling in a sweetgum plantation. Biogeochemistry 69: 379~403.

Li ZY, Xie Y, 2003. Invasive species in China (in Chinese). China Forest Publishing House.

Lichtenthaler HK, Wellburn AR,1983. Determination of total carotenoids and chlorophyll a and b of leaf extracts in different solvents. Biochem Soc Trans 603: 591~592.

Liu YJ, Oduor AMO, Zhang Z, Manea A, Tooth IM, Leishman MR, Xu XL, Van Kleunen M,2017. Do invasive alien plants benefit more from global environmental change than native plants? Global Change Biol 23: 3363~3370.

Liu YJ, Van Kleunen M,2017. Responses of common and rare aliens and natives to nutrient availability and fluctuations. J Ecol 105: 1111~1122.

Long SP, Ainsworth EA, Rogers A, Ort DR, 2004. Rising atmospheric carbon dioxide: Plants face the future. Annu Rev Plant Biol 55: 591~628.

Luo Y, Su B, Currie WS, Dukes JS, Finzi A, Hartwig U, Hungate B, McMurtrie RE, Oren M, Parton WJ, Patakai DE, Shaw R, Zak DR and Field CB,2004.Progressive nitrogen limitation of ecosystem responses to rising atmospheric carbon dioxide. BioScience 54: 731~739.

McGuire AD, Melillo JM, Kicklighter DW, Joyce LA,1995. Equilibrium responses of soil carbon to climate change: empirical and process-based estimates. J Biogeogr 22:785~796.

Medlyn BE, Badeck FW, de Pury DGG, Barton CVM, Broadmeadow M, Ceulemans R, de Angelis P, Forstreuter M, Jach ME, Kellomaki S, Laitat E, Marek M, Philippot S, Rey A, Strassemeyer J, Laitinen K, Liozon R, Portier B, Roberntz P, Wang K, Jarvis PG,1999. Effects of elevated CO_2 on photosynthesis in European forest species: a meta-analysis of model parameters. Plant Cell Environ 22: 1475~1495.

Meng GT, Li GX, He LP, Chai Y, Kong JJ, Lei YB,2013. Combined effects of CO_2 enrichment and drought stress on growth and energitic properties in the seedlings of a potential bioenergy crop Jatropha curcas. J Plant Growth Regul 32: 542~550.

Nagel JM, Griffin KL,2004. Can gas-exchange characteristics help explain the invasive success of *Lythrum salicaria*? Biol Invasions 6: 101~111.

Nagel JM, Wang XZ, Lewis JD, Fung HA, Tissue DT, Griffin KL,2005. Atmospheric CO_2 enrichment alters energy assimilation, investment and allocation in *Xanthium strumarium*. New Phytol 166: 513~523.

Nackley L, Hough-Snee N, Kim SH,2017. Competitive traits of the invasive grass *Arundo donax* are enhanced by carbon dioxide and nitrogen enrichment. Weed Res 57: 67~71.

Norby RJ, Warren JM, Iversen CM, Garten CT, Medlyn BE, McMurtrie RE,2010. CO_2 enhancement of forest productivity constrained by limited nitrogen availability. Proc Natl Acad Sci USA 107:19368~19373.

Onoda Y, Hikosaka K, Hirose T , 2004. Allocation of nitrogen to cell walls decreases photosynthetic nitrogen-use efficiency. Funct Ecol 18: 419~425.

Owensby CE, Ham JM, Knapp AK, Auen LM,1999. Biomass production and species composition change in a tallgrass prairie ecosystem after long-term exposure to elevated atmospheric CO_2. Global Change Biol 5: 497~506.

Pattison RR, Goldstein G, Ares A,1998. Growth, biomass allocation and photosynthesis of invasive and native Hawaiian rain-forest species. Oecologia 117: 449~459.

Penuelas J, Sardans J, Llusià J, Owen S, Carnicer JM, Giambelluca TW, Rezende EL, Waite M, NiinemetsÜ,2010. Faster returns on 'leaf economics' and different biogeochemical niche in invasive compared with native plant species. Global Change Biol 16: 2171~2185.

Poorter H, Remkes C,1990. Leaf area ratio and net assimilation rate of 24 wild species differing in relative growth rate. Oecologia 83: 553~559.

Qi SS, Dai ZC, Miao SL, Zhai DL, Si CC, Huang P, Wang RP, Du DL,2014. Light limitation and litter of an invasive clonal plant,*Wedelia trilobata*, inhibit its seedling recruitment. Ann Bot 114:425~433.

Raizada P, Singh A, Raghubanshi AS,2009. Comparative response of seedlings of selected native dry tropical and alien invasive species to CO_2 enrichment. J Plant Ecol 2: 69~75.

Rogers HH, Runion GB, Prior SA, Price AJ, Torbert HA, Gjerstad DH, 2008. Effects of elevated atmospheric CO_2 on invasive plants: Comparison of purple and yellow nutsedge (*Cyperus rotundus* L. and *C. esculentus* L.). J Environ Qual 37: 395~400.

Smith SD, Huxman TE, Zitzer SF, Charlet TN, Housman DC, Coleman JS, Fenstermaker LK, Seemann JR, Nowak RS, 2000. Elevated CO_2 increases productivity and invasive species success in an arid ecosystem. Nature 408: 79~82.

Song LY, Wu JR, Li CH, Li FR, Peng SL, Chen BM, 2009. Different responses of invasive and native species to elevated CO_2 concentration. Acta Oecol 35: 128~135.

Song LY, Li CH, Peng SL, 2010. Elevated CO_2 increases energy-use efficiecy of invasive *Wedelia trilobata* over its indigenous congener. Biol Invasions 12: 1221~1230.

Walch-Liu P, Neumann G, Engels C, 2001. Response of shoot and root growth to supply of different nitrogen forms is not related to carbohydrate and nitrogen status of tobacco plant. J Plant Nutr Soil Sci 164: 97~103.

Wilsey BJ, Polley HW, 2006. Aboveground productivity and root-shoot allocation differ between native and introduced grass species. Oecologia 150: 300~309.

Zhang L, Zou JW, Siemann E, 2017. Interactive effects of elevated CO_2 and nitrogen deposition accelerate litter decomposition cycles of invasive tree (*Triadica sebifera*). Forest Ecol Manag 385: 189~197.

Zheng YL, Feng YL, Liu WX, Liao ZY, 2009. Growth, biomass allocation, morphology, and photosynthesis of invasive *Eupatorium adenophorum* and its native congeners grown at four irradiances. Plant Ecol 203: 263~271.

Zou JW, Rogers WE, Siemann E, 2007. Differences in morphological and physiological traits between native and invasive populations of *Sapium sebiferum*. Funct Ecol 21: 721~730.

入侵植物蟛蜞菊和本地同属植物形态、生长状况、生物量分配和光合作用对CO_2富集的相似性反应

摘要：全球气候变化和生物入侵都威胁着全世界的生物多样性。然而，它们之间的相互作用和相关机制仍未得到很好的阐明。为了阐明导致入侵的潜在特征以及CO_2的持续增加是否会加剧入侵，我们比较了有害入侵植物三裂叶蟛蜞菊(*Wedelia trilobata*)和本地植物麻叶蟛蜞菊(*W. urticifolia*)和蟛蜞菊(*W. chinensis*)在正常CO_2浓度和CO_2浓度加倍情况下其生长状况、生物量、形态和生理特征。结果表明，入侵植物的叶片质量分数(LMF)和比叶面积(SLA)均高于本地植物，较高的叶片面积比例，加快了其生长速度，从而导致侵入性较强；根质量分数(RMF)越低，细根百分率越高，LMF值越高。另一方面，入侵植物将较高比例的氮分配给叶片光合装置，这与入侵植物较高的光合速率和资源利用效率有关。所有这些特点共同促成了它的入侵性。CO_2富集通过增加实际光合作用促进了所有研究物种的生长，但由于叶片和光合作用的减少而降低了光合能力。入侵植物和原生植物对CO_2浓度升高的反应并无显著差异，表明CO_2浓度的持续升高可能不会加剧生物入侵，与参考文献中的普遍结果不一致。因此，需要对相关的入侵植物和原生植物进行更多的比较研究，以阐明CO_2的富集是否会导致入侵加剧。

(本文发表于 *Plant Ecology*, 2018, 219)

第三篇 生物多样性保护研究

从高山到河谷:德钦藏药植物资源的多样性及利用研究

马建忠[1],庄会富[2]

(1. 云南省林业科学院,云南 昆明 650201;2. 中国科学院昆明植物研究所,云南 昆明 650201)

摘要:为探讨复杂生境条件下,藏药植物资源的多样性和利用现状,选取云南迪庆藏族自治州西北部澜沧江、金沙江河谷;梅里雪山、白马雪山山区和高寒山区展开了藏药植物资源调查。结果表明:该地区使用的藏药资源丰富,共调查到药用植物144种,隶属于63科126属;资源利用方式以野生采集为主(64%),人工栽培为辅(25%已开展栽培,11%正试验栽培);部分资源存在资源枯竭问题(26%)。统计分析表明:从河谷到山区、高寒山区,各海拔梯度的植物资源均有较多使用,但使用的资源类群不同;随海拔高度上升,植物资源易濒危,难栽培,栽培开展受恶劣环境的抑制,受资源枯竭推动,易形成先枯竭、后栽培的不合理开发模式;资源使用过度、植物种群恢复能力弱、采集伤害重是天然资源枯竭的相关因素。应针对上述问题,制定适合特殊生境条件下藏药资源特点的开发策略。

关键词:藏药;药用植物;重要性和濒危程度评价

From Valley to Alpine Mountain:Diversity and Utilization of Tibetan Medicinal Plants in Deqin

Man Jian-zhong[1], Zhuang Hui-fu[2]

(1. Yunnan Academy of Forestry, Kunming Yunnan 650201;
2. Kunming Institute of Botany, Chinese Academy of Sciences, Kunming Yunnan 650201)

Abstract:A field and interdisciplinary survey of Tibetan medicinal plant resources was undertaken in Deqin Country of Northwest Yunnan in 2008. Three typical habitats, river valley, mid-mountain and alpine regions were selected to document the use knowledge and species diversity used locally, and to evaluate the importance and vulnerability of medicinal plant species. 144 medicinal plant species of 63 families and 126 genera were recorded in this paper. Most of them (64%) are used through collecting wild plants, and the other 36% are supplied by pilot cultivation or conventional plantation. 26% of total 144 species are being threatened because of commercial harvesting and limited resources. Among the three surveyed plots along altitude gradients, the river valley has the least vulnerable plant species, while alpine region has the most plant species which are being threatened by overharvesting, limited by slow growth rate, and exacerbated by the difficulty of man-made cultivation. And reasonable strategies should be developed for plants resources in this and other special habitats.

Keyword: Tibetan medicine; Medicinal plant; Assessment of importance and vulnerability

藏医药是中国传统医药的重要组成部分,是藏族人民在与疾病的长期斗争中,结合传统药用经验、印度医学、中医等传统医学形成的具有青藏高原药物学特点的医药体系(李隆云和次仁巴珠,2001),其使用范围包括西藏、青海、甘肃甘南州、四川阿坝、甘孜、云南迪庆等广大藏区(罗达尚,1996)。目前针对藏医药的研究开发日渐增多,青海、西藏等地藏药企业发展迅速,已有100余家,藏医药产业现已成为西藏六大支柱产业之一。但是传统藏药使用的药源独特,主要是产于青藏高原的植物药材(占堆,2004)。由于青藏高原独特的自然生态条件,虽然孕育了种类丰富的植物资源,但是具体物种的资源储量有限、种群更新速度慢,资源采集后难以恢复(李隆云和次仁巴珠,2001)。产业开发相继带来了资源枯竭、生态环境破坏等不可持续利用问题,目前一些大宗藏药材,如红景天(*Rhodiola rosea*)、藏茵陈(*Swertia mussotii*)、独一味(*Lamiophlomis rotata*)、雪莲(*Saussurea* spp.)等已处于濒危或枯竭状态(许理刚,2003;杨青松等,2003)。为满足市场对部分药用植物的需求,当前藏药资源的栽培研究日渐增多,如波棱瓜(*Herpetospermum caudigerum*)、手参(*Crymnadenia conopsea*)、翼首花(*Pterocephalus bretschneider*)、藏木香(*Inula racemosa*)等传统藏药已开展人工栽培(甘玉伟等,2006;刘显福等,2006;陈灼等,2006;鲍隆友等,2008),但进展缓慢。

传统藏族医药在产业化开发、外来医药体系等因素的冲击下,藏药植物种类、资源储量和利用方式等正发生深刻的变化。本文选择藏药资源丰富的云南迪庆藏族自治州德钦县及周边进行实地调查,分析该区域传统藏药植物资源多样性与使用现状,旨在探讨复杂生境条件下藏药资源开发存在的问题,为合理开发提供决策依据。

1 研究地点与方法

1.1 研究地点

德钦县位于云南省西北部,与川藏交界,是著名的"三江"并流核心地区。境内高山河谷众多,其最低海拔仅1 486m,而最高海拔为云南省的最高峰——梅里雪山的卡瓦格博峰,高达6 740m。多样的地形地貌和巨大的海拔变化,形成了垂直分布的三种生态气候类型:高山带气候、中山带气候与河谷带气候。境内主要居民为藏族、汉族,其中藏族最多(占80%)。

德钦县所属的迪庆州是《迪庆藏药》的知识来源地(李玉娟等,2008),《迪庆藏药》记载了藏药598种,其中主要是植物药,占448种,多数是产自青藏高原的药材(杨竞生和初称江错,1987)。该县药用资源种类丰富,20世纪80年代的中药资源普查调查到的药用植物160科867种(勒安旺堆,2003)。

本研究调查的地区主要在德钦县,仅在河谷地区调查中涉及维西县。针对上述三类气候区,我们选取了具代表性和典型性的河谷地段、中山山区和高寒山区开展藏药植物资源与使用的调查,具体调查行程安排见表1。

1.2 野外调查

野外调查自2008年4月中旬至9月中旬,由当地知名的藏医药专家和民间组织德钦县藏医药研究会的成员,组成一个10人调查队,调查内容包括:①记录传统藏药资源的植物基原,采集植物凭证标本;②走访、调查药材资源分布、储量和枯竭状况(具体评估方法见1.3);③藏药资源的利用状况,包括利用部位、是否商品收购以及栽培开展情况;④传统药物使用方法、功效等。

表1 2008年迪庆藏区传统药用植物资源多样性调查行程
Table 1 Field survey itinerary in 2008

海拔梯度 Altitude gradients	野外工作时间 Field survey schedule	地点 Specific locations
河谷地区 River valley (alt. <2 500m)	4月19日至5月3日 April 19 –April 3	德钦县金沙江上游、维西县澜沧江河谷 Valley of upper Jinsha and Lancang River in Deqin and Weixi County
中山地区 Mid-mountain (alt. 2 500~3 500m)	6月17日至7月5日 June 17–July 5	德钦县梅里雪山、白马雪山中山地区 Mid-mountain of Meili Snow Mountain and Baima Snow Mountain
高寒山区 Alpine regions (alt. >3 500m)	9月3日至9月17日 September 3 –September 17	德钦县梅里雪山、白马雪山高寒山区 Alpine region of Meili Snow Mountain and Baima Snow Mountain

1.3 资料处理及分析方法

1.3.1 野外调查和标本鉴定

藏医药专家负责确定藏药对应的基原植物,请植物分类专家鉴定采集的标本。

1.3.2 开展重要程度和濒危程度评估

将某一植物物种的重要程度按照从低到高分为5级(1至5分,分值越高表示越重要),具体参照其他药用植物资源的评估方法(王雨华等,2003),结合药材的功效、使用范围,对某一特定物种的重要性进行打分;将濒危程度的评价同样分为5级,分值越低表示资源越丰富,越高表示资源越接近濒危,以实际调查中资源分布和储量的多少,结合当地藏医对资源分布变化的描述,评估濒危程度。由藏医药研究会的成员分别根据上述调查信息对每种药材重要程度和濒危程度评分,然后取平均值(四舍五入取整)。

1.3.3 整理资料,分析数据

统计调查地区的藏药资源多样性和药材的利用方式、产业开发情况;应用SPSS软件分析不同海拔地区使用的资源、采集方式、栽培开展之间的差异,包括重要程度、海拔梯度与濒危程度的对应分析(correspondence analysis),海拔梯度与药材生产方式的对应分析,并基于分析结果,探讨资源濒危的相关因素和栽培开展的制约因素。本次藏药调查的数据,如:濒危程度、重要程度、采集方式、区域分布等皆为属性数据(categorical data),不适合应用常用的统计分析方法。因此我们尝试使用对应分析这一社会学研究中常用的分析方法,它通过分析由定性变量构成的交互汇总表来揭示变量间的联系,并将联系的紧密程度通过类别的分布图直观展示出来,以代表各类别的散点间的距离揭示它们间的对应关系(郭志刚,1999)。

2 结 果

2.1 德钦、维西两县澜沧江、金沙江河谷至梅里、白马雪山高山的藏药资源

2008年在河谷、中山和高寒地区的调查,共采集和记录藏药植物144种,隶属于63科126属。生活型主要包括多年生草本(116种,占81%)和乔灌木(22种,占15%),其余藤本、寄生植物等共有6种。药用种数较多的科有菊科(Compositae),18种;毛茛科(Ranunculaceae)、百合科(Liliaceae)各8种;龙胆科(Gentianaceae)、唇形科(Labiatae),各7种;蔷薇科(Rosaceae)6种;十字花科(Cruciferae)、蝶形花科(Papilionaceae)和玄参科(Scrophulariaceae)各5种。

统计各海拔梯度使用的植物资源(表2),各海拔梯度的植物均使用较多。

表2 各区域使用的药用植物资源统计
Table 2 Statistics of Tibetan medicinal plants in three different places along altitude gradients

海拔梯度 Altitude gradients	药用植物数 Number of medicinal Plants	所属科数 Number of families	种类较多的科(>4种) Higher-frequency families (>4)
河谷地区 River valley	48	33	Compositae, Labiatae, Cruciferae, Rosaceae
中山地区 Mid-mountain	57	39	Liliaceae, Ranunculaceae, Papilionaceae, Gentianaceae
高寒山区 Alpine regions	41	22	Compositae, Ranunculaceae, Gentianaceae, Scrophulariaceae

注:当归和黄牡丹在多地区出现,所以三地区使用的物种总数会大于144。

菊科植物在各区域均使用较多,这与菊科物种数量丰富和分布广泛有关。统计显示有41个科仅在一个地区使用,17个科仅在河谷地区使用,仅中山地区使用的科也为17个,于高寒山区使用的为6个,其余22个科的植物分别在两类地区或三类地区有使用。

2.2 植物资源重要程度与濒危程度

统计重要程度评估分值,调查的144种植物较重要的(分值4~5)有金耳石斛(Dendrobium hookerianum)、红花(Carthamus tinctorius)、甘青乌头(Aconitum tanguticum)、刺红珠(Berberis dictyophylla)、藏木香、波棱瓜等50种(占35%),多数药用植物重要性分值为2~3(图1)。

144种植物中有各级别保护植物8种(国家一级、二级保护植物4种,CITES收录的4种)。植物濒危程度的评估分值统计显示:资源较濒危(分值4~5)的植物有紫檀(Pterocarpus indicus)、水柏枝(Myricaria germanica)、刺红珠(Berberis dictyophylla)、胡黄连(Picrorhiza scrophulariiflora)、水母雪莲花(Saussurea medusa)等37种(占26%);多数植物资源濒危分值为2~3(图1)。另外,调查发现存在规模化商业采集的植物种类有11种,其中9种因过度采集而使当地资源面临枯竭。

重要程度与濒危程度的SPSS对应分析显示,重要的植物资源濒危程度相应高,二者

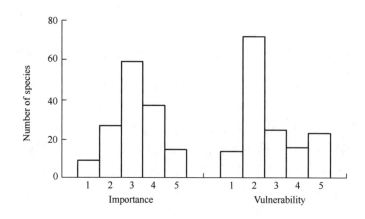

图 1　重要程度与濒危程度分值分布图

Figure 1　Distribution of importance and vulnerability of medicinal plants

表现很强的对应关系（图2）。

对三个海拔梯度区域的药用资源的濒危情况分析，发现呈不均匀分布的情况。区域与濒危程度的对应分析显示高寒山区的植物类群接近濒危分值较高，而中山和河谷地带植物类群接近中低濒危程度（2~3）（图3）。统计显示：高寒山区的植物资源中濒危类群（分值4~5）比例最高，18种，占该区域41种药用植物的44%；中山和河谷分别为11种（占19%）和8种（17%）。

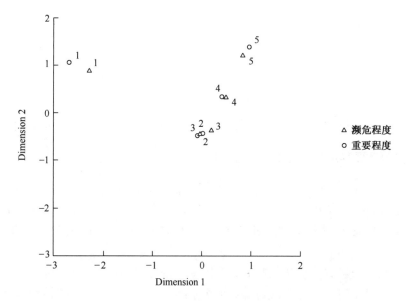

图 2　重要程度（○）与濒危程度（△）对应分析散点图

图中显示重要程度与濒危程度显著相关，重要程度高的类群濒危程度相应高

Figure 2　Correspondence analysis on importance (△) and vulnerability (○) This figure indicates that plant resources with high important degree are coincidently highly endangered

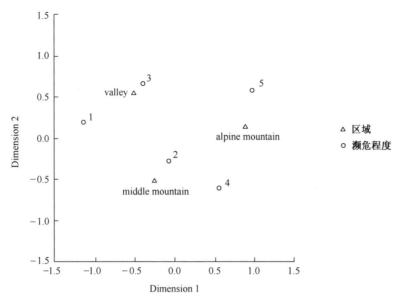

图 3　海拔梯度区域（△）与濒危程度（○）对应分析散点图
图中显示中山、河谷地带应用的药用植物多处于中低度濒危状态（濒危分值 2~3），
而与高山区域类群最接近的是濒危程度较高（4~5）的类群

Figure 3　Correspondence analysis on altitude gradients （△） and vulnerability （○）This figure indicates that the valley and mid-mountain regions have less vulnerability, but high vulnerability in alpine region

2.3 药材的生产

药材的生产有采集野生资源和人工栽培两种方式。在调查的 144 种植物中，完全使用野生资源的有 94 种，占 64%；已栽培的有 36 种，占 25%；有 16 种植物正进行试验性栽培，占 11%。对应分析结果显示：野生资源在各地区均有较多使用，但以中山山区的种类利用最多；栽培药材在河谷地带最多，其次是中山，高寒山区的种类栽培较少，但目前当地开展试验性栽培的药材主要是高山植物种类（表 3）。对应分析散点图较明显地显示以下对应关系：野生采集——山区；栽培——河谷地区；试验栽培——高寒山区（图 4）。

表 3　药材生产方式和生境区域的对应分析结果
Table 3 Correspondence analysis on production and habitat of medicinal plants

生产方式 Mode of production	海拔梯度 Elevation gradients			
	河谷 River valley	中山 Mid-mountain	高山 Alpine mountain	合计 Total
野生采集 Wild plants collection	28	41	25	94
实验栽培 Experimental cultivation	2	2	12	16
栽培 Conventional plantation	18	14	4	36
合计 Total	48	57	41	146

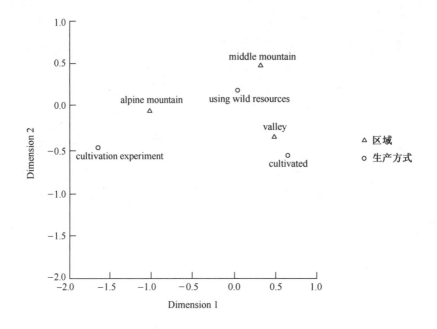

图 4 海拔梯度区域（△）与药材生产方式（○）的对应散点图

Figure 4　Correspondence analysis on production mode（○）and elevation gradients（△）

2.4 药材的利用部位

调查的 144 种藏药植物中，使用全草或地上部分的最多，占 63%；其次是根或根茎 32%，使用花的占 14%，其余果、种子、树皮等部位均有少数应用（表 4）。

表 4　各药用采集部位统计

Table 4　Percentage of different parts of collected medicinal plants

采集部位 Used Part	全草/地上部分 Herb/Above-ground Part	根/根茎 Root/Rhizome	花 Flower	果 Fruit	茎干 Stem	种子 Seed	树皮 Bark
数量 Number	63	46	20	14	9	9	5
比例 Percentage（%）	43	32	14	10	6	6	3

药材的采集部位与药材基原植物的伤害程度密切相关，我们依照药用采集部位将采集伤害分为 3 类，即营养伤害（采集枝、叶等），繁殖伤害（采集花、果、种子等繁殖器官）和致死伤害（采集根、根茎、全草等引起植株死亡的采集方式）。统计发现，致死采集是主要的采集方式（64%），其次是繁殖伤害（28%）和营养伤害（8%）。分析还发现：河谷地区采集伤害较轻，致死采集模式占 44%；中山、高寒山地采集伤害重，致死采集模式分别占 70%、78%（图 5）。

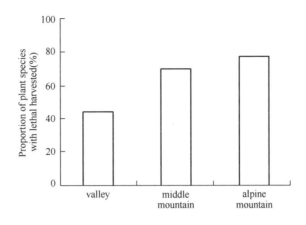

图 5　致死采集伤害在各区域类型药用资源中的比例

图中显示河谷地区的药材致死采集比例最小（44%），而中山、高寒山地比例都较高（70%和78%）

Figure 5　Proportion of species with lethal harvesting in each altitude gradients This figure shows that the lethal harvesting in river valley has a percentage of 44%, much lower than those in the mid-mountain and alpine regions

3　讨　论

3.1　从河谷到高山藏药植物资源及利用的多样性

迪庆州德钦县是藏区药用植物资源最丰富的地区之一，仅调查的样区就调查采集到63科144种药用植物。分析发现藏药使用的植物类群广泛，无论是河谷、中山还是高寒山区均有许多植物被利用，但三个海拔梯度所利用的种类有较大差异，多数植物仅在某一区域使用。从河谷到高山的药用植物数与科数之比分别为 48/33＝1.5、57/38＝1.5、41/22＝1.9，可见高寒山区的药用植物种类相对集中于少数的植物科属中，这与高山环境胁迫压力大，只有较少植物种类能适应高寒生境有关。

与传统中医药相比，藏药资源使用以全草和地上部分为主（表4），而中医药使用最多的部位是根或根茎；传统藏药对花的使用较频繁，调查中使用花的植物资源占14%，中医药对花使用很少。另外，党参（*Codonopsis* ssp.）、白芨［*Bletilla striata*（Thunb.）Reichb. f.］、秦艽（*Gentiana* spp.）、射干［*Belamcanda chinensis*（L.）DC］等药材中国药典中仅使用其根部（国家药典编委会，2005），藏药中全草都有应用，以此为依据，可以开展及拓展药源的研究。

3.2　藏药资源的濒危与相关因素

传统藏药资源种类丰富，但是资源量非常有限，调查中有37种（26%）资源较濒危。分析发现药材的濒危程度与重要程度显著相关（图2），即重要性越大其濒危程度越高，而药材重要程度高是药材使用量大和应用范围广的体现，因此药材的过度使用是资源濒危的因素之一；另一方面，规模化的商业收购是引起资源枯竭的原因，调查中存在商品收购植物资源多数面临严重枯竭（占88%）。

濒危的37种药用植物中，有18种（占49%）分布在高寒山区，中山地区次之，河谷地区最少;海拔梯度与濒危程度对应分析结果(图3)同样支持上述结果。从河谷到高山,随着环境胁迫压力的增大,药用植物资源越来越容易濒危,一方面缘于该高山地区植被多是灌丛草甸、高寒草甸等,植被中逆境耐受型植物(stress-tolerant strategy)占多数,该类型植物应对环境胁迫能力强,但应对采集干扰能力最弱(Grime,1977),种群恢复能力差,采集之后难以恢复;另一方面,采集伤害也是引起该区域资源濒危的因素之一,该区域的类群致死采集方式占78%,而河谷地区仅为44%(图5)。

3.3 从河谷到高山,传统藏药的生产方式

传统藏药以采集野生资源为主(占堆,2004)。在我们调查的藏药植物种类中,采集野生资源的植物占较高的比例,但栽培和试验性栽培的植物也占相当比例(25%,11%),随着传统医药的产业化发展,试验性和规模化栽培呈扩大趋势。分析各区域的药材生产方式发现,从河谷到高山,栽培的植物种类越来越少;但是正处于试验栽培的资源反而增多(表3,图4)。究其原因:环境胁迫压力在低海拔的河谷地区影响较小,容易开展栽培;高海拔地区生境特殊,植物生长缓慢,多数药材生长周期长,加之环境因素,不利于人们开展生产活动,造成栽培开展少;然而在资源枯竭的迫使下,人们不得不开展试验栽培,易形成先枯竭后栽培的模式。

3.4 应对藏药产区复杂生境,藏药资源开发的合理策略

在产业化开发背景下,传统藏医药正朝两个方向发展:一是传统医药的传统应用,现今在广大藏区传统藏药仍维持相当广泛的应用;二是传统藏药的商业化和产业化开发。尽管后者对带动藏区经济发展起到了积极作用(许理刚,2003),但应密切关注产业开发对资源、环境的破坏。青藏高原是一个特殊生态地理区域,其药用植物资源独特,应针对藏药资源的特点,制定相应的资源保护、监管以及栽培研究的策略,满足产业开发对资源持续性的需求。

(1)藏药资源种类丰富,有必要进一步调查、研究、发掘新的药源。

(2)调查中藏药资源使用仍以采集野生资源为主,恶劣的自然环境条件抑制人们开展栽培,而又在资源枯竭推动下被迫开展栽培,应改变这种先枯竭、后栽培的不合理模式,制定相应的鼓励措施,及时监测资源的变化,及早开展栽培研究。

(3)现在藏药资源的商业收购缺少监管,应采取相关措施,防止调查中出现的因商业收购引起的资源枯竭的现象,尤其是高山逆境中的植物类群。

致谢 德钦县藏医药研究会的仁钦旺学老师、次仁桑主医生在论文的形成过程中提供了大力的支持,昆明植物研究所杨永平研究员提出修改意见。

参考文献:

[1] 甘玉伟,陈灼,旦智草,等.藏药翼首草的人工栽培试验研究[J].甘肃科技纵横,2006(3):225.
[2] 李玉娟,杨梅,刘青,等.论迪庆藏药及应用特色[J].中国民族医药杂志,2008(3):39~40.
[3] 杨竞生,初称江错.迪庆藏药[M].昆明:云南民族出版社,1987.
[4] 陈灼,甘玉伟,杨勇,等.藏药藏木香的人工栽培试验[J].甘肃科技纵横,2006(3):227,187.
[5] 国家药典编委会.中华人民共和国药典[M].北京:化学工业出版社,2005.

[6] 郭志刚. 社会统计分析方法——SPSS软件应用[M]. 北京:中国人民大学出版社,1999.

[7] 勒安旺堆. 迪庆藏族自治州志[M]. 昆明:云南民族出版社,2003.

[8] Bao LY (鲍隆友), Yang XM (杨晓梅), Liu YJ (刘玉军), 2008. Study of the localization technology of the wild Tibetan indigenous medicinal herbs[J]. Forest By-Product and Speciality in China (中国林副特产), 92 (1):22~25.

[9] Gri me JP. Evidence for the existence of three pri mary strate-giesin plants andits relevance to ecological and evolutionarythe-ory[J]. The American Naturalist, 1977,111 (982):1169.

[10] Li LY (李隆云), Ci RBZ (次仁巴珠), 2001. Comprehensive uti-lization and develoment of traditional Tibetan medicine in China[J]. China Journal of Chinese Materia Medica (中国中药杂志), 26 (12):808~810.

[11] Liu XF (刘显福), Fang QM (方清茂), Liu DP (刘代品) et al., 2006. Research on the cultivation techniques of Herpetosper-mumcaudigerum Wall[J]. Journal of Liaoning University of TCM (辽宁中医药大学学报), 8 (4):131~132.

[12] Luo SD (罗达尚), 1996. Discussion of the strategies of Tibetanmedicine resources developing[J]. Journal of Shandong Colleege of Traditional Chinese Medicine (山东中医学院学报), 20 (6):417~418.

[13] Wang YH (王雨华), Xu JC (许建初), Li YH (李延辉) et al., 2003. Ethnobotanical valuation on medicinal plant re-sources in Ludian Administrative Village, Lijiang County, Yunnan Province[J]. Acta Botanica Yunnanica (云南植物研究), Suppl, XIV:41~50.

[14] Xu LG (许理刚), 2003. The sustainable development researehof Tibet's drugindustry[D]. Chengdu: Sichuan University.

[15] Yang QS (杨青松), Chen ST (陈绍田), Zhou ZK (周浙昆), 2003. Protection and sustainable utilization of traditional Tibetan medicine "Snow Lotuses" (Saussurea) in Diqing Autonomous Prefecture, Yunnan [J]. Acta Botanica Yun-nanica (云南植物研究), 25 (3):297~302.

(本文发表于《云南植物研究》,2010年)

野生中泰南五味子果实的形态及营养成分分析

陈伟,文进,袁莲珍,李江,陈绍安,刘际梅

(云南省林业科学院云南省森林植物培育与开发利用重点实验室,云南 昆明 650201)

摘要:以云南西双版纳野生中泰南五味子成熟果实为研究对象,对果实形态指标和主要营养成分进行测定和分析。结果表明,中泰南五味子果实为聚合浆果,球形,直径11.0~16.5cm,质量可达1kg以上。果实含水量90%,粗纤维、粗脂肪、还原糖、蛋白质、灰分、总酸、总碳水化合物、可溶性固形物含量分别为17.1mg·g^{-1}、11.3mg·g^{-1}、9.2mg·g^{-1}、5.6mg·g^{-1}、5.0mg·g^{-1}、2.3mg·g^{-1}、78.1mg·g^{-1}、82.8mg·g^{-1},能量值为1.84kJ·g^{-1};含有16种氨基酸,其中必需氨基酸有7种,必需氨基酸总含量为1.93mg·g^{-1}。钾、钙、锰、铁等矿物元素含量均较高,其中,钾元素含量高达2.22mg·g^{-1},钠元素含量较低。中泰南五味子果实体积大,营养丰富,具有高钾、低钠、低能量等特点。

关键词:中泰南五味子;果实;形态;营养成分

Analyses on morphology and nutrient component in fruit of wild *Kadsura ananosma*

Chen Wei, Wen Jin, Yuan Lian-zhen, Li Jiang, Chen Shao-an, Liu Ji-mei

(Yunnan Provincial key Laboratory of Cultivation and Exploitation of Forest Plants, Yunnan Academy of Forestry, Kunming Yunnan 650201, China)

Abstract: The morphological indexes and main nutritional components in nutrient component in mature fruit of wild *Kadsura ananosma* Kerr from Xishuangbanna of Yunnan were measured and analyzed. The results show that fruit of *K. ananosmais* aggregate berries and spherical with diameter of 11.0~16.5cm and weight of more than 1kg. Water content of fruit is 90%. Content of crude fiber, crude fat, reducing sugar, protein, ash, total acids, total carbohydrates and soluble solid are 17.1, 11.3, 9.2, 5.6, 5.0, 2.3, 78.1 and 82.8mg·g^{-1}, respectively, and energy value is 1.84kJ·g^{-1}. There are sixteen kinds of amino acids, in which, there are seven kinds of essential amino acids, and total content of essential amino acids is 1.93mg·g^{-1}. Contents of mineral elements such as K, Ca, Mn and Fe are all higher, while content of Na is lower, in which, content of K is as high as 2.22mg·g^{-1}. It is indicated that fruit of *K. ananosma* is big and contains rich nutrients with characteristics of high K, low Na and low energy.

Key words: *Kadsura ananosma* Kerr; fruit; morphological; nutritional component

中泰南五味子(*Kadsura ananosma* Kerr)为五味子科(Schisandraceae)南五味子属(*Kadsura*)攀缘植物[1],其根、藤茎、果实是珍贵的傣药,具有收敛固涩,益气生津,补肾宁心的作用。对中泰南五味子化学成分的研究结果表明:中泰南五味子体内含有较多的木脂素类和三萜类成分,其中有20多个成分为特有成分,这些化学成分具有抗氧化、抗HIV病毒

等药理活性[2~5]。中泰南五味子的果实具有果大、肉多酸甜可口等特点,好的可作为野生水果食用,但目前对中泰南五味子果实的相关研究尚未见报道;相关志书[1,6]对其果实的基本描述也不完整,甚至空缺。为此,作者对野生中泰南五味子果实的形态特征进行观察,并对其营养成分进行分析,以期完善中泰南五味子果实的基础数据,为其深入开发和利用提供依据。

1 材料与方法

1.1 材料

供试中泰南五味子果实于2014年10月采自云南省西双版纳傣族自治州景洪市普文镇的野生群体,地理坐标为22°25′41.0″N,101°05′32.0″E,海拔913m。选择5株样株,分别在各样株上采集无虫害、果形整齐、成熟度相近的果实。

1.2 方法

1.2.1 形态指标测定

分别在每株样株上随机选取5个果实(共25个),观察果实形状和果皮颜色;用游标卡尺(成都成量工具集团有限公司)测量聚合果、单果及种子的长度、宽度和厚度等指标;用XS4002S电子天平[精度0.01g,梅特勒-托利多(上海)有限公司]称量各聚合果、单果及种子质量。

1.2.2 营养成分含量测定

选取大小和成熟度相近的果实3个,用于营养成分测定。称取10g果实,参照GB 5009.3—2010的方法测定水分含量;称取1g果实参照GB 5009.5—2010的方法测定蛋白质含量;分别称取10g和3g果实,参照GB/T5009.6—2003的方法分别测定粗脂肪和粗纤维含量;称取25g果实,参照GB/T 12456—2008的方法测定总酸含量;称取3g果实,参照GB/T 5009.7—2008的方法测定还原糖含量;称取15g果实,参照GB/T 6195—1986的方法测定Vc含量;称取1g果实,参照NY/T 1295—2007的方法测定总黄酮含量;称取10g果实,参照GB 5009.4—2010的方法测定灰分含量;称取0.5g果实,参照GB/T5009.124—2003的方法测定氨基酸含量;称取3g果实,参照NY/T 1653—2008的方法测定矿质元素含量;称取250g果实,参照GB/T 8210—2011的方法测定可溶性固形物含量;根据参考文献[7]中的公式分别计算总碳水化合物含量和能量值。每个果实测定1组营养成分指标,重复测定3次,结果取平均值。

1.3 数据处理

用Excel 2003统计分析软件对相关实验数据进行统计和处理分析。

2 结果与分析

2.1 果实形态特征

中泰南五味子果实为聚合浆果,呈球形或椭圆形,直径11.0~16.5cm,重量1.05~

1.42kg,黄绿色,浆果聚合处为暗紫色;每个聚合果含12~38个单果。单果呈倒卵形,肉质,单果质量在16.2~44.1g;顶端宽厚,呈五边形或六边形,最宽处3.5~5.3cm,最窄处2.6~4.3cm;果皮3层,顶端外果皮革质,中部外果皮暗紫色,基部外果皮白色,中果皮肉质、内果皮胶质且紧密包裹种子,浆果基部插入果轴,整个浆果长5.0~6.1cm;每个单果含种子1或2枚,稀3枚。种子两侧压扁,心形或卵状心形,种皮褐色,光滑,长1.6~2.2cm,宽1.0~1.6cm,厚0.7~0.9cm、质量0.9~1.2g。在中国分布的10种南五味子属植物中,中泰南五味子的果实和种子均是最大的[6]。

2.2 果实营养成分分析

2.2.1 营养成分含量分析

测定结果表明:中泰南五味子果实中水分含量(质量分数)为90.00%,粗纤维、粗脂肪、还原糖、蛋白质、灰分、总酸、总碳水化合物和可溶性固形物含量分别为17.1、11.3、9.2、5.6、5.0、2.3、78.1和82.8 mg·g^{-1},Vc和总黄酮含量分别为20μg·g^{-1}和0.2μg·g^{-1},能量值1.84kJ·g^{-1}。

与常见水果[7~9]相比,中泰南五味子果实的水分含量超过苹果(*Malus pumila* Mill.)、甜橙[*Citrus sinensis* (Linn.) Osbeck]、桃(*Amygdalus persica* Linn.)、丰水梨[*Pyrus pyrifolia* (Burm. f.) Nakai]等,并高于同为热带水果的火龙果[*Hylocereus undulates* (Haw.) Britton et Rose];粗脂肪含量高于苹果(2.0mg·g^{-1})、甜橙(2.0mg·g^{-1})、桃(1.0mg·g^{-1})和火龙果(7.0mg·g^{-1}),因此,其熟透的果实具有独特的香味;其蛋白质含量高于苹果和丰水梨,但低于甜橙、桃和火龙果;其还原糖的含量相对较高,是丰水梨的1.3倍;此外,中泰南五味子果实的粗纤维含量也相对较高。果实中可溶性固形物含量相对较低,能量值低于苹果、甜橙、桃和火龙果。与同属植物南五味子(*Kadsura longipedunculata* Finet et Gagnep.)的果实[10]相比,中泰南五味子果实的蛋白质、Vc和粗脂肪含量低于前者,而总酸含量则高于前者。

2.2.2 氨基酸组成及含量分析

中泰南五味子果实的氨基酸组成及含量见表1。测定结果显示:其果实含有16种氨基酸,其中,谷氨酸含量最高(0.79mg·g^{-1}),天门冬氨酸和苯丙氨酸的含量也较高,而组氨酸和甲硫氨酸的含量相对较低。氨基酸总含量为4.70 mg·g^{-1},低于桃和甜橙,但明显高于苹果[7]。中泰南五味子果实含有7种人体必需的氨基酸(色氨酸未检测),总必需氨基酸含量占总氨基酸含量的41.1%,必需氨基酸含量与非必需氨基酸含量的比值为0.70。

表1 中泰南五味子果实中氨基酸组成和含量
Table1 Composition and content of amino acidsin fruit of *Kadsura ananosma*

氨基酸 Amino acid	含量(mg·g^{-1}) Content	氨基酸 Amino acid	含量(mg·g^{-1}) Content
天门冬氨酸(Asp)	0.45	亮氨酸(Leu*)	0.41
苏氨酸(Thr*)	0.21	酪氨酸(Tyr)	0.24
丝氨酸(Ser)	0.29	苯丙氨酸(Phe*)	0.45
谷氨酸(Glu)	0.79	赖氨酸(Lys*)	0.31

(续)

氨基酸 Amino acid	含量(mg·g^{-1}) Content	氨基酸 Amino acid	含量(mg·g^{-1}) Content
甘氨酸(Gly)	0.22	组氨酸(His)	0.11
丙氨酸(Ala)	0.31	精氨酸(Arg)	0.19
胱氨酸(Cys)	—	脯氨酸(Pro)	0.17
缬氨酸(Val*)	0.26	TAA	4.70
蛋氨酸(Met*)	0.06	EAA	1.93
异亮氨酸(Ile*)	0.23		

注:"*":人体必需氨基酸;"—":未检出;TAA:总氨基酸;EAA:总必须氨基酸。

2.2.3 矿物元素含量分析

中泰南五味子果实矿物质元素的含量见表2。从常量元素含量看,中泰南五味子果实的K元素含量最高,并且高于常见水果苹果、甜橙、桃和火龙果[7,8];Mg、Ca和P含量较高,其中,Mg和Ca含量高于苹果和桃;Na含量最低。从微量元素含量看,中泰南五味子果实的Mn含量相对最高,Fe、Cu和Zn含量也较高,其中,Fe含量是甜橙的2.2倍,是苹果的1.5倍,且锌铜比小于10、锌铁比小于1。

表2 中泰南五味子果实中矿物元素含量
Table2 Content of mineral elements in fruit of *Kadsura ananosma*

矿物元素 Element	含量(μg·g^{-1}) Content	矿物元素 Element	含量(μg·g^{-1}) Content
P	75.20	Ca	97.90
Zn	3.27	Cu	3.49
Fe	8.92	Na	2.81
Mn	9.64	K	2220.00
Mg	130.00		

3 讨论和结论

在水果资源的开发中,果实大、果形好、营养物质、风味佳的种类是鲜食和加工的理想原料,也是优良栽培品种的育种资源。中泰南五味子果实为球形,总质量可达1kg以上,果肉大,水分丰富且粗脂肪和粗纤维含量高;总必须氨基酸含量占总氨基酸含量的41.1%,必需氨基酸含量与非必需氨基酸含量的比值为0.70,这2项指标均略高于联合国粮食及农业组织(FAO)和世界卫生组织(WHO)建议的理想蛋白模式值(40%和0.6),表明其所含蛋白质营养价值较高。中泰南五味子果实能量值较低,属于低能量水果。其K含量高、Na含量低,且其锌铜比和锌铁比较为合理[11],为高钾低钠的健康水果。但中泰南五味子果实可溶性固形物含量相对较低,其成熟果实不适宜长时间贮存。

总之,中泰南五味子果实大、果味酸甜、营养丰富,并有低能量、高钾和低钠的特点,符合现代新型健康水果的选育方向,具有开发为保健食品的潜力。但中泰南五味子果实若要作为食用水果进入市场,还需确保其安全性,可借鉴相关的评估方法[12],高钾含量和低钠含量的特点。综合分析认为,中泰南五味子果实在保健食品开发方面极具潜力。同时,对其进行食品安全及毒理学分析。

参考文献:

[1] 中国科学院昆明植物研究所. 云南植物志:第十一卷[M]. 北京:科学出版社,2000.
[2] Yang J H, Zhang H Y, Wen J, et al. Dibenzocyclooctadiene Lignans with Antineurodegenerative Potential from *Kadsura ananosma*[J]. Journal of natural products,2011;74, 1028~1035.
[3] Yang J H, Pu J X, Wen J, et al. Cytotoxic Triterpene Dilactones from the Stems of *Kadsura ananosma*[J]. Journal of natural products,2010;73, 12~16.
[4] Yang J H, Wen J, Du X, et al. Triterpenoids from the stems of *Kadsura ananosma*[J]. Tetrahedron, 2010(66):8880~8887.
[5] Yang J H, Zhang H Y, Du X, et al. New dibenzocyclooctadiene lignans from the *Kadsura ananosma*[J]. Tetrahedron, 2011(67):4498~4504.
[6] 中国科学院中国植物志编辑委员会. 中国植物志:第三十卷(第一分册)[M]. 北京:科学出版社,1996.
[7] 中国疾病预防控制中心营养与食品安全所. 中国食物成分表2002[M]. 北京:北京大学医学出版社,2002.
[8] 蔡永强,向青云,陈家龙,等. 火龙果的营养成分分析[J]. 经济林研究,2008,26(4):53~56.
[9] 张朝飞,钟海雁,郑仕宏. 5种沙梨主要营养成分分析[J]. 食品与机械,2005,21(3):41~42,60.
[10] 卓雄标,林雄平,苏巧玲,等. 南五味子果实营养成分的研究[J]. 宁德师范学院学报(自然科学版),2015,27(4):418~420.
[11] 吴峰华,花雪梅,成纪予,等. 野芝麻的营养成分分析及评价[J]. 营养学报,2015,37(3):306~307.
[12] 李秀芬,张德顺,王少鸥,等. 木槿两变型花蕾的营养成分分析[J]. 植物资源与环境学报,2009,18(4):85~87.

(本文发表于《植物资源与环境学报》,2016年)

云南省藤黄属植物的地理分布及其区系特征

马婷,司马永康,马惠芬,陈少瑜

(云南省森林植物培育与开发利用重点实验室;国家林业局开放性重点实验室云南珍稀濒特森林植物保护和繁育实验室;云南省林业科学院,云南 昆明 650201)

摘要:云南省产藤黄属植物13种,占中国总种数的65.0%,是中国藤黄属植物种类最丰富的地区,因此是进行该属植物研究的典型区域。本文以文献和标本为基础资料,通过聚类和相似性分析,研究了云南省藤黄属植物的地理分布及其区系特征。结果表明:①藤黄属植物在云南的水平分布分区包括:滇东南至滇西南、滇南、滇东南偏滇南至滇西南、滇南偏滇西南和滇西南至滇西北独龙江等五个区。垂直分布分区为:丘陵至低山地带、低山至近低亚中山地带和亚中山地带等三个地带。②云南省藤黄属植物特有现象显著。其中,有8种(61.54%)属于中国特有分布区类型,5种(38.64%)属于热带亚洲(印度—马来西亚)分布区类型。③藤黄属主要分布范围是600~1500m的滇东南、滇南至滇西南的低山和近低亚中山地带,属于热带分布区类型。④藤黄属植物在云南的分布具有明显的地理替代现象。⑤与相邻省区的比较表明,云南省藤黄属植物区系与广西省联系最为密切。

关键词:藤黄属;分布区域;聚类分析;地理替代;相似性系数

Geographical Distribution and Floristic Characteristics of the Genus *Garcinia* Linn. from Yunnan Province

Ma Ting, Sima Yong-kang, Ma Hui-fen, Chen Shao-yu

(Yunnan Provincial Key Laboratory for Cultivation and Exploitation of Forest Plants; Yunnan Laboratory for Conservation of Rare, Endangered & Endemic Forest Plants, Public Key Laboratory of the State Forestry Administration; Yunnan Acaedmy of Forestry, Kunming, Yunnan 650201, China)

Abstract: Thirteen species of *Garcinia* Linn., accounting for 65.0% of the species from China, occour in Yunnan Province and Yunnan is the richest province of China in *Garcinia* species. Therefore, Yunnan is a key region for researches of the genus *Garcinia* Linn. Based on the data of herbarium specimens and literatures, the geographical distribution and floristic characteristics of the genus *Garcinia* Linn. from Yunnan province were studied by cluster analysis and similarity analysis. The results reveal that as follows: ①The horizontal distribution area of the genus *Garcinia* Linn. in Yunnan can be divided into southeastern to southwestern Yunnan region, southern Yunnan region, southeast-southern Yunnan to southwestern Yunnan region, southern-southwestern Yunnan region and southwestern to northwestern (Dulongjiang) Yunnan region. The vertical distribution area of the genus can be divided intohill to low-mountain zone, low-mountain to sub-mid-mountain near low-mountain zone and sub-mid-mountainzone. ②The *Garcinia* species from Yunnan are conspicuous in endemism. Among them, eight species (61.54%) belong to the endemic to China areal type, and five species (38.64%) belong to the tropical Asia (Indo-Malaysia) areal type. ③The most of the *Garcinia* species from Yunnan are distributed in the zones of the altitudes from 600 to 1500 meters in the Southeast, South and Southwest of Yunnan, and belong to the tropical areal type. ④The *Garcinia* species in Yunnan are obvious in geographical vicariance. ⑤Comparing with other neighboring regions, Guangxi plays the greatest connection in geographical relationships with Yunnan.

Key words: *Garcinia* Linn. ; Distribution areas; Cluster analysis; Geographical vicariance; Coefficient of similarity

藤黄属(*Garcinia* Linn.)是一类隶属藤黄科(Guttiferae/Clusiaceae)的中等大小的木本植物类群,全世界约450种,主要分布在热带亚洲、澳大利亚东北部、马达加斯加、热带非洲及非洲南部、波利尼西亚西部和热带美洲(Li et al., 2007),在分布类型上属于泛热带分布。该属植物中国有20种,产台湾南部、福建、广东、海南、广西南部、云南南部、西南部至西部、西藏东南部、贵州南部及湖南西南部(李锡文,1996;中国科学院昆明植物研究所,2006;Li et al., 2007)。本属多种的果实都可食用,种子富含油脂,黄色树脂供药用,多种的木材坚韧耐用,可供建筑和家具制品。近年来,藤黄属植物在经济上和医药上的重要价值,尤其是其食用、药用和保健功效越来越引起人们的重视(许本汉,1998;邹明宏,杜丽清等,2007;文定青,邹明宏,2008)。但就现有资料看,对该属植物的地理分布及区系特征的研究还鲜有报道。本文对云南省藤黄属植物资源的地理分布和区系特征进行分析和总结,以期为该属物种资源的开发利用和保护等提供基础资料。

1 资料来源与研究方法

1.1 资料来源

分析数据来源于中国科学院昆明植物研究所(KUN)、国家林业局云南珍稀濒特森林植物保护和繁育实验室(YCP)、云南省林业科学院(YAF)、中国科学院西双版纳热带植物园(HITBC)、广西中国科学院植物研究所(IBK)、中国科学院华南植物园(IBSC)、江西省中国科学院庐山植物园(LBG)、中国科学院西北高原生物研究所(HNWP)、中国科学院武汉植物园(HIB)、中国科学院植物研究所(PE)等植物标本馆(室)的标本资料,以及经核实和修订的有关植物志(李锡文,1996;中国科学院昆明植物研究所,2006;Li et al., 2007)、植被调查报告、地方资料、书籍、学术论文记载的藤黄属植物分布资料(LamphayInthakoun,Claudio O,2008;MARK N M,SOUNTHONE K,PHILIP T et al.,2007;U Kyaw Tun,U Pe Than,2006)。

1.2 研究方法

1.2.1 聚类分析
1.2.1.1 地理分布分区

以藤黄属植物在云南分布的33个县市为分类单位(Operational Taxonomic Unit,简写成"OTU"),以云南产13种藤黄属植物的有无为分类特性进行二态编码,即某县市有某种分布记为1,反之为0,分析云南省藤黄属植物的水平分布分区。

同时,以藤黄属植物在云南省分布的海拔为分类单位,海拔从200m为起始,每上升50m为一个梯度,以云南产13种藤黄属植物的有无为分类特性进行二态编码,即某海拔梯度有某种分布记为1,反之为0,研究云南省藤黄属植物的垂直分布分区。

1.2.1.2 植物分布区类型分类

以13种藤黄属植物为分类单位,以它们在云南省分布的33个县市和不同海拔的有无为分类特性分别进行二态编码,即某种在某县市或某海拔高度有分布记为1,反之为0,探讨云南省藤黄属植物水平和垂直分布区类型分类。

1.2.2 相似性分析

云南省藤黄属植物与相邻地区的关系,运用区系相似性系数(Coefficient of similarity)进行比较,采用张镱锂(1998)建议的Sørensen相似系数计算,其公式如下:$S=2c/A+B$,式中的S为相似性系数,A和B分别为两地的全部种类,c为两地共有种类(文定青,邹明宏,2008;王荷生,1992)。

1.2.3 数据处理

聚类分析数据用SPSS软件进行处理,先计算各OUT之间的平方欧式距离,作为表征相似程度的度量指标,然后用分类效果较好的Ward系统聚类法进行Q聚类分析,作出分类单位的树系图。聚类分析时,需要确定一条分类等级划分线,一般以类群聚合水平为纵坐标,聚合次数为横坐标,将全部聚合过程描绘成一条阶梯式折线,由结合线上聚合水平较大的跳跃位置来确定等级分界值(袁菊红,2010;刘春迎,王莲英,1995)。

2 结果与分析

2.1 地理分布分析

经文献查阅和标本鉴定,云南有藤黄属植物13种,特产5种,云南省种数占中国藤黄属植物总数的65.0%。该13种藤黄属植物分布于云南省的7个州33个县市,海拔分布范围在200~1 900m之间(表1)。

2.1.1 水平分布分区

通过对云南省藤黄属植物水平分布进行聚类分析,得到藤黄属植物水平分布聚类图(图1)。再根据该聚类图,作出聚类结合线(图2)。从结合线可以看出,类群的聚合在纵轴上分布是不均匀的,等级分界值取在结合线跳跃位置的中点,即$L_1=(19+25)/2=22$,$L_2=(10+14)/2=12$。

表1 云南省藤黄属植物分布表

Table 1 The distribution of thegarcinia species from Yunnan province

编号 Number	种名 Species	分布区域 Distribution	海拔(m) Altitude
1	大苞藤黄 Garcinia bracteata	云南(勐腊、勐海、景洪、沧源、富宁、西畴、麻栗坡、耿马、临沧、镇康、凤庆、金平、元江),广西	400~1 800
2	怒江藤黄 Garcinia nujiangensis	云南(贡山、盈江、陇川),西藏	800~1 700
3	大果藤黄 Garcinia pedunculata	云南(瑞丽、盈江、贡山),西藏。孟加拉国,印度	200~1 500

(续)

编号 Number	种名 Species	分布区域 Distribution	海拔(m) Altitude
4	山木瓜 *Garcinia esculenta*	云南(盈江、瑞丽、陇川、贡山)	900~1 650
5	大叶藤黄 *Garcinia xanthochymus*	云南(金平、绿春、勐腊、景洪、勐海、孟连、河口、西畴、景东、澜沧、景谷、云县、耿马、盈江、梁河),广西。越南,老挝,柬埔寨,泰国,缅甸,孟加拉国,印度,不丹,尼泊尔	200~1 400
6	云南藤黄 *Garcinia yunnanensis*	云南(勐腊、勐海、景洪、临沧、耿马、镇康、沧源、澜沧、潞西)	680~1 600
7	云树 *Garcinia cowa*	云南(金平、绿春、河口、屏边、勐海、勐腊、景洪、思茅、澜沧、耿马、沧源、临沧、瑞丽)。越南,老挝,柬埔寨,马来西亚,孟加拉国,印度	260~1 300
8	木竹子 *Garcinia multiflora*	云南(西畴、麻栗坡、马关、文山、金平、河口、元阳、绿春、屏边、双江、新平、勐海、勐腊、景洪、瑞丽)、湖南、江西、福建、贵州、广西、广东、香港、海南、台湾。越南	400~1 900
9	版纳藤黄 *Garcinia xipshuanbannaensis*	云南(勐腊、元阳、盈江)	600~1 660
10	金丝李 *Garcinia paucinervis*	云南(麻栗坡、盈江),广西	300~1 400
11	长裂藤黄 *Garcinia lancilimba*	云南(勐海、勐腊、景洪)。老挝	600~1 800
12	红萼藤黄 *Garcinia rubrisepala*	云南(盈江)	300~400
13	双籽藤黄 *Garcinia tetralata*	云南(景洪、沧源、耿马)	500~1 750

将L_1,L_2分别标在聚类图(图1)上。从图中可以看出,结合线L_1将云南省藤黄属植物分布的33个地区分为Ⅰ和Ⅱ两大类。在结合线L_2处将第Ⅰ大类分为A、B、C三组:A组包括云县、孟连、梁河、景东和景谷等五个县市;B组包括绿春、河口、屏边、思茅等十三个县市;C组为贡山、盈江和陇川等三个县。第Ⅱ大类分为D和E两组:D组为景洪、勐腊和勐海等三个县市;E组包括临沧、沧源、耿马等九个县市。

根据聚类分析的结果,以及各类群在云南省所处的地理位置,将藤黄属植物在云南的分布地划分为五个区,即:Ⅰ区为滇南偏滇西南区,Ⅱ区为滇东南偏滇南至滇西南区,Ⅲ区为滇西南至滇西北独龙江区,Ⅳ区为滇南区,Ⅴ区为滇东南至滇西南区。由藤黄属植物在云南的分布分区可知,藤黄属植物在云南主要分布于滇东南、滇南至滇西南的区域。

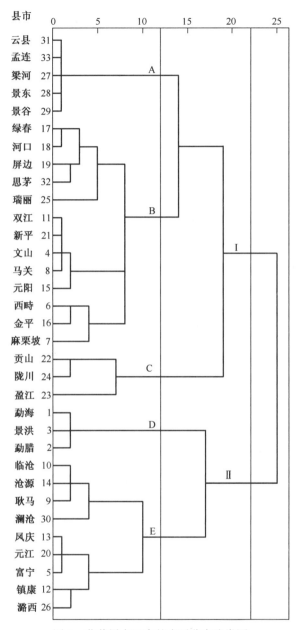

图 1 藤黄属在云南的水平分布聚类图

Figure 1　The dendrogram of the horizontal distribution of the genus *Garcinia* in Yunnan

2.1.2 垂直分布分区

对云南省藤黄属植物的垂直分布进行聚类分析,得到垂直分布聚类图(图3)。根据该聚类图作出垂直分布的聚类结合线(图4)。分界值取在结合线上最大跳跃位置的中点,即 $L=(9+21)/2=15$ 处。将 L 标在聚类分布图 3 上,则结合线 L 将云南省藤黄属植物分布的海拔区域划分为 Ⅰ、Ⅱ 和 Ⅲ 三类。其中,Ⅰ 大类为垂直分布范围在海拔 1 450~1 900m 之间的亚中山地带;Ⅱ 大类是海拔范围在 200~550m 之间的丘陵至低山地带;Ⅲ 大类为海拔在 600~1 500m 之间的低山至近低亚中山地带。

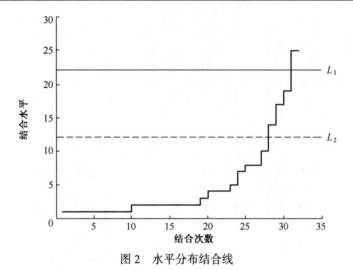

图 2　水平分布结合线

Figure 2　Combination line of the horizontal distribution

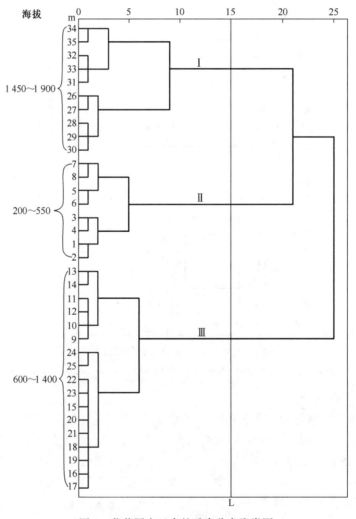

图 3　藤黄属在云南的垂直分布聚类图

Figure 3　The dendrogram of the vertical distribution of the genus *Garcinia* in Yunnan

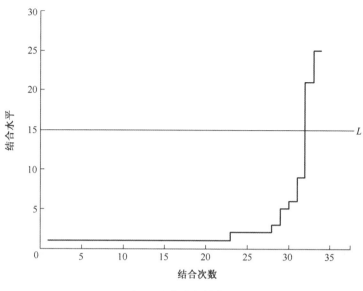

图 4 垂直分布结合线

Figure 4　Combination line of the vertical distribution

2.2　地理区系成分分析

2.2.1　属和种的分布区类型

按照吴征镒(1991)关于中国种子植物属分布区类型的划分标准,藤黄属植物属于泛热带分布(2)。而就在云南分布的种类而言,大苞藤黄(Garcinia bracteata C. Y. Wu ex Y. H. Li)、怒江藤黄(Garcinia nujiangensis C. Y. Wu et Y. H. Li)、山木瓜(Garcinia esculenta Y. H. Li)、云南藤黄(Garcinia yunnanensis H. H. Hu)、版纳藤黄(Garcinia xishuanbannaensis Y. H. Li)、金丝李(Garcinia paucinervis Chun et F. C. How)、红萼藤黄(Garcinia erythrosepala Y. H. Li)和双籽藤黄(Garcinia tetralata C. Y. Wu ex Y. H. Li)等8种属于中国特有分布区类型(15),占在云南分布种类总数的61.54%;大果藤黄(Garcinia pedunculata Roxburgh ex Buchanan-Hamilton)、大叶藤黄(Garcinia xanthochymus J. D. Hooker ex T. Anderson)、云树(Garcinia cowa Roxburgh)、木竹子(Garcinia multiflora Champion ex Bentham)和长裂藤黄(Garcinia lancilimba C. Y. Wu ex Y. H. Li)等5种属于热带亚洲(印度—马来西亚)分布区类型(7),占38.46%,其中大果藤黄属于热带印度至华南分布区亚型(7-2),木竹子和长裂藤黄属于越南(或中南半岛)至华南(或西南)分布区亚型(7-4)。

2.2.2　在云南的分布区类型

2.2.2.1　水平分布区类型

对藤黄属植物进行在云南的水平分布区类型聚类分析,得到水平分布区类型聚类图(图5)。根据该聚类图作出聚类结合线(图6),分界值取跳跃位置的中点,得出 $L_1 = (13+25)/2 = 19$, $L_2 = (10+13)/2 = 11.5$, $L_3 = (4+7)/2 = 5.5$。

将 L_1, L_2 和 L_3 标在聚类图(图7)上,可以看出:①L_1将藤黄属植物分为Ⅰ和Ⅱ两大类;②L_2将第Ⅱ大类划分为A和B两个亚类;③L_3将A亚类分成a、b、c三个小类。结合藤黄

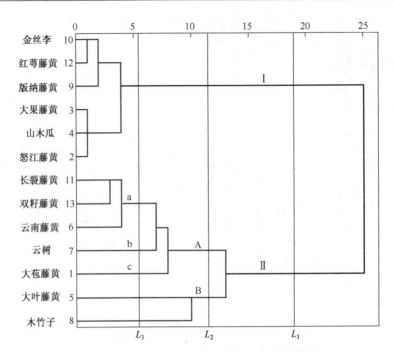

图 5 藤黄属在云南的水平分布区类型聚类图

Figure 5 The dendrogram of the horizontal areal-types of the genus *Garcinia* in Yunnan

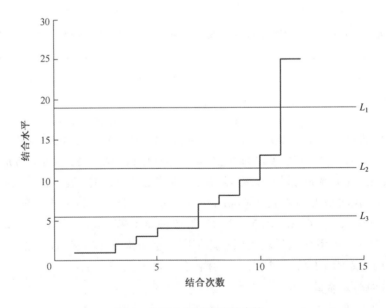

图 6 水平分布区类型结合线

Figure 6 Combination line of the horizontal areal-types

属植物在云南的分布分区看,第Ⅰ类包括金丝李、红萼藤黄、版纳藤黄、大果藤黄、山木瓜和怒江藤黄等 6 种,占在云南分布种类总数的 38.46%,分布于滇东南至滇西南和滇西北地区,为滇东南至滇西南和滇西北分布区类型;B 亚类包括大叶藤黄和木竹子等 2 种,占 15.38%,分布于滇东南至滇西南和滇中地区,为滇东南至滇西南和滇中分布区亚型;a 小类包括长裂藤黄、双籽藤黄和云南藤黄等 3 种,占 23.08%,分布于滇南至滇西南地区,为滇南至滇西南分布区亚型;b 小类包括云树 1 种,占 7.69%,分布于滇东南至滇西南地区,为滇东南至滇西南分布区亚型;c 小类包括大苞藤黄 1 种,占 7.69%,分布于滇东南至滇西南偏滇南和滇中地区,为滇东南至滇西南偏滇南和滇中分布区亚型。

2.2.2.2 垂直分布区类型

按照垂直分布对藤黄属植物进行在云南的垂直分布区类型聚类分析,得到垂直分布区类型聚类图(图 7)。根据该聚类图作出聚类结合线(图 8),分界值取跳跃位置的中点,得出 $L_1=(14+25)/2=19.5, L_2=(4+10)/2=7$。

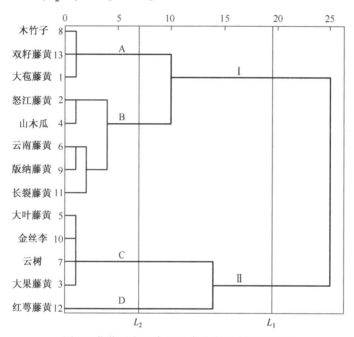

图 7 藤黄属在云南的垂直分布区类型聚类图

Figure 7 The dendrogram of the vertical areal-types of the genus *Garcinia* in Yunnan

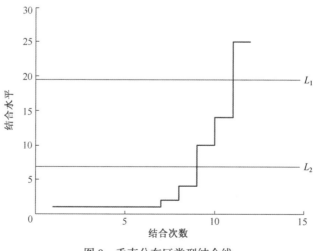

图 8 垂直分布区类型结合线

Figure 8 Combination line of the vertical areal-types

将 L_1,L_2 标在聚类分析图 7,则:①分界线 L_1 将 13 种藤黄属植物分为 Ⅰ 和 Ⅱ 两大类;② L_2 将第 Ⅰ 类分为 A 和 B 两个亚类,将第 Ⅱ 大类分为 C 和 D 两个亚类。结合藤黄属植物在云南的垂直分布区聚类分析的结果,可得出:A 大类包括木竹子、双籽藤黄和大苞藤黄等 3 种,占在云南分布种类总数的 23.08%,分布于海拔 400~1 900m 范围内,基本涵盖了整个藤黄属的垂直分布范围,属于丘陵至低山和亚中山分布区类型;B 类包括怒江藤黄、山木瓜、云南藤黄、版纳藤黄和长裂藤黄等 5 种,占 38.46%,海拔分布范围在 600~1 800m 之间,属低山至亚中山分布区类型;C 类有大叶藤黄、金丝李、云树和大果藤黄等 4 种,占 30.77%,海拔分布在 200~1 500m 之间,属丘陵至低山和近低亚中山分布区类型;D 类仅红萼藤黄 1 种,占 7.69%,海拔分布在 300~400m 之间,属于丘陵分布区类型。可见,藤黄属植物在云南主要分布于海拔 200~1 900m 范围内的丘陵、低山和亚中山,随着海拔的升高或降低,藤黄属植物种类总体上呈现逐渐减少的趋势。以 600~1 500m 海拔低山和近低亚中山分布的种类相对较多,分布较为集中。

2.3 地理替代现象

2.3.1 水平替代

藤黄属植物在云南的水平替代现象十分明显。大苞藤黄主要分布于滇东南,长裂藤黄和双籽藤黄分布于滇南,在西南部也偶有分布,云南藤黄分布于滇南到滇西南,怒江藤黄、大果藤黄和山木瓜分布于滇西,它们呈现出由东南至西南的水平替代规律。同时,分布于滇南的长裂藤黄、双籽藤黄、云树和大叶藤黄,分布于滇西南至滇西北的红萼藤黄、大果藤黄、山木瓜和怒江藤黄之间也构成了由南至北的地理替代分布。

2.3.2 垂直替代

红萼藤黄垂直分布范围较窄,分布于 300~400m 海拔范围内,随着海拔升高逐渐被分布范围在 200~1 500m 的大叶藤黄、金丝李、云树和大果藤黄取代;随着海拔的继续升高,这四种藤黄属植物又被垂直分布范围在 600~1 800m 间的怒江藤黄、山木瓜、云南藤黄、版纳藤黄和长裂藤黄取代;当海拔继续升高至 1 900m 时,只有木竹子、双籽藤黄和大苞藤黄还能生长。因此,这些种的分布呈现出海拔由低到高的地理替代规律。

2.4 与相邻地区藤黄属植物区系的关系

根据司马永康等(2000)的 7 级相似等级划分标准,结合种的相似性系数来看(表 2),云南藤黄属植物与广西的相似性系数最高,为 0.43,处于相似等级水平,地理关系最为密切;其次是中国西藏和老挝,与它们的相似性系数分别为 0.24 和 0.22,处于很近似水平,地理关系较为密切;而与缅甸、中国贵州和越南的相似性系数分别为 0.18、0.14 和 0.11,处于近似水平,地理关系较为疏远。这与中国植物区系图所示该分布区属于 ⅡG22 相一致(Wu Z Y,Wu S G,1998;吴征镒,1984;吴征镒,朱彦成,1987)。在中国,云南藤黄属植物与华南藤黄属植物分布最为接近,这与藤黄属植物主要分布于(7-2)热带印度至华南(尤其云南南部分布)相一致(吴征镒,1991)。但从相似性系数值的大小来看,其值多在 0.4 以下,说明云南藤黄属植物的区系特征与其他国家及省区有明显的区别。这与云南省特殊的地理位置和多样化的气候条件有着必然的联系。同时,古老的区系特征有利于孑遗物种的保存,奇特的地貌类型有利于新物种的形成。

表 2 云南省藤黄属植物区系与邻近地区比较
Table 2 The comparison of the floras of the genus *Garcinia* Linn. between Yunnan and other neighboring regions

地区	西藏	贵州	广西	缅甸	老挝	越南
总种数	4	1	10	20	14	22
与云南共有种数	2	1	5	3	3	2
相似性系数	0.24	0.14	0.43	0.18	0.22	0.11

3 讨 论

(1)该类文章的聚类分析中,距离系数大多采用欧式距离,聚类方法用非加权组平均法(UPGMA)(陈建明,2005;左家哺,1989)。刘凌燕(2008)认为采用哪一种聚类方法能得到最合理的分类结果,这在数量分类学理论上没有定论。目前尚无法完全从数学角度进行最优化分割,这需要我们从研究对象的实际属性等角度综合考查、比较来确定最合理的分类结果。因此本研究选取了多种距离系数及聚类方法对文中数据进行分析,发现用平方欧式距离作为表征相似程度的度量指标,然后用 Ward 系统聚类法进行聚类效果较好。

(2)藤黄属植物主要分布的滇东南、滇南至滇西南地区属于云南热带范围(吴征镒,1984;吴征镒,朱彦成,1987)。但云南省藤黄属植物中,有部分种分布至滇西北独龙江地区。这一地区海拔相差较大,立体气候明显,低海拔地带热量丰富,高海拔地区气候温凉(冯建孟,2010)。文中将这一地区仍归入热带范围,因为云南是一个基带为热带的大山原,云南的植被分布应视为一个大的沿海拔高度的垂直分布,在海拔高度 1 500m 以下地区均为热带(刘慎谔,1959;任美谔,1963)。而藤黄属植物在该区的垂直分布范围多在 1 500m 以下,因此本文将藤黄分布的滇西南至滇西北独龙江地区也归入热带范围。这一看法与朱华的观点,即:由于云南北高南低的地势,在云南北部特别是西北部,不论其纬度,只要是海拔低的区域都是热带区域相一致。

(3)云南省的藤黄属植物资源十分丰富,同时藤黄属植物具有十分重要的经济和生态价值。然而目前对该属植物资源的保护和利用研究还非常少,该属植物基本处于野生或者半野生状态。今后应该加强对该属的种子资源收集和保存工作。

致谢:感谢文中所提各单位标本馆及其工作人员在查阅标本过程中所提供的无私帮助!

参考文献:

[1] 李锡文,1990. 中国植物志 [M]. 北京:科学出版社.
[2] 王荷生,1992. 植物区系地理[M]. 北京:科学出版社,10~12.
[3] 文定青,邹明宏. 2008. 海南岛藤黄属植物的分布及其区系特征[J]. 科技创新导报,(7):245~246.
[4] 吴征镒,1984. 云南种子植物名录上册[M]. 昆明:云南人民出版社,附《云南省植物分区图》.
[5] 吴征镒,朱彦成,1987. 云南植被[M]. 北京:科学出版社,附《云南植被区划图》.
[6] 许本汉,1998. 山野佳果——云南藤黄[J]. 云南林业,19(6):22.
[7] 中国科学院昆明植物研究所. 2006. 云南植物志[M]. 北京:科学出版社.
[8]《云南植被》编写组,1987. 云南植被[M]. 北京:科学出版社.
[9] CHEN J M(陈建明),2005. Cluster analysis on geographical distribution of protective plants in Guizhou prov-

ince[J]. *Journal of Hangzhou Teachers College（Natural Science Edition）*杭州师范学院学报（自然科学版），4(5)：384~388.

[10] FENG J M(冯建孟)，XU C D(徐成东)，ZHA F S(查凤书)，et al，2010. Plant biodiversity and flora composition in north-west Yunnan[J]. *Resources and Environment in the Yangtze Basin*（长江流域资源与环境），19（1）：65~72.

[11] Lamphay Inthakoun，Claudio O，2008. Lao Flora A checklist of plants found in Lao PDR with scientific and vernacular names[M] Morrisville：Lulu Enterprises.

[12] LIU C Y(刘春迎)，WANG L Y(王莲英)，1995. The Numerical Classification of Some cultivars Florist's Chrysanthemum(I)[J]. *JOURNAL OF BEIJINGFORESTRYUNIVERSITY*（北京林业大学学报）1995, 17（2）：79~87.

[13] LIU L Y(刘凌燕)，ZHANG M L(张明理)，LI J Q(李建强)，et al.，2008. A Numerical Taxonomic Study of the Genus CyclobalanopsisOersted from China[J]. *Journal of Wuhan Botanical Research*（武汉植物学研究），26(5)：466~475.

[14] LIU S E(刘慎谔)，FENG Z W(冯宗纬)，ZHAO D C(赵大昌)，1959. Some principles of vegetational regionalization in China[J]. *Acta Botanica Sinica*（植物学报），8(2)：87~105.

[15] Li Xiwen（李锡文），Li Jie（李捷），Peter F Stevens，2007. Flora of China［M］. Beijing：Science Press. MARK N M，SOUNTHONE K，PHILIP T et al，2007. A Checklist of the Vascular Plants of Lao PDR［M］. Edinburgh：RoyalBotanic GardenEdinburgh.

[16] REN M E(任美锷)，XIANG R Z(相韧章)，1963. On some theoretical problems of the physical regionalization in China from a contradictory opinion& Re-discussion on the physical regionalization in China[J]. *Journal of Nanjing University*：Natural Sciences（南京大学学报：自然科学），3：1~12.

[17] SIMA Y K(司马永康)，WU X S(吴新盛)，LI K（李克明），2000. Similarity and Dissimilarity Percentage of Component Abundance. *Yunnan Forestry Science and Technology*（云南林业科技），2000，（2）：20~23.

[18] U Kyaw Tun，U Pe Than，2006. Medicinal Plants of Myanmar[EB/OL]. http://www.tuninst.net/MyanMedPlants/TIL/famH/Hypericaceae.htm#Garcinia-spp WU C Y(吴征镒)，1991. The areal-types of chinese genera of seed plants[J]. *Acta Botanica Yunnanica*（云南植物研究），IV：1~139.

[19] Wu Z Y，Wu S G，1998. A proposal for a new floristic kingdom (realm)-the E. Asiatic kingdom, its delineation and characteristics[M]. Beijing：China Higher Education Press.

[20] YUAN J H(袁菊红)，2010. Numerical Taxonomy and Principal Component Analysis of Lycoris from China[J]. *Subtropical plant science*（亚热带植物科学），39(3)：32~37.

[21] ZHANG Y L(张德锂)，1998. COEEFICIENT OF SIMILARITY—AN IMPORTANT PARAMETER IN FLORISTIC GEOGRAPHY[J]. *GEOGRAPHICALRESEARCH*（地理研究），17(4)：429~434.

[22] ZHU H（朱华)，2008. Distribution Patterns ofGeneraofYunnan Seed Plants withReferencesto Their Biogeographical Significances[J]. *ADVANCES IN EARTH SC IENCE*（地球科学进展），23(8)：830~839.

[23] ZOU M H(邹明宏)，DU L Q(杜丽清)，ZENG H(曾辉)，2007. Advances of Development and Utilization of Garcinia Resources[J]. *Chinese Journal of Tropical Crops*（热带作物学报），28(4)：122~127.

[24] ZUO J F(左家哺)，1989. Divisionofthe floristicelementsofmichelia ofguizhou by method of fuzzy clurtering analysis[J]. *Acta Botanica Yunnanica*（云南植物研究），11(4)：415~422.

（本文发表于《云南大学学报》，2013年）

铁力木苗木分级研究

杨斌,周凤林,史富强,徐玉梅,李玉媛

(云南省林业科学院,云南 昆明 650204)

摘要:采用逐步聚类分析方法,对铁力木容器苗苗木的分级标准进行了探讨。提出了以苗高和地径作为该树种苗木分级的质量指标,并得出铁力木苗木的3级分级标准,即:Ⅰ级苗:树高≥19.4cm,地径≥0.25cm;Ⅱ级苗:19.4cm>树高≥12.3cm, 0.25cm>地径≥0.16cm;Ⅲ级苗:树高<12.3cm,地径<0.16cm。

关键词:铁力木;苗木分级;逐步聚类

A Study of Seedling Grading of *Mesua ferrea* L.

Yang Bin, Zhou Feng-Lin, Shi Fu-qiang, Xu Yu-mei, Li Yu-yuan

(Yunnan Academy of Forestry, Kunming Yunnan 650201)

Abstract: The standard of seedling grading of *Mesua ferrea* L. was studied by using the methods of gradual cluster analysis. Seedling height and basal diameter were proposed as the main indexes of quality for seedling grading of this species. The following standards were produced for seedlings of the first to the third grades. The seedlings with the height of 19.4cm or above, basal diameter of 0.25cm or above were classified as grade one, those with height lower than 19.4cm but higher than or equal to 12.3cm, basal diameter lower than 0.25cm but higher than or equal to 0.16cm were classified as grade two, those with height lower than 12.3cm and basal diameter lower than 0.16cm were classified as grade three.

Keyword: *Mesua ferrea* L. ; seedling grading; gradual cluster analysis

铁力木(*Mesua ferrea* L.)属藤黄科,为热带珍贵用材树种,目前我国只在云南耿马县孟定海拔540~600m的低山尚有成片的小面积纯林。其木材为亚洲热带著名的硬木,铁力木枝密叶浓,新叶紫红,异常艳丽,花大香浓,是很好的观赏绿化树种。本文结合云南省"十五"攻关课题,对铁力木的苗木分级提出标准。

1 试验地概况

试验地位于西双版纳普文试验林场,22°24′~22°26′N,101°04′~101°06′E之间,属热带北缘季风气候类型,一年之中受湿润的西南季风和干暖的西风南支急流交替控制,年平均气温20.1℃,≥10℃积温7 459℃,最热月均温23.9℃,最冷月均温13.9℃,极端最高气温38.3℃,极端最低温-0.7℃。年降水量1 655.3mm,年平均相对湿度83%。土壤为赤红壤,呈酸性,pH值4.3~6.3,有机质含量低,缺氮,尤其少磷,而钾较丰富。

2 研究方法

在同一育苗点随机抽取 50 株苗龄 150d 的铁力木容器苗,测定苗高、地径、全株鲜重、地上重、地下重、根系长、高径比等 7 个生长量指标,经相关分析后确定质量指标,采用聚类分析的数学方法进行分级[1~4]。据此,提出铁力木苗木的分级标准。

3 苗木分级标准的计算

3.1 分级指标的确定

苗木的质量是由一系列苗木性质组成的,壮苗(或称合格苗)的标志一般是指:苗木通直、顶芽饱满、地径粗壮、枝叶繁茂、根系发达、有较多的侧根和须根、主根粗且较短、重量大、色泽正常、木质化程度高、无病虫害、无机械损伤,等等,这些标志反映了苗木质量指标的多样性。评定苗木质量时,由于各质量指标间相关极为密切,利用多个指标,其信息往往是重叠的[5~6],因而有必要对苗木质量指标的共性进行研究,提取既能反映苗木质量,又易于测量,便于应用的少数较直观指标,用以进行苗木分级。

为充分说明苗木各因子之间的相关性[7],进行相关分析。

表 1 铁力木容器苗各指标间的相关距阵
Table 1 Correlation matrix between different quality factors of *Mesua ferrea* L. container seedling

	苗高	地径	冠幅	全株鲜重	地上部分重	地下部分重	根系长	高径比
苗高	1.0							
地径	0.465 45	1						
冠幅	0.206 88	0.364 54	1					
全株鲜重	0.545 67	0.786 63	0.535 96	1				
地上部分重	0.547 81	0.697 43	0.554 16	0.955 29	1			
地下部分重	0.356 67	0.716 89	0.314 98	0.760 46	0.534 47	1		
根系长	-0.137 80	0.231 62	0.183 96	0.317 20	0.220 59	0.422 29	1	
高径比	0.673 59	-0.298 62	-0.087 9	-0.074 08	-0.012 32	-0.184 71	-0.334 8	1

从表 1 的相关距阵中可以清楚的看出:苗木全株鲜重为相关的中心,最能体现苗木质量,同时也说明在同一批苗木中,苗木重量越大,积累的干物质越多,其质量就越好,因而用全株鲜重来评价苗木的质量是最好的指标。

地径、苗高与全株鲜重相关性也较强,相关系数分别为 0.786 63 和 0.545 67,说明两者是评价苗木质量的次主要指标。根系长、高径比与全株鲜重的相关性均较差,而高径比仅是一个算术商,与苗木个体的大小无关,只能描述苗木个体的均衡度,因而不能用作苗木分级的评价指标。

苗木各指标间的相关性,一方面反映了苗木各器官之间相对均衡的生长作用,即其整体性,另一方面又说明了评价苗木质量时可选择较少的指标。

显然全株鲜重是评定苗木质量的一个比较可靠的指标,但在确定苗木质量的指标时,既要考虑到有足够多的信息量,也要考虑到在生产实践中的可操作性,因此本文选择苗高与地径作为铁力木容器苗苗木分级的指标。

3.2 分级的原则

按照生产实际,苗木一般分为Ⅲ级,其中Ⅰ、Ⅱ级苗为合格苗,可出圃上山造林,Ⅲ级苗为不合格苗,应留圃继续培养。

3.3 苗木分级标准的计算

聚类分析所依据的基本原则是直接比较属性中各样本反映出来的性质,将性质相近的属性(性状、变量)分在同一类,而将差别较大的分在不同的类,确定区分事物性质的聚类标志。而对于苗木分级来说,即是利用质量指标来划分苗木个体的相似程度。但如何确定苗木个体间的相似程度呢?在统计学上是以定义它们之间的距离来确定的,即距离越小则其相似程度越大。本文所使用的为欧氏距离公式:

$$d_{ij} = \sqrt{\sum_{k=1}^{n}(X_{ik} - X_{jk})} \quad ①$$

因仅采用苗高(H)和地径(D)两个分组指标,因此 $n=2$,从而①式变为:

$$d_{ij} = \sqrt{(H_I - H_J)^2 + (D_i - D_j)^2} \quad ②$$

3.3.1 数据标准化

为了使数据值能够在同一水平上进行比较和计算分析,必须对所有观测的数据进行标准化。公式如下:

$$Z_{ij} = \frac{x_{ij} - x_{i(\min)}}{x_{i(\min)} - x_{i(\max)}} \quad ③$$

式中:Z_{ij}:标准化值;i:苗高或地径;j:所观测的样苗号(1,2,3,……,50);$x_{i(\max)}$、$x_{i(\min)}$:所观测的总体样本中的苗高或地径的最大值、最小值。

用③式计算得苗高及地径的标准化值(表2)。

表2 苗高及地径标准化值
Table 2 Standardized values of seedling heights and base diameters

样苗号	苗高	地径	样苗号	苗高	地径	样苗号	苗高	地径
1	0.32	0.46	18	0.16	0.50	35	0.32	0.42
2	0.37	0.42	19	0.13	0.31	36	0.42	0.50
3	0.42	0.50	20	0.05	0	37	0.68	0.50
4	0.26	0.42	21	0.58	0.27	38	0.68	0.35
5	0.34	0.46	22	0.76	0.42	39	0.79	0.85
6	0.16	0.46	23	0.61	0.58	40	0.42	0.23
7	0.47	0.58	24	0.37	0.42	41	0.26	1.00
8	0.84	0.38	25	0.42	0.31	42	0.66	0.77
9	0.39	0.46	26	0.47	0.12	43	0.58	0.73
10	0.21	0.54	27	0.71	0.35	44	0.32	0.65
11	0.09	0.35	28	0.82	0.62	45	0.55	0.58
12	0.13	0.69	29	0.58	0.54	46	0.34	0.50
13	0.21	0.23	30	0.66	0.46	47	0.71	0.54
14	0	0.27	31	0.29	0.19	48	0.58	0.62
15	0.47	0.31	32	0.97	0.77	49	0.74	0.69
16	0	0.31	33	1.00	0.81	50	0.79	0.58
17	0.18	0.23	34	0.84	0.73			

3.3.2 初始分级

以苗高、地径值分级,苗木大体上可按"高大→矮小"的顺序排列。较高大的苗木,应有较粗的地径,由于苗木的分化现象,高而细、粗而矮的苗木是存在的,但仅为少量,因此对苗木质量的聚类就是对有序样本的聚类。经过简单的计算,样苗可按"$\Sigma_{标}$"由大到小排序($\Sigma_{标}=H_{标}+D_{标}$,即$\Sigma_{标}$为苗高和地径标准化值的和)。

根据各样苗的$\Sigma_{标}$值,在一维坐标上进行排序,在小群距离较明显的地方,将它们分为三群(即三级),完成苗木的初始分级(表3)。

表3 苗木初始分级结果
Table 3 Preliminary grading results of tested seedlings

	I级苗样苗号	II级苗样苗号	III级苗样苗号
初始分级	43、50、28、42、49、34、39、32、33	1、2、3、4、5、6、7、8、9、10、12、15、18、21、22、23、24、25、26、27、29、30、35、36、37、38、40、41、44、45、46、47、48	20、14、16、17、11、13、31、19

3.3.3 修改分级

由于初始分级是人为的,避免不了错误,因此必须利用数学公式进行修改以改正错误。其原理是按照初始分级结果,分别计算各级苗高、地径标准化值的平均数,并以此平均数作为该级的凝聚中心,计算各样苗与相邻凝聚中心的距离d,样苗距离那个凝聚中心的距离最近,即判为那一级(最短距离法)。每次修改后,如有变化,需按新的分级重新计算凝聚中心和距离。如此反复进行,直到完全没有变化,分级即结束。

开始分级修改时,按照表3的原始分级,将样苗的标准化值列入表4进行修改,并分

表4 欧氏距离修改表
Table 4 Grading results revised by "Minimum distance" method

样苗号	标准化值		第一次修改					第二次修改					第三次修改				
	H	D	X_I^1 (0.80 0.73)	X_{II}^1 (0.45 0.47)	X_{III}^1 (0.12 0.24)	判别 原级	变动	X_I^2 (0.80 0.79)	X_{II}^2 (0.45 0.47)	X_{III}^2 (0.13 0.29)	判别 原级	变动	X_I^3 (0.79 0.68)	X_{II}^3 (0.46 0.47)	X_{III}^3 (0.14 0.30)	判别 原级	变动
33	1	0.81															
50	0.79	0.58	0.15	0.36		I		0.1	0.36		I		0.1	0.35		I	
43	0.58	0.73	0.22	0.29		I		0.21	0.29		I		0.21	0.29		I	
47	0.71	0.54	0.21	0.27		II	I	0.16	0.27		I		0.16	0.26		I	
8	0.84	0.38	0.35	0.40		II	I	0.30	0.40		I		0.30	0.39		I	
41	0.26	1.0	0.60	0.56		II		0.61	0.56		II		0.61	0.57		II	
48	0.58	0.62	0.24	0.20		II		0.22	0.20		II		0.21	0.19		II	
⋮	⋮	⋮	⋮	⋮		⋮		⋮	⋮		⋮		⋮	⋮		⋮	
40	0.42	0.23	0.24	0.30		II		0.24	0.30		II		0.24	0.29		II	
26	0.47	0.12	0.35	0.37		II		0.35	0.38		II		0.35	0.38		II	
4	0.26	0.42	0.19	0.23		II		0.19	0.18		II	III	0.20	0.17		III	
6	0.16	0.46	0.29	0.22		II	III	0.29	0.17		III		0.30	0.16		III	
18	0.16	0.50	0.29	0.26		II	III	0.29	0.21		III		0.30	0.20		III	
31	0.29	0.19	0.32	0.18		III		0.32	0.18		III		0.33	0.19		III	
11	0.09	0.35	0.38	0.11		III		0.38	0.07		III		0.39	0.07		III	

注:┈┈代表初始分级;——代表修改后分级。

别计算各级苗高、地径的标准化值的平均值,作为该级的凝集中心。第一次修改时Ⅰ、Ⅱ、Ⅲ级的凝聚中心分别记为 $X_Ⅰ^1(0.80,0.73)$、$X_Ⅱ^1(0.45,0.47)$、$X_Ⅲ^1(0.12,0.24)$,括号内的数据分别是该级(初始分级)的苗高、地径标准化的平均值。进行第二次修改时记为 $X_Ⅰ^2$、$X_Ⅱ^2$、$X_Ⅲ^2$,余此类推。如第43号样苗其苗高、地径标准化值分别为0.58和0.73,与Ⅱ级苗的距离 $d_Ⅱ = \sqrt{(0.58-0.45)^2+(0.73-0.47)^2} = 0.29$,大于其与Ⅰ级苗的距离。$d_Ⅰ = \sqrt{(0.58-0.80)^2+(0.73-0.73)^2} = 0.22$,因而40号样苗应划为Ⅰ级苗。

第一次修改,Ⅰ、Ⅱ级苗临界处,原Ⅱ级苗中第47、8号两株样苗划归Ⅰ级苗;Ⅱ、Ⅲ级苗临界处,原Ⅱ级苗中第6、18号两株样苗划归Ⅲ级苗。利用第一次分级修改结果后计算所得的凝集中心分别为 $X_Ⅰ^2(0.79,0.68)$、$X_Ⅱ^2(0.45,0.47)$、$X_Ⅲ^2(0.13,0.29)$,再次计算各样苗与凝集中心的距离 d,经计算,第二次修改后Ⅰ、Ⅱ级苗临界处,样苗没有发生变化;Ⅱ、Ⅲ级苗临界处,原Ⅱ级苗中第4号样苗划归Ⅲ级苗。利用第二次分级修改结果,计算所得的凝集中心分别为 $X_Ⅰ^2(0.79,0.68)$、$X_Ⅱ^2(0.46,0.47)$、$X_Ⅲ^2(0.14,0.30)$,再次计算各样苗与凝集中心的距离 d,经计算,第三次修改后Ⅰ、Ⅱ级苗临界处,Ⅱ、Ⅲ级苗临界处,均没有变化。此时聚类分级结束。

3.3.4 临界值的确定

逐步聚类分级的结果是各级苗木聚集在以该级最终凝聚中心为圆心,以 d 为半径的圆内。过大的苗木在Ⅰ级苗的上方,过小的苗木则在Ⅱ级苗的下方,因此只要求出Ⅰ、Ⅱ级苗的下限,就可准确地确定各级别的界限。其计算方法是将最终修改分级的Ⅰ、Ⅱ级凝聚中心绘在方格纸上,求出半径 d,在图上即可读出Ⅰ、Ⅱ级苗的下限值(图1)。

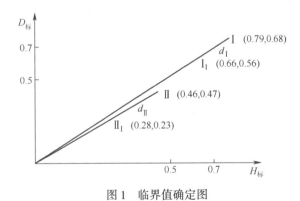

图1 临界值确定图

Figure 1 Critical values of grade Ⅰ and grade Ⅱ seedling

图中Ⅰ、Ⅱ分别为Ⅰ、Ⅱ级苗的最终凝聚中心,$d_Ⅰ$、$d_Ⅱ$ 分别为Ⅰ、Ⅱ级苗的最终聚类圆半径,$Ⅰ_1$、$Ⅱ_2$ 分别为欲求的Ⅰ、Ⅱ级苗的临界点。其中半径 $D = \sqrt{kS_{H标}^2 + S_{D标}^2}$,式中 $k=1$,$S_{H标}^2$、$S_{D标}^2$ 分别是苗高和地径的标准化值的标准差。Ⅰ、Ⅱ级苗的最终凝聚中心分别是 $X_Ⅰ^2(0.79,0.68)$、$X_Ⅱ^2(0.46,0.47)$,经计算 $d_Ⅰ=0.18$,$d_Ⅱ=0.23$,在方格纸中可读出Ⅰ、Ⅱ级苗的分级界限分别为 $Ⅰ_1(0.66,0.56)$、$Ⅱ_2(0.28,0.23)$,即两级间的临界处相应的苗高、地径的标准化值,把 $Ⅰ_1$、$Ⅱ_2$ 代回公式(3)中得出各级苗下限的值为:

Ⅰ级:$H \geq 9.4$ cm,$D \geq 0.25$ cm;

Ⅱ级:$H \geq 12.3$ cm,$D \geq 0.16$ cm。

凡苗高和地径均达以上标准的,即为合格苗,其中一项不达标者,则降一级(表5)。

表5　苗木分级结果
Table 5　Grading result of tested seedlings cm

样苗号	Ⅰ级苗 苗高	地径	样苗号	Ⅱ级苗 苗高	地径	样苗号	苗高	地径	样苗号	Ⅲ级苗 苗高	地径
28	22.5	0.26	1	13.0	0.22	29	18.0	0.24	4	12.0	0.21
32	25.5	0.30	2	14.0	0.21	30	19.5	0.22	6	10.0	0.22
33	26.0	0.31	3	15.0	0.23	35	13.0	0.21	10	11.0	0.24
34	23.0	0.29	5	13.0	0.23	36	15.0	0.23	11	8.7	0.19
39	22.0	0.32	7	16.0	0.25	37	20.0	0.23	12	9.5	0.28
42	19.5	0.30	8	23.0	0.20	38	20.0	0.19	13	11.0	0.16
49	21.0	0.28	9	14.5	0.22	40	15.0	0.16	14	7.0	0.17
50	22.0	0.25	15	16.0	0.18	43	18.0	0.29	16	7.0	0.18
			21	18.0	0.17	44	13.0	0.27	17	10.5	0.16
			22	21.5	0.21	45	17.5	0.25	18	10.0	0.23
			23	18.5	0.25	46	13.5	0.23	19	9.5	0.18
			24	14.0	0.21	47	20.5	0.24	20	8.0	0.10
			25	15.0	0.18	48	18.0	0.26	26	16.0	0.13
			27	20.5	0.19				31	12.5	0.15
41	12.0	0.36									

4　结　论

苗高和地径是反映苗木质量的最直观的指标。利用苗高和地径对铁力木苗木进行逐步聚类分级,其最终分级标准为Ⅰ级苗:$H \geq 19.4 cm$、$D \geq 0.25 cm$;Ⅱ级苗:$19.4 > H \geq 12.3 cm$、$0.25 > D \geq 0.16 cm$。除这两个评选指标外,在生产实践中,进行壮苗选择时还应考虑到壮苗还应具备生长充实、枝叶繁茂、色泽正常、木质化程度高、无病虫害、无机械损伤及饱满的顶芽,等等。由于各育苗地立地条件及育苗技术的差异性,造成苗木质量的参差不齐,故各地因应根据具体情况分析研究,以上分级标准仅供各地在确定铁力木苗木等级时作参照。

参考文献:

[1] 余远火昆. 苗木指标的提取及聚类分级法[J]. 热带林业科技, 1984(3):18~26.
[2] 梁及芝, 何爱华, 金建康. 枫香苗期密度试验及苗木分级指标探讨[J]. 湖南林业科技, 1997, 24(1):23~26.
[3] 杨斌, 赵文书, 陈建文, 等. 西南桦容器苗苗木分级研究[J]. 云南林业科技, 2003(2):17~21.
[4] 杨斌, 赵文书, 姜远标, 等. 思茅松容器苗苗木分级研究[J]. 西部林业科学, 2004(1):32~37.
[5] 刘勇. 我国苗木培育理论与技术进展[J]. 世界林业研究, 2000(5):43~49.
[6] 祁述雄. 中国桉树(第2版)[M]. 北京:中国林业出版社, 2002.
[7] 北京林学院. 数理统计[M]. 北京:中国林业出版社, 1980.

印度尼西亚结香植物种质资源及沉香
人工结香技术与借鉴

郑科,陈鹏,郭永清

(云南省林业科学院,云南 昆明 650201)

摘要:由于天然沉香资源的缺乏,人工结香技术得到了一定的发展,但仍需要进一步深入研究。文中介绍了印度尼西亚结香植物种质资源以及沉香的人工结香技术。印度尼西亚结香植物种质资源极其丰富,其沉香人工接菌诱导结香主要采用镰孢菌属系,可为我国天然香资源的扩大与生产提供借鉴。

关键词:结香植物,天然香,沉香属,拟沉香属,人工结香,镰刀菌属,印度尼西亚

"沉檀龙麝",沉香为众香之首。沉香为沉香属(*Aquilaria*)、拟沉香属(*Gyrinops*)等植物的树干受外界不良因素影响后留存的病理组织,一般为褐色至深黑色,且随颜色加深其品质也随之提高。沉香从古到今为珍品。明朝周嘉胄的《香乘》以及宋朝洪刍的《香谱》和陈敏的《陈氏香谱》都将沉香列为首香进行阐述。沉香用途广泛,可以用于香料、中药以及工艺品。近年来,随着沉香价格飙升,上佳品达到千金难求的地步。印度尼西亚是世界上沉香出产大国,也是世界沉香主要出口地之一,其种质资源丰富,野生沉香资源存量大。2011年我国已经与印度尼西亚签订了年进口沉香500t的协议。近年来,印度尼西亚有关科研机构与企业就沉香人工结香技术进行了探索,本文将对印度尼西亚沉香种质资源与结香技术研究状况进行简述与评价,以期为我国天然香资源扩大与沉香生产提供参考。

1 结香植物种质资源

1.1 种质资源

印度尼西亚结香植物种质资源丰富,资源优势十分明显。印度尼西亚能够产生沉香木的植物不仅仅包括瑞香科沉香属(*Aquilaria*)(15种)和拟沉香属(*Gyrinops*)(13种)。此外,*Aetoxylon*属植物也能结香,种质资源极其丰富。在印度尼西亚,沉香属包括马来西亚沉香(*A. malaccensis*)、*A. beccariana*、小果沉香(*A. microcarpa*)、*A. hirta*、卡氏沉香(*A. crassna*)、*A. cumingiana*、*A. filaria*、*A. tomentosa*、*A. audate*、*A. brachyantha*、*A. moszkowskii*、*A. borne*、*A. citrinaecarpa*、*A. urdanetensis*、*A. apiculata* 等,拟沉香属包括 *G. versteegii*、*G. landermanii*、*G. caudate*、*G. decipiens*、*G. acuminate*、*G. urdanentensis*、*G. citrinaecarpa*、*G. pubifolia*、*G. cumingiana*、*G. decemcostata*、*G. salicifolia*、*G. Podocarpus*、*G. moluccana* 等。

根据2013年统计数据,印度尼西亚野生沉香木 *Aquilaria fillaria* 的沉香产量达到515.8t,野生沉香木 *Aquilaria malaccensis* 为178.5t,*G. versteegii* 为156t,其余各种也相应有

数量不等的产量。印度尼西亚结香植物种质资源十分丰富,远远多于我国,其优良丰富的种质资源值得我国引进和研究。

1.2 主要品种资源介绍

在印度尼西亚沉香种质资源中,马来西亚沉香是当地的主要品种之一,主要分布于苏门答腊岛、邦加岛、加里曼丹岛和苏拉威西岛。分布比较多的还有沉香属卡氏沉香、小果沉香以及 A. fillaria,此外还有拟沉香属的 G. verstegii(分布于龙目岛、松巴哇岛、苏拉威西岛)、G. decipiens 和 G. salicifolia 等[1]。

在沉香属中,马来西亚沉香是印度尼西亚主要产生沉香的树种之一。其花绿色至暗黄色,在印度尼西亚当地,开花结果季节9~12月,同样是胸径10~30cm 的树木,在植物园中比在野外产种量高。在野外,胸径大于40cm 后,其单株种子产量会下降。同样是胸径20~60cm 的树木,马来西亚沉香的种子产量少于小果沉香,前者每棵树种子产量在 3 900~13 270粒。种子在播种 15 天后发芽。除印度尼西亚外,马来西亚沉香在世界上还主要分布于印度、缅甸、马来西亚和菲律宾等地。在原始林与次生林中也生长有马来西亚沉香,但在低地和砂性土上生长良好。马来西亚沉香树高可达 20~40m,胸径最大可以达到 60cm。其幼枝皮轻褐色披毛,老枝皮光滑白色。无结香的树木是白色的,结香后的木质黑色并且比较重。叶椭圆或者披针形,3~3.5cm 宽、6~8cm 长,12~16 对叶脉。果实绿色,每个果实种子2 粒,也有 1 粒的,种子重量为 1kg/1 500粒。印度尼西亚还分布有卡氏沉香(A. crassna),在印度尼西亚茂物种植园。但该沉香种主要分布于中南半岛,如越南、泰国(比较多)。其开花结实季节 8~12 月,种子发芽率比较高,在播种后 9~15d 后发芽率达到 92%,发芽率相对于其他沉香种要高。小果沉香(A. microcarpa)也是印度尼西亚的主要栽培品种之一,果实比马来西亚沉香小。在种植园中开花结果期为 9~12 月。在野外,当胸径大于 50cm 时种子产量会下降,胸径为 20~60cm 的树木种子产量达到 13 260~19 280 粒。拟沉香属的 G. verstegii 一般生长在海拔 400~800m 的低山区,与榕树、番樱桃属、红厚壳属等植物混生在一起。在松巴哇岛,该树种生长在棕色的、薄的腐殖土上。在印度尼西亚佛罗伊斯岛,该树种高度可达 10~17.5m,胸径 25~30cm。G. decipiens 也是拟沉香属的一种,在印度尼西亚分布比较广,包括在 Tembok Jerman 和 Lengke 山脉等地有分布。在目前发现的野生树中,其最高达到 17m,直径达到 30cm。拟沉香属还包括 G. salicifolia 等多种。

2 沉香结香技术

在印度尼西亚,沉香人工结香技术得到了一些发展,包括物理诱导结香技术、真菌接种诱导技术等。

2.1 使用机械工具伤害结香

使用刀砍与钉子钉入树干造成伤害而结香的方法已广泛使用。物理诱导的方法之一是将铁钉嵌入树体,等待结香。印度尼西亚通常的做法是,先用电钻在树体上钻出一个洞,然后将铁钉对准洞口钉入。另外,也可用凿子凿洞,短柄小斧削掉树皮,等待结香。这些研究表明,伤害方法影响沉香的形成,效果也受季节的影响,雨季相对于干旱季节更利于沉香的形成。

通过这种物理诱导的方法产生的沉香仅仅产生淡淡的香气和很小比例的香精油,沉香质量不高,不能满足市场需要,要得到高质量的沉香需要很多年的时间。

2.2 人工伤害与化学物质注射诱导结香

印度尼西亚的做法是将树打洞,在洞口接入塑料管,将一定浓度的化学物质通过塑料管导入树体刺激树木产生防御反应,从而产生沉香。这种化学诱导的方式在机制上可导致防御反应,开始韧皮部的细胞形成愈合组织,随即愈合组织形成被阻止,使树木结香。

在印度尼西亚巴布亚岛开展了不同化学物质使用效果的试验,这些化学物质主要包括甲酸和乙烯利等。不过,由于化学物质的污染作用,当地研究人员对这方面的开发应用还是很谨慎。

2.3 钻孔和真菌接种结香

沉香形成也可通过真菌侵染引起树木防御,从而使树木产生香料复合物的次生代谢物。在野外,香料累积的自然过程也是树木与菌体相互作用的过程,时间长达多年,时间越长,形成的香料越贵,价值更高。

2.3.1 结香诱导真菌的筛选

印度尼西亚林业部 FORDA 和 FNCRDC 这 2 个机构开发的人工沉香诱导技术比较多。他们从被侵染的沉香树中(样本采自印度尼西亚本地以及泰国的罗勇、尖竹汶府、那空沙旺、甲米府、董里府等地),分离出了酵母真菌(*Torula* sp.)、枝孢菌(*Cladosporium* sp.)、附球菌(*Epicoccum granulatum*)、桔青霉(*Pencillium citrinum*)、酱油曲霉(*Aspergillus tamari*)、腐皮镰刀菌(*Fusarium solani*)、球二孢属(*Botryodiplodia*)、毛色二孢菌属(*Lasiodiplodia*)、寄生瓶霉菌(*Philophora parasitica*)、附球孢菌(*Epicoccum* sp.)、球毛壳菌(*Chaetum globosum*)、*Fusarium xylarodes*、*Fusariumtricinctum*、柱孢属(*Cylindrocarpon*)、杧果蒂腐病菌(*Botryodiplodia theobromae*)、新月弯孢霉(*Curvularia lunata*)、尖孢镰刀菌(*Fusarium oxysporum*)、黑盘孢霉属、*Cercosporella* sp.、螺旋毛壳(*Chaetomium spirale*)、*Phialogeniculata* sp.、皮思霉属(*Pithomyces* sp.)、根霉属(*Rhizopus* sp.)、*Spiculostibella* sp. 和木霉菌(*Trichoderma* sp.)等。其中将酵母真菌接种于沉香树,发现有沉香形成,但是后来由于污染则停止了试验;接种枝孢属菌和附球菌,效果一般。研究发现,蜜香树管胞内螺旋形气蚀形成归因于寄生瓶霉菌,寄生瓶霉菌侵染后经常能产生较好质量的沉香。沉香的形成一般伴随着伤害,包括风与暴风雪对树的伤害、树体较衰弱等。试验表明,镰刀菌可成功侵染沉香树并产生沉香。

FORDA 微生物实验室在印度尼西亚各地,包括北苏门答腊、西苏门答腊、占碑、廖内、西加里曼丹、东加里曼丹等 17 个省,从被侵染后产生沉香的树体上分离,并根据基因特征最终确定了 36 个真菌家系,分离物均属于镰刀菌属,绝大多数属于腐皮镰刀菌(*Fusarium Solani*),另外几种真菌分别为西苏门答腊的 *F. ambrosium*、东努沙登加拉的 *F.* sp.(没有鉴别到种)、东加里曼丹的尖孢镰刀菌、西爪哇的 *F. falciforme* 和明古鲁的 *F.* sp.(没有鉴别到种)。印度尼西亚马塔兰大学农学院生物技术实验室的研究团队也开展了刺激沉香产生的接种技术。该团队发现砖红镰孢菌(引起树镰刀霉属癌肿病)可以在 PDA 液中培养,将该真菌注射接种到 4 年生的树上能形成沉香。

FORDA 在龙目岛马塔兰的 5~7 龄的拟沉香属树木 *G. versteegii* 和苏门答腊佩坎巴鲁

12种沉香属树木上进行人工诱导,分别在树上钻8个10cm深、1cm长和宽的洞,利用包括 *F. trifosfrium* 在内的5种镰刀细菌接种,结果表明,在钻孔周围形成了沉香,但未接种的伤害部位也形成了一定的沉香。此外,用2%糖溶液、*Acremonium* sp. 和茉莉酸甲酯混合液注射可生产一定等级的沉香。

印度尼西亚林业部森林研究与发展中心在西加里曼丹Sebadu村开展了人工诱导试验,对300棵8龄的小果沉香树注射了5种真菌:杧果蒂腐病菌、尖孢镰刀菌、*F. bulbigenium*、*F. laseritium* 和 *Phytium* sp.。在注射之前将棕榈糖和奶油与供试真菌混合,然后再注射。研究表明,加入棕榈糖与奶油可促进沉香的形成,其中3种镰刀菌注射效果比较好,在侵染的树体上形成了较大面积的沉香。该机构还与西加里曼丹省林业科研机构合作开展试验,将镰刀菌注入到9年生的马来西亚沉香树中,3个月以后树皮容易被撕破,沉香已形成,将被侵染的部分燃烧后发现有沉香味。

表1列出了Santoso等在印度尼西亚多地开展的利用镰刀菌属注入沉香树进行诱导结香的试验情况。

表1 印度尼西亚用钻孔法注入镰刀菌诱导沉香树结香试验

序号	地点	沉香树种	诱导树的数量(株)
1	邦加岛	小果沉香	160
2	苏门答腊岛北部巴哈洛克	马来西亚沉香	50
3	西苏门答腊	马来西亚沉香	200
4	万丹省的Carita	小果沉香	300
5	西加里曼丹、Bodok	小果沉香、马来西亚沉香、*A. beccarian*	200
6	坎当岸、南加里曼丹	马来西亚沉香、小果沉香	800
7	占碑省、Muaro	小果沉香	50
8	巴厘岛	拟沉香属	50
9	西努沙登加拉省	拟沉香属	200
10	西努沙登加拉省、Flores	拟沉香属	100
11	摩鹿加、伊拿岛	*A. cumminggiana*	150

资料来源:Santoso等的未公开资料。

经过试验,Santoso等认为,决定真菌诱导成功结香的重要因子包括注射方法、真菌系类型和真菌生长介质。

2.3.2 接菌结香的有效方法

印度尼西亚近几年来集中研究并确认了有效的结香方法:

①真菌的类型决定沉香的形成,因此真菌的筛选十分重要,目前印度尼西亚使用比较多的是腐皮镰刀菌(*F. solani*)。②用混合器稀释诱导剂,注射诱导剂前,先用电钻在树干上打孔(大小5mm左右),注射时将针头顺孔插入。③专用注射器针头为小尺寸孔(孔直径3mm),在诱导剂注射完后不久树干注射孔就会自然封闭。注射孔自然封闭过程对刺激沉香形成是很重要的。④将诱导剂以液体的形式用注射器注入,每个孔注射1mL。⑤注射孔之间的距离要足够大,以阻止孔之间的疾病交叉感染。野外接菌刺激诱导沉香,使用自制竹

梯围绕树木用于攀登以完成全树注射。随着接种时间的增加，所得沉香的质量会提高。用这种方法结香3年时收获的产品比 Kemedangan 等级的质量要高（印度尼西亚沉香分级从高到低分别为：Doublesuper，Super，AB，Kcangan，Teri，Kemedangan，Abuk。如果结香时间比较短的话，沉香质量相当于 Kemedangan 等级。

从总体上来说，印度尼西亚发展人工结香技术多年，取得了一些效果，包括方法与配套工具（野外驱动电机、电线、混合器、专用注射器、电钻以及木工工具等）相对较规范。但是，人工结香存在着一些不确定因素，同样的方法、同样地点、同样品种以及同样年龄的树，人工结香的效果都不一样，这还需要从树木个体差异（包括遗传分子水平）、个体生长小环境方面去研究探讨。

3 启示与建议

沉香用途广泛，可用于香料、中药、工艺品以及宗教（宗教视之为灵物）等，因而扩大我国结香种质资源植物基础，培育沉香树，引进、交流和研究有效的结香技术有利于我国沉香产业可持续发展。

（1）由于资源丰富，印度尼西亚野生天然沉香存量大；但是随着沉香价格提高，开采强度增大而导致天然香资源越来越少，人工结香技术需求强烈。印度尼西亚发展了物理诱导结香、化学结香以及人工接菌结香技术，取得了一定的成效。其中，接菌诱导与我国采用黄绿墨耳菌[2]、毛色可可二孢子菌不同的是，印度尼西亚沉香接菌诱导结香采用的是镰刀菌属，使用时不是采用输液的方法，而是采用针注射方法（印度尼西亚研究人员认为输液方法菌液量大，容易伤害树，在没能结香的情形下树木即已死亡）。根据2014年在印度尼西亚实地考察，有些树木得到了沉香，有些树木由于菌液少而没有结香，并且结香的质量与数量都有待提高。目前人工结香技术需要进一步深入研究与探索。

（2）目前人工速成结香产品的质量、数量都还存在一定的不足，应致力于相关机理与技术的研究。我国采用的人工结香方法包括砍伤法、半段干法、凿洞法等，人工接菌结香。而适生地香农在找香的同时，采用物理伤害沉香成年树木以期给后人以可利用的资源（形成好的天然沉香的时间是比较长的）。目前，尽管对沉香结香机理做了一些研究[3~6]，但尚不十分清楚。近年来，愈来愈多的实验证明，激发子能够激活植物的信号传递，调节植物次生代谢途径。沉香结香激发子包括真菌及其组分、伤害、伤害信号分子茉莉酸甲酯（MeJA）等。有研究者以白木香根悬浮培养细胞为材料，以黄绿墨耳真菌提取物为真菌激发子，在组织培养物中成功诱导产生了2-（2-苯乙基）色酮化合物（沉香特征性活性成分，而未经诱导的悬浮培养细胞中均不能检测到色酮类化合物[7~8]。我国目前应加强沉香结香机理的研究，包括宏观形成规律与微观内部生理机制两个方面。在加强人工结香研究的同时（时间比较长），应加强其特定种源的枝叶利用、复合种植经营研究，解决时间长短结合问题，促进沉香产业的发展。

（3）我国大陆沉香属种质资源主要有2种，一种是白木香（A. sinensis），另一种是云南土沉香（A. yunnanensis），种质资源较单一。印度尼西亚属于热带雨林地区，物种资源丰富，能够结香的植物除了沉香属外，还有拟沉香属和 Aetoxylon 属等，各个属包括的种也比较多。相比较之下，我国结香植物资源较匮乏，仅有1~2种[9~10]，留存的古树也濒临灭绝。为了促进天然沉香产业的发展，应从印度尼西亚乃至其他东南亚国家引进结香植物种质资源

(不仅仅是沉香属植物)于我国的适生区域,包括云南、海南、广西、广东、福建等地,以扩大和丰富生物天然香植物资源,推动结香产业发展。

参考文献:

[1] SANTOSO E,TURJAMAN M. Fragrant wood gaharu: when the wildcan no longer provide? [M]. Bogor,2010.

[2] 林峰,戴好富,王辉,等. 两批接菌法所产沉香挥发油化学成分的气相色谱-质谱联用分析[J]. 时珍国医国药,2010,21(8):1901~1902.

[3] NG L T,CHANG Y S,AZIZOL L K. A review on agar(gaharu)producing Aquilaria species[J]. Journal of Tropical Forest Products,1997,2(2):272~285.

[4] RAHMAN M A,BASAK A C. Agar production in agar trees byartificial inoculation and wounding[J]. Bano-BiganPatrika,1980,9(1/2):87~93.

[5] NOBUCHI T,SOMKID S. Preliminary observation of Aquliariacrassnawood associated with the formation of aloes wood bult[J]. Kyoto University Forests,1991,63(2):226~235.

[6] BLANCHETTE R A,HEUVELING V B H. Cultivated agar wood: EUWO02094002[P]. 2001:11~28.

[7] QI S Y,HE M L,LIN L D,*et al*. Production of 2 - (2 - phenyl -ethyl) chromones in cell suspension cultures of Aquilariasinensis[J]. Plant Cell,Tissue and Organ Culture,2005,83(2):217~221.

[8] 何梦玲,戚树源,胡兰娟. 白木香悬浮培养细胞中 2 - (2 -苯乙基)色酮化合物的诱导形成[J]. 广西植物,2007,27(4):627~632.

[9] 梅全喜,吴惠妃,梁食,等. 中山沉香资源调查与开发利用建议[J]. 今日药学,2011,21(8):487~490.

[10] 刘军民,徐鸿华. 国产沉香研究进展[J]. 中药材,2005,28(7):627~632.

(本文发表于《世界林业研究》,2015 年)

云南珍稀濒危植物五裂黄连种群现状及生态习性初报

段宗亮,胡青

(云南省林业科学院,云南 昆明 650204)

摘要:五裂黄连(Coptis quinquesecta W. T. Wang)为我国特有种,也是大陆地区黄连属叶片五裂的唯一种。为深入研究五裂黄连濒危机理并开展种群的有效保护,详细调查了五裂黄连种群状况并初步分析了其生态学习性。结果显示:五裂黄连仅存2丛(1丛22株、1丛32株,共计54株),物种濒临灭绝;五裂黄连分布区狭窄,对生境要求独特,是其种群规模极小而呈现濒危状态的生物生态学因素,而当地少数民族的过度采集是导致五裂黄连濒临灭绝的人为因素。根据实地种群调查的结果,提出五裂黄连紧急保护的措施。

关键词:五裂黄连;极小种群;濒危;保护

Primary Study on Present Situation and Ecological Habits of Rare and Endangered Species *Coptis quinquesecta* W. T. Wang in Yunnan Province

Duan Zong-liang, Hu Qing

(Yunnan Academy of Forestry, Kunming Yunnan 650204, P. R. China)

Abstract: *Coptis quinquesecta* W. T. Wang is endemic to China, and it is also the only species of the five-lobed-leaf *Coptis* the mainland. In this study, the condition of *Coptis quinquesecta* W. T. Wang population was comprehensively investigated and its ecological habit was analyzed to figure out the endangered mechanism and protect the population effectively. The results showed that there were only 2 clumps of *C. quinquesecta* W. T. Wang left (22 strains in one clump and 32 strains in the other, 54 in total). The species was on the verge of extinction. The narrowness of area where *Coptis quin quesecta* W. T. Wang grows and its special requirement for habitats constitute its bio-ecological factors leading to its extremely small population size, whereas the excessive collecting by the local people because of its high medicinal value is the main human factor leading to its extinction. Based on the results of the field survey, relevant measures for protecting *Coptis quinquesecta* W. T. Wang were proposed.

Key words: *Coptis quinquesecta* W. T. Wang; plant with extremely small population; endangered species; conservation

五裂黄连(Coptis quinquesecta W. T. Wang)为毛茛科(Ranunculaceae)黄连属(Coptis)多年生草本植物,是我国特有种,也是大陆地区黄连属叶片五裂的唯一种。属药用植物,亦为国家二级保护植物。五裂黄连在1956年,由中苏联合云南考察团于云南金平苗族瑶族傣族自治县首次发现,并采集标本保存于中国科学院植物研究所[1]。本种记载于《云南植物志:第27卷》中,到目前为止,其相关研究报道较少,仅有系统分类和形态特征研究[2~4]。基础性资料的欠缺严重影响了对极度濒危植物五裂黄连的保护,鉴于此,于2016年开展了五裂黄连种群调查，以期了解五裂黄连资源分布状况及生态习性，为五裂黄连的有效保护提供基础资料和科学依据。

1 研究方法

1.1 资料收集

资料由标本资料和文献资料2个部分组成。标本资料通过查阅CVHI中国数字植物标本馆（http://www.cvh.ac.cn）、NSII中国国家资源标本平台（http://www.nsii.org.cn）、Global plants（http://www.plants.jstor.org）、Tropicos（http://www.tropicos.org）等多个国内外标本馆进行网络资料收集，同时实地查阅昆明植物研究所标本馆标本信息。文献资料主要从Flora of China、《中国植物志》[1,3]、《云南植物志》[2,4]及《云南国家重点保护野生植物》[5]等文献中收集。

1.2 野外调查

在查阅五裂黄连的相关文献资料、标本信息的基础上，依据五裂黄连的生物学生态学及生境习性特征，划定出可能有野生种群存在的区域，携带标本图片、信息资料，深入目的地，与当地林业部门、自然保护区管理部门及周边社区进行沟通、寻访，在整合所有信息资料后确定位于金平苗族瑶族傣族自治县马鞍底乡的金平分水岭国家级自然保护区五台山保护站区域内开展实地调查。五裂黄连野生种群调查，通过使用全球定位系统（GPS）定位，按照30m×30m的规格布设样方。调查内容包括五裂黄连分布的地理位置（分布坐标、海拔、坡度、坡向、坡位等）、生境特征、种群状况（数量、高度）、更新情况及植物群落特征。

2 云南五裂黄连的资源分布

2.1 云南五裂黄连的标本信息

通过标本信息查询，云南的五裂黄连在标本馆目前藏有标本3份，均采自云南省金平苗族瑶族傣族自治县，从馆藏标本的信息看，中苏联合云南考察团于1956年采集了本种；其余2份标本均由杨增宏于1988年10月27日采集，分别馆藏于中国科学院昆明植物研究所、云南珍稀植物引种繁育中心。从以上信息可以推测，20世纪80年代，五裂黄连普遍分布于云南省金平苗族瑶族傣族自治县。

表1 五裂黄连的标本信息
Table 1 The specimens of *Coptis quinquesecta* W. T. Wang

采集号	采集人	采集日期	采集地	标本馆
2471*	中苏联合考察团	1956年	云南省金平苗族瑶族傣族自治县	中国科学院植物研究所
88-1210*	杨增宏	1988年10月27日	云南省金平苗族瑶族傣族自治县	中国科学院昆明植物研究所
88-1710*	杨增宏	1988年10月27日	云南省金平苗族瑶族傣族自治县	云南珍稀植物引种繁育中心

*：引自国家标本资源共享平台。

2.2 云南五裂黄连的种群现状

在开展云南的五裂黄连的种群调查中,共调查了红河哈尼族彝族自治州的金平苗族瑶族傣族自治县、屏边苗族自治县、河口瑶族自治县的10个乡镇,仅在金平苗族瑶族傣族自治县的金平分水岭国家级自然保护区内的林间沟谷溪边发现2丛(一丛32株,一丛22株,共计54株)。详见表2云南五裂黄连种群样方表。

表2 云南五裂黄连群落样方表
Table 2 The location of *Coptis quinquesecta* W. T. Wang community in Yunnnan province

生境与总的特点	
地点:金平县马鞍底五台山国家级自然保护区774界碑旁 海拔高度:1 822m 地貌:深切割中山峡谷 位置:山中上部坡面 土壤类型:黄色砖红壤 土层厚度:20~30cm 母岩:千枚岩坡向:东北坡度:70° 群落高度:12m 总覆盖度:65%	样地面积:900m^2 地理坐标:0343129,2518572 人为活动:属保护区,但有人在林下种植草果,生境受到一定的影响 采伐情况:无 更新情况:因周边草果茂盛林下更新受到限制,大树以木兰科的红花木莲、云南拟单性木兰和山茶科的岗柃(*Eurya groffii*)为主

乔木1层 层盖度20%					
植物名称	株数	树高(m)		胸径(cm)	
		平均	最高	平均	最高
红花木莲[*Manglietia insignis*(Wal.)Blume]	3	26	36	68	76
云南拟单性木兰(*Parakmeria yunnanensis*)	2	25	35	45	50
木莲(*Manglietia fordiana* Oliv.)	2	16	25	25	25
红花荷(*Rhodoleia parvipetala* Tong)	1	16	15	16	16

乔木2层			
植物名称	层高度(m)	层盖度(%)	备注
岗柃 *Eurya groffii* Merr. 红花荷 *Rhodoleia parvipetala* Tong 大果藤黄 *Garcinia pedunculata* Roxb. 中华桫椤 *Alsophila costularis* 香竹 *Chimonocalamus delicates* Hsueh et Yi 云南七叶树 *Aesculus wangii* Hu ex Fang var. *wangii* 长叶野桐 *Mallotus decipiens* Muell-Arg.	10~15	15	

灌木层			
植物名称	层高度(m)	层盖度(%)	备注
中华尖药花 *Acranthera sinensis* 伞形紫金牛 *Ardisia corgmbifera*	1~5	30	

(续)

灌木层			
植物名称	层高度(m)	层盖度(%)	备注
瑞丽紫金牛 *Ardisia shueliesis*	1~5	30	
紫金牛 *Ardisia japonica*			
齿叶紫金牛 *Ardisia sellatce*			
柏那参 *Brassaiopsis glomerulata*			
土密树 *Bridelia monoica*			
文山山柑 *Capparis fangii*			
锯叶竹节树 *Carallia diplopetala*			
总序山柑 *Capparis assamica*			
双籽棕 *Didgmosperma caudatum*			
点叶麻 *Diospyros punctilimba*			
狗牙花 *Ervatamia* sp.			
细弱杜茎山 *Maesa macilenta*			
小叶杜茎山 *Maesa parvifolia*			
长花腺萼木 *Mycelia lougiflora*			
细柄腺萼木 *Mycetia gracicis*			
垂花密脉木 *Myzioueuzon nutans*			
尖子木 *Oxyspera paniculata*			
水锦树 *Wendlandia uvarifolia*			
糙叶水锦树 *Wendlandia scabra*			
宿苞山矾 *Symplocos persistens*			
短柱肖拔葜 *Heterosmilax yunnanensis*			

草本层			
植物名称	层高度(m)	层盖度(%)	备注
五裂黄连 *Coptis quinquesecta*	0.3~1.0	40	调查样方内的五裂黄连呈丛状生长,共计两丛,第一丛32株,第二丛22株
球兰 *Hoya carnosa*			
滇南星 *Arisaema yunnanense*			
大花蜘蛛抱蛋 *Aspididstra tonkinensis*			
铁角蕨属一种 *Asplomium* sp.			
秋海棠 *Begonia grnadis*			
蜂出巢 *Cenlroslemma multiflorum*			
大青 *Clerodendrum cyrotophy*			
野芋 *Colocasia antiguorum*			
黄花球兰 *Hoya fusca*			
糙叶火焰花 *Phlogacanthus vitellinus*			
庐山楼梯草 *Elatostema stewardii*			
箭根薯 *Tacca chantrieri*			
闭鞘姜 *Costus speciosus*			
莲座蕨一种 *Angiopteris* sp.			

3 云南五裂黄连的生境特征

3.1 气候特征

五裂黄连在云南的分布区域较为狭窄,仅见于云南省东南部红河哈尼族彝族自治州的金平苗族瑶族傣族自治县,分布地属滇南低纬高原地区,南亚热带和热带季风气候类型,雨量充沛,干湿分明。通过分析五裂黄连分布点的气候因子(表3),得出适于五裂黄连生长的气候条件,即年平均气温大于21℃,年降水量不低于1 900mm,年平均相对湿度在64%以上,极端最低温不低于-4℃,极端最高温不高于42℃,积温不低于7 600℃,干湿季节明显,且要求冬春多雾,以弥补旱季缺水。

表3 五裂黄连模式标本采集地气候因子
Table 3 Climatic factors of *C. quinquesecta* W. T. Wangspecimen collection sites

地点	年均气温 (℃)	年均降水量 (mm)	相对湿度 (%)	极端最高气温 (℃)	极端最低气温 (℃)	≥10℃积温 (℃)
勐桥乡*	23	1 900	69	42	-2	8 163.8
马鞍底乡*	21.5	2 200	64	39	-4	7 638.6

注:此表数据为云南红河州气象局提供数据。

3.2 地形地貌特征

云南的五裂黄连主要分布于森林生态系统完整优良的自然保护区内的林间沟谷溪边潮湿处,通常沿着溪流分布,分布地海拔为1 700~2 500m[5],本次调查野生种群的分布地海拔为1 800m。野生五裂黄连分布于林间溪边常年潮湿的环境,本种的人工培育应充分考虑相对湿度的控制。在原生地恢复种群时,水分条件应是选择恢复地点的关键因子,相对湿度大的环境有利于五裂黄连生长,选择地森林覆盖度高的溪边,并沿溪谷栽植可有效提高植株的成活率[6~7]。

3.3 土壤特征

云南五裂黄连分布地母岩为千枚岩,土壤为表层浅黄色的砖红壤,生境地土体厚度在山中上部坡面为20~30cm,土壤呈酸性,pH值为5.2~5.4,有机质质量分数高,达8.3%,缺磷,缺钾,土壤具有典型的季雨林土壤特征,即土壤养分指标不高,湿热的条件致生物小循环旺盛,有利于植物的生长[5,8]。

表4 云南五裂黄连分布地土壤营养测定结果
Table 4 Soil nutrition in the distribution area of *C. quinquesecta* W. T. Wang

土壤类型	土样深度	pH值	全氮(g/kg)	全磷(g/kg)	全钾(g/kg)
砖红壤	25(h)/cm	5.2	2.7	1.15	11.23

3.4 群落特征

五裂黄连群落特征结构复杂,可分为乔木层、灌木层和草本层。

乔木层可分为2层,乔木层高度大于15m,盖度约为20%,主要植物有红花木莲[*Manglietia insignis* (Wal.) Blume]、云南拟单性木兰(*Parakmeria yunnanensis* Hu)、木莲(*Manglietia fordiana* Oliv.)、红花荷(*Rhodoleia parvipetala* Tong)。乔木2层高度在10~15m,为群落的主林冠层,盖度约80%,主要植物有岗柃(*Eurya groffii* Merr.)、红花荷(*Rhodoleia parvipetala* Tong)、大果藤黄(*Garcinia pedunculata* Roxb.)、中华桫椤(*Alsophila costularis*)、香竹(*Chimonocalamus delicates* Hsueh et Yi)、云南七叶树(*Aesculus wangii* Hu ex Fang var. *wangii*)、长叶野桐(*Mallotus decipiens* Muell-Arg.)等。

灌木层高度为1~5m,植物种类丰富,主要有中华尖药花(*Acranthera sinensis*)、伞形紫金牛(*Ardisia corgmbifera*)、瑞丽紫金牛(*Ardisia shueliesis*)、紫金牛(*Ardisia japonica*)、齿叶紫金牛(*Ardisia sellatce*)、柏那参(*Brassaiopsis glomerulata*)、土密树(*Bridelia monoica*)、文山山柑(*Capparis fangii*)、锯叶竹节树(*Carallia diplopetala*)、总序山柑(*Capparis assamica*)、双籽棕(*Didgmosperma caudatum*)、点叶麻(*Diospyros punctilimba*)、狗牙花(*Ervatamia* sp.)、细弱杜茎山(*Maesa macilenta*)、小叶杜茎山(*Maesa parvifolia*)、长花腺萼木(*Mycelia lougiflora*)、细柄腺萼木(*Mycetia gracicis*)、垂花密脉木(*Myzioueuzon nutans*)、尖子木(*Oxyspera paniculata*)、水锦树(*Wendlandia uvarifolia*)、宿苞山矾(*Symplocos persistens*)、短柱肖拔葜(*Heterosmilax yunnanensis* Gagnep.)等。

草本层高度为0.3~1m,盖度为40%,主要植物有五裂黄连(*Coptis quinquesecta* W. T. Wang)、球兰(*Hoya carnosa*)、滇南星(*Arisaema yunnanense*)、大花蜘蛛抱蛋(*Aspididstra tonkinensis*)、秋海棠(*Begonia grnadis*)、蜂出巢(*Cenlroslemma multiflorum*)、大青(*Clerodendrum cyrotophy*)、野芋(*Colocasia antiguorum*)、黄花球兰(*Hoya fusca*)、糙叶火焰花(*Phlogacanthus vitellinus*)、庐山楼梯草(*Phrygnium capitalum*)、箭根薯(*Tacca chantrieri*)、闭鞘姜(*Costus speciosus*)等。

4 结论与讨论

经实地调查,云南的五裂黄连极度濒危,目前仅于金平分水岭国家级自然保护区内发现2丛,共54株,五裂黄连的野生种群现在已很难找到,加之个体数量极少,已经低于稳定存活界限,濒临灭绝,属于极小种群物种和极度濒危物种,亟待紧急救护。

五裂黄连在云南的分布范围非常狭窄,迄今为止,本种的记录仅限于云南省东南部红河哈尼族彝族自治州的金平苗族瑶族傣族自治县。分布地气候类型属于低纬高原地区的热带季风气候,从黄连属植物分布于北温带的生物学特征,五裂黄连的生境处于热带的北缘,植物群落的组成物种热带性质显著,其分布区已为黄连属植物的最南端。因此,五裂黄连狭窄的适宜分布区是导致其种群规模极小的生物地理学因素。五裂黄连对生境的要求独特,其生境海拔在1 700~2 500m之间,自然分布于常年潮湿的林中沟谷溪边,通常沿溪边生长,母岩为千枚岩,土壤为浅黄色砖红壤,酸性强。对自然环境因子如水分、土壤、相对湿度等的特殊要求也限制了五裂黄连种群向山地森林的繁衍[9]。

五裂黄连在当地哈尼族、瑶族、傣族、苗族等少数民族传统药材中是清热、解毒等的上等药材,在当地少数民族药方中是解蛇毒特效药的主要成分。当地少数民族每当见到五裂黄连都会连根采集到家中种植待用,又或连根采集晒干后卖到市场,由于当地少数民族对其药用价值的高度认可,加之过度的采集使得其在原生境内野生种群数量急剧减少。信守着"靠山吃山"的当地少数民族,缺乏相关的人工繁育技术,为满足药用的需求,野生五裂黄连已采集殆尽,随时濒临灭绝。目前,五裂黄连已被列为云南省极小种群拯救保护规划纲要(2010~2020年)保护物种。根据其种群现状及生态习性,应紧急开展保护工作,建议采取如下措施:①就地保护。本次调查的五裂黄连已在保护区内,适生环境已有保障,应严禁当地人采集,为其种群的自然繁衍提供良好的条件,同时保护、保存现有野生种质资源[10]。②引种保育。在不破坏原生种群结构的情况下,采集五裂黄连到人工适生环境中人工栽培,掌握其人工繁育技术,扩大物种数量,逐步在原生地恢复、重建种群。③建议将五裂黄连种质基因保存到"中国西南野生生物种子资源库"中,为该种的种子生物学、植物基因组学和保护生物学等科学提供研究基础。

参考文献:

[1] 中国科学院中国植物志编辑委员会. 中国植物志[M]. 北京:科学出版社, 1984:183.

[2] 中国科学院昆明植物研究所. 云南植物志(第27卷)[M]. 北京:科学出版社, 2003:737.

[3] 吴征镒, 洪德元. 中国植物志(第19卷)[M]. 北京:科学出版社, 1999:294.

[4] 云南省植物研究所. 云南植物志(第二卷)[M]. 北京:科学出版社, 1979:175.

[5] 李玉媛. 云南国家重点保护野生植物[M]. 昆明:云南科学技术出版社, 2005:230~231.

[6] JANE H. Distribution, habitat and red list status of the New Caledonian endemic treeCanacomyricamonticola (Myricaceae)[J]. Biodiversity and Conservation, 2006, 15:1459~1466.

[7] 李发根, 夏念和. 水松地理分布及其濒危原因[J]. 热带亚热带植物学报, 2004, 12(1):13~20.

[8] Sun Wei bang, Zhou Yuan, Han Chunyuan, *et al.*, status and conservation of Trigonobalanusdoichangensis (Fagaceae)[J]. Biodiversity and Conservation, 2006, 15:1303~1318.

[9] Luis N, Javier G. Seed germination and seedling survival of two threatened endemic species of the northwest Iberian peninsula[J]. Bio-logical Conservation, 2003, 109:313~320.

[10] 费永俊, 雷泽湘, 余昌均, 等. 中国红豆杉属植物的濒危原因及可持续利用对策[J]. 自然资源, 1997(5):59~63.

(本文发表于《西部林业科学》,2019年)

Phenotypic Plasticity of Lianas in Response to Altered Light Environment

YUAN Chun-ming[1,2,3], WU Tao[1,2,3], GENG Yun-fen[1,2,3], CHAI Yong[3], HAO Jia-bo[1,2,3]

[1] Key Laboratory for Conservation of Rare, Endangered and Endemic Forest Plants in Yunnan, State Forestry Administration, Kunming Yunnan 650201, China;

[2] Yunnan Provincial Key Laboratory for Cultivation and Utilization of Forest Plants, Kunming Yunnan 650201, China;

[3] Yunnan Academy of Forestry, Kunming Yunnan 650201, China

Abstract: The growth, morphology and biomass allocation of 11 liana species (six light-demanding and five shade-tolerant) were investigated on growing plants under three contrasting light environments (i.e., field, forest edge and forest interior). Our objectives were to determine: (1) changes in plant traits at the species level; and (2) differences in light-demanding and shade-tolerant species in response to altered light environment. We found that all seedlings of liana species increased in total biomass, total leaf area, relative growth rate (RGR), net assimilation rate (NAR), height, basal diameter, root length, leaf number, root mass / total plant mass (RMR) and root-to-shoot dry biomass (R/S ratio), and decreased in leaf area ratio (LAR), specific leaf area (SLA), leaf size, stem mass-to-total plant mass ratio (SMR) and leaf mass-to-total plant mass ratio (LMR) with increasing light availability. Under the three light environments, the two types of species differeded significantly in total biomass, total leaf area, RGR, NAR, LAR, SLA and leaf number, and not in leaf area. Only light-demanding species differed significantly in height, root length, basal diameter, RMR, SMR, LMR and R/S ratio. The mean plasticity index of growth and biomass allocation were relatively higher than the morphological variables, with significant differences between the two groups. Our results showed that liana species responded differently to changing light environments and that light-demanding species exhibited higher plasticity. Such differences may affect the relative success of liana species in forest dynamics.

Key words: liana seedling; functional traits; phenotypic plasticity; biomass allocation; changes in light environment

Introduction

Light is one of the most important environmental factors, providing plants with a source of energy and controlled growth and development (Lambers *et al.*, 2008). Rain forest species are often classified into two functional groups, based on seed germination and seedling establishment (Swaine and Whitmore, 1988). Shade-tolerant species germinate, grow and survive in low light (e.g., forest interior), whereas light-demanding species need high levels of light (e.g., treefall gap, forest edge, and other disturbed forests) for their establishment. Light-demanding species grow under exposed conditions of the canopy. As the amount of light is not limited they maximize their photosynthetic capacity and usually exhibit high growth rate (Kitajima, 1994; Poorter and Bongers, 2006). Shade-tolerant speciestend to have thicker leaves that are tougher and live longer and therefore have a life-time light acquisition and carbon gain comparable to the shorter lived but more productive leaves of light-demanding species (Selaya and Anten, 2010). The two groups of species also differ in response to altered light environment. Light-demanding species manifest higher plasticity in growth, morphology and physiology than shade-tolerant species, because they grow in a more variable environment (Bazzaz, 1979; Huante and Rincon, 1998; Valladares et al., 2000). However, no general consensus exists as greater, similar, and even lower plasticity has been found in pioneers compared with shade-tolerant species (Rozendaal et al., 2006).

Phenotypic plasticity is the ability of a genotype to produce distinct phenotypes under changing environmental conditions. It often involves ecologically relevant behavioral, physiological, morphological and life-history traits (Miner et al., 2005). Such plasticity may be of paramount importance for species to adjust to temporal and spatial variation in resource availability. Shade-grown plants typically invest in high aboveground biomass and also have thin leaves to optimize light capture and utilization. In contrast, plants grown under high light allocate relatively less biomass to leaves and more to roots to capture water and nutrients to sustain the high transpiration and growth rates (Brouwer, 1962). The diversity of species may be partly explained by their potential for plastic response to the environment. Strong evidence suggests that plant species may differ remarkably in the extent of their plastic responses to comparable environmental challenges (Valladares et al., 2007).

Lianas (woody vines) are diverse and abundant in many tropical forests, especially in disturbed sites, such as treefall gaps (Putz, 1984; Schnitzer and Carson, 2001), forest margins (Laurance et al., 2001; Zhu et al., 2004; Londre and Schnitzer, 2006), and other disturbed forests (Hegarty and Caballe, 1991; De Walt et al., 2000; Yuan et al., 2009). As the forest ecosystem is increasingly disrupted worldwide the relative importance of lianas increased. The lianas play a role in many aspects of forest dynamics (Schnitzer and Bongers, 2002): driving floristic changes, suppressing tree regeneration, exacerbating tree mortality, and decreasing whole-forest biomass and carbon sequestration (Schnitzer and Bongers, 2011; Schnitzer et al., 2011). Understanding the response of lianas to environmental changes (such as light) is therefore important in predicting their impact on forests.

Lianas are usually considered light-demanding because of their rapid growth under high-light conditions (Richards, 1996). Nevertheless, contrasting evidence supporting the regeneration of lianas also exists (Gerwing, 2004). Shade-tolerant species of liana that germinate and survive in deeply shaded forests also exist (Putz, 1984; Carter and Teramura, 1988; Nabe-Nielsen, 2002; Sanches and Vállo, 2002a; Gilbert, 2006; Yuan et al., 2015). Furthermore, lianas rely on the surrounding plants for their structural support, and therefore, may invest less biomass in stems and more in leaves (Putz, 1983; Castellanos et al., 1989; Niklas, 1994). However, studies fail to support this hypothesis, e.g., Kaneko and Homma (2006) showed that liana species (*Hydrangea petiolaris*) did not invest more in leaves and reproductive organs than the three Hydrangea shrub species. Cai et al., (2007) investigated the differences in growth patterns, biomass allocation and leaf traits in five closely related liana and tree species of the genus *Bauhinia* species. They found that the faster growth of light-demanding lianas compared with light-demanding trees is based on functional traits (i. e. , specific leaf area, leaf mass ratio, and leaf area ratio), and cannot be attributed to higher photosynthetic rates at the leaf level. Cai *et al.*, (2008) further analyzed the responsiveness to light and nutrient availability of the five *Bauhinia* species, and suggested that lianas were no more responsive to variation in light and nitrogen availability than trees. However, all the above studies compared only a few liana species with non-climbing species (i. e. , trees or shrubs), and it might be difficult to generalize the results to all lianas. Sanches and Válio (2002b) studied the initial growth of a few seedlings of liana and herbaceous vine species in the forest margins and under forest canopy. They found that the climbers showed high rates of growth in sunlight when compared with those under canopy, but with a diverse response of morphological and physiological traits (e. g. , leaf area, specific leaf area (SLA), leaf mass ratio (LMR), chlorophyll a and b, as well as chlorophyll a/b ratio).

In this study, the growth, morphology and biomass allocation of 11 liana species were investigated by growing plants in three contrasting light environments, i. e. , open field (high light), forest edge (intermediate light) and forest interior (low light). Six of the species' seedlings are principally found in gaps or forest edge environments, whereas the other five species are principally found in shaded understory. Our objectives were to determine: (1) how plant traits change under altered light environment at the species level; (2) differences in light-demanding species from shade-tolerant species in response to altered light environment. We found that plants used their plasticity to invest in organs that captured the most limiting resources (e. g. , light and water). High plasticity occurred in light-demanding plants rather than shade-tolerant species.

Materials and Methods

Study site

The experiment was carried out in the Ailao Mountains Subtropical Forest Ecosystem Research Station (24°32′N, 101°01′E), the Chinese Academy of Sciences. The study area exhibits a typical mountainous monsoon climate. Annual mean precipitation in 1991—1995 averaged 1931. 1 mm, of which, 85. 0% occurred during the rainy season from May to October. The average annual temperature is 11. 3°C with an average of 5. 4°C in January to 16. 4°C in July (Qiu and Xie, 1998). The predominant vegetation is mid-montane moist evergreen broad-leaved primary forest, which accounts for nearly 80 % of the total area, along with secondary patches within the forest. The forest flora consists of a combination of tropical and temperate species, including those endemic to the region (Wu and Fan, 1990). Canopy tree species are mainly composed of *Castanopsis wattii*, *Lithocarpus xylocarpus*, *Schima noronhae* and *Lithocarpus jingdongensis*, and evenly distributed within the forest (Liu *et al.*, 2001). The forest reaches 20~25 m in height, with a closed canopy (> 90%).

Species

Eleven liana species (Table 1), which are native to the studied region (Yuan et al., 2009, 2015) were selected. Species selection was based on differences in shade tolerance or regeneration, including six light-demanding species and five shade-tolerant species (Yuan et al., 2008, 2015; species guilds see Table 1). The adults of all species climb to the forest canopy, except *Embelia procumbens*, which is a typical understory liana. Due to lack of seedlings for the three light-demanding species (i.e., *Rosa longicuspis*, *Actinidia callosa* and *Celastrus angulatus*) in the forest, we collected fresh seeds for germination in a greenhouse in March, 2013. The seedlings of the other eight species were transplanted from the forest.

Table 1　Summary of the species studied: climbing type (H, hook climbers; S, stem twiners; R, root climbers; T, tendril climbers), life form (E, evergreen species; D, deciduous species), regeneration habitats (G, gap; E, edge; U, understory), source of plant material (S, seed; T, transplanted seedling from forest). Classification of the species in light-demanding and shade-tolerant species was based on differences in regeneration requirements (L, light-demanding; S, shade-tolerant).

Species	Family	Species code	Climbing type	Life form	Regeneration habitat	Guild	Adult stature	Source
Rosa longicuspis	Rosaceae	RL	H	E	G; E	L	Canopy	S
Actinidia callosa	Actinidiaceae	AC	S	D	G; E	L	Canopy	S
Celastrus angulatus	Celastraceae	CA	S	D	G; E	L	Canopy	S
Holboellia latifolia	Lardizabalaceae	HL	S	E	G; E	L	Canopy	T
Kadsura coccinea	Schisandraceae	KC	S	E	G; E	L	Canopy	T
Parthenocissus himalayana	Vitaceae	PH	R	D	G; E	L	Canopy	T
Euonymus vagans	Celastraceae	EV	R	E	U	S	Canopy	T
Jasminum urophyllum	Oleaceae	JU	S	E	U	S	Canopy	T
Hydrangea anomala	Saxifragaceae	HA	R	D	U	S	Canopy	T
Heterosmilax japonica	Smilacaceae	HJ	T	E	U	S	Canopy	T
Embelia procumbens	Myrsinaceae	EP	S	E	U	S	Understory	T

Experimental design

We designed three light environments, i.e., field, forest edge and forest interior (50 meters from edge into forest interior). During June 2013, we transplanted the germinated seedlings and wild types into 20cm × 30cm plastic pots (one plant per pot and 40 plots in total, for each species) containing topsoil from the nearby forest and then moved into a shade house for material preparation (10% of full sun light). All plants were watered on days without rain to maintain the soil near field capacity.

At the end of April 2014, 24 pots per species with similar plant size (i.e., height or length) were selected for the experiment. Of these, six plants (pots) per species for each treatment (i. e., field, forest edge and forest interior) were moved for growth experiment (each plant was tagged and supported with a dry bamboo shoot), which lasted six months (at the end of October, totaled 184 days). The remaining six plants were used for initial measurement. During this experimental period, the fallen leaves of deciduous species were collected for each plant species and treatment and retained in paper bags for the final measurements (i. e., leaf number, area, and biomass).

During the preparation period of the seedlings, there was no death. During the experimental period, all the plants survived, except three plants of Rosa longicuspis, in the forest interior, and one *Embelia procumbens*, which died in the field.

Plant functional traits

At the initial and final harvest, basal diameter, height, root length, leaf number (and leaflets counted in compound leaves), leaf area and leaf (including petiole), stem and root mass were determined. Liana's height was defined as either actual height of upright individuals or the length of climbing individuals, measured from the plant stem base to the apex.

After harvest, plants were separated into leaves (including petioles), stems and roots. All leaves in each species were photocopied by EPSON V700 scanner and analyzed with an image analysis system (ImageTool Version 2.0) to calculate leaf area. Roots were washed in tap water. All the tissues were dried to a constant weight at 70℃ for 48 h.

From the primary data the following variables were derived: root mass ratio (RMR; root mass/ total plant mass, in $g \cdot g^{-1}$), stem mass ratio (SMR; stem mass / total plant mass, in $g \cdot g^{-1}$), leaf mass ratio (LMR; leaf + petiole mass / total plant mass, in $g \cdot g^{-1}$), R/S ratio (root-to-shoot dry biomass), specific leaf area (SLA; leaf area / leaf mass, in $cm^2 \cdot g^{-1}$), leaf area ratio (LAR; leaf area / total plant mass, in $cm^2 \cdot g^{-1}$), and mean leaf size (total leaf area / total leaf number, in cm^2) (Poorter, 1999; Cornelissen et al., 2003; Pérez-Harguindeguy et al., 2013). The relative growth rate (RGR, dry biomass increment per unit total plant biomass per unit time) for each treatment was calculated as: $RGR = (\ln W_2 - \ln W_1)/t$, the net assimilation rate (NAR, the rate of dry matter production per unit leaf area) as: $NAR = [(W_2 - W_1)/t] \times [(\ln A_2 - \ln A_1)/(A_2 - A_1)]$, where W is the total plant dry biomass in grams, A is the total leaf area in cm^2 and t is time in days (Hunt, 1978).

Statistical analysis

The differences in growth, morphology and biomass allocation variables of each species in the three light environments (field, forest edge and forest interior) were tested using a one-way ANOVA ($P=0.05$). The Fisher LSD test was used for post-hoc analysis. Plant responses were analyzed using a two-way ANCOVA, with light and species as fixed factors. Plant biomass may differ between light environments and species at the end of the experiment, and biomass was therefore included as a covariable in the analysis. Data were checked for normality and homogeneity of variances, and an ln- or square-root transformation was used when necessary to satisfy the assumptions of ANCOVA.

To compare the plasticity in growth, morphology and biomass allocation, we calculated a plasticity index for each measured variable in each species, following Valladares et al., (2000). The index ranged from zero to one and represented the difference between the maximum and minimum mean value of a variable among treatments divided by the maximum value. Finally, a mean plasticity index was calculated for growth, morphology and biomass allocation variables, respectively. Independent-sample t-test was used to compare the differences of mean plasticity index of growth, morphology and biomass allocation between the two functional group species. Statistical analyses were done using SPSS 19.0 (SPSS, Chicago, IL, USA).

Results

Growth, morphology and biomass allocation

Total biomass, total leaf area, RGR and NAR of all species were significantly higher in the field and at forest edge than in the forest interior ($P<0.05$), and most of them showed the highest value in the field (Fig. 1a, b, c, d). All these growth variables differed significantly under the three light environments in both light-demanding and shade-tolerant species.

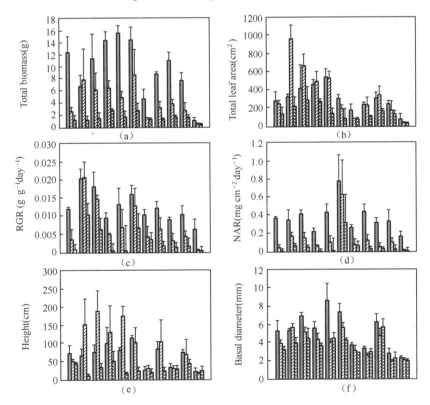

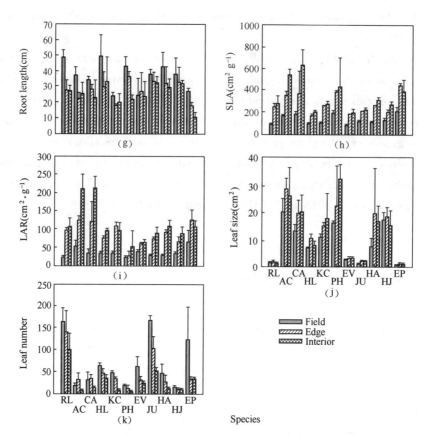

Figure 1 Liana functional traits in response to altered light environment
(a) total biomass; (b) total leaf area; (c) RGR; (d) NAR; (e) height; (f) basal diameter; (g) root length;
(h) SLA; (i) LAR; (j) leaf size; (k) leaf number. RGR, relative growth rate; NAR, net assimilation rate;
SLA, specific leaf area; LAR, leaf area ratio. Species codes are as defined in Table 1.
Data are means ±SD.

Among the morphology variables, seedlings of liana species increased in height, basal diameter, root length, leaf number, and decreased in LAR, SLA and leaf size with increasing light availability. In the forest interior, the height of all the species was the lowest under the three light environments, except Embelia procumbens (Fig. 1e). Light-demanding species showed significant differences under different light environments, with *Rosa longicuspis* and *Parthenocissus himalayana* reaching the highest altitude in the field, and *Actinidia callosa*, *Celastrus angulatus*, *Holboellia latifolia* and *Kadsura coccinea* at the forest edge. On the contrary, there was no significant difference among the shade-tolerant species, except *Jasminum urophyllum*, which was significantly taller at the forest edge and in the field than in the forest interior. Basal diameter (Fig. 1f) and root length (Fig. 1g) were greater in the field than at the forest edge and in the forest interior. Except for a few cases, light-demanding species differed significantly among the three light environments, unlike shade-tolerant species.

The SLA (Fig. 1h) and LAR (Fig. 1i) of all species were significantly lower in the field than at forest edge and in the forest interior ($P<0.05$). All of them reached the highest value in the forest interior, excluding the LAR of Kadsura coccinea, and LAR and SLA of *Embelia procumbens*. LAR and SLA differed significantly under the three light environments in both light-demanding and shade-tolerant species.

The leaf size in all species (Fig. 1j) was smaller in the field than at the forest edge and in the forest interior, without any significant differences under the three light environments among both light-demanding and shade-tolerant species, excluding *Holboellia latifolia* and *Jasminum urophyllum*. The leaf number in all the species (Fig. 1k) was higher in the field than at the forest edge and in the forest interior, except *Actinidia callosa* and *Celastrus angulatus*, which reached the highest value at forest edge. Contrary to leaf size, both light-demanding and shade-tolerant species showed significant differences under the three light environments, except *Heterosmilax japonica*.

In general, RMR and R/S ratios were higher in the field than at the forest edge and in the forest interior (Fig. 2a, d), whereas LMR followed the opposite pattern (Fig. 2c) while SMR was higher at forest edge than in the field and in the forest interior (Fig. 2b). RMR, SMR, LMR and R/S ratio differed significantly under the three light environments for light-demanding species, unlike shade-tolerant species, except for *Jasminum urophyllum* (Fig. 2).

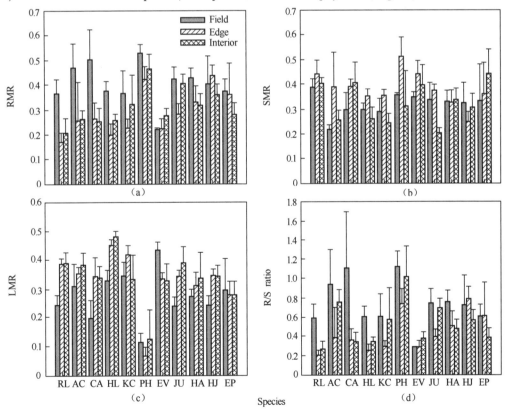

Figure 2 Liana biomass allocation under the three light environments (field, forest edge and forest interior): (a) RMR; (b) SMR; (c) LMR; (d) R/S ratio; RMR, root mass ratio; SMR, stem mass ratio; LMR, leaf mass ratio; and R/S ratio, root-to-shoot dry biomass ratio. Species codes are as defined in Table 1. Data are means ±SD.

Interaction between light and species: relative significance

The two-way ANCOVA explained much of the variation in variable values, with a mean R^2 of 0.84 (range 0.65~0.91, Table 2). Biomass at the final harvest had no effect on biomass allocation, but showed a strong effect on seedling growth and morphological variables, except root length and leaf size. Both light and species had significant effects on the 15 variables of growth, morphology and biomass allocation. There were significant light vs. species interactions among all the variables, except RGR, NAR, root length and leaf size.

Table 2 Results of a two-way ANCOVA with light ($N = 3$) and species ($N = 11$) as fixed factors and ln (1 + biomass) as covariable. F-values, levels of significance (P), regression coefficient (b) and coefficient of determination (R^2) are shown. RGR, relative growth rate; NAR, net assimilation rate; SLA, specific leaf area; LAR, leaf area ratio; RMR, root mass ratio; SMR, stem mass ratio; LMR, leaf mass ratio; R/S ratio, root-to-shoot dry biomass.

Levels of significance: ns, not significant; *, $P < 0.05$; **, $P < 0.01$; ***, $P < 0.001$

Group	Variable	Light		Species		Light×Species		Biomass		b	R^2
		F	P	F	P	F	P	F	P		
Growth	Total biomass	184.4	***	16.0	***	5.4	***				0.91
	Total leaf area	40.6	***	15.0	***	3.2	***	69.7	***	357.02	0.89
	RGR	5.3	**	34.9	***	0.3	ns	57.9	***	0.01	0.93
	NAR	4.8	**	13.6	***	0.7	ns	27.4	***	0.24	0.87
Morphology	Height	40.8	***	4.4	***	3.7	***	16.2	***	50.15	0.84
	Root length	5.0	**	4.5	***	1.3	ns	3.5	ns	6.58	0.69
	Basal diameter	6.5	**	22.7	***	4.9	***	10.5	**	0.93	0.92
	SLA	4.7	**	16.0	***	2.5	**	8.4	**	-93.75	0.86
	LAR	8.1	**	21.5	***	5.5	***	12.4	**	-31.21	0.91
	Leaf size	5.1	**	18.5	***	1.1	ns	0.6	ns	2.01	0.82
	Leaf number	5.1	**	45.2	***	4.0	***	16.3	***	35.48	0.91
Biomass allocation	RMR	18.8	***	10.6	***	2.9	**	0.5	ns	0.02	0.77
	SMR	10.0	***	5.3	***	1.8	*	1.3	ns	0.04	0.65
	LMR	4.8	**	22.3	***	2.7	**	0.3	ns	-0.01	0.83
	R/S ratio	13.4	***	8.5	***	2.4	**	0.4	ns	-0.06	0.73

Phenotypic plasticity

The plasticity index was calculated for each species under the three light environments (Table 3). RGR and NAR showed the highest mean plasticity index, while basal diameter was the lowest. The mean plasticity index of growth and biomass allocation were relatively higher than that of morphology. However, the leaf number and R/S ratio showed a relatively higher mean plasticity index, when compared with the other morphological and biomass allocation variables.

Table 3 Phenotypic plasticity index, (maximum- minimum)/maximum, for 15 variables of eleven lianas in response to changes in light environment. RGR, relative growth rate; NAR, net assimilation rate; SLA, specific leaf area; LAR, leaf area ratio; RMR, root mass ratio; SMR, stem mass ratio; LMR, leaf mass ratio; R/S ratio, root-to-shoot dry biomass. Species codes are as defined in Table 1.

Group	Variable	Species											
		RL	AC	CA	HL	KC	PH	EV	JU	HA	HJ	EP	Mean
Growth	Total biomass	0.89	0.85	0.87	0.81	0.89	0.81	0.72	0.85	0.85	0.79	0.64	0.82
	Total leaf area	0.49	0.76	0.56	0.45	0.73	0.70	0.53	0.54	0.50	0.46	0.53	0.57
	RGR	0.96	0.85	0.91	0.98	0.95	0.85	0.86	0.96	0.94	0.93	0.93	0.92
	NAR	0.98	0.81	0.89	0.98	0.98	0.96	0.75	0.93	0.92	0.92	0.95	0.92
	Mean	0.83	0.82	0.81	0.81	0.89	0.83	0.72	0.82	0.80	0.78	0.76	0.81
Morphology	Heights	0.39	0.93	0.82	0.61	0.91	0.79	0.38	0.77	0.12	0.61	0.35	0.61
	Root length	0.44	0.41	0.33	0.40	0.30	0.50	0.12	0.15	0.31	0.16	0.59	0.33
	Basal diameter	0.40	0.32	0.35	0.35	0.48	0.43	0.24	0.24	0.25	0.32	0.09	0.32
	SLA	0.67	0.69	0.72	0.51	0.64	0.55	0.57	0.48	0.64	0.53	0.53	0.59
	LAR	0.79	0.75	0.84	0.66	0.69	0.57	0.40	0.69	0.75	0.62	0.49	0.66
	Leaf size	0.30	0.30	0.35	0.32	0.38	0.49	0.22	0.35	0.62	0.19	0.57	0.37
	Leaf number	0.77	0.87	0.74	0.68	0.88	0.67	0.61	0.83	0.91	0.87	0.77	0.77
	Mean	0.54	0.61	0.59	0.50	0.61	0.57	0.36	0.50	0.51	0.47	0.48	0.52
Biomass allocation	RMR	0.67	0.81	0.70	0.59	0.64	0.42	0.39	0.50	0.39	0.43	0.50	0.55
	SMR	0.28	0.75	0.54	0.42	0.41	0.77	0.38	0.56	0.31	0.51	0.58	0.50
	LMR	0.51	0.47	0.69	0.42	0.38	0.83	0.40	0.57	0.46	0.43	0.53	0.52
	R/S ratio	0.79	0.90	0.86	0.71	0.77	0.65	0.48	0.66	0.54	0.61	0.67	0.86
	Mean	0.56	0.73	0.70	0.54	0.55	0.67	0.41	0.57	0.43	0.50	0.57	0.61
	Total mean	0.66	0.68	0.65	0.54	0.66	0.66	0.59	0.66	0.64	0.60	0.61	0.63

The mean plasticity index of growth, morphology and biomass allocation differed significantly between light-demanding and shade-tolerant species (growth: $t = 2.70$, $P = 0.024$; morphology: $t = 3.40$, $P = 0.008$; biomass allocation: $t = 2.64$, $P = 0.027$; Fig. 3). No significant differences were found in the mean plasticity index of total variables between the two functional group species ($t = 0.84$, $P = 0.422$).

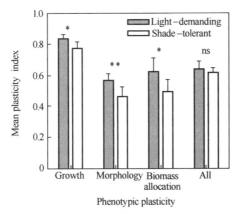

Figure 3 Mean phenotypic plasticity index of the growth, morphology, biomass allocation and all variables combined, of light-demanding (closed bars) and shade-tolerant (open bars) species. Asterisks indicate a significant difference: *, $P<0.05$; **, $P<0.01$; ns, not significant. Data are means ±SD.

Discussion

Response to altered light environment

This study demonstrated that both light-demanding and shade-tolerant liana species differed significantly in growth variables among the three light environments. The total biomass, total leaf area, RGR and NAR of the eleven liana species were significantly higher in the field and at forest edge than in the forest interior, and most of the species exhibited the highest value in the field (Figure 1a,b,c,d). However, under low light conditions (i.e., shade house and forest interior), nearly all trial seedlings survived, with considerably smaller and shorter stature compared with seedling growth at the forest edge and in the field. This indicated that lianas prefered high-light environments to forest interior, and fully captured the high irradiance for their growth and development. The finding also explains the rapid growth under high-light conditions, such as treefall gaps, forest edge, and disturbed forest ecosystems.

Plant response includes increases or reductions in certain traits due to phenotype plasticity. Seedlings of liana species increased in height, basal diameter, root length, leaf number, and decreased in LAR, SLA and leaf size with increasing light availability. Plants that grow in a shady environment invest relatively more in photosysthesis and other resources in leaf area, with a high LAR and SLA (Lambers et al., 2008). A high SLA is advantageous in a low-light environment, where photoreception is of primary importance (Poorter, 1999) as seen in the current study.

RMR and R/S ratio were higher in the field than at the forest edge and in the forest interior. The LMR and SMR followed the opposite pattern by investing biomass in roots in the field, and the shade lianas invest in leaves suggesting that plants use their plasticity to allocate resources to areas that capture the most limiting resource (e.g., light, water, and nutrients). The altered pattern of this study is consistent with other studies for trees (Gyimah and Nakao, 2007). Poorter and Nagel (2000) carried out a meta-analysis of the studies, and found that the responses to light, nutrients and water were consistent with the prediction of the 'functional equilibrium' theory stating that plants respond to a decrease in aboveground resources with increased allocation of biomass to shoots (leaves), whereas they respond to a decrease in underground resources with increased allocation to roots. Furthermore, nearly all the R/S ratio values were lower than 1, suggesting that plants generally allocated higher mass to shoots than to roots, which is favorable for accumulation of photosynthetic products, and met the needs of plant growth and development. All liana species, excluding the understory liana, *Embelia procumbens*, reached the greatest height at the forest edge and in the field, and allocated more resources to canopy development, and stem elongation, which enabled rapid growth and a competitive advantage (Wright, 2002).

Functional response tochanges in light environment

The results presented in this study show that light-demanding species differed from shade-tolerant species in response to altered light environment. The results are in accordance with the response to edge effects of lianas in the forests studied (Yuan et al., 2016), where liana species responded differently to edge effects, including species present exclusively at or near the edges (within 20 m of the edge). The species density decreased with increasing distance from the edge, and was insensitive to edge effects, which showed minor variation or random fluctuation throughout the gradient.

Among the three light environments, significant differences between light-demanding and shade-tolerant species were found in the plasticity of seven of the fifteen variables. The two functional group species differed significantly in total biomass, total leaf area, RGR, NAR, LAR, SLA and leaf number, and not in leaf area. Only light-demanding species differed significantly in height, root length, basal diameter, RMR, SMR, LMR and R/S ratio. The plasticity response of height and root length may reflect an important growth strategy for light-demanding lianas, in terms of enhanced light access and water use, respectively.

Phenotypic plasticity

Our data demonstrate that the mean plasticity index of growth, morphology and biomass allocation variables was significantly greater for light-demanding than for shade-tolerant species. The result was consistent with the hypothesis of Bazzaz (1979), and other studies (Huante and Rincon, 1998; Valladares et al., 2000) demonstrating a greater phenotypic plasticity in pioneer species. Plasticity is generally thought to be the greatest for pioneer species, as they occur in variable, heterogeneous environments with high resource availability. Nevertheless, this hypothesis was rejected by Rozendaal et al., (2006), as short-lived pioneers showed the lowest plasticity to irradiance. They hypothesized that plasticity was the largest for tall species that experience large ontogenetic changes in irradiance during their life cycle.

In this study, growth and morphological variables changed with the plant size (Table 2). The regression slope shows altered plant variables with biomass. The SLA, LAR, LMR and R/S ratio declined with biomass, whilst the other traits increased with biomass.

Conclusions

In summary, the growth variables of all liana species were significantly higher in the field and at the forest edge than in the forest interior. However, seedlings of liana species increased in height, basal diameter, root length, leaf number, and decreased in LAR, SLA and leaf size with increased light availability. This result indicates that light is the most important factor explaining the success of lianas in high light environments. Plants use their plasticity to invest in areas that capture the most limiting resource (e.g., light, water, and nutrients). Furthermore, our results demonstrated that only light-demanding species differed significantly in height, root length, basal diameter, RMR, SMR, LMR and R/S ratio among the three light environments, and light-demanding species have a higher plasticity. Thus, lianas respond differently to altered environments with a diverse role in forest dynamics.

Acknowledgements

We would like to thank the Ailao Mountains Subtropical Forest Ecosystem Research Station, Chinese Academy of Sciences, for providing the logistics and facility management support. We are also indebted to Mr. Li Dawen for managing the experimental materials during the whole research process. We especially thank the editors of Ecological Research and two anonymous reviewers for their constructive suggestions, which have considerably improved this paper. This work was financially supported by a grant (No. 31160136) from the National Natural Science Foundation in China.

References

Bazzaz FA,1979. The physiological ecology of plant succession. Annu Rev Ecol Evol Syst, 10: 351~371.

Brouwer R,1962. Distribution of dry matter in the plant. Neth J Agr Sci, 10: 361~376.

Cai ZQ, Poorter L, Han Q, Bongers F,2008. Effects of light and nutrients on seedlings of tropical *Bauhinia* lianas and trees. Tree Physiol, 28: 1277~1285.

Cai ZQ, Poorter L, Cao KF, Bongers F,2007. Seedling growth strategies in Bauhinia species: comparing lianas and trees. Ann Bot, 100: 831~838.

Carter GA, Teramura AH,1988. Vines photosynthesis and relationships to climbing mechanics in a forest understory. Am J Bot, 75: 1011~1018.

Castellanos A, Mooney HA, Bullock SH, Jones C, Robichaux R,1989. Leaf, stem and metamer characteristics of vines in a tropical deciduous forest in Jalisco, Mexico. Biotropica, 21: 41~49.

Cornelissen JHC, Lavorel S, Garnier E, Díaz S, Buchmann N, Gurvich DE, Reich PB, ter Steege H, Morgan HD, van der Heijden MGA, Pausas JG, Poorter H,2003. A handbook of protocols for standardized and easy measurement of plant functional traits worldwide. Aust J Bot, 51: 335~380.

DeWalt SJ, Schnitzer SA, Julie S, Denslow JS,2000. Density and diversity of lianas along a chronosequence in a central Panamanian lowland forest. J Trop Eco, 16: 1~19.

Gerwing JJ,2004. Life history diversity among six species of canopy lianas in an old-growth forest of the eastern Brazilian Amazon. For Ecol Manage, 190: 57~72.

Gilbert B, Wright SJ, Muller-Landau HC, Kitajima K, Hernandez A,2006. Life history trade-offs in tropical trees and lianas. Ecology, 87: 1281~1288.

Gyimah R, Nakao T,2007. Early growth and photosynthetic responses to light in seedlings of three tropical species differing in successional strategies. New Forests, 33: 217~236.

Hegarty EE, Caballe G,1991. Distribution and abundance of vines in forest communities. In: Putz FE and Mooney HA (eds) The Biology of Vines. Cambridge University Press, Cambridge, pp 263~282.

Huante P, Rincon E,1998. Responses to light changes in tropical deciduous woody seedlings with contrasting growth rates. Oecologia, 113: 53~66.

Hunt R (1978) Plant Growth Analysis. Edward Amold, London.

Kaneko Y, Homma K,2006. Differences in the allocation patterns between liana and shrub Hydrangea species. Plant Spec Biol, 21: 147~153.

Kitajima K,1994. Relative importance of photosynthetic and allocation traits as correlates of seedling shade tolerance of 15 tropical tree species. Oecologia, 98: 419~428.

Lambers H, Chapin FS, Pons TL,2008. Plant Physiological Ecology. Springer, New York.

Laurance WF, Perez-Salicrup DR, Delamonica P, Fearnside PM, Angelo SD, Jerozolinski A, Pohl L, Lovejoy TE, 2001. Rain forest fragmentation and the structure of Amazonian liana communities. Ecology, 82: 105~116.

Liu WY, Fox JED, Xu ZF, 2001. Community characteristics, species diversity and management of middle-mountain moist evergreen broad-leaved forest in the Ailao Mountains, Southwestern China. Pac Conserv Biol, 7: 34~44.

Londre RA, Schnitzer SA, 2006. The distribution of lianas and their change in abundance in temperate forests over the past 45 years. Ecology, 87: 2973~2978.

Miner BG, Sultan SE, Morgan SG, Padilla DK, Relyea RA, 2005. Ecological consequences of phenotypic plasticity. Trends Ecol Evol, 20: 685~692.

Nabe-Nielsen J, 2002. Growth and mortality rates of the liana Machaerium cuspidatum in relation to light and topographic position. Biotropica, 34: 319~322.

Niklas KJ, 1994. Comparisons among biomass allocation and spatial distribution patterns of some vine, pteridophyte, and gymnosperm shoots. Am J Bot, 81: 1416~1421.

Pérez-Harguindeguy N, Díaz S, Garnier E, Lavorel S, Poorter H, Jaureguiberry P, Bret-Harte MS, Cornwell WK, Craine JM, Gurvich DE, Urcelay C, Veneklaas EJ, Reich PB, Poorter L, Wright IJ, Ray P, Enrico L, Pausas JG, de Vos AC, Buchmann N, Funes G, Quétier F, Hodgson JG, Thompson K, Morgan HD, ter Steege H, van der Heijden MGA, Sack L, Blonder B, Poschlod P, Vaieretti MV, Conti G, Staver AC, Aquino S, Cornelissen HC, 2013. New handbook for standardised measurement of plant functional traits worldwide. Aust J Bot, 61: 167~234.

Poorter H, Nagel O, 2000. The role of biomass allocation in the growth response of plants to different levels of light, CO2, nutrients and water: a quantitative review. Funct Plant Biol, 27: 1191.

Poorter L, 1999. Growth responses of 15 rain-forest tree species to a light gradient: the relative importance of morphological and physiological traits. Funct Ecol, 13: 396~410.

Poorter L, Bongers F, 2006. Leaf traits are good predictors of plant performance across 53 rain forest species. Ecology, 87: 1733~1743.

Putz FE, 1983. Liana biomass and leaf area of a 'tierra firme' forest in the Rio Negro basin, Venezuela. Biotropica, 15: 185~189.

Putz FE, 1984. The natural history of lianas on Barro Colorado Island, Panama. Ecology, 65: 1713~1724.

Qiu XZ and Xie SC, 1998. Studies on the Forest Ecosystem in Ailao Mountains. Yunnan Science and Technology Press, Kunming (in Chinese)

Richards PW, 1996. The Tropical Rain Forest. Cambridge University Press, Cambridge.

Rozendaal DMA, Hurtado VH, Poorter L, 2006. Plasticity in leaf traits of 38 tropical tree species in response to light; relationships with light demand and adult stature. Funct Ecol, 20: 207~216.

Sanches MC, Válio IFM, 2002a. Seed and seedling survival of some climber species in a southeast Brazilian tropical forest. Biotropica, 34: 323~327.

Sanches MC, Válio IFM, 2002b. Seedling growth of climbing species from a southeast Brazilian tropical forest. Plant Ecol, 154:51~59.

Schnitzer SA and Bongers F, 2002. The ecology of lianas and their role in forests. Trends Ecol Evol, 17: 223~230.

Schnitzer SA., Carson WP, 2001. Treefall gaps and the maintainance of species diversity in a tropical forest. Ecology, 82: 913~919.

Schnitzer SA, Bongers F, Wright SJ, 2011. Community and ecosystem ramifications of increasing lianas in neotropical forests. Plant Signal Behav, 6: 598~600.

Schnitzer SA and Bongers F, 2011. Increasing liana abundance and biomass in tropical forests: emerging patterns and putative mechanisms. Ecol Lett, 14: 397~406.

Selaya NG, Anten PR, 2010. Leaves of pioneer and later-successional trees have similar lifetime carbon gain in tropical secondary forest. Ecology, 91: 1102~1113.

Swaine MD, Whitmore TC, 1988. On the definition of ecological species groups in tropical rain forests. Vegetatio, 75: 81~86.

Valladares F, Wright SJ, Lasso E, Kitajima K, Pearcy RW, 2000. Plastic phenotypic response to light of 16 congeneric shrubs from a Panamanian rainforest. Ecology, 81: 1925~1936.

Valladares F, Gianoli E, Gómez JM, 2007. Ecological limits to plant phenotypic plasticity. New Phytol, 176: 749~763.

Wright SJ, 2002. Plant diversity in tropical forests: a review of mechanisms of species coexistence. Oecologia, 130: 1~14.

Wu BX, Fan JR, 1990. Floristic structure of mid-montain moist evergreen broad-leaved forest at Xujiaba in Ailao Mountains (in Chinese with English abstract). Scientia Silvae Sinicae, 26: 396~401.

Yuan CM, Liu WY, Yang GP, 2008. Species composition and diversity of lianas in forest gaps of montane moist evergreen broadleaved forest in Ailao Mountains, Yunnan, China (in Chinese with English abstract). J Mt Sci, 26: 29~35.

Yuan CM, Liu WY, Tang C. Q, Li XS, 2009. Species composition, diversity and abundance of lianas in different secondary and primary forests in a subtropical mountainous area, SW China. Ecol Res, 24: 1361~1370.

Yuan CM, Liu WY, Yang GP, 2015. Diversity and spatial distribution of lianas in a mid-montane moist evergreen broad-leaved forest in the Ailao Mountains, SW China (in Chinese with English abstract). Biodivers Sci, 23: 332~340.

Yuan CM, Geng YF, Chai Y, Hao JB, Wu T, 2016. Response of lianas to edge effects in mid-montane moist evergreen broad- leaved forests in the Ailao Mountains, SW China (in Chinese with English abstract). Biodivers Sci, 24: 40~47.

Zhu H, Xu ZF, Wang H, Li BG, 2004. Tropical rain forest fragmentation and its ecological and species diversity changes in southern Yunnan. Biodivers Conserv, 13: 1355~1372.

木质藤本植物的表型可塑性对光环境变化的响应

摘要：采用田间实验的方法,对林内、林缘和林外3种不同光环境条件下11种木质藤本(6种喜光5种耐阴)幼苗的生长、形态和生物量配置进行了研究。目的是探讨以下两个科学问题:①物种水平上木质藤本植物功能性状对光环境变化是如何响应的?②喜光种和耐阴种对光环境变化的响应是否存在差异?结果表明:①在从林内到林缘、林外的光环境梯度上,所有木质藤本幼苗的总生物量、总叶面积、相对生长速率、净同化速率、高度、基径、根长、叶片数量、根生物量比和地上/地下生物量比增加,而叶面积比、比叶面积、叶片大小、茎生物量比、叶生物量比降低。②所有木质藤本物种的总生物量、总叶面积、相对生长速率、净同化速率、叶面积比、比叶面积、叶片数量在林内、林缘和林外3种光环境下存在显著差异;3种光环境下,喜光物种的高度、根长、基面积、根生物量比、茎生物量比、叶生物量比和根茎比存在显著差异,而耐阴种则没有显著差异。③木质藤本植物生长和生物量配置的可塑性指数均较高,而形态可塑性指数均较低;木质藤本植物生长、形态和生物量配置的可塑性指数在喜光和耐阴的物种之间存在显著的差异,且喜光物种的可塑性指数显著地高于耐阴的物种。以上研究结果说明木质藤本植物对光环境变化的响应在不同物种间存在差异,喜光木质藤本表现出较高的可塑性。这种差异性可能会影响木质藤本植物物种在森林动态的相对作用。

[本文发表于 Ecological research,2016,31(6)]

A New System for the Family Magnoliaceae

SIMA Yong-Kang[1,2], and LU Shu-Gang[1]*

1. Institute of Ecology and Geobotany, Yunnan University, Kunming 650091, China;

2. Yunnan Academy of Forestry; Yunnan Laboratory for Conservation of Rare, Endangered & Endemic Forest Plants, State Forestry Administration; Yunnan Provincial Key Laboratory for Cultivation And Exploitation of Forest Plants, Kunming 650204, China.

Abstract: A new system for the family Magnoliaceae is proposed on the basis of the latest data on DNA and the observations of morphological characters especially in living plants. In the new system, a total of 2 subfamilies, 2 tribes and 15 genera are recognized, of which, two genera are described as new. A key to the subdivision of the family and 52 new combinations are presented.

Key words: Magnoliaceae; Paramagnolia; Metamagnolia; new genus; new combination

1 Introduction

As it is well known, the family Magnoliaceae is one of the most primitive taxa in angiosperms (Wu *et al.* 2003). Southeastern Asia is very rich in species, but the greatest concentration of species and highest diversity occur in southern and southwestern China (Law 1984; Liu *et al.* 1995; Sima *et al.* 2001). Magnoliaceous plants are of great values in botanical studies; theoretically, they as the classic representatives of primitive taxa are the key materials for the research of the origin and evolution of angiosperms, and for the reconstruction of the natural system of angiosperms; practically, they are the main components of evergreen broad-leaved forests and deciduous broad-leaved forests from tropical to temperate zones as well as famous trees for ornamental, timber, medicinal and perfume (Sima *et al.* 2001).

The taxonomic studies on the family Magnoliaceae (sensu stricto) has a long history; since the early part of 18th century many systems of the family have been published, of which, Dandy's system was followed by most later authours for about half a century and Law's system is very popular in China, but none of the systems at generic level were generally accepted. Not only the genus number but also the conception of the genera are quite different in different systems (de Candolle, 1817; Benthan & Hooker, 1862; Baillon, 1866; Engler & Gilg, 1924; Dandy, 1927, 1964; Law, 1984; Nooteboom, 1985, 1993, 2000; Liu, 2000; Figlar & Nooteboom, 2004; Sima, 2005; Xia *et al.*, 2008; Sima & Lu, 2009). All present studies reveal that the generic delimitation and classification of the family Magnoliaceae are controversial, and further researches to reconstruct a more objective and natural system and solve evolutionary and phylogenetic problems in Magnoliaceae are necessary.

Based on the data on DNA (Chase *et al.* 1993; Jin *et al.* 1999; Azuma *et al.* 1999, 2001, 2004; Shi *et al.* 2000; Kim *et al.* 2001; Wang *et al.* 2006; Nie *et al.* 2008) and the observations of morphological characters especially in living plants (Fig. 1-2) (Tiffney 1977; Zhang *et al.* 1996; Li 1997; Figlar 2000; Xu *et al.* 2000; Sima *et al.* 2001; Xu & Wu 2002; Gong *et al.* 2003; Li & Conran 2003), the present authors proposed a new system for the family Magnoliaceae. In the new system, a total of 2 subfamilies, 2 tribes and 15 generas are recognized, of which, two genera are described as new. Of course, it is a start to table a hypothesis on Magnoliaceae system. In the future, this system will be tested to determine whether the hypothesis is correct or incorrect with refining the knowledge of phylogenetic relationships in Magnoliaceae. It is believed that the day to solve the evolutionary and phylogenetic problems in Magnoliaceae is coming soon.

2 Taxonomical System

2.1 Morphological charaters

On the basis of taxonomic theories and evolutionary principles, the following morphological charaters are selected for taxonomy and their evolutionary tendency is presented.

1. Plants pubescent → glabrous;
2. Branching sympodial → monopodial;
3. Branches produced by syllepsis → by prolepsis;
4. Leaves evergreen → deciduous;
5. Young leaf blades in bud conduplicate → open;
6. Young leaves in bud erect → pendant;
7. Leaves on the branch arranged spirally → distichously;
8. Leaves at the base cuneate to rounded → cordate to auriculate;
9. Leaf blades unlobed → lobed;
10. Leaf margin thin and not sclerophyllous → thick and sclerophyllous;
11. Stipules adnate to → free from the petiole;
12. The type of leaf stomatal apparatus paracytic → anomocytic;
13. Mixed bud scales or bracts foliaceous → spathaceous;
14. Axillary buds in the mixed bud sprouted → never sprouted;
15. Floral branches by mixed bud after flowering or fruiting shed partly → completely;
16. Flowers bisexual → androdioecious → unisexual monoecious → unisexual dioecious;
17. Pseudophyllaries foliaceous → spathaceous;
18. Tepals less than 9 → 9 to more;
19. Big parasepals present → absent;
20. Tepals coloured only on the abaxial surface → on both surfaces;
21. Stamens caducous → persistent;
22. Anthers dehiscent introrsely (→ sublatrorsely to latrorsely)→extrorsely
23. Anther connective appendages shorter → longer than the anther cells;
24. Gynoecium sessile → stipitate;
25. Gynoecium in diameter 8mm to more → less than 8mm;
26. Mature carpels not samariod →samariod;
27. Fruits not spicate → spicate;
28. Mature carpels on torus sparse → dense → concrescent;

29. Mature carpels not follicular → follicular;

30. Mature carpels with a dorsal suture groove → without a dorsal suture groove;

31. Mature carpels without a dorsal suture ridge → with a dorsal suture ridge;

32. Mature carpels dehiscing along the dorsal and/or ventral suture → circumscissile → indehiscent;

33. Mature carpels not falling → falling off from fruit axis;

34. Mature fruit axes not split → split;

35. Placentation marginal → apical;

36. Ovules 2 to more → 2 in each carpel;

37. Testae free from → adnate to the endocarp;

38. The morphological charater of chalazal region on endotesta of seed belongs to the pore type → the tube type.

2.2 Taxonomical system

I. Magnoliaceae subfam. Magnolioideae

i. Tribe Magnolieae

1. *Manglietia* Blume

2. *Lirianthe* Spach

3. *Magnolia* Linn.

4. *Dugandiodendron* Lozano-Contreras

5. *Talauma* Juss.

6. *Houpoëa* N. H. Xia et C. Y. Wu

7. *Oyama* (Nakai) N. H. Xia et C. Y. Wu

8. *Kmeria* (Pierre) Dandy

9. *Pachylarnax* Dandy

10. *Paramagnolia* Sima et S. G. Lu, gen. nov.

11. *Metamagnolia* Sima et S. G. Lu, gen. nov.

ii. Tribe Michelieae Y. W. Law

12. *Aromadendron* Blume

13. *Yulania* Spach

14. *Michelia* Linn.

II. Magnoliaceae subfam. Liriodendroideae (Nurk.) Y. W. Law

15. *Liriodendron* Linn.

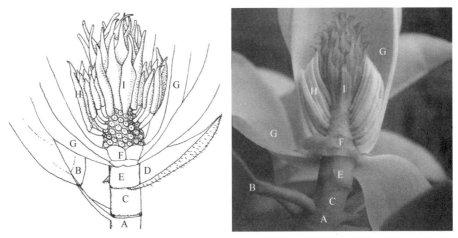

Figure 1   A flower of Magnoliaceae: A. Internode; B. pseudophyllary (foliaceous scale or leaf); C. peduncle; D. bract (spathaceous); E. pedicle; F. receptacle; G. tepals; H. stamens; I. carpels.

Figure 2   The positions of floral branches of Magnoliaceae:
A. *Michelia lacei* W. W. Smith (terminal); B. *Yulania liliiflora* (Desr.) D. L. Fu (terminal);
C. *Michelia yunnanensis* Franch. ex Finet et Gagnep. (axillary);
D. *Michelia macclurei* Dandy (axillary, 2-branched).

2.3 Key to the subfamilies, tribes and genera of Magnoliaceae:

1 (28) Leaf blades unlobed or rarely 2-lobed at the apex; anthers dehiscent introrsely or sublatrorsely to latrorsely; placentation marginal; mature carpels not samaroid, dehiscent; testae fleshy, free from the endocarp (I. Subfam. Magnolioideae).

2 (23) Fruits ovoid, ellipsoid or globose; mature carpels with a dorsal suture groove or ridge, not follicular even if not concrescent; branches produced only by syllepsis or rarely by prollepsis; stamens caducous or rarely persistent (i. Tribe Magnolieae).

3 (20) Leaves cuneate to rounded or rarely subcordate at the base, and arranged spirally or distichously on the branch, evergreen or deciduous.

4 (19) Plants with more or less trichomes; young leaf blades conduplicate in bud.

5 (18) Plants with bisexual flowers or rarely with bisexual and unisexual flowers; tepals 9 or 9 to more, the flowers with less than 9 tepals can be never seen.

6 (9) Ovules unconstant, 2 to more in each carpel.

7 (8) Pseudophyllaries foliaceous or rarely spathaceous and caducous; all mixed buds with spathaceous scales and foliaceous scales (normal leaves); leaf margin thin and not sclerophyllous .... 1. *Manglietia*

8 (7) Pseudophyllaries spathaceous and caducous or rarely foliaceous; most mixed buds only with spathaceous scales or without foliaceous scales (normal leaves); leaf margin thick and sclerophyllous .... 2. *Lirianthe*

9 (6) Ovules constant, 2 in each carpel.

10 (15) Plants evergreen.

11 (14) All mixed buds with spathaceous scales and foliaceous scales (normal leaves); pseudophyllaries foliaceous, persistent.

12 (13) Anther connective appendages not elongated, triangular or semicircular .... 3. *Magnolia*

13 (12) Anther connective appendages elongated, bristlelike .... 4. *Dugandiodendron*

14 (11) Most mixed buds only with spathaceous scales or without foliaceous scales (normal leaves); pseudophyllaries spathaceous, caduceus .... 5. *Talauma*

15 (10) Plants deciduous.

16 (17) Branches produced only by syllepsis; leaves arranged spirally on the branch; peduncles and pedicles robust, erect; stamens caduceus .... 6. *Houpoëa*

17 (16) Branches produced by prollepsis or rarely by prollepsis and syllepsis; leaves arranged distichously on the branch; peduncles and pedicles slender, pendent; stamens persistent .... 7. *Oyama*

18 (5) Plants only with unisexual flowers; tepals 2 to 13, the flowers with less than 9 tepals can be seen .... 8. *Kmeria*

19 (4) Plants entirely glabrous; young leaf blades open in bud .... 9. *Pachylarnax*

20 (3) Leaves cordate to auriculate at the base, and arranged spirally on the branch, deciduous.

21 (22) Plants entirely glabrous; tepals coloured only on the abaxial surface, not blotched at the base of adaxial surface ··· 10. *Paramagnolia*

22 (21) Plants with more or less trichomes; tepals coloured on both surfaces, blotched at the base of adaxial surface ··· 11. *Metamagnolia*

23 (2) Fruits cylindrical, ovoid or globose; mature carpels without a dorsal suture groove or ridge, follicular when not concrescent; branches produced only by prollepsis or rarely by prollepsis and syllepsis; stamens persistent (ii. Tribe Michelieae Y. W. Law).

24 (27) All axillary buds or some upper axillary buds in mixed bud developed and sprouted; only the parts, from flowers or fruits to pseudophyllaries, of the floral branches into which mixed buds formed shed after flowering or fruiting; branching sympodial and monopodial.

25 (26) Plants evergreen; peduncles slender, more than 2.5cm long; Anthers dehiscent introrsely ··· 12. *Aromadendron*

26 (25) Plants deciduous; peduncles shorter, less than 2.0cm long; Anthers dehiscent sublatrorsely to latrorsely ··· 13. *Yulania*

27 (24) All axillary buds in mixed bud undeveloped and not sprouted, or the basal axillary bud in mixed bud developed, sprouted and formed into a scorpioid cyme of 2 to 4 floral branches; all of the floral branches into which mixed buds formed shed after flowering or fruiting; branching only monopodial ··· 14. *Michelia*

28 (1) Leaf blades 4- to 10-lobed; anthers dehiscent extrorsely; placentation apical; mature carpels samaroid, indehiscent; testae thin and dry, adnate to the endocarp (II. subfam. Liriodendroideae (Nurk.) Y. W. Law) ··· 15. *Liriodendron*

3 Taxonomic Treatment

I. Magnoliaceae subfam. Magnolioideae

i. Tribe Magnolieae

1. *Manglietia* Blume, Verh. Batav. Genootsch. Kunsten 9: 149. 1823. = *Magnolia* sect. Manglietia (Blume) Baill., Adansonia 7: 66. 1866. – Type: *Manglietia glauca* Blume

–*Paramanglietia* Hu et W. C. Cheng, Acta Phytotax. Sin. 1 (3–4): 255. 1951. – Type: *Paramangietia aromatica* (Dandy) Hu et W. C. Cheng = *Manglietia aromatica* Dandy

–*Sinomanglietia* Z. X. Yu, Acta Agric. Univ. Jiangxiensis 16 (2): 202. 1994. – Type: *Sinomanglietia glauca* Z. X. Yu et Q. Y. Zheng = *Manglietia decidua* Q. Y. Zheng

Description. Trees, evergreen, or rarely semievergreen or deciduous, hairy or glabrescent. Branching sympodial and monopodial; branches produced by only syllepsis. Leaves conduplicate, erect in the bud when young, arranged spirally, rarely fascicled and pseudowhorled on the shoot; leaf blades unlobed, cuneate to rounded at the base. Stipules adnate to the petiole. All axillary buds or some upper axillary buds in the mixed bud developed and sprouted; only the parts, from flowers or fruits to pseudophyllaries, of the floral branches into which mixed buds formed shed after flowering or fruiting. Flowers terminal, solitary, bisexual. Pseudophyllaries solitary, foliaceous, persistent or rarely spathaceous, caducous. Bracts solitary, spathaceous, caducous. Peduncles robust or slender. Pedicles present or absent, visible or invisible. Tepals 9 to 18, 3-merous or 4-merous, subequal, coloured only on the abaxial surface. Stamens caducous; anthers dehiscent introrsely, anther connective appendages shorter than the anther cells. Gynoecium sessile. Fruits apocarpous to syncarpous; mature carpels with or rarely without a dorsal suture groove, dehiscing along the dorsal and/or ventral suture. Placentation marginal; ovules 2 to more in each carpel. Testae free from the endocarp. The morphological charater of chalazal region on endotesta of seed belonging to the pore type.

About 40 species distributed in tropical and subtropical Asia.

2. *Lirianthe* Spach, Hist. Nat. Vég., Phan. 7: 485. 1839. = *Magnolia* sect. *Lirianthe* (Spach) Dandy in Roy. Hort. Soc., Camellias and Magnolias, Conf. Rep.: 68. 1950. – Type: *Lirianthe grandiflora* (Roxb.) Spach – *Lirianthe pterocarpa* (Roxb.) Sima et S. G. Lu

—*Magnolia* sect. *Gwillimia* DC., Syst. Nat. 1: 455, 548. 1817. – Type: *Magnolia pumila* Andr. – *Lirianthe coco* (Lour.) N. H. Xia et C. Y. Wu

—*Blumia* Nees ex Blume, Verh. Batav. Genootsch. Kunst. 9: 147. 1823, nom. rejec., non *Blumea* DC. (1833), nom. cons. = *Magnolia* sect. *Blumia* (Nees ex Blume) Baill., Adansonia 7: 2. 1866. – Type: *Blumia candollei* (Blume) Nees – *Lirianthe liliifera* (Linn.) Sima et S. G. Lu

—*Talauma* sect. *Blumiana* Blume, Fl. Javae 19-20: 32. 1829. = *Magnolia* subsect. *Blumiana* (Blume) Figlar et Noot., Blumea 49 (1): 90. 2004. – Type: *Talauma candollei* Blume – *Lirianthe liliifera* (Linn.) Sima et S. G. Lu

Description. Trees or shrubs, evergreen, hairy or glabrescent. Branching sympodial and monopodial; branches produced by only syllepsis. Leaves conduplicate, erect in the bud when young, arranged spirally on the shoot; leaf blades unlobed, cuneate to rounded at the base. Stipules adnate to the petiole. All axillary buds or some upper axillary buds in the mixed bud developed and sprouted; only the parts, from flowers or fruits to leaves or the first developed axillary buds, of the floral branches into which mixed buds formed shed after flowering or fruiting. Flowers terminal, solitary, bisexual. Pseudophyllaries solitary, spathaceous, caducous or rarely foliaceous, persistent. Bracts solitary, spathaceous, caducous. Peduncles robust or slender. Pedicles absent or present, invisible or rarely visible. Tepals 9 or 9 to 10, 3-merous, subequal, coloured only on the abaxial surface. Stamens caducous; anthers dehiscent introrsely, anther connective appendages shorter than the anther cells. Gynoecium sessile. Fruits apocarpous or synocarpous; mature carpels not samariod, with or without a dorsal suture groove, dehiscing circumscissile, or along the dorsal and/or ventral suture. Placentation marginal; ovules 2 or 2 to more in each carpel. Testae free from the endocarp. The morphological charater of chalazal region on endotesta of seed belonging to the pore type.

About 20 species distributed in tropic and subtropical SE Asia.

Taxonomic combinations:

Lirianthe clemensiorum (Dandy) Sima et S. G. Lu, comb. nov. = *Magnolia clemensiorum*

Dandy, J. Bot. 68: 207. 1930.

Lirianthe gigantifolia (Miq.) Sima et S. G. Lu, comb. nov. = *Talauma gigantifolia* Miq., Fl. Ned. Ind. 1 (2): 15. 1858.

Lirianthe lasia (Noot.) Sima et S. G. Lu, comb. nov. = *Magnolia lasia* Noot., Blumea 32 (2): 377. 1987.

Lirianthe liliifera var. *angatensis* (Blanco) Sima et S. G. Lu, comb. nov. = *Magnolia angatensis* Blanco, Fl. Filip. : 859. 1837. = *Magnolia liliifera* var. *angatensis* (Blanco) Govaerts, World Checklist Bibliogr. Magnoliaceae: 71. 1996.

Lirianthe liliifera var. *beccarii* (Ridl.) Sima et S. G. Lu, comb. nov. = *Talauma beccarii* Ridl., Bull. Misc. Inform. Kew 1912: 381. 1912. = *Magnolia liliifera* var. *beccarii* (Ridl.) Govaerts, World Checklist Bibliogr. Magnoliaceae: 71. 1996.

Lirianthe liliifera var. *obovata* (Korth.) Sima et S. G. Lu, comb. nov. = *Talauma obovata* Korth., Ned. Kruidk. Arch. 2 (2): 89. 1851. = *Magnolia liliifera* var. *obovata* (Korth.) Govaerts, World Checklist Bibliogr. Magnoliaceae: 71. 1996.

Lirianthe liliifera var. *singapurensis* (Ridl.) Sima et S. G. Lu, comb. nov. = *Talauma singapurensis* Ridl., Bull. Misc. Inform. Kew 1914: 323. 1914. = *Magnolia liliifera* var. *singapurensis* (Ridl.) Govaerts, World Checklist Bibliogr. Magnoliaceae: 71. 1996

Lirianthe mariusjacobsia (Noot.) Sima et S. G. Lu, comb. nov. = *Magnolia mariusjacobsia* Noot., Blumea 32 (2): 381. 1987.

Lirianthe nana (Dandy) Sima et S. G. Lu, comb. nov. = *Magnolia nana* Dandy, J. Bot. 68: 207. 1930.

Lirianthe persuaveolens (Dandy) Sima et S. G. Lu, comb. nov. = *Magnolia persuaveolens* Dandy, Bull. Misc. Inform. Kew 1928: 186. 1928.

Lirianthe persuaveolens var. *pubescens* (Noot.) Sima et S. G. Lu, comb. nov. = *Magnolia persuaveolens* var. *pubescens* Noot., Blumea 32 (2): 379. 1987.

Lirianthe persuaveolens subsp. *rigida* (Noot.) Sima et S. G. Lu, comb. nov. = *Magnolia persuaveolens* subsp. *rigida* Noot., Blumea 32 (2): 379. 1987.

Lirianthe poilanei (Dandy ex Gagnep.) Sima et S. G. Lu, comb. nov. = *Magnolia poilanei* Dandy ex Gagnep. in P. H. Lecomte, Fl. Indo-Chine, Suppl. 1: 40. 1938.

Lirianthe pulgarensis (Elmer) Sima et S. G. Lu, comb. nov. = *Talauma pulgarensis* Elmer, Leafl. Philipp. Bot. 5: 1809. 1913.

Lirianthe sarawakensis (A. Agostini) Sima et S. G. Lu, comb. nov. = *Talauma sarawakensis* A. Agostini, Atti Reale Accad. Fisiocrit. Siena, X, 1: 190. 1926.

Lirianthe villosa (Miq.) Sima et S. G. Lu, comb. nov. = *Talauma villosa* Miq., Fl. Ned. Ind. Eerste Bijv. : 366. 1861.

3. *Magnolia* Linn., Sp. Pl. : 535. 1753. – Type: *Magnolia virginiana* Linn.

Description. Trees, evergreen or semievergreen, hairy. Branching sympodial and monopodial; branches produced by only syllepsis. Leaves conduplicate, erect in the bud when young, arranged spirally on the shoot, or spirally on the terminal shoot and distichously on the axillary shoot; leaf blades unlobed, cuneate to rounded at the base. Stipules adnate to or free from the petiole. All axillary buds or some upper axillary buds in the mixed bud developed and sprouted; only the parts, from flowers or fruits to pseudophyllaries, of the floral branches into which mixed buds formed shed after flowering or fruiting. Flowers terminal, solitary, bisexual. Pseudophyllaries solitary, foliaceous, persistent. Bracts solitary, spathaceous, caducous. Peduncles robust. Pedicles present or absent, visible or invisible. Tepals 9 to 12, 3-merous, subequal, coloured only on the abaxial surface. Stamens caducous; anthers dehiscent introrsely, anther connective appendages shorter than the anther cells. Gynoecium sessile. Fruits apocarpous; mature carpels with a dorsal suture groove, dehiscing along the dorsal and/or ventral suture. Placentation marginal; ovules 2 in each carpel. Testae free from the endocarp. The morphological charater of chalazal region on endotesta of seed belonging to the tube type.

Sixteen species distributed in southeast North America and Central America.

4. *Dugandiodendron* Lozano – Contreras, Caldasia 11 (53): 33. 1975. = *Magnolia* subsect. *Dugandiodendron* (Lozano-Contreras) Figlar et Noot., Blumea 49 (1): 90. 2004. – Type: *Dugandiodendron mahechae* Lozano-Contreras

–*Magnolia* subsect. *Cubenses* Imkhan., Novosti Sist. Vyssh. Rast. 28: 60. 1991. – Type: *Magnolia cubensis* Urb. = *Dugandiodendron cubense* (Urb.) Sima et S. G. Lu

–*Magnolia* sect. *Splendentes* Dandy ex A. Vázquez, Brittonia 46: 4. 1994. = *Magnolia* subsect. *Splendentes* (Dandy ex A. Vázquez) Figlar et Noot., Blumea 49 (1): 91. 2004. – Type: *Magnolia splendens* Urb. = *Dugandiodendron spendens* (Urb.) Sima et S. G. Lu

Description. Trees, evergreen, hairy. Branching sympodial and monopodial; branches produced by only syllepsis. Leaves conduplicate, erect in the bud when young, arranged spirally on the shoot; leaf blades unlobed, cuneate to rounded at the base. Stipules free from the petiole. All axillary buds or some upper axillary buds in the mixed bud developed and sprouted; only the parts, from flowers or fruits to pseudophyllaries, of the floral branches into which mixed buds formed shed after flowering or fruiting. Flowers terminal, solitary, bisexual. Pseudophyllaries solitary, foliaceous, persistent. Bracts solitary, spathaceous, caducous. Peduncles robust or slender. Pedicles present or absent, visible or invisible. Tepals 9 to 15, 3-merous, subequal, coloured only on the abaxial surface. Stamens caducous; anthers dehiscent introrsely, anther connective appendages longer than the anther cells, embedded to the gynoecium. Gynoecium sessile. Fruits apocarpous to synocarpous; mature carpels with or without a dorsal suture groove, dehiscing circumscissile, or along the dorsal and/or ventral suture. Placentation marginal; ovules 2 in each carpel. Testae free from the endocarp. The morphological charater of chalazal region on endotesta of seed belonging to the tube type.

About 25 species distributed in tropic Central and South America.

Taxonomic combinations:

Dugandiodendron cacuminoides (Bisse) Sima et S. G. Lu, comb. nov. = *Magnolia cacuminoides* Bisse, Repert. Spec. Nov. Regni Veg. 85 (9-10): 587. 1974.

Dugandiodendron cristalense (Bisse) Sima et S. G. Lu, comb. nov. = *Magnolia cristalensis*

Bisse, Repert. Spec. Nov. Regni Veg. 85 (9-10): 588. 1974.

Dugandiodendron cubense (Urb.) Sima et S. G. Lu, comb. nov. = *Magnolia cubensis* Urb., Symb. Antill. 1 (2): 307. 1899.

Dugandiodendron cubense subsp. *cacuminicolum* (Bisse) Sima et S. G. Lu, comb. nov. = *Magnolia cacuminicola* Bisse, Repert. Spec. Nov. Regni Veg. 85 (9-10): 587. 1974. = *Magnolia cubensis* subsp. *cacuminicola* (Bisse) G. Klotz, Wiss. Zeitschr. Friedrich-Schiller-Univ. Jena, Math. -Naturwiss. Reihe 29 (4): 464. 1980.

Dugandiodendron domingense (Urb.) Sima et S. G. Lu, comb. nov. = *Magnolia domingensis* Urb., Repert. Spec. Nov. Regni Veg. 13: 447. 1914.

Dugandiodendron ekmanii (Urb.) Sima et S. G. Lu, comb. nov. = *Magnolia ekmanii* Urb., Ark. Bot. 23A (11): 12. 1931.

Dugandiodendron emarginatum (Urb. et Ekman) Sima et S. G. Lu, comb. nov. = *Magnolia emarginata* Urb. et Ekman, Ark. Bot. 23A (11): 11. 1931.

Dugandiodendron hamori (R. A. Howard) Sima et S. G. Lu, comb. nov. = *Magnolia hamori* R. A. Howard, Bull. Torrey Bot. Club 75: 351. 1948.

Dugandiodendron pallescens (Urb. et Ekman) Sima et S. G. Lu, comb. nov. = *Magnolia pallescens* Urb. et Ekman, Ark. Bot. 23A (11): 10. 1931.

Dugandiodendron portoricense (Bello) Sima et S. G. Lu, comb. nov. = *Magnolia portoricensis* Bello, Anales Soc. Esp. Hist. Nat. 10: 233. 1880.

Dugandiodendron splendens (Urb.) Sima et S. G. Lu, comb. nov. = *Magnolia splendens* Urb., Symb. Antill. 1 (2): 306. 1899.

5. *Talauma* Juss., Gen. Pl.: 281. 1789. = *Magnolia* sect. *Talauma* Baill., Adansonia 7: 3, 66. 1866. = *Magnolia* subgen. *Talauma* (Juss.) Pierre, Fl. Forest. Cochinch.: sub. t. 1. 1880. - Type: *Talauma plumieri* (Sw.) DC. - *Talauma dodecapetala* (Lam.) Urb.

—*Svenhedinia* Urb., Repert. Spec. Nov. Regni Veg. 24: 3. 1927. = *Talauma* sect. *Svenhedinia* (Urb.) Imkhan., Novosti Sist. Vyssh. Rast. 29: 74. 1993. - Type: *Svenhedinia minor* (Urb.) Urb. = *Talauma minor* Urb.

Description. Trees, evergreen, hairy. Branching sympodial and monopodial; branches produced by only syllepsis. Leaves conduplicate, erect in the bud when young, arranged spirally on the shoot; leaf blades unlobed, cuneate to rounded at the base. Stipules adnate to the petiole. All axillary buds or some upper axillary buds in the mixed bud developed and sprouted; only the parts, from flowers or fruits to leaves or the first developed axillary buds, of the floral branches into which mixed buds formed shed after flowering or fruiting. Flowers terminal, solitary, bisexual. Pseudophyllaries solitary, spathaceous, caducous or rarely foliaceous, persistent. Bracts solitary, spathaceous, caducous. Peduncles robust or slender. Pedicles present or absent, visible or invisible. Tepals 9 to 15, 3-merous, subequal, coloured only on the abaxial surface. Stamens caducous; anthers dehiscent introrsely, anther connective appendages shorter than the anther cells. Gynoecium sessile. Fruits synocarpous; mature carpels with or without a dorsal suture groove, dehiscing circumscissile. Placentation marginal; ovules 2 in each carpel. Testae free from the endocarp. The morphological charater of chalazal region on endotesta of seed belonging to the tube type.

About 30 species distributed in tropic southeast – central North America and Central and South America.

6. *Houpoëa* N. H. Xia et C. Y. Wu, Fl. China 7: 64. 2008. = *Magnolia* sect. *Rytidospermum* Spach, Hist. Nat. Vég. Phan. 7: 474. 1839. – Type: *Magnolia umbrella* Desr. – *Houpoëa tripetala* (Linn.) Sima et S. G. Lu

Description. – Trees, deciduous, hairy. Branching sympodial and monopodial; branches produced by only syllepsis. Leaves conduplicate, erect in the bud when young, arranged spirally, often fascicled and pseudowhorled on the shoot; leaf blades unlobed or rarely 2 – lobed at the apex, cuneate to rounded or rarely subcordate at the base. Stipules adnate to the petiole. All axillary buds or some upper axillary buds in the mixed bud developed and sprouted; only the parts, from flowers or fruits to pseudophyllaries, of the floral branches into which mixed buds formed shed after flowering or fruiting. Flowers terminal, solitary, bisexual. Pseudophyllaries solitary, foliaceous, persistent. Bracts solitary, spathaceous, caducous. Peduncles robust. Pedicles present, visible. Tepals 9 to 17, 3 – to 5 – merous, subequal, coloured only on the abaxial surface. Stamens caducous; anthers dehiscent introrsely, anther connective appendages shorter than the anther cells. Gynoecium sessile. Fruits apocarpous; mature carpels with a dorsal suture groove, dehiscing along the dorsal and/or ventral suture. Placentation marginal; ovules 2 in each carpel. Testae free from the endocarp. The morphological charater of chalazal region on endotesta of seed belonging to the pore type.

Four species distributed in temperate eastern North America and temperate E and SE Asia.

Taxonomic combinations:

Houpoëa tripetala (Linn.) Sima et S. G. Lu, comb. nov. = *Magnolia virginiana* Linn. var. *tripetala* Linn., Sp. Pl.: 536. 1753. = *Magnolia tripetala* (Linn.) Linn., Syst. Nat., ed. 10: 1082. 1759.

7. *Oyama* (Nakai) N. H. Xia et C. Y. Wu, Fl. China 7: 66. 2008. = *Magnolia* sect. *Oyama* Nakai, Fl. Sylv. Koreana 20: 117. 1933. – Type: *Magnolia parviflora* Siebold et Zucc. – *Oyama sieboldii* (K. Koch) N. H. Xia et C. Y. Wu

= *Magnolia* sect. *Cophantera* Dandy, Curtis's Bot. Mag. 159: sub. t. 9467. 1936. – Type: *Magnolia sieboldii* K. Koch = *Oyama sieboldii* (K. Koch) N. H. Xia et C. Y. Wu

Description. Trees or sbrubs, deciduous, hairy. Branching sympodial and monopodial; branches produced by prolepsis. Leaves conduplicate, erect in the bud when young, arranged distichously on the shoot; leaf blades unlobed, cuneate to rounded at the base. Stipules adnate to the petiole. All axillary buds or some upper axillary buds in the mixed bud developed and sprouted; only the parts, from flowers or fruits to pseudophyllaries, of the floral branches into which mixed buds formed shed after flowering or fruiting. Flowers terminal, solitary, bisexual. Pseudophyllaries solitary, foliaceous, persistent. Bracts solitary, spathaceous, caducous. Peduncles slender. Pedicles present or rarely absent, visible or invisible. Tepals 9 to 10, 3 – merous, subequal, coloured only on the abaxial surface. Stamens persistent; anthers dehiscent introrsely, anther connective appendages absent or shorter than the anther cells. Gynoecium sessile or rarely stipitate. Fruits apocarpous; mature carpels without a dorsal suture groove, dehiscing along the dorsal and/or ventral suture. Placentation marginal; ovules 2 in each carpel. Testae free from the endocarp. The morphological charater of chalazal region on endotesta of seed belonging to the pore type.

Three species distributed in E and SE Asia.

8. *Kmeria* (Pierre) Dandy, Bull. Misc. Inform. Kew 1927: 262. 1927. = *Magnolia* subgen. *Kmeria* Pierre, Fl. Forest. Cochinch. : sub. t. 1. 1880. = *Magnolia* sect. Kmeria (Pierre) Figlar et Noot., Blumea 49 (1): 91. 2004. − Type: *Magnolia duperreana* Pierre = *Kmeria duperreana* (Pierre) Dandy

−*Woonyoungia* Y. W. Law, Bull. Bot. Res., Harbin 17 (4): 354. 1997. − Type: *Woonyoungia septentrionalis* (Dandy) Y. W. Law = *Kmeria septentrionalis* Dandy

Description. Trees, evergreen, hairy or glabrescent. Branching sympodial and monopodial; branches produced by only syllepsis. Leaves conduplicate, erect in the bud when young, arranged spirally on the shoot, or spirally on the terminal shoot and distichously on the axillary shoot; leaf blades unlobed, cuneate at the base. Stipules adnate to the petiole. All axillary buds or some upper axillary buds in the mixed bud developed and sprouted; only the parts, from flowers or fruits to pseudophyllaries, of the floral branches into which mixed buds formed shed after flowering or fruiting. Flowers terminal, solitary, unisexual monoecious or dioecious. Pseudophyllaries solitary, foliaceous, persistent or rarely spathaceous, caducous. Bracts solitary, spathaceous, caducous. Peduncles slender. Pedicles present or absent, invisible. Tepals 2 to 17, 2− to 5−merous, subequal, coloured only on the abaxial surface. Stamens caducous; anthers dehiscent introrsely or sublatrorsely to latrorsely, anther connective appendages shorter than the anther cells. Gynoecium sessile. Fruits apocarpous or synocarpous; mature carpels without a dorsal suture groove, dehiscing along the dorsal and/or ventral suture. Placentation marginal; ovules 2 in each carpel. Testae free from the endocarp. The morphological charater of chalazal region on endotesta of seed belonging to the tube type.

Three species distributed in subtropical SE Asia.

9. *Pachylarnax* Dandy, Bull. Misc. Inform. Kew 1927: 260. 1927. − Type: *Pachylarnax praecalva* Dandy

−*Magnolia* sect. *Gynopodium* Dandy, Curtis's Bot. Mag. 165: t. 16. 1948. = *Magnolia* subgen. *Gynopodium* (Dandy) Figlar et Noot., Blumea 49 (1): 94. 2004. − Type: *Magnolia nitida* W. W. Sm. = *Pachylarnax nitida* (W. W. Sm.) Sima et S. G. Lu

−*Parakmeria* Hu et W. C. Cheng, Acta Phytotax. Sin. 1 (1): 1. 1951. − Type: *Parakmeria omeiensis* W. C. Cheng = *Pachylarnax omeiensis* (W. C. Cheng) Sima et S. G. Lu

−*Micheliopsis* H. Keng, Quart. J. Taiwan Mus. 8: 207. 1955. − Type: *Micheliopsis kachirachirai* (Kaneh. et Yamam.) H. Keng = *Pachylarnax kachirachirai* (Kaneh. et Yamam.) Sima et S. G. Lu

—*Manglietiastrum* Y. W. Law, Acta Phytotax. Sin. 17 (4): 72. 1979. = *Magnolia* sect. *Manglietiastrum* (Y. W. Law) Noot., Blumea 31 (1): 91. 1985. = *Manglietia* sect. *Manglietiastrum* (Y. W. Law) Noot., Ann. Missouri Bot. Gard. 80 (4): 1051. 1993. – Type: *Manglietiastrum sinicum* Y. W. Law = *Pachylarnax sinica* (Y. W. Law) N. H. Xia et C. Y. Wu

Description. Trees, evergreen, glabrous. Branching sympodial and monopodial; branches produced by only syllepsis. Leaves open, erect in the bud when young, arranged spirally on the shoot, or spirally on the terminal shoot and distichously on the axillary shoot; leaf blades unlobed, cuneate to rounded at the base. Stipules free from the petiole. All axillary buds or some upper axillary buds in the mixed bud developed and sprouted; only the parts, from flowers or fruits to pseudophyllaries, of the floral branches into which mixed buds formed shed after flowering or fruiting. Flowers terminal, solitary, bisexual or androdioecious. Pseudophyllaries solitary, spathaceous, caducous. Bracts solitary, spathaceous, caducous. Peduncles robust. Pedicles present or absent, visible or invisible. Tepals 9 to 11, 3-merous, subequal, coloured only on the abaxial surface. Stamens caducous; anthers dehiscent introrsely, anther connective appendages shorter than the anther cells. Gynoecium stipitate or sessile. Fruits apocarpous to synocarpous; mature carpels with or without a dorsal suture groove, dehiscing along the dorsal and/or ventral suture. Placentation marginal; ovules 2 or 2 to 8 in each carpel. Testae free from the endocarp. The morphological charater of chalazal region on endotesta of seed belonging to the tube type or rarely to the pore type.

Seven species distributed in tropical and subtropical SE Asia.

10. *Paramagnolia* Sima et S. G. Lu, gen. nov. – Type: *Paramagnolia fraseri* (Walter) Sima et S. G. Lu = *Magnolia* sect. *Auriculatae* Figlar et Noot., Blumea 49 (1): 92. 2004 ['*Auriculata*']. – Type: *Magnolia fraseri* Walter = *Paramagnolia fraseri* (Walter) Sima et S. G. Lu

Diagnosis. – Folia decidua, glabra, pseudoverticillata, basi auriculata. Tepala non nisi subtus colorata.

Description. Trees, deciduous, glabrous. Branching sympodial and monopodial; branches produced by only syllepsis. Leaves conduplicate, erect in the bud when young, arranged spirally, often fascicled and pseudowhorled on the shoot; leaf blades unlobed, auriculate at the base. Stipules adnate to the petiole. All axillary buds or some upper axillary buds in the mixed bud developed and sprouted; only the parts, from flowers or fruits to pseudophyllaries, of the floral branches into which mixed buds formed shed after flowering or fruiting. Flowers terminal, solitary, bisexual. Pseudophyllaries solitary, foliaceous, persistent. Bracts solitary, spathaceous, caducous. Peduncles robust. Pedicles present, visible. Tepals 9, 3-merous, subequal, coloured only on the abaxial surface. Stamens caducous; anthers dehiscent introrsely, anther connective appendages shorter than the anther cells. Gynoecium sessile. Fruits apocarpous; mature carpels with a dorsal suture groove, dehiscing along the dorsal and/or ventral suture. Placentation marginal; ovules 2 in each carpel. Testae free from the endocarp. The morphological charater of chalazal region on endotesta of seed belonging to the pore type.

One species and one variety distributed in SE North America.

Taxonomic combinations:

Paramagnolia fraseri (Walter) Sima et S. G. Lu, comb. nov. = *Magnolia fraseri* Walter, Fl. Carol.: 159. 1788.

Paramagnolia fraseri var. *pyramidata* (Bartram) Sima et S. G. Lu, comb. nov. = *Magnolia pyramidata* Bartram, Travels Carolina: 408. 1791. = *Magnolia fraseri* var. *pyramidata* (Bartram) Torr. et A. Gray, Fl. N. Amer. 1: 43. 1838.

11. *Metamagnolia* Sima et S. G. Lu, gen. nov. – Type: *Metamagnolia macrophylla* (Michx.) Sima et S. G. Lu

= *Magnolia* sect. *Macrophyllae* Figlar et Noot., Blumea 49 (1): 92. 2004 ['*Macrophylla*']. – Type: *Magnolia macrophylla* Michx. = *Metamagnolia macrophylla* (Michx.) Sima et S. G. Lu

Diagnosis. – Folia decidua, pubescentia, pseudoverticillata, basi auriculata vel cordata. Tepala utrinque colorata.

Description. Trees, deciduous, pubescent. Branching sympodial and monopodial; branches produced by only syllepsis. Leaves conduplicate, erect in the bud when young, arranged spirally, often fascicled and pseudowhorled on the shoot; leaf blades unlobed, deeply cordate to auriculate at the base. Stipules adnate to the petiole. All axillary buds or some upper axillary buds in the mixed bud developed and sprouted; only the parts, from flowers or fruits to pseudophyllaries, of the floral branches into which mixed buds formed shed after flowering or fruiting. Flowers terminal, solitary, bisexual. Pseudophyllaries solitary, foliaceous, persistent. Bracts solitary, spathaceous, caducous. Peduncles robust. Pedicles present, visible. Tepals 9, 3-merous, subequal, coloured on both surfaces, blotched at the base of adaxial surface. Stamens caducous; anthers dehiscent introrsely, anther connective appendages shorter than the anther cells. Gynoecium sessile. Fruits apocarpous; mature carpels with a dorsal suture groove, dehiscing along the dorsal and/or ventral suture. Placentation marginal; ovules 2 in each carpel. Testae free from the endocarp. The morphological charater of chalazal region on endotesta of seed belonging to the tube type.

Two species and one subspecies distributed in SE North America.

Taxonomic combinations:

Metamagnolia macrophylla (Michx.) Sima et S. G. Lu, comb. nov. = *Magnolia macrophylla* Michx., Fl. Bor.-Amer. 1: 327. 1803.

Metamagnolia macrophylla subsp. *ashei* (Weath.) Sima et S. G. Lu, comb. nov. = *Magnolia ashei* Weath., Rhodora 28: 35. 1926. = *Magnolia macrophylla* subsp. *ashei* (Weath.) Spongberg, J. Arnold Arbor. 57: 268. 1976.

Metamagnolia dealbata (Zucc.) Sima et S. G. Lu, comb. nov. = *Magnolia dealbata* Zucc., Abh. Math.-Phys. Cl. Königl. Bayer. Akad. Wiss. 2: 373. 1836.

ii. Tribe Michelieae Y. W. Law

12. *Aromadendron* Blume, Bijdr.: 10. 1825. = *Talauma* sect. *Aromadendron* Miq., Ann. Mus. Bot. Lugduno-Batavi 4: 70. 1868. = *Magnolia* sect. *Aromadendron* (Blume) Noot.,

Blumea 31 (1): 89. 1985. = *Magnolia* subsect. *Aromadendron* (Blume) Figlar et Noot., Blumea 49 (1): 94. 2004. - Type: *Aromadendron elegans* Blume

—*Alcimandra* Dandy, Bull. Misc. Inform. Kew 1927: 260. 1927. = *Magnolia* sect. *Alcimandra* (Dandy) Noot., Blumea 31 (1): 88. 1985. - Type: *Alcimandra cathcartii* (J. D. Hook. et Thomson) Dandy = *Aromadendron cathcartii* (J. D. Hook. et Thomson) Sima et S. G. Lu

—*Magnolia* sect. *Maingola* Dandy, Curtis's Bot. Mag. 165: t. 16. 1948. = *Magnolia* subsect. *Maingola* (Dandy) Figlar et Noot., Blumea 49 (1): 93. 2004. - Type: *Magnolia maingayi* King = *Aromadendron maingayi* (King) Sima et S. G. Lu

Description. Trees, evergreen, hairy or glabrescent. Branching sympodial and monopodial; branches produced by prolepsis, or rarely by prolepsis and syllepsis. Leaves conduplicate, erect in the bud when young, arranged distichously on the shoot, or spirally on the terminal shoot and distichously on the axillary shoot; leaf blades unlobed, cuneate to rounded at the base. Stipules free from the petiole. All axillary buds or some upper axillary buds in the mixed bud developed and sprouted; only the parts, from flowers or fruits to pseudophyllaries, of the floral branches into which mixed buds formed shed after flowering or fruiting. Flowers terminal, solitary, bisexual. Pseudophyllaries solitary, foliaceous, persistent. Bracts solitary, spathaceous, caducous. Peduncles short or slender. Pedicles absent or rarely present, invisible or visible. Tepals 9 or 9 to 45, 3 - to 5-merous, subequal, coloured only on the abaxial surface. Stamens persistent; anthers dehiscent introrsely, anther connective appendages absent, or shorter to longer than the anther cells, not embedded to the gynoecium. Gynoecium sessile or stipitate. Fruits apocarpous to synocarpous; mature carpels without a dorsal suture groove, dehiscing along the dorsal and/or ventral suture, or circumscissile. Placentation marginal; ovules 2 or 2 to 9 in each carpel. Testae free from the endocarp. The morphological charater of chalazal region on endotesta of seed belonging to the pore type or the tube type.

Thirteen species distributed in tropical and subtropical SE Asia.

Taxonomic combinations:

Aromadendron annamense (Dandy) Sima et S. G. Lu, comb. nov. = *Magnolia annamensis* Dandy, J. Bot. 68: 209. 1930.

Aromadendron ashtonii (Dandy ex Noot.) Sima et S. G. Lu, comb. nov. = *Magnolia ashtonii* Dandy ex. Noot., Blumea 32 (2): 363. 1987.

Aromadendron bintuluense (A. Agostini) Sima et S. G. Lu, comb. nov. = *Talauma bintuluensis* A. Agostini, Atti Reale Accad. Fisiocrit. Siena, X, 1: 187. 1926.

Aromadendron borneense (Noot.) Sima et S. G. Lu, comb. nov. = *Magnolia borneensis* Noot., Blumea 32 (2): 366. 1987.

Aromadendron carsonii (Dandy ex Noot.) Sima et S. G. Lu, comb. nov. = *Magnolia carsonii* Dandy ex Noot., Blumea 32 (2): 348. 1987.

Aromadendron carsonii var. *drymifolium* (Noot.) Sima et S. G. Lu, comb. nov. = *Magnolia carsonii* var. *drymifolia* Noot., Blumea 32 (2): 351. 1987.

Aromadendron carsonii var. *phaulantum* (Noot.) Sima et S. G. Lu, comb. nov. = *Magnolia phaulanta* Dandy ex Noot., Blumea 32 (2): 359. 1987. = *Magnolia carsonii* var. *phaulanta* (Dandy ex Noot.) S. Kim et Noot., Blumea 47 (2): 332. 2002.

Aromadendron griffithii (J. D. Hook. et Thomson) Sima et S. G. Lu, comb. nov. = *Michelia griffithii* J. D. Hook. et Thomson in J. D. Hooker, Fl. Brit. Ind. 1: 41. 1872.

Aromadendron gustavii (King) Sima et S. G. Lu, comb. nov. = *Magnolia gustavii* King, Ann. Roy. Bot. Gard. (Calcutta) 3 (2): 209. 1891.

Aromadendron macklottii (Korth.) Sima et S. G. Lu, comb. nov. = *Manglietia macklottii* Korth., Ned. Kruidk. Arch. 2 (2): 97. 1851.

Aromadendron macklottii var. *beccarianum* (A. Agostini) Sima et S. G. Lu, comb. nov. = *Micheliabeccariana* A. Agostini, Atti Reale Accad. Fisiocrit. Siena, X, 1: 184. 1926. = *Magnolia macklottii* var. *beccariana* (A. Agostini) Noot., Blumea 32 (2): 348. 1987.

Aromadendron pahangense (Noot.) Sima et S. G. Lu, comb. nov. = *Magnolia pahangensis* Noot., Blumea 32 (2): 367. 1987.

Aromadendron pealianum (King) Sima et S. G. Lu, comb. nov. = *Magnolia pealiana* King, Ann. Roy. Bot. Gard. (Calcutta) 3 (2): 210. 1891.

13. *Yulania* Spach, Hist. Nat. Vég., Phan. 7: 462. 1839, nom. cons. propos. = *Magnolia* subgen. *Yulania* (Spach) Reichenbach, Der Deutsch. Bot. 1: 192. 1841. - Type: *Yulania conspicua* (Salisb.) Spach - *Yulania denudata* (Desr.) D. L. Fu

= *Lassonia* Buc'hoz, Pl. Nouv. Découv.: t. 19. 1779, nom. rejec. propos. Type: *Lassonia heptapeta* Buc'hoz, nom. rejec. - *Yulania denudata* (Desr.) D. L. Fu

- *Tulipastrum* Spach, Hist. Nat. Vég., Phan. 7: 461. 1839. = *Magnolia* sect. *Tulipastrum* (Spach) Dandy in Roy. Hort. Soc., Camellias and Magnolias, Conf. Rep.: 74. 1950. = *Yulania* sect. *Tulipastrum* (Spach) D. L. Fu, J. Wuhan Bot. Res. 19 (3): 198. 2001. = *Magnolia* subsect. *Tulipastrum* (Spach) Figlar et Noot., Blumea 49 (1): 92. 2004. - Type: *Tulipastrum americanum* Spach - *Yulania acuminata* (Linn.) D. L. Fu

- *Buergeria* Siebold et Zucc., Abh. Math.-Phys. Cl. Königl. Bayer. Akad. Wiss. 4: 186. 1845. = *Magnolia* sect. *Buergeria* (Siebold et Zucc.) Dandy in Roy. Hort. Soc., Camellias and Magnolias, Conf. Rep.: 73. 1950. = *Yulania* sect. *Buergeria* (Siebold et Zucc.) D. L. Fu, J. Wuhan Bot. Res. 19 (3): 198. 2001. - Type: *Buergeria stellata* Siebold et Zucc. = *Yulania stellata* (Siebold et Zucc.) Sima et S. G. Lu

- *Magnolia* subgen. *Pleurochasma* Dandy, J. Roy. Hort. Soc. 75: 161. 1950. - Type: *Magnolia campbellii* J. D. Hook. et Thomson = *Yulania campbellii* (J. D. Hook. et Thomson) D. L. Fu

- *Magnolia* sect. *Axilliflora* B. C. Ding et T. B. Chao, Acta Agric. Univ. Henan. 19 (4): 360. 1985. = *Yulania* sect. *Axilliflora* (B. C. Ding et T. B. Chao) D. L. Fu, J. Wuhan Bot. Res. 19 (3): 198. 2001. - Type: *Magnolia axilliflora* (T. B. Chao, T. X. Zhang et J. T. Gao) T. B. Chao - *Yulania biondii* (Pamp.) D. L. Fu

—*Magnolia* sect. *Trimorphaflora* B. C. Ding et T. B. Chao, Acta Agric. Univ. Henan. 19
(4): 359. 1985. − Type: *Magnolia henanensis* B. C. Ding et T. B. Chao − *Yulania biondii* (Pamp.) D. L. Fu

—*Magnolia* sect. × *Zhushayulania* W. B. Sun et T. B. Chao, J. Centr. South Forest. Univ. 19 (2): 27. 1999. = *Yulania* sect. × Zhushayulania (W. B. Sun et T. B. Chao) D. L. Fu, J. Wuhan Bot. Res. 19 (3): 198. 2001. − Type: *Magnolia* × *soulangeana* Soul. -Bod. = *Yulania* × *soulangeana* (Soul. -Bod.) D. L. Fu

Description. Trees or shrubs, deciduous, hairy. Branching sympodial and monopodial; branches produced by prolepsis. Leaves conduplicate, erect in the bud when young, arranged distichously on the shoot; leaf blades unlobed or rarely 2-lobed at the apex, cuneate to rounded or rarely subcordate at the base. Stipules adnate to the petiole. All axillary buds or some upper axillary buds in the mixed bud developed and sprouted; only the parts, from flowers or fruits to pseudophyllaries, of the floral branches into which mixed buds formed shed after flowering or fruiting. Flowers terminal, solitary, bisexual. Pseudophyllaries solitary, spathaceous or foliaceous, caducous or rarely persistent. Bracts solitary, spathaceous, caducous. Peduncles short. Pedicles absent or rarely present, invisible or visible. Tepals 9 to 38, 3- to 5-merous, subequal, or unequal, ones in outmost whorl smaller than 1/2 of ones in other inner whorls, coloured only on the abaxial surface. Stamens persistent; anthers dehiscent sublatrorsely to latrorsely, anther connective appendages shorter than the anther cells. Gynoecium sessile. Fruits apocarpous to synocarpous; mature carpels without a dorsal suture groove, dehiscing along the dorsal and/or ventral suture, or circumscissile. Placentation marginal; ovules 2 or 2 to 3 in each carpel. Testae free from the endocarp. The morphological charater of chalazal region on endotesta of seed belonging to the tube type.

Eleven species distributed in temperate E Asia and eastern North America.

Taxonomic combinations:

Yulania stellata (Siebold et Zucc.) Sima et S. G. Lu, comb. nov. = *Buergeria stellata* Siebold et Zucc., Abh. Math. -Phys. Cl. Königl. Bayer. Akad. Wiss. 4 (2): 186. 1845.

14. *Michelia* Linn., Sp. Pl.: 536. 1753. = *Magnolia* sect. Michelia (Linn.) Baill., Adansonia 7: 66. 1866. − Type: *Michelia champaca* Linn.

=*Champaca* Adans., Fam. Pl. 2: 365. 1763. − Type: *Champaca michelia* Noronha − *Michelia champaca* Linn.

−*Liriopsis* Spach, Hist. Nat. Vég., Phan. 7: 460. 1839. − Type: *Liriopsis fuscata* (Andr.) Spach − *Michelia figo* (Lour.) Spreng.

−*Magnolia* sect. *Micheliopsis* Baill., Adansonia 7: 4, 66. 1866. = *Michelia* sect. Micheliopsis (Baill.) Dandy in J. Praglowski, World Pollen Spore Fl. 3: 5. 1974. − Type: *Magnolia figo* (Lour.) DC. = *Michelia figo* (Lour.) Spreng.

=*Sampaca* Kuntze, Revis. Gen. Pl.: 6. 1891. − Type: *Sampacca euonymoides* Kuntze = *Michelia champaca* Linn.

−*Elmerrillia* Dandy, Bull. Misc. Inform. Kew 1927: 261. 1927. = *Magnolia* subsect.

Elmerrillia (Dandy) Figlar et Noot., Blumea 49 (1): 93. 2004. – Type: *Elmerrillia papuana* (Schltr.) Dandy = *Michelia tsiampacca* Linn.

–*Paramichelia* Hu, Sunyatsenia 4: 142. 1940. = *Michelia* sect. *Paramichelia* (Hu) Noot. et B. L. Chen, Ann. Missouri Bot. Gard. 80 (4): 1087. 1993. – Type: *Paramichelia baillonii* (Pierre) Hu = *Michelia baillonii* (Pierre) Finet et Gagnep.

–*Tsoongiodendron* Chun, Acta Phytotax. Sin. 8 (4): 281. 1963. = *Michelia* sect. *Tsoongiodendron* (Chun) Noot. et B. L. Chen, Ann. Missouri Bot. Gard. 80 (4): 1086. 1993. – Type: *Tsoongiodendron odorum* Chun = *Michelia odora* (Chun) Noot. et B. L. Chen

Description. Trees or shrubs, deciduous, hairy. Branching only monopodial; branches produced by prolepsis, or rarely by prolepsis and syllepsis. Leaves conduplicate, erect in the bud when young, arranged distichously on the shoot, or spirally on the terminal shoot and distichously on the axillary shoot; leaf blades unlobed, cuneate to rounded at the base. Stipules adnate to or free from the petiole. All axillary buds in mixed bud undeveloped and not sprouted, or the basal axillary bud in mixed bud developed, sprouted and formed into a scorpioid cyme of 2 to 4 floral branches; all of the floral branches into which mixed buds formed shed after flowering or fruiting. Flowers terminal, solitary, bisexual. Pseudophyllaries solitary, spathaceous or rarely foliaceous, caducous. Bracts solitary, spathaceous, caducous. Peduncles short. Pedicles absent or rarely present, invisible or visible. Tepals 4 to 23, 3– to 5–merous, subequal, or rarely unequal, ones in outmost whorl smaller than 1/2 of ones in other inner whorls, coloured only on the abaxial surface. Stamens persistent; anthers dehiscent sublatrorsely to latrorsely, anther connective appendages shorter than the anther cells. Gynoecium stipitate or sessile. Fruits apocarpous to synocarpous; mature carpels without a dorsal suture groove, dehiscing along the dorsal and/or ventral suture, or circumscissile. Placentation marginal; ovules 2 to 36 in each carpel. Testae free from the endocarp. The morphological charater of chalazal region on endotesta of seed belonging to the tube type or the pore type.

About 60 species distributed in tropical and subtropical Asia.

Taxonomic combinations:

Michelia citrata (Noot. et Chalermglin) Sima et S. G. Lu, comb. nov. = *Magnolia citrata* Noot. et Chalermglin, Blumea 52 (3): 559. 2007.

Michelia ovalis (Miq.) Sima et S. G. Lu, comb. nov. = *Talauma ovalis* Miq., Ann. Mus. Bot. Lugduno-Batavi 4: 69. 1868.

Michelia pubescens (Merr.) Sima et S. G. Lu, comb. nov. = *Talauma pubescens* Merr., Philipp. J. Sci. 3: 133. 1908.

Michelia tsiampacca var. *glaberrima* (Dandy) Sima et S. G. Lu, comb. nov. = *Elmerrillia papuana* var. *glaberrima* Dandy, Bull. Misc. Inform. Kew 1928: 185. 1928.

Michelia tsiampacca subsp. *mollis* (Dandy) Sima et S. G. Lu, comb. nov. = *Elmerrillia mollis* Dandy, Bull. Misc. Inform. Kew 1928: 184. 1928. = *Elmerrillia tsiampacca* subsp. *mollis* (Dandy) Noot., Blumea 31(1): 108. 1985.

II. Magnoliaceae subfam. Liriodendroideae (Nurk.) Y. W. Law

15. *Liriodendron* Linn. , Sp. Pl. : 535. 1753. – Type: *Liriodendron tulipifera* Linn.

= *Tulipifera* Mill. , Gard. Dict. Abr. , ed. 4. 1754. – Type: *Tulipifera liriodendrum* Mill.

= *Liriodendron tulipifera* Linn.

Description. Trees, deciduous, hairy and glabrescent or glabrous. Branching sympodial and monopodial; branches produced by only syllepsis. Leaves conduplicate, pendant in the bud when young, arranged spirally on the shoot; leaf blades 4- to 10-lobed, rounded, truncate or slightly cordate at the base. Stipules free from the petiole. All axillary buds or some upper axillary buds in the mixed bud developed and sprouted; only the parts, from flowers or fruits to pseudophyllaries, of the floral branches into which mixed buds formed shed after flowering or fruiting. Flowers terminal, solitary, bisexual. Pseudophyllaries solitary, foliaceous, persistent. Bracts solitary, spathaceous, caducous. Peduncles short or slender. Pedicles present or absent, invisible or visible. Tepals 9 to 12, 3-merous, subequal, coloured on both surfaces. Stamens persistent; anthers dehiscent extrorsely, anther connective appendages shorter than the anther cells. Gynoecium sessile. Fruits apocarpous; mature carpels samariod, without a dorsal suture groove, indehiscent. Placentation apical; ovules 2 in each carpel. Testae adnate to the endocarp. The morphological charater of chalazal region on endotesta of seed belonging to the pore type.

Two species distributed in SE Asia and SE North America.

Acknowledgements

The work was supported by the National Natural Science Foundation of China (30660154, 31060096), the Foundation of The Magnolia Society International and the Foundation of Yunnan Provincial Key Laboratory for Cultivation and Exploitation of Forest Plants, China.

References

Azuma H, Thien L B, and Kawano S, 1999. Molecular phylogeny of *Magnolia* (Magnoliaceae) inferred from cpDNA sequences and floral scents. *Journal of Plant Research* 112: 291~306.

Azuma H, Garcia-Franco J G, Rico-Gray V, and Thien L B, 2001. Molecular phylogeny of the Magnoliaceae: The biogeography of tropical and temperate disjunctions. *American Journal of Botany* 88 (12): 2275~2285.

Azuma H, Rico-Gray V, Garcia-Franco J G, Toyota M, Asakawa Y, and Thien L B, 2004. Close relationship between Mexican and ChineseMagnolia (subtropical disjunct of Magnoliaceae) inferred from molecular and floral scent analyses. *Acta Phytotaxomica et Geobotanica* 55 (3): 167~180.

Baillon H E, 1866. Mémoire sur la famille des Magnoliacées. *Adansonia* 7: 1~16, 65~69.

Bentham G, and Hooker J D, 1862. Genera plantarum 1. Reeve & Co. , London.

Chase M W, Soltis D E, Olmstead R G, Morgan D, Les D H, Mishler B D, Duvall M R, Price R A, Hills H G, Qiu Y L, Kron K A, Rettig J H, Conti E, Palmer J D, Manhart J R, Sytsma K J, Michaels H J, Kress W J, Karol K G, Clark W D, Hedren M, Gaut B S, Jansen R K, Kim K J, Wimpee C F, Smith J F, Furnier G R, Strauss S H, Xiang Q Y, Plunkett G M, Soltis P S, Swensen S, Williams S E, Gadek P A, Quinn C J, Eguiarte L E, Golenberg E, Learn G H Jr. , Graham S W, Barrett S C H, Dayanandan S, and Albbert V A, 1993. Phylogenetics of seed plants: An analysis of nucleotide sequences from the plastid generbcL. *Annals of the Missouri Botanical garden* 80: 528~580.

Dandy J E, 1927. The genera of Magnolieae. *Bulletin of Miscellaneous Information* (Royal Botanic Gardens, Kew) 1927 (7): 257~264.

Dandy J E, 1964. Magnoliaceae. In: J Hutchinson. *The Genera of Flowering Plants*, Angiospermae I. Clarendron Press, Oxford. 50~57.

De Candolle A P, 1817. *Regni vegetabilis systema naturale* 1. Treuttel & Würtz, Paris. pp. 449~560.

Engler A, and Gilg E, 1924. *Syllabus der Pflanzenfamilien*. 2nd ed. Berlin.

Figlar R B, 2000. Proleptic branch initiation in *Michelia* and *Magnolia* subgenus *Yulania* provides basis for combinations in *subfamily Magnolioideae*. In: Liu Y H, Fan H M, Chen Z Y, Wu Q G, and Zeng Q W (eds.). *Proceeding of the international symposium on the family Magnoliaceae*. Science Press, Beijing. pp. 14~25.

Figlar R B, and Nooteboom H P, 2004. Notes on Magnoliaceae IV. *Blumea* 49 (1): 87~100.

Gong X, Shi S H, Pan Y Z, Huang Y L, and Yin Q, 2003. An observation on the main taxonomic characters of subfamily Magnolioideae in China. *Acta Botanica Yunnanica* 25 (4): 447~456.

Law Y W, 1984. A preliminary study on the taxonomy of the family Magnoliaceae. *Acta Phytotaxonomica Sinica* 22 (2): 89~109.

Li J, 1997. A cladistic analysis of Magnoliaceae. *Acta Botanica Yunnanica* 19 (4): 342~356.

Li J, and Conran J G, 2003. Phylogenetic relationships in Magnoliaceae subfam. Magnolioideae: a morphological cladistic analysis. *Plant Systematics and Evolution* 242: 33~44.

Liu Y H, 2000. Studies on the phylogeny ofMagnoliaceae In: Liu Y H, Fan H M, Chen Z Y, Wu Q G, and Zeng Q W (eds.). *Proceeding of the international symposium on the family Magnoliaceae*. Science Press, Beijing. pp. 3~13.

Liu Y H, Xia N H, and Yang H Q, 1995. The origin, evolution and phytogeography of Magnoliaceae. *Journal of Tropical and Subtropical Botany* 3 (4): 1~12.

Nie Z L, Wen J, Azuma H, Qiu Y L, Sun H, Meng Y, and Sun W B, 2008. Phylogenetic and biographic complexity of Magnoliaceae in the Northern Hemisphere inferred from three nuclear data sets. *Molecular Phylogenetics and Evolution* 48: 1027~1040.

Noteboom H P, 1985. Notes onMagnoliaceae with a revision of *Pachylarnax* and *Elmerrillia* and the Malesian species of *Manglietia* and *Michelia*. *Blumea* 31 (1): 65~121.

Noteboom H P, 1993. Magnoliaceae. In: Kubitzki K (ed). *The Families and Genera of Vascular Plants* 2. Springer-Verlag, Berlin. pp. 391~401.

Noteboom H P, 2000. Different looks at the classification of theMagnoliaceae. In: Liu Y H, Fan H M, Chen Z Y, Wu Q G, and Zeng Q W (eds.). *Proceedings of the International Symposium on the Family Magnoliaceae*. Science Press, Beijing. pp. 26~37.

Jin H, Shi S H, Pan H C, Huang Y L, and Zhang H D, 1999. Phylogenetic relationships betweenMichelia (Magnoliaceae) and its related genera based on the matK gene sequence. *Acta Scientiarum Naturalium Universitatis Sunyatseni* 38 (1): 93~97.

Kim S, Park C W, Kim Y D, and Suh Y, 2001. Phylogenetic relationship in family *Magnoliaceae* inferred from *ndh*F sequences. *American Journal of Botany* 88 (4): 717~728.

Shi S, Jin H, Zhong Y, He X, Huang Y, Tan F, and Boufford D E, 2000. Phylogenetic relationships of theMagnoliaceae inferred from cpDNA matK sequences. *Theoretical and Applied Genetics* 101: 925~930.

Sima Y K, 2005. Magnoliaceae. In: Li Y Y (ed). *National Protected Wild Plants in Yunnan Province, China*. Yunnan Science and Technology Press, Kunming. pp. 199~244.

SimaY-K, and Lu S-G, 2009. Magnoliaceae. In: Shui Y-M, Sima Y-K, Wen J, Chen W-H (eds.). *Vouchered Flora of Southeast Yunnan*, 1. Yunnan Publishing Group Corporation; Yunnan Science and Technology

Press, Kunming. pp. 16~67.

Sima Y K, Wang Q, Cao L M, Wang B Y, and Wang Y H, 2001. Prefoliation features of the Magnoliaceae and their systematic significance. *Journal of Yunnan University* (Natural Sciences Edition) 23 (Suppl.): 71~78.

Tiffney B H, 1977. Fruits and seeds of the Brandon Lignite: Magnoliaceae. *Botanical Journal of the Linnean Society* 75: 299~323.

Wang Y L, Li Y, Zhang S Z, and Yu X S, 2006. The utility of matK gene in the phylogenetic analysis of the genus *Magnolia*. *Acta Phytotaxonomica Sinica* 44 (2): 135~147.

Wu Z Y, Lu A M, Tang Y C, Chen Z D, and Li D Z, 2003. *The Families and Genera of Angiosperms in China: A Ccomprehensive Analysis*. Science Press, Beijing. pp. 57~68.

Xia N H, Liu Y H (Law Y W), and H P Nooteboom, 2008. Magnoliaceae. In: Wu Z Y, Raven P H, and Hong D Y (eds). *Flora of China* 7. Science Press, Beijing; Missouri Botanical Garden Press, St. Louis. pp. 48~91.

Xu F X, Chen Z Y, and Zhang D X, 2000. A Cladistic Analysis of Magnoliaceae. *Journal of Tropical and Subtropical Botany* 8 (3): 207~214.

Xu F X, and Wu Q G, 2002. Chalazal region morphology on the endotesta of Magnoliaceous seeds and its systematic significance. *Acta Phytotaxonomica Sinica* 40 (3): 260~270.

Zhang B, Huang Y H, Su Y J, and Wang T, 1996. Observation on the morphology of seed endotesta at chalazal region of Magnoliaceae. *Ecologic Science* 1996 (1): 30~34.

木兰科植物一新分类系统

摘要:根据对木兰科尤其活植物性状认真仔细的观察和研究,结合最新的DNA数据,本文提出了一个木兰科植物新分类系统。该系统将木兰科划分为2亚科2族15属,其中2属为新属。同时,还提供了一个科下分类检索表和52个新组合。

In: Xia N-H, Zeng Q-W, Xu F-X, Wu Q-G (eds). Proceedings of the Second International Symposium on the Family Magnoliaceae [C]. Wuhan: Huazhong University of Science & Technology Press, 2012: 55-71.

In Vitro Germination and Low-temperature Seed Storage of *Cypripedium lentiginosum* P. J. Cribb & S. C. Chen, A Rare and Endangered Lady's Slipper Orchid

JIANG Hong[1], CHEN Ming-chuan[2], Yung-I Lee[3,4]*

[1] Yunnan Academy of Forestry, Kunming Yunnan, 650201

[2] Department of Biology, National Museum of Natural Science, Taichung Taiwan 40453

[3] Department of Biology, National Museum of Natural Science, Taichung Taiwan 40453

[4] Department of Life Sciences, National Chung Hsing University, Taichung Taiwan 40227

Abstract: Procedures for asymbiotic seed germination and seed storage were established for *Cypripedium lentiginosum*, an endangered lady's slipper orchid. Based on a defined time frame, the optimum germination was recorded for immature seeds collected at 90-105 days after pollination (DAP), at which time early globular to globular embryos can be observed. After 120 DAP, as seeds matured, the germination decreased sharply. At maturity, two distinct layers of seed coats enclosed the embryos tightly. Histochemical staining suggests that the lignified seed coats may cause the coat-imposed dormancy. Pretreatments with 0.5 or 1% NaOCl for 45 min were effective in increasing the permeability of seed coats and improving the germination of mature seeds. Among different MS salt concentrations examined, the germination and protocorm formation were higher for seeds cultured on 1/4 and 1/2 MS media. Among cytokinins tested, only 2iP showed a stimulatory effect on germination, while both 2iP and TDZ enhanced the formation of multiple protocorm-like bodies. For seeds dehydrated to 13.5 % of initial water content and stored at 5℃, both germination and viability decreased slightly after 12 months of storage, while the extended storage to 24 months resulted in a sharp decrease of germination and viability. Using this protocol, seedlings with numerous roots were readily acclimatized to greenhouse conditions after 6 months of culture.

Keywords: Conservation; Basal media; Embryology; Seed pretreatment; Seed storage.

1 Introduction

The genus Cypripedium consists of approximately 50 species, distributed from subtropical to temperate latitudes of the Northern Hemisphere (Cribb, 1997). Many Cypripedium species have attractive flowers that are popular in ornamental market as the potted and garden plants. *Cypripedium lentiginosum* (section Trigonopedia) is an endangered terrestrial species restricted to southeastern Yunnan province in China and Ha Giang province in Vietnam (Rankou and Averyanov, 2014). The species in section Trigonopedia is characterized by two black-spotted hairy leaves with a short flower stalk at the time of anthesis (Cribb, 1997; Li *et al*., 2011). According to the study on pollination biology of *C. fargesii* (a relative species to *C. lentiginosum* in section Trigonopedia), these black spots on leaves mimic black mold spots to attract mycophagous flies to disperse its pollen with no reward (Ren *et al*., 2011). Because of its high value in the ornamental market, many plants have been illegally collected from the wild population, making this species endangered (Rankou and Averyanov, 2014). According to the long-term exploration of *C. len-*

tiginosum in southeastern Yunnan province by Hong Jiang (*unpublished data*), the sizes of many wild populations have decreased rapidly as a consequence of over-collection. Therefore, the establishment of a reliable protocol for seed germination of this species is crucial for ex situ conservation works (Ramsay and Dixon, 2003) and commercial market demands.

For the establishment of plantlets from seeds in orchids, germination is the limiting step. In most orchids, a capsule usually holds a large number of tiny seeds after successful pollination events (Arditti, 1967). At the time of capsule maturation, the tiny seed has a thin seed coat covering a globular embryo without an endosperm. Though numerous seeds are released from the capsule, only a few seeds will germinate into protocorms, and even fewer protocorms will develop into mature plants. Due to limited food reserves, the germination of orchid seeds is dependent on the formation of an association with the compatible mycorrhizal fungi. During the germination and the early stages of seedling establishment, orchids reply on nutrients supplied from the mycorrhizal association (Rasmussen, 1995). Since the establishment of asymbiotic seed germination by Lewis Knudson in 1922, this method has become a useful propagation technique for orchids (Arditti, 1967). Nonetheless, as compared to the epiphytic orchids in the tropical area, seed germination of terrestrial orchids in the temperate region is more complicated, such as Calanthe (Lee *et al.*, 2007), Cypripedium (Rasmussen, 1995) and Epipactis (Rasmussen, 1992). There are some causes of poor germination of mature seeds in terrestrial orchids. The mature seeds of Cypripedium species generally have two tough layers of seed coats, i.e. the inner and outer seed coats covering the embryos (Rasmussen, 1995). The impermeable character of seed coats may result in the poor germination. In addition, the accumulation of inhibitory substances, such as phenolic compounds and abscisic acid may induce the physiological dormancy in embryos as the seeds matured (Kako, 1976; Lee *et al.*, 2007; Lee *et al.*, 2015; Van der Kinderen, 1987; Van Waes and Debergh, 1986). Different methods have been applied to optimize the germination of Cypripedium species, including the culture of immature seeds (De Pauw and Remphrey, 1993; Lee *et al.*, 2005; Zhang *et al.*, 2013), pre-chilling treatments (Masanori and Tomita, 1997; Miyoshi and Mii, 1998; Shimura and Koda, 2005), scarification treatments (Bae *et al.*, 2010; Miyoshi and Mii, 1998), the liquid suspension culture (Chu and Mudge, 1994), and the addition of plant growth regulators to the culture medium (De Pauw *et al.*, 1995; Miyoshi and Mii, 1998).

Although protocols for seed germination of *Cypripedium* species have been reported in section Acaulie, e.g. *C. acaule* (Lauzer *et al.*, 1994; St. Arnaud *et al.*, 1992); in section Bifolia, e.g. *C. guttatum* (Bae *et al.*, 2009); in section Cypripedium, e.g. *C. calceolus* (Ramsay and Stewart, 1998), *C. calceolus* var. *pubescens* (Chu and Mudge, 1994; De Pauw and Remphrey, 1993; Light and Mac Conaill, 1990), *C. macranthos* (Bae *et al.*, 2010; Miyoshi and Mii, 1998; Zhang *et al.*, 2013) and *C. candidum* (De Pauw and Remphrey, 1993); in section Flabellinervia, e.g. *C. formosanum* (Lee *et al.*, 2005) and *C. japonicum* (Masanori and Tomita, 1997); in section Obtusipetala, e.g. *C. regina* (De Marie *et al.*, 1991; Harvais, 1973; Harvais, 1982; Steele, 1996); in section Retinervia, e.g. *C. debile* (Hsu and Lee, 2012), the critical period for immature seed culture and the culture requirements vary from species to species (Zeng *et al.*, 2014). Nevertheless, to date, no study has been made to establish a protocol for the asymbiotic seed germination of Cypripedium species in section Trigonopedia. The aims of this study are to establish an effective propagation system by asymbiotic germination of *C. lentiginosum*. For culturing immature seed in vitro, we investigated the histological and histochemical changes of seed development based on a precise time frame and determined the critical stage for seed collection. We also examined the effects of basal salt strength and cytokinins on asymbiotic seed germination. For improving the germination of mature seeds, we examined the effects of various strengths and durations of pretreatment conditions using sodium hypochlorite solution on seed germination. Besides, the storage of seeds is one of the most important ways of preserving genetic resources of orchid species (Pritchard et al., 1999; Seaton and Pritchard, 2003; Thornhill and Koopowitz, 1992). Here, we examined the effects of seed initial water content and storage temperature on seed viability. Such studies would benefit the mass propagation to meet commercial needs and to conserve this endangered orchid species.

2 Materials and methods

2.1 Plant material and seed collection

Developing capsules of *C. lentiginosum* P. J. Cribb & S. C. Chen were collected from the natural population at Malipo county, Yunnan province, People's Republic of China (Fig. 1A and B). Anthesis generally occurs in the early July, each year. For good capsule setting, flowers were hand-pollinated by transferring the pollinia onto the stigma of the same flower. After capsule setting, at least three capsules were collected continuously every 30 days from July to November for the morphological measurement, the histological observation and the asymbiotic seed germination experiments. In the middle of November, capsules began to mature and split after 135 days after pollination (DAP).

Figure 1 *C. lentiginosum* and its natural habitat in Malipo county, Yunnan, China.
(A) A typical habitat of *C. lentiginosum* at the limestone mountain.   (B) The flower at the time of anthesis.
(C) Significant elongation of stalk after a successful pollination event.
Scale bar = 10cm.

2.2 Histological and histochemical studies

Seeds of different developing stages were collected and fixed in 2.5% glutaraldehyde and 1.6% paraformaldehyde buffered with 0.1M phosphate buffer (pH 6.8) for 24 h at room temperature. After fixation, the samples were dehydrated through the ethanol series, and then were infiltrated gradually (3 : 1, 1 : 1, and 1 : 3 100% ethanol: Technovit 7100, 24 hours each) with Technovit 7100 (Kulzer & Co., Germany), followed by three changes of pure Technovit 7100. The samples were then embedded in Technovit 7100 as described by Yeung (1999). A Reichert-Jung 2040 Autocut rotary microtome was employed for the sectioning (3μm thick). Sections were stained with Periodic Acid-Schiff's (PAS) reaction for the structural carbohydrate and then counter-stained with either 1% (w/v) amido black 10B in 7% acetic acid for protein (Sigma-Aldrich Co., St. Louis, Mo., USA), or 0.05% (w/v) toluidine blue O (TBO, Sigma-Aldrich Co.) in benzoate buffer for general histological staining (Yeung, 1984). For detecting the deposition of cuticular material in seeds, sections were stained with 1 μg · mL^{-1} of Nile red (Sigma-Aldrich Co.) for 3 min, and were washed in running tap water for 1 min. Then, sections were mounted in distilled water containing 0.1% n-propyl gallate (Sigma-Aldrich Co.). The fluorescence pattern of Nile red was viewed using an epifluorescence microscope (Axioskop 2, Carl Zeiss AG, Germany) equipped with the Zeiss filter set 15 (546/12 nm excitation filter and 590 emission barrier filter). All images were taken using a CCD camera attached to the Carl Zeiss microscope.

2.3 Effect of collection timing on asymbiotic germination

The capsules of different developmental stages were collected and taken to the laboratory, and were surface sterilized with a 1% sodium hypochlorite (NaOCl) solution with one drop of a wetting agent (Tween 20, Sigma-Aldrich Co.) for 20 min. After surface sterilization, the capsules were cut open, and then the seeds were scooped out with forceps onto the culture medium. To ensure the seed quality and developmental stages of each capsule, the remaining seeds of each capsule were fixed and examined under a microscope. Approximately 90% of seeds had fully developed embryos (as confirmed by a compound microscopy). The culture medium used in this experiment was the 1/4 Murashige and Skoog (MS) medium (Murashige and Skoog, 1962), which was supplemented with 2 mg · L^{-1} glycine (Sigma-Aldrich Co.), 0.5 mg · L^{-1} niacin (Sigma-Aldrich Co.), 0.5 mg · L^{-1} pyridoxine HCl (Sigma-Aldrich Co.), 0.1 mg · L^{-1} thiamine (Sigma-Aldrich Co.), 1 · g · L^{-1} tryptone (Sigma-Aldrich Co.), 20 g · L^{-1} sucrose (Sigma-Aldrich Co.), 100 ml · L^{-1} coconut water, 0.5 g · L^{-1} activated charcoal (Sigma-Aldrich Co.), and solidified with 7 g · L^{-1} agar (Plant cell culture, tested, powder, Sigma-Aldrich Co.). The pH of the medium was adjusted to 5.6 before autoclaving at 121℃ for 15 min. Ten milliliters of medium were placed into each culture tube (20×100mm). After sowing, the cultures were incubated in a growth room in the dark at 25 ± 1℃.

2.4 Effect of MS basal salt strength on asymbiotic germination

Given that the seeds collected at 105 DAP showed good germination percentages, for investigating the influence of basal salt strength on asymbiotic germination, the immature seeds (105 DAP) were inoculated on 1/2, 1/4, 1/8 and 1/10 strength of macroelements of MS basal salt. These culture media were supplemented with 2 mg · L^{-1} glycine, 0.5 mg · L^{-1} niacin, 0.5 mg · L^{-1} pyridoxine HCl, 0.1 mg · L^{-1} thiamine, 0.5 g · L^{-1} tryptone, 20 g · L^{-1} sucrose, 100 ml · L^{-1} coconut water, 0.5 g · L^{-1} activated charcoal, and solidified with 7 g · L^{-1} agar. The pH of media was adjusted to 5.6 before autoclaving at 121℃ for 15 min. Ten milliliters of medium were placed into each culture tubes. After sowing, the cultures were incubated in a growth room in the dark at 25±1℃.

2.5 Effect of cytokinins on asymbiotic germination

For investigating the influence of cytokinins on seed germination in vitro, seeds collected at 105 DAP were used in this experiment. The basal medium used was 1/4 MS medium, supplemented with 1 μm of either 6-(γ,γ-Dimethylallylamino)purine (2iP, Sigma-Aldrich Co.), 6-benzylaminopurine (BA, Sigma-Aldrich Co.), kinetin or thidiazuron (TDZ, Sigma-Aldrich Co.). These media were modified by supplemented with 2 mg · L^{-1} glycine, 0.5 mg · L^{-1} niacin, 0.5 mg · L^{-1} pyridoxine HCl, 0.1 mg · L^{-1} thiamine, 0.5 g · L^{-1} tryptone, 20 g · L^{-1} sucrose, and solidified with 7 g · L^{-1} agar. The pH of media was adjusted to 5.6 before autoclaving at 121℃ for 15 min. After sowing, the cultures were incubated in a growth room in the dark at 25±1℃.

2.6 Effect of seed pretreatments on asymbiotic germination

For investigating the effectiveness of different concentrations and durations of NaOCl pretreatments in promoting germination, the mature seeds at 135 DAP were collected for this experiment. The procedure of pretreatments included soaking the seeds in 0.5% or 1% NaOCl solution with one drop of Tween 20, for 15, 30, 45, or 60 min, respectively. In the control, seeds were soaked only in water. After the pretreatments, the seeds were washed three times with sterilized distill water, and then were inoculated on 1/4 MS medium as described above.

2.7 Seed storage

In this experiment, the mature seeds (135 DAP) were collected and placed in the sample tubes, and were equilibrated with relative humidity (RH) in sealed jars controlled by saturated solutions of calcium chloride (ca. 30% RH) or lithium chloride (ca. 11% RH) at 25℃ for 3 days as described by Seaton and Ramsay (2005). After equilibration, seeds were dried at 70℃ oven for 48 h. The water content was estimated as the percentage of water loss: fresh weight minus dry weight, to its fresh weight. After equilibration, the seeds in sample tubes were directly sealed, then were stored at 5°C or 25°C for 1, 6, 12 or 24 months. For checking seed viability, the seeds from each treatment were evaluated using the 2,3,5-triphenyltetrazolium chloride (TTC, Sigma-Aldrich Co.) test. Seeds were incubated in 1% TTC solution at 27 ± 2℃ in darkness for 3 days. Staining experiment was performed a minimum of three times with 100–150 seeds. Embryos remaining yellow were considered as unstained, and those turning orange to red were classified as stained. For checking seed germinability, the seeds from each treatment were taken out of sample tubes and were surface-sterilized with 1% NaOCl solution with one drop of Tween 20 for 30 min. After sterilization, the seeds were inoculated on 1/4 MS medium as described above.

2.8 Experimental design and data analysis

In each experiment, three capsules were collected for inoculation. The seeds from each capsule were evenly sown into the replicates of each treatment. Each treatment comprised twelve culture tubes (replicates) and was conducted three times. Experiments were performed in a completely randomized design. Each culture tube was examined at 15-day intervals for 180 days in culture by using a stereomicroscope(Carl Zeiss AG). Germination was considered as emergence of the embryo from the testa. The germination percentage was calculated as the percentage of the number of seeds germinated among the total countered number of seeds. The data were statistically analyzed using analysis of variance (ANOVA). The means were separated using Fisher's protected least significant difference test at $P = 0.05$ (SAS statistical software, version 8.2, SAS Institute, Cary, NC).

2.9 Seedling growth medium and greenhouse acclimatization

After 4 months of culture, young protocorms (Fig. 6F) were transferred onto the seedling growth medium: 1/4MS medium supplemented with 20 g · L^{-1} sucrose, 1 g · L^{-1} activated charcoal, 20 g · L^{-1} potato homogenate, and 7 g · L^{-1} agar for the growth of seedling described by Lee (2010). The potato was peeled and cut into about 1cm^3 sections, and then the potato cubes were boiled for 10 min and homogenized with a kitchen blender. After mixing the potato homogenate, the pH was adjusted to 5.6 before autoclaving at 121℃ for 15 min. One hundred milliliters of medium was placed into a 500-mL culture flask. After transferring the young protocorms, the flasks were incubated in a growth room at 25 ± 2℃ in constant darkness. After 3 months of culture in the dark, plantlets (about 2 cm in height) were washed three times with tap water and transplanted to pots containing akadama soil. The seedlings were grown in a greenhouse with sunshade nets under the natural light no more than 200 μmol · m^{-2} · s^{-1}, average temperatures ranged from 15 to 25℃ and the RH ranged from 50 to 80%.

3 Results

3.1 Capsule and stalk development

After successful pollination, the ovary began to enlarge and the stalk began to elongate rapidly. The main structural changes occurring within the ovary from anthesis until maturity at 135 DAP are summarized in Table 1. The length of the ovary increased steadily until the maximum size (54.3±4.2 mm) reached at 75 DAP, and the diameter of the ovary also increased steadily until the maximum size (21.2 ± 2.4 mm) reached at 75 DAP (Fig. 2). At anthesis, the stalk was short (38.5 ± 1.8 mm) (Fig. 1B). Following pollination, the length of the stalk increased rapidly (approximately 4.1 mm per day) by 60 DAP, and then grew slowly, until the maximum size (325.4 ± 47.2 mm) reached at 75 DAP (Fig. 1C; Fig. 2).

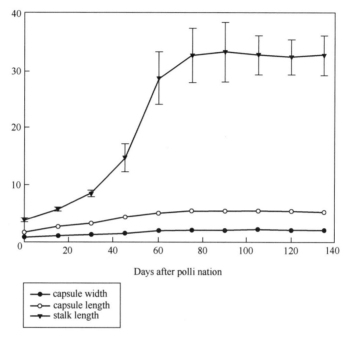

Figure 2 Changes in capsule length and width, and stalk length of *C. lentiginosum* during development. Error bars represent SE (n=3).

Table 1 Major microscopic structural events occurring in the developing fruits of *C. lentiginosum* after hand pollination.

DAP[z]	Developmental Stage	Seed color
0~30	Ovule development	—
45	Fertilization and the formation of zygote	White
60	Proembryo	White
75	Proembryo and the developing of early globular embryo	White
90	Early globular embryo with a single-celled suspensor	Yellowish white
105	Globular embryo	A mixture of yellow and light brown seeds
120	Late globular embryo and the suspensor has degenerated	Light brown seeds
135	Dry and mature seed	Brown

[z] DAP = days after pollination.

3.2 Seed development

After fertilization (45 DAP), the zygote had an elongated shape with a prominent nucleus located toward the chalazal end and a large vacuole situated toward the micropylar end (Fig. 3A). At 60 DAP, a three-celled proembryo could be observed. The basal cell was larger than the other two cells at the terminus (Fig. 3B). At 75 DAP, the cells toward the terminus divided further, giving rise to a six-celled proembryo (Fig. 3C). As development progressed (90 DAP), more cell divisions occurred within embryo proper cells, resulting in the formation of an early globular embryo (Fig. 3D). This species had a single-celled suspensor that connected the embryo proper to the embryo sac, and the suspensor did not divide or enlarge further and eventually degenerated as seed matured. As the embryo developed to the globular stage at 105 DAP (Fig. 3E), the single-celled suspensor had degenerated, and there were several starch grains accumulated near the nuclei within embryo proper cells. As the seed approached maturity (120 DAP), cell divisions within embryo proper had ceased and the embryo proper cells expanded as storage products accumulated (Fig. 3F). At maturity (135 DAP), the embryo was only six cells long and five cells wide without differentiation of shoot apical meristem and cotyledon (Fig. 3G). At this stage, the cytoplasm of embryo proper cell was filled with a large quantity of storage products, e.g. protein bodies and lipid bodies, and starch grains had disappeared. By 135 DAP, the capsule had fully matured and desiccated. Subsequently, the capsule split and the mature seeds were released.

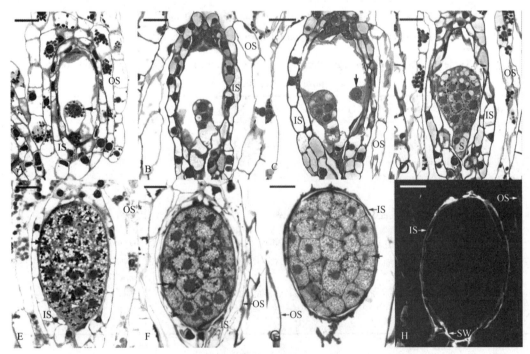

Figure 3 The embryo development of *C. lentiginosum*.

(A) At 45 DAP, light micrograph showing a longitudinal section through a zygote. The nucleus is located toward the chalazal end and there are several tiny starch grains (arrow) surrounding the nucleus. IS = the inner seed coat. OS = the outer seed coat. Scale bar = 30 μm.

(B) At 60 DAP, light micrograph showing a longitudinal section through a 3 - celled embryo. Scale bar = 30 μm.

(C) At 75 DAP, light micrograph showing a longitudinal section through a proembryo. The endosperm fails to develop in this species and the endosperm nuclei (arrow) would finally degenerate. Scale bar = 30 μm.

(D) At 90 DAP, light micrograph showing a longitudinal section through an early globular embryo with a single-celled suspensor (S). Scale bar = 30 μm.

(E) At 105 DAP, light micrograph showing a longitudinal section through a globular embryo. At this stage, the suspensor has degenerated. Many starch grains (arrow) and protein bodies (arrowhead) are present in the cells of embryo proper. Scale bar = 50 μm.

(F) At 120 DAP, light micrograph showing a longitudinal section through a globular embryo. At this stage, the inner seed coat (IS) is dehydrating and compressing. Starch grains (arrow) are decreasing and numerous protein bodies (arrowhead) occur within the cells of embryo proper. Scale bar = 50 μm.

(G) At 135 DAP, light micrograph showing a longitudinal section through a mature seed. At maturity, the embryo is enveloped by the shriveled innerseed coat (IS) and the outer seed coat (OS). Starch grains has disappeared and protein bodies (arrow) of various sizes can be observed within the embryo proper. Though lipid could not be preserved in historesin, plentiful translucent vesicles within the cytoplasm of embryo proper indicate the deposition of lipid bodies. Scale bar = 50 μm.

(F) Light micrograph showing a fluorescence outline of a mature seed at the stage similar to Figure 3G after Nile red staining. The surface wall (SW) of the embryo proper, and the shriveled inner seed coat (IS) and outer seed coat (OS) reacted positively. Scale bar = 50μm.

In the mature seed of C. lentiginosum, the embryo were enveloped by two distinct layers of seed coats, i.e. the inner and outer seed coats derived from the inner and outer integuments of the ovule separately (Fig. 3G). Both the inner and outer seed coats were two cells thick. The cells of outer seed coat were highly vacuolated and contained numerous starch grains, while no starch grains occurred within the cells of inner seed coat (Fig. 3D). As the seed approached maturity, the cells of the inner and outer seed coats became dehydrated and compressed into a thin layer covering the embryo tightly (Fig. 3F). Using TBO staining, the walls of inner and outer seed coats stained greenish blue, indicating the presence of phenolic compounds in the walls. Besides, the surface wall of embryo proper and the inner seed coat reacted positively to Nile red, suggesting the possible accumulation of cuticular substance in the walls (Fig. 3H). In the outer seed coat, the fluorescence intensity of Nile red staining is weak and could be easily quenched by pre-staining of sections with TBO, suggesting that a distinct cuticular substance might be absent (Fig. 3H).

3.3 Influence of seed maturity on asymbiotic germination

The timing of seed collection clearly affected on germination percentage of *C. lentiginosum*

(Fig. 4). No germination was recorded in the seeds collected at 45 DAP. The optimum germination percentage (54.9%–57.8%) was recorded at 90–105 DAP. During this period, the seeds were moist and yellowish white to light brown in color. By 120 DAP, a remarkable decrease in

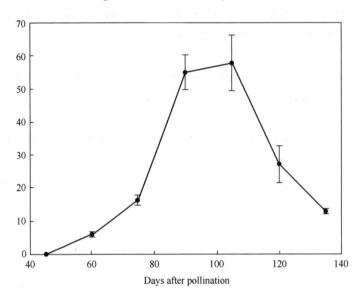

Figure 4 Mean percent germination of *C. lentiginosum* seeds collected at each successive 15 days after pollination on 1/4 MS medium. Data were scored after 180 days of culture. Error bars represent SE (n=3).

germination percentage (27.2%) was observed. At this stage, the majority of seeds had turned light brown and dehydrated. After 135 DAP, the germination percentage had declined to 12.8%, and the mature seeds were dry and brown in color.

3.4 Influence of medium composition on asymbiotic germination

Results of this experiment demonstrated that germination and protocorm formation were significantly affected by MS salt concentrations. Among different MS salt concentrations tested, the germination percentage was higher for seeds cultured on 1/4 MS (54.1%) and 1/2 MS (52.8%) as compared to 1/8 MS (45.1%) and 1/10 MS (43.8%) (Table 2). In addition, the higher protocorm formation was observed in the cultures on 1/4 MS (51.2%) and 1/2 MS (51.7%).

Table 2 Effect of different strengths of MS basal salts on seed germination in vitro and protocrom formation of *C. lentiginosum*. Data were recorded after 180 days of culture

MS basal salts strength	Germination/%	Protocorm formation/%
1/2	52.8a	51.2a
1/4	54.1a	51.7a
1/8	45.1b	37.2b
1/10	43.8b	36.5b

Mean within a column followed by the same letter are not significantly different at $P = 0.05$ by Fischer's protected LSD test.

3.5 Influence of cytokinins on asymbiotic germination and protocorm morphology

The germination percentage was affected by the addition of different types of cytokinins to the media (Table 3). Among the cytokinins tested, only 2iP showed a stimulatory effect on seed germination compared to the control with no cytokinin, while the addition of BA, kinetin and TDZ suppressed seedgermination. After germination, the presence of cytokinins also had an effect on protocorm morphology (Table 3). Without cytokinins (control) and with the addition of kinetin, the higher percentages of single protocorm bodies (SPBs) were observed. With the addition of 2iP and TDZ, the higher percentages of multiple protocorm bodies (MPBs) were observed. The addition of TDZ enhanced the formation of more amorphous protocorm bodies (APBs) than the others.

Table 3 Effects of various cytokinins on seed germination and protocorm development of *C. lentiginosum* after 180 days of culture in vitro

Cytokinins/1 μm	Germination/%	Protocorm development type/%		
		Single	Multiple	Amorphous
Control	51.2b	78.2a	12.2d	9.6d
2iP	62.1a	60.5b	28.7ab	10.8cd
BA	44.2c	54.8c	24.4b	20.8b
Kinetin	33.5d	68.1ab	18.4c	13.5c
TDZ	35.4d	34.6d	33.7a	31.7a

Mean within a column followed by the same letter are not significantly different at $P = 0.05$ by Fischer's protected LSD test.

3.6 Influence of pretreatments on asymbiotic germination

The pretreatment with NaOCl solutions significantly improved germination percentage (Fig. 5). The soaking with 0.5% NaOCl solution improved germination percentage with the increase in soaking duration, causing the optimum germination (56.9 ± 6.4%) with a 45 min pretreatment, while pretreatment for 60 min reduced germination slightly to 47.2 ± 9.2%. Extending the soaking duration at higher concentration (1% NaOCl solution) from 30 to 45 min enhanced germination significantly from 58.6 ± 5.3% to 60.3 ± 6.1%. Prolonged soaking duration for 60 min reduced germination remarkably to 29.5 ± 7.6%.

3.7 Seed storage

In two storage temperatures (at 5 or 25℃), the germination and viability were greatly affected by the initial water content of seeds (Table 4). Viability examined using TTC staining indicated that the viability was regularly higher than the germinability (Table 4). For the seeds dehydrated to 13.5% of initial water content, the germination percentage decreased slightly from 57.2 ± 3.7% to 52.9 ± 4.7% during 12 months of storage at 5℃, and the extended storage to 24 months resulted in a rapid decrease of germination to 34.3 ± 5.6%. While, for the seeds dehydrated to 13.5% of initial water content, the germination percentage decreased from 58.3 ±

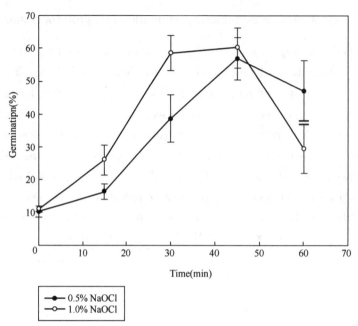

Figure 5 Effect of NaOCl concentrations and durations of pretreatment on seed germination in vitro of *C. lentiginosum*. Data were recorded after 180 days of culture. Error bars represent SE (n=3).

2.9% to 27.2 ± 1.6% during 12 months of storage at 25℃, and no germination occurred as extended storage to 24 months. On the contrary, for the seeds dehydrated to 31.4% of initial water content, the germination percentage decreased dramatically to 45.4 ± 3.5% and 28.4 ± 4.2% after one month of storage at 5℃ and 25℃, respectively.

Table 4 Effects of initial water content of mature seeds and storage temperature on the asymbiotic germination and TTC staining (inparentheses) percentages of *C. lentiginosum*.

Initial water content /%	Storage temperature /℃	Storage period/month			
		1	6	12	24
13.5	5	57.2 ± 3.7 (61.3 ± 4.5)	58.5 ± 2.2 (53.2 ± 7.4)	42.9 ± 4.7 (50.5 ± 5.2)	14.3 ± 5.6 (18.5 ± 2.2)
	25	58.3 ± 2.9 (66.5 ± 4.7)	32.1 ± 3.2 (35.4 ± 2.1)	27.2 ± 1.6 (32.5 ± 7.6)	0 ± 0 (0 ± 0)
31.4	5	45.4 ± 8.5 (52.2 ± 3.3)	20.5 ± 5.8 (22.7 ± 2.6)	8.2 ± 1.2 (10.9 ± 3.1)	0 ± 0 (0 ± 0)
	25	38.4 ± 4.2 (44.2 ± 6.4)	11.9 ± 1.1 (15.4 ± 7.1)	0 ± 0 (0 ± 0)	0 ± 0 (0 ± 0)

Data were scored after 180 days of culture. Error bars represent SE (n=3)

3.8 Protocorm formation and seedling growth

By 30 days of culture on 1/4 MS medium, most embryos were still enveloped by the seed coat (Fig. 6A), while a few seeds had become swollen (Fig. 6B). By 45 days of culture, seed germination (as emergence of the embryo from the testa) could be observed (Fig. 6C). After germination, the young protocorm enlarged and the shoot apical meristem had differentiated in top region of protocorm (Fig. 6D). By 60 days of culture, the formation of first root could be observed (Fig. 6E). By 90 days of culture, the protocorm elongated further and the formation of numerous rhizoids could be observed at its basal region (Fig. 6F). The developing protocorms were transferred onto the seedling growth medium supplemented with 20 g · L^{-1} potato homogenate as described by Lee (2010), and the seedlings with several healthy shoots and roots could be observed after 3 months of culture (Fig. 6G). After taking out of flasks, the leaves of seedlings had expanded with the appearance of black spots (Fig. 6H).

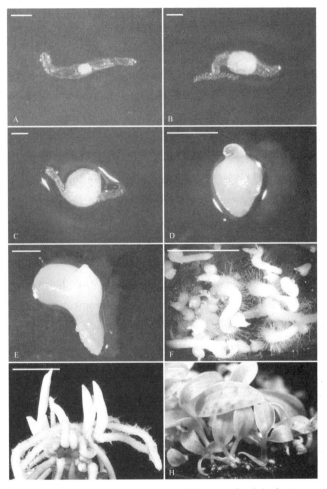

Figure 6 Protocorm development and seedling growth of *C. lentiginosum*.
(A) Before germination, the embryo is enveloped tightly by the seed coat. Scale bar = 200μm.
(B) The embryo has become swollen due to the absorption of water and nutrients. Scale bar = 200μm.
(C) The embryo continues to swell and results in rupture of the seed coat. Scale bar = 200μm.
(D) The shoot apical meristem has appeared in the young protocorm. Scale bar = 200μm.
(E) The protocorm has enlarged and the formation of first root could be observed. Scale bar = 1mm.
(F) The protocorm has elongated further and numerous rhizoids occurred at its basal part. Scale bar = 1cm.
(G) The development of seedling with several shoots and roots after 3 months of culture. Scale bar = 1cm.
(H) After taking out of flasks, the leaves of seedlings have expanded with the appearance of black spots. Scale bar = 1cm.

4 Discussion

Based on the precise time frame, the general overview of capsule, stalk and seed development with emphasis on main structural changes in *C. lentiginosum* was investigated in this study (Table 1; Fig. 2). It is notable that a successful pollination event not only triggers the development of ovary but also stimulates the significant elongation of stalk (Fig. 1 B and C; Fig. 2). Among Cypripedium species, the rapid elongation of stalk after pollination occurs primarily in section Trigonopedia. The rapid elongation of stalk after pollination has been observed in a number of dwarf orchids, such as Auxopus, Corybas and Gastrodia, suggesting to be a benefit to the seed dispersal (Segerback, 1983; Pellow et al., 2009; Li et al., 2016). *C. lentiginosum* grows on the limestone rocks in the thicket floor, the significant elongation of stalk would be helpful to seed dispersal as the capsule matured.

Along the timeline, we investigated the effect of seed maturity on asymbiotic germination from fertilization to seed mature (Fig. 3), and observed that the timing of seed collection is critical in asymbiotic germination of *C. lentiginosum* (Fig. 4). In terrestrial orchids, the culture of immature seeds usually results in a higher germination percentage than the culture of mature seeds (De Pauw and Remphrey, 1993; St-Arnaud et al., 1992). In *C. lentiginosum*, the immature seeds collected at 90 and 105 DAP were the optimal stages for culture in vitro (Fig. 4). As we compared the profile of optimum timing for seed collection to maximize germination in different Cypripedium species, the optimal duration for seed collection in *C. lentiginosum* (90–105 DAP) is similar to *C. formosanum* (90–105 DAP), but is much longer than *C. macranthos* (42 DAP). From histological investigations of these species, the embryos at the early globular to globular stages (Fig. 3D and E) are the optimum developmental stages for the culture of immature seeds. Both *C. lentiginosum* and *C. formosanum* occur in the subtropical mountain area, in which the growth season is longer than those of *C. macranthos* in the north temperate area. The various time spans of these Cypripedium species to reach the early globular stage may attribute to the length of growing seasons in different natural habitats.

In this study, it is noteworthy that the germination percentage of *C. lentiginosum* mature seeds decreases sharply to 12.8%, but it is much higher than values (near zero) obtained from the mature seeds of *C. formosanum* (Lee at al., 2005) and *C. macranthos* (Zhang et al., 2013). In *C. debile*, the seed coat has a less hydrophobic nature than those of *C. formosanum* and *C. macranthos*, making mature seeds of *C. debile* easier to germinate (Hsu and Lee, 2012). As compared to the results of histochemical staining in mature seeds of *C. formosanum* and *C. macranthos*, the wall of outer seed coat of *C. lentiginosum* fluoresces weakly after Nile red stain. Besides, the weak fluorescence can be easily quenched by pre-staining with TBO, suggesting the absence of distinct cuticular material in the wall of outer seed coat (see Holloway, 1982; Yeung et al., 1996). These results suggest that different permeability of seed coat may cause mature seeds of Cypripedium species with varying degrees of coat-imposed dormancy.

An interesting feature of this study demonstrated that the optimum germination and protocorm development were observed in the relatively high concentration of mineral salts, i. e. 1/2 and 1/4 MS(Table 2). The results observed here are unexpected because terrestrial orchids usually prefer the medium with a low concentration of mineral salts for seed germination (Rasmussen, 1995). In previous reports, seed germination of terrestrial orchids was significantly inhibited with the increasing concentrations of mineral salts (Pierik et al., 1988; Zhang et al., 2015; Huh et al., 2016). In *C. lentiginosum*, the tolerance to higher concentrations of mineral salts reveals that the nutritional requirement for seed germination in some subtropical terrestrial orchids may be different from temperate terrestrial orchids.

Besides the outer seed coat, seeds of many terrestrial orchids contain a well-developed inner seed coat (known as 'carapace'), causing the coat-imposed dormancy(Carlson, 1940; Lee et al., 2005; Rasmussen, 1995; Yamazaki and Miyoshi, 2006). In order to make the seed coats more permeable, seed pretreatments using ultrasound or bleaching solutions are critical to improve the germination of mature seeds (Linden, 1980; Miyoshi and Mii, 1988; Van Waes and Debergh, 1986; Steele, 1996). In this study, the germination was greatly enhanced when mature seeds were pretreated with 0.5 and 1% NaOCl solution (Fig. 5). Different hypochlorite solutions, such as $Ca(OCl)_2$ and NaOCl have been used for the surface-sterilization of orchid seeds and demonstrated the effect of enhancement in germination (Rasmussen, 1995; Miyoshi and Mii, 1998). As a strong oxidizing agent, NaOCl are commonly used for the degradation of lignin from the wood pulp (Holik, 2006). After pretreating with the NaOCl solution, the seed coat were scratched, and the oxidation of cell wall increased the hydrophilic character of the seed coat (Lee, 2011). In addition to the scarification of seed coat, hypochlorite solutions may accelerate the leaching of inhibitory substances, such as the endogenous ABA from orchid seeds (Van Waes and Debergh, 1986; Lee et al., 2007; Lee et al., 2015).

Among the cytokinins tested here, only 2iP improved seed germination of *C. lentiginosum* (Table 3). The presence of cytokinins in the medium has been demonstrated stimulatory effect on seed germination of several terrestrial orchids (Arditti and Ernst, 1993; Stewart and Kane, 2006). Nevertheless, the response to various cytokinins is species-specific. For example, in *C. candidum*, BA and 2iP improved seed germination, but not kinetin (De Pauw et al., 1995). Meanwhile, kinetin significantly promoted seed germination of *C. macranthos* (Miyoshi and Mii, 1998). In *Cypripedium*, the addition of cytokinin does not only enhance seed germination, but also promote proliferation of protocorms (Barabe et al., 1993; De Pauw et al., 1995). In this study, both 2iP and TDZ resulted in higher ratios of multiple protocorms. Besides, TDZ also increased the formation of amorphous protocorms (Table 3). For multiplication of Cypripedium seedlings, cytokinins alone or with combination of auxins had a significant effect on protocorm multiplication or callus induction (Lee and Lee, 2003; Shimura and Koda, 2004; Yan et al., 2006). The addition of 2iP and TDZ could be useful for the development of a protocol of micropropagation of *C. lentiginosum*.

Our results demonstrate that the initial water content in mature seeds of *C. lentiginosum* and the storage temperature significantly influence the seed germination and viability after storage (Table 4). In this study, after 24 months of storage at 5℃, the seeds with a initial water content of 13.5% only retained 14.3 ± 5.6% of germination, while the seeds stored in the other conditions completely lost their viability. In *Phaius tankervilleae*, the germination of seeds with a initial water content of 5% decreased drastically (9%) after 6 months of storage at 4℃ (Hirano et al., 2009). In *Calanthe tricarinata*, the storage of dry seeds at 5℃ for 26 months resulted in the loss of seed viability (more than 50%) (Godo et al., 2010). TTC staining has been successfully applied to evaluate the longevity of orchid seeds (Van Waes and Debergh, 1986; Vujanovic et al., 2000). In this study, TTC stainability was higher than the germinability recorded in *C. lentiginosum* (Table 4). However, TTC stainability still provides a practical means for understanding relative germination performance after storage with a calibration based on germination testing (Van Waes and Debergh, 1986). The overestimation of seed viability by TTC staining has been reported in some orchids (Lauzer et al., 1994; Hu et al., 2013). In *C. acaule*, after the treatment of sodium hypochlorite, the release of dehydrogenase from injured embryos may cause higher TTC stainability of embryos (Lauzer et al., 1994). In this study, the low temperature storage (4-5℃) may be not effective for preserving seed viability longer than one year. Further development of a cryopreservation protocol is required for long-term storage of *C. lentiginosum* seeds.

5 Conclusions

In this report, we provide reliable protocols for seedling production of *C. lentiginosum* through asymbiotic seed germination. For the culture of immature seeds, the optimum germination could be obtained from collecting the immature seeds at 90-105 DAP (the early globular to globular embryo stages) on 1/4 or 1/2 MS media. Moreover, the addition of 2iP enhances the germination and the formation of multiple protocorm bodies. For the culture of mature seeds, pretreatments with 0.5% or 1% NaOCl solution for 45 min are required for releasing the seed coat-imposed dormancy and improving the germination. For seed storage, mature seeds dehydrated to 13.5% of initial water content could be preserved for 12 months at 5℃.

Acknowledgements

This work was supported by the grants from the Yunnan Academy of Forestry, People's Republic of China to JIANG Hong and grants from National Museum of Natural Science, Taiwan to Yung-I Lee.

References

Arditti J, 1967. Factors affecting the germination of orchid seeds. Bot. Rev. 33, 1~97.

Arditti J, Ernst R, 1993. Micropropagation of orchids. Wiley, New York.

Bae K H, Kwon H K, Choi Y E, 2009. In vitro germination and plantlet conversion from the culture of fully mature seeds of *Cypripedium guttatum* Swartz. Propag. Ornam. Plants, 9, 160~165.

Bae K H, Kim C H, Sun B Y, Choi Y E, 2010. Structural changes of seed coats and stimulation of in vitro germination of fully mature seeds of *Cypripedium macranthos* Swartz (Orchidaceae) by NaOCl pretreatment. Propag. Ornam. Plants 10:107~113.

Barabe D, Saint-Arnaud M, Lauzer D,1993. Sur la nature des protocormes d'Orchidées (Orchidaceae). Comptes Rendus de l'Académie des Sciences. Série 3, Sciences de la vie 316, 139~144.

Carlson M C,1940. Formation of the seed of *Cypripedium parviflorum*. Bot. Gaz. 102, 295~301.

Chu C C, Mudge K W,1994. Effects of prechilling and liquid suspension culture on seed germination of the yellow lady's slipper orchid (*Cypripedium calceolus* var. *pubescens*). Lindleyana 9, 153~159.

Cribb P J. 1997. The Genus *Cypripedium*. Timber Press, Portland, Oregon.

DeMarie E, Weimer M, Mudge W, 1991. *In vitro* germination and development of 'Showy Lady' slipper orchid (*Cypripedium reginae* Walt.) seeds. HortScience26, 272.

De Pauw M A, Remphrey W R,1993. In vitro germination of three *Cypripedium* species in relation to time of seed collection, media and cold treatment. Can. J. Bot. 71, 879~885.

De Pauw M A, Remphrey W R, Palmer, C E,1995. The cytokinin preference for in vitro germination and protocorm growth of *Cypripedium candidium*. Ann. Bot. 75, 267~275.

Godo T, Komori M, Nakaoki E, Yukawa T, Miyoshi, K,2010. Germination of mature seeds of *Calanthe tricarinata* Lindl., an endangered terrestrial orchid, by asymbiotic culture in vitro. In Vitro Cell. Dev. Biol. -Plant 46, 323~328.

Harvais G,1973. Growth requirements and development of *Cypripedium* reginae in axenic culture. Can. J. Bot. 51, 327~332.

Harvais, G,1982. An improved culture medium for growing the orchid *Cypripedium reginae* axenically. Can. J. Bot. 60, 2547~2555.

Hirano T, Goto T, Miyoshi K, Ishikawa K, Ishikawa M, Mii, M,2009. Cryopreservation and low-temperature storage of seeds of *Phaius tankervilleae*. Plant Biotechnol. Rep. 3, 103~109.

Holik H,2006. Handbook of paper and board. Wiley, Weinheim, Germany.

Holloway P J,1982. Structure and histochemistry of plant cuticular membranes: an overview. In: Cutler, D. F., Alvin, K. L., Price, C. E. (Eds.) The plant cuticle. Academic Press, London, pp. 1~32.

Hsu R C C, Lee Y I,2012. Seed development of *Cypripedium debile* Rchb. f. in relation to asymbiotic germination. HortScience 47, 1495~1498.

Hu W H, Yang Y H, Liaw S I, Chang C,2013. Cryopreservation the seeds of a Taiwanese terrestrial orchid, Bletilla formosana (Hayata) Schltr. by vitrification. Bot. Stud. 54, 33.

Huh Y S, Lee J K, Nam S Y, Hong E Y, Paek K Y, Son S W,2016. Effects of altering medium strength and sucrose concentration on in vitro germination and seedling growth of *Cypripedium macranthos* Sw. J. Plant Biotechnol. 43, 132~137.

Kako S,1976. Study on the germination of seeds of Cymbidium goeringii. In: Torigata, H. (Ed.), Seed formation and sterile culture of orchids. Tokyo Seibundoshinkosha, pp. 174~237 (in Japanese).

Knudson L,1922. Nonsymbiotic germination of orchid seeds. Bot. Gaz. 73, 1~25.

Lauzer D, St-Arnaud M, Barabe D,1994. Tetrazolium staining and in vitro germination of mature seeds of *Cypripedium acaule*(Orchidaceae). Lindleyana 9, 197~204.

Lee Y I, 2010. Micropropagation of *Cypripedium formosanum* Hayata through axillary buds from mature plants. HortScience 45, 1369~1372.

Lee Y I,2003. Growth periodicity, changes of endogenous abscisic acid during embryogenesis, and in vitro propagation of *Cypripedium formosanum* Hay. Ph. D. dissertation. National Taiwan University, Taipei, Taiwan.

Lee Y I, Lee N,2003. Plant regeneration from protocorm-derived callus of *Cypripedium formosanum*. In Vitro

Cell. Dev. Biol. -Plant 39, 475~479.

Lee Y I, Lee N, Yeung E C, Chung M C, 2005. Embryo Development ofCypripedium formosanum in relation to seed germination *in vitro*. J. Am. Soc. Hortic. Sci. 130, 747~753.

Lee Y I, Lu C F, Chung M C, Yeung E C, Lee N, 2007. Developmental changes in endogenous abscisic acid concentrations and asymbiotic seed germination of a terrestrial orchid, *Calanthe tricarinata* Lindl. J. Am. Soc. Hortic. Sci. 132, 246~252.

Lee Y I, 2011. In vitro culture and germination of terrestrial Asian orchid seeds. In: Thorpe, T. A., Yeung, E. C. (Eds.), Plant embryo culture; methods and protocols. Methods in Molecular Biology. Humana Press, New York, pp. 53~62.

Lee Y I, Chung M C., Yeung E C, Lee N, 2015. Dynamic distribution and the role of abscisic acid during seed development of a lady's slipper orchid, *Cypripedium formosanum*. Ann. Bot. 116, 403~411.

Li J H, Liu, Z. J, Salazar G A, Bernhardt P, Perner, H., Yukawa T, Jin X H, Chung S W, Luo, Y. B, 2011. Molecular phylogeny of *Cypripedium*(Orchidaceae: Cypripedioideae) inferred from multiple nuclear and chloroplast regions. Mol. Phylogenet. Evol. 61, 308~320.

Li Y Y, Chen X M., Guo S X, Lee Y I, 2016. Embryology of two mycoheterotrophic orchid species, *Gastrodia elata* and *Gastrodia nantoensis*: ovule and embryo development. Bot. Stud. 57, 18.

Light M H S, MacConaill M, 1990. Characterization of the optimal capsule development stage for embryo culture of *Cypripedium calceolus var. pubescens*. In: North American Native Terrestrial Orchid Propagation and Production Conference Proceedings, Pennsylvania, pp. 92.

Linden B, 1980. Aseptic germination of seeds of Northern terrestrial orchids. Ann. Bot. Fennici 17, 174~182.

Masanori T, Tomita M, 1997. Effect of culture media and cold treatment on germination in asymbiotic culture of *Cypripedium macranthos* and *Cypripedium japonicum*. Lindleyana 12, 208~213.

Miyoshi K. and M Mii, 1988. Ultrasonic treatment for enhancing seed germination of terrestrial orchid, *Calanthe discolor*, in asymbiotic culture. Sci. Hortic. 35, 127~130.

Miyoshi K, Mii M, 1998. Stimulatory effects of sodium and calcium hypochlorite, pre-chilling and cytokinins on the germination of *Cypripedium macranthos* seed in vitro. Physiol. Plant. 102, 481~486.

Murashige T, Skoog F, 1962. A revised medium for rapid growth and bioassays with tobacco tissue cultures. Physiol. Plant. 15, 473~497.

Pellow B J, Henwood M J, Carolin R C, 2009. Flora of the Sydney Region. Sydney University Press, Sydney.

Pierik R L M, Sprenkels P A, van der Harst B, van der Meys Q G, 1988. Seed germination and further development of plantlets ofPaphiopedilum ciliolare Pfitz. in vitro. Sci. Hortic. 34, 139~153.

Pritchard H W, Poyner A L C, Seaton P T, 1999. Interspecific variation in orchid seed longevity in relation to ultra-dry storage and cryopreservation. Lindleyana 14, 92~101.

Ramsay M, Stewart J, 1998. Re-establishment of the lady's slipper orchid (*Cypripedium calceolus* L.) in Britain. Bot. J. Linn. Soc. 126, 173~181.

Ramsay M M, Dixon K W, 2003. Propagation science, recovery and translocation of terrestrial orchids. In: Dixon, K. W., Kell, S. P., Barrett, R. L., Cribb, P. J. (Eds.), Orchid Conservation. Natural History Publications (Borneo), Kota Kinabalu, pp. 259~288.

Rankou H, Averyanov L, 2014. *Cypripedium lentiginosum*. The IUCN Red List of Threatened Species 2014: e. T201846A2722359. http://dx.doi.org/10.2305/ IUCN. UK. 2014-1. RLTS. T201846A2722359. en.

Rasmussen H N, 1992. Seed dormancy patterns in *Epipactis palustris*(Orchidaceae): requirements for germination and establishment of mycorrhiza. Physiol. Plant. 86, 161~167.

Rasmussen H N, 1995. Terrestrial Orchids, From Seed to Mycotrophic Plant. Cambridge University Press, Cambridge.

Segerback L B, 1983. Orchids of Nigeria. CRC Press, Boca Raton, Florida.

Shimura H, Koda Y, 2004. Micropropagation of *Cypripedium macranthos* var *rebunense* through protocorm-like bodies derived from mature seeds. Plant Cell Tiss. Org. Cult. 78, 273~276.

Shimura H, Koda Y, 2005. Enhanced symbiotic seed germination of *Cypripediummacranthos* var. *rebunense* following inoculation after cold treatment. Physiol. Plant. 123, 281~287.

Seaton P, Ramsay M, 2005. Growing orchids from seed. Royal Botanical Gardens, Kew, Richmond, UK.

Seaton P, Pritchard H W, 2003. Orchid germplasm collection, storage and exchange. In: Dixon, K. W., Kell, S. P., Barrett, R. L., Cribb, P. J. (Eds.), Orchid Conservation. Natural History Publications (Borneo), Kota Kinabalu, pp. 227~258.

St-Arnaud M, Lauzer D, Barabe D, 1992. In vitro germination and early growth of seedling of *Cypripedium acaule* (Orchidaceae). Lindleyana 7, 22~27.

Steele W K, 1996. Large scale seedling production of North American *Cypripedium* species. In Allen, C. (Ed.), North American native terrestrial orchids. Propagation and production. North American Native Terrestrial Orchid Conference, Germantown, Maryland, pp. 11~26.

Stewart S L, Kane M E, 2006. Asymbiotic seed germination and in vitro seedling development of *Habenaria macroceratitis* (Orchidaceae), a rare Florida terrestrial orchid. Plant Cell Tiss. Org. Cult. 86, 147-158.

Thornhill A, Koopowitz, H, 1992. Viability of *Disa nuiflora* Berg (Orchidaceae) seeds under variable storage conditions: Is orchid gene-banking possible? Biol. Conserv. 62, 21~27.

Van der Kinderen G, 1987. Abscisic acid in terrestrial orchid seeds: a possible impact on their germination. Lindleyana 2, 84~87.

Van Waes J M, Debergh P C, 1986. In vitro germination of some Western European orchids. Physiol. Plant. 67, 253~261.

Vujanovic V, St-Arnaud M, Barabe D, Thibeault G, 2000. Viability testing of orchid seed and the promotion of colouration and germination. Ann. Bot. 86, 79~86.

Yamazaki J, Miyoshi K, 2006. In vitro asymbiotic germination of immature seed and formation of protocorm by *Cephalanthera falcata* (Orchidaceae). Ann. Bot. 98, 1197~1206.

Yan N, Hu H, Huang J L, Xu K, Wang H, Zhou Z K, 2006. Micropropagation of *Cypripedium flavum* through multiple shoots of seedlings derived from mature seeds. Plant Cell, Tissue Organ Cult, 84, 113~117.

Yeung, E C, 1984. Histological and histochemical staining procedures. In: Vasil, I. K. (Ed.), Cell culture and somatic cell genetics of plants. vol. 1. Laboratory procedures and their applications. Academic Press, Orlando, Florida, pp. 689~697.

Yeung E C, 1999. The use of histology in the study of plant tissue culture systems - some practical comments. In Vitro Cell. Dev. Biol. Plant 35, 137~143.

Yeung E C, Zee S Y, Ye X L, 1996. Embryology of *Cymbidium sinense*: embryo development. Ann. Bot. 78, 105~110.

Zhang Y Y, Wu K L, Zhang J X, Deng R F, Duan J, Teixeira da Silva J A, Huang W C, Zeng S J, 2015. Embryo development in association with asymbiotic seed germination in vitro of *Paphiopedilum armeniacum* S. C. Chen et F. Y. Liu. Sci. Reports. 5:16356.

Zhang Y, Lee Y I, Deng L, Zhao S, 2013. Asymbiotic germination of immature seeds and the seedling development of *Cypripedium macranthos* Sw., an endangered lady's slipper orchid. Sci. Hortic. 164, 130-136.

Zeng S, Zhang Y, Teixeira da Silva J A, Wu K, Zhang J, Duan J, 2014. Seed biology and *in vitro* seed germination of *Cypripedium*. Crit. Rev. Biotechnol. 34, 358~371.

濒危兰科植物长瓣杓兰离体萌发和低温种子贮藏

摘要：本研究建立了濒危兰科植物——长瓣杓兰（*Cypripedium lentiginosum*）的无菌萌发与种子保藏技术。授粉90~105天后采收的长瓣杓兰未成熟种子发芽率最高，此时胚发育至早期球型胚与球型胚阶段；授粉120天后，种子接近成熟阶段，其发芽率急剧下降；当种子完全成熟时，其具有两层种皮紧实的包住球型胚。组织化学染色表明木质化的种皮或许是造成强制性休眠的因素。以0.5%或1%的次氯酸钠溶液处理成熟种子45分钟，可增加种皮的通透性，促进成熟种子发芽。不同MS盐类浓度试验结果显示，1/4和1/2MS培养基上种子的萌发和原球茎形成较高。在被测的细胞分裂素中，只有2 ip对萌发有促进作用，而2 ip和tdz都能促进多个原球茎样体的形成。当成熟种子脱水至初始含水量的13.5%，在5°C贮藏时，种子的萌发率和活力在贮藏12个月后略有下降，而将贮藏期延长到24个月，种子的萌发率和活力急剧下降。采用这种方法，经过6个月的培养，幼苗根系健壮，可移植到温室驯化种植。

关键词：长瓣杓兰；培养基；胚胎学；种子预处理；种子贮藏

（本文发表于 *Scientia Horticulturae*, 2017, 225）

Climate Change-induced Water Stress Suppresses the Regeneration of the Critically Endangered Forest Tree *Nyssa yunnanensis*

ZHANG Shan-shan, KANG Hong-mei, YANG Wen-zhong

Yunnan Key Laboratory of Forest Plant Cultivation and Utilization, State Forestry Administration Key Laboratory of Yunnan Rare and Endangered Species Conservation and Propagation, Yunnan Academy of Forestry, Kunming Yunnan 650201, China

Abstract: Climatic change-induced water stress has been found to threaten the viability of trees, especially endangered species, through inhibiting their recruitment. *Nyssa yunnanensis*, a plant species with extremely small populations (PSESP), consists of only two small populations of eight mature individuals remaining in southwestern China. In order to determine the barriers to regeneration, both *in situ* and laboratory experiments were performed to examine the critical factors hindering seed germination and seedling establishment. The results of *in situ* field experiments demonstrated that soil water potentials lower than -5.40 MPa (experienced in December) had significant inhibitory effects on seedling survival, and all seedlings perished at a soil water potential of -5.60 MPa (January). Laboratory experiments verified that *N. yunnanensis* seedlings could not survive at a 20% PEG 6000 concentration (-5.34 MPa) or 1/5 water-holding capacity (WHC; -5.64 MPa), and seed germination was inhibited in the field from September (-1.10 MPa) to November (-4.30 MPa). Our results suggested that soil water potentials between -5.34 and -5.64 MPa constituted the range of soil water potentials in which *N. yunnanensis* seedlings could not survive. In addition to water deficit, intensified autotoxicity, which is concentration-dependent, resulted in lower seed germination and seedling survival. Thus, seed establishment was probably simultaneously impacted by water deficit and aggravated autotoxicity. Meteorological records from the natural distribution areas of *N. yunnanensis* indicated that mean annual rainfall and relative humidity have declined by 21.7% and 6.3% respectively over past 55 years, while the temperature has increased by 6.0%. Climate change-induced drought, along with a poor resistance and adaptability to drought stress, has severely impacted the natural regeneration of *N. yunnanensis*. In conclusion, climate change-induced drought has been implicated as a regulating factor in the natural regeneration of *N. yunnanensis* through suppressing seed germination and screening out seedlings in the dry season. Based on the experimental findings, habitat restoration and microclimate improvement should both be highlighted in the conservation of this particular plant species.

Keywords: Climatic change-type drought; Endangered plant species; Forest regeneration; Plant conservation; *Nyssa yunnanensis*

Introduction

Global climatic change-induced drought stress has manifested significant influences on forest regeneration in many regions of the world[1~4]. In recent decades, changes in the global climate as a result of increasing greenhouse gas emissions have resulted in drought and/or higher temperatures[5]. Furthermore, the frequency and intensity of droughts is increasing[6]. In fact, increased drought has already resulted in ecological change in many forest communities[7~8], and drought-related forest regeneration failure has been well documented[1,9~13]. As suggested by Bartholomeus et al.[14], water-related stresses have more severe effects on the future distributions of endangered plant species, as these species usually have restricted distributions and do not possess the physiological and morphological adaptations to prevent excessive water loss. Furthermore, plants in early life stages are more sensitive to changes in climate, which can present a major bottleneck to recruitment[2,15~16].

Seed germination and seedling establishment are two critical early life stages in whichdrought stress affects forest regeneration[17~20]. Increases in temperature as well as moisture limitation may make species particularly susceptible to climate change during these life stages. For many plant species, constraints experienced during the phases of seed germination and seedling establishment severely limit regeneration[15~16,21]. The fate of seedlings has been found to be closely associated with precipitation, which is often highly seasonal[22]. Studies on woody plants have indicated that the length of the dry season affects seedling growth and survival, and continuous precipitation is also favorable for seedling growth[22~24]. Additionally, seed germination and early seedling establishment depend primarily on water availability, and increased germination rates are correlated with higher rainfall in some species[25~26]. Thus, studies on the effects of drought on plant regeneration may provide insights into population dynamics, mechanisms of survival as well as inform conservation strategies[27,28].

Nyssa yunnanensis is a critically endangered and range-restricted tree species known only from two small populations, consisting of five and three mature individuals respectively. It is distributed in ravine rainforest and remained in the tropical forest region of southern Yunnan Province, southwestern China[29,30]. It has been delineated on the China's Flora Red List[31] and the IUCN Red List as a Critically Endangered Species. Although *N. yunnanensis* naturally produces abundant seeds, no seedlings were found surrounding the mother trees upon inspection of the locality. Chen et al.[32] suggested that difficulties in seed germination and seedling growth in wild populations of *N. yunnanensis* hamper its regeneration. Studies on the taxonomy, morphology, seed germination characteristics and reproductive biologyof *N. yunnanensis*[33~36] have not identified any physiological constraints to regeneration. However, Zhang et al.[30] proposed that regeneration in *N. yunnanensis* may be negatively impacted by its autotoxicity. Moreover, autotoxic effects were found to be concentration-dependent, such that rainfall and soil moisture availability becomes a significant influencing factor. Significantly, meteorological records since 1960 have indicated that annual precipitation in the natural distribution areas of *N. yunnanensis* has decreased significantly during the past half century[30]. However, the impacts of drought stress on the suppression of *N. yunnanensis* regeneration and survival under climate change are not currently known.

Using field surveys and observations, we designed crosschecking experiments in order to examine the impact of drought as a regulating factor in the natural regeneration of *N. yunnanensis*. *In situ* experiments were designed to examine whether water acted as a vital factor in both seed germination and seedling survival. Laboratory experiments were then performed to verify the observations from the *in situ* field experiments. In the laboratory experiments, the seedlings were measured at two development stages, i. e., early seedling and post-seedling, in order to examine how drought stress influences the entire seedling establishment process.

Material and methods

Study site and climatic data

Nyssa yunnanensis occurs in the natural forests of Puwen Forest Farm (22°25'N, 101°05'E), Yunnan Province, southwestern China. In accordance with the seasonal changes in temperature and rainfall, three seasons in a year were defined for the study site, i. e., rainy season, fog cool season and hot season. The rainy season is from May to October and experiences a mean annual rainfall of 1 167mm; the fog cool season lasts from November to February and experiences less rainfall (110mm per annum); the hot season is from March to April and is associated with elevated temperatures and the least rainfall (73mm per annum)[37]. Seed germination and seedling establishment of *N. yunnanensis* occurs from September to February, during the latter part of the rainy season and the entire fog cool season.

We obtained monthly climatic data including mean temperature, precipitation and relative humidity from meteorological stations located near the study sites for the period of 1960~2014 (S1 Fig and S2 Fig).

Materials

Mature seeds of *Nyssa yunnanensis* were harvested from both remaining populations in August 2013 from their native habitat in Puwen Forest Farm, Yunnan Academy of Forestry. They were stored in a cold room (4℃) after drying at room temperature (20℃). In order to exclude the autotoxic effects of its pericarp, a seed-washing treatment was employed to break dormancy[30]. Before sowing, the seeds were selected based on size homogeneity and surface-sterilized to prevent decay by soaking in 5% Na-hypochlorite for five min, followed by washing five times with distilled water. We used 50 post seedlings in the follow-up experiments. They were cultivated for five weeks in a growth chamber (Safe Experimental Instrument Company, Haishu, Ningbo, China) at 20/15℃ (day/night), under a 14-h photoperiod and a light intensity of 100-110μmol · m^{-2} · s^{-1} and 90% relative humidity. The seedlings were then transplanted into pots containing sterilized bed soil (60% vermiculite, 20% cocopeat and 20% of other additives such as zeolite, loess and peat moss), and were placed into regular microcosm pots (10 cm dia. × 7 cm depth) in a greenhouse.

Methods

In situ field experiment

The aim of the field experiment was to examine seed germination and seedling emergence of *N. yunnanensis* in *in situ* soil cultures within their natural environment. A 1-factorial experiment

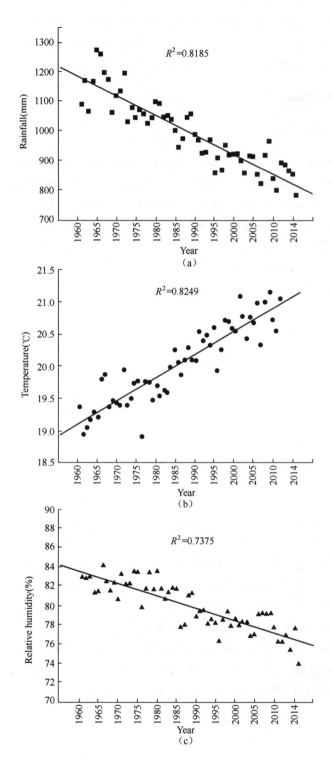

S1 Figure　Trend in annual mean rainfall
(a) temperature, (b) and relative humidity, (c) over the past 55 years at Puwen Experimental Forest Farm.

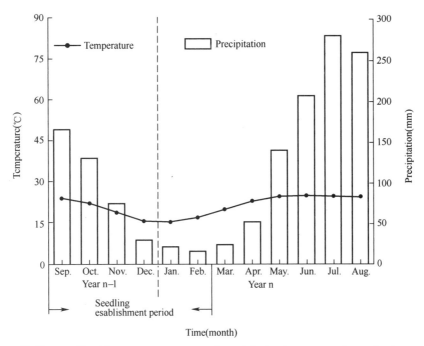

S2 Figure Monthly mean temperature and precipitation in a year at the study site.

was designed consisting of two treatments: sowing in pots, and sowing directly in the soil in the field. In the "pot" treatment, five microcosm pots (30 cm dia. ×25 cm dep.) containing 5 kg of *in situ* soil were placed under the canopy of each target mother tree and treated with sufficient moisture artificially in the morning every two days. A 0.8 m tall fence was installed around the perimeter of the plot to deter entry and subsequent trampling. In the "field" treatment, five furrows were dug (45 cm long × 20 cm wide) in the field close to the pots, and received only natural moisture. Three *N. yunnanensis* mother trees bearing fruits were used for the study. In total, there were 15 replicates for the three target mother trees in each treatment. No additional nutrients were added to either treatment. One hundred seeds were sown in each plot. Seed germination rate, seedling emergence, seedling survival and soil water potential (using a soil moisture sensor: Watermark, Spectrum, USA) were recorded monthly during the experimental period. The equation to calculate germination rate is: GR= seeds germinated /total seeds × 100.

Laboratory experiments

Experiment 1: Effects of simulated drought stress (SDS) on seed germination and early seedling growth

This experiment was carried out to determine the effects of simulated drought stress on germination and early seedling growth. Osmotic solutions of polyethylene glycol (PEG 6000) were used to induce drought stress reproducibly under *in vitro* conditions, and to maintain consistent water potential throughout the experimental period. The solutions were prepared by dissolving the required amount of PEG 6 000 in distilled water and shaking in a shaker bed (at 25℃) for 12 h.

Seed germination and early seedling growth were evaluated under five concentrations of PEG 6000 treatment (5%, 10%, 15%, 18% and 20%), corresponding to water potentials of -0.58, -1.66, -3.25, -4.29 and -5.34 MPa, respectively, while distilled water was used as the control (-0.01 MPa). The experiment was performed in Petri dishes (9cm diameter) in a greenhouse. A 1 - factorial seed germination experiment was designed with five PEG 6000 concentrations including six replicates, resulting in a total of 30 Petri dishes.

The seeds were sterilized by soaking in a 5% solution of H_2O_2 for five min. After the treatment, the seeds were washed several times with distilled water. Thirty seeds based on size homogeneitywere placed on filter paper moistened with the respective PEG 6000 solutions in each 9cm diameter Petri dish. The Petri dishes were covered with parafilm to prevent the loss of moisture by evaporation. The dishes were then placed in an incubator for 40 days at 25°C. After 40 days of incubation, seed germination and early seedling growth indicators including shoot length, root length, seedling length and seedling fresh weight were measured. Seeds were considered germinated when the emergent radical reached 2mm in length. Time to 50% germination (time taken to reach 50% germination) was calculated referred to Cavallaro et al. [38] Germination rate (GR) was calculated using the formula: GR = seeds germinated/total seeds × 100.

Experiment 2: Effects of watering treatments on post-seedling growth

A water-controlled experiment was conducted to study the effects of simulated water conditions corresponding to various soil moistures on seedling growth. In this experiment, one-year-old seedlings were subsequently subjected to five soil moisture levels: 1 WHC (water holding capacity), 1/2 WHC, 1/3 WHC, 1/4 WHC and 1/5 WHC encompassing the range of soil moistures experienced in the field. The corresponding soil water potentials were determined using the filter paper technique[39], and were determined to be -0.01, -1.18, -2.30, -4.46 and -5.64 MPa, respectively. Indicators of seedling growth were measured at the end of experiment.

Intact forest soils from the natural habitat of *N. yunnanensis* were used in this experiment to assess seedling performance across a gradient of soil moisture contents. At the field site, 30 soil cores (10 cm in diameter and 15 cm in height) in three sampling areas (2 m × 2 m) were collected from the rhizospheres of *N. yunnanensis*. Soil cores were transported to the laboratory immediately. Each soil core was transferred into a microcosm pot (10 cm diameter × 7 cm depth).

The pots were arranged in a greenhouse and positioned in a way that they would have likely received uniform amounts of sunlight, amount of shading, etc. Seedlings were irrigated daily with deionized water. Annual seedlings used in the experiment were propagated. All of the seedlings received watering for approximately one month prior to the initiation of experiments in order to help them acclimatize, as well as to minimize initial mortality. Desired soil moisture levels were obtained by allowing the soil to dry until it approached the selected moisture level, as determined gravimetrically on a parallel set of identical pots maintained for this purpose in the glasshouse (Khurana and Singh, 2004). The pots were weighed on alternate days and the required amount of water was added in order to maintain the respective moisture levels as accurately as possible. The plants were harvested following 90 days of growth. The aboveground organ of each plant was cut at the soil level and dried at 65℃ until a constant mass was achieved, and then biomass was determined by weighing. Pots were placed at 4°C until roots could be washed (within one week). Belowground mass was separated from the growing media by washing with water, and dried at 65℃ until a constant mass was reached. Growth measurements were determined as follows: leaf area ratio (LAR) = total leaf area/ total biomass; specific leaf area (SLA) = total leaf area/ total leaf biomass.

Statistical analyses

A repeated-measure one-way ANOVA (SPSS V. 16.0) was performed to determine the effect of sowing treatments in *in situ* experiments. Data obtained from the laboratory experiments, including seed water absorption, germination rate, early seedling growth and post seedling growth were evaluated using one-way ANOVA. If ANOVAs were significant, LSD was used to separate means at the 5% level. As the early seedlings in Experiment 1 did not survive under the treatment with 20% PEG 6000 solution (−5.34 MPa), and similarly, the post-seedlings in Experiment 2 did not survive under the 1/5 WHC (−5.64 MPa) treatment, they were excluded from the above analyses.

Results

Seed germination

Results of the *in situ* field experiments revealed that the rate of seed germination in *N. yunnanensis* was significantly different between the sowing treatments (pot and field) and months (from September to January; Table 1). Seed germination began one week after sowing in September and was determined to be 25.81% in the pots and 6.64% in the field (Fig 1a). It increased with time and reached its highest value in November. The highest germination rate in the pots was 66.61%, while in the field it was 40.69%. During the following two months there were no significant changes in both treatments. This suggests that seed germination terminates at the end of November under conditions of both sufficient and natural moisture. Seed germination in the pots was significantly higher than that in the field during the germination period ($P < 0.01$); soil water potential in the pots (−0.01 MPa) was kept in saturation state during the entire period of seed germination, while in the field it decreased progressively, i.e., −1.10 MPa in Sept., −2.50 MPa in Oct., −4.30 MPa in Nov., −5.40 MPa in Dec. and −5.60 MPa in Jan.

Table 1 Significance level of the effects of factors and factor interactions on variables based on repeated-measure one-way ANOVA

Variables	df	Seed germination	Seedling emergence	Seedlings survived	Soil water potential
Sowing treatment	1	** ($F=895.957$)	** ($F=628.606$)	** ($F=1.636E3$)	** ($F=3.995E3$)
Month	4	** ($F=349.581$)	** ($F=82.360$)	** ($F=98.175$)	** ($F=210.688$)
Sowing treatment×Month	4	** ($F=6.025$)	* ($F=4.011$)	** ($F=29.037$)	** ($F=209.446$)

*: $P< 0.05$; **: $P< 0.001$.

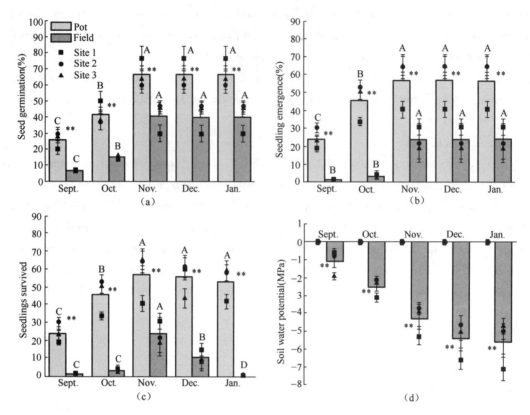

Figure 1 Seed germination (a), seedling emergence (b), seedling survival (c), soil water content (d) and relative soil water content (e) for *Nyssa yunnanensis* in pots with sufficient moisture, and in field with natural moisture at three sites during the entire seedling establishment period. Squares, circles and triangles represent site means; values are means ± SD; * * indicates a significant difference between treatments in "Pot" and "Field" at $P< 0.01$; For the soil with sufficient moisture (Pot) or natural moisture (Field), means with different uppercase are significantly different from September to January at $P< 0.05$.

The SDS experiment on *N. yunnanensis* seed germination indicated that the final germination rate declined significantly with the increase in PEG 6000 concentration. It decreased from 54.5% in distilled water (control) to zero at a PEG 6000 concentration of 20% (−5.34MPa; Fig 2a, $P < 0.05$). Furthermore, the time to 50% germination increased significantly with the increase in concentration of PEG 6000, from 9.67 days in distilled water to 38.2 days at a PEG 6000 concentration of 20% (−5.34MPa; Fig 2b, $P < 0.05$). Unfortunately, seeds began to germinate and then all died at a PEG 6000 concentration of 20%.

Seedling survival

Earlyseedlings survived for the entire experimental period (40 d) up to a water stress of 15% PEG 6000 (−3.25MPa). At a PEG 6000 concentration of 20% (−5.34MPa), seedlings perished. Post-seedlings survived for the entire experimental period (90 d) up to a water stress of 1/4 WHC (−4.46MPa), but perished under the 1/5 WHC (−5.64MPa) treatment.

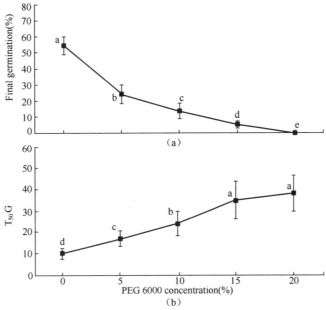

Figure 2 Effects of simulated drought stress on final seed germination rate (a) and time to 50% germination (b) of *Nyssa yunnanensis* seeds

Seedling growth

Results of the *in situ* field experiment revealed that seedling emergence was related to seed germination and was found to be significantly different between sowing treatments and months (Table 1). Seedling emergence lasted from September to November and the emergence rate 23.89% in pots and 0.85% in field during the first month (Fig. 1b). The highest seedling emergence rate reached in November, which corresponded to 56.61% in the pots and 23.69% in the field. Seedling emergence stopped from the end of November. As found in the germination experiments, the potted treatment had significantly higher seedling emergence than the field treatment.

Seedling survival in the pots was significantly different to that in the field across the different months (Table 1). The seedlings survived in both the pot and field experiments decreased from November (Fig. 1c). In the field treatment, the survival rate was 23.69% in November, dropping to 10.50% in December and reaching 0 in January, indicating that all seedlings perished under natural conditions. However, the seedling survival rates in the pot treatment only decreased slightly, retaining 53.00% survival in January.

The SDS experiment on early seedling growth revealed that all growth indices (root length, shoot length and fresh weight) decreased significantly with an increase in PEG 6000 concentrations (Table 2). Compared to the distilled water control, the growth of seedlings decreased by more than four times and fresh weight decreased by seven times at the PEG 6000 concentration of 10% (−1.66MPa). No significant differences were observed in any of the parameters under PEG 6000 concentrations of 10% and 15% ($P > 0.05$). However, at a PEG 6000 concentration of 20% (−5.34MPa), all seedlings perished by the end of the experiment.

Table 2 Effects of simulated drought stress on early seedling growth of *Nyssa yunnanensis*

PEG 6000 concentration (%)	Root length (cm)	Shoot length (cm)	Seedling length (cm)	Seedling fresh weight (g)
0	3.04±0.20a	3.12±0.27a	6.16±0.45a	0.07±0.018a
5	0.97±0.22b	1.45±0.19b	2.41±0.23b	0.03±0.010b
10	0.57±0.13c	0.72±0.08c	1.49±0.14c	0.01±0.006c
15	0.29±0.04d	0.28±0.07d	0.57±0.09d	0.01±0.001c
20	0.26±0.06d	0.27±0.04d	0.53±0.06d	0.01±0.001c

Data are mean±SD (n=6). Values marked by different letters (a, b, c, d) within a column are significantly different at P<0.05. Same as below.

The water-controlled experiment indicated that relative soil water content significantly influenced the post seedling growth of *N. yunnanensis*. All growth indices declined with the decrease in relative soil water content (Table 3).

Table 3 Effects of water condition on the growth parameters of *Nyssa yunnanensis* seedlings

Soil moisture level	Relative growth rate	Total leaf area (cm^2)	Number of leaves	Height (cm)	Aboveground biomass (g)	Belowground biomass (g)	Total biomass (g · plant^{-1})	Root/Shoot ratio	LAR	SLA
1 WHC	17.79±0.93a	111.66±7.97a	20.60±2.19a	51.22±4.22a	7.61±0.88a	4.05±0.54a	11.66±0.98a	0.53±0.05a	197.21±12.58a	302.27±4.101a
1/2 WHC	13.45±1.10b	50.96±5.04b	16.25±1.98b	30.25±3.35bc	3.16±0.51b	1.48±0.19bc	4.64±0.29b	0.47±0.04a	178.62±16.38a	262.38±17.13a
1/3 WHC	10.05±1.53c	36.81±2.38bc	11.20±1.11c	27.75±2.92c	2.07±0.21bc	0.84±0.07c	2.91±0.26b	0.41±0.04a	165.45±10.86a	199.18±12.36b
1/4 WHC	9.19±0.82c	23.15±3.17c	6.75±0.87d	23.66±2.20c	2.04±0.07c	0.76±0.11c	2.80±0.22b	0.37±0.05a	101.81±8.61b	206.30±18.05b
1/5 WHC	8.59±0.49c	21.75±1.96c	5.45±0.59d	20.66±1.85c	1.84±0.24c	0.53±0.07c	2.41±0.26b	0.32±0.03a	98.44±10.87b	195.76±16.87b

Discussion

Seed germination and seedling establishment depend primarily on moisture availability[40], soil salinity[17] and phytotoxic molecules[41]. Poor water availability is considered to be one of the principal causes of unsuccessful seedling establishment[42]. Soil moisture is especially critical for seedlings, which are prone to die-off in the dry season[43]. In our *in situ* field experiment, seed germination and seedling establishment were significantly lower under field conditions with naturally available moisture than under potted conditions with a sufficient moisture supply, across all three sites. Furthermore, seed germination, seedling emergence and seedling survival continued to increase until November, corresponding to a soil water potential of −4.30 MPa. Thereafter, seed germination and seedling emergence terminated physiologically in both the potted and field experiments. Seedlings began to perish in Dec. (-5.40 MPa), and had all died -off by Jan. at a soil water potential of -5.60 MPa in the field treatment. This implied that a soil water potential of -5.40 MPa severely limited *N. yunnanensis* seedlings. These results support our hypothesis that drought stress hinders seed germination and seedling survival. However, the exact mechanism by which drought stress suppresses *N. yunnanensis* regeneration and threatens its survival remains elusive.

Numerous studies have demonstrated that drought stress inhibits seed germination and seedling growth[44~49]. For many plants, the stages of seed germination and early seedling growth are highly sensitive to environmental stresses[50]. The strongest decline in germination has been observed at a high PEG concentration in several studies[51~53]. Similarly, seed germination in *N. yunnanensis* was progressively inhibited with increasing PEG 6000 concentrations (-0.01 ~ -5.34 MPa) in our simulated drought stress (SDS) experiment. Our results indicated that the soil water potentials experienced between Sept. and Jan. (-1.10 ~ -5.60 MPa) in the field result in water stress, suppressing seed germination.

The early seedling growth parameters were also significantly affected by simulated drought stress. Root length, shoot length and fresh weight decreased significantly with an increase in PEG 6000 concentration (-0.01 ~ -5.34 MPa), and all the seedlings perished at a PEG 6000 concentration of 20% (-5.34 MPa). Our results support the idea that seed germination and early seedling growth are impacted by simulated drought stress.[54~56] Early seedlings were significantly inhibited when soil water potential was at a PEG 6000 concentration of 18% (-4.29 MPa), corresponding to that observed in the field in Nov. (-4.30 MPa). Results of the water-controlled experiment confirmed these observations and indicated poor drought resistance in *N. yunnanensis*. Ten growth indicators of *N. yunnanensis* seedlings were significantly inhibited by drought stress (-0.01 ~ -4.46 MPa) and seedlings perished at 1/5 WHC (-5.64 MPa). The fact that root/shoot ratio declined with the decrease in relative soil water content is not in accordance with the normal response of plants, which is to increase belowground organ biomass to encourage nutrient and water update under drought conditions. This suggests that *N. yunnanensis* seedlings are not able to dynamically adapt to water deficit via responses in morphology. Hence, the implication is that seedlings in field had also been suppressed by drought stress, and a range of soil water potentials from -5.34 MPa to -5.64 MPa are fatal for *N. yunnanensis* seedlings. These results also indicated that seed germination and seedling establishment in the field had been suppressed during the seedling establishment period from Sept. to Feb. In accordance with the findings of other plant studies[60~62], low water availability significantly inhibited the growth of *N. yunnanensis* seedlings. Conversely, higher soil moisture content culminated in an increase in leaves and larger leaf areas of seedlings, which resulted in a greater photosynthetic rate as well as a higher biomass. When experiencing drought stress, most plants will reduce their aboveground biomass and allocate more resources to their belowground biomass, so as to improve the likelihood of survival by increasing access to water and nutrients[2,14,57~59]. Interestingly, the root/shoot ratio of *N. yunnanensis* seedlings declined with the decrease of soil moisture. Therefore, susceptibility to drought stress along with low phenotypic plasticity in a changing environment might be a major factor in the regeneration failure of *N. yunnanensis*.

Climate change-induced drought is the most frequent cause of forest regeneration failure[19,42,50]. Global climate change is projected to produce more frequent droughts in many regions of the world[6,12], and can trigger regeneration failure in natural forests[19]. According to the meteorological records of Puwen Forest Farm over the past 55 years, the climate in the natural distribution area of *N. yunnanensis* has changed dramatically. There has been a clear decreasing trend in annual mean rainfall and relative humidity, declining by 21.7% and 6.3% respectively from 1960 to 2014, while annual mean temperature has increased by 6.0% (S1 Fig.). In the context of decreased precipitation, higher temperatures are associated with an increased occurrence of drought due to higher rates of evapotranspiration. Importantly, decreasing precipitation and increasing temperature from September to February (S2 Fig) intensifies the soil moisture deficit during the key growth phrases of seedling establishment of *N. yunnanensis*. It appears that water stress as a result of decreasing rainfall severely influences the integrity of *N. yunnanensis* seedlings, especially when autotoxicity is concerned. According to Zhang et al.[30], the autotoxic effects of *N. yunnanensis* might be concentrated under lower water potential conditions, thereby further hindering seed germination and seedling establishment. Thus, climatic change-induced drought may be primarily responsible for the regeneration failure of *N. yunnanensis*, resulting in a decline in population size. In addition to the changes in global climate trends, the microclimate changes in natural habitat of *N. yunnanensis* are influenced by local production activities. For instance, many local natural forests have been replaced by rubber, coffee, tea and other plantations (Zhang et al., 2015). Therefore, the conservation strategies for *N. yunnanensis* should encompass physiological or adaptive obstacles in seedling establishment, preventing habitat loss as a result of human disturbances, as well as the maintenance of microclimate stability.

Conclusion

Results from our field and laboratory experiments revealed that climate change-induced drought stress impacts the natural forest habitat of *N. yunnanensis*, suppressing seed germination and seedling growth in the process of natural regeneration, threatening the survival of the species. It was discovered that soil water potentials ranging from -5.34 MPa to -5.64 MPa are fatal for *N. yunnanensis* seedlings, and can probably be attributed to the simultaneous impacts of water deficit and aggravated autotoxicity. Effective conservation measures, such as habitat restoration and microclimate improvement, should be considered in the conservation strategy.

Acknowledgements

We gratefully acknowledge funding for this project provided by the National Natural Science Foundation of China (NSFC-31460119 and NSFC-31660164), State Forestry Administration of China (2014YB1004, 2015YB1021, 2016YB1038), Yunnan Provincial Natural Science Foundation (2013FD074), and Yunnan Forestry Department (2015SX1001, 2016SX1018 and

2017SX1012).

References

[1] Allen CD, Macalady AK. Chenchouni H, Bachelet D, McDowell N, Vennetier M, et al. A global overview of drought and heat-induced tree mortality reveals emerging climate change risks for forests. Forest Ecol Manag. 2010; 259: 660~684.

[2] Walck JL, Hidayati SN, Dixon KW, Thompsons K, Poschlod P. Climate change and plant regeneration from seed. Global Change Biol. 2011; 17: 2145~2161.

[3] Stone C, Penman T, Turner R. Managing drought-induced mortality in Pinus radiata plantations under climate change conditions: A local approach using digital camera data. Forest Ecol Manag. 2012; 265: 94~101.

[4] Kouba Y, Camarero JJ, Alados CL. Roles of land-use and climate change on the establishment and regeneration dynamics of Mediterranean semi-deciduous oak forests. Forest Ecol Manag. 2012; 274: 143~150.

[5] IPCC. Climate change 2007: the physical science basis. In: Solomon, S., Qin, D., Manning, M., Chen, Z., Marquis, M., Averyt, K. B., Tignor, M., Miller, H. L. (Eds.), Contribution of Working Group I to the Fourth Assessment. Report of the Intergovernmental Panel on Climate Change. Cambridge University Press, Cambridge, United Kingdom/New York, NY, USA. 2007a; 996.

[6] IPCC. Climate change 2007: impacts, adaptation and vulnerability. Contribution of Working Group II to the fourth assessment report of the Intergovernmental Panel on Climate Change. New York, NY: Cambridge University Press, Press, Cambridge, UK. 2007b; 976.

[7] Van Mantgem PJ, Stephenson NL, Byrne JC, Daniels LD, Franklin JF, Fule PZ, et al. Widespread increase of tree mortality rates in the western United States. Science. 2009; 323: 521~524.

[8] Williams AP, Allen CD, Macalady AK, Griffin D, Woodhouse CA, Meko DM, et al. Temperature as a potent driver of regional forest drought stress and tree mortality. Nat Clim Change. 2013; 3: 292~297.

[9] Peñuelas, J, Prieto P, Beier C, Cesaraccio C, Paolo de Angelis, Giovanbattista de Dato, et al. Response of plant species richness and primary productivity in shrublands along a north-south gradient in Europe to seven years of experimental warming and drought: reductions in primary productivity in the heat and drought year of 2003. Global Change Biol. 2007; 13: 2563~2581.

[10] Peñuelas J, Lloret F, Montoya R. Severe drought effects on Mediterranean woody flora in Spain. Forest Sci. 2001; 47: 214~218.

[11] Mueller RC, Scudder CM, Porter ME, Trotter RT, Gehring CA, Whitham TG. Differential tree mortality in response to severe drought: evidence for long-term vegetation shifts. J Ecol. 2005; 93: 1085~1093.

[12] Breshears DD, Myers OB, Meyer CW, Barnes FJ, Zou CB, Allen CD, et al. Tree die-off in response to global change- type drought: mortality insights from a decade of plant water potential measurements. Front Ecol Environ. 2009; 7: 185~189.

[13] Huang CY, Anderegg WRL. Large drought-induced aboveground live biomass losses in southern Rocky Mountain aspen forests. Global Change Biol. 2012; 18: 1016~1027.

[14] Bartholomeus RP, Witte JPM, van Bodegom PM, van Dam JC, Aerts, R. Climate change threatens endangered plant species by stronger and interacting water-related stresses. J Geophys Res. 2011; 116: G4.

[15] Fay PA, Schultz MJ. Germination, survival, and growth of grass and forb seedlings: effects of soil moisture variability. Acta Oecol. 2009; 35: 679~684.

[16] Dalgleish HJ, Koons DN, Adler PB. Can lifehistory traits predict the response of forb populations to changes

in climate variability. J Ecol. 2010; 98: 209~217.

[17] Almansouri M, Kinet JM, Lutts S. Effect of salt and osmotic stresses on germination in durum wheat (Triticum durum Desf.). Plant Soil. 2001; 231: 243~254.

[18] Cochrane AM, Daws I, Hay FR. Seed-based approach for identifying flora at risk from climate warming. Austral Ecol. 2011; 36: 923~935.

[19] Carón MM, Frenne PD, Brunet J, Chabrerie O, Cousins SAO. Interacting effects of warming and drought on regeneration and early growth of *Acer pseudoplatanus* and *A. platanoides*. Plant Biology. 2015; 17: 52~62.

[20] Cavallaro V, Barbera AC, Maucieri C, Gimma G, Scalisi C, Patanè C. Evaluation of variability to drought and saline stress through the germination of different ecotypes of carob (*Ceratonia siliqua* L.) using a hydro-time model. Ecol Eng. 2016; 95: 557~566.

[21] Lloret F, Penuelas J, Estiarte M. Effects of vegetation canopy and climate on seedling establishment in Mediterranean shrubland. J Veg Sci. 2005; 16: 67~76.

[22] Poorter L, Hayashida-Oliver Y. Effects of seasonal drought on gap and understorey seedlings in a Bolivian moist forest. J Trop Ecol. 2000; 16: 481~498.

[23] Engelbrecht BMJ, Kursar TA. Comparative droughtresistance of seedlings of 28 species of co-occurring tropical woody plants. Oecologia. 2003; 136: 383~393.

[24] Khurana E, Singh JS. Ecology of seed and seedling growth for conservation and restoration of tropical dry forest: a review. Environ Conser. 2000; 28: 39~52.

[25] Rawal DS, Kasel S, Keatley MR, Nitschke CR. Environmental effects on germination phenology of co-occurring eucalypts: implications for regeneration under climate change. Int J Biometeorol. 2015; 59: 1237~1252.

[26] Camarero JJ, Gazol A, Sangüesa-Barreda G, Oliva J, Vicente-Serrano SM. To die or not to die: early warnings of tree dieback in response to a severe drought. J Ecol. 2015; 103: 44~57.

[27] Jeltsch F, Moloney KA, Schurr FM, Kochy M, Schwager M. The state of plant population modelling in light of environmental change. Perspect. Plant Ecol. 2008; 9: 171~189.

[28] Milbau A, Graae BJ, Shevtsova A, Nijs I. Effects of a warmer climate on seed germination in the subarctic. Ann Bot. 2009; 104: 287~296.

[29] Ma YP, Chen G, Grumbine RE, Dao ZL, Sun WB, Guo HJ. Conserving plant species with extremely small populations (PSESP) in China. Biodivers Conserv. 2013; 3: 803~809.

[30] Zhang SS, Shi FQ, Yang WZ, Xiang ZY, Kang HM, Duan ZL. Autotoxicity as a cause for natural regeneration failure in *Nyssa yunnanensis* and its implications for conservation. Isr J Plant Sci. 2015; 62: 187~197.

[31] AAVV, Lista roja de la flora vascular española (valoración según categorías UICN). Conservation and Vegetation. 2000; 6: 1~40.

[32] Chen W, Shi FQ, Yang WZ, Zhou Y, Chen HW. Population Status and Ecological Characteristics of *Nyssa yunnanensis*. J Northeast For Univ. 2011; 39:17~19, 61.

[33] Sun BL, Zhang CQ. A Revised Description of *Nyssa yunnanensis* (Nyssaceae). Acta Bot. Yunnanica. 2007a; 29: 173~175.

[34] Sun BL, Zhang CQ, Zhou FL, Shi FQ, Wu ZK. Seed morphology and effects of different treatments on germination of the critically endangered *Nyssa yunnanensis* (Nyssaceae). Acta Botany of Yunnanica. 2007b; 29: 351~354.

[35] Sun BL, Zhang CQ, Porter PL, Wen J. Cryptic Dioecy in *Nyssa yunnanensis* (Nyssaceae), A Critically Endangered Species from Tropical Eastern Asia. Ann Mo Bot Gard. 2009; 96: 672~684.

[36] Yuan RL, Xiang ZY, Yang WZ, Zhang SS. Seed dormancy and germination traits of *Nyssa yunnanensis*. Forest Res. 2013; 26: 384~388.

[37] Liu WJ, Li HM. Tourism climatic resources in Xishuangbanna. Nat Resour. 1997; 2: 62~66.

[38] Cavallaro V, Maucieri C, Barbera AC. *Lolium multiflorum* Lam. cvs germination under simulated olive mill wastewater salinity and pH stress. Ecol Eng. 2014; 71: 113~117.

[39] Deka RN, Wairiu M, Mtakwa PW, Mullin CE, Veenendaalel EM, Townend J. Use and accuracy of filter-paper technique for measurement of soil matric potential. Eur J Soil Sci. 1995; 46:233~238.

[40] Mohammad ME, Benbella M, Talouizete A. Effect of sodium chloride on sunflower (*Helianthus annuus* L.) seed germination. Helia. 2002; 37: 51~58.

[41] Barbera AC, Maucieri C, Ioppolo A, Milani M, Cavallaro V. Effects of olive mill wastewater physico-chemical treatments on polyphenol abatement and Italian ryegrass (*Lolium multiflorum* Lam.) germinability. Water Res. 2014; 52 (4): 275~281.

[42] Springer TL. Germination and early seedling growth of shaffy-seeded grasses at negative water potentials. Crop Sci. 2005; 45: 2075~2080.

[43] McDowell N, Pockman WT, Allen CD, Breshears DD, Cobb N, Kolb T, et al. Mechanisms of plant survival and mortality during drought: why dosome plants survive while others succumb to drought? Tansley review. New Phytol. 2008; 178: 719~739.

[44] Nepstad DC, Moutinho P, Dias MB, Davidson E, Cardinot G, Markewitz,D. et al. The effects of rainfall exclusion on canopy processes and biogeochemistry of an Amazon forest. J Geophys Res. 2002; 107: 1~18.

[45] Clark DB, Clark DA, Oberbauer SF. Annual wood production in a tropical rain forest in NE Costa Rica linked to climatic variation but not to increasing CO_2. Global Change Biol. 2010; 16: 747~759.

[46] Clark DA, Piper SC, Keeling CD, Clark DB. Tropical rain forest tree growth and atmospheric carbon dynamics linked to interannual temperature variation during 1984-2000. PNAS. 2003; 100: 5852~5857.

[47] Rolim SG, Jesus RM, Nascimento HEM, do Couto HTZ, Chambers JQ. Biomass change in an Atlantic tropical moist forest: the ENSO effect in permanent sample plots over a 22-year period. Oecologia. 2005; 142: 238~246.

[48] Phillips OL, Aragão LEOC, Lewis SL, Fisher JB, Lloyd J, López-González G, et al. Drought sensitivity of the Amazon rainforest. Science. 2009; 323: 1344~1347.

[49] Barbeta A, Ogaya R, Peñuelas J. Dampening effects of long-term experimental drought on growth and mortality rates of a Holm oak forest. Global Change Biol. 2013; 19: 3133~3144.

[50] McLaren KP, McDonald MA. The effects of moisture and shade on seed germination and seedling survival in a tropical dry forest in Jamaica. Forest Ecol Manag. 2003; 183: 61~75.

[51] Gamze OKCU, Kaya MD, Atak M. Effects of Salt and Drought Stresses on Germination and seedling growth of Pea (*Pisum sativum* L.). Turk Entomol Derg-Tu. 2005; 29: 237~242.

[52] Khazaie H, Earl H, Sabzevari S, Yanegh J, Bannayan M. Effects of Osmo-Hydropriming and Drought Stress on Seed Germination and Seedling Growth of Rye (*Secale Montanum*). ProEnviron. 2013; 6: 496~507.

[53] Khodadad M. A Study Effects of Drought Stress on Germination and Early Seedling Growth of Flax (*Linum Usitatissimum* L.) Cultivars. Adv Environ Biol. 2011; 5: 3307~3311.

[54] Sadeghian SY, Fasli H, Mohammadian R, Taleghani DF, Mesbah,M. Genetic variation for drought stress in sugar beet. J Sugar Beet Res. 2000; 37: 55~77.

[55] Murillo-Amador B, López-Aguilar R, Kaya C, Larrinaga-Mayoral JA, Flores-Hernández A. Comparative

effects of NaCl and polyethylene glycol on germination, emergence and seedling growth of cowpea. Crop Sci. 2002; 188: 235~247.

[56] Sadeghian SY, Yavari N. Effect of water-deficit stress on germination and early seedling growth in sugar beet. J. Agron. Crop Sci. 2004; 190: 138~144.

[57] Wang ML, Feng YL. Effects of soil nitrogen levels on morphology, biomass allocation and photosynthesis *Ageratina adenophora* and *Chromoleana odorata*. Acta Phytoecologica Sinica. 2005; 9: 697~705.

[58] Jump AS, Peñuelas J. Running to stand still: adaptation and the response of plants to rapid climate change. Ecol Lett. 2005; 8: 1010~1020.

[59] Mainiero R, Kazda M. Depth-related fine root dynamics of Fagus sylvatica during exceptional drought. Forest Ecol Manag. 2006; 237: 135~142.

[60] Knipe OD. Western wheatgrass germination as related to temperature, light, and moisture stress. J Range Manag. 1973; 26: 68~69.

[61] Sharma ML. Simulation of drought and its effect on germination of five pasture grasses. Agro J. 1973; 65: 982~987.

[62] Sharma ML. Interaction of water potential and temperature effects on germination of three semi-arid plant species. Agron J. 1976; 68: 390~394.

气候变化导致的水分胁迫是抑制极度濒危植物云南蓝果树的主要原因

摘要：气候变化引起的水分胁迫已经威胁到树木的生存能力，尤其威胁到了濒危物种。云南蓝果树现仅存2个天然种群，野外个体数量为8株，天然更新困难，属典型的极小种群物种。本研究采用野外原位生态学实验和人工控制实验相结合的手段，探究了干旱导致的水分需求限制对极小种群野生植物云南蓝果树天然更新过程中种子萌发和幼苗生长2个关键阶段的影响。原位试验结果表明，土壤水势低于-5.40 MPa时对幼苗的存活率和幼苗均有显著的抑制作用。幼苗在-5.60 MPa的土壤水势(一月)条件下就会死亡。室内受控试验证明云南蓝果树幼苗在20% PEG 6000浓度(土壤水势为-5.34 MPa)或1/5的持水(WHC; -5.64 MPa)条件下是无法存活的；野外种子的萌发率从九月(-1.10 MPa)到十一月(-4.30 MPa)呈降低趋势。我们的研究结果表明，-5.34~5.64 MPa的土壤水势范围是云南蓝果树幼苗不能存活的水势。除了干旱导致的水分缺失，还加剧了浓度依赖性的自毒作用，导致了更低的种子种子萌发与幼苗存活。因此，幼苗的建立可能是受干旱胁迫导致的水分缺失和自毒效应加剧同时影响的。自然气象记录表明，云南分布区平均降水量和相对降雨量在过去的55年里，湿度分别下降了21.7%和6.3%，温度增加了6%。气候变化引起的干旱，以及对干旱胁迫的较低适应性，严重影响了云南蓝果树的自然更新。总之，气候变化引起的干旱胁迫被认为是主要通过种子萌发和幼苗生长两种方抑制云南蓝果树天然更新的调控因子。根据试验结果，生境恢复和小气候的改善都可以在保护这种特殊的植物物种上有重大作用。

（本文发表于 *PLOS ONE*, 2017 年）

A Sophisticated Species Conservation Strategy for *Nyssa yunnanensis*, A Species with Extremely Small Populations in China

YANG Wen-Zhong, ZHANG Shan-Shan, WANG[1] Wei-Bin[1,2] *, KANG Hong-Mei[1], MA Na[1,3]

1. Key Laboratory of Rare and Endangered Forest Plants of State Forestry Administration, Yunnan Academy of Forestry, Kunming 650201, China
2. Puwen Forest Farm, Yunnan Academy of Forestry, Xishuangbanna 666102, China
3. School of Landscape Architecture, Southwest Forestry University, Kunming 650224, China

Abstract: Conservation of plant species with extremely small populations (PSESP) is a focus of wild plant conservation in China at present. A relevant strategy for PSESP conservation requires improvement from previous programs for rare and endangered plants and national key protected plants. An integrated strategy for PSESP conservation of the *Nyssa yunnanensis* was initiated and applied over a 7-year period (2009-2015). Here, we reviewed the processes to implement the strategy: resource inventory, formulation of conservation action plan (CAP), *in situ* conservation, seedling propagation, *near situ* conservation, *ex situ* conservation and scientific research. Major concerns and technical requirements for each action are described and further analyzed within a broad scope to conserve PSESPs. A detailed resource inventory that highlights both the change in population status and the participation of local residents is recommended before the formulation of the CAP. Techniques for determination of the area of a mini-reserve is developed for *in situ* conservation of *N. yunnanensis*. Near situ conservation is a novel approach whereby establishing new viable population in contiguous areas with a similar climate, habitat and community, in which techniques for seedling preparation, soil preparation, and early management are introduced. A population-based species conservation strategy for *N. yunnanensis* may aid additional PSESP conservation, so as to contribute to overall wild plant conservation.

Key words: Wild plant conservation; plant species with extremely small populations (PSESP); species conservation strategy (SCS); near situ conservation; *Nyssa yunnanensis*

1 Introduction

Wild plant conservation in China has experienced three stages with different protected plant lists. A rare and endangered plant list covering 354 species was issued in 1984 (Environment Protection Commission of China, 1984; State Environment Protection Bureau and Chinese Academy of Sciences, 1987). The national key protected plant list with 246 species and eight speciose groups was issued in 1999, two years after the promulgation of the National Wild Plant Protection Regulation (Yu, 1999). Most recently, a list of plant species with extremely small populations (PSESPs), covering 120 species, was launched in 2012 (Yang *et al.*, 2015). The concept of PSESPs was proposed to highlight the most endangered species in China according to the results of the first inventory of national key protected plants and other thematic surveys of rare and endangered plants (Gu, 2003; Li *et al.*, 2003a). The term PSESP applies to a plant species with a narrow geographical distribution, that have been subject to anthropogenic disturbance over a long period, and those in which the population number and size are smaller than the minimum required to prevent extinction (Yunnan Forestry Department and Yunnan Science and Technology Department, 2009a; State Forestry Administration of China, 2012; Ren *et al.*, 2012; Volis, 2016). A National Conservation Program for PSESPs has operated since its approval by the State Council of China in 2012 (Guo, 2012; Ma *et al.*, 2013; Sun and Yang, 2013).

Nyssa yunnanensis is one of the focal species for exploration of the methodologies for PSESP conservation. *N. yunnanensis*, a canopy tree up to 30 m in height, is a cryptically dioecious species consisting of individuals bearing staminate flowers and other individuals bearing morphologically perfect flowers but with inaperturate and inviable pollen grains (Sun and Zhang, 2007; Sun *et al.*, 2009). Flowers are in anthesis from February to April and drupes mature from August to September. According to previous field surveys only three small populations are known, with 37 trees growing in ravine rainforest at the northern margin of the tropical zone in Southwestern China (Sun, 2008). The species is classified as Critically Endangered (CR) in the IUCN Red List of Threatened Species™ (IUCN, 2012), and is a national key protected species under grade I protection in China. Studies on *N. yunnanensis*, a basal taxon of the Nyssa clade, are indispensable for resolution of phylogenetic relationships of the genus Nyssa and floristic linkages of the disjunct distribution between East Asia and North America (Sun *et al.*, 2009). Therefore, *N. yunnanensis* is included in the PSESP list. Several national and provincial pilot projects on *N. yunnanensis* conservation were conducted even before the inception of the National PSESP Conservation Program (Yunnan Forestry Department and Yunnan Science and Technology Department, 2009b).

An integrated strategy for *N. yunnanensis* conservation has been implemented. As a pilot project of PSESP conservation, no guidance or effective method was available beforehand to guide implementation of *N. yunnanensis* conservation. Owing to continuous project supports from protection authorities at both national and provincial levels, our efforts to conserve *N. yunnanensis* were assisted. These projects covered, for example, resource investigation, seedling propagation, mini-reserve establishment, *near situ* and *ex situ* conservation, reinforcement, and research on endangerment mechanism. As conservation actions were taken, various methods were tested and relevant techniques were developed (Yang *et al.*, 2015). An integrated strategy for *N. yunnanensis* conservation was formed through implementation of pilot projects. Here, we summarize the strategy with the aim to share knowledge and accumulated experiencesfor PSESP conservation.

2 Resource inventory and population status

Understanding the status of focal species, and in particular natural populations, is crucial for conservation programs. To determine the current status of *N. yunnanensis*, a framework was generated for establishment of a resource inventory (Fig 1). The framework involved three steps: preparation, field survey and data analysis (Yang *et al.* 2010).

Preparation began with a literature review, examination of herbarium specimens and consultation of experts. Based on previous field surveys (Li et al., 2003b; Sun, 2008), specimens collected from 1957 to 2007, consultation of experts and relevant literatures, we gained a preliminary understanding of the biological and ecological characteristics of *N. yunnanensis* and its possible geographical distribution. A pre-survey was carried out at known sites of *N. yunnanensis* to learn morphological characters for its identification, habitat features and community structure. Subsequently, the preparation phase involved identification of a filed survey area, formulation of survey forms, development of photographs taken during the pre-survey and purchase of investigatory tools.

A field survey was initiated with fresh specimens collected from the known sites. We assumed that local residents are better acquainted with local plant distributions than botanical experts; hence the field survey was highly reliant on local communities. Fresh specimens and photographs were helpful aids when interviewing local residents in the field. Once a location was suspected from interviews conducted in local communities, the site would be visited to search for the focal species. A detailed investigation on plant community, population and disturbances was carried out at the verified distribution site. A 7-month intensive field survey covering 10 townships in four counties in the Xishuangbanna and Pu'er Prefectures was completed in October 2010.

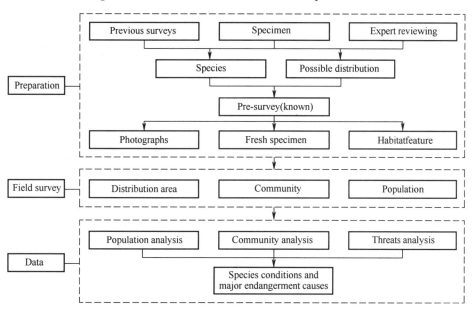

Figure1   Framework for resource inventory of *Nyssa yunnanensis*

Data analysis focused on changes in distribution area, population structure, community composition and habitat features through comparison of historical and current conditions. Detailed assessment of the population status and ecological characteristics of *N. yunnanensis* has been concluded (Chen et al., 2011), of which notable results are the following. ①Three confusable species, *Litsea glutinosa*, *Elaeocarpus prunifolioides* and *Nothapodytes nimmoniana*, were casually recognized as *N. yunnanensis* and properly excluded in this resource inventory. ②The distribution area of *N. yunnanensis* has decreased by 95% over the past 50 years, and only two small populations containing eight individuals remain in the wild. ③The population adjacent to an old dam described by Sun *et al.* (2009) was verified to be of artificial origin rather than a natural population. ④The remnant natural population I consists of five individuals with diameter at breast height (DBH) of 48.0, 44.7, 23.3, 10.1, and 9.1cm, respectively, while the population II comprises three trees with DBH of 90.4, 85.0, and 33.3cm, respectively. ⑤Three mature trees in total bear fruits but no seedling or sapling was observed in the forest. ⑥Two remnant populations of *N. yunnanensis* occur near water in valleys, implying that the species may prefer wet habitats. ⑦Most of historical habitats, including the type localities, have been replaced by coffee, tea and rubber plantations.

3 Conservation action plan

An action plan has been formulated for *N. yunnanensis* conservation. Though resource inventory gave the background of *N. yunnanensis* and indicated an urgent need to take action, the first requirement was to determine what actions and their sequence were needed. A planning team consisting of botanists, ecologists, forestry planners and conservation practitioners was formed to formulate a conservation action plan (CAP) for *N. yunnanensis*. The formulation went through the stages of status review, problem analysis, plan drafting and panel assessment (Yang *et al.* 2011b). A final CAP with assessments on its relevance, rationality and feasibility was submitted to the provincial protection authority for approval.

Six targets with 12 actions are contained in the CAP for *N. yunnanensis*. Based on problem analysis, actions to conserve remnant populations were first proposed, which included establishment of a mini-reserve and recovery of the remnant populations and habitats (Table 1). To prevent extinction in the wild, conservation of germplasm resources, including seed preservation and *ex situ* collections, was suggested. Once the remnant populations and the extant germplasm are safely conserved, further actions would be taken to increase the population number and size, for example, through seedling propagation, reinforcement and establishment of new populations. To overcome the threats and endangerment causes, two targets with four actions were suggested in the CAP. One was to strengthen awareness building in local communities because the resources inventory indicated that plantations of cash crops were the major threat to *N. yunnanensis*. The other was to elucidate biological and ecological mechanisms underlying the endangerment through scientific research and population monitoring. To implement the CAP and effectively conserve *N. yunnanensis*, capacity building of related stakeholders should also be taken into account.

The duration of the CAP is 10 years from 2011 to 2020, which was divided into two phases. The first phase proceeded from 2011 to 2015 with a focus on natural population protection, germplasm preservation, seedling propagation, population enhancement, population establishment, and purchase of equipment and facilities. The second phase will operate from 2016 to 2020 and will highlight population maintenance, awareness building and personnel training. Scientific research and population monitoring will be conducted throughout the duration of the CAP. The total budget of CAP is RMB 8 million, which will be mainly invested in establishment and management of a mini-reserve, recovery of remnant populations, establishment of new populations, monitoring of populations and their habitats, scientific research, and technical development.

Table 1 Targets and actions incorporated in the conservation action plan for *Nyssa yunnanensis*

Targets	Actions	Phase *
I: Natural population protection and restoration	1. Establishment of mini-reserve	I
	2. Natural population recovery	I
II: Germplasm resources preservation and seedling propagation	3. Germplasm preservation	I
	4. Nursery construction and management	I
III: Establishment of new populations	5. Near situ conservation	I
	6. Ex situ conservation	II
IV: Capacity building	7. Equipment and facilities for management and research	I&II
	8. Technical trainings and information sharing	II
V: Scientific research and monitoring	9. Scientific research on *N. yunnanensis*	I&II
	10. Population and habitat monitoring	I&II
VI: Awareness building	11. Propaganda-related equipment, facilities and materials	I
	12. Awareness building activities	II

* Phase I proceeded from 2011 to 2015; Phase II will operate from 2016 to 2020.

4 Implementation of the CAP

4.1 Mini-reserve establishment

Protection of remnant populations is urgent, in particular for those PSESP species outside nature reserves. In accordance with the CAP, we established a mini-reserve to protect natural populations of *N. yunnanensis* in Puwen Forest Farm. Establishment of the mini-reserve consisted of three stages of planning, construction and management (Yang et al., 2011a). The planning included five steps: team formation, field survey, preliminary design, action confirmation, and plan evaluation. The construction covered boundary demarcation, institutional setup, equipment and facility purchase, clearance of patrol pathways, and signpost construction. Management of the mini-reserve included formulation of rules and regulations, monitoring and patrolling, awareness building, facility maintenance and archive management (Yang et al., 2016).

The total area of the *N. yunnanensis* mini-reserve is 49.46 ha. It was divided into two functional zones in accordance with changes in landform, site conditions, vegetation and anthropogenic disturbances. The core zone comprises an area of 10.00 ha, accounting for 20.22% of the total area of the mini-reserve, while the remaining area is a buffer zone. The core zone was demarcated by 10 boundary poles coded with red paint. The buffer zone was demarcated by 18 boundary poles coded with blue paint, which was also the mini-reserve boundary. The function of the core zone is to protect remnant individuals and provide suitable habitats for future expansion of the population, whereas the buffer zone functions mainly to maintain a stable environment for the population.

In addition to infrastructure construction, we established a mini-reserve management office in the administrative building of Puwen Forest Farm. Two staff members of the management office were assigned with responsibilities for monitoring and patrolling, awareness building, facility maintenance and assistances with scientific research.

4.2 Propagation and seedling cultivation

As soon as the remnant populations were formally protected, propagation was essential for enhancement of the remnant populations and establishment of new populations. A nursery with a total area of 380 m^2, as planned in the CAP, was constructed adjacent to the natural populations in Puwen Forest Farm, so as to meet the climatic and edaphic requirements of *N. yunnanensis* seedlings.

Techniques for both sexual and asexual propagations were developed to raise *N. yunnanensis* seedlings. Sexual propagation techniques included seed collection and treatment, germination facilitation and seedling management (Yuan *et al.* 2013). Fruits were collected in September and soaked in tap water for 24 h to enable removal of the flesh and thereby remedy autotoxicity effects (Zhang *et al.* 2015). Seeds were soaked in 200 mg · L^{-1} gibberellic acid (GA$_3$) for 24 h, and then sowed in plastic trays and seedbeds that were sterilized with carbendazim (800×). No obvious insect or disease damage was observed during seedling growth, but seedlings were sensitive to water deficit. Therefore, irrigation is a key consideration for raising *N. yunnanensis* seedlings.

Asexual propagation was mainly focused on cutting techniques, which included cutting collection and treatment, rooting facilitation and seedling management (Qiu *et al.*, 2013). Nursery beds contained river sand 20–25 cm thick as growing medium, which was sterilized with 0.5% potassium permanganate 3 days before the operation. Cuttings were collected from first-year shoots in March and April and cut to a length of 15 cm. Cuttings were sterilized with 0.1% carbendazim for 10 min, washed three times with purified water and soaked in 500 mg · L^{-1} ABT-1 rooting powder for 5 s. Successful cutting propagation requires consistent soil moisture and relative humidity (~80%).

Up to December 2015, more than 3000 saplings in five batches were raised, of which 2500 were propagated from seeds and about 500 were propagated from cuttings. Saplings produced were used for reinforcement in remnant populations in the mini-reserve and establishment of new populations in *near situ* conservation sites.

4.3 Reinforcement in remnant populations

Reinforcement, or augmentation, is the addition of individuals to an existing population, with the aim of increasing the population size or diversity and thereby improving population viability (Falk *et al.*, 1996). Given the existence of only two remnant natural populations, our emphasis was to reinforce population I in the first phase of the CAP, whereas population II comprising three old trees (DBH 90.4, 85.0, and 33.3 cm, respectively) would be reinforced in the second phase.

Techniques for reinforcement cover seedling preparation, soil preparation, field planting, and early management. Saplings of different ages used in reinforcement included seedlings raised and cuttings propagated from staminiferous plants. (Sun *et al.*, 2009). Block land clearance and pit soil preparation were carried out in canopy gaps, so as to minimize damage to microenvironments (Fig. 2b). The density of pits depended upon the scale of canopy gaps and the size of the seedlings. Field planting was carried out at the same time as soil preparation, so as to increase the survival rate. Furthermore, saplings were pruned and transplanted with clay root balls when DBH exceeded 5cm (Fig. 2a). The main aim of early management was to prevent water loss by plants. Both ground irrigation and trunk injection of water were applied, especially for big saplings (Fig. 2c).

a. Sapling with clay root ball   b. Pit soil preparation with minimum impact on habitat   c. Planted sapling with trunk injection

Figure 2 Reinforcement in population I of *Nyssa yunnanensis*

Reinforcements of *N. yunnanensis* were carried out in the vicinity of natural population I in the buffer zone of the mini-reserve. Four reinforcements have been conducted within an area of 0.67 ha. The first reinforcement was carried out in 1979, and nine trees remained. The second reinforcement was carried out in 1995, and five trees remained. These two reinforcements were not intended, though they contributed to natural population enhancement. The individuals were cultivated as part of a tree species collection in the past, and no monitoring record was found. The third batch, consisting of 85 1-year-old seedlings, was planted in 2009. The last reinforcement of 204 4-year-old saplings was carried out in 2013. A total of 303 individuals remained in the natural population I after the four reinforcements.

4.4 *Near situ* conservation

Based on our understanding of *near situ* conservation, we established three new populations of *N. yunnanensis* in natural and semi-natural environments. At the time of formulating the CAP, we learned from *near situ* conservation only that the site for establishment of a new population should be geographically close and have a similar climate to the native habitat. Although available knowledge was limited, it guided site selection for establishment of new populations. In the sites selected, land clearance and soil preparation were required to minimize impact on the original habitats, as that in reinforcement. Field planting was carried out in the wet season (from June to

August), so as to increase survival rate and reduce maintenance cost.

The first *near situ* population was established in Guanping in June 2009 through cooperation with the local forestry administration. Guanping is 30 km southwest of the natural population. A total of 300 1-year-old seedlings were planted along the bottom of the Beiliaoqing Valley, where the natural environment remains in a sound condition. Except for watering at the time of field planting, no further management measures were applied. Monitoring of the site in January 2013 showed that over 200 individuals remained, and the average DBH was 2.41 cm and average height was 3.65 m. The death of individuals was primarily due to shading from other trees in the community.

The second *near situ* population was established in Caiyanghe Forest Park in July 2010. The park is located in 18 km north of the natural population. A total of 170 1-year-old seedlings were planted in abandoned farmland in Datianqing Valley, a semi-natural environment in an early stage of recovery. As the first *near situ* population in Guanping, no further management measure was applied other than watering at the time of field planting. Monitoring of the population in January 2013 showed that 110 trees remained, and the average DBH was 3.70 cm and average height was 2.45 m. The death of seedlings was mainly caused by drought, particularly in the dry season (from October to May of the following year).

The third *near situ* population was established in Puwen Forest Farm in August 2014. The site is located at the bottom of the Houziqing Valley, 1 km from the natural populations. The natural vegetation comprised grasses and bushes. First, grasses and bushes were cleared but trees exceeding 3 m in height were retained. Then pit soil preparation was carried out at a density of approximately 3 m × 4 m. A total of 500 1-year-old seedlings were planted. In the first 3 years, it was planned to carry out weeding, scarifications, and fertilizations twice a year. So far, two-thirds of the management has been completed and the survival rate is about 90%. From the start of the fourth year (September 2017), all management measures will cease, so as to leave planted trees as a part of natural community.

4.5 *Ex situ* conservation

Ex situ conservation includes two approaches to preserve genetic resources. One strategy is to preserve germplasm resources in a seed bank, and the other approach is to maintain living collections in botanic gardens or arboreta. For *N. yunnanensis*, the former has not been carried out yet. The latter has been completed through establishment of four *ex situ* conservation sites in the Kunming Botanic Garden, the Kunming Arboretum, the Xishuangbanna Tropical Botanic Garden and the Xishuangbanna Botanic Garden for Rare and Endangered Plants. The number of seedlings planted in these sites ranged from 50 to 100. The management of *ex situ* collections includes watering, weeding, scarification, and fertilization. However, frost damage to collections, in particular to the top shoots, has been observed every winter in Kunming, which experiences a subtropical climate, whereas collections have shown vigorous growth in Xishuangbanna, where the climate is tropical.

To further facilitate *ex situ* conservation and thereby preserve germplasm resources of *N. yunnanensis in additional locations, we donated to landscaping and gardening companies approximately* 1000 seedlings, and encouraged use of the species as an ornamental tree in artificial landscape environments.

5 Discussion

5.1 Differences to previous wild plant conservation programs

Prior to establishment of PSESP conservation programs, China's wild plants, including rare and endangered plants and national key protected plants, were conserved in a traditional manner of natural resources management. Four major actions were incorporated in the previous strategy. The first was an attempt to ease pressures on wild protected plants via laws, regulations and administrations. The second was to preserve germplasm resources through *ex situ* conservation in botanic gardens. The third was to advocate public involvement in plant conservation via awareness building (Wu *et al.*, 2004). The fourth action was to conserve germplasm resources through establishment of seed banks. There are two obvious shortcomings in this strategy. One is that wild plant protection was generally attempted through creation of a favorable social atmosphere rather than implementation of a specific strategy for certain species. The other is that the population, based on which species persist and evolve, was seldom taken into account (Yang *et al.*, 2015).

The concept of population management was introduced into PSESP conservation because strategies are species-specific. The essence of species conservation is to manage and adjust the number, size, structure and dynamics of populations. In other words, long-term persistence of a species is achievable only through managing the structure and dynamics of existing populations and establishing new populations. In this sense, theories and methods of population ecology are inevitably highlighted in species conservation, through which the integration of scientific research and conservation practices are facilitated, thereby enhancing the effectiveness of species conservation. A population-based strategy for PSESP conservation represents progress in China's wild plant conservation program.

5.2 Significance of resource inventory

A resource inventory is imperative in PSESP conservation. Wild plant species were treated as a collective entity rather than specific species in the past in China, resulting in general conservation strategies. Presently, resource inventories of national protected wild plants have been conducted twice, but the data obtained are inadequate to generate population-based conservation strategy for a specific species. It is therefore necessary to carry out a resource inventory for a focal species, so as to satisfy the requirements for generation of species conservation strategy (SCS). A detailed resource inventory also helps to corroborate the rationality of a protected plant list, in which species were suggested for inclusion by botanical experts.

A resource inventory consists of two parts: distribution assessment and population demographics. The distribution investigation informs the native range and population number of the species. The current method of resource inventory for national key protected species relies on professionals and lacks the participation of local inhabitants. This approach usually results in omission of distribution sites or populations and thereby misleads the SCS. Demographic changes in recruitment, growth, death and dispersal rate of individuals comprising a population will manifest as local extinction or colonization events (Hansen et al., 2001). Studies focused on plant population demography have demonstrated that the largest loss of reproductive potential occurs mostly between seed and seedling establishment (Schupp and Fuentes, 1995; Clark et al., 1999; Nathan and Muller-Landau, 2000). Therefore, emphasis on the demographic changes during these crucial life-history stages is essential (Qian et al., 2016). Unfortunately, only individuals with DBH ≥ 5cm are measured and those with DBH < 5cm are simply counted as saplings or seedlings in conventional methods (Yunnan Forestry Department, 2012). This introduces difficulties in analyzing population structure and dynamics, suggesting that the method of resource inventory requires improvement in accordance with population demography, so as to satisfy PSESP conservation.

Internationally, there is also a lack of a resource inventory prior to planning actions. Technical manuals on species conservation emphasize the process of generating the SCS, whilst a resource inventory is mentioned in passing as a "status review" in the preparation period (IUCN, 2008a). The SCS formulated might be generic without fully understanding the current status of a focal species. For this reason, SCSs attracted widespread admiration, but their relevance was often not clear to practical conservation programs (IUCN, 2008b). A resource inventory carried out with an improved methodology to reflect demographic changes should be an essential component of the SCS.

5.3 Area of a mini-reserve

A mini-reserve represents *in situ* conservation of natural populations of PSESPs, in particular those outside nature reserves. However, questions pertaining to mini-reserve establishment remain unanswered, of which the most crucial one is how the scale of the mini-reserve is determined.

It is not the case that the mini-reserve with bigger area is better. For range-restricted species, in particular those that occur in an area subject to intensive human activities, an overly large mini-reserve area will increase costs and lead to wastage of limited conservation resources. Plant micro-reserve, as an accepted model for the effective protection of the endemic, rare and threatened flora in the Mediterranean regions of Europe, is defined with an area of up to 20 ha (Laguna et al., 2004; Kargiolaki et al., 2007; Laguna 2014). There is, however, a lack of proper method to determine the area of a mini-reserve. We suggest using two parameters to determine the area in a simple manner: the area of the focal natural population and the protection goal. For instance, if the area of the focal population is 5 ha and the goal is to triple the size of the focal population, then the projected area of the mini-reserve is 15 ha.

More rationally, three parameters, i.e., number of individuals at genetic saturation point, population density, and protection goal, should be taken into account simultaneously. The number of individuals at genetic saturation point is obtained through genetic diversity analysis. A total of 64 *N. yunnanensis* accessions, including eight wild individuals and their progenies, were analyzed using inter-simple sequence repeat markers (Xiang et al., 2015). The results showed that Shannon's information index attained 95.4% of overall diversity when random samples comprised 24 individuals, from which no significant change in genetic diversity was observed as the number of samples increased (Fig. 3). The population density is obtained from a field survey and GIS-based area measurement. This approach yielded six mature individuals per hectare in natural populations of *N. yunnanensis*. The protection goal is set in accordance with the investment available and planned actions, such as reinforcement and habitat restoration. In the case of *N. yunnanensis*, it was supposed to have tripled the number of individuals at genetic saturation point, so that a total of 72 individuals (24 × 3) should be contained in the mini-reserve. Furthermore, to maintain this many individuals at a natural density requires 12 ha (72 ÷ 6) of analogous habitats, which is the area of the *N. yunnanensis* mini-reserve.

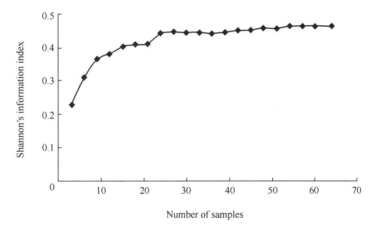

Figure 3 Number of individuals at genetic saturation point for *Nyssa yunnanensis*
The "S" on the genetic diversity curve is the genetic saturation point. The corresponding value on the x-axis, 24 individuals, is the number of individuals at genetic saturation point.

A 50 ha mini-reserve was established for *N. yunnanensis* because it occurs in natural forests subject to fewer human disturbances. As all remnant individuals are located in the bottom of the valley, the effective area of the mini-reserve is limited to the core zone that covers an area of 10 ha. There is a 2-ha deficit to attaining the goal of tripling the number of individuals in the core zone. Therefore, good management of the buffer zone is required to make up for the deficit, through restoration of habitats for colonization by *N. yunnanensis*.

5.4 *Near situ* conservation

Near situ conservation is a novel approach initiated specifically for PSESPs that require exacting habitats and environment. Based on studies of identically sourced plants grown in different botanic gardens, the *near situ* conservation approach, whereby growing and preserving focal species in contiguous locations, such as a botanic garden, arboretum, forest farm, village margin, domestic garden, roadside and beside water bodies, was originally advocated to highlight climatic similarity in plant conservation (Guo, 2011). However, its fatal limitation is that humanized site conditions may result in domestication risks, especially when measures to maintain cultivated plants and facilitate their growth are encouraged. The concept of *quasi in situ* conservation encompassed *ex situ* collections maintained in a natural or semi-natural environment to preserve neutral and adaptive genetic diversities (Volis and Blecher, 2010; Heywood, 2015), subsequently, near situ conservation was redefined as the growth and preservation of focal species with viable populations in natural or semi-natural areas with a similar climate, habitat (in particular soil and water) and community to its native environments (Sun and Yang, 2013; Xu and Guo, 2014). There is, however, no explicit requirement for the maintenance and management of *near situ* populations.

The approach of *near situ* conservation is promoted through practices of *N. yunnanensis* conservation. First, both neutral and adaptive genetic diversities have been taken into account for *N. yunnanensis* conservation. Whilst varied site conditions from natural to semi-natural environments within a range of 1-30 km from the remnant populations were selected for preservation and examination of ecological adaptability, adequate neutral genetic diversity was maintained in the newly established populations through using saplings propagated from all remnant individuals. It is anticipated that newly cultivated plants will form the basis of viable populations in the wild.

Second, different management measures performed at three *near situ* sites, i.e., 3-year intensive management at Houziqing and close-to-nature management at Guanping and Caiyanghe, would help to identify requirements for maintenance of *near situ* populations. This would improve the current *near situ* approach if the tradeoff between the failure of natural occupancy and the risk of domestication of newly established populations is attained through proper human assistance.

Third, *near situ* conservation may supplement the portfolio of techniques for plant species conservation. Prior to promulgation of *near situ* conservation, reintroduction and assisted colonization seemingly covered the full range from within to outside indigenous areas for intentional movement and release of an organism (IUCN, 2013). Nevertheless, it is difficult to distinguish reintroduction and assisted colonization from one another in contiguous areas, because in reality the boundary of the range of disappearing PSESPs are changing and cannot be accurately defined. Hence, we suggest that the term *near situ* conservation is used for establishment of new viable population in contiguous areas with a similar climate, habitat and community.

6 Conclusion

An integrated strategy for PSESP conservation has been gradually developed through implementation of pilot projects on *N. yunnanensis* conservation. The strategy, centered on remnant populations, aims to improve population viability, increase population number and expand the species distribution range. Actions are taken in the sequence of resource inventory, formulation of a CAP, remnant population protection via a mini-reserve, seedling propagation, natural population enhancement via reinforcement, new population establishmentvia *near situ* conservation, and *ex situ* germplasm preservation. Implementation of the strategy has achieved secure protection of natural populations, development of propagation technologies and production of vigorous seedlings, and establishment of three new populations and four *ex situ* germplasm collections of *N. yunnanensis*. The conservation outlook for *N. yunnanensis* has been greatly improved, and the strategy developed may be a valuable reference for conservation of other PSESPs.

Acknowledgements

The work was continuously supported by the State Forestry Administration of China (2012YB1001, 2014YB1004, 2015YB1021, 2015 – LY – 193) and Yunnan Forestry Department (201003005, 2013YB1005, 2015SX1001, 2016SX1018). Studies to unveil the endangerment causes of *N. yunnanensis* are mainly funded by the National Natural Science Foundation of China (NSFC – 31460119 and NSFC – 31660164) and Yunnan Science and Technology Department (2013FD075 and 2016FD097). We are grateful to three anonymous reviewers for their comments.

References

Chen W, Shi FQ, Yang WZ, Zhou Y, Chen HW (2011) Population status and ecological characteristics of *Nyssa yunnanensis*. J Northwest For Univ39(9):17~19, 61.

Clark JS, Beckage B, Camill P, Cleveland B, HilleRisLambers J, Lichter J, McLachlan J, Mohan J, Wyckoff P (1999) Interpreting recruitment limitation in forests. Amer J Bot 86: 1~16.

Environment Protection Commission of the State Council of China (1984) The protected rare and endangered plants of China (I). Environment Protection Commission of the State Council of China, Beijing

Falk DA, Millar CI, Olwell M (1996) Restoring diversity: strategies for reintroduction of endangered plants. Island Press, Washington, DC, pp xx.

Gu YC (2003) Status quo of China's state priority protected wild plants. Centr South For Inven and Plann 22(4): 1~7.

Guo HJ (2011) Protecting wild plant with extremely small populations at the front of houses. China Green Times, Beijing, March 10, pp 4.

Guo HJ (2012) Initiating a new phase to conserve wild plant with extremely small populations. Yunnan For33(6): 16~17.

Hansen AJ, Neilson RP, Dale VH, Flather CH, Iverson LR, Currie DJ, Shafer S, Cook R, Bartlein PJ (2001) Global change in forests: responses of species, communities, and biomes: interactions between climate change and land use are projected to cause large shifts in biodiversity. Bioscience 51:765~779.

Heywood VH (2015) In situ conservation of plant species – an unattainable goal? Israel J Plant Sci, DOI: 10. 1080/07929978. 2015. 1035605

International Union for Conservation of Nature (IUCN) (2008a) Strategic planning for species conservation: a handbook (version 1.0). IUCN Species Survival Commission, Gland, Switzerland, pp 104.

International Union for Conservation of Nature (IUCN) (2008b) Strategic planning for species conservation: an overview (version 1.0). IUCN Species Survival Commission, Gland, Switzerland, pp 22.

International Union for Conservation of Nature (IUCN) (2012)IUCN red list categories and criteria (version 3.1, 2nd edn). Gland, Switzerland and Cambridge, UK, pp 32.

International Union for Conservation of Nature (IUCN) (2013) Guidelines for reintroductions and other conservation translocations (version 1.0). IUCN Species Survival Commission, Gland, Switzerland, pp 57.

Kargiolaki H, Thanos CA, Fournaraki C, Maria EA, Karpathaki H (2007) Plant micro-reserves (a pilot project implemented in western Crete) & Samaria Biosphere Reserve. In: Guziova Z comp. Proceedings of the Conference on Priorities for Conservation of Biodiversity in Biosphere Reserves in Changing Conditions (2-6 June), Bratislava, Slovakia, pp 17~23.

LagunaE, DeltoroVI, Pèrez-BotellaJ, Pèrez-RoviraP, Serra L, OlivaresA, FabregatC (2004) The role of small reserves in plant conservation in a region of high diversity in eastern Spain. Biol Conserv 119(3):421~426.

Laguna E (2014) Origin, concept and evolution of plant micro-reserves: the pilot network of the Valencian Community (Spain). In Vladimirov V eds. A pilot network of small protected sites for conservation of rare plants in Bulgaria. Institute of Biodiversity and Ecosystem Research, Bulgarian Academy of Sciences and Ministry of Environment and Water ofBulgaria, Sofia, pp 14~24.

Li YY, SiMa YK, Fang B, Guo LQ, Jiang H, Zhao WS (2003a) Current situation and evaluation of natural resources of the priority protection wild plants in Yunnan Province of China. Acta Bot Yunnan25(2):181~191.

Li YY, Zhao WS, Chai Y, Fang B, Jiang H, Yang WZ (2003b) National key protected wild plants in Yunnan Province. Yunnan Univ Press, Kunming, China, pp 265~266.

Ma YP, Chen G, Grumbine RE, Dao ZL, Sun WB, Guo HJ (2013) Conserving plant species with extremely small populations (PSESP) in China. Biodivers Conserv 22:803~809.

Nathan R, Muller-Landau HC (2000) Spatial patterns of seed dispersal, their determinants and consequences for recruitment. Trend Ecol Evolut 15:278~285.

Qian SH, Yang YC, Tang CQ, Momohara A, Yi SR, Ohsawa M (2016) Effective conservation measures are needed for wild *Cathaya argyrophylla* populations in China: insights from the population structure and regeneration characteristics. Forest Ecol Manage 361:358~367.

Qiu Q, Yang DJ, Wang L, Zhong P, Zhang SX (2013) Test on cutting propagation of *Nyssa yunnanensis*. J West China For Sci 42(5):105~108.

Ren H, Zhang QM, Lu HF, Liu HX, Guo QF, Wang J, Jian SG, Bao HO (2012) Wild plant species with extremely small populations require conservation and reintroduction in China. Ambio 41:913~917.

Schupp EW, Fuentes M (1995) Spatial patterns of seed dispersal and the unification of plant population ecology. Ecoscience2:267~275.

State Environment Protection Bureau of China, Chinese Academy of Sciences (1987) A list of protected rare and endangered plants of China. Bull Biol 7:23~28.

State Forestry Administration of China (2012) National conservation program for plant species with extremely small populations in China. State Forestry Administration of China, Beijing.

Sun BL (2008) Taxonomy ofNyssa L. (Nyssaceae) and conservation biology of *Nyssa yunnanensis*, a critically endangered species (Ph. D. Thesis). Kunming Botany Institute of Chinese Academy of Sciences, Kunming, pp 48~50.

Sun BL, Zhang CQ (2007) A revised description of *Nyssa yunnanensis* (Nyssaceae). Acta Bot Yunnan29:73~175.

Sun BL, Zhang CQ, Porter PL, Wen J (2009) Cryptic dioecy in *Nyssa yunnanensis* (Nyssaceae), a critically endangered species from tropical eastern Asia. Ann Mo Bot Gard96:672~684.

Sun WB, Yang WZ (2013) Conservation of wild plant species with extremely small populations in Yunnan: practice & exploration. Yunnan Science and Technology Press, Kunming, pp 8~94.

Volis S (2016) How to conserve threatened China's wild plants with extremely small populations? Plant Divers1:

53~62.

Volis S, Blecher M (2010) Quasi *in situ*: a bridge between *ex situ* and *in situ* conservation of plants. Biodivers Conserv 19:2441~2454.

Wu XQ, Huang BL, Ding YL (2004) The advance on the study of protection of rare and endangered plants in China. J Nanjing For Univ (Nat Sci Edn)28(2):72~76.

Xiang ZY, Zhang SS, Yang WZ, Kang HM (2015) Conservation of *Nyssa yunnanensis* based on genetic diversity analysis using ISSR markers. J Plant Genet Resour 16(3):664~668.

Xu ZF, Guo HJ (2014) *Near situ* conservation for wild plant species with extremely small populations. Plant Divers and Resour36:533~536.

Yang WZ, Kang HM, Xiang ZY, Zhang SS (2014) Methods and techniques for conserving wild plant species with extremely small populations. J West China For Sci43(5):24~29.

Yang WZ, Li YJ, Yu CY (2011a) Mini-reserve establishment plan for *Nyssa yunnanensis*, a species with extremely small population (2011-2015). Key Laboratory of Rare and Endangered Forest Plant of State Forestry Administration, Kunming, pp 1~34.

Yang WZ, Li YJ, Zhang SS, Yu CY, Kang HM, Shi FQ, Chen Y, Zhang KF (2016) Mini-reserve of *Nyssa yunnanensis*: the first practice of mini-reserve construction for plant species with extremely small populations (PSESP) in China. J West China For Sci45(3):149~154.

Yang WZ, Xiang ZY, Zhang SS, Kang HM, Shi FQ (2015) Plant species with extremely small populations (PSESP) and their significance in China's national plant conservation strategy. Biodivers Sci 23:419~425.

Yang WZ, Zhang SS, Xiang ZY (2011b) Conservation action plan for *Nyssa yunnanensis*, a species with extremely small population (2011-2020). Key Laboratory of Rare and Endangered Forest Plant of State Forestry Administration, Kunming, pp 1~20.

Yang WZ, Zhou Y, Jiang H, Xiang ZY, Zhang SS (2010) Resource inventory and population analysis of *Nyssa yunnanensis*. Key Laboratory of Rare and Endangered Forest Plant of State Forestry Administration, Kunming, pp 1~29.

Yu YF (1999) A list of national key protected wild plant (I) released: a milestone of China's wild plant conservation. J Plant5:3~11.

Yuan RL, Xiang ZY, Yang WZ, Zhang SS (2013) Seed dormancy and germination traits of *Nyssa Yunnanensis*. Forest Res 26:384~388.

Yunnan Forestry Department (2012) Technical guidance for the second inventory of Yunnan's key protected wild plant resources. Yunnan Forestry Department, Kunming, pp 3~17.

Yunnan Forestry Department, Yunnan Science and Technology Department (2009a) Planning outline of rescuing and conserving Yunnan's plant species with extremely small populations (2010-2020). Yunnan Forestry Department, Yunnan Science and Technology Department, Kunming, pp 4.

Yunnan Forestry Department, Yunnan Science and Technology Department (2009b)Emergency action plan for rescuing and conserving Yunnan's plant species with extremely small populations (2010-2015). Yunnan Forestry Department, Yunnan Science and Technology Department, Kunming, pp 1~51.

Zhang SS, Shi FQ, Yang WZ, Xiang ZY, Kang HM, Duan ZL (2015) Autotoxicity as a cause for natural regeneration failure in *Nyssa yunnanensis* and its implications for conservation, Israel J Plant Sci 62(3):187~197.

以中国代表性物种云南蓝果树为例的极小种群野生植物保护策略研究

摘要: 极小种群野生植物物种保护是目前中国野生植物保护的重点。极小种群植物保护不同于以往的珍稀濒危植物和国家重点保护植物保护,其保护策略需要在此基础上改进。关于云南蓝果树的综合保护策略于 2009 年被提出(2009–2015)。在此,我们回顾了保护行动计划实施的过程:资源调查、保护计划行动计划制定、就地保护、种苗繁育、近地保护、迁地保护和科学研究。文章详细介绍了每个保护行动的重点和技术要求,在此基础上进一步分析了如何保护极小种群野生植物。在制定保护行动计划之前,建议先进行详细的资源调查,以了解此物种的种群动态和和当地居民的参与情况。基于云南蓝果树就地保护,我们提出了保护小区建立的技术要求。近就地保护是一种新型的保护方法,通常在目的物种的邻近区内,选择与天然种群分布点地貌、小气候、土壤、水文、生物等生境因子相似的自然或半自然地段作为近地保护点;同时介绍了种苗培育、整地和早期管护的方法。对云南蓝果树开展的基于种群保护的物种保护策略,可为其他极小种群野生植物的拯救保护提供示范,同时带动其他野生植物的保护。

[本文发表于 *Biodivers Conserv*, 2017, 26(4)]

第四篇　林木遗传育种研究

5个乡土绿化阔叶树种不同育苗基质的当年育苗效应试验

刘云彩

(云南省林业科学院,云南　昆明　650201)

摘要:在昆明树木园所进行的连香树、多果槭、榉树、枫杨和川滇桤木等5个乡土绿化阔叶树种不同育苗基质的当年育苗效应试验的结果表明:在其当年生苗木的苗高、地径和冠幅3个性状上,与对照红壤育苗基质相比,不同育苗基质对此5个乡土绿化阔叶树种当年生苗木的苗高、地径和冠幅生长的作用不一,红壤4+福贝菌2育苗基质(处理A)和(复合肥2kg+钙镁磷肥2kg)/100kg红壤育苗基质(处理C)对5个乡土绿化阔叶树种当年生苗木的生长均具有促进作用,对连香树、多果槭、榉树和川滇桤木等4个阔叶树种的当年生苗木生长具较佳促进作用的育苗基质是处理A,其连香树当年生苗木的苗高、地径和冠幅分别比对照高34.04%、15.00%和15.38%;多果槭当年生苗木的苗高、地径和冠幅分别比对照高70.55%、23.81%和39.18%;榉树当年生苗木的苗高、地径和冠幅分别比对照高31.93%、35.33%和33.33%;川滇桤木当年生苗木的苗高、地径和冠幅分别比对照高101.65%、38.27%和50.63%。而促进枫杨当年生苗木生长的较佳育苗基质是处理C,其当年生苗木的苗高、地径和冠幅分别比对照提高44.30%、23.55%和30.93%。

关键词:阔叶树种;育苗基质;当年育苗效应

Seedling Medium Experiment for Five Broad-leaved Tree Species

Liu Yun-cai

(Iunnan Academy of Forestry, Kunming Iunnan 650201)

Abstract: The experiments of nursery medium of five broad-leaved tree species namely Cercidiphyllum japonicum, Acer prolificum, Zelkova schneideriana, Pterocaryastenoptera and Alnuscremastogynecv. yanshan, Cercidiphyllum japonicum, were carried out in Kunming arboretum. The experimental results showed that compared with the control (red soil as medium), tested medium formulation had different effects on the performance of seedling on height, basal diameter and crown diameter. Treatment A (red soil and microbial strain Faby of the proportion of 4 to 2) and treatment C (compound fertilizer 2kg + calcium magnesium phosphate fertilizer 2kg) /100 kg red soil can promote the growth of tested seedlings. Formulation A is the optimum medium for Cercidiphyllum japonicum, Acer prolificum, Zelkova schneideriana and Alnuscremastogynecv. yanshan, Cercidiphyllum japonicum grew in medium A had the height, basal diameter and crown diameter 34.04%, 15.00% and 15.38% higher than the control respectively. And that of Acer prolificum in medium A were 70.55%, 23.81% and 39.18% higher than the control. The values for Zelkova schneideriana were 31.93%, 35.33% and 33.33% in order, and 101.65%, 38.27% and 50.63% respectively for Alnuscremastogynecv. yanshan. The optimum formulation for Pterocaryastenoptera was treatment C, the seedling height, basal diameter and crown diameter were 44.30%, 23.55% and 30.93% higher than the control.

Keyword:broad-leaved tree species; seedling raising; medium formulation

为增强云南乡土绿化造林树种的产苗性能，2008年在云南省林业科学院的昆明树木园进行了连香树（*Cercidiphyllum japonicum*）、多果槭（*Acer prolificum*）、榉树（*Zelkova schneideriana*）、枫杨（*Pterocarya stenoptera*）和川滇桤木（*Alnus cremastogyne cv. yanshan*）5个乡土绿化阔叶树种不同育苗基质的当年育苗效应试验，现将试验情况报道于后。

1 试验方法

连香树、多果槭、榉树、枫杨和川滇桤木5个乡土绿化阔叶树种不同育苗基质的当年育苗效应试验在云南省林业科学院的昆明大马山树木园的苗圃内进行[1~9]。其地理位置为102°45′E，25°09′N，海拔高度1 950 m，属北亚热带半湿润高原季风气候类型。

试验采用单因素随机区组试验设计，设置了红壤与不同肥料配比的4种育苗基质，以不配比肥料的红壤作为对照育苗基质共5个处理，3次重复，每一重复的小区面积为1 m²。其参试的各育苗基质处理为:A处理（红壤4+福贝菌2），B处理（红壤4+福贝菌1），C处理（复合肥2 kg+钙镁磷肥2 kg）/100 kg红壤，D处理（复合肥1 kg+钙镁磷肥1 kg）/100 kg红壤，E处理（红壤即对照）。2008年4月初对连香树、多果槭、榉树、枫杨和川滇桤木的种子做消毒处理，冷水浸种后播种进行苗床育苗，待其苗高达3 cm左右时，于6月移入装好不同育苗基质的塑料育苗袋中作容器苗继续培育。其容器袋的规格为10 cm×14 cm。在其育苗试验地上搭建荫棚，上覆70%的遮阴网。2008年年末，当苗木生长停止后，每个小区机械抽取30株苗木观测苗高、地径和冠幅，计算出各小区当年生苗木的苗高、地径和冠幅平均值，以此进行5个阔叶树种不同育苗基质的当年育苗效应评估。

2 试验结果与分析

2.1 不同育苗基质的当年生连香树苗木的生长效应

不同育苗基质的当年生连香树苗木生长状况及各生长性状的方差分析结果见表1和表2。由表可见:不同育苗基质处理间当年生连香树苗木的生长差异十分明显，在苗高、地径和冠幅3个性状上，各重复间差异均不显著，而各处理间差异均达极显著水平，表明育苗基质对当年生连香树苗木生长的促进作用十分明显。

不同育苗基质处理对当年生连香树苗木的高、径和冠幅生长的促进作用不一，处理A、C和D对其当年生苗木的生长具有促进作用，处理B作用不明显，其当年生苗木的高和冠幅的生长不如对照。各参试育苗基质对连香树当年生苗木的苗高生长的促进作用依次为C>A>D>E>B，对地径生长的促进作用依次为A>C>D>E≥B，对苗木冠幅生长的促进作用依次为A>D>C>E>B。与对照红壤育苗基质相比较，在几种育苗基质处理中，以处理A和处理C促进苗木生长的效果较好，处理A对连香树当年生苗木的地径和苗木冠幅的生长有较好的促进作用,处理 C 对连香树当年生苗木高生长的促进作用较好。

表1 不同育苗基质处理的当年生连香树苗木生长状况

Table 1 Growth performance of *Cercidiphyllum japonicum* of different medium formulations

生长性状	处理	A	B	C	D	E(对照)
苗高	重复Ⅰ(cm)	25.0	18.5	26.7	24.3	20.2
	重复Ⅱ(cm)	23.7	20.2	32.2	24.2	17.3
	重复Ⅲ(cm)	26.8	16.7	24.6	22.6	18.8
	均值(cm)	25.2	18.5	27.8	23.7	18.8
	>对照值(%)	34.04	-1.60	47.87	26.06	
地径	重复Ⅰ(cm)	0.332	0.304	0.317	0.308	0.303
	重复Ⅱ(cm)	0.344	0.311	0.319	0.295	0.298
	重复Ⅲ(cm)	0.360	0.293	0.325	0.305	0.298
	均值(cm)	0.345	0.303	0.320	0.303	0.300
	>对照值(%)	15.00	1.00	6.67	1.00	
冠幅	重复Ⅰ(cm)	11.6	10.3	11.2	11.8	10.3
	重复Ⅱ(cm)	11.9	9.6	11.3	11.5	10.1
	重复Ⅲ(cm)	12.4	9.2	11.1	11.4	10.9
	均值(cm)	12.0	9.7	11.2	11.6	10.4
	>对照值(%)	15.38	-6.73	7.69	11.54	

表2 不同处理的连香树当年生苗木各生长性状的方差分析

Table 2 Variance analysis on growth performance of *Cercidiphyllum japonicum* under different treatments

生长性状	SU	SS	df	MS	F
苗高	重复	6.737 3	2	3.368 7	0.656 2
	处理	200.370 7	4	50.092 7	9.757 7**
	误差	41.069 3	8	5.133 7	
	总计	248.177 3	14		
地径	重复	0.000 03	2	0.000 02	0.196 5
	处理	0.004 45	4	0.001 11	13.282 8**
	误差	0.000 67	8	0.000 08	
	总计	0.005 16	14		
冠幅	重复	0.069 3	2	0.034 7	0.208 4
	处理	9.909 3	4	2.477 3	14.893 8**
	误差	1.330 7	8	0.166 3	
	总计	11.309 3	14		

注:**表示在0.01水平上差异显著。

2.2 不同育苗基质的当年生多果槭苗的生长效应

不同育苗基质的当年生多果槭苗木生长状况及各生长性状的方差分析结果见表3和表

4。由表可见：不同育苗基质处理间当年生多果槭苗木的生长差异十分明显，在苗高、地径和苗木冠幅3个生长性状上，各重复间差异均不显著，而在苗高和苗木冠幅2个性状上各处理间差异均极显著，在地径上各处理间差异显著，表明不同育苗基质对当年生多果槭苗木生长的促进作用明显。

表3 不同育苗基质处理的当年生多果槭苗木生长状况
Table 3 Growth performance of *Acer prolificum* of different medium formulations

生长性状	处理	A	B	C	D	E（对照）
苗高	重复Ⅰ(cm)	26.9	21.3	22.5	13.9	13.5
	重复Ⅱ(cm)	25.9	25.2	24.8	17.3	13.4
	重复Ⅲ(cm)	21.8	24.6	23.9	17.4	16.8
	均值(cm)	24.9	23.7	23.7	16.2	14.6
	>对照值(%)	70.55	62.33	62.33	10.96	
地径	重复Ⅰ(cm)	0.435	0.345	0.353	0.297	0.314
	重复Ⅱ(cm)	0.393	0.349	0.344	0.325	0.300
	重复Ⅲ(cm)	0.342	0.356	0.347	0.287	0.331
	均值(cm)	0.390	0.350	0.348	0.303	0.315
	>对照值(%)	23.81	11.11	10.48	-3.81	
冠幅	重复Ⅰ(cm)	14.2	12.8	12.9	9.9	9.6
	重复Ⅱ(cm)	13.4	13.6	11.9	11.3	9.0
	重复Ⅲ(cm)	12.8	13.3	12.7	10.9	10.6
	均值(cm)	13.5	13.2	12.5	10.7	9.7
	>对照值(%)	39.18	36.08	28.87	10.31	

表4 不同处理的多果槭当年生苗木各生长性状的方差分析
Table 4 Variance analysis on growth performance of *Acer prolificum* under different treatments

生长性状	苗高				地径				冠幅			
	重复	处理	误差	总计	重复	处理	误差	总计	重复	处理	误差	总计
SS	7.841 3	280.177 3	33.698 7	321.717 3	0.000 7	0.014 0	0.005 0	0.019 7	0.137 3	32.169 3	4.082 7	36.389 3
df	2	4	8	14	2	4	8	14	2	4	8	14
MS	3.920 7	70.044 3	4.212 3		0.000 3	0.003 5	0.000 6		0.068 7	8.042 3	0.510 3	
F	0.930 8	16.628 4**			0.527 0	5.540 6*			0.134 6	15.759 0**		

注：**表示在0.01水平上差异显著，*表示在0.05水平上差异显著。

参试的几种育苗基质处理对当年生多果槭苗木的生长一般都有促进作用（除处理D的苗木地径生长小于对照外），其中以处理A、B和C对其当年生苗木生长的促进作用较明显，其对苗木苗高生长的促进作用依次为A>B≥C>D>E，对地径的生长促进作用依次为A>B>C>E>D，对苗木冠幅生长的促进作用依次为A>B>C>D>E。与对照红壤育苗基质相比较,在几种育苗基质处理中,以处理 A 对当年生多果榄苗木生长的促进作用最大,其次是处理 B 和处理 C。

2.3 不同育苗基质的当年生枫杨苗的生长效应

不同育苗基质的当年生枫杨苗木的生长状况及各生长性状的方差分析结果见表 5 和表 6。由表可见:不同育苗基质处理间当年生枫杨苗木的生长差异十分明显,在苗高、地径和冠幅 3 个生长性状上,各重复间差异均不显著,在苗高上各处理间差异达极显著水平,而在地径和冠幅 2 个性状上各处理间的差异显著,表明不同育苗基质处理对当年生枫杨苗木生长的促进作用明显。

表 5 不同育苗基质处理的当年生枫杨苗木生长状况

Table 5 Growth performance of *Pterocaryas tenoptera* of different medium formulations

生长性状	处理	A	B	C	D	E(对照)
苗高	重复Ⅰ(cm)	19.5	15.6	20.4	15.7	14.6
	重复Ⅱ(cm)	16.5	15.0	21.5	17.3	14.9
	重复Ⅲ(cm)	16.1	12.8	22.5	16.9	15.1
	均值(cm)	17.4	14.5	21.5	16.6	14.9
	>对照值(%)	16.78	-2.68	44.30	11.41	
地径	重复Ⅰ(cm)	0.356	0.348	0.337	0.308	0.295
	重复Ⅱ(cm)	0.328	0.317	0.383	0.329	0.273
	重复Ⅲ(cm)	0.368	0.334	0.367	0.338	0.31
	均值(cm)	0.351	0.333	0.362	0.325	0.293
	>对照值(%)	19.80	13.65	23.55	10.92	
冠幅	重复Ⅰ(cm)	12.1	8.7	11.6	9.5	9.2
	重复Ⅱ(cm)	10.0	9.9	12.4	11.0	10
	重复Ⅲ(cm)	10.4	9.4	14.2	11.3	9.9
	均值(cm)	10.8	9.3	12.7	10.6	9.7
	>对照值(%)	11.34	-4.12	30.93	9.28	

表 6 不同处理的枫杨当年生苗木各生长性状的方差分析

Table 6 Variance analysis on growth performance of *Pterocarya stenoptera* under different treatments

生长性状	SU	SS	df	MS	F
苗高	重复	0.624 0	2	0.312 0	0.173 9
	处理	93.542 7	4	23.385 7	13.037 9**
	误差	14.349 3	8	1.793 7	
	总计	108.516 0	14		
地径	重复	0.000 9	2	0.000 4	1.288 9
	处理	0.008 6	4	0.002 1	6.340 9*
	误差	0.002 7	8	0.000 3	
	总计	0.012 2	14		

(续)

生长性状	SU	SS	df	MS	F
冠幅	重复	1.684 0	2	0.842 0	0.920 7
	处理	21.036 0	4	5.259 0	5.750 7*
	误差	7.316 0	8	0.914 5	
	总计	30.036 0	14		

注：** 表示在 0.01 水平上差异显著，* 表示在 0.05 水平上差异显著。

除处理 B 外，其他育苗基质处理对当年生枫杨苗木的生长都具有促进作用。对当年生枫杨苗木的苗高和冠幅生长具促进作用的育苗基质处理依次为 C>A>D>E>B，对地径生长具促进作用的育苗基质处理依次为 C>A>B>D>E。与对照红壤育苗基质相比较，在几种育苗基质处理中，以处理 C 的促进当年生枫杨苗木生长的效果最好，其次是处理 A 和处理 D。

2.4 不同育苗基质的当年生榉树苗的生长效应

不同育苗基质的当年生榉树苗木生长状况及各生长性状的方差分析结果见表 7 和表 8，由表可见：不同育苗基质处理间当年生榉树苗木的生长差异十分明显，在苗高、地径和冠幅 3 个生长性状上，各重复间差异均不显著，在苗高和冠幅 2 个性状上各处理间差异均显著，在地径上各处理间差异极显著，表明不同育苗基质对当年生枫杨苗木生长的促进作用明显。

参试的几种育苗基质处理对当年生榉树苗木的生长都具有促进作用，对其苗高生长具促进作用的育苗基质处理依次为 A>C>B>D>E，对其苗木地径生长有促进作用的育苗基质处理依次为 A>B>C>D>E，对其苗木冠幅生长有促进作用的育苗基质处理依次为 A>C>D>B>E。与对照红壤育苗基质相比，在几种育苗基质处理中，对榉树当年生苗木的苗高、地径、冠幅生长具促进作用的以处理 A 的效果最好，其次是处理 C 和处理 B。

表 7 不同育苗基质处理的当年生榉树苗木生长状况
Table 7 Growth performance of *Zelkova schneideriana* of different medium formulations

生长性状	处理	A	B	C	D	E(对照)
苗高	重复Ⅰ(cm)	21.3	20.4	20.0	18.2	18.3
	重复Ⅱ(cm)	20.3	19.0	22.8	18.0	16.4
	重复Ⅲ(cm)	24.0	22.5	21.9	20.3	15.0
	均值(cm)	21.9	20.6	21.6	18.8	16.6
	>对照值(%)	31.93	24.10	30.12	13.25	
地径	重复Ⅰ(cm)	0.250	0.241	0.221	0.176	0.180
	重复Ⅱ(cm)	0.220	0.228	0.238	0.180	0.193
	重复Ⅲ(cm)	0.276	0.274	0.227	0.213	0.178
	均值(cm)	0.249	0.248	0.229	0.190	0.184
	>对照值(%)	35.33	34.78	24.46	3.26	

生长性状	处理	A	B	C	D	E(对照)
冠幅	重复Ⅰ(cm)	9.0	7.7	7.9	7.0	7.2
	重复Ⅱ(cm)	8.7	7.3	9.1	7.1	6.3
	重复Ⅲ(cm)	10.0	7.5	9.1	9.3	7.3
	均值(cm)	9.2	7.5	8.7	7.8	6.9
	>对照值(%)	33.33	8.70	26.09	13.04	

表8 不同处理的榉树当年生苗木各生长性状的方差分析

Table 8 Variance analysis on growth performance of Zelkova schneideriana under different treatments

生长性状	SU	SS	df	MS	F
苗高	重复	5.665 3	2	2.832 7	1.095 4
	处理	58.296 0	4	14.574 0	5.635 7*
	误差	20.688 0	8	2.586 0	
	总计	84.649 3	14		
地径	重复	0.001 5	2	0.000 7	2.505 8
	处理	0.011 7	4	0.002 9	10.017 1**
	误差	0.002 3	8	0.000 3	
	总计	0.015 5	14		
冠幅	重复	2.769 3	2	1.384 7	3.479 12
	处理	10.300 0	4	2.575 0	6.469 8*
	误差	3.184 0	8	0.398 0	
	总计	16.253 3	14		

注:** 表示在0.01水平上差异显著,* 表示在0.05水平上差异显著。

2.5 不同育苗基质的当年生川滇桤木苗的生长效应

不同育苗基质的当年生川滇桤木苗木的生长状况及各生长性状的方差分析结果见表9和表10。由表可见:不同育苗基质处理间当年生川滇桤木苗木的生长状况差异十分明显,在苗高、地径和冠幅3个性状上,各重复间差异均不显著,而各处理间差异均极显著,表明不同的育苗基质处理对当年生川滇桤木苗木生长的促进作用十分明显。

几种育苗基质处理对当年生川滇桤木苗木的各生长性状都具有促进作用,对其苗高和冠幅生长具有促进作用的处理依次为A>B>C>D>E,对其苗木地径生长有促进作用的处理依次为A>B>D>C>E。与对照红壤育苗基质相比,在几种育苗基质处理中,对当年生川滇桤木苗木的苗高、地径及冠幅生长的促进作用以处理 A 的效果最好,其次是处理 B。

表9 不同育苗基质处理的当年生川滇桤木苗木生长状况
Table 9 Growth performance of *Alnus cremastogyne* cv. *yanshan* of different medium formulations

生长性状	处理	A	B	C	D	E(对照)
苗高	重复Ⅰ(cm)	23.4	20.8	19.1	15.4	12.6
	重复Ⅱ(cm)	26.5	19.8	12.1	14.6	13.9
	重复Ⅲ(cm)	23.4	18.2	16.7	15.8	9.7
	均值(cm)	24.4	19.6	16.0	15.3	12.1
	>对照值(%)	101.65	61.98	32.23	26.45	
地径	重复Ⅰ(cm)	0.346	0.331	0.289	0.287	0.236
	重复Ⅱ(cm)	0.335	0.299	0.262	0.301	0.282
	重复Ⅲ(cm)	0.326	0.298	0.25	0.296	0.212
	均值(cm)	0.336	0.309	0.267	0.295	0.243
	>对照值(%)	38.27	27.16	9.88	21.40	
冠幅	重复Ⅰ(cm)	11.8	11.5	11.3	8.9	7.6
	重复Ⅱ(cm)	11.9	11.5	7.2	7.9	8.8
	重复Ⅲ(cm)	11.9	10.8	8.8	9.3	7.4
	均值(cm)	11.9	11.3	9.1	8.7	7.9
	>对照值(%)	50.63	43.04	15.19	10.13	

表10 不同处理的川滇桤木当年生苗木各生长性状的方差分析
Table 10 Variance analysis on growth performance of *Alnus cremastogyne* cv. *yanshan* under different treatments

生长性状	SU	SS	df	MS	F
苗高	重复	5.6813	2	2.8407	0.5758
	处理	268.0067	4	67.0017	13.5819**
	误差	39.4653	8	4.9332	
	总计	313.1533	14		
地径	重复	0.0014	2	0.0007	1.9026
	处理	0.0156	4	0.0039	10.5964**
	误差	0.0029	8	0.0004	
	总计	0.0199	14		
冠幅	重复	1.5773	2	0.7887	0.6654
	处理	34.8093	4	8.7023	7.3417**
	误差	9.4827	8	1.1853	
	总计	45.8693	14		

注:**表示在0.01水平上差异显著。

3　结论

（1）不同育苗基质（处理）对连香树、多果械、榉树、枫杨和川滇桤木等5个树种当年生苗木的苗高、地径、冠幅生长的方差分析结果表明，在苗高、地径和冠幅3个性状上，各重复间差异均不显著，而各处理间的差异均达极显著或显著水平，表明参试5种育苗基质（含对照红壤基质）对这5个乡土绿化阔叶树种的当年生苗木生长的促进作用十分明显。

（2）不同育苗基质对5个乡土绿化阔叶树种当年生苗木的苗高、地径和冠幅生长的促进作用不一，红壤4+福贝菌2育苗基质（处理A）和（复合肥2 kg+钙镁磷肥2 kg）/100 kg红壤育苗基质（处理C）对连香树、多果械、枫杨、榉树和川滇桤木等5个乡土绿化阔叶树种当年生苗木的苗高、地径和冠幅生长均具有促进作用；红壤4+福贝菌1育苗基质（处理B）对多果械、榉树和川滇桤木3个乡土绿化阔叶树种当年生苗木的苗高、地径和冠幅的生长具有促进作用；而（复合肥1kg+钙镁磷肥1kg）/100kg红壤育苗基质（处理D）对连香树、枫杨、榉树和川滇桤木4个阔叶树种当年生苗木的苗高、地径和冠幅生长具有促进作用。

（3）促进连香树当年生苗木生长的较佳育苗基质是红壤4+福贝菌2育苗基质（处理A），其苗高、地径和冠幅分别达25.2cm、0.345cm和12.0cm，而红壤基质（对照）的苗高、地径和冠幅分别为18.8cm、0.300cm和10.4cm，分别比对照高34.04%、15.00%和15.38%。促进多果械当年生苗木生长较佳的育苗基质亦是红壤4+福贝菌2育苗基质（处理A），其苗高、地径和冠幅分别达24.9cm、0.390cm和13.5cm，而红壤基质（对照）的苗高、地径和冠幅分别为14.6cm、0.315cm和9.7cm，分别比对照高70.55%、23.81%和39.18%。促进枫杨当年生苗木生长的较佳育苗基质是（复合肥2kg+钙镁磷肥2kg）/100kg红壤育苗基质（处理C），其苗高、地径和冠幅分别达21.5cm、0.362cm和12.7cm，而红壤基质（对照）的苗高、地径和冠幅分别为14.9cm、0.293cm和9.7cm，分别比对照高44.30%、23.55%和30.93%。促进榉树当年生苗木生长的较佳育苗基质仍是红壤4+福贝菌2育苗基质（处理A），其苗高、地径和苗木冠幅分别达21.9cm、0.249cm和9.2cm，而红壤基质（对照）的苗高、地径和冠幅分别为16.6cm、0.184cm和6.9cm，分别比对照高31.93%、35.33%和33.33%。促进川滇桤木当年生苗木生长较佳的育苗基质还是红壤4+福贝菌2育苗基质（处理A），其苗高、地径和冠幅分别达24.4cm、0.336cm和11.9cm，而红壤基质（对照）的苗高、地径和冠幅分别为12.1cm、0.243cm和7.9cm，分别比对照高101.65%、38.27%和50.63%。

参考文献：

[1] 云南树木图志编委会. 云南树木图志[M]. 昆明：云南科学技术出版社，1988.
[2] 李淑琴. 榉树的育苗方法[J]. 江苏林业科技，2000，27（3）：39~41.
[3] 徐有明，邹明宏，史玉虎，等. 枫杨的生物学特性及其资源利用的研究进展[J]. 东北林业大学学报，2002，30（3）：42~48.
[4] 中国树木志编委会. 中国主要树种造林技术[M]. 北京：中国林业出版社，1993.
[5] 赵培仙，赵悦仙，赵各琼，等. 金沙江干热河谷3种阔叶育苗试验[J]. 西南林学院学报，2006，73（3）：40~43.
[6] 吴朝斌，杨汉远，龙舞，等. 伯乐树种子育苗试验[J]. 林业实用技术，2007，62（2）：17~18.

[7] 赵秋玲,刘林英. 连香树育苗试验研究[J]. 甘肃科技,2007,23(7):224~226.
[8] 曲良谱,喻方圆,张新. 枫杨容器苗育苗技术研究[J]. 江苏林业科技,2008,35(2):9~12.
[9] 陈代喜,陈健波. 尾叶桉幼林施肥效应初步研究[J]. 广西林业科学,1995(2):169~172.

(本文发表于《西部林业科学》,2009年)

高松香思茅松无性系的选育

李思广¹, 付玉嫔¹, 张快富¹, 赵永红¹, 蒋云东¹, 姚志琼², 李明²

(1. 云南省林业科学院, 云南 昆明 650204; 2. 景谷林业股份有限公司, 云南 思茅 665002)

摘要: 对40个思茅松(*Pinus kesiya* var. *langbianensis*)高产脂无性系的松香产量进行了测定分析。结果表明: 所有无性系的松香产量均大于对照, 较对照松香产量平均提高192.4%; 无性系间松香产量存在极显著差异。无性系松香产量的遗传力为0.68。初步选择出20个高产脂的优良无性系。入选无性系松香产量的对照遗传增益为178.8%。

关键词: 思茅松; 嫁接无性系; 松香

Study on Superior Clone Selection of High-Rosin-Yield *Pinus kesiya var langbianensis*

Li Sig-uang¹ Fu Yu-pin¹ Zhang Kuai-fu¹ Zhao Yong-hong¹
Jiang Yundong¹ Yao Zhiqiong² Li Ming²

(1 Yunnan Academy of Forestry, Kunming 650204; 2 Jinggu Forestry Co. Ltd., Simao 665002)

Abstracts: Based on the measurement results of the rosin-production of 40 high-resin-yield clones and a control, it showed that the rosin production of every clone was higher than that of the control, and the average rosin production was 192.4% higher than the control. The rosin production among the clones had a significant deference. The family heritability was 0.68. 20 superior clones with high-rosin-yield were selected, and the genetic gain to the control was 178.8%.

Key words: *Pinus kesiya* var.*langbianensis*, grafting clone, rosin

思茅松(*Pinus kesiya* var. *langbianensis*)是材、脂兼用树种, 具有速生、优质、高产脂和生态适应性强等特点, 以大面积纯林或针阔叶混交林的形式集中分布于云南省思茅市的翠云、普洱、景谷、景东、镇沅、江城、墨江等县以及临沧地区和红河州的部分县。据调查, 目前云南省思茅松的林地面积约102.5万 hm², 立木蓄积量约1亿 m³, 松香年蓄积量为11.5万 t[1~2]。到2004年云南省松香产量已达7.3万 t(含小厂的松香), 在全国排第3位, 其中思茅松松香的产量占云南省松香产量的90%以上。思茅松松脂中的化学组分可分为松香和松节油两大类, 现在生产中对松脂利用最多的是生产松香。近年来, 思茅松天然林面积逐年下降, 采脂树亦随之减少, 已不能适应林产化工业的需要, 因此选择思茅松天然林中的高产脂、高松香含量的基因资源并加以繁殖利用, 营造人工高产脂、高松香的原料林已迫在眉睫, 是林业可持续发展的必然趋势。2007年, 笔者对已建立的5年生思茅松高产脂嫁接林进行了产脂力及松脂化学成分的测定, 以期从中选出高产脂、高松香产量的无性系, 进行繁殖, 应用于生产, 同时也可为下一轮回选择提供基本群体, 并为早期选择提供科学依据。

1 材料与方法

1.1 田间试验设计及试验材料

2002年,在云南省思茅市景谷县建立高产脂无性系嫁接试验林,试验采用随机区组设计,共设40个处理,6次重复,6株小区,以普通未嫁接思茅松作为对照,株行距3m×3m,试验面积为2hm²。

2007年4月份进行产脂力测定,对1~3重复每重复取1株平均木进行产脂量及松脂化学成分测定分析。采脂方式用常规的"V"形下降采脂法,割脂高度为1m左右。测沟夹角70°~90°,采割深入木质部0.3~0.4cm,割面负荷率45%~50%。每2d加割1刀,连续割4刀后测定产脂量。产脂力=产脂量/采割沟水平长。为了消除各单株之间因割沟夹角、直径大小及割面负荷率不完全一致而带来的误差,将树木的产脂力进行校正,校正公式:校正产脂力=产脂力/单株胸径。

1.2 松脂化学组分

将试管内松脂搅拌均匀,称取0.5g松脂溶于无水乙醇中,以酚酞为指示剂,用四甲基氢氧化铵-乙醇溶液滴至微红,对滴定好的松脂溶液进行气相色谱和质谱分析。

气相色谱分析仪器为美国安捷伦6890型气相色谱仪。质谱分析仪器为英国VG公司的FISONS MD 800GC/MS/DS联用仪。

GC条件:使用HP-5毛细管柱(30m×0.32mm×0.25μm),以高纯氮为载气,二阶程序升温80℃→240℃→280℃,升温速率分别为3℃·m^{-1}和2℃·m^{-1},汽化室温度290℃,检测器温度300℃,柱前压为50kPa,分流比50∶1,进样量0.3μL。

MS条件:EI-MS,电子能量70eV;离子源温度200℃;灯丝电流4.1A;质量扫描范围35~600u;扫描周期1s;数据处理采用LAB-BASE系统,使用美国国家标准局NBS谱库检索。

1.3 统计分析方法

松香产量:因为思茅松个体的松香含量与松香产量并不一定呈正比关系,所以会出现松脂中松香含量很高但松树产脂力并不高的情况,所以为了选择出真正高松香产量的无性系,本研究以松香产量作为选择标准。松香产量=校正产脂力×松脂中松香质量。

遗传方差和环境方差[3-6]: $\delta_g^2 = \frac{1}{r}(M_1 - M_2)$, $\delta_e^2 = M_2$。式中:M_1无性系均方差;M_2为环境均方差;r为重复数。

表型方差和广义遗传力(重复力)[3-6]: $\delta_P^2 = \delta_g^2 + \frac{1}{r}\delta_e^2$; $h^2 = \frac{\delta_g^2}{\delta_P^2} \times 100 = \frac{M_1 - M_2}{M_1} \times 100$。

遗传增益:$\Delta G = h^2 \cdot S/\overline{X}$。式中:$S$为选择差;$\overline{X}$为性状平均值。

实际增益:$G实 = (\overline{x}_i - \overline{x})/\overline{x}$。式中:$\overline{x}_i$为各无性系平均值;$\overline{x}$为对照平均值。

2 结果与分析

2.1 松香产量及方差分析

根据收集到的各无性系及对照的松脂产脂力,计算出其校正产脂力,并对从试验林中收集到的思茅松松脂样品进行松香、松节油等松脂特征组分的分析测定,最后计算出各重复的松香产量。结果表明,高产脂无性系的平均松香产量($2.865g \cdot cm^{-1}$)远高于对照($0.980g \cdot cm^{-1}$),约为普通思茅松产脂力2.9倍。各高产脂思茅松无性系松香产量变化极大,变幅为1.175~5.043$g \cdot cm^{-1}$。所有无性系的松香产量均大于对照的产量。

为比较无性系间松香产量的差异显著性,将松香产量调查数据,进行方差分析,分析结果见(表1)。由表1可见,无性系间松香产量差异极显著($P<0.01$),表明思茅松无性系的松香产量存在着丰富的变异,这些变异主要由遗传特性决定的,因此,定向选择具有很大的潜力。

表1 思茅松无性系松香产量的方差分析结果

变异来源	平方和	自由度	方差	F	$F_{0.01}$	方差组成
无性系间	89.218 2	39	2.287 6	3.09	1.86	$\delta_e^2 + r\delta_g^2$
无性系内	59.305 1	80	0.741 3			δ_e^2
总变异	148.523 3	119				

2.2 松香产量遗传参数

由表1根据1.3中的统计方法可以估算出松香产量的遗传参数:环境方差$\delta_e^2 = 0.741\ 3$,遗传方差$\delta_g^2 = 0.515\ 4$,表型方差$\delta_p^2 = 0.762\ 5$,无性系遗传力$h^2 = 0.68$,可见松香产量差异中遗传性因素占较大比例。说明通过一定强度的选择,松香产量能获得很高的遗传增益,但考虑到以后的遗传基础不能过于狭窄,且林分处于幼林阶段,性状没有完全稳定,故选择强度不能太大。

为了能选出最优的思茅松高松香产量无性系,可以根据各个参试无性系松香产量与对照进行选择差、实际增益和遗传增益的估算,估算结果见表2。从表2可以看出:40个参试无性系的松香产量实际增益变幅为19.9%~414.6%,总平均达192.4%;理论遗传增益的变幅为13.5%~281.9%,总平均达130.8%,其增益效果极为显著。

表2 思茅松无性系松香产量增益

序号	无性系号	松香产量 ($g \cdot cm^{-1}$)	实际增益 (%)	理论遗传增益 (%)	松香产量增益
1	238	5.043	414.6	281.9	4.063**
2	212	4.481	357.3	243.0	3.501**
3	118	4.336	342.4	232.9	3.356**
4	252	3.971	305.2	207.6	2.991**
5	122	3.931	301.2	204.8	2.951**
6	220	3.631	270.5	183.9	2.651**

(续)

序号	无性系号	松香产量 (g·cm^{-1})	实际增益 (%)	理论遗传增益 (%)	松香产量增益
7	234	3.629	270.3	183.8	2.649**
8	184	3.534	260.6	177.2	2.554**
9	134	3.519	259.1	176.2	2.539**
10	282	3.484	255.5	173.7	2.504**
11	150	3.482	255.4	173.6	2.502**
12	154	3.384	245.3	166.8	2.404**
13	288	3.382	245.1	166.7	2.402**
14	242	3.304	237.2	161.3	2.324**
15	110	3.186	225.1	153.1	2.206**
16	160	3.150	221.4	150.6	2.170**
17	162	3.046	210.8	143.3	2.066**
18	128	2.922	198.2	134.8	1.942**
19	274	2.896	195.5	132.9	1.916**
20	136	2.830	188.8	128.4	1.850**
21	222	2.802	185.9	126.4	1.822*
22	210	2.775	183.2	124.6	1.795*
23	278	2.728	178.4	121.3	1.748*
24	130	2.706	176.1	119.8	1.726*
25	266	2.704	175.9	119.6	1.724*
26	284	2.624	167.7	114.0	1.644*
27	106	2.383	143.2	97.4	1.403*
28	198	2.359	140.7	95.7	1.379
29	196	2.284	133.1	90.5	1.304
30	194	2.261	130.7	88.9	1.281
31	138	2.198	124.2	84.5	1.218
32	214	2.192	123.7	84.1	1.212
33	294	2.076	111.8	76.0	1.096
34	208	2.052	109.3	74.4	1.072
35	286	1.838	87.5	59.5	0.858
36	200	1.755	79.1	53.8	0.775
37	268	1.644	67.8	46.1	0.664
38	218	1.511	54.2	36.9	0.531
39	108	1.41	43.9	29.9	0.430
40	182	1.175	19.9	13.5	0.195
41	ck	0.980			
无性系均值		2.865	192.4	130.8	

注:松香产量增益为各无性系松香产量-对照无性系松香产量。

2.3 优良无性系的评选

无性系评选采用最小显著差数法(LSD)[7-9]。评选指标按评选公式大于标准正态 $\alpha = 0.01$ 水平时单侧临界值 t 时的 LSD 就可以认为该无性系极显著大于对照,即可入选。经计算,$L_{SD0.05} = 1.385$,$L_{SD0.01} = 1.838$,经比较有 20 个无性系与对照相比差异达到极显著水平(见表2),这 20 个无性系平均松香产量达 $3.56 g \cdot cm^{-1}$,其松香产量的平均实际增益为 263.0%,平均理论遗传增益为 178.8%,增益效果异常显著。这 20 个无性系可作为入选无性系进行推广造林,实现早期增益。

3 小结与讨论

对 5 年生的 40 个无性系及对照进行松香产量的测定,结果表明,各无性系间松香产量差异极显著。与对照相比,所有参试思茅松无性系的松香产量均大于对照;20 个入选无性系的松香产量平均为 $3.56 g \cdot cm^{-1}$,约为对照产量($0.980 g \cdot cm^{-1}$)的 3.6 倍;其松香产量的平均实际增益为 263.0%,平均理论遗传增益为 178.8%,说明这 20 个思茅松无性系的松香产量的增益效果异常显著。通过计算松香产量性状的遗传参数得出,松香产量的无性系遗传力为 0.68,表明思茅松松香产量在无性系间存在着很大的遗传差异,并且这种差异受较强的遗传控制,这种差异性为筛选优良无性系提供了科学依据。因此,可以继续扩大高产脂高松香产量的思茅松的优树选择,并建立初级无性系种子园,开展多层次的遗传改良工作,进一步选择出更高松香产量的思茅松无性系。考虑到思茅松林分林龄只有 5a 左右,尚处于幼林阶段,性状没有完全稳定,故选择强度不能太大。通过多重比较(LSD)法,选择出松香产量与对照差异达极显著水平的 20 个无性系,入选率为 50%。对于初步选择出的无性系,还需进行进一步的观测及试验,并进行多点试验,以便找出适合不同立地条件下的优良无性系,但为尽快实现高松香产量思茅松培育的目标,可用这 20 个优良无性系进行适当地推广种植。

参考文献:

[1] 陈少瑜,赵文书,王炯. 思茅松天然种群及其种子园的遗传多样性[J]. 福建林业科技,2002,29(3):1~5.

[2] 蒋云东,李思广,杨忠元,等. 土壤化学性质对思茅松人工幼林生长的影响[J]. 东北林业大学学报,2006,34(1):25~27.

[3] 朱之悌. 林木的遗传学基础[M]. 北京:中国林业出版社,1990:176~178.

[4] 王明麻. 林木遗传育种学[M]. 北京:中国林业出版社,2001:174~176.

[5] 李会平,黄大庄,杨梅生,等. 高抗光肩星天牛优良杨树无性系的选择[J]. 东北林业大学学报,2003,31(05):30~32.

[6] 刘月蓉. 高产脂马尾松半同胞的产脂力优良单株的选择[J]. 林业科技,2006(3):1~4.

[7] 李春喜,王志和,王文林. 生物统计学[M]. 北京:科学出版社. 2000:87~97.

[8] 曾令海,王以珊,阮梓材,等. 高脂马尾松产脂力和遗传稳定性分析[J]. 广东林业科技,1998(2):1~6.

[9] 许兴华,李霞,孟宪伟,等. 毛白杨优良无性系选育研究[J]. 山东林业科技,2006(2):30~32.

(本文发表于《西南林业大学学报》,2018 年)

不同育苗基质对长蕊甜菜树苗木生长的影响

李娅[1], 李恩良[2], 毛云玲[1], 周江[2], 刘永国[3],
常恩福[1], 李勇鹏[1], 景跃波[1]

(1. 云南省林业科学院,云南 昆明 650201; 2. 云南省墨江西歧桫椤省级自然保护区管理局,
云南 墨江 654800; 3. 云南 省文山州林业科学研究所,云南 文山 666300)

摘要:以咖啡壳沤制物、腐殖土及森林土为原料配制成5种基质,通过对基质的理化性状及长蕊甜菜树苗木的苗高、地径、主根长等7个生长指标的测定,分析不同基质对长蕊甜菜树苗木生长的影响。结果表明:不同育苗基质苗木的各项生长指标均存在显著($P<0.05$)或极显著($P<0.01$)差异,适宜的基质能有效促进苗木的生长及生物量的累积,苗木的生长差异是不同育苗基质理化性状综合作用的结果。4种配方基质苗木的苗高、地径分别较对照高38.2%、22.6%、1.3%、4.4%和49.0%、23.3%、27.2%、15.0%;1#、2#和3#基质苗木的主根长、根幅、一级侧根数分别较对照高10.9%~17.3%、9.2%~27.1%和34.3%~68.1%。以理想基质适宜指标为标准,基于苗木生长指标测定结果,以不同占比的咖啡壳沤制物与腐殖土配制的1#和2#基质适宜用作长蕊甜菜树的育苗基质,苗龄6个月的苗木苗高为15.31cm 和13.58cm,地径为3.07mm 和2.54mm。

关键词:长蕊甜菜树;咖啡壳沤制物;理化性状;生长指标

Effects of Growing Media on Seedling Performance of *Melientha longistaminea*

Li Ya[1], Li En-liang[2], Mao Yun-ling[1], Zhou Jiang[2],
Liu Yong-guo[3], Chang En-fu[1], Li Yong-peng[1], Jing Yue-bo[1]

(1. Yunnan Academy of Forestry, Kunming Yunnan 650201, P. R. China;
2. Yunnan Provincial Nature Reserve Management Bureau of Xiqi *Alsophila spinulosa*
in Mojiang, Mojiang Yunnan 654800, P. R. China;
3. Forest Institute of Wenshan Prefecture, Wenshan Yunnan 666300, P. R. China)

Abstract: Using coffee husk compost, humus and forest soil as the components, five kinds of growing media were prepared, eight physical-chemical properties of the tested growing media were examined, and seven growing parameters of *Melientha longistaminea* were determined, to analyzed the effects of growing media on seedling. The results showed that in different growing media, there were significant ($P<0.05$) or extreme significant differences ($P<0.01$) on the growing parameters, suitable media could effectively promoted the seedling growth and biomass accumulation. The differences on seedling growing were the integrated results of all the physical-chemical properties of growing media. For the growing media with different bulk weight, the root system development and biomass accumulation were affected by porosity and water-air ratio. Taking the indexes of quality growing media as the standard, based on the growing performance, two medium formula with different proportion of coffee husk compost and humus, got the six month seedling height of 15.31cm and 13.58cm respectively, and the basal diameter of 3.07mm and 2.54mm respectively.

Key words：growing medium；*Melientha longistaminea*；coffee husk compost seedling

长蕊甜菜树(*Melientha longistaminea*)是山柚子科(Opiliaceae)甜菜树属(*Melientha*)植物,常绿小乔木或灌木[1],主要分布在云南热区,其嫩茎、嫩芽叶含有P、S、K、Ca、Mn、Fe、Zn等多种矿质元素[2],对人体健康有益,是热区有名的天然、无污染的绿色木本蔬菜之一[3]。近年来,随着人们生活水平的提高,森林蔬菜因其污染少、风味独特、营养丰富等特点而广受人们的关注,森林蔬菜已成为市场上的新宠和人们餐桌上的时尚食品[4]。长蕊甜菜树野生种群由于对生境要求极为苛刻,天然下种更新困难,加之人类活动的加剧及无序采摘,导致其种源数量日益减少,其野生资源不仅难以满足市场需求,而且有濒临灭绝的危险。就目前的技术手段而言,种群数量的增加必须依靠种子繁殖和人工种植方式来实现[5],而传统的长蕊甜菜树容器育苗化以森林土作为育苗基质,出苗率低、苗木生长缓慢是苗木繁育中普遍存在的问题,这不仅造成了种子的巨大浪费,同时也制约其人工规模化栽培。目前,国内虽有学者开展其育苗技术及栽培技术方面的研究[6-9],但也仅限于简单的介绍,尚缺乏系统、深入的研究与探讨,种苗繁育技术仍是长蕊甜菜树资源保护与开发利用中亟须解决的关键技术。

育苗基质是木本植物育苗的基础和关键因素[10-11],而基质的筛选不仅是容器育苗、移栽成功与否的关键[12-13],同时也是工厂化育苗的关键技术之一[14-15]。国内多位学者利用草炭土、蛭石和珍珠岩等材料配制为育苗基质代替以天然土配制而成的传统育苗基质,开展了多个树种[14-20]的育苗试验,取得了较好效果。邱琼等[21]及常恩福等[22]则利用经堆肥发酵的咖啡壳与蛭石、珍珠岩等材料配制成育苗基质,开展了印度紫檀(*Pterocarpus indicus*)、铁橡栎(*Quercus cocciferoides*)、乌桕(*Sapium sebiferum*)等树种苗木培育的研究,筛选出了适宜的育苗基质。但不同的育苗基质以及同一种育苗基质的育苗效果,往往会因培育树种、育苗地区及育苗设施的不同而发生变化[23]。因此,本项研究以咖啡(*Coffea Arabica*)壳沤制物、腐殖土及森林土为基质原料,按体积比配制成不同基质,通过不同育苗基质理化性状测定及长蕊甜菜树苗木生长状况的分析,探究基质理化性状与长蕊甜菜树苗木各项生长指标间的相互关系,以期筛选出适宜的、能培育壮苗的育苗基质,为其苗木繁育技术的研发提供科学依据。

1 试验区自然概况

试验区位于云南墨江县新安镇挖岩村,海拔888m。属南亚热带半湿润山地季风气候,干湿季明显；年降水量1 338mm,主要集中在5~10月份,蒸发量1 696.7mm,全年平均相对湿度80%；年均气温17.8℃,最冷月(1月)平均气温11.5℃,最热月(6月)平均气温22.1℃；全年总日照时间2 161.2h,年日照率为50%,月均日照时间149h。土壤为赤红壤。

2 材料与方法

2.1 供试种子

供试长蕊甜菜的种子于2017年6月下旬采自墨江县新抚镇车沙村生长健壮的成熟母

树。采集到种子后立即去皮,用500~800倍的多菌灵液浸泡消毒30min,清水漂洗干净后,用洁净河沙层积催芽。每层河沙铺设厚度8cm左右,播种后覆盖厚度2cm左右,播种量3 300粒·m^{-2},铺设层数视催芽器具大小而定,不宜超过3层。催芽期间注意淋水,以保持河沙湿润。

2.2 育苗基质

以咖啡壳沤制物、腐殖土及森林土为基质原料,按体积比配制成5种育苗基质。
基质1为70%咖啡壳沤制物(70%咖啡壳、15%牛粪、15%锯末):30%腐殖土;
基质2为70%咖啡壳沤制物(65%咖啡壳、20%牛粪、15%锯末):30%腐殖土;
基质3为70%咖啡壳沤制物(60%咖啡壳、20%牛粪、20%锯末):30%腐殖土;
基质4为10%咖啡壳沤制物(70%咖啡壳、15%牛粪、15%锯末):90%森林土;
对照(CK)为传统育苗基质(94%森林土、4%羊粪、2%复合肥)。

其中,咖啡壳沤制物按堆肥标准化工艺堆肥发酵后使用,羊粪经充分发酵后施用。5种基质的理化性状详见表1。

表1 不同基质的理化学性状

基质	密度 (g·cm^{-3})	孔隙度 (%)	碱解氮质量分数 (mg·kg^{-1})	速效磷质量分数 (mg·kg^{-1})	速效钾质量分数 (mg·kg^{-1})	有机质(%)	pH值	EC值 (ms·cm^{-1})
1#	0.68	70.47	413.8	130.3	241.4	9.85	6.72	0.554
2#	0.66	70.42	467.5	129.6	251.1	11.16	6.69	0.528
3#	0.71	66.26	405.9	127.5	233.6	9.48	6.82	0.505
4#	1.19	48.23	150.3	121.1	111.5	3.66	7.55	0.290
对照	1.25	46.85	168.3	245.3	87.4	5.7	7.62	0.478

2.3 试验方法

将配制好的5种育苗基质分别装于24cm×40cm的无纺布育苗容器内,基质与容器口持平。每种育苗基质作为1个处理,在苗床上采用单因素随机区组设计排列试验小区,每处理3次重复。共设置15个试验小区,每小区参试的苗木株数为200株。

2017年7月下旬,把催芽处理后长出胚根(0.2~0.5mm)的种子将其胚根向下点播于无纺布容器内。每袋点播种子1粒,种子距基质表面0.8~1cm左右,浇透定根水,苗床上用透光度65%~75%的遮阳网搭建遮阴篷,防止阳光直射苗木,同时保持无纺布容器内的基质湿润。苗木出土后的1~2个月,采取少量多次的方法进行浇水,同时每隔10d喷施500~800倍的多菌灵液1次,以防止病害的发生。移栽或造林前1~2个月,逐渐揭除遮阳网炼苗,促进苗木的木质化及苗木的适应性,以提高造林成活率。

2018年3月19日,参考常恩福等[22]的苗木调查方法,在每个小区中进行抽样调查。

2.4 数据处理与分析

应用Excel 2010进行试验数据处理,DPS 7.05软件进行差分析、多重比较及相关分析。

3 结果与分析

3.1 不同基质对苗高及地径的影响

从表2可以看出,1#、2#基质与3#、4#基质和对照间的苗高差异达显著或极显著水平,1#基质与其余4种基质间及2#、3#、4#基质与对照间的地径差异达显著或极显著水平。4种配方基质的苗高、地径均高于对照,分别较对照高38.2%、22.6%、1.3%、4.4%和49.0%、23.3%、27.2%、15.0%。苗高从大到小的顺序为1#>2#>4#>3#>对照,地径则为1#>3#>2#>4#>对照。可见,不同育苗基质对长蕊甜菜树的苗高、地径均有着明显的影响,适宜的配方基质对苗高、地径生长具有明显的促进作用。相较而言,1#基质最适宜其苗高、地径的生长,苗龄6个月的苗木苗高、地径达15.31cm和3.07mm,其次是2#基质,苗高、地径为13.58cm和2.54mm。

表2 不同育苗基质苗高和地径的差异性比较

基质类型	苗高(cm)	地径(mm)
1#	(15.31±0.34)Aa	(3.07±0.09)Aa
2#	(13.58±0.21)ABa	(2.54±0.18)Bb
3#	(11.22±0.80)Bb	(2.62±0.16)Bb
4#	(11.57±1.84)Bb	(2.37±0.05)BCb
对照	(11.08±0.38)Bb	(2.06±0.07)Cc

注:同列不同的小写字母表示在0.05水平上差异显著,同列不同大写字母表示在0.01水平上差异极显著。

3.2 不同育苗基质对根系生长发育的影响

从表3可以看出,1#、2#和3#基质与4#基质和对照间的主根长、一级侧根数的差异达显著或极显著水平,且高于对照。根幅仅1#基质与对照间的差异达显著水平,但2#、3#和4#基质的根幅仍高于对照。可见,以咖啡壳沤制物和腐殖土配制成的1#、2#和3#基质由于具有容重小、孔隙度大的特点及较好的通气透水性能和较强的持水能力,从而有效地促进了根系的生长发育。综合3个根系性状指标的测定结果,1#、2#和3#基质均有利于其根系的生长发育,主根长、根幅、一级侧根数分别较对照高10.9%~17.3%、9.2%~27.1%和34.3%~68.1%。

表3 不同育苗基质根系生长的差异性比较

基质类型	主根长(cm)	根幅(cm)	一级侧根数(根)
1#	(21.70±0.46)Aa	(6.23±0.25)Aa	(21.53±0.83)ABa
2#	(20.73±0.91)ABab	(5.73±0.15)Aab	(18.53±1.00)Bb
3#	(21.93±2.40)Aa	(5.87±0.76)Aab	(23.20±1.41)Aa
4#	(17.40±0.66)Bc	(5.17±0.21)Aab	(14.60±0.53)Cc
对照	(18.70±2.00)ABbc	(4.90±0.60)Ab	(13.80±1.56)BCc

注:同列不同的小写字母表示在0.05水平上差异显著,同列不同大写字母表示在0.01水平上差异极显著。

3.3 不同基质对生物量的影响

从表4可以看出,1#基质与其余4种基质间、2#、3#基质与4#基质和对照间及4#基质与对照间的地上部分生物量差异达显著或极显著水平,3#基质与1#、4#基质间地下部分生3量差异达显著或极显著水平,1#、2#、3#基质与4#基质和对照间全株生物量差异达极显著水平。这说明,不同基质对苗木地上、地下及全株生物量的累积有不同的影响,不同基质对其苗高、地径及根系生长发育的影响直接影响其地上、地下及全株生物量的累积,影响程度不同是其生物量产生差异的原因。

从根冠比来看,4种配方基质与对照间及1#基质与4#基质间的根冠比差异达显著或极显著水平,说明不同基质对苗木地上、地下生物量的分配有一定影响,但主要还是与其生态适应性有关。5种供试基质的根冠比在1.46~4.48之间变化,说明其地下部分生长明显大于地上部分的生长,主要是为适应其要求极为苛刻的生境,苗期主要以扩展根系,增强根系吸收功能为主。就生物量的累积而言,还是1#、2#和3#基质更为适宜。

表4 不同育苗基质生物量的差异性比较

基质类型	地上部分生物量(g)	地下部分生物量(g)	全株生物量(g)	根冠比
1#	(2.15±0.09)Aa	(3.14±0.10)Bb	(5.29±0.18)Aa	(1.46±0.02)Bc
2#	(1.65±0.07)Bb	(3.44±0.23)ABab	(5.09±0.22)Aa	(2.09±0.18)Bbc
3#	(1.68±0.03)Bb	(3.68±0.12)Aa	(5.36±0.12)Aa	(2.19±0.09)Bbc
4#	(1.09±0.25)Cc	(3.22±0.07)ABb	(4.30±0.28)Bb	(3.05±0.60)ABb
对照	(0.78±0.18)Cd	(3.37±0.10)ABab	(4.16±0.21)Bb	(4.48±1.15)Aa

注:同列不同的小写字母表示在0.05水平上差异显著,同列不同大写字母表示在0.01水平上差异极显著。

3.4 基质理化性状与苗木生长性状间的相关性分析

从表5可以看出,苗高、地径、地下生物量与基质各理化状无显著的相关性,仅地径与总孔隙度存在显著的相关性,主根长、根幅、一级侧根数、地上部分生物量和全株生物量与容重、总孔隙度、碱解N、速效K、pH值则存在显著或极显著的相关性。速效P与苗木各生长性状无显著的相关性,有机质与主根长、全株生物量存存显著的相关性,EC值与主根长存在显著的相关性。地下生部分物量与基质理化性状则无显著的相关性。

表5 基质理化性状与苗木生长性状的相关性

性状	密度	总孔隙度	碱解N	速效P	速效K	有机质	pH值	EC值
苗高	-0.64	0.71	0.62	-0.40	0.65	0.60	-0.67	0.53
地径	-0.80	0.82*	0.72	-0.67	0.80	0.63	-0.81	0.48
主根长	-0.91*	0.90*	0.91*	-0.34	0.89*	0.91*	-0.91*	0.86*
根幅	-0.93**	0.93**	0.87*	-0.66	0.93**	0.80	-0.93**	0.63
一级侧根数	-0.89*	0.86*	0.83*	-0.57	0.88*	0.77	-0.87*	0.63
地上部分生物量	-0.92**	0.93**	0.86*	-0.67	0.92**	0.79	-0.92**	0.61
地下部分生物量	-0.28	0.21	0.33	0.02	0.27	0.35	-0.25	0.27
全株生物量	-0.97**	0.96**	0.94**	-0.63	0.97**	0.88*	-0.96**	0.68

注:*表示显著相关($P<0.05$),**表示极显著相关($P<0.01$)。

4 结论与讨论

基质的密度、总孔隙度、pH 值、EC 值及有机质、各种营养元素的含量直接影响植物栽培的效果[24],目前还未见作物栽培基质的标准化性状参数[25]。马太源等[26]及杨延杰等[27]的研究表明,理想栽培基质的密度为 $0.1 \sim 0.8 g \cdot cm^{-3}$,总孔隙度在 70%~90%之间,透气性良好,性质稳定,pH 值以弱酸性或中性为宜,EC 值小于 $2.6 ms \cdot cm^{-1}$。从 5 个供试基质理化性状测定结果来看,以咖啡壳沤制物和腐殖土配制成的 1#和 2#基质完全符合这些要求,这两种基质具有密度小、孔隙度大的特点及较好的通气透水性能和较强的持水能力,丰富且均衡的养分则为苗木的生长提供所需的有机质营养及 N、P、K,适宜的 pH 值及 EC 值则确保苗木对必需元素的吸收及转化。

不同育苗基质苗木的各项生长指标均存在显著或极显著的差异,说明不同育苗基质对长蕊甜菜树苗木的各项生长指标有着不同的影响,其苗木对基质表现敏感,适宜的基质能有效促进苗木苗高、地径、根系的生长及生物量的累积。不同育苗基质各理化指标含量的变化与苗木大多数生长指标的优劣排序相吻合,表明苗木的生长差异是不同育苗基质理化性状综合作用的结果,而秦爱丽等[14]在不同育苗基质对圃地崖柏出苗率和苗木生长的影响的研究中也得出了相同的结论。

育苗基质的理化性质对苗木的生长起决定性的作用,一般认为基质的物理性质是影响苗木生长的主要性质,而密度和孔隙度是影响苗木长势的主要物理性质[28]。王春荣等[29]的研究也表明,基质密度并非直接影响苗木的生长,而是不同密度的基质存在孔隙度和水汽比的差异,进而影响了苗木生长。相关分析结果显示,基质的密度与苗高、地径无显著的相关性,而与主根长、根幅、一级侧根数、地上部分生物量及全株生物量存在显著或极显著的相关性,说明不同密度的基质是通过孔隙度和水汽比的差异影响根系发育及生物量累积来影响苗木的长势,这与前人的研究结论基本吻合。苗高、地径与基质的各化学性状间无显著的相关性,可能与基质的养分含量较为丰富且满足了苗木生长对养分需求有关。速效 P 与苗木的各项生长指标无显著的相关性,可能是在传统育苗基质森林土中(对照)添加了一定比例的羊粪和复合肥致使速效 P 含量大幅提高所致。EC 值与除主根长外的苗木生长指标无显著的相关性,说明 EC 值处于植物生长的安全范围内。地下部分生物量与基质理化性状间无显著的相关性,其原因还有待进一步的研究。

以理想基质适宜指标为标准,基于苗木生长指标测定结果,以 70%咖啡壳沤制物(70%咖啡壳、15%牛粪、15%锯末):30%腐殖土配制的 1#基质及以 70%咖啡壳沤制物(65%咖啡壳、20%牛粪、15%锯末):30%腐殖土配制的 2#基质适宜用作长蕊甜菜的育苗基质。但鉴于目前人工育苗及栽培的现状,综合考虑影响种子萌发和苗木生长的多种因子,应进一步开展种子萌发和苗木生长的生理机制方面的研究,为其人工育苗及栽培的规模化提供科学依据和技术支撑。

参考文献:

[1] 吴征镒,李德铢.甜菜树属——我国云南产山柚子科一原始新属及其植物地理学意义[J].云南植物研究,2000,22(3):248~250.

[2] 吴志霜,王跃华.野生植物甜菜树嫩茎叶的营养成分分析[J].植物资源与环境学报,2005,14(1):

60~61.
[3] 李莲芳,孟梦,温琼文,等.云南热区的5种木本森林蔬菜及其培育技术[J].西部林业科学,2005,34(1):9~14.
[4] 温琼文.元江县6种森林蔬菜的育苗及修剪试验[J].西部林业科学,2012,41(3):106~109.
[5] 杨超本.开发木本蔬菜长蕊甜菜树大有可为[J].云南林业,2015,(2):61.
[6] 温琼文,方福生,李桥安,等.元江6种热区森林蔬菜的驯种培育及示范种植成效[J].西部林业科学,2006,35(2):108~112.
[7] 袁莲珍,史富强,童清,等.不同基质对长蕊甜菜苗木生长的影响[J].陕西林业科技,2015(3):8~9,28.
[8] 普玉明.长蕊甜菜栽培技术初探[J].热带农业科技,2014,37(2):40~42,46.
[9] 杨超本.优良的木本蔬菜-长蕊甜菜树栽培技术[J].云南林业,2013,34(4):65.
[10] 刘士哲.现代实用无土栽培技术[M].北京:中国农业出版社,2001.
[11] 中国科学院南京土壤研究所.土壤理化分析[M].上海:上海科学技术出版社,1978.
[12] 李晓玲.不同基质配比的育苗基质对黄瓜苗木生长发育及其质量的影响[J].山西农业科学,2009,37(7):34~36.
[13] 任杰,崔世茂,刘杰才,等.不同基质配比对黄瓜穴盘育苗质量的影响[J].华北农学报,2013,28(2):128~132.
[14] 秦爱丽,郭泉水,简尊吉,等.不同育苗基质对圃地崖柏出苗率和苗木生长的影响[J].林业科学,2015,51(9):9~16.
[15] 燕丽萍,邵伟,刘翠兰,等.不同基质对日本赤松和日本柳杉播种出苗与苗木生长的影响[J].山西农业科学,2017,45(11):1806~1809.
[16] 刘帅成,何洪城,曾琴.国内外育苗基质研究进展[J].北方园艺,2014(15):205~208.
[17] 李新利,李向军,支恩波,等.不同育苗基质对白榆长根苗生长的影响[J].河北林业科技,2012,(4):9~11.
[18] 彭邵锋,陈永忠,杨小胡.不同育苗基质对油茶良种容器苗生长的影响[J].中南林业科技大学学报,2009,29(1):25~31.
[19] 王因花,刘翠兰,孔雨光,等.不同育苗基质处理对日本落叶松出苗及苗木生长的影响[J].山东农业科学,2017,49(6):68~70.
[20] 蓝肖,程琳,陈琴,等.不同育苗基质及移苗时间对杉木苗期生长的影响[J].广西林业科学,2016,45(4):359~363.
[21] 邱琼,陈勇,杨德军,等.云南热区印度紫檀容器育苗基质选择[J].林业科技开发,2015,29(3):16~19.
[22] 常恩福,李娅,李品荣,等.不同育基质对铁橡栎和乌桕苗木生长的影响[J].西部林业科学,2018,47(3):56~62.
[23] 邓煜,刘志峰.温室容器育苗基质及苗木生长规律的研究[J].林业科学,2003,6(5):33~40.
[24] 王欣,高文瑞,徐刚,等.以猪发酵床废弃垫料为主要原料的无土栽培基质理化性状分析[J].江苏农业科学,2017,45(24):251~254.
[25] 董晓宇,蔡晓红,翟春峰,等.新型有机栽培基质的研究进展及展望[J].陕西农业科学,2007,(4):88~90.
[26] 马太源,蓝炎阳,洪志方,等.花卉栽培介质不同配方理化性状比较研究[J].福建热作科技,2010,35(1):1~5.
[27] 杨延杰,赵康,林多,等.基质理化性状与番茄壮苗指标的通径分析[J].华北农学报,2013,28(6):104~110.

[28] 余萍,丁志彬,程龙霞,等. 不同基质对欧洲鹅耳枥1年生播种苗生长及生理特性影响[J]. 中南林业科技大学学报,2016,36(9):44~50,56.

[29] 王春荣,毕君,杨静宇. 菇渣基质的理化性状及其对油松苗木生长的影响[J]. 林业科技,2016,41(4):37~41.

(本文发表于《东北林业大学学报》,2019年)

辣木叶提取物的抗氧化活性研究

赵一鹤[1],李沁[1],夏菁[2],施蕊[2],
陈薇[2],邹洁[2],张静美[3]*

(1.云南省林业科学院,云南 昆明 650021;
2.西南林业大学轻工与食品学院,云南 昆明 650024;
3.云南林业职业技术学院,云南 昆明 650021)

摘要:[目的]研究辣木叶提取物的抗氧化活性。[方法]对辣木叶乙醇初提物用不同极性的溶剂进行萃取,得到辣木叶石油醚提取物、二氯甲烷提取物、乙酸乙酯提取物,同时对各萃取物的生物活性进行研究。[结果]辣木叶的石油醚,二氯甲烷,乙酸乙酯3个提取层对DPPH自由基均有清除作用,在相等的浓度条件下,乙酸乙酯层>二氯甲烷层>石油醚层。虽然抗氧化活性较Vc弱,但浓度达到0.5mg/ml以上时,辣木叶各层提取物均能清除90%左右的DPPH自由基。[结论]为辣木叶中具有抗氧化活性的化合物的进一步分离奠定基础,并为辣木叶开发成天然抗氧化剂或天然日化产品提供参考依据。

关键词:辣木;提取物;抗氧化活性

Study on Antioxidant Activity of Extracts fromMoringa oleifera Leaves

(Zhao Yi-he[1], Li Qin[1], Xia Jing[2], Shi Kui[2],
Chen Wei[2], Zou jie[2], Zhang Jing-mei[3])

(1 Yunnan Academy of Forestry, Kunming Yunnan 650021, China;
2 School of Light Industry and Food, Southwest Forestry University, Kunming Yunnan 650024, China;3 Yunnan Forestry Technological College, Kunming Yunnan 650021, China)

Abstract:[Objective]Study on Antioxidant Activity of Extracts from *M. oleifera* Leaves. [Method]The ethanol extracts of *M. oleifera* leaves were extracted with different organic solvents to get petroleumether, dichloromethaneand ethyl acetateextracts. Theantioxidant activity study were carried out of these different polarity extracts. [Result]The results showed the petroleum ether, dichloromethane and ethyl acetate extracts had a certain degree of scavenging activities on DPPH radical in the same concentrations, the antioxidant activity of ethyl acetate extract>dichloromethane extracts>petroleum ether extract. Although their antioxidant activity were weaker than Vc, the extract could clear approximately 90% DPPH radical when the concentration reaches more than 0.5 mg/ml. [Conclusion]This study provided a foundation for the further separation of compounds with antioxidant activity in *M. oleifera* leaves and provides a reference for the development of *M. oleifera* leaves into natural antioxidants or natural daily chemical products.

Key words:*Moringa oleifera*;Extacts;Antioxidant activity

辣木(*Moringa oleifera*)又称为鼓槌树(Drum stick tree),因其具有辛辣味的根而得名[1,2],是辣木亚目(Moringineae)辣木科(Moringaceae),辣木属(*Moringa* Adans.)植物的统称[3],原产于热带及南亚热带的干旱或半干旱区域,在我国未见自然分布,目前我国已将其作为经济型植物引进种植。

自由基是一类具有不成对电子的原子和基团,是机体氧化反应中产生的有害物质,具有强的氧化性,可损害机体组织和细胞,进而引起慢性毒性和细胞衰亡作用,存在于植物和动物细胞中,是生命体细胞凋亡的一个重要诱因。体内活性氧自由基包括超氧阴离子自由基、羟自由基、脂氧自由基、二氧化氮和一氧化氮自由基等,本身具有一定的功能,但过多的活性氧自由基就会有破坏行为,导致人体正常细胞和组织的损坏,从而引起多种疾病。如心脏病、老年痴呆症、帕金森病和肿瘤。外界诱因会使人体产生更多活性氧自由基,使核酸突变,这是人类衰老和患病的根源。因此,抗氧化作用成为维护健康的关注热点,各类抗氧化剂和抗氧化保健日化品成为热销产品。Sholapu等[5]研究结果表明辣木能够抑制周缘组织内由地塞米松诱导的胰岛素抗性。Jaiswa等[6]通过体外和体内试验证实辣木叶片具有显著的抗氧化活性,能保护正常人及糖尿病患者免受氧化胁迫的损伤。辣木籽油因含辣木多糖具有一定的抗氧化活性,辣木叶抗氧化活性的研究未见报道。由于国内引进时间较短,所以国内对辣木的研究主要集中于在辣木的引种栽培[7,8],辣木的营养成分[9]以及辣木油的提取工艺[10]等方面,但是关于辣木生物活性研究较少[11~14]。本实验对辣木叶乙醇初提物用不同极性的溶剂进行萃取,得到辣木叶石油醚提取物、二氯甲烷提取物、乙酸乙酯提取物,采用DPPH和苯三酚自氧化法对辣木叶的抗氧化活性进行测定,目的是通过不同的抗氧化活性测定,明确辣木叶中抗氧化活性部位及其抗氧化作用位点,为辣木叶中具有抗氧化活性的化合物的进一步分离奠定基础,也希望能为辣木叶开发成天然抗氧化剂或天然日化产品提供参考依据。

1 材料与方法

1.1 辣木叶

辣木鲜叶样品采自云南省大理白族自治州宾川县,将其阴干,备用。

1.2 试验仪器

主要仪器包括:旋转蒸发仪、UV紫外分光光度计、循环水式多用真空泵、电炉、电子天平、移液管、移液枪、胶头滴管、试管、培养皿、镊子。

1.3 试验试剂

石油醚、乙酸乙酯、二氯甲烷、乙醇、蒸馏水、DPPH、无水乙醇、抗坏血酸。

1.4 原材料处理

将辣木叶用75%的乙醇冷浸3次,合并3次滤液,于55℃下旋蒸浓缩,再依次用等体积石油醚、二氯甲烷、乙酸乙酯对所得浸膏进行萃取,要求每种试剂都萃取3次,以达到萃取层基本无杂色。合并每种溶剂所得萃取液,减压浓缩至干。萃取结束后,得到3层浓缩浸膏,即石油醚层浸膏、二氯甲烷层浸膏及乙酸乙酯层浸膏。

1.5 抗氧化活性测定

1.5.1 溶液的配置

DPPH 抗氧化试验:精密称取辣木各提取层 0.1g,用 95% 乙醇溶解并定容于 50ml 量瓶,摇匀,超声 20min,作为母液。精密称取 2mgDPPH,加无水乙醇 40ml,超声 5min。用分光光度计在 517nm 处测得分光度值。

邻苯三酚自氧化法:样品溶液:精密称取 0.300g 样品,加 10ml 缓冲液,备用。$0.1mol·L^{-1}$ Tris:精密称取 1.21gTris,0.0372gEDTA 用蒸馏水溶解并定容于 100ml 容量瓶中;$0.1mol·L^{-1}$ HCl:0.85ml 36%HCl 溶液,用蒸馏水溶解并定容于 100ml 容量瓶中;pH8.2—Tris-HCl 缓冲液:50ml $0.1mol·L^{-1}$ Tris,22.9ml $0.1mol·L^{-1}$ HCl 蒸馏水定容于 100ml 蒸馏水;邻苯三酚:精密称取 0.6306g 邻苯三酚加 $10mmol·L^{-1}$ 定容于 100ml 容量瓶中。

1.5.2 DPPH 抗氧化试验方法

按表 1 分别取 2ml 不同浓度(100,200,300,400,500μg·ml^{-1})的溶液,加入 2ml DPPH 溶液,混合均匀。室温放置 30min 后于 517nm 处测吸光值。用 Vc(抗坏血酸)作为阳性对照。

表 1 DPPH 法试验加样表
Table 1 DPPH sample table

	受试组 A1 Subjects groupA1	空白组 A2 Blank group A2	对照组 A3 Contrast Group A3	阳性对照组 AcPositive control groupAc
DPPH	2	2	0	2
样品溶液 Sample solution	2	0	2	0
无水乙醇 Anhydrous ethanol	0	2	2	0
抗坏血酸 Ascorbic acid	0	0	0	2

计算样品对 DPPH 自由基的清除率公式

DPPH 清除率 = [1-(A1-A3)/A2] ×100%

A1: 2ml 样品溶液 + 2ml DPPH 溶液的吸光值;
A2: 2ml 无水乙醇 + 2ml DPPH 溶液的吸光值;
A3: 2ml 样品溶液 + 2ml 无水乙醇的吸光值。

1.5.3 邻苯三酚自氧化法

按表 2 取 $50mmol·L^{-1}$ Tris-HCl 缓冲液(pH8.2)4.5ml,置 25℃ 水浴中保温 20min,按表依次加入不同浓度(100,200,300,400,500ug·ml^{-1})样品溶液、邻苯三酚溶液和蒸馏水,混匀后于 25℃ 水浴中反应 5min,每 30 秒在 325nm 处测定吸光度。

表2 邻苯三酚实验试剂加入量
Table 2　Addition of pyrogallol experimental reagent　　　　　ml

组别/试剂 Group/reagentml	受试组 A1 Subjects groupA1	对照组 A2 Contrast Group A2	空白组 A3 Blank group A3
Tris-HCl	4.5	4.5	4.5
样品溶液 Sample solution	0.3	0	0.3
邻苯三酚 Pyrogallol	0.3	0.3	0
蒸馏水 distilled water	3.9	4.2	4.2

计算公式:按下式计算 O_2^- 抑制率:

邻苯三酚自氧化率=(第4min 的 D_{325nm} 值-第1min 的 D_{325nm} 值)/3

抑制率=(邻苯三酚自氧化率-加样后邻苯三酚自氧化率)/邻苯三酚自氧化率×100%

2　结果与分析

2.1　辣木叶提取物对 DPPH 自由基的清除作用

不同质量浓度辣木叶提取物对 DPPH 自由基的清除作用结果见图1。

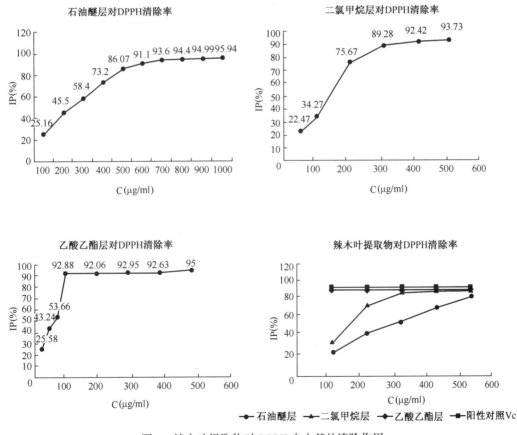

图1　辣木叶提取物对 DPPH 自由基的清除作用

Figure 1　DPPH radical scavenging effect of the extract from the leaves of *Moringa oleifera*

由图1可以看出,辣木叶石油醚层具有抗氧化活性,并且随着质量浓度的增大,辣木叶石油醚层对DPPH自由基的清除能力逐渐增强。在浓度为600μg·ml^{-1}时趋于平缓,清除率达91.1%;辣木叶二氯甲烷层对DPPH自由基的清除能力随着质量浓度的增大逐渐增强,在浓度为300μg·ml^{-1}时趋于平缓,清除率达89.32%;辣木叶乙酸乙酯层对DPPH自由基的清除能力逐渐增强。在浓度为200μg·ml^{-1}时趋于平缓,清除率达92.06%。由上图可知,辣木提取物的石油醚层,二氯甲烷层和乙酸乙酯层都对DPPH自由基均具有一定的清除作用。同时,乙酸乙酯层清除DPPH自由基的能力强于二氯甲烷层,而二氯甲烷层的清除能力强于石油醚层。并且在相同浓度梯度下,辣木叶各个提取层清除DPPH自由基的能力低于阳性对照组Vc。

2.2 辣木叶各提取层对超氧阴离子的清除作用

辣木叶提取物对超氧阴离子的清除作用如图2所示。

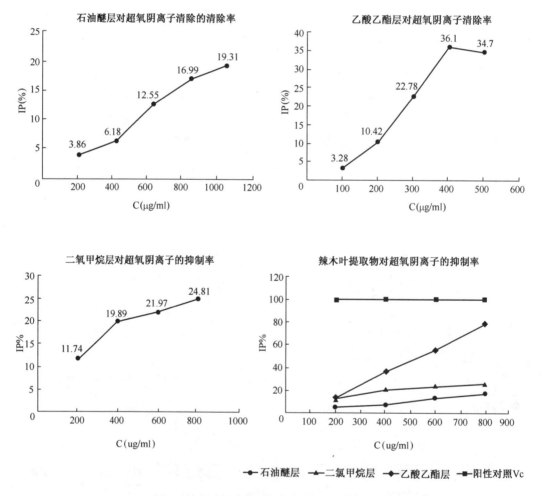

图2 辣木叶提取物对超氧阴离子的清除作用

Figure 2　Scavenging effect of the extract from the leaves of *Moringa oleifera* on superoxide anion

由图2可知,辣木叶石油醚层提取物对超氧阴离子的清除率较低,在浓度达到 1mg·ml^{-1}时为19.31%,而二氯甲烷层提取物在浓度为0.8mg·ml^{-1}时IP%为24.8%,相对乙酸乙酯层的超氧阴离子清除率较低,相同实验条件下,乙酸乙酯层提取物在浓度为0.4mg·ml^{-1}时,对超氧阴离子的清除率达到36.1%,和其他提取物相比,其抗氧化作用较强,虽然和阳性对照物Vc相比,对超氧阴离子的抑制率还是较低,但本实验采用的材料是辣木叶提取物而不是单体,而Vc为单体化合物,因此,下一步将考虑把辣木叶乙酸乙酯层进行分离纯化,获得活性单体化合物再进行抗氧化活性研究。

辣木叶提取物的石油醚层,二氯甲烷层和乙酸乙酯层都对DPPH自由基和超氧阴离子具有一定的清除作用。同时,乙酸乙酯层清除DPPH自由基的能力强于二氯甲烷层,而二氯甲烷层的清除能力强于石油醚层。虽然在相同浓度梯度下,辣木叶各个提取层清除DPPH自由基和超氧阴离子的能力均低于阳性对照组Vc,但是,当浓度达到0.5mg·ml^{-1}以上时,辣木叶各层提取物能清除90%左右的DPPH自由基。

3 结论与讨论

本研究对辣木叶乙醇初提物用不同极性的溶剂进行萃取[15,16],得到辣木叶石油醚提取物、二氯甲烷提取物、乙酸乙酯提取物,并同时进行抗氧化活性实验。

植物抗氧化测定方法主要集中在脂质体过氧化,自由基清除率和抗过氧化物歧化酶等方面,其中DPPH是最常用的抗氧化测定方法,DPPH法是基于DPPH(二苯代苦味酰基的有机溶剂在520处有最大吸收波长,如抗氧剂同时存在时与其故对电子配对而吸收减弱的原理检测抗氧化剂对自由基的清除效率,是评价抗氧化剂最常用的方法。邻苯三酚法则属于抗过氧化物歧化酶活性测定范围,利用邻苯三酚在自氧化过程中产生有色中间体和氧阴离子,利用SOD分解而间接推算酶活力的方法。本研究通过比较不同的抗氧化活性测定,明确辣木叶中的抗氧化活性部位,通过两种抗氧化方法的比较,确定辣木叶提取物中的乙酸乙酯提取物的抗氧化作用主要为对自由基的清除作用,为下一步开发辣木叶天然抗氧剂提供理论基础[16]。

在抗氧化实验中,测定辣木叶的石油醚、二氯甲烷、乙酸乙酯3个提取层的DPPH自由基清除实验结果显示,3个提取层均有清除作用,且随浓度的增大而增大,同时在相等的浓度条件下,乙酸乙酯层抗氧化效果最强,其次是二氯甲烷层,再次是石油醚层。植物的抗氧化能力多与其多酚类物质和黄酮类物质的含量有一定的关系,乙酸乙酯层比其他两层抗氧化效果好可能是由于其主要萃取植物中黄酮和多酚成分。虽然各提取层清除DPPH自由基的能力均低于Vc,但是在浓度达到0.5mg·ml^{-1}以上时,各层对DPPH自由基清除率都能达到90%以上。

由此可见,辣木叶具有良好的抗氧化性,具有潜在的开发价值。抗氧剂是食品、医药和化工领域常用和必须使用的辅料和添加剂,化学合成抗氧剂对身体有一定的毒性和致突变作用,随着消费者对天然健康的需求,天然抗氧剂将是未来抗氧剂研究和开发的重点,通过本研究的结果表明,辣木叶是一种极具开发潜力的天然抗氧化剂,其在2012年列入国家卫生部批准的新资源食品目录后,使得辣木叶的综合开发利用成为可能,作为食品级原料,其安全性已经得到认可,如能在此基础上对其功能性进行开发利用,无疑将扩大辣木叶的价值和利用度。因此,辣木叶可作为一种新型抗氧化剂或抗衰老保健品,也可以进一步对其单体化合物的抗氧化功能上作进一步研究,以期望开发出相关医药产品。

参考文献：

[1] 中国科学院中国植物志编辑委员会. 中国植物志:第三十四卷第一分册[M]. 北京:科学出版社,1984.

[2] 陈惠明. 植物中的钻石——高经济价值的辣木[M]. 台湾:商周出版社,2003.

[3] 彭兴民,郑益兴,段琼芬,等. 印度传统辣木引种栽培研究[J]. 热带亚热带植物学报,2008,16(6):579~585.

[4] NDHLALA AR, MULAUDZI R, NCUBE B, et al. Antioxidant, antimicrobial and phytochemical variations in thirteen Moringa oleifera Lam cultivars. [J]. Molecules,2014,19(7):10480~10494.

[5] SHOLAPUR H N, PATIL B M. Effect of Moringa Oleifera Bark Extracts on Dexamethasone induced Insulin resistance in rats[J]. Drug res(stuttg),2013,63(10):527~531.

[6] JAISWAL D, RAI P K, MEHTA S, et al. Role of Moringa oleifera in regulation of diabetes-induced oxidative stress[J]. Asian Pac J Trop Med, 2013, 6(6):426~432.

[7] OGUNSINA B S, INDIRA T N, BHATNAGAR A S, et al. Quality characteristics and stability of Moringa oleifera seed oil of Indian origin[J]. J Food Sci Technol,2014,51(3):503~510.

[8] 郑毅,解培惠,伍斌,等. 辣木在金沙江干热河谷造林试验研究[J]. 中国农村小康科技,2011(2):52~54.

[9] 周明强,班秀文,刘清国,等. 贵州辣木的引种栽培技术及特征特性研究[J]. 安徽农业科学,2010,38(8):4086~4088.

[10] 杨东顺,樊建麟,邵金良,等. 辣木不同部位主要营养成分及氨基酸含量比较分析[J]. 山西农业科学,2015,43(9):1110~1115.

[11] 苏瑶,赵一鹤,冯武,等. 云南引种辣木籽营养成分分析与评价[J]. 西部林业科学,2015,44(4):142~145.

[12] 白旭华,黎小清,刘昌芬,等. 超临界CO_2流体萃取辣木籽油工艺研究初报[J]. 热带农业科技,2007,30(3):29~31.

[13] 马李一,余建兴,张重权,等. 水酶法提取辣木油的工艺研究[J]. 林产化学与工业,2010,30(3):53~56.

[14] 陈玫,张海德,陈敏,等. 几种中药不同溶剂组分的抗氧化活性研究[J]. 中山大学学报,2006,45(6):131~133.

[15] WATERMAN C, CHENG DM, P. Stable, water extractable isothiocyanates from Moringa oleifera leaves attenuate inflammation in vitro[J]. Phytochemistry,2014,103:114~122

[16] ZHAO J H, ZHANG Y L, WANG L W, et al. Bioactive secondary metabolites from Nigrospora sp. LL-GLM003, an endophytic fungus of the medicinal plant Moringa oleifera Lam [J]. World J Microbiol Biotechnol,2012,28(5):2107~2112.

（本文发表于《安徽农业科学》,2019 年）

七彩红竹二氢黄酮醇4-还原酶基因 *IhDFR*1 的克隆及表达分析

缪福俊[1]，陈剑[1]，孙浩[2]，王毅[1]，王晨晨[3]，原晓龙[1]，杨宇明[1]，王娟[1]

(1. 云南省林业科学院，国家林业局云南珍稀濒特森林植物保护和繁育重点实验室，
云南省森林植物培育与开发利用重点实验室，云南 昆明 650201;
2. 中国科学院沈阳应用生态研究所，辽宁 沈阳 110016;
3. 西南林业大学林学院，云南 昆明 650224)

摘要：二氢黄酮醇4-还原酶(DFR)是植物花色素苷合成途径中的关键酶，在植物花色的形成过程中起重要作用。依据七彩红竹转录组数据设计特异引物，采用RT-PCR技术从七彩红竹中克隆获得了一个新的DFR基因cDNA全长，命名为*IhDFR*1(登录号为KF728205)。序列分析结果表明，*IhDFR*1基因cDNA全长945 bp，编码314个氨基酸。生物信息学预测显示，该基因编码的蛋白具有典型的DFR蛋白功能结构域，存在2个特异结合位点，属于非Asn/Asp型DFR酶，与禾本科植物中的DFR具有较高的相似性。对不同发育时期七彩红竹的*IhDFR*1基因进行时空表达的结果显示，只有在竹秆颜色呈现红紫色时，*IhDFR*1基因才有表达。以上结果初步显示*IhDFR*1蛋白可能作为一个重要的酶参与竹秆花色素苷的代谢调控，同时为进一步研究七彩红竹花色素苷产生的分子机理和综合开发利用奠定了基础。

关键词：七彩红竹；花色素苷；二氢黄酮醇4-还原酶；基因表达

Cloning and Expression Analysis of Dihydroflavonol 4-Reductase Gene *IhDFR*1 from *Indosa sahispida* cv. 'Rainbow'

(Miao Fu-Jun[1], Chen Jian[1], Sun Hao[2], Wang Yi[1], Wang Chen-Chen[3],
Yuan Xiao-Long[1], Yang Yu-Ming[1], Wang Juan[1])

(1. Key Laboratory for Conservation of Rare, Endangered & Endemic Forest Plants,
State Forestry Administration, Yunnan Provincial Key Laboratory of Cultivation and Exploitation
of Forest Plants, Yunnan Academy of Forestry, Kunming Yunnan 650201, China; 2. Institute of Applied
Ecology, Chinese Academy of Sciences, Liaoning Shenyang 110016, China;
3. College of Forestry, Southwest Forestry University, Kunming Yunnan 650224, China)

Abstract: Dihydroflavonol 4-reductase (DFR) is a key enzyme in the anthocyanins biosynthesis pathway, and plays a critical role in flower pigmentation. The gene-specific primers were designed according to the transcriptome sequencing data, and the full-length cDNA of a novel DFR gene was cloned from *Indos asahispida* cv. 'Rainbow' with the method of reverse transcription PCR. This novel gene was named as *IhDFR*1 (GenBank accession No. KF728205). Sequence analysis indicated that *IhDFR*1 was 945 bp in length and encoded a protein with 314 amino acids. Bioinformatics analysis showed that IhDFR1 had the typical functional domains of DFR protein, containing two specific binding sites and belonging to the non-Asn/Asp DFR. The IhDFR1 was homologous with the DFRs from the gramineous plants. The temporal-spatial expression analysis based on different growth stages indicated that the IhDFR1was expressed only in the reddish violet culms. The results above preliminarily suggested that the IhDFR1might be an important enzyme governing anthocyanin metabolism, and lay a theoretical basis for further exploration of molecular mechanism of anthocyanins and for the comprehensive exploitation and utilization of *I. hispida* cv. 'Rainbow'.

Key words: *Indosasa hispida* cv. 'Rainbow'; anthocyanins; dihydroflavonol 4-reductase; gene expression

　　七彩红竹(Indosasa hispida cv.'Rainbow')为禾本科大节竹属植物,是浦竹仔的产花色素苷变种,主要分布于云南南部景洪、勐腊、普洱、江城等地,海拔在1 700m以下(王娟等 2012;徐永椿 1991)。我们于2009年在云南普洱发现当地自然分布的蒲竹仔群落中偶见竹秆中下部出现不同程度的红色至红紫色,经鉴定,在分类上应属大节竹属的浦竹仔变种,具有重要的观赏价值和潜在的商业开发价值。而出现红色秆性状的居群,并非是种属差异导致,是因样本收集地环境因子诱发基因突变所致,其诱发因子目前未知(王娟等 2012)。花色素苷是植物次生代谢过程中产生的一类水溶性天然食用色素,是构成花瓣、果实和茎秆颜色的主要色素之一,还具一定的营养和药理作用(Singh等 2009;卢钰等 2004)。然而,在花色素苷合成过程中,二氢黄酮醇4-还原酶(dihydroflavonol 4-reductase,DFR)对花色素苷的最终形成起决定性作用,是一个重要的调控点(符红艳等 2012;Dick等 2011;陈大志等 2010)。DFR是依赖二氢堪非醇(dihydrokaempferol,DHK)、二氢栎皮黄酮(dihydroquercetin,DHQ)和二氢杨梅黄酮(dihydromyrivetin,DHM)3种底物,分别生成无色花葵素、无色花青素和无色翠雀素,在花色素苷合成酶的作用下又分别合成花葵素(橙色—砖红色)、花青素(红色—粉色)和翠雀素(蓝色—紫色)(张龙等 2008)。研究表明,不同物种的DFR对底物选择性不同,合成了不同的花色素苷,呈现出各异的花色,如矮牵牛(*Petunia hybrida*)的DFR缺乏还原底物DHK活性,所以其花瓣缺少橙色,而大丁草(*Gerbera anandria*)DFR能还原DHK,其花瓣就能产生橙色(潘丽晶等 2010;Polashock等 2002)。依据DFR结合区中的134位与145位的氨基酸可直接影响酶的底物特异性,依据第134位氨基酸残基的不同将DFR分为3类(李春雷等 2009;Johnson等 2001)。第一类为Asn型DFR,在绝大多种植物DFR中第134位氨基酸残基为N;第二类为Asp型DFR,其在134位上存在一个D,此类DFR不能有效地将DHK还原成无色花葵素,导致缺少橙色,如矮牵牛的花瓣;第三类是DFR的第134位氨基酸残基既不是D和N,又称为非Asn/Asp型。研究表明,非Asn/Asp型DFR酶的作用底物仅限于底物DHQ,合成红色的花色素苷,如蔓越橘(*Vaccinium macrocarpon*)的第134位氨基酸残基为V,属非Asn/Asp型DFR,致使其花和果实为红色(刘娟等 2005)。

　　目前,除本课题组发表了一篇在高山箭竹——棉花竹(*Fargesia fungosa*)中发现花色素苷的论文(彭桂莎等,2011)外,尚无其他有关竹子产花色素苷的报道。七彩红竹的发现为竹类植物中研究花色素苷代谢流提供了极好的试验材料。本研究以七彩红竹为材料,采用RT-PCR方法,克隆了七彩红竹DFR基因的cDNA全长,并对其编码蛋白的结构特征及其在不同生长时期竹秆中的表达特性进行了分析,为进一步研究IhDFR1基因功能和探讨七彩红竹红紫色秆形成的分子机制提供理论依据。

1 材料与方法

1.1 材料

七彩红竹(*Indosasa hispida* cv. 'Rainbow')采自云南省林业科学院温室棚,根据其竹秆颜色变化情况,分为 7 个不同的生长时期:第 1 和第 2 时期为幼嫩顶端时期,基本组织和表皮都不显颜色;第 3 时期为中间秆段,基本组织为红紫色,表皮不显色;第 4 时期为中间秆段,基本组织和表皮都显红紫色;第 5 时期为中间秆段,基本组织略带红紫色,表皮显红紫色;第 6 时期为成熟老秆,基本组织不显色,表皮显红紫色;第 7 时期为成熟老秆,基本组织和表皮都不显红紫色(图 1)。分别采集 7 个时期的竹秆组织,经液氮速冻后于 $-80℃$ 保存备用。

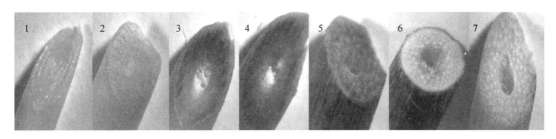

图 1 七彩红竹不同生长时期的竹秆横切面图

Figure 1 The cross profile of *I. hispida* cv. 'Rainbow' in different growth stages

数字 1~7 分别代表 7 个不同的生长时期。

1.2 方法

1.2.1 cDNA 克隆和基因表达

采用 TRNzol Reagent 试剂盒(TIANGEN 公司)方法提取各样品中的总 RNA。用 PrimeScript TM II 1st Strand cDNA Synthesis Kit (TaKaRa 公司)试剂盒方法扩增 cDNA。

RT-PCR 采用第 4 时期竹秆的 RNA 为模板扩增 cDNA,上游引物 F 为:5′-ATGTGTC-CGGCGCCGGCGGCATCGGGTAAG-3′,下游引物 R 为:5′-CTAATCCAACTCTCCATATATTC-CACTGGC-3′。反应体系为:1μL cDNA、1μL 上游引物 F、1μL 下游引物 R、9.5μL ddH$_2$O、12.5μLPrimeSTAR Max DNA Premix(2×)。反应条件为:98℃ 10s,55℃ 5s,72℃ 10s;30 个循环。PCR 产物经 1.2%琼脂糖凝胶电泳检测后,将预期片段经 DNA 回收试剂盒(上海生工公司)切胶回收后,连接载体 pEASY-Blunt 转入 Trans1-T1 感受态细胞中,挑取阳性克隆,用菌液 PCR 法(M13F、M13R 引物)鉴定重组子,确认包含重组子的克隆,送上海生工公司用 M13F、M13R 引物测序。

半定量 RT-PCR 分别采用 7 个时期的 RNA 为模板扩增 cDNA,上游引物 F 为:5′-GG-GAGAGGAGGTGAA-3′,下游引物 R 为:5′-CGGGGTCTTTGGATT-3′。Actin 上游引物 F 为:5′-ATGGCTGAAGAGGATATCCAGC-3′;下游引物 R 为:5′-TYCCATGCCAATA-AAAGATGGCTG-3′。反应体系为:1μL cDNA、1μL 上游引物 F、1μL 下游引物 R、9.5μL ddH$_2$O、12.5μLPrimeSTAR Max DNA Premix(2×)。反应条件为:98℃ 10s,57℃ 5s,72℃ 5s;30 个循环。PCR 产物经 1.2%琼脂糖凝胶电泳检测。

1.2.2 生物信息学分析

测序结果用 DNAman 去除载体后拼接得到 *IhDFR*1 基因全长。使用 NCBI (http://www.ncbi.nih.gov)中 Conserved Domain Database 数据库搜索 IhDFR1 蛋白的结构功能域；理化性质的预测借助 ProtParam (http://www.expasy.ch/tools/protparam.html)在线工具完成的；利用 MEGA3 软件进行聚类分析的；信号肽预测由 SignalP (http://www.cbs.dtu.dk/services/signalp/)完成；亚细胞定位由 WoLF PSORT (http://wolfpsort.seq.cbrc.jp/)完成；利用 SWISS-MODEL (http://swissmodel.expasy.org//swiss-model.html)在线工具分析蛋白质的三维结构，并用 PROCHECK (http://nihserver.mbi.ucla.edu/saves/)在线工具构建拉氏图。

2 结果与分析

2.1 七彩红竹 *IhDFR*1 基因全长 cDNA 的克隆

七彩红竹 *IhDFR*1 基因扩增结果如图 2A 所示，阳性克隆检测结果如图 2B 所示，经测序 *IhDFR*1 基因 cDNA 全长为 945bp，编码一个长 314 个氨基酸的蛋白质，后续的生物信息学预测分析表明，克隆的 cDNA 序列编码 DFR 蛋白，因此将该基因命名为 *IhDFR*1，GenBank 登录号为 KF728205。

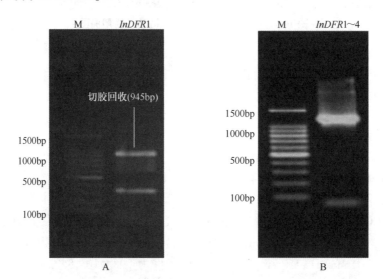

图 2 *IhDFR*1 基因 PCR 扩增的电泳分析

Figure 2 Agarose gel electrophoresis analysis of *IhDFR*1 PCR amplification products

A: RT-PCR 扩增产物；B: 阳性克隆扩增产物

2.2 *IhDFR*1 蛋白生物信息学分析

2.2.1 *IhDFR*1 基因序列和蛋白结构域预测分析

*IhDFR*1 基因 cDNA 全长及推测的氨基酸序列见图 3。蛋白结构功能域的预测分析表明，*IhDFR*1 蛋白有 1 个典型的 NADB-Rossmann 超家族(superfamily)保守结构域，存在 2 个结合位点(图3)：一个是 NADP 特异性结合位点(VMDASGPLGHALVDRLLRRGY)，而该位点是 NADP 依赖型 DFR 关键酶的特征性结构域，在已知的 DFR 氨基酸序列中均有发现；

另一个是底物特异性结合位(TMERVVFTSSVTAVVWKENHKLVDAF),属于短链的脱氢酶及还原酶(SDR)家族(图4)。DFR 对不同底物的结合是由其分子中底物结合区的氨基酸序列所决定,这个序列在不同物种中也是高度保守的。IhDFR1 基因编码的氨基酸第 134 位氨基酸残基为 R,属于非 Asn/Asp 型 DFR 酶。

```
  1 ATGTGTCCGGCGCCGGCGGCATCAGGTAAGAGCGTGTGCGTCATGGACGCCTCCGGCCCGCTGGGCCACGCCCTC
  1 M  C  P  A  P  A  A  S  G  K  S  V  C  V  M  D  A  S  G  P  L  G  H  A  L
 76 GTCGACCGGCTCCTCCGCCGCGGCTACACCGTCCACGCCGCCACCTACACCCACCACGCAGACCAAGACGACCAA
 26 V  D  R  L  L  R  R  G  Y  T  V  H  A  A  T  Y  T  H  H  A  D  Q  D  D  Q
151 GACTCCGAGTCGCTGCTGAGGCAGCTGTCCTCGTCGTGCAGCGGCGACAAACAGCAGCGGCTCAAGGTGTTCCAG
 51 D  S  E  S  L  L  R  Q  L  S  S  S  C  S  G  D  K  Q  Q  R  L  K  V  F  Q
226 GCCGACCCGTTCGACTACCACACCATCGCCGGCGCCGTCCGCGGCTGCTCCGGGCTTCTGCTGCATGTTCAGCACG
 76 A  D  P  F  D  Y  H  T  I  A  G  A  V  R  G  C  S  G  L  F  C  M  F  S  T
301 CCCCACGACCAGGCCACCTGCGACGAGGCGATGGCTGAGATGGAGGTGCGCGCGGCGCACAACGTGCTGGAGGCG
101 P  H  D  Q  A  T  C  D  E  A  M  A  E  M  E  V  R  A  A  H  N  V  L  E  A
376 TGCGCGCAGACGGAGACCATGGAGAGGGTCGTCTTCACCTCCTCCGTCACCGCCGTCGTCTGGAAGGAGAACCAC
126 C  A  Q  T  E  T  M  E  R  V  V  F  T  S  S  V  T  A  V  V  W  K  E  N  H
451 AAGCTCGTCGACGCCTTCGACGAGAAGAACTGGAGCGAGCTCAGCTTCTGCAGGAAATTCAAGCTGTGGCATGCT
151 K  L  V  D  A  F  D  E  K  N  W  S  E  L  S  F  C  R  K  F  K  L  W  H  A
526 CTGGCCAAGACACTGTCGGAGAAGACCGCCTGGGCGTTGGCCATGGACAGAGGAGTGGACATGGTGGCCATCAAC
176 L  A  K  T  L  S  E  K  T  A  W  A  L  A  M  D  R  G  V  D  M  V  A  I  N
601 GCCGGCCTGCTCACCGGGCCGGGGCTCACCGCCGCCCACCCCTACCTCAAGGGAGCCCCCGACATGTACGAGGAC
201 A  G  L  L  T  G  P  G  L  T  A  A  H  P  Y  L  K  G  A  P  D  M  Y  E  D
676 GGCGTCCTCGTCACCGTCGACGTCGACTTCCTCGCCGACGCCCACGTGGCCGTCTACGAGTCCCCGACGGTAC
226 G  V  L  V  T  V  D  V  D  F  L  A  D  A  H  V  A  V  Y  E  S  P  T  A  Y
751 GGCCGCTACCTCTGCTTCAACAATGCCGTGTGCGGGCCCGAGGATGCAGTGAAGCTGGCCCAGATGCTCTCCCCA
251 G  R  Y  L  C  F  N  N  A  V  C  G  P  E  D  A  V  K  L  A  Q  M  L  S  P
826 TCCGCTCCGCGCTCCCCACCAAGTGATGAGCTGAAGGTGATCCCACAGAGGATTCAGGACAAGAAGCTGAACAAG
276 S  A  P  R  S  P  P  S  D  E  L  K  V  I  P  Q  R  I  Q  D  K  K  L  N  K
901 CTCATGGTGGAGTTTGCCAGTGGAATATATGGAGAGTTGGATTAG
301 L  M  V  E  F  A  S  G  I  Y  G  E  L  D  *
```

图 3 IhDFR1 基因 cDNA 全长及推测的氨基酸序列

Fugyre 3 Complete cDNA and deduced amino acid sequences of InDFR1

下划线表示 NADP 结合位点;双下划线表示底物结合特异性位点。

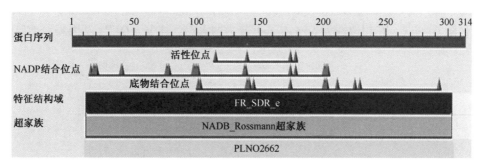

图 4 IhDFR1 蛋白保守结构域的预测分析

Figure 4 The analysis of conserved domain prediction of IhDFR1 protein

2.2.2 IhDFR1 蛋白质的理化性质预测

利用 ProtParam 在线软件预测 IhDFR1 蛋白的理化性质,结果表明其分子式为 $C_{1505}H_{2349}N_{413}O_{462}S_{21}$,相对分子量 34 293.9 Da,等电点 5.35(酸性蛋白),脂肪系数 78.66,半衰期 30 h,不稳定参数 31.25。根据不稳定参数值在 40 以下才是稳定蛋白的标准(薛庆中 2010),IhDFR1 蛋白是一种较稳定的蛋白质。SignalP 软件分析结果表明,该蛋白不存在信号肽,为非分泌蛋白,可能在细胞质中合成,不进行蛋白转运。进而采用 WoLF PSORT 软件进行亚细胞定位，结果显示 *IhDFR*1 蛋白在细胞质中合成。

2.2.3 IhDFR1 蛋白质三维结构预测

利用 SWISS-MODEL 在线软件预测 IhDFR1 蛋白质的三级结构，并用 PyMOL 软件对建模结果进行处理，结果显示，三级结构的构象呈现致密球状结构（图 5A）。另外，利用 PROCHECK 在线软件分析同源建模结果，可见该蛋白质残基的二面角位于黄色核心区域（图 5B），表明其空间结构稳定，建模结果可靠。

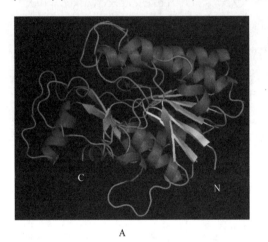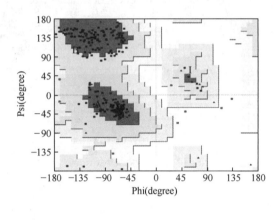

图 5　IhDFR1 蛋白的三维结构模型（A）及拉氏构象图（B）

Figure 5　The putative 3D structure (A) and ramachandram (B) of IhDFR1 protein

2.2.4 IhDFR1 基因编码氨基酸的同源比对分析

将 *IhDFR*1 基因编码的氨基酸在 NCBI 数据库中 Blastp 后，选取二穗短柄草（*Brachypodium distachyon*）、玉米（*Zea mays*）、葡萄（*Vitis vinifera*）、粗山羊草（*Aegilops tauschii*）、大豆（*Glycine max*）、乌拉尔图小麦（*Triticum urartu*）、蒺藜苜蓿（*Medicago truncatula*）7 个物种构建聚类树（图 6）可见，在相似性为 54% 水平上分为两大类，其中 *IhDFR*1 基因编码的氨基酸

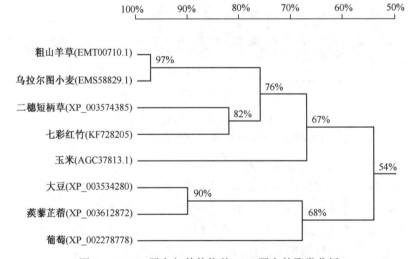

图 6　IhDFR1 蛋白与其他物种 DFR 蛋白的聚类分析

Figure 6　Phylogenetic tree analysis of the IhDFR1 and DFR proteins from other species

序列与玉米、粗山羊草、二穗短柄草、乌拉尔图小麦等单子叶禾本科植物 DFR 的氨基酸序列亲缘关系最近，属 Asn 型 DFR，与二穗短柄草 DFR 相似性高达 82%，与大豆、蒺藜苜蓿、葡萄等亲缘关系较远，相似性为 68%。

2.3 *IhDFR*1 基因的表达特性分析

以 Actin 作为内参，通过半定量 RT-PCR 检测七彩红竹 7 个不同生长时期竹秆组织中 *IhDFR*1 基因的表达，结果表明，*IhDFR*1 基因在竹秆不显红紫色的幼嫩时期（即第 1、2、7 时期）不表达，在第 3～6 时期都有表达，其中在第 3 和 4 时期表达最强（图 7）。可见，*IhDFR*1 基因的表达与七彩红竹 7 个时期的竹秆颜色变化相对应，竹秆颜色呈现红紫色时基因表达，反之，基因不表达，表明 *IhDFR*1 基因与竹秆呈红紫色的变化紧密相关。

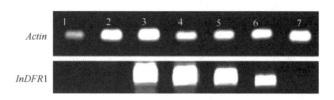

图 7 *IhDFR*1 基因在七彩红竹茎部的时空表达分析

Figure 7 Temporal-spatial expression of *IhDFR*1 gene in the culms of *I. hispida* cv. 'Rainbow'

数字 1~7 分别代表 7 个不同的生长时期。

3 讨论

本研究采用 RT-PCR 和生物信息学方法对七彩红竹的花色素苷途径中的关键酶 DFR 的基因进行克隆和序列分析，IhDFR1 蛋白属于酸性稳定的非分泌蛋白，不存在信号肽。研究结果与樊云芳等（2011）对宁夏枸杞（*Lycium barbarum*）LbDFR1 和周琳等（2011）对牡丹（*Paeonia suffruticosa*）PsDFR1 基因的研究结果一致，说明植物的 DFR 蛋白不具有信号肽，是非分泌蛋白。从而推断 DFR 蛋白可能在细胞质中合成，不进行蛋白转运，牡丹的 PsDFR1 蛋白定位于质膜。IhDFR1 蛋白具有典型的 DFR 蛋白功能结构域，存在 2 个结合位点，不同植物的 DFR 氨基酸序列在 2 个特异区域有较高的同源性，DFR 氨基酸序列中的 NADP 结合位点是高度保守的，由 26 个氨基酸序列组成的底物特异性结合区也是高度保守的（覃建兵等 2012；Petit 等 2007）。IhDFR1 蛋白第 134 位氨基酸残基为 R，属于非 Asn/Asp 型 DFR 酶，前人的研究表明非 Asn/Asp 型 DFR 酶的作用底物仅限于底物 DHQ，合成红色的花色素苷，由此可以从底物选择特异性结合位点的角度解释七彩红竹红色秆个体产生的原因，可能是由于编码第 134 位氨基酸的碱基发生突变而导致浦竹仔产生红色秆。这说明七彩红竹出现花色素苷性状并非种属差异导致，而是由于样本收集地环境因子诱发基因突变所致（王娟等 2012），相比来自其他种源地的植株，七彩红竹 IhDFR1 基因是否发生变异（包括启动子和编码区域），还有待后续进一步的研究。

IhDFR1 蛋白三级结构的构象呈现致密球状结构，空间结构稳定。陈大志等（2010）研究唇形亚纲植物 DFR 基因的结果表明，DFR 基因的空间结构分为松散 C-末端和致密球状结构两个部分。DFR 与 NADPH 辅助因子的底物结合区域都在致密球状结构中。IhDFR1

蛋白与单子叶禾本科植物(玉米、粗山羊草、二穗短柄草、乌拉尔图小麦)的 DFR 亲缘关系最近,说明该基因在进化上比较保守,而单子叶植物的 DFR 均为 Asn 型 DFR,进一步证实非 Asn/Asp 型 DFR 极有可能是由于地理隔离等原因由 Asn 型进化而来的说法(张龙等 2008)。另外,七彩红竹 *IhDFR*1 基因的在竹秆不同时期的表达情况不同,且与竹秆颜色的变化有关联,但其分子机制还有待研究。以上研究结果为后期七彩红竹花色素苷的开发利用和培育具有较高观赏价值的竹种提供了理论依据。

参考文献

[1] 陈大志,周嘉裕,李萍(2010). 二氢黄酮醇 4-还原酶的生物信息学分析[J]. 生物技术通报,(12):206~212.

[2] 符红艳,于晓英,廖祯妮(2012). 观赏植物 DFR 基因的研究进展[J]. 天津农业科学,18(6):14~19.

[3] 李春雷,崔国新,许志茹,李玉花(2009). 植物二氢黄酮醇 4-还原酶基因的研究进展[J]. 生物技术通讯,20(3):442~445.

[4] 刘娟,冯群芳,张杰(2005). 二氢黄酮醇 4-还原酶基因(*DFR*)与花色的修饰[J]. 植物生理学报,41(6):715~719.

[5] 卢钰,董现义,杜景平,李永强,王明林(2004). 花色苷研究进展[J]. 山东农业大学学报,35(2):315~320.

[6] 樊云芳,陈晓军,李彦龙,戴国礼,曹有龙(2011). 宁夏枸杞 DFR 基因的克隆与序列分析[J]. 西北植物学报,31(12):2373~2379.

[7] 潘丽晶,张妙彬,范干群,陈伟庭,曹友培(2010). 石斛兰 *dfr* 基因的克隆、序列分析及原核表达[J]. 园艺学报,37(1):129~134.

[8] 彭桂莎,杨宇明,孙浩,张怀璧,王娟(2011). 棉花竹 *FsMyb*1 基因的生物信息学研究[J]. 西部林业科学,40(4):6~11.

[9] 覃建兵,唐亚萍,曾卫军(2012). 新疆雪莲 DFR 基因的分离及同源遗传转化[J]. 植物研究,32(6):695~700.

[10] 王娟,孙浩,彭桂莎,王明月,熊智,张兴波,孙茂盛,杨宇明(2012). 浦竹仔红色秆变异个体的 rDNA ITS 序列及系统发育研究[J]. 西部林业科学,41(1):1~6.

[11] 徐永椿(1991). 云南树木图志(下册)[M]. 昆明:云南科学技术出版社,1482~1488.

[12] 薛庆中(2010). DNA 和蛋白质序列数据分析工具[M]. 北京:科学出版社,75~106.

[13] 张龙,李卫华,姜淑梅,朱根发,王碧青,李洪清(2008). 花色素苷生物合成与分子调控研究进展[J]. 园艺学报,35(6):909~916.

[14] 周琳,王雁,任磊,彭镇华(2011). 牡丹二氢黄酮 4-还原酶基因 PsDFR1 的克隆及表达分析[J]. 植物生理学报,47(9):885~892.

[15] Dick C A, Buenrostro J, Butler T, Carlson M L, Kliebenstein D J, Whittall J B (2011). Arctic mustard flower color polymorphism controlled by petal-specific down regulation at the threshold of the anthocyanin biosynthetic pathway. PLoS One, 6 (4): e18230.

[16] Johnson E T, Ryu S, Yi H, Shin B, Cheong H, Choi G (2001). Alteration in a single amino acid changes the substrate specificity of dihydroflavonol 4-reductase. Plant J, 25 (3): 325~333.

[17] Petit P, Granier T, Estaintot BL, Manigand C, Bathany K, Schmitter J M, Lauvergeat V, Hamdi S, Gallois B (2007). Crystal structure of grape dihydroflavonol 4-reductase, a key enzyme in flavonoid biosynthesis. J Mol Biol, 368 (5): 1345~1357.

[18] Polashock J J, Griesbach R J, Sullivan R F, Vorsa N (2002). Cloning of a cDNA encoding the cranberry

dihydroflavonol-4-reductase(DFR) and expression in transgenic tobacco. Plant Sci, 163 (2): 241~251.

[19] Singh K, Kumar S, Yadav S K, Ahuja P S (2009). Characterization of dihydroflavonol 4-reductase cDNA in tea [*Camellia sinensis*(L.) O. Kuntze]. Plant Biotechnol Rep, 3 (1): 95~101.

(本文发表于《植物生理学报》,2014 年)

蓝桉 6×6 全双列交配生长性状的遗传效应分析

李淡清[1],刘永平[2],曾德贤[2],钟开鹏[3],范林元[2]

(1. 云南省林业科学院,云南 昆明 650204;2. 云南省林木种苗站,云南 昆明 650215;
3. 云南省禄丰县一平浪林场,云南 禄丰 651214)

摘要:以蓝桉 6×6 完全双列交配的子代测定林为试验材料,对树高、胸径、材积、枝下高、冠幅、弯曲度和枝径比 7 个性状的遗传效应进行分析。结果表明,这些性状主要受一般配合力(GCA)、特殊配合力(SCA)和特殊正反交效应(SRE)的控制,母本效应(FEM)对弯曲度有较小作用,而父本效应(MAL)对所有性状影响甚微。有 3 个亲本对改良数量性状或树干品质有显著的作用,有 8 个组合对提高数量性状遗传增益的贡献极显著或显著。蓝桉的数量性状既有自交衰退,也有自交优势。对蓝桉育种的策略和方法也提出了建议

关键词:蓝桉;双列交配;遗传效应;GCA;SCA;生长性状

Analysis of Genetic Effects for Growth Traits of *Eucalyptus globulus* Labill. in a 6×6 Diallel Design

Li Dan-qing[1], Liu Yong-Ping[2], Zeng De-xian[2], Zhong Kai-peng[3], Fang Lin-yuan[2]

(1. Yunnan Academy of Forestry, kunming Yunnan 650204, China; 2. Tree Seed
and Seedling Station of Yunnan Province, Kunming Yunnall 650215, China;
3. Yiping lang Forest Farm, Lufeng Country, Yunnar 651214, China)

Abstract: The genetic effects of general and specific combining ability, male, female, and specific reciprocal cross were estimated for height, diameter of breast height (dbh), volume, crown width, under branch height, crook degree and ratio of branch with shoot in Eucalyptus globulus Labill. . Three quantitative characters, i.e. tree height, dbh, volume were focused, and their effects were discussed in detail. It was found that the GCA, SCA and specific reciprocal cross effects (SRE) were most important, and female effect (FEM) and male effect (MAL) were less important. Three parents with high GCA were good materials used for establishing seed orchard. Eight combinations significantly improved quantitative characters. One of them was improved on quantitative character and trunk quality at the same time. The breeding strategy and methods were also suggested for E. globulus.

Keyword: Eucalyptus globulus; complete diallel; genetic effects; combining ability; growth traits

蓝桉(*Eucalyptus globulus* Labill.)原产于澳大利亚的塔斯马尼亚的东海岸,维多利亚南部沿海及其附近岛屿,大约在南纬 38.5°~43.5°。原产地气候温凉湿润,多为冬雨型[1]。它最先引种到法国、英国、意大利等国家,现在欧洲、美洲、亚洲、非洲等的许多国家都有种植,以西班牙、葡萄牙、中国、巴西、印度、埃塞俄比亚等国栽培较多[2]。蓝桉生长迅速,适应性强,已成为世界上重要的工业纤维树种之一,主要用于制浆和造纸。西班牙就是靠栽培蓝桉,在短短的 10 年间由一个纸浆进口国变为出口国。我国引种蓝桉的历史已有 110 年以上,在云南表现最好,已成为重要的造林树种。因此,早在 15 年前就从本地的

基本群体中选出优树,建立了实生种子园[3,4]。为了解此树种的遗传效应、进一步提高育种水平,1992~1993年在种子园中作了6×6完全双列交配。

双列交配方法,可以估算有关群体的遗传和环境参数,具有统计上的稳健性,较高的统计效率和精度。它在作物育种中应用相当普遍[5~7],现已应用到林木育种中来,国内以杉木做得最多[8,9],国外主要是用在一些针叶树上。有关桉树遗传效应的研究还很罕见。

1 材料和方法

1.1 供试材料

选择种子园中生产区内生长量大、长势好、干形优、当年花蕾多的蓝桉6株,按完全双列交配设计进行控制授粉,种子成熟后采种育苗,然后上山定植。获得种子及苗木情况见表1。除G-89×G-59仅得到少量种子所育苗木甚少未参与实验外,其余组合都参加了实验。蓝桉的供试材料比直干桉好得多。

表1 蓝桉6×6完全双列交配亲本及F1代号
Table 1 Code of parents and F1 of 6×6 diallel mating in *Eucalyptus globulus*

♂ ♀	G-7	G-9	G-41	G-59	G-60′	G-89
G-7	1	7	13	19	25	31
G-9	2	8	14	20	26	32
G-41	3	9	15	21	27	33
G-59	4	10	16	22	28	34(a)
G-60′	5	11	17	23	29	35
G-89	6	12	18	24	30	36

注:(a) 得到少量种子,但无苗造林。

1.2 田间试验

1996年在一平浪林场大平地用完全随机区组设计,5株单行小区造林,株行距2m×3m,共8个区组,以种子园混合种子作对照。由于某些组合苗木不足,部分区组内有组合缺失。试验地位于101°55′E,25°09′N,海拔1 860m,土壤为发育在砂岩上的山地红壤。

每年1月,在桉树停止生长时,每木实测树高、胸径、冠幅、冠长、弯曲度、枝径比,并按 $V=4-1\times\pi fHD^2$ 计算单株材积(f为形数,取0.44,H为树高,D为胸径)。以树木最弯处的弓形高与弦长之比评估弯曲度(为计算方便,所得数值乘以100),用最粗枝条靠近树干处的直径与该处分枝上的树干直径的比值作为枝径比。

1.3 统计分析方法

1.3.1 不均衡资料的方差分析
用SAS System的广义线性模型完成不平衡数据的方差分析。

1.3.2 遗传效应分析
用南京林业大学SPQG软件包中的DAILL程序对双列交配设计进行方差分量及效应值估算[10]。

2 结果与分析

蓝桉是工业纤维林,收获的对象主要是木材。因此,本文重点研究树高、胸径和材积3个数量性状。枝下高、冠幅、弯曲度和枝径比与木材的品质及桉叶的产量有关,必要时也加以讨论。

2.1 组合各性状的方差分析

试验林在3年5个月时,各性状的平均值为:树高10.76m;胸径10.18cm;材积0.041 77m^3;枝下高5.50m;冠幅1.76m;弯曲度3.18;枝径比0.15。对它们作方差分析(表2)看出,7个性状组合间的差异都在0.10水平以上显著,3个数量性状显著或极显著。区组差异显著或极显著是试验林的立地条件差异较大。

表2 蓝桉各性状方差分析结果(F值)
Table 2 Analysis of variance for growth traits of E. globulus (F)

变异来源 VS	树高 Height	胸径 dbh	材积 Volume	枝下高 Under branch height	冠幅 Crown-width	弯曲度 Crook degree	枝径比 Ratio of branch with shoot
组合 Combinations	2.04**	1.543*	1.499*	4.290**	1.478*	1.385+	1.378+
区组 Blocks	2.99**	2.707**	2.802**	8.330**	3.419**	2.357*	2.509*

注:**,*,+表示在0.01,0.05,0.10水平上显著,下同。

2.2 各性状的遗传效应分析

Griffing(1956)将双列交配设计的组合均值分解为一般配合力(GCA)、特殊配合力(SCA)、正反交效应(REC)和随机误差。Keuls(1977)和Garretson(1978)在此基础上,又进一步将正反交效应分解为一般正反交效应(GRE),包括父本效应(MAL)和母本效应(FEM),以及特殊正反交效应(SRE),并提供了分析模型[8,11,12]。该模型用最小二乘法原理,对数据有缺失的双列交配设计进行分析,仍具有相当高的精度和可靠性。

在本研究中,由最小二乘法所得的期望均方为:

GCA:$Ve+Vsre+2.944\ 7Vmal+2.944\ 7Vfem+1.666Vsca+11.726Vgca$

SCA:$Ve+Vsre+0.027\ 96Vmal+0.027\ 96Vfem+1.669Vsca$

FEM:$Ve+0.999\ 999\ 7Vsre+0.028\ 57Vmal+5.829Vfem+0.874Vsca+5.926Vgca$

MAL:$Ve+0.999\ 999\ 7Vsre+5.828Vmal+0.028\ 57Vfem+0.874Vsca+5.926Vgca$

SRE:$Ve+Vsre-0.046\ 59Vmal-0.046\ 59Vfem-0.02.107Vsca$

误差:Ve

式中的 Ve、$Vsre$、$Vmal$、$Vfem$、$Vsca$ 和 $Vgca$ 分别为误差、特殊正反交、父本、母本、特殊配合力和加性效应的方差分量,其相应的自由度分别为168,9,5,5,15和5。

用随机模型分析各效应的方差分量和比例(表3)看出,GCA对各性状都有影响,以对枝径比、枝下高和弯曲度最强,均在60%以上;冠幅最弱,仅5.9%。在数量性状中,对树高的控制最强(53.5%),对材积(32.2%)和胸径(20.0%)也有一定作用。SCA对冠幅的控制最强(72.0%),对树高也有相当大的作用(39.7%),对材积和枝径比几乎没有什么作用,对其余性状的控制也很低。FEM仅对弯曲度和枝下高有影响,而MAL对所有性状的贡献都微乎其微。

表3 蓝桉各遗传效应的方差分量及百分比
Table 3 Variance component and its percentage for various genetic effects

遗传效应 Genetic effect	树高 Height	胸径 dbh	材积 Volume	枝下高 Under branch height	冠幅 Crown-width	弯曲度 Crook degree	枝径比 Ratio of branch with shoot
2GCA	0.175 8 (53.5%)	0.164 2 (20.0%)	0.000 023 (32.2%)	0.315 7 (69.0%)	0.000 884 (5.9%)	0.449 0 (63.1%)	0.000 137 (75.3%)
SCA	0.130 2 (39.7%)	0.125 3 (15.2%)	- (0)	0.137 7 (30.1%)	0.010 795 (72.0%)	0.117 2 (16.5%)	-
FEM	-	-	-	0.004 4 (1.0%)	-	0.145 8 (20.5%)	-
MAL	-	-	-	-	-	-	-
SRE	0.022 3 (6.8%)	0.532 3 (64.8%)	0.000 049 (67.8%)	-	0.003 318 (22.1%)	0	0.000 045 (24.7%)
h_N^2 N2	0.32	0.12	0.18	0.56	0.03	0.35	0.40
h_B^2 B2	0.59	0.60	0.56	0.81	0.48	0.55	0.53

注:-表示方差分量为负值,按约定视为0,括号内为各遗传效应方差分量占总遗传方差分量的百分数

对蓝桉的数量性状进行遗传改良时,除要利用GCA外,要特别注意SRE的作用,它对材积和胸径的贡献率分别达到67.8%和64.8%。供试材料为全同胞家系,按有关文献[13]估算出狭义遗传力(h_N^2)和广义遗传力(h_B^2)。从中看出,树高、枝下高、弯曲度和枝径比的h_N^2在0.3~0.6之间,选择育种效果较好。对材积、胸径等性状则不太理想,h_N^2都在0.2以下,而它们的h_B^2是h_N^2的4~5倍,用其他方式改良会取得更好的效果。这些都是今后制定育种策略的依据。

2.3 亲本的一般配合力分析

对亲本各性状所作的一般配合力分析表明(表4),在选择育种中,G-9和G-89对改良数量性状将起到很重要的作用,G-89尤其优良,它可同时使数量性状和树干品质得到改良。G-7对改良树干品质有着明显的优势,同时也不会对数量性状产生明显的负面影响。这3个是优良亲。其余3个亲本的数量性状和枝下高的加性效应值都是负数,并且大多数都达到0.01水平,不宜将它们作为选择育种的亲本,但并不是说它们就不可能成为优良的杂交亲本。

表4 6个亲本各性状的加性效应估算值
Table 4 Estimates of GCA for 6 parents

亲本	树高	胸径	材积	枝下高	冠幅	弯曲度	枝径比
G-7	0.078 4	-0.127 3	-0.001 1	0.375 1**	-0.102 8**	-0.406 8**	-0.004 6**
G-9	0.390 7**	0.181 0*	0.003 3**	0.165 5**	-0.013 7	0.904 1**	-0.002 2
G-41	-0.138 4**	-0.272 1**	-0.002 3**	-0.138 3**	-0.009 6	0.117 4	0.015 4*
G-59	-0.195 9**	-0.038 8	-0.001 6	-0.174 7**	0.043 5**	-0.126 0	0.002 7
G-60′	-0.472 4**	-0.313 2**	-0.003 9**	-0.676 0**	0.053 9**	0.100 9	-0.009 1**
G-89	0.350 3**	0.618 8**	0.005 9**	0.473 4**	0.035 3*	-0.654 6**	-0.002 2

2.4 各组合的特殊配合力效应分析

从这些组合的树高、胸径和材积的特殊配合力效应值（表5）看出，蓝桉有些自交组合严重衰退，如2×2、3×3、6×6的SCA值都是负值，绝大多数都达到极显著。但有些组合，如1×1的胸径和材积、4×4的树高、5×5的树高、胸径和材积均为正值，且极显著，具有明显的自交优势。

表5 各组合的特殊配合力效应估算值
Table 5 Estimates of SCA for different combinations

组合	树高	胸径	材积	枝下高	冠幅	弯曲度	枝径比
1×1	-0.001 2	0.369 9**	0.003 8**	-0.299 4**	0.124 6**	0.485 4**	-0.007 7**
2×2	-1.197 5**	-1.693 5**	-0.013 2**	-0.266 8**	-0.107 9**	-0.060 2	0.000 7
3×3	-0.837 1**	-2.429 7**	-0.018 9**	0.602 2**	-0.485 7**	-0.566 9**	-0.010 8**
4×4	0.443 2**	0.137 0	-0.000 5	0.216 8**	-0.009 6	0.247 5	0.008 0**
5×5	0.871 2**	0.484 1**	0.004 8**	0.923 3**	-0.068 4**	1.268 7**	0.001 7
6×6	-1.044 2**	-1.682 0	-0.016 5**	-0.241 2**	-0.158 8**	0.273 2**	-0.018 8**
1×2	0.331 4**	-0.032 0	-0.001 4	0.257 5**	-0.032 1	-0.144 5	-0.012 3**
1×3	0.374 1**	0.663 1**	0.006 0**	0.189 0**	0.065 3**	-0.074 2	0.004 0*
1×4	-0.271 4**	-0.347 1**	-0.005 1**	-0.334 2**	-0.035 3	0.187 2	0.009 2**
1×5	-0.250 0**	-0.200 1**	0.001 1	0.133 7**	-0.040 5*	-0.375 8**	-0.002 7
1×6	-0.182 6**	-0.500 6**	-0.005 4**	0.041 4	-0.087 8**	-0.032 3	0.008 5**
2×3	0.085 0	0.586 0**	0.002 8**	-0.406 4**	0.063 1**	1.470 6**	0.017 1**
2×4	0.073 4	0.567 3**	0.005 3**	0.276 9**	0.064 8**	-0.965 6**	-0.000 8
2×5	-0.013 6	-0.061 3	-0.000 7	-0.406 9**	0.070 0**	-0.316 2**	-0.004 7**
2×6	0.694 7**	0.587 0**	0.006 5**	0.533 7**	-0.063 8**	0.061 6	-0.000 9
3×4	-0.048 9	-0.123 1	-0.001 1	-0.240 3**	-0.016 0	-0.153 8	-0.007 5**
3×5	-0.108 9	0.085 2	-0.001 3	-0.236 5**	0.054 8**	-0.588 9**	-0.000 02
3×6	0.536 1**	1.171 8**	0.011 5**	0.079 9	0.312 6**	-0.051 1	-0.003 7**

(续)

组合	树高	胸径	材积	枝下高	冠幅	弯曲度	枝径比
4×5	-0.399 0**	-0.573 4**	-0.004 2**	0.029 5	-0.033 4*	0.452 8**	-0.009 2**
4×6	0.431 7**	0.732 0**	0.012 5**	0.140 1	0.039 0	0.242 2	0.004 6
5×6	-0.126 4*	0.218 8*	-0.000 3	-0.455 21**	0.011 6	-0.404 7**	0.014 0**

注：1~6分别代表编号为 G-7、G-9、G-41、G-59、G-60′、G-89 的亲本。

应当说明，并非有自交优势的组合就有优良的基因型。如 5×5 的 3 个数量性状 SCA 为正值，且极显著，但这 3 个性状的表型值却没有一个超过总体平均值。因为亲本 5 的 3 个数量性状的 GCA 值均为负，且极显著，这样 GCA、SCA、还有其他遗传效应（如 SRE）共同作用，再加上环境的影响，5×5 这个组合表现就很差。但有的组合不仅有自交优势，基因型值也高，如 1×1，其树高、胸径和材积都超过了总体平均值，可在育种中加以利用。

从 3 个数量性状上看，杂交组合中的 2×6、3×6、4×6、2×3、2×4、1×3、1×2 的 SCA 值至少在 1、2 个，甚至 3 个性状上都极显著。这些优良组合中都有一个亲本数量性状的 GCA 值极显著（如亲本 2 和 6），或对改良树干品质效果显著（如亲本 1）。因此，有可能选出 GCA 高的亲本、其后代的 SCA 也高的组合。但 GCA 和 SCA 并没有必然联系，如 1×6 和 5×6 这两个组合数量性状的 SCA 均为负值，且显著或极显著，而 1×3 的 SCA 是正值且极显著。

2.5 特殊正反交效应

特殊配合力没有考察亲本作为父本或母本时对后代的影响，即特殊正反交效应（SRE），但在林木育种实践中，这个问题必须加以考虑。现将有关数量性状 SRE 显著和极显著的正交组合的效应值列在下面：

树高：G-9×G-41，-0.648 4*；胸径：G-7×G-9，-0.647 5*；G-9×G-41，0.972 0**；G-9×G-59，-1.372 0**；材积：G-7×G-9，-0.006 4**；G-9×G-41，0.009 1**；G-9×G-59，-0.012 8**。

由上看出，SRE 对胸径和材积的影响远远大于树高，这与随机模型的分析结果是一致的。G-9×G-41、G-9×G-7、G-59×G-9，这 3 个组合的 SRE 显著或极显著，具有明显的杂种优势。尤其是 G-59×G-9 在胸径和材积上的杂种优势最为显著，不仅 SRE 效应值最大，而且其表型值分别超过总体平均值的 25.5% 和 63.6%，是所有组合中最大的。G-9×G-41 这个组合，材积增加，弯曲度减小，能在改良数量性状的同时，也使树干品质得到改良，这是选育用材林希望达到的目标。

3 讨论与建议

蓝桉无性繁殖困难，采用有性繁殖，建立种子园是目前切实可行的办法。从此研究结果看，除冠幅外，GCA 对各性状的作用都显著，说明用选择育种来改良蓝桉可取得良好的效果。建立第二代种子园，既要从已大大增加的基本群体中选优树以扩大遗传基础，也应从已作过加性选择的育种和生产群体挑选育种材料，以提高遗传增益。选择优良组合不仅以 SCA 大小，同时也考虑 GCA 的作用（即亲本之一的 GCA 值大），才能提高选择育种的效果。按树高、胸径和材积的表型值，选出了 8 个优良组合，它们是 2、10、12、13、14、18、

32和33号，其树高比总体平均值超出5.6%～14.8%，胸径超出4.9%～25.5%，材积超出14.1%～63.6%。如不考虑遗传基础狭窄的风险，用G-7建立单系种子园，用G-9和G-59、G-41和G-89建杂交（双系）种子园，用G-89与G-9、G-41、G-59等建立种子园，都可以做实验。

材积是数量性状的集中体现，胸径对材积的影响占绝对优势，但这两个性状GCA的方差分量仅20%～30%，而SRE的方差分量却高达65%～68%，后者是前者的2～3倍。若要大幅度提高这两个性状的遗传增益，还得走无性系林业的道路，即先进行杂交育种，再选择优良源株进行无性繁殖。从长远考虑，研究蓝桉大量无性繁殖技术是值得的。优良源株可以从最具杂种优势的G-59×G-9等8个组合以及上面所提的单系、杂交等类种子园的后代群体中选择。

致谢：云南省澄江林业局张必福、高永能同志参与控制授粉，谨致谢忱。

参考文献：

[1] Boland D J et al. Forest Tress of Australia. Four the dition [M]. Published by Thomas Nelson Australia, 1984, 452～453.

[2] Eldridge K G et al. Genetic Improvement of Eucalyptus globules and E. nitens—are view of the worlds cene in blue-gum breeding and relevance to China. Paper for International Academic Eucalyptus Symposium. Zhanjiang, China, 20～30 November, 1990.

[3] 李淡清. 蓝桉、直干桉优树选择研究[J]. 林业科学，1990，26（3）：167～174.

[4] 李淡清，张必福，高永能等. 蓝桉、直干桉优树子代的遗传参数和预期增益[J]. 云南林业科技，1993，（1）：1～9.

[5] 1982，391～410. 马育华. 植物育种的数量遗传学基础[M]. 南京：江苏科学技术出版社，1982，391～410.

[6] 高之仁. 数量遗传学[M]. 成都：四川大学出版社，1986，310～367.

[7] Zhu J. Analysis Method for Genetic Models. Beijing: China Agric. Press. 1997, 66～74. 朱军. 遗传模型分析方法[M]. 北京：中国农业出版社，1997，66～74.

[8] 叶志宏，施季森，翁玉榛等. 杉木十一个亲本双列交配遗传分析[J]. 林业科学研究，1991，4（4）：380～385.

[9] 胡德活，阮梓材，吴青等. 杉木生长性状配合力分析[J]. 广东林业科技，1998，14（2）：7～13.

[10] 叶志宏. SPQG软件包用户指南[M]. 南京：南京林业大学林木遗传育种教研室，1990，178～188，303～308.

[11] Keuls M et al. A General Method for the Analysis of Genetic Variation in Complete and In complete diallels and North Carolina II. Designs, Part II. proce dures and general formulas for the random model, Euphitica, 1977, 26: 537～551.

[12] Carreston F et al. A General Method for the Analysis of Genetic Variation in Complete and Incomplete diallels and North Carolina II. Designs, Part II. proce dures and general for mulas for the fixed model. Euphitica, 1978, 27: 49～68.

[13] 王明庥. 林木遗传育种学[M]. 北京：中国林业出版社. 1989，230～244.

（本文发表于《遗传学报》，2002年）

Temporal Changes in Wetland Plant Communities with Decades of Cumulative Water Pollution in Two Plateau Lakes in China's Yunnan Province Based on Literature Review

WANG Si-hai[1,2,3], WU Chao[2], XIAO De-rong[2],
WANG Juan[2,4]*, CHENG Xi-ping[5], GUO Fang-bin[5],

[1] Key Laboratory of the State Forestry Administration on Conservation of Rare,
Endangered and Endemic Forest Plants, Yunnan Academy of Forestry, Kunming 650201, China

[2] National Plateau Wetlands Research Center, Southwest Forestry University, Kunming 650224, China

[3] State Key laboratory of Phytochemistry and Plant Resources in West China,
Kunming Institute of Botany, Chinese Academy of Sciences, Kunming 650201, China

[4] Yunnan Provincial Key Laboratory of Cultivation and Exploitation of Forest Plants,
Yunnan Academy of Forestry, Kunming 650201, China

[5] Faculty of Ecotourism, Southwest Forestry University, Kunming 650224, China

Abstract: Wetland plant communities in the plateau lakes of Yunnan Province, China, have decreased significantly over the past decades. To better understand this degradation, we analyzed the processes and characteristics of changes in wetland plant communities in two of the largest lakes in Yunnan Province, Dianchi and Erhai lakes. We collected records of native and alien plant communities in the two lakes from literature published from the 1950s to current period. We calculated plant community types and their area in some historical periods when related data were reported, and analyzed the relationship between changes in plant communities and water pollution. In Dianchi Lake, 12 community types of native plant communities, covering over 80% of the surface in the 1950s and 1960s, were reduced to four types covering 2.4% by the late 2000s. Alien plant communities started to appear in the lake in the late 1970s, and have since come to cover 4.9% of the lake surface, thereby becoming dominant. In Erhai Lake, 16 types of native plant communities, covering 47.1% of the lake surface in the late 1970s, declined to 10 community types, covering 9.3% of the surface, by the late 2000s. Alien plant communities appeared in the middle 1980s, and at present cover 0.7% of the surface area. It was indicated that changes in plant communities were significant related to water eutrophication. The area occupied by native and alien plant communities was, respectively, negatively and positively related to the content of nutrients in water. This showed lacustrine pollution played an important role in native plant loss and alien plant invasion in the two plateau lakes.

Keywords: Dianchi Lake; Erhai Lake; Wetland plant community; Eutrophication; Alien plant

Introduction

Plateau lakes are water towers for downstream river basins, but they are ecologically very fragile and prone to human disturbances, for example, land reclamation and cumulative pollution, in terms of the lacustrine environment due to the geological formations and patterns of human habitation. Most of the lakes in Yunnan Province in Southwest China are sag ponds formed by land sinking along fault lines. In general the lakes are surrounded by mountains and forests, and have relatively steady water levels. These are lakes are rich in plant communities and biodiversity (Yang et al. ,2008). Around some of these large and fertile lakes, humans have been living for over 2000 years, and hence have been affecting the lacustrine environment. However, the effects of early human habitation, mostly resulting from traditional and small-scale agricultural activities, were minimal in comparison to the effects in recent decades, which are caused by industrial pollution and urban development. Lacustrine environments have experienced different magnitudes of change, and biodiversity has correspondingly degraded over the past decades (Wang et al., 2013).Plant communities are the basis of maintaining the structures and functions of plateau lakes (Li et al., 2014) and their changes may be subject to the dynamics in the lacustrine environment. The cumulative effects of the lacustrine environment on plant communities are rarely immediate and often require long-term regular monitoring to detect the changes. Processes of long-term changes (several decades) in the plant communities were rarely known owing to the lack of regular monitoring. This paper endeavors to study the processes and causes of plant community changes resulting from the cumulative impact of water pollution based on reviewing and analysis of historical data in the past decades.

Dianchi and Erhai lakes are the two largest plateau lakes in Yunnan province. Their lacustrine environment has rapidly changed in recent decades, most likely owing to human disturbance. In the 1950s there was still a high diversity of species, and many types of stable plant communities existed (Li et al.,1963; Dai,1986; Li and Shang,1989). After 1970s, great changes occurred in the species composition and community structure as a result of the deterioration of the lacustrine environment (Yu et al.,2000; Li et al.,2011; Lu et al.,2012; Xiang et al.,2013; Wang et al.,2013). Previous studies on the two lakes focused mainly on the short-term (several years) changes of plant composition and community structure. It is unknown how the lacustrine environment affects changes in plant communities. In this paper, we summarize the distribution patterns of plant communities and the changing water conditions from 1950s to 2000s, and analyze the relationship of plant community change with regard to water pollution.

1 Methods

1.1 Study sites

The drainage basin of Dianchi Lake has a northern subtropical monsoon climate with an annual rainfall of about 1 000 mm. More than 20 large and small rivers run into the lake, of which, Panlong River, the largest and longest, traverses through Kunming City, the capital of Yunnan Province. Similarly, Erhai Lake is located in the subtropical monsoon climate zone with an annual precipitation of 1 060 mm and there are about 25 main rivulets injecting into the lake. The human population has multiplied ten-fold in Dianchi Lake basin (Lian 2011), and three-fold in the Erhai Lake basin (Liao & Zhou,1993; Liao et al.,2003) in the last 60 years. Rapid population growth was accompanied by fast agricultural expansion and intensification, industrialization and urbanization, posing enormous pressure on the water quality in both lakes. Table 1 gives brief information about the two lakes (Wang et al.,2013; Fu et al.,2013).

Table 1 Characteristics of Dianchi and Erhai lakes

Character	Dianchi Lake	Erhai Lake
Co-ordinates	24°29′–25°28′ N 102°29′–103°01′ E	25°35′–25°58′ N 100°05′–100°17′E
River basin	Yangtze River	Mekong River
Lake-surface area	297.9 km^2	249.8 km^2
Catchment area	2920 km^2	2565 km^2
Lake elevation	1886 m	1974 m
Average water depth	4.4 m	10.5 m
Water volume	1.17×10^9 m^3	2.88×10^9 m^3

1.2 Data Collection

We collected records of native and alien plant communities from literature from the 1950s to the present (Table 2). The data for Erhai Lake were only available from the 1970s. The maps of community distribution from previous publications were scanned, imported into, and rectified in Auto CAD 2013. All maps of different scales from literature were scaled up or down so as to match the boundary of all maps. In the historical data of plant community distribution, the areas of individual plant communities in the maps were represented by the distribution of a single dominant species, we thus adopted the same mapping scheme. The boundary of each plant community was then redrawn on the basis of the background distribution maps and characteristics of plant distribution, and their areas were computed in the graphic software. We focused on the temporal changes of the plant communities represented by the single dominant species (Appendices I, II) in different time periods, but companions of individual communities were not considered, as the communities are mostly dominated by single species.

As our data were from literatures published in various journals, rather than from direct long-term regular monitoring, the criteria for assessing community area were not consistent, and deviation existed among different publications. Furthermore, we redrew the boundary of the plant communities by graphic software based on small-scale illustration of community distribution published in literatures, so area measurement was not precise. In contrast to long-term and large-amplitude changes in plant communities, the effect of this inaccuracy was acceptable on our conclusion. Appendices I and II show the estimated areas of plant communities over the past decades. In certain time periods, the areas of some communities were non-existent as the researchers at that time had not assessed it as reaching the level of a plant community due to scattered or insignificant distribution area.

We chose the easily available data of water pollution, the total nitrogen (TN) and total phosphorus (TP), as the indicators of water quality. Historical data of water quality were acquired from the collected literature (Table 2). Process of water pollution, and content of TN and TP were showed in Appendix III. These water pollution data were matched with area of plant communities in the same period, in order to correlate the analysis.

Table 2 Literature assessed to estimate community distribution and water quality

Type	Dianchi Lake	Erhai Lake
Community distribution	Ley et al. 1963; Li 1980; Dai 1985, 1986; Zhao et al. 1999; Yu. 2000; Shen et al. 2010; Xiang et al. 2013; Xie et al. 2013	Ley et al. ,1963; Li. ,1980, 1989; Li and Shang,1989; Qian,1989; Dai,1989; Dong et al. ,1996; Hu et al. ,2005; Li et al. ,2001; Fu et al. ,2013
Water quality	Ley et al. 1963; Wang et al. 2004; Su 2011; Lu et al. 2012	Wang et al. 1999; Li 2001; Ni 2003; Zhao 2003; Han 2005; Fu et al. 2013

1.3 Data analysis

We summed up the area of each plant community and the number of community types in each period (Appendix I and II) in which distribution of plant communities was reported by literature. Four quantified indicators were used to assess (1) the total area of all plant communities types; (2) changes in the number of plant community types; (3) the correlation coefficient between total area of all native plant community types, and TN and TP. (4) the correlation coefficient between total area of all alien plant community types, and TN and TP. All statistical analyses were conducted in Microsoft Excel 2010.

2 Results

2.1 Changes in area of native and alien plant communities

Most of the surface area of Dianchi Lake was covered by native plant communities in the 1950s and 1960s, reaching over 80%. In the late 1970s, the coverage of plant communities rapidly shrank to 20.9% of the lake surface. In the late 2000s, native plant communities covered 2.4% of the lake surface only (Figure 1). In Erhai Lake, 47.1% of the lake surface was covered by native plant communities in the late 1970s. The coverage subsequently decreased to 9.3% in the late 2000s (Figure 1). Alien plant communities were initially recorded in Dianchi and Erhai Lake in the late 1970s and the middle 1980s, respectively. The spread of these alien communities was rapid. In the late 2000s, 4.9% of the surface of Dianchi Lake was covered by alien communities, equaling twice the area of native communities, while in Erhai Lake only 0.7% of its surface was covered by alien communities (Figure 1).

2.2 Changes in the types of plant communities

The types of plant communities were reduced in recent decades. In Dianchi Lake there were 12 and 13 types of native plant communities respectively in the 1950s and the 1960s, and 12 types still remained in the late 1970s, while the area of communities substantially decreased during the 1970s (Figure1 & Figure 2), and the number of community types had decreased to 4 by the late 2000s. In Erhai Lake the process of degradation followed the same pattern, with the

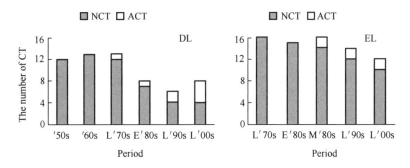

Figure 1 Changes in the total area of vascular plant communities in Dianchi Lake (DL) and Erhai Lake (EL) over the past decades. NPC, Native plant communities; APC, Alien plant communities; '50, 1950s; L'70s, the late 1970s; E'80s, the early 1980s; M'80s, the mid 1980s, and so forth.

number of community types decreasing from 16 in the late 1970s to 10 in the late 2000s (Figure 2). The types of alien plant communities increased from one in the late 1970s to four in the late 2000s in Dianchi Lake, while in Erhai Lake only two alien communities were present from the middle 1980s (Figure 2). Even though there were few alien plants at present, these invasive species had occupied much of the lake surface (Figure 1), and had posed serious ecological impacts on lakes.

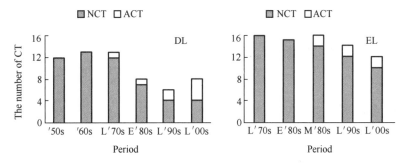

Figure 2 Changes in number of the vascular plant community types in Dianchi Lake (DL) and Erhai Lake (EL) over the past decades. NCT, native plant community type; ACT, alien plant community types; CT, community types; '50, 1950s; L'70s, the late 1970s; E'80s, the early 1980s; M'80s, the middle 1980s, and so forth.

2.3 Relationship between area of native plant communities and nutrient contents s in water

The area of native plant communities was significantly affected by nutrient load (TN and TP) in the two lakes. In Dianchi Lake the negative correlation was significant for both TN ($R^2 = 0.9546$, $P<0.05$) and TP ($R^2 = 0.9158$, $P<0.1$); similarly, in Erhai Lake negative correlation occurred with TN ($R^2 = 0.9845$, $P<0.05$) and TP ($R^2 = 0.5836$, $P<0.5$) (Figure 3).

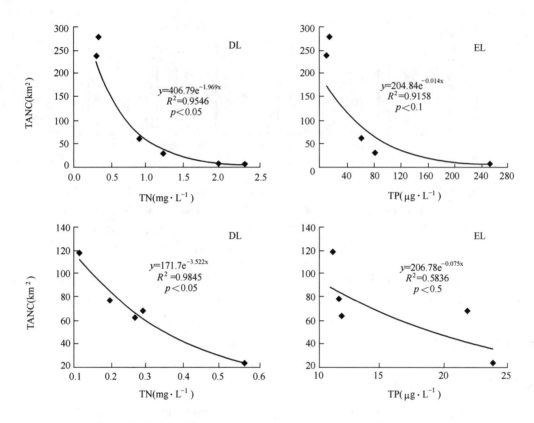

Figure 3 Relationship between area of native plant communities and content of nutrients in water in Dianchi Lake (DL) and Erhai Lake (EL). TANC, total area of native plant communities; TN, total nitrogen; TP, total phosphorus.

2.4 Relationship between area of alien plant communities and content of nutrients

The spread of alien plant communities was positively correlated with the nutrient load (TN and TP) based on the observation of the time span of 30–40 years. In Dianchi Lake the area covered by communities was strongly correlated with TN ($R^2 = 0.8818$, $P<0.01$), and with TP ($R^2 = 0.755$, $P<0.05$); similarly, in Erhai Lake, the area is strongly correlated with TN and TP ($R^2 = 0.7055$ and $R^2 = 0.9956$ respectively, $P<0.01$) (Figure 4).

3 Discussion

3.1 Impacts of cumulative pollution on native plant communities

Many abiotic and biotic factors interact to influence the performance and abundance of species in the plant community. Anthropogenic pollutant enrichment is a specific stressor on wetland systems that alter these abiotic and biotic interactions, potentially altering species composition (Mahaney et al., 2005; Porter et al., 2013). Dianchi and Erhai Lake experienced a large amount of cumulating pollutant input over the past decades. Dianchi Lake has undergone a

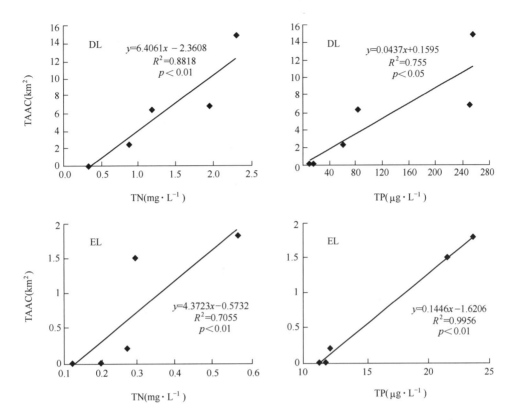

Figure 4 Relationship between area of alien plant communities and content of nutrients in water in Dianchi Lake (DL) and Erhai Lake (EL). TAAC, total area of alien plant communities; TN, total nitrogen; TP, total phosphorus.

process of rapid eutrophication since the 1960s, deteriorating from being oligotrophic to mesotrophic by the 1970s, and reaching a hypertrophication level by the middle 1990s, which continued until today (Wang et al., 2013). Similarly, Erhai Lake received increasing nutrient input since the middle 1970s, and became mesotrophic by the late 1980s. Subsequently, eutrophication continued to worsen, and while Erhai Lake is still mesotrophic, it is very close to becoming eutrophic (Wang et al., 2013). A significant negative correlation exists between lake area covered by plant communities, and the nutrient load of the water implies eutrophication is a key factor for shrinking native plant communities. Area and types of native plant communities diminished rapidly during the period in which the two lakes changing from mesotrophic to eutrophic. This perhaps indicates that it was difficult for plants living in oligotrophic water to cope with a eutrophic environment. For example, *Ottelia acuminata* covered 40% of the Dianchi Lake surface in the 1960s, but disappeared completely in the late 1970s (Dai 1986). The areas of native plant communities shrunk rapidly from 1960s to 1970s in Dianchi Lake from over 80% to 20.9%, and from 1970s to 1980s in Erhai Lake from 47.1% to 24.9%, whereas the types of native plant communities remained relatively stable, maintaining at 12 to 13 types in Dianchi Lake, and 14 to 16 types in Erhai Lake, during the corresponding time periods. The drastic loss of large plant community areas preceded the rapid disappearance of plant community types. The period of imminent disappearance of a large number of plant communities is critical in terms of interventions for plant community restoration, as it would be very difficult to restore the plant community after it has disappeared completely.

3.2 Role of alien plants in deteriorating lacustrine environment

Wetland is extremely vulnerable to plant invasions. Many wetland invaders form monotypes, which alter habitat structure, reduce biodiversity, disrupt nutrient cycling and productivity, and modify food webs (Zedler & Kercher, 2004). Human impact on global nutrient cycles can benefit invasive species; many native plant communities are susceptible to invasion by undesirable species under elevated nutrients (Ostertag & Verville, 2002; Perry et al., 2004). Increased nutrient availability may change the outcome of competitive interactions to favor invaders, resulting in their rapid spread (Kettenring et al., 2011). Pollution-tolerant alien plants, e.g. *Eichhornia crassipes*, *Alternanthera philoxeroides*, and *Pistia stratiotes*, could better adapt to eutrophic water, so these alien plants could quickly increase in abundance in the two lakes since the 1970s. The area of their spread had significant correlation with water pollution (Figure 1 & Figure 4). The area of alien plant communities has surpassed that of native plant communities in Dianchi Lake (Figure 1), and alien plants have become dominant. Nutrient inflow contributed to invasion success, and the expansion of invasive plants further suppressed the native species (Currie et al., 2013). This showed lacustrine pollution played an important role for these pollution-tolerant alien plant species in invasive success.

3.3 Possible trend of change in plant community

China has witnessed rapid economic growth driven by industrialization and urbanization since the 1950s, especially since the 1970s. Accompanying this fast development, eutrophication has become an ubiquitous problem for lakes in China (Jin et al., 1990; Xiong et al., 2010). Such a trend of degrading lacustrine ecosystems was paid more attention under the pressure of water pollution. It is more difficult to control the pollution of plateau lakes with smaller catchment area and long cycle of water replacement in the event of increasing water pollution. Great changes (lacustrine eutrophication, extinction of native plants and increasing abundance of alien plants) have occurred in Dianchi Lake in the recent five decades. With the deteriorating water quality, it can be expected that alien plant species will continue to increase in abundance in Dianchi Lake and the coverage of their respective communities will further expand.

Although Erhai Lake has remained mesotrophic until today, the water quality is gradually deteriorating. If this problem is not dealt with in time, Erhai Lake will likely face the same destiny as Dianchi Lake, and alien plant species will come to dominate the flora in the lake.

4 Conclusions

In both investigated lakes, the surface area covered by native plant communities shrank rapidly over the recent three decades. Furthermore, many native plant community types were lost, while alien plant species became more abundant rapidly. As the date of publication, alien plant species have mostly displaced native species and have become the dominant flora in Dianchi Lake.

Changes in plant communities were more drastic in Dianchi Lake than those in Erhai Lake, which was probably due to differences in water pollution levels of the two lakes. Changes in plant communities have significant correlation with water pollution. The surface area covered by native plants in the two lakes is negatively correlated with the eutrophication level, while the surface area covered by alien species is significantly positively correlated.

Considering the ongoing deterioration of the environmental conditions of the two lakes, we have to assume that native plant will continue to decrease, and that alien species will continue to expand their area of coverage in the two lakes. Erhai Lake is under serious threat of alien plant, which in the worst case scenario will eventually dominate the lake in the future.

Acknowledgements

This study was funded by the National Natural Science Foundation of China (Grant No. 31560092, U0933601), National Scientific and Technological Basic Work of China (No. 2012FY110300), Science Fund of China's Yunnan Government (Grant No. 2015BB018, 2009CC024), and the State Key Laboratory of Phytochemistry and Plant Resources in West China (Grant No. P2015-KF11). Dr. Tobias Marczewski contributed to the first revision of the original manuscript and Prof. LI Mao-biao helped with the final revisions and proof-reading of the paper.

References

Carvalho P, Thomaz SM, Kobayashi JT, *et al.* (2013) Species richness increases the resilience of wetland plant communities in a tropical floodplain. Austral Ecology 38(5): 592~598. DOI: 10. 1111/aec. 12003

Dai QY (1989) A preliminary study of the aquatic vegetation in Erhai Lake. In: Erhai Lake Management Bureau and the Science and Technology Commission of Dali Bai Autonomous Prefecture of Yunnan (eds), Collected Scientific Papers on Erhai Lake in Yunnan. The Nationalities Publishing House of Yunnan, Kunming, China. pp 235~243. (In Chinese).

Dai QY (1985) The ecological characteristics of the aquatic vegetation in the lake of Fuxian, Erhai and Dianchi in Yunnan plateau. Acta Ecologica Sinica 5(4):324~335. (In Chinese).

Dai QY (1986) Observation and analysis of aquatic vegetation in Kunminghu Lake. Transactions of Oceanology and Limnology 2: 65~75. (In Chinese).

Dong YX, Xie JP, Wang SF, *et al.* (1996) Vegetation resources of Erhai Lake and sustainable utilization. Ecological Economy 5: 15~19. (In Chinese).

Fu H, Yuan GX, Cao T, *et al.* (2013) Succession of submerged macrophyte communities in relation to environmental change in Lake Erhai over the past 50 years. Journal of Lake Sciences 25(6): 854~861. (In Chinese).

Han T, Peng WQ, Li HE, *et al.* (2005) Evolution of eutrophication in the Erhai Lake and its relevant research progress. Journal of China Institute of Water Resources and Hydropower Research 3(1): 71~78. (In

Chinese).

Hu XZ, Jin XC, Du BH, *et al.* (2005) Submerged macrophyte of Lake Erhai and its dynamic change. Research of Environment Science 18(1): 1~5. (In Chinese).

Jin XC, Liu HL, Tu QY, *et al.* (1990) Eutrophication of lakes in China. Chinese Research Academy of Environmental Sciences, Beijing, China. (In Chinese).

Kettenring KM, McCormick MK, Baron HM (2011) Mechanisms of Phragmites australis invasion: feedbacks among genetic diversity, nutrients, and sexual reproduction. Journal of Applied Ecology 48(5): 1305-1313. DOI: 10.1111/j.1365-2664.2011.02024.

Ley SH, Yu MK, Li KC, *et al.* (1963) Limnological survey of the lakes of Yunnan plateau. Oceanologia Et Limnologia Sinica 5(2):87~114. (In Chinese).

Li EH, Wang XL, Cai XB, *et al.* (2011) Features of aquatic vegetation and the influence factors in Erhai lakeshore wetland. Journal of Lake Sciences 23(5): 738~746. (In Chinese).

Li H (1980) A study on the lake vegetation of Yunnan plateau. Acta Botanica Yunnanica 2(2): 113~141. (In Chinese).

Li H (1989) A recall of aquatic vegetation of Erhai Lake. In: Erhai Lake Management Bureau and the Science and Technology Commission of Dali Bai Autonomous Prefecture of Yunnan (eds), Collected Scientific Papers on Erhai Lake in Yunnan. The Nationalities Publishing House of Yunnan, Kunming, China. pp 31~44. (In Chinese).

Li H, Shang YM (1989) Aquatic vegetation in Lake Erhai, Yunnan. Mountain Research 7(3): 166~174. (In Chinese).

Li JJ (2001) Research and counter measures for Erhai Lake eutrophication. Journal of Lake Sciences 13(2):187~192. (In Chinese).

Li W, Tan R, Yang YM, Wang J (2014) Plant diversity as a good indicator of vegetation stability in a typical plateau wetland. Journal of Mountain Science 11(2): 464~474. DOI: 10.1007/s11629-013-2864-5

Lian F (2011) Pulsation of Dianchi Lake. Yunnan People's Publishing House, Kunming, China. pp 78~79. (In Chinese).

Liao HZ, Zhang TM, Shang YM, Song LQ (2003) Research of regional synthesis exploitation and environment management programming of Erhai Lake area, Dali, P. R. China. In: Bai JK, Shang YM, Kui LX (eds), Dali Erhai Lake Scientific Research. The Ethnic Publishing House, Beijing, China. pp 521~529. (In Chinese).

Liao HZ, Zhou KM (1993) The forecast and analysis on the development of Population in Erhai Lake region. Journal of Yunnan University 15(3): 228~235. (In Chinese).

Lu J, Wang HB, Pan M, *et al.* (2012) Using sediment seed banks and historical vegetation change data to develop restoration criteria for a eutrophic lake in China. Ecological Engineering 39: 95~103. DOI: 10.1016/j.ecoleng.2011.11.006

Mahaney WM, Wardrop DH, Brooks RP (2005) Impacts of sedimentation and nitrogen enrichment on wetland plant community development. Plant Ecology 175(2): 227~243. DOI: 10.1007/s11258-005-0011-2

Ni CJ (2003) Gray Prediction on the development trend of nourishment state in Erhai Lake. In: Bai JK, Shang YM, Kui LX (eds), Dali Erhai Lake Scientific Research. The Ethnic Publishing House, Beijing, China. pp 189~191. (In Chinese).

Ostertag R, Verville JH (2002) Fertilization with nitrogen and phosphorus increases abundance of non-native species in Hawaiian montane forests. Plant Ecology 162: 77~90. DOI: 10.1023/A:1020332824836

Perry LG, Galatowitsch SM, Rosen CJ (2004) Competitive control of invasive vegetation: a native wetland sedge suppresses Phalaris arundinacea in carbon-enriched soil. Journal of Applied Ecology 41: 151~162. DOI: 10.1111/j.1365-2664.2004.00871.

Porter EM, Bowman WD, Clark CM, et al. (2013) Interactive effects of anthropogenic nitrogen enrichment and climate change on terrestrial and aquatic biodiversity. Biogeochemistry 114: 93~120. DOI: 10.1007/s10533-012-9803-3

Qian DR (1989) Investigation on aquatic vegetation in Erhai Lake. In: Erhai Lake Management Bureau and the Science and Technology Commission of Dali Bai Autonomous Prefecture of Yunnan (eds), Collected Scientific Papers on Erhai Lake in Yunnan. The Nationalities Publishing House of Yunnan, Kunming, China. pp 45~67. (In Chinese).

Shen YQ, Wang HJ, Liu XQ (2010) Aquatic flora and assemblage characteristics of submerged macrophytes in five lakes of the central Yunnan Province. Resources and Environment in the Yangtze Basin 19(Z1): 111~119. (In Chinese).

Su T (2011) Change trend and reasons of water quality of Dianchi Lake during the Eleventh Five-year Plan period. Environmental Science Survey 30(5): 33~36. (In Chinese).

Wang SH, Wang J, Li MB, et al. (2013) Six decades of changes in vascular hydrophyte and fish species in three plateau lakes in Yunnan, China. Biodiversity and Conservation 22: 3197~3221. DOI: 10.1007/s10531-013-0579-0

Wang YF, Pan HX, Wu QL, Huang Q (1999) Impacts of human activity on Erhai Lake and countermeasures. Journal of Lake Sciences 11(2): 123~128. (In Chinese).

Wang YZ, Peng YA, Li YM (2004) The characteristic of water pollution and engineering-oriented prevention on Dianchi. Areal Research and Development 23(1): 88~92. (In Chinese).

Xiang XX, Wu ZL, Luo K, et al. (2013) Impacts of human disturbance on the species composition of higher plants in the wetlands around Dianchi Lake, Yunnan Province of Southwest China. Chinese Journal of Applied Ecology 24(9): 2457~2463. (In Chinese).

Xie J, Wu DY, Chen XC, et al. (2013) Relationship between aquatic vegetation and water quality in littoral zones of Lake Dianchi and Lake Erhai. Environmental Science & Technology 36(2): 55~59. (In Chinese). DOI: 10.3969/j.issn.1003-6504.2013.02.012

Xiong YQ, Wu FC, Fang JD (2010) Organic geochemical record of environmental changes in Lake Dianchi, China. Journal of Paleolimnol 44: 217~231. DOI:10.1007/s10933-009-9398-4

Yang YM, Tian K, Hao JM, et al. (2004) Biodiversity and biodiversity conservation in Yunnan, China. Biodiversity and Conservation 13: 813~826. DOI: 10.1023/B:BIOC.0000011728.46362.3c

Yu GY, Liu YD, Qiu CQ, Xu XQ (2000) Macrophyte succession in Dianchi Lake and relations with the environment. Journal of Lake Sciences 12(1): 73~80. (In Chinese).

Zedler JB, Kercher S (2004) Causes and consequences of invasive plants in wetlands: opportunites, opportunists and outcomes. Critical Reviews in Plant Sciences 23(5): 431~452. DOI: 10.1080/07352680490514673

Zhao FQ (2003) The water quality analysis and synthesis evaluation of Erhai Lake (1980-1990). In: Bai JK, Shang YM, Kui LX (eds.), Dali Erhai Lake Scientific Research. The Ethnic Publishing House, Beijing, China. pp 313~316. (In Chinese).

Zhao S, Wu XC, Xia F (1999) Aquatic plants in Dianchi Lake. Environmental Science of Yunnan 18(3): 3~8. (In Chinese)

几十年的累积性水污染对云南两个
高原湖泊植物群落的影响

摘要：水生植物群落是维持高原湖泊生态系统稳定的基础，植物群落的变化将对湖泊生态环境造成多方面的影响。云南的滇池和洱海由于受人类的强度影响，水生植物群落在近几十年间发生了很大变化。为了查明两个湖泊水生植物群落变化的程度，文章分析了近几十年植物群落变化情况，以及与水污染的关系。滇池在 20 世纪 50 和 60 年代本地植物群落占湖泊总面积的 90% 以上，目前仅为 2.4%，群落类型从 12 种减少到 4 种；外来植物在 70 年代被引入滇池，目前已发展成为湖泊的优势种，占湖泊总面积的 4.9%。在 70 年代，洱海本地植物群落占湖泊总面积的 47.1%，目前仅为 8.6%，群落类型从 16 种减少到 10 种；外来植物群落表现出快速发展趋势，已占湖泊总面积的 0.7%。污染更为严重的滇池比污染相对较轻的洱海植物群落变化表现更为迅速。从近几十年的时段表现看，水生植物群落面积的变化与湖泊污染程度紧密相关。本地水生植物群落总面积与污染物（TN 和 TP）浓度呈显著的负相关关系，外来水生入侵植物群落总面积与污染物（TN 和 TP）浓度呈显著的正相关关系。从研究结果分析，由于这两个湖泊的环境污染状况没有得到有效的改善，在未来很有可能本地植物群落继续减少，外来植物群落继续快速扩张。特别对于洱海，如果水环境污染趋势得不到控制，外来入侵植物群落会逐步取代本地中，发展成为优势种，在湖泊中占统治地位。

[本文发表于 *Journal of Mount Science*, 2017, 14(7)]

Somatic Embryogenesis in Mature Zygotic Embryos of *Picea likiangensis* (Franch.) Pritz

CHEN Shao-yu[1], CHEN Shan-na[1], CHEN Fang[2], WU Tao[2],
WANG Yin-bin[3] & YI Shan-Jun[3]

[1] College of Life Science, Yunnan Universty, Kunming Yunnan 650091, China

[2] Laboratory of Forest Plant Cultivation and Utilization, Yunnan Academy of Forestry, Kunming Yunnan 650204, China

[3] Resource Department, Southwest Forestry College, Kunming Yunnan 650224, China

Abstract: Somatic embryogenesis (SE) was successfully induced from mature zygotic embryos of seven families of *Picea likiangensis* (Franch.) Pritz after 20 weeks culture on initiation, proliferation and maturation media. The medium for embryogenic cultures (EC) initiation is one-half strength LM medium supplemented with combination of 2, 4 - dichlorophenoxyacetic acid (2, 4 - D) and 6 - benzyladenine (6-BA). The initiation frequencies of EC varied greatly from different families when culturing on the same initiation medium. The highest frequency (41.3%) was induced from one of the families on one-half strength LM medium supplemented with 3 mg $\cdot$ L^{-1} 2,4-D and 1.5 mg L^{-1} 6-BA and 15.37% on average for seven families. EC were subcultured and proliferated on the same medium as the initiation one every two weeks. Cotyledonary-stage embryos were observed after EC were transferred to maturation media of one-half strength LM medium containing 20-80 mg $\cdot$ L^{-1} abscisic acid and 7.5% polyethylene glycol. Most cotyledonary-stage embryos germinated normally on DCR medium containing 0.2% activated carbon. The success on SE induction of the species has provided an effective clonal propagation method for this important tree's genetic improvement.

Key words: *Picea likiangensis* (Franch.) Pritz; somatic embryogenesis; Embryogenic cultures; Embryo suspensor masses; plant growth regulators.

Abbreviations: 2,4-D, 2,4-dichlorophenoxyacetic acid; 6-BA, 6-benzyladenine; ABA, abscisic acid; PEG, polyethylene glycol; AC, activated carbon; SE, somatic embryogenesis; EC, embryogenic cultures; ESMs, embryo suspensor masses; PGRs, plant growth regulators.

Introduction

Distributed widely in Yunnan Province of China, *Picea likiangensis* (Franch.) Pritz is the only tree species in genus of Picea growing in the Province. Although it is one of the most important conifers used as timber and garden trees, little progress on its genetic improvement has been made because of its slow growth and low rate of seed maturation in nature. Sexual propagation has been the normal reproduction for this species, which leads to a great variation in commercially interested traits. Due to the long flowering time and difficulty in controlled pollination, it is difficult to produce sufficient seeds for genetic improvement. If clonal propagation is possible, however, we can capture the benefits of genetic variation or genetic engineering programs to improve this species in many commercially interested traits, including growth, wood quality and uniformity.

SE is one of the most practical ways in large scale vegetative propagation. In some cases, SE is favoured over other methods of vegetative propagation because of the possibility to scale up the propagation by using bioreactors. In addition, the somatic embryos or the embryogenic cultures can be cryopreserved, which makes it possible to establish gene banks (von Arnold et al. ,2002). Embryogenic cultures are also an attractive target for genetic modification. Although more and more conifer species have been successful in SE induction (Stasolla et al. ,2002a, 2003), it still remains limited in practical application because of low regeneration frequency, EC initiation genotype-dependent and explant limited (always mature or immature zygotic embryo), especially for some of conifer tree species.

Conifer SE has been demonstrated for many genera (Park Y. S. ,2002, Sutton B. ,2002, Tautorus T. E. et al. ,1991). SE proceeds through initiation, proliferation, maturation and germination. Since the first description of somatic embryogenesis in *Picea abies* in 1985 (Chalupa, 1985, Hakman & von Arnold, 1985), somatic embryogensis has been reported in some *Picea* trees, including *Picea abies* (Egertsdotter & von Arnold, 1998), *Picea glauca* (Tremblay 1990) and *Picea meyeri* (Yang et al. ,1997), but besides *Picea meyeri*, this has not been done in any other *Picea* species in China. This study was the first in China to establish protocols for somatic embryogenesis in Picea *likiangensis* (Franch.) Pritz.

Material and methods

Plant material

Mature seeds were collected from open-pollinated cones of seven *Picea likiangensis* (Franch.) Pritz families in October 2005 in Lijiang Prefecture of Yunnan. The seeds from different families were stored in plastic bags seperately at 4℃ before they were used. Mature zygotic embyos were excised and used as explants.

Seeds were disinfested by immersion in 70% v/v ethanol for 30 s and in 0.1% mercuric chloride for 30 min, rinsed 3 times in sterile, distilled water and soaked overnight in sterile water. Mature zygotic embryos were aseptically removed from disinfested seeds and placed on solidified callus induction medium. Each treatment consisted of about 150 explants, and the experiment was repeated twice.

Induction of EC—explants, media and culture conditions

Two groups of experiment were arranged for initiation of embryogenic cultures (EC). One group was to explore effects of basal media and plant growth regulators (PGRs) combinations on EC initiation. Three basal media were used in the experiments and they were one-half strength LM medium(1/2 LM) (Litway et al. ,1985), one-half strength LP medium (1/2 LP) (von Arnold & Eriksson 1981) and improved LP medium (decreasing concentration of NH_4NO_3 from 1200 mg · L^{-1} to 600 mg · L^{-1}), containing various combinations of 2,4-dichlorophenoxyacetic acid (2,4-D) and 6-benzyladenine (BA) concentration in each basal media (Table 1). In this group experiment, explants were mature zygotic embryos picked randomly from mixed seeds of seven families.

Based on the first group experiment, another group experiment was set to investigate EC initiation from different families. This was to put mature zygotic embryos of seven families respectively on two kinds of media selected from former experiments (Table 2).

The concentrations of sucrose, glutamine, casein hydrolysate, inositol and agar in all media were 20 g · L^{-1}, 500 mg · L^{-1}, 1 g · L^{-1}, 100 mg · L^{-1} and 7 g · L^{-1}, respectively. The pH was adjusted to 5.8 before phytagel was added. The media were then sterilized by autoclaving. Stock solutions of glutamine and casein hydrolysate were filter-sterilized and added to the media after they cooled to about 50℃. 15 excised mature embryos were laid on culture media in each petri dishes, which grew in darkness at 21℃.

The EC initiation frequency was assessed after 12 weeks culture.

EC proliferation, somatic embryos mature and germination

EC was subcultured on proliferation medium that was same as initiation medium. After several times of subculture, they were transferred to maturation medium that was one-half strength LM medium containing 20~80 mg · L^{-1} abscisic acid (ABA) and 7.5% polyethylene glycol(PEG). For maturation, cultures were maintained in darkness at 21℃ for about 4 weeks when cotyledonary-stage embryos formed. And then the cotyledonary-stage embryos were transferred on the medium for germination. The medium for germination is DCR medium (Gupta & Durzan, 1985), containing 0.2% activated carbon (AC).

Table 1 —EC induction (%) of *Picea likiangensis* (Franch.) Pritz maturezygotic embryos on different basal media and PGRs combinations—Explants: mature zygotic embryos from mixed seeds of seven families

Basal media	PGRs combinations		EC induction/%
	2,4-D/mg · l^{-1}	BA/mg · l^{-1}	
1/2 LM	1	0.5	0.74 e
		1.0	2.20 d
		1.5	1.48 d
	2	0.5	8.14 b
		1.0	3.33 c
		1.5	4.76 c
	3	0.5	7.50 b
		1.0	9.60 b
		1.5	15.37 a
			Average for all above 5.90
1/2 LP	Same as above		0
Improved LP	Same as above		0

- 1/2 LM: one-half strength LM medium; 1/2 LP: one-half strength LP medium; Improved LP: decreasing concentration of NH$_4$NO$_3$ from 1 200 mg · L^{-1} to 600 mg · L^{-1}.
- 2,4-D: 2,4-dichlorophenoxyacetic acid; BA: 6-benzyladenine.
- EC: embryogenic cultures; PGRs: plant growth regulators.
- Percentages followed with the same letter are not significantly different at the 0.05 level of confidence.

Table 2 —EC induction (%) from seven families of *Picea likiangensis* (Franch.) Pritz on two basal media and PGRs combinations——Explants: mature zygotic embryos from seeds of seven families respectively

Families / EC induction/%	Media & PGRs combinations		Media & PGRs average
	1/2 LM + 2,4-D_3 + $BA_{1.5}$	1/2 LM + 2,4-D_2 + BA_1	
AYL-226	16.3 c	2.7 b	9.5
BYL-632	2.9 e	0 d	1.45
CYL-1102	7.4 d	2.5 b	4.95
DBS-98	41.3 a	0.6 c	20.95
EBS-690	20.0 c	10.0 a	15
FBS-802	0.7 e	0.2 c	0.45
GBS-1406	29.2 b	0.6 c	14.9
Family average	16.83	2.38	9.6

- 1/2 LM: one-half strength LM medium.
- 2,4-D: 2,4-dichlorophenoxyacetic acid; BA: 6-benzyladenine. 2,4-D_3: the concentration of 2,4-D is 3 mg l^{-1}; BA_1: the concentration of BA is 1 mg l^{-1}.
- EC: embryogenic cultures; PGRs: plant growth regulators.
- Percentages followed with the same letter are not significantly different at the 0.05 level of confidence

Results and discussion

The basal media and PGRs affected EC initiation of *Picea likiangensis* (Franch.) Pritz greatly. The typical structure of 10–12 weeks of EC initiation culture, embryo suspensor masses (ESMs) were induced, but only from the basal media of 1/2 LM with combinations of 2,4-D and BA. None were induced from 1/2 LP and improved LP. However, the frequency of EC initiation was different from the different 2,4-D and BA concentrations (Table 1). Most ECs emerged firstly from the part of hypocotyl of zygotic embryo [Fig. 1(a, b)] and Fig. 1(c) shows the typical structure of ESMs for conifers. The cultures on 1/2 LM with 3 mg · L^{-1} 2,4-D and 1.5 mg · L^{-1} BA showed the highest initiation frequency of 15.37% and 1/2 LM with 1 mg · L^{-1} 2,4-D and 0.5 mg · L^{-1} BA gave the lowest frequency of only 0.74%. The average frequency for 3 group PGR combinations on the basal medium of 1/2 LM is 5.9%.

Initiation of embryogenic cultures is very important for the application of somatic embrygenesis in conifers (Li X. Y. *et al.*, 1998). There were more than 10 species reported to undergo SE process out of all about 40 species in Picea spp. to date. And most frequently used basal media for their EC initiation were LP medium, LM medium and MS medium. Among them, LP or 1/2 LP medium was the optimal selected. However, our studies showed that 1/2 LM medium was suitable for EC initiation of *Picea likiangensis* (Franch.) Pritz. A relatively high concentration of auxins (0.4–2 mg · L^{-1}) and low concentration of cytokinins (0.1–1 mg · L^{-1}) were always needed for initiation and proliferation of spruce EC (Yang *et al.*, 1999). In our studies,

3 mg·L^{-1} 2,4-D combined with 1.5 mg·L^{-1} BA gave the highest frequency of EC initiation (15.37% for seven families on average, and 41.3% for family DBS-98). Generally, EC initiation frequencies in *Picea likiangensis*(Franch.) Pritz were lower than those reported from mature embryos of *Picea abies* (50%) (von Arnold 1987) and *Picea glauca* (10% - 50%) (Tremblay 1990).

EC initiation frequencies of mature zygotic embryos in *Picea likiangensis*(Franch.) Pritz varied greatly among seven families (Table 2). On 1/2 LM media supplemented with 3 mg·L^{-1} 2,4-D and 1.5 mg·L^{-1} BA, the frequency was from 0.7% (Family FBS-802) to 41.3% (Family DBS-98), and the average of seven families was 16.83%. While on the media of 1/2 LM supplemented with 2 mg·L^{-1} 2,4-D and 1.0 mg·L^{-1} BA, the induction frequency was from 0% (Family BYL-632) to 10% (Family EBS-690), and the average value was only 2.38%. The results suggested that EC initiation in *Picea likiangensis*(Franch.) Pritz was influenced greatly by genotypes. The reason why the differences in EC induction frequency existed among *Picea likiangensis*(Franch.) Pritz genotypes is unknown, it is possible that these differences are related to genetic characteristic of genotyps and due to the complex interaction of genes controlling adaptability in genotypes and the culture protocols.

The capacity for somatic embryogenesis is genetically determined. There are major genotype differences for this trait (von Arnold et al.,2002). The pattern of developmental response of cultured tissue is epigenetically determined and is influenced by the stage of development of the plant, the nature of the explant etc. (Litz & Gray,1995). It has been agreed generally that genotype is one of the internal factors affecting somatic embryogenesis. And many reports show the results that the SE induction varies greatly among different genotypes under completely same culture protocals (Chelial & Klimaszewska,1991, Jain et al.,1995, Tang W. et al.,2001). Somatic embryogenesis can probably be achieved for all plant species provided that the appropriate explant, culture media and environmental conditions are employed (von Arnold et al.,2002). On this point, it is possible to get SE or relatively high induction frequency for specific genotype of all plant species, especially those of elite genotypes of great commercial value tree species. In our study, the EC initiation protocols are not appropriate for some families, such as family BYL-632, CYL-1102 and FBS-802. So, for them, further explores on explant, culture media and environmental conditions need to be made.

EC of *Picea likiangensis*(Franch.) Pritz was proliferated on the same medium as the EC initiation medium with a better initiation, that is 1/2 LM with 3 mg·L^{-1} 2,4-D and 1.5 mg·L^{-1} BA. They were subcultured for interval of 2 weeks and also could be cryopreserved. Fig. 1(d) shows vigorously growing of ECs that have been subcultured for more than half a year. And they still have capacity of embryo maturation.

The EC developed to cotyledonary-stage somatic embryos after about 5 weeks of culture on media of 1/2 LM containing 20-80 mg·L^{-1} ABA and 7.5% PEG. However, maximum numbers of cotyledonary-stage somatic embryos were observed at 40-60 mg·L^{-1} ABA [Fig. 1(e, f)]. Although all the EC lines looked morphologically the same, they showed a different maturation ef-

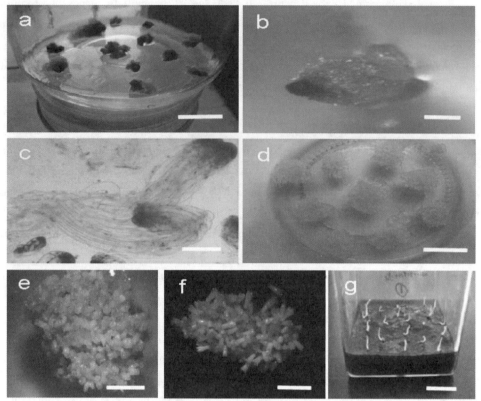

Figure 1 Somatic embrygenesis in mature zygotic embryos of *Picea likiangensis* showing SE initiation, proliferation, muturation and germination. a, b-initiation of EC from mature zygotic embryos, scale bar=5 mm in a and 1mm in b, c-embryo suspensor masses, scale bar=150 μm, d-proliferation of EC, scale bar=10 mm, e, f-maturation of SE, scale bar=2 mm, g-germination of SE, scale bar=6mm.

feciency when cultured under the same conditions. Most somatic embryos germinated normally on DCR medium, containing 0.2% activated carbon [Fig. 1(g)].

Conclusions

Somatic embryogenesis has the potential for eventual mass propagation of superior genotypes of forest trees, and for genetically engineering genotypes in both conifers and hardwood species, for most trees, especially conifers, limitations due to low initiation frequency and the genetic specificity of explants are important problems associated with embryogenesis. The present work has successfully established protocol for induction of somatic embryogenesis and regeneration of plantlets in *Picea likiangensis*(Franch.) Pritz. Further studies are needed to improve frequency of EC initiation and efficiency of somatic embryogenesis maturation and germination, especially for those of the specific elite genotypes, before this species can be propagated in large scales and be of commercially practical importance.

Acknowledgements

This work was financially supported by the Forestry Ministry of People's Republic of China (2003-4-44). We would like to express our thanks to Professor Bailian Li in North Carolina State University for his helpduring the progress of this work and critical reading on the manuscript.

References

Chalupa V. 1985. Somatic embryogenesis and plant regeneration from cultured immature and mature embryos of *Picea abies* (L.) Karst. Comm Inst For Cech, 14: 57~63.

Cheliak W. M. & klimaszewska K. 1991. Genetic variation in somatic embryogenic response in open-pollinated families of black spruce. Theor Appl Genet, 82:185~190.

Egertsdotter U. & von Arnold S. 1998. Development of somatic embryos in Norway spruce. J Expe. Bot. 49: 155~162.

GuptaP. K & DurzanD. J. 1985. Shoot multiplication from mature Douglas fir and sugar pine. Plant Cell Rep. 4: 177~179.

HakmanI & von Arnold S. 1985. Plantlet regeneration through somatic embryogenesis in *Picea abies* (Norway spruce). J Plant physiol. 121: 149~158.

Jain S M, Gupta P K & Newton R J. 1995. Somatic embryogenesis in woody plants. Volum 3: Gymnosperms Dordrecht Kluwer Academic Publishers.

Li X. Y., Huang F. H. & Gbur E. E. 1998. Effect of basal medium, growth regulators and phytagel concentration on initiation of embryogenic cultures from immature zygotic embryos of loblolly pine (*Pinus taeda* L.). Plant cell Rep. 17: 298~301

LitwayJ. D, Verma D. C & JohnsonM. A. 1985. Influence of loblolly pine (*Pinus taeda*) culture medium and its components on growth and somatic embryogenesis of the wild carrot (*Daucus carota*). Plant Cell Rep. 4: 325~328.

Litz RE & Gray DJ. 1995. Somatic embryogenesis for agricultural improvement. World journal of mocrobiology and biotechnology 11:416~425.

Park Y. S. 2002. Implementation of somatic embryogenesis in clonal forestry: technical requirements and deployment strategies. Ann. For. Sci. 59:651~656.

Stasolla C, Kong L, Yeung E C & Thorpe T A. 2002a. Maturation of somatic embryos in conifers: morphogenesis, physiology, biochemistry and molecular biology. In vitro cell dev pl, 38:93~105

Stasolla C & Yeung E C. 2003. Recent advances in conifer somatic embryogenesis: improving somatic embryo quality. Plant Cell Tiss Org. 74:15~35.

Sutton B. 2002. Commercial delivery of genetic improvement to conifer plantations using somatic embyogensis. Ann. For. Sci. 59: 657~661.

Tang Wei, Ross Whetten & Ron Sederoff. 2001. Genotypic control of high-frequency adventitious shoot regeneration via somatic organogenesis in loblolly pine. Plant science 161:267~272.

Tautorus T. E., Fowke L. C. & Dunstan D. I. 1991. Somatic embryogenesis in conifers. Can. J. Bot. 69: 1873~1899.

TremblayF. M. 1990. Somatic embryogenesis and plantlet regeneration from embryos isolated from stored seeds of Picea glauca. Can. J. Bot. 68: 236~242.

von Arnold S & ErikssonT. 1981. *In vitro* studies of adventitious shoot formation in *Pinus contorta*. Can. J. Bot. 59: 870~874.

von Arnold S. 1987. Improvement efficiency of somatic embryogenesis in mature embryos of *Picea abies* (L.) Karst. J Plant Physiol 128: 297~302.

von Arnold S. Sabala I & Bozhkov P. 2002. Developmental pathways of somatic embryogenesis. Plant cell Tiss. Org. Cult. 69:233~249.

Yang J. L, Gui Y. L, Yang Y. G, Ding Q. X & Guo Z. C. 1997. Somatic embryogenesis and plantlet regeneration in mature zygotic embryos of *Picea meyeri*. Acta Bot. Sinica. 39:315~321.

Yang J. L, Gui Y. L & Guo Z. C. 1999. Somatic embryogenesis of *Picea* Species. Chinese bulletin of Botany 16 (1): 59~66.

丽江云杉成熟合子胚的体细胞胚胎发生

摘要:经20周的诱导培养成功地由7个家系丽江云杉[*Picea likiangensis* (Franch.) Pritz]的成熟合子胚诱导产生体细胞胚胎。研究进行了添加不同浓度2,4-D和6-BA的3种基本培养基(1/2 LM培养基、1/2 LP培养基和改良LP培养基)的胚性愈伤诱导试验,结果只有在添加2,4-D和6-BA的1/2 LM基本培养基成功诱导胚性愈伤。不同家系在相同的诱导培养基中胚性愈伤的诱导率差异较大,7个家系在1/2 LM + 2,4-D 3 mg·L^{-1} + 6-BA 1.5 mg·L^{-1}培养基中胚性愈伤的平均诱导率为16.83%,其中诱导率最高的1个家系达41.3%。胚性愈伤每10天在与诱导培养基相同的培养基中继代增殖1次。之后将同一个家系诱导产生的胚性愈伤的3个细胞系用于体胚成熟培养,胚性愈伤转入成熟培养基1/2 LM + ABA 20~80 mg·L^{-1} + PEG 7.5%中培养后可观察到子叶胚的形成,其中1/2 LM + ABA 40 mg·L^{-1} + PEG 7.5%和1/2 LM + ABA 60 mg·L^{-1} + PEG7.5%培养基效果最好。超过80%的子叶胚在DCR + 0.2%活性炭的培养基中可以正常萌发。丽江云杉体细胞胚的成功诱导为这一树种的遗传改良提供了一条有效的无性繁殖途径。

(本文发表于*Biologia*,2010,65115)

Determination of Anthocyanins and Flavonols in *Paeonia delavayi* by High-Performance Liquid Chromatography with Diode Array and Mass Spectrometric Detection

HUA Mei[1,2], MA Hui-fen[1,2], Tan Rui[1,2], Yuan Xiaolong[1,2],
CHEN Jian[1,2], YANG Wei[1,2], WANG Yi[1,2], KONG Jijum[1,2],
Hu Yanli[1,2], YANG Yuming[1], WANG. Juan[1,2]

[1] Yunnan Academy of Forestry, Kunming Yunnan 650201, China
[2] Key Laboratory for Conservation of Rare, Endangered & Endemic Forest Plants,
State Forestry Administration/Yunnan Provincial Key Laboratory of Cultivation and
Exploitation of Forest Plants, Kunming Yunnan 650201, China

Abstract: The colors of the leaves in *Paeonia delavayi* change from pure green, green-red to dark red according to different habitats. The aim of this study was to compare the type and the total content of anthocyanins and flavonols in red and greenpaeonia delavayi. The constituents and concentration of anthocyanins and flavonols were identified and determined by using high-performance liquid chromatography, diode array detection and mass spectrometry. Cyanidin-3, 5-di-O-glucoside, cyanidin-3-O-glucoside, and peonidin-3-O-glucoside were first analyzed in redpaeonia delavayi. There were no anthocyanin in green paeonia delavayi. 7 flavonol compounds were identified both in red and green paeonia delavayi, only with content difference. The content of total anthocyanins and flavonoids were respectively calculated at 525 nm and 360 nm, using external standard method. The total anthocyanins of red paeonia delavayi reached 152.24 mg · 100g^{-1} in leaves and 78.92 mg · 100 g^{-1} in stems. And the total flavonoids were 805.4 mg · 100g^{-1} in leaves and 438.3 mg · 100 g^{-1} in stems, which were much higher than in green paeonia delavayi (both leaves and stems). The research will help to make clear the composition and content differences in red and green paeonia delavayi from different ecological habitats.

Keywords: Anthocyanin; Flavonol; *Paeonia delavayi*; High-performance liquid chromatography; Electrospray ionization-tandem mass spectrometry

Introduction

Paeonia delavayi is distributed mainly in the northwest Yunnan province, southwest Sichuan province and southeast Tibet, China. It is one of the wild species of arbor peony, with rich colors including white, yellow, yellow-green, orange, red, dark red, purple-red and purple in petals. *Paeonia delavayi* is very special due to its extreme variability of morphological characters in leaves and flowers at populations level, especially the diversity of flower color is very valuable for research (Hong, 1998; Jing-ji, 1999; Li, 2011).

The previous wild studies found that the leaves' color of *Paeonia delavayi* were different in diffrent environmental condition. The leaves' color of *Paeonia delavayi*is pure green from stems to leaves in low altitude areas such as Liangwang mountain(Qujing) and Xishan park(Kunming). In some regions, the colors of leaves change from green, green-red, to pure red (Erhai, Dali and Lijiang Snow Mountain). In higher altitude areas, the red leaves of *Paeonia delavayi* account for a much higher proportion than other colors in paeonia delavayi's natrual communities (Shangri-la). The pure red *Paeonia delavayi* still keep stable red traits after introducing and cultivating. Some researches also reported that the meteorological conditions and environmental factors influence the accumulation of flavonols and anthocyanins, which resulted in color diversity (Rodrigues,2011; Pérezgregorio,2014).

A lot of researches about flavonoids and anthocyanidins in peony organs have been reported. Chalcones, flavones and flavonols (Wang, 2005; Zhang, 2007) were the main components, mainly including isosalipurposide, kaempferol, quercetin, isorhamnetin, chrysoeriol and apigenin-glycopyranoside (Li,2009; Zhou,2011; Yuan,2013). There were only 6 anthocyanidins in Peony, including cyanidin-3,5-di-O-glucoside, cyanidin-3-O-glucoside, peonidin-3,5-di-O-glucoside, peonidin-3-O-glucoside, pelargonidin-3,5-di-O-glucoside and pelargonidin-3-O-glucoside (Hosoki,1991; Fan,2007; Li,2012; Gao,2012; Izquierdo-Hernández,2016). Flavonoids and anthocyanidins are synthesized through the phenylpropanoid biosynthesis pathway (Shi,2011; Luo,2017). Each compound was a very important metabolic product in phenylpropanoidbiosynthetic pathway and directly resulted in color's change (Wang,2014; Shi, 2015; Zhao,2016).

In order to help to explore the internal reasons of the leaves' turning from green to red, the chemical constituents in red and green *Paeonia delavayi* were analyzed and identified by using high-performance liquid chromatography-diode array detectionand mass spectrometry. The total content of anthocyanins and flavonols were detected both in leaves and stems in red and green *Paeonia delavayi*.

Materials and methods

Standards and solvents

Cyanidin 3-O-glucopyranoside chloride was purchased from sigma aldrich. The rutin trihydrate was obtained from Tauto biotech (Shanghai, China). Acetonitrile used for high-performance liquid chromatographyanalysis were of chromatographic grade and purchased from J&K Scientific Limited Company (Beijing, China). Methanol, acetic acid and hydrochloric acid were of analytical grade and purchased from Beijing Chemical Works (Beijing, China). Ultra-pure water was obtained from a Milli-Q advantage A10 system (Millipore, USA).

Plant materials

Red and green *Paeonia delavayi* were collected from Malong (Qujing, Yunnan, China) cultivation base of *Paeonia delavayi*, which were introduced from Liangwang mountain of Qujing city (green *Paeonia delavayi*) and Shangri-la (red *Paeonia delavayi*), respectively. The materials were collected on October 30, 2017 and stored in -20℃ condition for later analysis.

Preparation of standard solutions

Standards of Cyanidin 3-O-glucopyranoside chloride and rutintrihydrate were accurately weighted, dissolved in 1% hydrochloric acid-methanol and methanol, respectively, and then diluted to appropriate concentrations, containing 0.133 4-1.001 mg·mL^{-1} for Cyanidin 3-O-glucopyranoside chloride and 0.09~0.45 mg · mL^{-1} for rutin trihydrate.

Extraction of anthocyanins and flavonols

Approximately 2.00 g red and green *Paeonia delavayi*, whose stems and leaves were separated for detection respectively, were cut into pieces and extracted with 10 mL of 70% methanol aqueous solution (including 1% hydrochloric acid) in 4℃ refrigerator without light and over night. The extracts were filtered and passed through 0.45 μm reinforced nylon membrane filters before high-performance liquid chromatography analysis. Three replicates were performed for each sample.

High-performance liquid chromatography analysis condition and procedures

High-performance liquid chromatography analyses were performed by using Agilent 1260 and mass spectrometry analyses were performed by using Q-TOF LC/MS 6540 series. The related high-performance liquid chromatographic conditions were as follows: Eclipse plus C18 column (4.6× 150 mm, 5 μm), with 1% acetic acid (A) and acetonitrile (B) as mobile phase in gradient mode. The Eluting condition: 0~10min, 0~20% B; 10~20 min, 20%~60% B. The detective wavelength was 360 nm for flavonoidand 525nm for anthocyanidin, the flow rate was 0.8 mL · min^{-1}, the column temperature was 35℃ and injection volume was 20 μL. Absorption spectra were scanned inthe range of 200-600 nm.

Mass spectrometry analyses were carried out in both positive and negative ion mode. Anthocyanins were analyzed only in positive ion mode and flavonoids were analyzed in negative ion mode. Mass spectrometric studies were carried out on a quadrupole time-of-flight high-resolution mass spectrometer (Q-TOF LC/MS 6540 series, Agilent Technologies, Santa Clara, CA, USA) coupled with electrospray ionization. The data was acquired by using mass hunter workstation software. The MS parameters were optimized as follows: the fragmentor voltage was set at 135 V; the capillary was set at 3500 V; the skimmer was set at 65 V; and nitrogen was used as the drying (350℃, 6 L · min^{-1}) and nebulizing (25 psi) gas.

Anthocyanins and flavonols were primarily identified according to their high-performance liquid chromatographyretention times, elution order, ultraviolet absorptionspectra and mass spectrometry fragmentation pattern, and by comparison with published data.

Semi-quantitative analysis of anthocyanins and flavonols

The amount of total anthocyanins was calculated at 525 nm, in accordance with a linear regression equation generated from a standard curve of reference concentrations of cyanidin 3-O-glucopyranoside chloride. The rutin trihydratestandard curve was used for the semi-quantification of total flavonols at 360 nm.

Results and discussion

The samples of red and green *Paeonia delavayi*

The pictures of red and green *Paeonia delavayi* were listed in Fig. 1. The red *Paeonia delavayi* were full dark red in stems and leaves, comparing with pure green in green *Paeonia delavayi* (Fig. 1 b). The red *Paeonia delavayi* was introduced from Shangri-la with the natural conditon of higher altitude and stronger ultraviolet rays comparing with Liangwang mountain (green *Paeonia delavayi*). After introducing and artificial subculture analysis, we found that strong ultraviolet rays might be the main reason for color turning red. The reported research also found that the natural levels of flavonoids in fresh-cut onion slices significantly increased at 35% for flavonols, and at 29% for anthocyanins after UV irradiation (Pérez-Gregorio, 2011). The cultivated red *Paeonia delavayi* still maintain the red character in introducing region.

Figure 1　The picture of red and green *Paeonia delavayi* (a. red *Paeonia delavayi*; b. green *Paeonia delavayi*)

Table 1　The constituents identified in red and green *Paeonia delavayi*

Retention time (min)	The maximum absorption wavelength (nm)	Identification of chromatographic peaks	Electrospray ionization-tandem mass spectrometry (positive ion mode) m/z	Electrospray ionization-tandem mass spectrometry (negative ion mode) m/z
7.602	514, 278	Cyanidin-3,5-di-O-glucoside	611$[M+H]^+$, $[Y_0+H]^+$	
9.400	514, 280	Cyanidin-3-O-glucoside	449$[M+H]^+$, $[Y_0+H]^+$	
9.839	515, 282	Uncertain	431$[M+H]^+$, $[Y_0+H]^+$	
10.701	517, 279	Peonidin-3-O-glucoside	463$[M+H]^+$, $[Y_0+H]^+$	
14.724	521, 281	Uncertain	523$[M+H]^+$, $[Y_0+H]^+$	
15.216	518, 279	Uncertain	597$[M+H]^+$, $[Y_0+H]$	
11.062	254, 352	Isorhamnetin di-hexoside		639$[M-H]^-$, 477, 315 $[Y_0-H]^-$
13.341	254, 354	Salipurposide		433$[M-H]^-$, 271$[Y_0-H]^-$
13.691	262, 346	Quercetin		301$[M-H]^-$, 145$[Y_0-H]^-$
14.008	254, 348	Luteolin hexoside		447$[M-H]^-$, 285$[Y_0-H]^-$
14.398	266, 337	Apigenin-glucoside		431$[M-H]^-$, 268$[Y_0-H]^-$
14.572	254, 346	Isosalipurposide		433$[M-H]^-$, 271$[Y_0-H]^-$
14.784	254, 346	Uncertain		599$[M-H]^-$, 285$[Y_0-H]^-$
14.861	254, 338	Uncertain		629$[M-H]^-$, 13,169$[Y_0-H]^-$

(续)

Retention time (min)	The maximum absorption wavelength (nm)	Identification of chromatographic peaks	Electrospray ionization-tandem mass spectrometry (positive ion mode) m/z	Electrospray ionization-tandem mass spectrometry (negative ion mode) m/z
15.026	254,348	Uncertain		545[M-H]$^-$, 469,169[Y$_0$-H]$^-$
15.226	254,345	Uncertain		609[M-H]$^-$, 447 [Y$_0$-H]$^-$
15.552	254,348	Apigenin		271[M-H]$^-$, 151[Y$_0$-H]$^-$

Identification of anthocyanins and flavonoids in red and green *Paeonia delavayi*

Fig. 2 shows the high-performance liquid chromatograms of extracts in red *Paeonia Delavayi* at 525 nm. There were no chromatographic peaks observed in green *paeonia delavayi*, which indicated that no anthocyanin existed in green *Paeonia delavayi*.

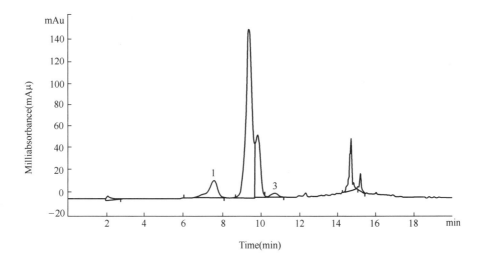

Figure 2 High-performance liquid chromatograms of red *Paeonia delavayi* at 525 nm
Peak identification: (1) Cyanidin-3,5-di-O-glucoside; (2) Cyanidin-3-O-glucoside;
(3) Peonidin-3-O-glucoside; Eclipse plus C18 column (4.6×150 mm, 5 μm), with 1%
acetic acid (A) and acetonitrile (B) as mobile phase; Gradient mode: 0-10 min, 0-20% B;
10-20 min, 20%-60% B; Detective wavelength: 525 nm (anthocyanidin); Flow rate: 0.8mL · min^{-1};
column temperature: 35℃; Injection volume: 20 μL.

Fig. 3 is the high-performance liquid chromatograms of green *Paeonia delavayi* at 360 nm. Baseline separation was achieved between each peak after using optional high-performance liquid chromatography analysis method. Flavonoids were analyzed at 360 nm, and were both detected in red and green *Paeonia delavayi* (including stems and leaves), only with content difference.

According to the high-performance liquid chromatography retention times, mass spectrometry

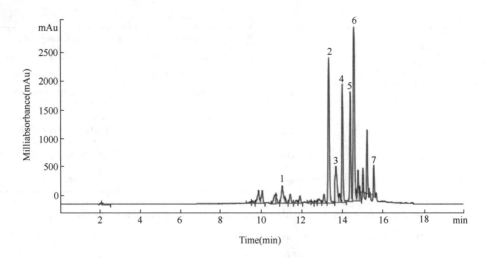

Figure 3 High-performance liquid chromatograms of green *Paeonia delavayi* at 360 nm
Peak identification: (1) Salipurposide; (2) Quercetin; (3) Luteolin hexoside; (4) Apigenin-glucoside;
(5) Isosalipurposide; (6) Apigenin; Eclipse plus C18 column (4.6×150 mm, 5 μm), with 1% acetic
acid (A) and acetonitrile (B) as mobile phase; Gradient mode: 0-10 min, 0-20% B; 10-20 min, 20-60% B;
Detective wavelength: 360 nm (flavonoid); Flow rate: 0.8mL·min^{-1}; column temperature: 35℃; Injection volume: 20 μL.

spectra, 17 compounds were detected and 10 compounds were identified from the red and green *Paeonia delavayi*, including 3 anthocyanins and 7 flavonoids. The constituents identified in red and green *Paeonia delavayi* shows in Table 1. The 3 anthocyanins were cyanidin-3,5-di-O-glucoside, cyanidin-3-O-glucoside and peonidin-3-O-glucoside respectively. The mainly anthocyanin was cyanidin-3-O-glucoside in red *Paeonia delavayi*. After serious and careful analysis, we found that all the flavonoids in red and green *Paeonia delavayi* were almost the same in composition, and the unique difference was. the content. The identified flavonoids are as follows: isorhamnetin di-hexoside, salipurposide, quercetin, luteolin hexoside, apigenin-glucoside, isosalipurposide and apigenin (according to the retention times).

The amount of total anthocyanins in red and green *Paeonia delavayi*

The amount of total anthocyanins was calculated at 525 nm, in accordance with a linear regression equation generated from a standard curve of cyanidin 3-O-glucopyranoside chloride. The linear regression equation of cyanidin 3-O-glucopyranoside chloride is as follows: A = 1 295 7.439 6 C (mg·mL^{-1})-131.325 4, R^2 = 0.998 8. The standard curve of cyanidin 3-O-glucopyranoside chloride are shown in Fig. 4. The total anthocyanins in red and green *Paeonia delavayi* are listed in Table 2. From the table, the red *Paeonia delavayi*, both in leaves and stems, have anthocyanins, and the average total anthocyanins are 152.24mg·100g^{-1} in red *Paeonia delavayi*(leaves), and 78.92mg·100g^{-1} in stems. The amount of anthocyanins in leaves and stems nearly doubled.

Eclipse plus C18 column (4.6×150 mm, 5 μm), with 1% acetic acid (A) and acetonitrile (B) as mobile phase; Gradient mode: 0-10 min, 0-20% B; 10-20 min, 20%-60% B; Detective wave length: 525 nm (anthocyanidin); Flow rate: 0.8 mL·min^{-1}; Column temperature:

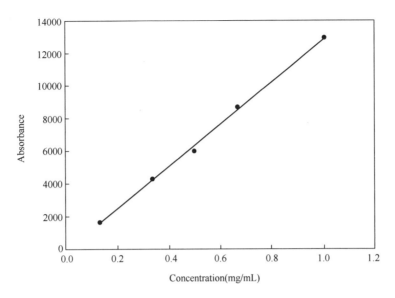

Figure 4 The standard curve of cyanidin 3-O-glucopyranoside chloride

35℃; Injection volume: 20 μL.

Table 2 The total content of anthocyanins in red and green *Paeonia delavayi*

Name of sample	Total content of anthocyanins /mg · 100g^{-1}	mean ± standard deviation/mg · 100g^{-1}
red *Paeonia delavayi* (leaves)	151.95 152.23 152.54	152.24 ±0.30
red *Paeonia delavayi* (stems)	78.45 79.30 79.02	78.92±0.43
green *Paeonia delavayi* (leaves)	— — —	—
green *Paeonia delavayi* (stems)	— — —	—

The amount of total flavonoids in red and green *Paeonia delavayi*

The rutin trihydrate were used for the semi-quantification of total flavonols at 360nm. The linear regression equation of rutin trihydrate standard as follows: A = 51 498.454 6 C (mg · mL^{-1}) + 817.175 6, R^2 = 0.993 1 (Fig. 5). The amount of total flavonoids in red and green *Paeonia delavayi* were listed in Table 3. From the table, red and green *Paeonia delavayi* both contain flavonoids. The average total flavonoids were much higher in red *Paeonia delavayi* than in green *Paeonia delavayi* (both in leaves and stems). The results are consistent with those reported in the literature.

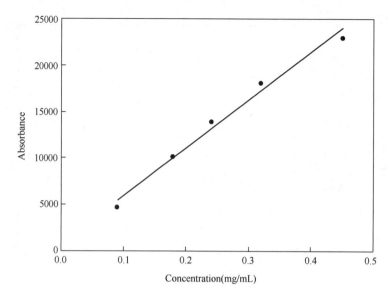

Figure 5　The standard curve of rutin trihydrate

The reported research showed that the red cultivar was about twice the flavonol level than the white cultivar in portuguese onion (Rodrigues 2012).

Eclipse plus C18 column (4.6×150 mm, 5 μm), with 1% acetic acid (A) and acetonitrile (B) as mobile phase; Gradient mode: 0-10min, 0-20% B; 10-20 min, 20%-60% B; Detective wavelength: 360 nm (flavonoid); Flow rate: 0.8 mL/min; Column temperature: 35℃; Injection volume: 20 μL.

Table 3　The total content of flavonoids in red and green *Paeonia delavayi*

Name of sample	Total content of flavonoids (mg·100g^{-1})	mean ± standard deviation (mg·100g^{-1})
red *Paeonia delavayi* (leaves)	805.8 807.6 802.9	805.4 ±2.37
red *Paeonia delavayi* (stems)	438.5 436.6 439.9	438.3±1.66
green *Paeonia delavayi* (leaves)	607.8 609.9 612.6	610.1±2.41
green *Paeonia delavayi* (stems)	245.1 247.2 248.2	246.8±1.58

Conclusions

In this study, the chemical constituents in red and green *Paeonia delavayi* were fist analyzed and identified by using high-performance liquid chromatography-diode array detection and mass spectrometry. The content of total anthocyanins and flavonoids were almost twice in red *Paeonia delavayi* than in green *Paeonia delavayi*, which implied the high altitude and strong ultraviolet rays had stimulated the accumulation of anthocyanins and flavonoids. The results were significant in exploring the reasons of the color's change from green to red in leaves and stems of *Paeonia delavayi* in different altitude areas.

Acknowledgements

This research was supported by Applied Basic Research Project of Yunnan, Grant No. 2014FD069 and Yunnan Science and Technology Program (international cooperation), Grant No. 2015IA005. Thanks Jianhong Yang for analysis and Testing Center of Kunming Institute of botany for spectra analysis of mass spectrometry.

References

Fan, J. L., W. X. Zhu, G. J. Tao, and J. W. Shen. 2007. Analysis of anthocyanins in tree peony flower by high performance liquid chromatography-electrospray ionization mass spectroscopy. Food Sci. 28: 367~371.

Gao, Y. H., S. X. Zhang, and S. Wan. 2012. Research progress in extraction and performance of peony pigment. Science and technology of food industry, 33: 431~435.

Hosoki, T., M. Hamada, T. Kando, R. Moriwaki, and K. Inaba. 2008. Comparative study of anthocyanins in tree peony flowers. Engei Gakkai Zasshi, 60: 395~403.

Hong, D. Y., K. Y. Pan, and H. Yu. 1998. Taxonomy of thepaeonia delavayi complex (Paeoniaceae). Annals of the Missouri Botanical Garden. 85: 554~564.

Izquierdo-hernández, A., Álvaro PeñaNeira, R. Lópezsolís, and E. Obrequeslier, 2016. Comparative determination of anthocyanins, low molecular weight phenols, and flavanol fractions in vitis vinifera L. cv carménère skins and seeds by differential solvent extraction and high-performance liquid chromatography. Analytical Letters, 49: 1127~1142.

Li, J. J. 1999. Chinese tree peony and herbaceous peony. Beijing: China forestry publishing house.

Li, C. H., H. Du, and L. S. Wang. 2009. Flavonoid composition and antioxidant activity of tree peony (Paeonia section Moutan) yellow flowers. Journal of agricultural and food chemistry, 57: 8496~8503.

Li, K., Y. Wang, B. Q. Zheng, X. T. Zhu, H. Z. Wu, and Q. Q. Shi. 2011. Pollen morphology of 40paeonia delavayi (Paeoniaceae) populations. Journal of Beijing forestry university, 33: 94~103.

Li, K., N. Zhou, and H. Y. Li. 2012. Composition and function research of peony flowers and peony seeds. Food research and development, 33: 228~230.

Luo, J., Q. Shi, L. Niu, and Y. Zhang. 2017. Transcriptomic analysis of leaf in tree peony reveals differentially expressed pigments genes. Molecules, 22: 324.

Pérezgregorio M. R., J. Regueiro, J. Simalgándara, A. S. Rodrigues, and D. P. Almeida. 2014. Increasing the added-value of onions as a source of antioxidant flavonoids: a critical review. Critical reviews in food science & nutrition, 54: 1050~1062.

Pérez-Gregorio M. R., C. González-Barreiro, R. Rial-Otero, and J. Simal-Gándara. 2011. Comparison of sanitizing technologies on the quality appearance and antioxidant levels in onion slices. Food control, 22: 2052~2058.

Rodrigues A. S., M. R. Pérezgregorio, M. S. Garcíafalcón, J. Simalgándara, and A. Dpf. 2011. Effect of meteorological conditions on antioxidant flavonoids in Portuguese cultivars of white and red onions. Food chemistry, 124: 303~308.

Rodrigues A. S., D. P. F. Almeida, M. S. Garcíafalcón, J. Simalgándara, and M. R. Pérezgregorio. 2012. Postharvest storage systems affect phytochemical content and quality of traditional Portuguese onion cultivars. Acta horticulturae, 934: 1327~1334.

Shi, S. C., Y. K. Gao, X. H. Zhang, J. Q. Sun, L. L. Zhao, and Y. Wang. 2011. Progress on plant genes involved in biosynthetic pathway of anthocyanins. Bulletin of Botanical Research, 31: 633~640.

Shi, Q., L. Zhou and Y. Wang. 2015. Transcriptomic analysis of paeonia delavayi wild population flowers to identify differentially expressed genes involved in purple-red and yellow petal pigmentation. PloS one, 10: e0135038.

Wang, X., C. G. Cheng, Q. L. Sun, F. Li, J. Liu, and C. C. Zheng. 2005. Isolation and purification of four flavonoid constituents from the flower of Paeonia suffruticosa by high-speed counter-current chromatography. Journal of Chromatography A, 1075: 127~131.

Wang, X., H. Fan, Y. Li, and X. Sun. 2014. Analysis of genetic relationships in tree peony of different colors using conserved DNA-derived polymorphism markers. Scientia Horticulturae, 175: 68~73.

Yuan, Y. G., C. C. Wang, and H. Y. Rao. 2013. Research progress of flavonoids in peony flowers. Journal of Shandong Institute of Light Industry (Natural Science Edition), 27: 31~34.

Zhang, J., L. Wang, Q. Shu, Z. Liu, C. Li, and J. Zhang. 2007. Comparison of anthocyanins in non-blotches and blotches of the petals of Xibei tree peony. Scientia Horticulturae, 114: 104~111.

Zhou, L., Y. Wang, C. Y. Lü, and Z. H. Peng. 2011. Identification of components of flower pigments in petals of Paeonia lutea wild population in Yunnan. Journal of northeast forestry university, 39: 52~54.

Zhao, D. Q., M. R. Wei, D. Liu and J. Tao. 2016. Anatomical and biochemical analysis reveal the role of anthocyanins in flower coloration of herbaceous peony. Plant physiology and biochemistry, 102: 97~106.

HPLC/MS 对滇牡丹叶片中花青素和黄酮的分析测定

摘要：不同的生境条件下，滇牡丹叶片的颜色有纯绿色、红绿色和深红色。为了分析比较滇牡丹红色和绿色叶片中花青素和黄酮的类型和总含量，采 HPLC/MS 技术分析检测花青素和黄酮的组成和含量。结果表明：在滇牡丹绿色叶片中并未检测到花青素，在滇牡丹红色叶片中分析得到矢车菊素-3,5-双葡糖苷、矢车菊素-3-O-葡萄糖苷和芍药素-3-O-葡糖苷三种花青素。在滇牡丹红色和绿色叶片中都分析得到 7 种黄酮化合物，只是在含量上有区别。用 HPLC 外标法，分别在 525nm 和 360nm 波长下测定花青素和黄酮的总含量。滇牡丹红色叶片中叶子和茎中总花青素含量分别为 152.24mg·100g^{-1} 和 78.92mg·100g^{-1}，总黄酮分别为 805.4mg·100g^{-1} 和 438.3mg·100g^{-1}，比滇牡丹绿色叶片和茎中的总黄酮要高很多。这一结果能帮助我们更加清楚不同的生态栖息地下滇牡丹红色和绿色叶片居群中化学组成和含量差异，进一步研究气候条件和环境因子对黄酮和花青素积累的影响。

（本文发表于 Analytical Letters, 2018, 51(15)）

Identification of A Putative Polyketide Synthase Gene Involved in Usnic Acidbiosynthesis in the Lichen *Nephromopsis pallescens*

WANG Yi[1,2], GENG Chang-an[3], YUAN Xiao-long[1], HUA Mei[1], TIAN Feng-hua[2], Li Chang-tian[2]*

[1] Key Laboratory of Forest Plant Cultivation and Utilization, Yunnan Academy of Forestry, Kunming Yunnan 650201. China;

[2] Engineering Research Center of Chinese Ministry of Education for Edible and Medicinal Fungi, Jilin Agricultural University, Changchun 130188, China

[3] State Key Laboratory of Phytochemistry and Plant Resources in West China, Kunming Institute of Botany, Chinese Academy of Sciences, Kunming 65021, China

Abstract: Usnic acid is a unique polyketide produced by lichens. To reveal usnic acid biosynthesis, the transcriptome of the usnic acid producing lichen forming fungus Nephromopsis pallescens was sequenced with Illumina NextSeq. Seven complete non-reducing polyketide synthase genes and nine highly-reducing polyketide synthase genes were obtained through transcriptome analysis. Results from gene expression assessed with qPCR and usnic acid detection by LCMS-IT-TOF showed. that NpPKS7 was involved in usnic acid biosynthesis in N. pallescens. NpPKS7 is a non-reducing polyketide synthesis gene with a MeT domain, which possessed β-ketoacyl synthase, acyl transferase, product template domain, acyl carrier protein, C-methyltransferase and claisen cyclase domain. Phylogenetic analysis showed that NpPKS7 and other polyketide synthases from lichens formed a unique monophyletic clade. Taken together, our data indicate that NpPKS7 is a novel PKS of *N. pallescens* involved in usnic acid biosynthesis.

Keywords: Lichen forming-fungi (LFF); polyketide synthase (PKS); Usnic acid; Transcriptomics; qRT-PCR; *Nephromopsis pallescens*

Introduction

Usnic acid is a unique natural compound produced by lichens, which has antibacterial (Francolini et al., 2004), antiviral (Campanella et al., 2002), and antitumor (Bačkorová et al., 2012; Song et al., 2012) bioactive properties. Several reviews about usnic acid bioactive properties show that usnic acid is a valuable pharmaceutical (Cocchietto et al., 2002; Ingólfsdóttir, 2002; Mayer et al., 2005; Guo et al., 2008).

Chemoenzymatic synthesis demonstrated that methylphloracetophenone is the precursor of usnic acid (Hawranik et al., 2009), and it was hypothesized that usnic acid biosynthesis involved a non-reducing PKS with a MeT domain. However, the biosynthesis mechanism of usnic acid remains unclear. Like other lichen substances, there are several challenges limiting the application of usual techniques, such as gene knockout or heterologous expression, to reveal mechanisms of lichen metabolite biosynthesis (Abdel-Hameed et al., 2016). Lichens are stable and self-supporting symbioses between fungi (lichen forming fungi, LFF) and photoautotrophic algal partners (Honegger, 1984). However, it has been shown to be the LFF that synthesize usnic acid and other interesting bioactive substances in lichens, rather than the algal partner (Stocker, 2008). Consequently, many LFF have been isolated, yet under laboratory conditions these LFF have failed to produce bioactive compounds detected in the lichen thallus (Molina et al., 2003; Abdel-Hameed et al., 2016). Further, there is no universal and effective transformation method of LFF (Park et al., 2013a; Wang et al., 2015).

Although several PKS genes have been cloned from LFF (Chooi et al., 2008; Wang et al., 2014; Hametner and Stocker, 2015), functional characterizations of PKS were limited to bioinformaticsapproaches. Some researchers have attempted heterologous expression of lichen PKS (Chooi et al., 2008; Gagunashvili et al., 2009; Wang et al., 2016) but were unable to demonstratede novo biosynthesis of a lichen metabolite. However, two research teams have used qRT-PCR and HPLC techniques to show that putative PKS related genes are associated with the biosynthesis of target lichen metabolites (Armaleo et al., 2011; Abdel-Hameed et al., 2016). Specifically, Abdel-Hameed et al. (2016) first reported putative an usnic acid biosynthetic gene cluster in *Cladonia uncialis*, which was followed by other methods to identify biosynthesis pathways of lichen metabolites.

Although most LFF couldn't produce lichen metabolites produced in natural lichen thallus, a few LFF can produce usnic acid in laboratory conditions, such as *Neuropogon sphacelatus* in LB (Stocker et al., 2013) and *Nephromopsis pallescens* in MY (Luo et al., 2011). Wei et al., (2008) first reported that an extract of N. pallescens exhibited antifungal activity. Luo et al., (2011) then reported an anti-*helicobacter pylorus* bioactivity and confirmed that the bioactive compound was usnic acid. In this study, the transcriptome data of the usnic acid producing N. pallescens was obtained by RNA-seq. Putative PKS genes involved in usnic acid biosynthesis obtained by bioinformatics analysis were then confirmed by the combination analysis of qRT-PCR and LCMS-IT-TOF.

Material and Methods

Lichen forming fungi

Lichen forming fungi (LFF) *Nephromopsis pallescens* (KOLRI-040516) was granted by Jae-Seoun Hur from Korean Lichen Research Institute (KoLRI), Sunchon National University.

Transcriptome sequence and analysis

Mycelium of LFF *N. pallescens*was cultured in 1.5% Malt-Yeast (MY) liquid medium at 15℃. After 2 months of culturing, the mycelia was sampled and immediately submerged in liquid nitrogen and preserved for RNA extraction. Other culture medium was collected by filtration and extracted with the same volume of Ethyl acetate. The extract was then evaporated with a Buchi Rotavapor (Flawil, Switzerland). The crude extracts were redissolved in 2 mL of MeOH, after centrifugation a 12,000 g, and filtratation through a 0.22 μm filter membrane. A 2 μl crude extract solvent from each sample was injected into a LCMS-IT-TOF for analysis. For LFF samples where usnic acid production was confirmed through LCMS-IT-TOF analyses, the frozen mycelia were ground into powder for RNA extraction. The usnic acid producing *N. pallescens* transcriptome was *de novo* sequenced by Personal Biotechnology Co., Ltd. (Shanghai, China), with an Illumina Next Seq protocol. To seek putative PKS genes, located BLAST searches with conserved KS domains were made in Nr, Egg NOGGO, KO, and Swissprot databases.

PKS Cloning

Through transcriptome analysis, the sequences of seven complete non-reducing PKS genes from *N. pallescens* were obtained. Specific primers for gene cloning were designed base on the transcriptome sequences of seven PKS genes (S1 Table 1). The RNeasy Plant Mini Kit (Qiagen) was used to isolate RNA from *N. pallescens*. cDNA was generated with SuperScript II reverse transcriptase (Invitrogen, Carlsbad, USA). The full PKS gene of *N. pallescens*was cloned using cDNA as template and pEASY-Uni Seamless Cloning and Assembly Kits (TransGen, Beijing, China).

Phylogenetic analyses

A total of 40 fungal PKS sequences were retrieved from GenBankto analyze relationships between PKS obtained from *N. pallescens* and known fungal PKSs. (S2 Table 2). These PKS protein sequences were aligned with ClustalW, implemented in MEGA 7.0.14. Phylogenetic trees were constructed using the minimum evolution method in MEGA 7.0.14 with 1000 bootstrap replicate analyses.

Different culture of LFF *N. pallescens* and HPLC-MS analysis

In previous study, we found that media composition influenced the production of antibacterial compounds by *N. pallescens*. Therefore, mycelia of *N. pallescens* weretransferredinto 100 mL of various broths including: malt-yeast (MY) (Difco, Lawrence, USA); MYM (MY+2% mannitol); PDB (potato dextrose broth); MS (Murashige and Skoog medium); CMG (10 g·L^{-1} casein peptone; 5 g·L^{-1} maltose, 10 g·L^{-1} glucose);SMG (10 g·L^{-1} soya peptone, 5 g·L^{-1} maltose, 10 g·L^{-1} glucose);TMG (10 g·L^{-1} tomato extract, 5 g·L^{-1} maltose, 10 g·L^{-1} glucose). After 2 months of culture, the mycelia were harvested for q-PCR, and the culture liquid was extracted with EtOAc to obtain crude extracts for LCMS-IT-TOF detection, respectively.

qPCR detection of expression of non-reducing PKSs

The expression of seven putative PKS genes in different media were detected by q-PCR with special primers (S3 Table 3). RNA from different samples was extracted with the RNeasy Plant Mini Kit (Qiagen). cDNA synthesis was performed with 1 μg total RNA and a reverse transcriptase kit (TaKaRa Super RT Kit) according to the manufacturer's instructions. SYBR Green was used for detection of PCR products one StepOnePlus Real-Time PCR system (Applied Biosystems). The tubulin gene was used as an internal standard for the normalization of gene expression. At least two independent biological replicates and three technical replicates for each biological replicate was analyzed through q-PCR for each sample to ensure reproducibility and reliability.

Results

Transcriptome de novo assembly

To obtain thetranscriptome of usnic acidproducing *N. pallescens*, an RNAseq library was constructed from mycelia, and sequenced using IlluminaNextSeq. We obtained 9 683 470 092 bases from 67 422 278 cleaned reads. Finally, 9 636 unigenes were identified from assembled transcripts. The mean unigene size was 2 419 bp with lengths ranging 200 from 15 815 bp, and the mean N50 was 3 393 bp. The unigene length distribution showed that approximately 75.5% of the unigenes contained more than 1 000 bp. Sequence data generated in this study have been deposited into NCBI (accession number: SRP091413).

Putative polyketide synthase from *N. pallescens*

For unigene annotation, sequence similarity searches were conducted against the NCBI non-redundant protein, Swiss-Prot protein databases, gene ontology (GO), and KEGG Ontology (KO) using the BLASTX algorithm. Nintey-nine putative polyketide synthase unigenes were obtained by searching annotation files with polyketide synthases and local BLAST transcriptome assemble files having a KS domain. The results from ORF and domain analyses showed that only 16 unigenes were complete cDNA sequence and had basic domain of polyketide synthase (Table 1). The 16 complete PKSs included seven non-reducing PKSs and nine highly-reducing PKSs.

Gene clone and domain organization

Seven non-reducing PKS genes were identified from the transcriptome of *N. pallescens*: NpPKS1 (c31_g1_i1), NpPKS2 (c4521_g1_i2), NpPKS3 (c478_g1_i2), NpPKS4 (c7957_g1_i1), NpPKS5 (c6270_g1_i3), NpPKS6 (c388_g1_i1), NpPKS7 (c7776_g1_i11). NCBI ORFfinder and FGENESH software were used to analyze open reading frames (ORF). Specific primers including initiation and termination codes were designed for seven PKS sequences, for which complete cDNA were obtained by seamlesscloning with special primers. Domain analysis (http://nrps.igs.umaryland.edu/index.html and https://blast.ncbi.nlm.nih.gov) showed that there are two non-reducing PKS including C-methyltransferase (MeT) domain (NpPKS6 and NpPKS7). Although both NpPKS6 and NpPKS7 had ketosynthase (KS), acyltransferase (AT), product template (PT), acyl carrier protein (ACP) and C-methyltransferase (MeT) domains, NpPKS6 had one more acyl carrier protein domain than NpPKS7. What's more, NpPKS7 had the Claisen cyclase (CLC) domain but NpPKS6 had the thioesterase (TE) domain. Two non-reducing PKSs (NpPKS5 and NpPKS4) possessed the same domain organization (KS-AT-PT-ACP). NpPKS2 and NpPKS3 had similar domain organization with KS-AT-PT-ACP-ACP. However, their last domain was different, NpPKS2 had the thioesterase (TE) domain and NpPKS3 had the short-chain dehydrogenase/reductase (SDR). Comparing NpPKS2 and NpPKS2, NpPKS1 had only one acyl carrier protein (ACP) domain, and the domain organization of NpPKS1 was KS-AT-PT-ACP-TE (Table 1).

Phylogenetic analysis of non-reducing PKS

The amino acid sequence of the KS domain of seven non-reducing PKSs from *N. pallescens* and 41 fungal non-reducing PKSs were used to generate multiple alignments and phylogenetic trees (Figure 1). The phylogenetic tree showed that they can be divided into five main groups according to their domain organization: Group I (KS-AT-PT-ACP), Group II (KS-AT-PT-ACP-ACP-SDR), Group III (KS-AT-PT-ACP-ACP-TE), Group IV (KS-AT-PT-ACP-TE), Group V (KS-AT-PT-ACP-(ACP)-MeT-TE/CLC). The known PKSs in group I usually were involved in anthraquinone biosynthesis; NpPKS4 and NpPKS5 belonged into group I. The known

Identification of A Putative Polyketide Synthase Gene Involved in Usnic Acidbiosynthesis in the Lichen *Nephromopsis pallescens*

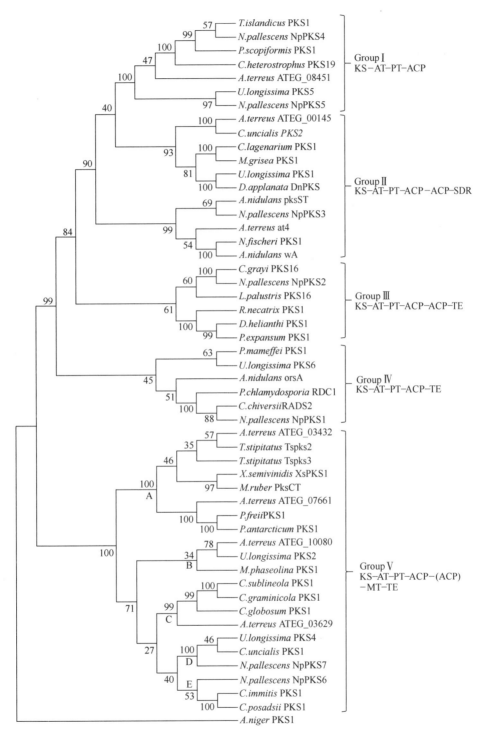

Figure 1 Phylogenetic relationships between *Nephromopsis pallescens* polyktetide synthase (PKS) genes and other fungal PKSs. Deduced PKS proteins were aligned with fungal PKS sequences retrieved from GenBank. PKSs of the *R. conduplicans* clade is marked in bold.

PKSs in group II usually were involved in pigment biosynthesis in fungi; NpPKS3 belonged into group II. The known PKSs in Group III were also related to fungal pigment biosynthesis and had different cylcization method, comparing with group II. NpPKS2 belonged to Group III. The known PKSs in group IV were involved in orsellinic acid biosynthesis, such as *A. nidulansors* A, *C. chiversii* RADS2 and *P. chlamydosporia* RDC1. NpPKS1 belonged to this group. Group V, the non-reducing PKS with Met domain is complex and varied. For example, there are five non-reducing PKS with Met domain in *Aspergillus terreus* (Chiang et al. 2013). Heterologous expression of these five PKSs gene from *A. terreus* showed four PKS with Met domain produced different polyketide and other one didn't detect product. Phylogenetic analysis shows that Group V can be divided into five sub-group. NpPKS6 and NpPKS7 were grouped into different sub-group. NpPKS7, *C. uncialis* PKS1 and *U. longissima* PKS4 formed one clade (sub-group D). *Cladonia uncialis* PKS1 was supposed to be involved in usnic acid (Abdel-Hameed et al. 2016).

Detection of PKS gene expression and usnic acid

According to a previous study, the growth medium influences the production of usnic acid in *N. pallescens* (Luo et al., 2011). Consequently, we incubated the mycelia of *N. pallescens* in seven different liquid media. After 2 months of culture, mycelia were harvested for gene expression and usnic acid production. Usnic acid was detected with LCMS-IT-TOF in extracts from MY, MYM and TMG cultures (Fig. 2). According to a previous study, the PKS related with

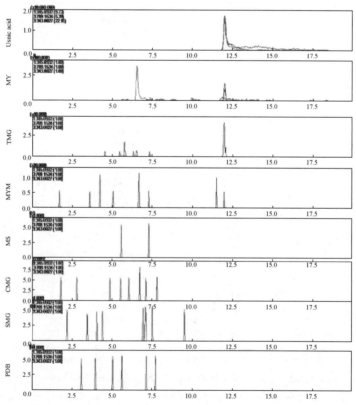

Figure 2 Detection of usnic acid in extracts of different *N. pallescens* cultures.
MY, TMG, MYM, MS, CMG, SMG, PDB showed different extracts from different medium, respectively.

usnic acid biosynthesis should be non-reducing. Therefore, we assessed through q-PCR the expression of seven PKS genes in transcriptome data from usnic acid producing strains (Fig. 3). q-PCR results show that seven non-reducing PKSs can be highly expressed in MYM medium. NpPKS1 was highly expressed in TMG and NpPKS3 was highly expressed in MY. However, only NpPKS7 was highly expressed in MY, MYM and TMG media and weakly expressed in SMG, MS, CMG and PDB medium. Therefore, NpPKS7 appears to be critical for usnic acid biosynthesis in N. pallescens.

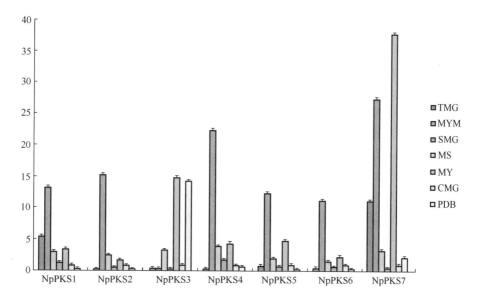

Figure 3 Seven non-reducing PKS gene expression profile in seven media.
TMG, MYM, SMG, MS, MY, CMG, PDB showed the different cDNA from mycelia of
N. pallescens cultured in TMG, MYM, SMG, MS, MY, CMG, PDB medium, respectively.

Discussion

Many clinical drugs, such as antibiotics, immunosuppressants, cytotoxins, and cholesterol-lowering substances, are derived from polyketides (Helfrich et al. ,2014). Lichens can produce diverse and unique polyketides, thus have a high potential pharmaceutical value. However, the slow growth of lichens and lichen forming fungi limits their direct application in biotechnology. However, since Miao (2001) proposed harvesting of lichen products through genetic approaches, several PKS genes were isolated from lichens, and lichen forming fungi, following the homologous clone method (Chooi et al. ,2008; Brunauer et al. ,2009; Wang et al. ,2011; Wang et al. , 2014). It is difficult to obtain many genes at one time using homologous clone methods, and it is also difficult to find new type genes. Next-generation sequencing (NGS) is a powerful and cost-effective tool that be used to obtain genetic information and large amounts of sequence data quickly (Mardis 2013). So far, several genomes of lichen forming fungi have been sequenced on NGS platforms (Abdel-Hameed et al. ,2016; Park et al. ,2013b; Park et al. ,2014). Compared to genome sequencing, RNA sequencing (RNA-Seq) is less expensive, easier and there are more efficient tools to obtain gene expression information from RNA (Ozsolak and Milos, 2011). In this study, RNA sequencing was used to release transcriptome of LFF N. pallescens, and 16 complete PKS genes were obtained. These sixteen PKSs included nine reducing PKS sand seven non-reducing PKSs. Genome analysis of lichen forming fungi indicated that lichen forming fungi had 13-34 complete PKS genes in their genome, such as Cladonia macilenta with 34 PKS genes, Cladonia metacorallifera with 31 PKS genes, Umbilicaria muehlenbergii with 20 PKS genes, Endocarpon pusillum with 15 PKS genes, Caloplaca flavorubescens with 13 PKS genes, Cladonia uncialis with 32 PKS genes (Abdel-Hameed et al., 2016; Park et al., 2013b; Park et al., 2014). The different genera of lichen forming fungi had different PKS content in their genome. One transcriptome released 16 compete PKSs from N. pallescens, indicating that there are abundant PKS genes in the N. pallescens genome.

Although many PKS genes were found in lichen forming fungi, the function of PKSs from lichen forming fungi is still unclear. Especially, the non-reducing PKSs having a MeT domain (Group V in phylogeny), which is complex and variable. So far, the function of seven non-reducing PKS with MeT domain were confirmed by heterologous expression (Fig 4). It was firstly confirmed that PksCT with KS-AT-PT-ACP-MeT-R domain is involved in citrinin biosynthesis (chooi et al., 2008). Next, five non-reducing PKS genes with MeT domain fromA. terreus were heterologous expressed in the engineered A. nidulans host (Chiang et al., 2013). These results show that ATEG-10080 with KS-AT-PT-ACP-MT-TE domain organization is involved in 3,5-dimethylorsellinic acid biosynthesis, ATEG-03629 with KS-AT-PT-ACP-ACP-MT-TE domain organization is involved in 5-methylorsellinic acid. Although ATEG-10080 and ATEG-03629 consisted of very similar domain organization, their products were different because ATEG-03629 has one more ACP domain than ATEG-10080. The number of domains will affect the final PKS product, and differences in releasing domains also will lead to variety of their product. ATEG-03432 and ATEG-10080 possessed a similar domain organization. ATEG-10080 had TE domain and ATEG-03432 had R domain. The product of ATEG-03432 was 6-acetyl-2,7-dihydroxy-3-methylnaphthalene-1,4-dione According to known non-reducing PKS with MeT domain, we can know that the number of domain, releasing domain, and PT domain decided will influence the structure of their product. The transcriptome of N. pallescens producing usnic acid showed that there are two non-reducing PKS with MeT (NpPKS6 and NpPKS7). NpPKS6 possessed KS-AT-PT-ACP-ACP-MeT-TE domain organization. NpPKS6 and ATEG-03629 had very similar domain organization. The phylogenetic analysis shows that NpPKS6 and ATEG-03629 belonged to different clades. Lichen or lichen forming fungi can produce large amounts of βorsellinic acid, and ATEG-03629 produce 5-methylorsellinic acid. It is possible that NpPKS6 is involved in βorsellinic acid biosynthesis. NpPKS7 possesses a KS-AT-PT-ACP-MeT-CLE domain organization. Comparing known non-reducing PKS, NpPKS7 took claisen cyclase domain as releasing domain. Recently, a non-reducing PKS (CuPKS1) from C. uncialis with a similar domain to NpPKS7 was predicted to be involved in usnic acid based upon genome cluster and gene expres-

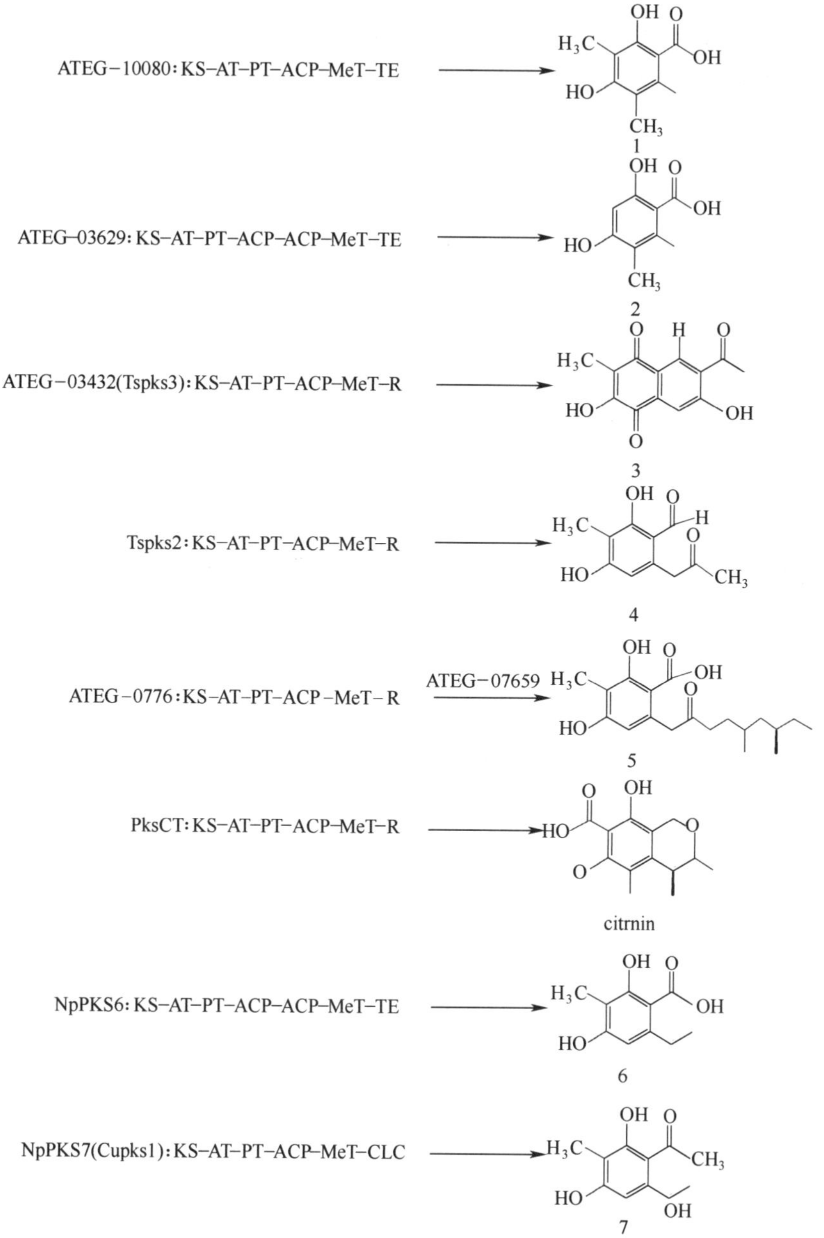

Figure 4 Known non-reducing PKS with MeT domain and their products. 1: 3,5-dimethylorsellinic acid; 2: 5-methylorsellinic acid; 3: 6-acetyl-2,7-dihydroxy-3-methylnaphthalene-1,4-dione; 4: 2,4-dihydroxy-6-(5,7-dimethyl-2-oxo-trans-3-trans-5-nonadienyl)-3-methylbenzaldehyde; 5: βorsellinic acid; 6: methylphloracetophenone. KS: b-ketoacyl synthase, AT: acyl transferase, PT: product template domain, ACP: acyl carrier protein, MeT: Cmethyltransferase, R: releasing domain, TE: thioesterase domain, CLC: Claisen cyclase domain.

sion analyses (Abdel-Hameed et al. 2016). However, the lichen forming fungi of C. uncialis in lab culture did not produce usnic acid. They didn't confirm the function of CuPKS1 by association analysis with gene expression and usnic acid production. In this study, the relevance analysis of gene expression by q-PCR detection and usnic acid by LCMS-IT-TOF detection demonstrates that gene expression of NpPKS7 is consistent with the production of usnic acid in different media. And the phylogenetic tree shows that NpPKS7, CuPKS1, UlPKS4 group into a single monophyletic clade. Therefore, all of results showed that NpPKS7 was involved in usnic acid biosynthesis in *N. pallescens* (Fig 4).

Conclusions

This work presents the first *de novo* transcriptome sequencing analysis of usnic acid producing *N. pallescens*. Illumina RNA-Seq was used to complete PKS genes obtained from transcriptome analyses of sixteen isolates, and seven non-reducing PKS genes were cloned by seamless Cloning with specific primers. Among these, two non-reducing PKS with MeT domain were detected. Gene expression with q-PCR was used to detect these non-reducing PKS gene expression and LCMS-IT-TOF was used to detect usnic acid production in *N. pallescens* growing in different culture media. The combined analyses of gene expression and usnic acid production show that NpPKS7 is involved in usnic acid biosynthesis. Domain organization and phylogenetic analyses also indicate it is possible that NpPKS7 produces methylphloracetophenone, which is the precursor of usnic acid.

Acknowledgements

Sequencing service were provided by Personal Biotechnology Co., Ltd. Shanghai, China.

Fundiong

This work was supported by a grant from National Natural Science Foundation of China (31400488), by the Applied Basic Research Project of Yunnan Province Grant (2016FB055) and by the Yunnan Academy of Forestry Innovation Fund Project, Grant (QN2018-01).

References

Abdel-Hameed M, Bertrand RL, Piercey-Normore MD, Sorensen JL. 2016. Putative identification of the usnic acid biosynthetic gene cluster by de novo whole-genome sequencing of a lichen-forming fungus. Fungal Biology 120:306~316.

Armaleo D, Sun X, Culberson C. 2011. Insights from the first putative biosynthetic gene cluster for a lichen depside and depsidone. Mycologia 103:741~754.

Bačkorová M, JendželovskýR, Kello M, Bačkor M, Mikeš J, Fedoročko P. 2012. Lichen secondary metabolites are responsible for induction of apoptosis in HT-29 and A2780 human cancer cell lines. Toxicology in Vitro An International Journal Published in Association with Bibra 26:462~468.

Brunauer G, Muggia L, Stocker WE, Grube M. 2009. A transcribed polyketide synthase gene from *Xanthoria elegans*. Mycological Research 113:82~92.

Campanella L, Delfini M, Ercole P, Iacoangeli A, Risuleo G. 2002. Molecular characterization and action of usnic acid: a drug that inhibits proliferation of mouse polyomavirus in vitro and whose main target is RNA transcription. Biochimie 84:329~334.

Chiang YM, Oakley CE, Ahuja M, Entwistle R, Schultz A, Chang SL, Sung CT, Wang CCC, Oakley BR. 2013. An efficient system for heterologous expression of secondary metabolite genes in *Aspergillus nidulans*. Journal of the American Chemical Society 135:7720~7731.

Chooi YH, Stalker DM, Davis MA, Fujii I, Elix JA, Louwhoff SH, Lawrie AC. 2008. Cloning and sequence characterization of a non-reducing polyketide synthase gene from the lichen *Xanthoparmelia semiviridis*. Mycological Research 112:147~161.

Cocchietto M, Skert N, Nimis PL, Sava G. 2002. A review on usnic acid, an interesting natural compound. Die Naturwissenschaften 89:137~146.

Francolini I, Norris P, Piozzi A, Stoodley GD. 2004. Usnic acid, a natural antimicrobial agent able to inhibit bacterial biofilm formation on polymer surfaces. Antimicrobial Agents & Chemotherapy 48:4360~4365.

Gagunashvili AN, Davídsson SP, Jónsson ZO, Andrésson OS. 2009. Cloning and heterologous transcription of a polyketide synthase gene from the lichen *Solorina crocea*. Mycological Research 113:354~363.

Guo L, Shi Q, Fang JL, Mei N, Ali AA, Lewis SM, Leakey JE, Frankos VH. 2008. Review of usnic acid and *Usnea barbata* toxicity. Journal of Environmental Science & Health Part C Environmental Carcinogenesis & Ecotoxicology Reviews 26:317~338.

Hametner C, Stocker WE. 2015. Type I NR-PKS gene characterization of the cultured lichen mycobiont *Xanthoparmelia Substrigosa* (Ascomycota). In: Upreti DK, Shukla V, Divakar PK, Bajpai R, eds. Recent advances in lichenology. Springer, New Delhi, India, p. 95~110.

Hawranik DJ, Anderson KS, Simmonds R, Sorensen JL. 2009. The chemoenzymatic synthesis of usnic acid. Bioorganic & Medicinal Chemistry Letters 19:2383~2385.

Helfrich EJ, Reiter S, Piel J. 2014. Recent advances in genome-based polyketide discovery. Current Opinion in Biotechnology 29:107~115.

Honegger R. 1984. Cytological aspects of the mycobiont-phycobiont relationship in lichens. Haustorial types, phycobiont cell wall types, and the ultrastructure of the cell surface layers in some cultured and symbiotic myco- and phycobionts. Lichenologist 16:111~127.

Ingólfsdóttir K. 2002. Usnic acid. Phytochemistry 61:729~736.

Luo H, Yamamoto Y, Jeon HS, Liu YP, Jung JS, Koh YJ, Hur JS. 2011. Production of anti-Helicobacter pylori metabolite by the lichen-forming fungus *Nephromopsis pallescens*. Journal of Microbiology 49:66~70.

Mardis ER. 2013. Next-generation sequencing platforms. Annual Review of Analytical Chemistry 6:287~303.

Mayer M, O'Neill MA, Murray KE, Santos MNS, Carneiro LAM, Thompson AM, Appleyard VC. 2005. Usnic acid: a non-genotoxic compound with anti-cancer properties. Anti-cancer drugs 16:805~809.

Miao V, Coëffetlegal MF, Brown D, Sinnemann S, Donaldson G, Davies J. 2001. Genetic approaches to harvesting lichen products. Trends in Biotechnology 19:349~355.

Molina MC, Crespo A, Vicente C, Elix JA. 2003. Differences in the composition of phenolics and fatty acids of cultured mycobiont and thallus of *Physconia distorta*. Plant Physiology & Biochemistry 41:175~180.

Ozsolak F, Milos PM. 2011. RNA sequencing: advances, challenges and opportunities. Nature Reviews Genetics 12:87~98.

Park SY, Jeong MH, Wang HY, Kim JA, Yu NH, Kim S, Yong HC, Kang S, Lee YH, Hur JS. 2013a. Agrobacterium tumefaciens-mediated transformation of the lichen fungus, *Umbilicaria muehlenbergii*. Plos One 8:e83896.

Park SY, Choi J, Kim JA, Yu NH, Kim S, Kondratyuk SY, Lee YH, Hur JS. 2013b. Draft genome sequence of lichen-forming fungus *Caloplaca flavorubescens* strain KoLRI002931. Genome Announcements 1:e00678~00613.

Park SY, Choi J, Lee GW, Kim JA, Oh SO, Jeong MH, Yu NH, Kim S, Lee YH, Hur JS. 2014. Draft Genome sequence of lichen-forming fungus *Cladonia metacorallifera* strain KoLRI002260. Genome Announc 2: e01065~01013.

Sakai K, Kinoshita H, Shimizu T, Nihira T. 2008. Construction of a citrinin gene cluster expression system in heterologous *Aspergillus oryzae*. Journal of Bioscience & Bioengineering 106:466~472.

Song Y, Dai F, Dong Z, Dong Y, Zhang J, Lu B, Luo J, Liu M, Yi Z. 2012. Usnic acid inhibits breast tumor angiogenesis and growth by suppressing VEGFR2-mediated AKT and ERK1/2 signaling pathways. Angiogenesis 15:421~432.

Stocker WE. 2008. Metabolic diversity of lichen-forming ascomycetous fungi: culturing, polyketide and shikimate metabolite production, and PKS genes. Natural Product Reports 25:188~200.

StockerWE, Cordeiro LMC, Iacomini M. 2013. Chapter 10 - Accumulation of potential pharmaceutically relevant lichen metabolites in lichens and cultured lichen symbionts. In: Atta-ur-Rahman FRS, ed. Studies in natural products chemistry. Elsevier, Karachi, Pakistan, p. 337~380.

Wang Y, Kim JA, Cheong YH, Koh YJ, Hur JS. 2011. Isolation and characterization of a non-reducing polyketide synthase gene from the lichen-forming fungus *Usnea longissima*. Mycological Progress 11:75~83.

鉴定皮革肾岛衣中一个参与松萝酸生物合成的聚酮合酶基因

摘要: 松萝酸是地衣产生的独特化合物,未来确定参与松萝酸生物合成的基因,完成了一个产生松萝酸时期的皮革肾岛衣地衣型真菌的转录组测序。从转录组中我们分析获得7个非还原型PKS基因和9个高度还原型PKS基因。我们用荧光定量PCR检测基因表达量,同时用LCMS-IT-TOF检测松萝酸含量,通过关联分析确定NpPKS7可能参与松萝酸的合成。NpPKS7是具有MeT结构域的非还原型PKS,含有6不同的结果域,系统进化显示NpPKS7和其他来源于地衣的聚酮合酶形成一个独特的分支。结合这些分析显示NpPKS7是一种新的聚酮合酶类型参与了松萝酸生物合成。

(本文发表于 *PLOS ONE*,2018年)

第五篇 森林病虫害防治研究

不同日龄和吊飞过程中桔小实蝇成虫飞行肌能量代谢相关酶活性的变化

袁瑞玲[1,2]，王晓渭[3]，杨珊[4]，陈鹏[2,*]

(1. 云南省森林植物培育与开发利用重点实验室，昆明 650201；
2. 云南省林业科学院，昆明 650201；3. 西南林业大学林学院，昆明 650224；
4. 西南林业大学生命科学学院，昆明 650224)

摘要：【目的】明确桔小实蝇 *Bactrocera dorsalis* (Hendel) 飞行肌对能源物质的利用。【方法】通过生化方法测定了能源物质代谢相关 5 种酶[3-磷酸甘油醛脱氢酶（GAPDH）、3-磷酸甘油脱氢酶（GDH）、乳酸脱氢酶（LDH）、柠檬酸合酶（CS）和 3-羟酰辅酶 A 脱氢酶（HOAD）]活性的变化。【结果】桔小实蝇成虫中所测的 5 种酶活性随日龄的变化而变化，4 日龄 GAPDH、GDH、LDH 和 CS 活性最高，20 日龄 HOAD 活性最高。吊飞过程中，GAPDH，GDH 和 CS 的活性变化基本一致，随吊飞时间的延长活性逐渐升高；LDH 和 HOAD 的活性变化雌、雄虫完全不同。雄虫 LDH 活性除吊飞 2 h 外其他时间均高于静息状态，雌虫则始终低于静息状态；雄虫 HOAD 活性只有吊飞 24 h 低于静息状态水平，而雌虫吊飞后 HOAD 活性一直在静息状态水平及以下波动。【结论】桔小实蝇飞行所利用的能源物质包括糖类和脂肪，以糖类能源为主。吊飞过程中，雄虫除可以进行高速有氧代谢以外，还具备一定的无氧代谢能力，而雌虫只进行有氧代谢；雄虫能利用脂肪供给能量，雌虫则几乎不动用脂肪。研究结果为进一步阐明桔小实蝇的迁飞行为机制提供了依据。

关键词：桔小实蝇；飞行肌；吊飞；能量代谢；酶活性

Changes in the activities of enzymes related to energy metabolism in flight muscles of adult *Bactrocera dorsalis* (Diptera: Tephritidae) at different ages and during tethered flight

Yuan Rui-Ling[1,2], Wang Xiao-Wei[3], Yang Shan[3], Chen Peng[2,*]

(1. Key Laboratory of Forest Plant Cultivation & Utilization, Kunming 650201, China;
2. Yunnan Academy of Forestry, Kunming 650201, China; 3. School of Forestry,
Southwest Forestry University, Kunming 650224, China; 4. School of Life Sciences,
Southwest Forestry University, Kunming 650224, China)

Abstract:【Aim】This study aims to clarify the utilization of energy substances in flight muscles of *Bactrocera dorsalis* (Hendel).【Methods】The activities of five key enzymes related to energy metabolism in flight muscles of *B. dorsalis*, i.e., glyceraldehyde phosphate dehydrogenase (GAPDH), glycerol-3-phosphate dehydrogenase (GDH), lactate dehydrogenase (LDH), citrate synthase (CS) and 3-

hydroxyacyl-CoA dehydrogenase (HOAD), were assayed and analyzed by tethered flight testing and biochemical methods in the laboratory.【Results】The activities of the five enzymes assayed changed withage (day-old) of *B. dorsalis* adults. The activities of GAPDH, GDH, LDH and CS peaked in the 4-day-old adults, while the HOAD activity peaked in the 20-day-old adults. During tethered flight, the activities of GAPDH, GDH and CS showed similar change trend, gradually increasing with tethered flight duration, while the activities of LDH and HOAD showed distinctively different change trend between male and female adults. The LDH activityin female adults during tethered flight was lower than that of female adults in resting state, while that in male adultswas just the oppositeexcept the 2 h tethered flight treatment. The HOAD activityin male adults fluctuated above the resting state level except the 24 htethered flighttreatment, while that in female adults fluctuated between the resting state level and below.【Conclusion】Energy substrates in flight muscles of *B. dorsalis* include carbohydrate and lipid, but carbohydrate metabolism provides primaryenergy. During tethered flight, significantly different energy metabolismsexist between male and female adults. Both aerobic and anaerobic metabolismsare presentin flightmuscles of male adults, but only aerobic metabolism is present in female adults. Male adults can utilize lipids as energy substrate for flight, but female adults hardly utilize lipids as energy substrate for flight. Our study provided a scientific basisfor further illuminatingthe mechanisms of the migratory behavior of *B. dorsalis*.

Key words:*Bactrocera dorsalis*; flightmuscle; tethered flight; energy metabolism; enzyme activity

昆虫的迁飞伴随着较高的代谢速率和能源物质的消耗,主要的能源物质一般有碳水化合物、脂类化合物、氨基酸等其中一种或两种以上(Rankin and Burchsted,1992)。判断昆虫能源利用类型的重要依据是代谢糖类和脂肪的关键酶3-磷酸甘油醛脱氢酶(GAPDH)和3-羟酰辅酶A脱氢酶(HOAD)的活性之比。东亚飞蝗(*Locusta migratoria*)属于糖类和脂肪都能利用的混合型(GAPDH:HOAD比值接近于1.0),小柏天蛾(*Philosa miacynthia*)主要利用脂肪(0.2),而红头丽蝇(*Calliphora erythrocephala*)及大多数双翅目昆虫主要利用糖类为飞行提供能源(>100)(Beenakkers,1969;韩兰芝等,2005;李克斌等,2005a)。

桔小实蝇(*Bactrocera dorsalis* Hendel)又名东方果实蝇,属双翅目(Diptera),实蝇科(Tephritida),果实蝇属(*Bactrocera*),是一种世界性检疫害虫,危害40多科250多种水果和蔬菜(Drew and Hancock,1994;Chen and Ye,2008;Suganya et al.,2010;Arévalo-Galarza and Follett,2011)。已有很多文献报道桔小实蝇有极强的远距离飞行能力,这是造成其分布和发生区不断扩张,以及根除后又再次发生的重要原因之一(陈鹏等,2007;李伟丰等,2007;Shi et al.,2010)。但目前国内外涉及桔小实蝇迁飞行为发生机理方面的相关研究甚少。袁瑞玲等(2015)研究了不同日龄桔小实蝇干重和糖原的累积,以及在室内静风且无补充营养条件下,飞翔后虫体重下降和糖原消耗的情况,分析表明,糖原是桔小实蝇飞行的能源物质之一。但能源物质代谢酶的活性研究还没有报道。3-磷酸甘油醛脱氢酶(GAPDH)是糖酵解途径中的一种关键酶,催化3-磷酸甘油醛转变为1,3-二磷酸甘油醛,同时以NAD^+为受氢体生成NADH,其活性水平往往代表糖酵解循环活性的高低。3-磷酸甘油脱氢酶(GDH)是昆虫体内一种较特殊的酶,它能维持糖类代谢的高度有氧状态,催化NADH,将氢和电子转移到磷酸二羟丙酮(DHAP,糖酵解的中间产物)上,氧化成NAD^+,磷酸二羟丙酮则被还原成3-磷酸甘油酯(G-3-P)。在以糖类作为主要能源的昆虫飞行肌中,G-3-P穿梭循环是有氧酵解供能的主要途径(王荫长,2004)。乳酸脱氢酶(LDH)是糖无氧酵解及糖异生的一种重要酶,催化丙酮酸与乳酸之间的还原与氧化反应,其活性可以反映动物体无氧代谢的能力(Beenakkers et al., 1984)。柠檬酸合成酶(CS)是一个调控酶,在三羧酸循环中,催化乙酰辅酶A与草酰乙酸释放出辅酶A,得到柠檬酸。三羧酸循环是糖、脂肪和蛋白质3种主要有机物在体内彻底氧化供能的共同通路,也是能源物质相互转化的核心生化过程,其活性的高低是制约该循环运转的关键因素之一(王荫长,2004)。3-羟酰辅酶A脱氢酶(HOAD)活性是衡量脂肪代谢水平高低的关键酶之一,其活性水平比肉毒碱酰基转移酶更能反映昆虫飞行肌动用脂肪的能力(Gunn and Gatehouse,1988)。GAPDH:HOAD比值是判断昆虫飞行肌能源物质利用类型的重要指标(Beenakkers,1969)。本文拟通过测定不同日龄桔小实蝇这5种能源物质代谢相关酶的活性动态以及最强飞行日龄的桔小实蝇在飞行过程中能源物质代谢相关酶活性的变化,为明确桔小实蝇飞行肌对能源物质的利用情况,进一步阐明桔小实蝇的迁飞行为机制提供依据。

1 材料与方法

1.1 供试虫源

试验所用虫源为云南省林业科学院森林保护研究所饲养的桔小实蝇试验种群,经室内连续继代饲养约10代的成虫。饲养条件为:温度25~28℃,湿度为60%~80%,光周期为16L:8D。同一时期羽化的成虫聚养在同一养虫笼(75cm×60cm×45cm)内,每24h补充一次饲料(豆渣45.0g、麦胚7g、蔗糖14g、酵母粉7g、盐酸0.3mL、对羟基苯甲酸甲酯0.3g,加以适量的蒸馏水配制成糊状饲料,pH 3.6,现配现用)和自来水。

1.2 供试药剂

还原型辅酶Ⅰ(NADH)、丙酮酸钠(sodium pyruvate acid)和腺苷三磷酸(ATP)均为AM-RESCO产品;3-磷酸甘油酸(3-phosphoglycerate)、3-磷酸甘油激酶(3-phosphoglycero-kinase)、磷酸二羟丙酮(dihydrox-yaceton-phosphate)、苹果酸脱氢酶(malate dehydrogenase)、乙酰辅酶A(acetyl-CoA)和乙酰乙酰辅酶A(acetoacetyl-CoA)、砷酸钾(sodium arsenate)、盐酸胱氨酸(L-cystine HCl)、氟化钠(sodium fluoride)、3-磷酸甘油醛(glyceraldehyde-3-phosphate)、MOPS半钠盐均为SIGMA公司产品;考马斯亮蓝G-250(coomassie brilliant blue G-250),Fluka公司进口分装;三羟甲基氨基甲烷(Tris),Solarbio公司;还原型谷胱甘肽(reduced glutathione),BBI公司进口分装;三乙醇胺(triethanolamine),分析纯,上海凌峰化学试剂有限公司生产;乙二胺四乙酸二钠(EDTANa$_2$),分析纯,上海试剂一厂生产;L-苹果酸(L-malic acid),南京化学试剂公司产品;氧化型辅酶Ⅱ(NAD),华美生物工程有限公司生产;牛血清白蛋白(albumin fraction),国药集团公司产品。

1.3 试虫处理与取样

1.3.1 不同日龄成虫飞行肌代谢酶活性的测定:选择1,4,7,10,15,20和25日龄的桔小实蝇雌、雄成虫供测试。

1.3.2 静风吊飞对成虫代谢酶活性的影响:选择最强飞行日龄15日龄(袁瑞玲等,2014)的桔小实蝇雌、雄成虫,用乙醚轻微麻醉,在飞行磨(即昆虫飞行数据微机采集系统)吊臂的一端直接粘取少量市售502胶水,小心地粘于成虫的前胸背板上,确认桔小实蝇头胸部、腹部和翅基部未受到影响,将带有桔小实蝇的吊臂放回飞行磨上待测,吊臂的另一端与吊虫端需保持水平。温度25~28℃,湿度为60%~80%,光照为自然光照,晚间补给光照,吊飞时间分别为1,2,5,10和24h,整个飞行过程中不补充水分和营养。设同一日龄的不飞行个体为对照,供测试。以上每处理雌、雄成虫各12头,每处理重复4次。所有试虫均立即置于液氮中处死,去头、足、翅及腹部,仅留胸部组织,电子天平称重并记录,放入-70℃低温冰箱中保存备用。

1.4 酶活性及蛋白质含量的测定

1.4.1 酶液的提取:将虫体组织从-70℃低温冰箱中取出,加入液氮,快速用研磨棒碾碎组织成粉末状,并按1∶9(m/v)比例加入冰冷的磷酸钾缓冲溶液(0.1mol·L^{-1},pH7.3),其中含2mmol·L^{-1}的EDTA,涡旋混匀器上剧烈混匀30s,间隔2min,重复操作2~3次,制备待测液。将待测液在4℃条件下以14 000×g冷冻离心20min,小心吸取上清液转移到新的1.5ml离心管进行酶活性测定。

1.4.2 酶活性测定:酶活性用北京普朗DNM-9602酶标仪测定,并进行数据处理。具体操作步骤参照Beenakkers(1969)的方法进行。

1.4.3 蛋白质含量的测定:以牛血清白蛋白为标准蛋白,制作标准曲线,用考马斯亮蓝G-250方法进行测定(Bradford,1976)。

1.5 数据分析处理

所获数据用SPSS17.0统计软件进行方差分析,对各处理间方差进行齐性Levene检验,各组方差齐时($P>0.05$)或组间方差不能达到齐性,但对数据进行转换后,组间方差齐时,经正态检验后,用Duncan氏新复极差法进行多重比较;组间方差经数据转换后仍不齐时($P<0.05$),采用Games-Howell法进行多重比较。

2 结果

2.1 不同日龄桔小实蝇成虫代谢酶活性的变化

2.1.1 糖类代谢相关酶的活性变化:不同日龄桔小实蝇成虫飞行肌GAPDH活性不同(图1)。就雄虫而言,1日龄GAPDH活性较高,显著高于除4日龄外的其他日龄($P<0.05$),4日龄GAPDH活性最高,极显著高于其他日龄($P<0.01$),之后GAPDH活性极显著下降,至20日龄降到最低,25日龄略有回升,10、20日龄GAPDH活性显著低于25日龄($P<0.05$),7,15和25日龄间GAPDH活性差异不显著。雌虫GAPDH活性变化趋势与雄虫相似,4日龄GAPDH活性最高,极显著高于其他日龄($P<0.01$),其他日龄中,15日龄GAPDH活性最低,显著低于7和20日龄,1,7,10,20和25日龄间没有显著差异。表明桔小实蝇在4日龄时糖酵解最活跃。

桔小实蝇雌、雄虫的GDH活性变化趋势基本一致(图1),类似GAPDH活性的变化。1日龄成虫GDH活性较高,显著高于除4日龄外的其他日龄($P<0.05$),到4日龄GDH活性达最大值,极显著($P<0.01$)高于其他日龄[雄虫4日龄与1日龄间差异只达到显著水平(P

<0.05)除外],以后随日龄的增加 GDH 活性逐渐下降,各日龄间 GDH 活性除 15 和 20 日龄没有显著差异外,差异达到显著水平($P<0.05$)。说明 4 日龄前与糖酵解有关的有氧代谢较为旺盛。

与 GAPDH 和 GDH 相比,桔小实蝇的 LDH 活性较低。随日龄的增加,雌、雄虫 LDH 活性均在 4 日龄达最高,以后,雄虫 LDH 活性先降低,至 10 日龄达最小值后逐渐升高,到 25 日龄 LDH 活性回升至 4 日龄水平,雌虫则一直在降低,10 日龄后差异不显著。说明桔小实蝇有一定的无氧代谢能力,但这种能力较弱。

2.1.2 脂肪酸 β-氧化的关键酶 HOAD 的活性变化:不同日龄桔小实蝇雌、雄虫 HOAD 的活性变化趋势基本一致。1 日龄时 HOAD 活性最低,极显著低于其他日龄($P<0.01$)。随日龄的增加,HOAD 活性升高,至 10 日龄已上升至较高值,之前,各处理 HOAD 活性差异极显著($P<0.01$),之后,HOAD 活性缓慢上升,到 20 日龄 HOAD 活性达最高。25 日龄雄虫 HOAD 活性与 20 日龄相当,而雌虫 HOAD 活性则是显著降低($P<0.05$)。这些结果表明,随日龄的增加桔小实蝇成虫动用脂肪的能力迅速增强,10 日龄后 HOAD 活性一直维持在较高的水平,这对桔小实蝇高效快速地动用脂肪相当重要。不同日龄的桔小实蝇飞行能力表现为:15 日龄最强,15~25 日龄具备较强的飞行能力。这说明桔小实蝇的 HOAD 活性与飞行能力存在一定的正相关关系。结合糖类代谢相关酶的活性变化以及这两种酶活性的绝对值来看,说明在桔小实蝇低龄阶段主要利用糖类作为能源物质,以后主要是通过脂类和糖类代谢共同提供能量。

2.1.3 三羧酸循环关键酶的活性变化:研究结果显示(图 1),桔小实蝇雌、雄虫不同日龄的 CS 活性动态基本一致。1 和 4 日龄低龄期 CS 活性最高,极显著高于其他日龄($P<0.01$)。之后 CS 活性随日龄的增加而显著下降($P<0.05$),雌虫至 20 日龄以后 CS 活性维持在最低状态,雄虫到 10 日龄已降到最低。说明低龄的桔小实蝇成虫中存在活跃的三羧酸循环。

2.1.4 GAPDH:HOAD 比值的变化:从表 1 可以看出,1 和 4 日龄雌、雄虫 GAPDH:HOAD 活性之比远远大于 1,说明低龄期所利用的能源物质以糖类为主。到 7 日龄比值下降到 1.2,此时成虫利用糖的能力比利用脂肪的能力稍强。从 10 日龄以后,比值维持在 1.0 左右,表明桔小实蝇具备远距离飞行能力(袁瑞玲等,2015)以后,成虫利用脂肪和糖的能力相当。

图 1 中误差线为标准差;柱上不同大小写字母分别表示在 5% 和 1% 水平上不同日龄成虫同一酶活性差异显著(Games-Howell 检验,雌虫 GDH 为 Duncan 氏检验)。Error bars are standard deviations. Different small and capital letters above bars indicate significant differencein enzyme activity among different day-old adults at the 5% and 1% level, respectively (Games-Howell test, GDH of the female by Duncan's test).

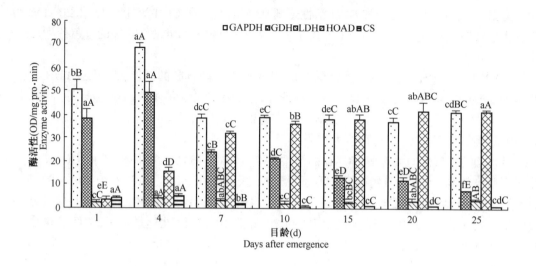

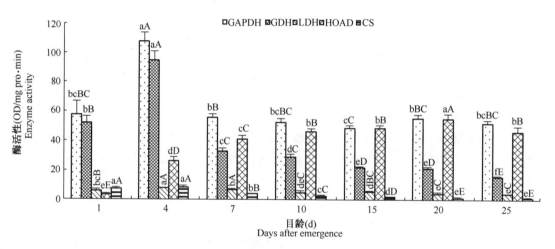

图 1　不同日龄桔小实蝇雌(B)、雄(A)成虫飞行肌代谢酶的活性

Figure 1　Enzyme activities in flight muscles of male(A) and female(B) adults of *Bactrocera dorsalis* at different ages (day-old)

表 1　不同日龄桔小实蝇成虫 GAPDH:HOAD 比值

Table 1　Ratios of GAPDH:HOAD in *Bactrocera dorsali* sadults at different ages (day-old)

日龄 Age (day-old)	雄性 Male	雌性 Female
1	14.66	18.08
4	4.36	4.08
7	1.20	1.36
10	1.08	1.13
15	1.00	1.00
20	0.89	1.00
25	1.00	1.13

2.2 飞行过程中桔小实蝇成虫代谢酶活性的变化

静风吊飞不同的时间后,桔小实蝇代谢酶活性的变化测试结果如图2。

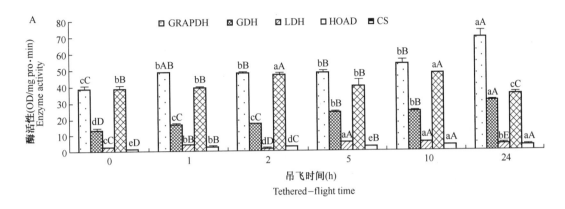

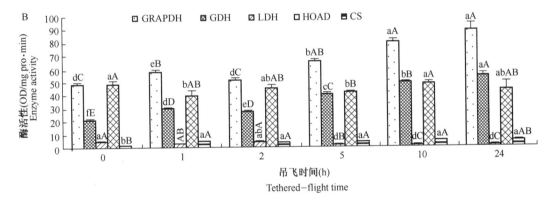

图 2 桔小实蝇雌(B)、雄(A)成虫吊飞不同时间飞行肌代谢酶的活性变化

Figure 2 Changes in enzyme activities in flight muscles of male (A) and female (B) adults of *Bactrocera dorsalis* at different tethered-flight time

图 2 中误差线为标准差;柱上不同大小写字母分别表示在5%和1%水平上不同吊飞时间同一酶活性差异显著(Games-Howell 检验)。Error bars are standard deviations. Different small and capital letters above bars indicate significant differencein enzyme activity among different tethered-flight time at the 5% and 1% level, respectively (Games-Howell test).

表 2 15 日龄桔小实蝇成虫吊飞不同时间 GAPDH:HOAD 比值

Table 2 Ratios of GAPDH:HOAD in 15-day old adults of *Bactrocera dorsalis* at different tethered-flight time

吊飞时间(h) Tethered-flight time	雄性 Male	雌性 Female
0	1.00	1.00
1	1.25	1.47
2	1.02	1.14

(续)

吊飞时间(h) Tethered-flight time	雄性 Male	雌性 Female
5	1.20	1.55
10	1.12	1.64
24	2.01	2.03

2.2.1 糖类代谢相关酶的活性变化：从结果可以看出，GAPDH 活性雄虫吊飞 1h 后升高极为明显（$P<0.01$），吊飞 2~5h 与 1h 相比变化不明显，吊飞 10h 以后迅速上升，吊飞 24h 达最高点，显著高于其他吊飞时间的活性（$P<0.05$）。雌虫吊飞 1h 后 GAPDH 活性极显著升高（$P<0.01$），吊飞 2h 时该酶活性急剧下降至吊飞前的水平，吊飞 5h 后 GAPDH 活性随吊飞时间的延长急剧上升，吊飞达 24h 时最高，10~24h 该酶活性显著高于其他吊飞时间的活性（$P<0.05$）。雌、雄虫整个吊飞过程（0~24h）中 GAPDH 的酶活性始终高于静息状态（0h，下同）的活性，表明糖原在桔小实蝇的飞行过程中一直处于动用状态。

GDH 的活性变化类似 GAPDH，随吊飞时间的延长 GDH 的活性逐渐升高，高活性的 GDH 保证了桔小实蝇飞行中旺盛的糖类有氧代谢。

雌、雄虫 LDH 的活性变化完全不同。雄虫吊飞 1h，LDH 活性急剧上升，吊飞 2h 又急剧下降，然后再度迅速回升，吊飞 5~10h 达最高点，极显著高于其他任何吊飞时间的活性（$P<0.01$）。雌虫吊飞 1h，LDH 活性明显下降，吊飞 2h 回升至静息状态水平，之后持续下降。说明桔小实蝇在飞行过程中，雄虫除了可以进行高速有氧代谢以外，还具备一定的无氧代谢能力，而雌虫只进行有氧代谢。

2.2.2 脂肪酸 β-氧化的关键酶 HOAD 的活性变化：桔小实蝇雌、雄虫在飞行过程中 HOAD 的活性变化也不一样。吊飞后雄虫 HOAD 活性缓慢上升，吊飞达 2h 时上升到较高值，吊飞 5 h 下降到静息状态水平，然后又迅速上升，在吊飞 10h 时达到最高值，稍高于吊飞 2h 时的水平，两处理间差异不大。吊飞 24h 后该酶活性明显下降，显著低于静息状态水平。雌虫吊飞后 HOAD 活性则一直在静息状态水平及以下波动。说明桔小实蝇在飞行过程中，雄虫偶尔利用脂肪供给能量，雌虫则几乎不动用脂肪。

2.2.3 三羧酸循环关键酶的活性变化：桔小实蝇雌、雄虫飞行过程中 CS 的活性变化基本一致。吊飞开始后 1h，CS 活性急剧升高，吊飞 2h 稍有下降，然后缓慢上升。飞行后的 CS 活性一直高于静息状态水平，说明桔小实蝇的飞行过程中存在活跃的三羧酸循环，这保证了飞行肌对能量的大量需求。

2.2.4 GAPDH：HOAD 比值的变化：从表 2 可以看出，桔小实蝇雄虫飞行过程中，吊飞时间少于 24h 的情况下，GAPDH：HOAD 比值略高于 1；吊飞时间大于 24h，该比值超过 2。这些说明桔小实蝇雄虫飞行过程中的能源物质利用属于混合型，既能利用糖类又能利用脂肪提供飞行能量，但长时间飞行主要利用糖类供能。雌虫不同吊飞时间的 GAPDH：HOAD 比值大于雄虫，且随吊飞时间的增加该比值增大。结合飞行过程 HOAD 的活性变化情况，结果说明桔小实蝇雌虫主要利用糖类为飞行提供能量。

3 讨论

3.1 不同日龄桔小实蝇飞行肌能源物质的变化

已有研究表明,在昆虫飞行肌的能量供应中,糖类和脂肪被认为是两类主要的能源物质,大多数的双翅目昆虫、膜翅目昆虫及某些鳞翅目昆虫,糖类是唯一的可以利用的能源物质(李克斌等,2005b;韩兰芝等,2005)。本研究证实了糖类和脂肪是桔小实蝇飞行所利用的主要能源物质,其中以糖类为主。在室内饲养条件下,桔小实蝇飞行肌的能量代谢有以下特点:不同日龄的桔小实蝇利用不同能源物质的比例不同。在低龄期(4日龄以前)主要以糖酵解提供能量;随着日龄的增加,脂肪动用增加,10日龄时动用脂肪的能力接近利用糖类的能力;15~25日龄,飞行肌中糖酵解及脂肪酸β-氧化能力相当。根据测定结果,桔小实蝇成虫GAPDH:HOAD活性之比大于等于1,说明桔小实蝇虽然既能利用糖类又能利用脂肪为飞行提供能量,但糖类为飞行提供的能量高于脂类所提供的能量。

在以糖类作为主要能源的昆虫飞行肌中,3-磷酸甘油酯穿梭(G-3-P shuttle)循环是有氧酵解供能的主要途径(王荫长,2004)。Beenakkers等(1975)对直翅目、鳞翅目和双翅目的研究报道,飞行肌中存在着高浓度的GAPDH,没有或只有很低浓度的乳酸脱氢酶(LDH)。桔小实蝇飞行肌中含有高活性的GAPDH,而LDH活性比较低,说明桔小实蝇飞行肌代谢中可能存在活跃的G-3-P,且主要以有氧代谢为主,极少进行无氧代谢,本文结论与前人研究结论相符。

3.2 桔小实蝇吊飞过程中的能量代谢

对桔小实蝇吊飞过程中酶活性的研究结果表明,桔小实蝇在室内静风迫飞过程中其能量代谢有以下特点:糖类代谢所产生的能量是桔小实蝇飞行过程中的主要能源物质。雌、雄虫的能量代谢差异很大,雄虫除可以进行高速有氧代谢以外,还具备一定的无氧代谢能力,而雌虫只进行有氧代谢。

有氧代谢的特点是强度低、有节奏、不中断,持续时间长,而且方便易行,容易坚持。桔小实蝇雌成虫飞行过程中只进行有氧代谢,更利于其在有限的能源供给条件下,以便进行更远距离的飞行,为下一代种群繁衍寻找到更宜适的寄主,为下一代种群生存寻找到更适宜的环境,因此,桔小实蝇雌成虫飞行过程中具备的有氧代谢,很可能是桔小实蝇雌成虫生存策略的一种表现。

无氧代谢是肌肉剧烈运动时氧供应满足不了需要,肌肉即利用三磷酸腺苷(ATP)、磷酸肌酸(CP)的无氧分解和糖的无氧酵解生成乳酸,释放出能量,再合成三磷酸腺苷供给肌肉需要的一种代谢过程,这同样与桔小实蝇雄成虫的生存策略有关。桔小实蝇雄成虫不仅需要具备远程飞行能力,以获取寄主和适宜的生存环境;同时雄虫与雄虫之间,在获取与雌虫的交配权方面,存在强烈的竞争关系,具备无氧代谢能力,使桔小实蝇雄虫具备了有别于雌虫的飞行爆发能力,有利于雄成虫寻找到中意的雌虫,以获得交配的主动权。

测试表明,13h吊飞,桔小实蝇雌成虫的平均累计飞行距离明显大于雄虫(♀3.57km;♂3.26km),体现了桔小实蝇雌成虫具备有氧代谢能力的优势;尽管吊飞13h雌成虫的平均累计飞行距离明显大于雄虫,但雄虫的平均累计飞行时间、平均飞行速度和平均最大飞行速度(分别是 1.82h,1.05m·s⁻¹和 1.65m·s⁻¹)均大于雌成虫(分别是 1.30h,0.98m·s⁻¹和 1.58m·s⁻¹),可以看出,雄虫相比雌虫具有更多的飞行技巧,这是桔小实蝇雄成虫即可以进行高速有氧代谢,也还具备一定的无氧代谢能力的体现(袁瑞玲等,2014)。

本研究表明,在飞行过程中桔小实蝇雄虫能利用脂肪供给能量,雌虫则几乎不动用脂肪。昆虫脂肪体从表面上看似乎只是类脂物的贮存,但实际上是昆虫生长、发育、变态和生殖等代谢活动的中心组织,桔小实蝇雌成虫远程飞行过程中不动用脂肪,这可能与雌虫的卵子发育有关,也可能是雌虫集聚能量为了孕育后代的一种生存策略的表现。

3.3 能源物质代谢酶活性与能源物质动用的关系

本研究结果表明,能源物质代谢酶活性的高低反映了能源物质动用的多与少。研究报道,棕煌蜂鸟(*Selasphorus rufus*)在休息时燃烧脂肪,当觅食花蜜时很快转为消耗碳水化合物(Suarez et al., 1990;Dingle, 1996),桔小实蝇飞行肌能量代谢类型与之有类同之处。不同日龄的桔小实蝇试虫分别圈养在不同的 75cm×60cm×45cm 养虫笼中,实验过程中发现,由于空间有限,养虫笼中的桔小实蝇很少进行飞行活动,多数时间处于休息或爬行状态。没进行室内静风吊飞的桔小实蝇,随着日龄的增加,脂肪代谢相关酶 HOAD 活性增加,说明桔小实蝇在不飞行的状态下动用了脂肪作为能量,而在吊飞的情况下,HOAD 活性变化很小,糖代谢相关酶 GAPDH 活性随吊飞时间的延长迅速上升,说明桔小实蝇飞行过程中消耗的能源物质主要为糖类。对桔小实蝇飞行过程中能源物质消耗的研究结果为,随飞行时间的延长糖原含量显著下降(袁瑞玲等, 2015),本文结论与之相符。

3.4 桔小实蝇生殖器官发育进程中的能源物质变化

本研究中,桔小实蝇成虫在其生长发育的不同阶段利用的能源物质种类不同。GAPDH 酶活性随成虫的性成熟(陈敏等, 2014)逐渐降低,HOAD 酶活性随成虫的性成熟逐渐增高,但 GAPDH:HOAD 活性之比始终大于等于 1,说明,糖类和脂类是桔小实蝇成虫生长发育所需要的能源物质,且糖类为主要能源物质,但在其生殖器官发育进程中,糖类的消耗相对减少,脂肪的消耗相对增加,可能的原因之一是糖原是昆虫卵黄的一个重要组成部分,它为胚胎发育提供合成几丁质的葡萄糖单位(王荫长, 2004;陈敏等, 2014)。然而,桔小实蝇在飞行期间雄虫能一定程度利用脂肪提供能量,雌虫几乎不动用脂肪。本文结论与昆虫成虫卵巢成熟的程度或多或少地影响了能源物质的含量,即飞行和生殖所需的能源物质是相似的(Rankin and Burchsted, 1992)的结论略有不同。雌、雄虫在能量代谢上存在差异的原因有待进一步研究。

参考文献:

[1] Arévalo-Galarza L. Follett PA. Response of *Ceratitiscapitata*, *Bactroceradorsalis*, and *Bactroceracucurbitae* (Diptera: Tephritidae) to metabolic stress disinfection and disinfestation treatment. *Journal of Economic Entomology*, 2011,104(1): 75~80.

[2] Beenakkers AMT. Carbohydrate and fat as a fuel for insect flight: a comparative study. *Journal of Insect Physiology*, 1969,15: 353~361.

[3] Beenakkers AMT. Van den Broek ATM, De Ronde TJA, Development of catabolic pathways in insect flight

muscle: a comparative study. *Journal of Insect Physiology*, 1975, 21: 849~859.

[4] Beenakkers AMT, Van der Horst DJ, Van Marrewijk WJA. Insect flight muscle metabolism. *Insect Biochem.*, 1984, 14: 243~260.

[5] Bradford MM. A rapid and sensitive method for quantization of microgram quantities of protein utilizing the principle of protein-dye binding. *Analytical Biochemistry*, 1976, 72: 248~254.

[6] 陈敏, 陈鹏, 叶辉, 季清娥, 黎剑平, 袁瑞玲, 杨珊. 桔小实蝇卵巢形态结构及发育特征研究. 环境昆虫学报, 2014, 36(1): 83~89.

[8] Chen P, Ye H. Relationship among five populations of Bactrocera dorsalis based on mitochondrial DNA sequences in western Yunnan, China. *Journal of Applied Entomology*, 2008, 132 (7): 530~537.

[9] 陈鹏, 叶辉, 母其爱. 基于荧光标记的怒江流域桔小实蝇(*Bactrocera dorsalis*)的迁移扩散. 生态学报, 2007, 27(6): 2468~2476.

[10] Dingle H. Migration: The Biology of Life on the Move. Oxford University Press, New York. 1996, 164.

[11] Drew RAI, Hancock DL. The *Bactrocera dorsalis* complex of fruit flies (Diptera: Tephritidae: Dacinae) in Asia. *Bulletin of Entomological Research*, Supplement 1994, 2: 1~68.

[12] Gunn A, Gatehouse AG. The development of enzymes involved in flight muscle metabolism in *Spodopteraexempta* and *Mythimnaseparata*. *Comp. Biochem. Physiol.*, 1988, 91B(2): 315~324.

[13] 韩兰芝, 翟保平, 张孝羲, 刘培磊. 甜菜夜蛾飞行肌中与能量代谢有关的酶活性. 生态学报, 2005, 25(5): 1101~1106

[14] 李克斌, 高希武, 曹雅忠, 罗礼智, 江幸福. 甜菜夜蛾能源物质积累及其飞行能耗与动态. 植物保护学报, 2005a, 32(1): 13~17.

[15] 李克斌, 高希武, 罗礼智, 尹姣, 曹雅忠. 黏虫飞行过程中四种相关酶的活性变化. 昆虫学报, 2005b, 48(4): 643~647.

[16] 李伟丰, 杨朗, 唐侃, 曾玲, 梁广文. 中国桔小实蝇种群的微卫星多态性分析. 昆虫学报, 2007, 50(12): 1255~1262.

[17] Rankin MA, Burchsted JCA. The cost of migration in insects. *Annu. Rev. Entomol.*, 1992, 37: 533~559.

[18] Shi W, Kerdelhué C, Ye H. Population genetic structure of the oriental fruit fly, *Bactrocera dorsalis*(Hendel) (Diptera: Tephritidae) from Yunnan province (China) and nearby sites across the border. *Genetica*, 2010, 138(3): 377~385.

[19] Suarez, RK, Lighten JRB, Moyes CD, Brown GS, Gass CL, Hochachka PW. Fuel selection in rufous hummingbirds: ecological implications of metabolic biochemistry. *Proc. Natl. Acad. Sci.* USA, 1990, 87: 9207~9210.

[20] Suganya, R, Chen SL, Lu KH. Target of rapamycin in the oriental fruit fly *Bactrocera dorsalis*(Hendel): its cloning and effect on yolk protein expression. *Archives of Insect Biochemistry and Physiology*, 2010, 75: 45~56.

[21] 王荫长. 昆虫生理学. 北京: 中国农业出版社. 2004, 88~123.

[22] 袁瑞玲, 杨珊, 王晓渭, 陈鹏. 桔小实蝇飞行能力测试. 西部林业科学, 2014, 43(6): 66~71.

[23] 袁瑞玲, 杨珊, 王晓渭, 陈鹏. 桔小实蝇体重和糖原的积累及其飞行消耗. 环境昆虫学报, 2015, 37(2): 305~311.

(本文发表于《昆虫学报》, 2015 年)

松实小卷蛾在云南生活史及其对思茅松的危害

胡光辉,雷玮,槐可跃,张俊波,陈宏伟,周云

(云南省林业科学院,云南昆明 650204)

摘要:松实小卷蛾(*Retiniacristata walsingham*)在云南危害思茅松是个新发现。2001~2003年对其生活史进行了观察。松实小卷蛾在思茅地区1 a发生3代,以幼虫在受害梢中越冬。该虫主要危害2~3 a生幼林,同一株树以向阳面受害最重,受害枝梢的直径范围为4.7~13.0 mm。

关键词:关键词:思茅松;松实小卷蛾;生活史;危害

Life history of Retiniacristata in Yunnan Province and its damage on Pinus *kesiya var. angbianensis*

Hu Guang-hui, Lei Wei, Huai Ke-yue, Zhang Jun-bo, Chen Hong-wei, Zhou Yun

(Forestry Academy of Yunnan Province, Kunming 650204, China)

Abstract: *Retiniacristata* was newly found attacking *Pinus kesiya var. langbianensis* in Yunnan Province. It has 3 generations a year in Simao District and overwinters as larva in the infested branch tips. *Retiniacristata* mainly attacks young trees at the age of 2 or 3 years. The sunny side of the tree suffered severer damage. The diameters of the infested branches were from 4.7 mm to 13.0 mm.

Key words: *Pinus kesiya var. langbianensis*; *Retiniacristata*; life history; damage

思茅松(*Pinus kesiya* var. *langbianensis*)是云南省主要用材及采脂树种。广泛分布于哀牢山以西的南亚热带山地(思茅、西双版纳、德宏、红河、文山、临沧等地州),垂直分布海拔700~1 800m,集中分布在海拔1 000~1 700m地区,现有森林面积102.5万hm²;活立木蓄积量10 011.1万m³。松香年产量11.5万t。在云南省林业产业结构中占有重要地位[1],思茅松是云南省目前最有发展前景的速生造林树种之一,其人工林正以每年约7 000hm²的速度增长[2]。思茅松人工幼林的食叶害虫主要有思茅松毛虫 *Dendrolimus kikuchii*、云南松毛虫 *D. houi*、祥云新松叶蜂 *Neodiprion xiangyunicus*、文山松叶蜂 *Diprion wenshanicus*、云南松镰象 *Drepanoderus leucofasciatus*;枝干害虫有松梢螟 *Diorytria splendidella* 和松实小卷蛾 *Retinia cristata*。其中对思茅松人工幼林生长影响最大、危害最严重的是松实小卷蛾。松实小卷蛾已知在国内分布于北京、黑龙江、辽宁、河北、山西、山东、江苏、浙江、广东、广西、云南、湖南、江西、四川、河南、安徽等省(自治区、直辖市);国外分布于日本。寄主有马尾松 *Pinus massoniana*、黑松 *P. thunbergii*、黄山松 *P. faiwanensis*、油松 *P. tabulaeformis*、湿地松 *P. elliottii*、赤松 *P. densiflora* 等。其生物学特性、幼虫和蛹的雌雄区分及蛹发育起点温度和产卵量已有报道[3~6],该虫云南分布和为害思茅松未见报道。2001~2003年,课题组对松实小卷蛾生活史进行了观察,并调查了其危害思茅松情况,现报道如下。

1 材料与方法

1.1 试验地概况

试验地设在思茅市翠云区锦屏镇清水河"现代林业资源培育产业化试验与示"项目试验林。该处山势起伏,沟壑纵横,属横断山脉,海拔1 080~1 280m,年均气温18.2℃,年均降水量1 524mm,为1999~2000年7月定植的思茅松纯林。

1.2 试验方法

1.2.1 生活史观察

每月上、中、下旬各观察1次,每次剪取受害枝梢100个,解剖并分别记录幼虫、蛹和成虫等情况,统计幼虫、蛹和成虫占总虫数的比率,绘制各虫态所占比率与时间变化曲线图,分析其年发生世代数。

1.2.2 危害与树高的关系

调查松实小卷蛾与树高关系,对分析该虫入侵时间以及利用其性信息素进行虫情监测和综合防治都有重要意义。在受松实小卷蛾危害较重林分,各随机调查50株树,平均树高为195cm,重复3次,分距离地面0~50cm、51~100cm、101~150cm、151~200cm、200cm以上5个区段调查受害枝梢数量,统计各区段受害率,分析该虫对不同林龄危害情况。

1.2.3 危害与方向的关系

在受松实小卷蛾危害较重的同龄林分,随机调查150株树,3个重复,分东、南、西、北4个方位记录每株树受害侧枝数量,计算各方位受害梢占总被害梢的百分率,分析不同方位受害情况。

1.2.4 危害与枝梢粗细的关系

在调查受害枝梢不同方向分布同时,记录每个受害枝梢的枝径,分析松实小卷蛾对枝梢粗细的选择。

2 结果与分析

2.1 松实小卷蛾年发生世代

鉴于松实小卷蛾虫龄不整齐、世代重叠的现象,本文未采取常用的以卵、幼虫、蛹、成虫等来表示其生活史,区划其世代数。而选用不同时间幼虫或蛹占所调查幼虫、蛹和成虫总虫数的比例,绘制出时间—幼虫(或蛹)比例曲线图(图1、图2),根据幼虫出现峰的多少,判定松实小卷蛾的世代数。图上明显有3个峰,是幼虫数量的高峰期,即上年11月上旬~2月下旬,5月中旬~6月上旬,7月上旬~8月中旬;蛹高峰期是:3月上旬~4月中旬,6月上旬~6月下旬,9月下旬~10月中旬。由此说明其在思茅1a发生3代,以幼虫在受害枝梢内越冬。南京,松实小卷蛾1a发生4代,以蛹在枝梢内越冬[3]。由思茅与南京两地6~9月平均气温比较(思茅:21.7℃、21.5℃、21.2℃、17.1℃;南京:29.5℃、32.5℃、33.0℃、17.5℃);6~9月最高气温(思茅:26.6℃、26.1℃、26.2℃、20.5℃;南京:29.5℃、32.5℃、33.6℃、28.3℃),看出南京有效积温高于思茅,因此松实小卷蛾南京每年比思茅多发生1代。

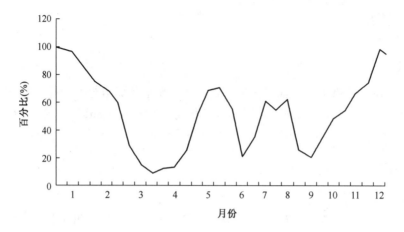

图 1　松实小卷蛾幼虫年发生世代

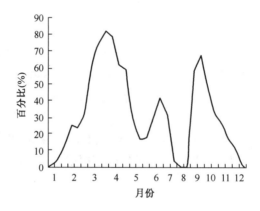

图 2　松实小卷蛾幼虫蛹年发生世代

2.2　松实小卷蛾危害与树高关系

调查结果详见表 1。经方差分析不同高度之间危害差异显著,其中侧梢 101~150cm 与其他 4 个区段有明显差异;主梢 101~150cm 和 151~200cm 两者间差异不明显,但与其他 3 个区段差异显著,说明以 101~200cm 区间主梢危害最重。思茅松树高年生长量是 82~125cm[1],侧枝和主梢受害情况显示,思茅松幼林受松实小卷蛾危害的林龄为 2~3a。由此得出结论:思茅松人工林松实小卷蛾的防治最佳时间是 2~3a 生幼林。诱捕器最佳设置高度为 100~200cm。

表 1　松实小卷蛾危害与树高关系

危害部位	受害枝梢										合计
	0~50cm		51~100cm		101~150cm		151~200cm		>200cm		
	枝梢数	%	枝梢数	%	枝梢数	%	枝梢数	%	枝梢数	%	
侧梢	0	0	29	13.24	160	73.06	28	12.79	2	0.91	219
主梢	0	0	1	2.70	17	45.95	15	40.54	4	10.81	37

2.3 松实小卷蛾危害与方向关系

经调查,150 株树共有 188 枝受害枝,其中东面有 62 枝,占 33.58%;南面为 58 枝,占 30.34%;西面为 43 枝,23.18%;北面为 25 枝,占 12.90%。方差分析显示:东面、南面与西面三者之间差异不明显;北面和其他 3 个方向之间有明显差异,证明不同方位松实小卷蛾危害差异显著。东面、南面和西面等向阳面受害比例高于阴面,3 个方向占 86.90%,说明松实小卷蛾危害有趋光性。该虫危害特点是,阳面受害大于阴面,危害多在郁闭度小、阳光充足的林分,以东南方向最重。

2.4 松实小卷蛾危害与枝径关系

枝梢粗细严重影响蛀梢害虫取食活动。粗大枝梢木质化程度高,有机养分含量少;而枝梢太细,蛀食后极易枯萎,不利于筑坑取食[7]。对 139 个受害枝梢的调查表明,对思茅松而言,该虫主要蛀害枝径为 5.2~8.7 mm 的枝梢,其蛀害率为全部受害枝梢的 70% 以上,受害枝梢的直径范围为 4.7~13.0 mm(表 2)。

表 2 不同梢径枝梢占总受害枝梢百分比

枝径(mm)	<4.7	5.2	5.7	6.2	6.7	7.2	7.7	8.2	8.7	9.2	9.7	10.0	11.0	12.0
枝数	7	10	17	11	20	12	11	12	9	5	8	2	6	4
%	5.1	7.2	12.3	8.0	14.5	8.7	8.0	8.7	6.5	3.6	5.8	1.4	4.3	2.9

3 结论与讨论

松实小卷蛾是思茅松人工幼林最主要的虫害之,1 年发生 3 代,以幼虫在枯梢中越冬,南京、浙江以蛹越冬[3,4]。该虫对 2~3 a 生思茅松幼林危害严重,同一株树阳面受害大于阴面,受害部位垂直高度 100~200 cm,受害枝梢的直径范围为 4.7~13.0 mm。所以,思茅松人工林松实小卷蛾防治关键在 2~3 a 生幼林。

松实小卷蛾不仅危害马尾松主侧梢,还取食其球果[3,4],近年研究主要针对思茅松幼林,思茅松种子园尚属空白,随着云南省思茅市思茅松种子园建设面积不断扩大,思茅松种子园松实小卷蛾防治研究是今后作的重点。

致谢:松实小卷蛾标本承蒙中国科学院北京动物研所宋士美先生、武春生先生鉴定,在此表示感谢!

参考文献:

[1] 赵文书,唐社云,李莲芳,等.思茅松无性系种子园营建技术[J].云南林业科技,1999(增刊):38-45.
[2] 王达明,李莲芳.思茅松速生丰产林培育的关键技术[J].云南林业科技,1999(4):6-8.
[3] 刘友樵,李广武.中国动物志昆虫纲第二十七卷鳞翅目卷蛾科[M].北京:科学出版社,2002:350.
[4] 萧刚柔.中国森林昆虫[M].北京:中国林业出版社,1992:836-837.
[5] 中国林木种子公司.林木种实病虫害防治手册[M].北京:中国林业出版社,1988:20.

[6] 宣家发,万东,何俊旭,等.松实小卷蛾幼虫和蛹的雌雄区分及蛹发育起点温度和产卵量的研究[J].森林病虫通讯,1996(4):28-29.
[7] 叶辉.纵坑切梢小蠹蛀梢期生物学研究[J].昆虫学报,1996,39(1):58-61.

(本文发表于《中国森林昆虫》,2005年)

广南油茶虫害调查及糖醋液诱捕试验

陈福[1],郭晓春[1],胡光辉[2],郑畹[2],闫争亮[2]

(1 云南省林业科学院油茶研究所,云南 广南 663300;2 云南省林业科学院,云南 昆明 650201)

摘要:对广南县油茶园的主要害虫进行了踏查和调查,发现油茶象、油茶金花虫、茶小绿叶蝉和小青花金龟为重度危害,茶梢蛾、绿鳞象甲、油茶肖叶甲、无斑弧丽金龟、龟蜡蚧、油茶绵蚧、茶梨蚧、茶蚜、茶黄蓟马中度危害,其余害虫危害较轻或零星发生。利用糖醋液进行了对油茶害虫的诱捕试验。糖醋液对油茶金花虫诱捕效果最好,同时也诱捕到大量的茶梢蛾、茶枝镰蛾、小青花金龟、茶小绿叶蝉等。试验证明,糖醋液可用于油茶害虫的监测和诱杀。最后,对试验结果进行了讨论并对油茶病虫害防控提出了建议。

关键词:油茶;害虫;调查;诱集;糖醋液

Investigation and trap catches by sweet and sour liquid of insect pestsin Guangnan oil-tea orchards

Chen Fu[1], Guo Xiaochun[1], Hu Guanghui[2], Zheng Wan[2], Yan Zhengliang[2]*

(1. Institwte of Oil Tea, Yunnan Academg of Forestry, Guangnan Yunnan 663300; 2. Yunnan Academg of Forestry, kunming Yunnan 650201)

Abstract: Investigation and trap catches of insect pests were conducted in Guangnan oil-tea orchards in 2017, 2018. In the oil-tea orchards, *Curculio chinensis*, *Colaspoides opaca*, *Oxycetoniajucunda*, andEmpoascaflavescens were the most dangerous pests; and *Parametriatesthea*, *Hypomecessquamosus*, *Demotinathei*, *Poecilocoris latus*, *Popilliamutans*, *Metaceronema japonica*, *Pinnaspistheae*, *Toscopteraaurantii*, and *Scirtothrips dorsalis* were moderately harmful pests. In trap catch tests by sweet and sour liquid, a large number of Colaspoides opaca was catched, and the trapped number of *Parametriatesthea*, *Casmarapatrona*, *Poecilocoris latus*, *Oxycetoniajucunda*, andEmpoascaflavescens were significantly more than those by water (CK). The results were discussed and some suggestion was proposed.

Keywords: oil-tea; insect pests; investigation; trap catch; sweet and sour liquid

油茶(*Camellia oleifer*)是我国重要的经济林树种,具有极高的经济价值。据统计,我国油茶林面积 6 000 多万亩,油茶籽产量超过 200 万 t,茶油产量达到 50 多万 t,产值超过 550 亿元。云南省有油茶面积接近 300 万亩,油茶成林达到 60 余万亩,已成为山区农民重要的收入来源之一。随着油茶开发力度的日益加大,油茶的病虫害发也日趋严重。油茶病虫害不仅导致大量落花落叶落果,甚至可致全株枯死,严重影响油茶产量与质量[1,2]。某些虫害对油茶的取食,造成枝叶损伤,还加速病害的传播,如油茶金花虫(又名刺股沟臀肖叶甲,*Colaspoides opaca*)的取食可加速传播油茶毁灭性病害炭疽病的病原菌,导致茶果减产;又如油茶棉蚧(*Metaceronema japonica*)、龟蜡蚧(*Ceroplastes floridensis*)和黑胶粉虱(*Aleurotrachelus camellia*)吸食油茶树叶子和枝干的汁液,分泌蜜露,导致煤污病发生[3,4],使树势衰弱,引起落花落果，重者全株枯死。

由于油茶害虫的隐蔽性，调查和监测难度较大，当发现已经有明显症状时，可能已经错过了最佳的防治时机。作为一种传统的虫害防治措施，糖醋液是一种具有广谱性质的昆虫引诱物，对鳞翅目、双翅目、鞘翅目害虫等都有明显的诱集效果，被广泛运用于害虫的监测和诱杀[5,6]。我们曾用糖醋酒液诱杀严重为害油橄榄的陈齿爪鳃金龟(*Holotrichia cheni*)，取得良好效果[7]。为了研究油茶害虫监测和诱杀的方法，我们开展了糖醋液的诱集试验。

1 材料与方法

1.1 调查方法

以踏查和抽样调查方法调查油茶主要虫害的分布和危害程度。选择位于文山壮族苗族自治州广南县旧莫乡的板茂村和莲城镇老龙村油茶标准化栽培基地调查样点，调查主要虫害的发生程度及特点，详细记录发生地、危害部位等有关数据，并对虫种进行鉴定。在板茂村进行重点调查，分别选择具代表性相对均匀一致的 10m×10m 样方 3 个（共 9 个样方），调查主要虫害的发生程度。

1.2 供试材料

红糖（元宝红糖 500g 装，昆明古凤食品有限公司）；白砂糖（380g 装，云南大理市八千方食品厂）；食用醋（七醋，433mL，富宁金泰得剥隘七醋有限公司，总酸≥3.5g/100mL）；550mL 云南山泉矿泉水瓶；细铁丝（市售）。

1.3 糖醋液配制

糖醋液有 2 个配方。配方 1：红糖 50g，食用醋 50mL，清水 100mL；配方 2：白砂糖 50g，食用醋 50mL，清水 100mL。

1.4 诱捕装置

在 550mL 无色透明塑料矿泉水瓶的三分之一处，以剪刀裁剪出 3 个直径约为 5.0cm 的圆形小口，3 个圆形小口之间距离相等。在矿泉水瓶的底部放入拇指大小蘸有溴氰菊酯乳油的 100 倍液的医用棉球；然后加入约 200mL 的糖醋液。试验时，将装有杀虫剂和糖醋液的矿泉水瓶以细铁丝悬挂于油茶树的侧枝上。

1.5 试验地概况

试验地位于文山壮族苗族自治州广南县旧莫乡的板茂村。该地基地海拔 1 200m，属于中亚热带高原季风气候，年平均气温 16℃，年平均日照数 1 860h。建有东昌农林产品开发公司和云南省林业科学院油茶研究所共建的高原山地油茶丰产栽培技术示范基地，油茶树树龄 10a。基地周边为云南松纯林以及新栽的油茶幼树。

1.6 诱捕试验设置

将由矿泉水瓶制成的诱杀器悬挂于树高约 2.0m 的油茶植株侧枝外缘，高度约 1.5m 的位置,正好与冠幅持平,诱杀器以避免树叶遮蔽或遭受阳光直晒为宜。每个糖醋液配方共设置诱杀装置5个,另设1个空白诱杀装置(内盛清水)为对照;每个诱杀器之间的距离约为30m。每5天统计一次数据,统计诱捕到的害虫种类和数量。诱捕时间从4月27日至6月1日。

1.7 昆虫种类鉴别

尽量在试验地现场当场鉴别诱捕到的昆虫。一时难以确定的昆虫种类,则标明采集时间、地点、诱捕装置编号,带回实验室后查对《油茶病虫害诊断与防治原色生态图谱》《茶树病虫害防治彩色图册》等文献确定鉴定。

2 试验结果

2.1 广南油茶要虫害名录、与危害

广南县距北回归线及太平洋、印度洋较近,随着海拔高度的不同,呈现亚热带高原立体气候和季风气候的特点。

油茶主要虫害种类及发生程度见表1所示。从调查结果看,茶梢蛾、绿鳞象甲、油茶肖叶甲、油茶宽盾蝽、无斑弧丽金龟、油茶棉蚜、茶梨蚧、茶蚜、茶黄蓟马等属于中度为害的害虫,需要给予适当关注。危害较为严重的害虫有油茶象、油茶金花虫、小青花金龟、茶小绿叶蝉等,需要及时采取的防治措施。其余害虫则危害较轻。

表1 广南县油茶主要虫害及其危害(2017~2018)
Table 1 Insect pests and damge in oil-tea orchards in Guangnan county(2017~2018)

种名 species	危害部位 Damage parts	危害程度 Damage degree
相思拟木蠹蛾 Arbela bailbarana	幼虫蛀食枝干	+
茶蓑蛾 Cryptothelea minuscula	危害叶片	+
白囊蓑蛾 Chalioides kondonis	幼虫为害树叶	+
大蓑蛾 Cryptothelea vaiegata	危害叶片	+
茶梢蛾 Parametriates thea	幼虫叶片,翌年春转移危害春梢	+ +
茶蚕 Andraca bipunctata	危害叶片	+
茶枝镰蛾 Casmara patrona	幼虫蛀食枝干	+
丽绿刺蛾 Parasa lepida	取食树叶	+
褐边绿刺蛾 Parasa consocia	取食树叶	+
油茶尺蠖 Biston marginata	油茶危害叶片	+
茶天牛 Aeolesthes induta	幼虫蛀食枝干	+
茶材小蠹 Xyleborus rornicatus	成虫、幼虫为害枝干	+
油茶象 Curculio chinensis	危害嫩枝和嫩果,油茶蛀食茶果	+ + +
绿鳞象甲 Hypomeces squamosus	成虫为害嫩叶	+ +
油茶金花虫 Colaspoides dentata	成虫取食嫩枝和嫩叶	+ + +
油茶肖叶甲 Demotina thei	危害嫩梢、嫩叶	+ +
油茶宽盾蝽 Poecilocoris latus	1龄若虫取食叶面,后为害茶果	+ +
小青花金龟 Oxycetonia jucunda	食害花瓣、花蕊、芽及嫩叶	+ + +

（续）

种名 species	危害部位 Damage parts	危害程度 Damage degree
无斑弧丽金龟 Popillia mutans	成虫食叶	++
茶小绿叶蝉 Empoasca flavescens	成、若虫吸汁液	+++
龟蜡蚧 Ceroplastes floridensis	吸食嫩枝叶汁液,分泌蜜露引发煤污病	+
油茶棉蚧 Metaceronema japonica	同上	++
茶梨蚧 Pinnaspis theae	若虫和雌虫附着叶脉取食	++
茶蚜 Toscoptera aurantii	吸食嫩梢嫩叶	++
茶黄蓟马 Scirtothrips dorsalis	成虫、若虫锉吸为害新梢嫩叶	++

注:"+++"表示发生严重;"++"表示中度发生;"+"表示发生轻微或偶尔发生。The symbol "+++" means severe damage; "++" means moderate damage; "+" means slight damage or occasional damage.

2.2 糖醋液诱捕结果

利用糖醋液作为食诱剂,共诱捕到13种油茶害虫。其中,鳞翅目害虫6种,有相思拟木蠹蛾(*Arbela bailbarana*)、茶蓑蛾(*Cryptothelea minuscula*)、茶梢蛾(*Parametriates thea*)、大蓑蛾(*Cryptothelea vaiegata*)、茶蚕(*Andraca bipunctata*)、茶枝镰蛾(*Casmara patrona*)等。鞘翅目害虫共7种:茶天牛(*Aeolesthes induta*)、油茶金花虫(*Colaspoides dentata*)、油茶宽盾蝽(*Poecilocoris latus*)、绿鳞象甲(*Hypomeces squamosus*)、茶材小蠹(*Xyleborus rornicatus*)、小青花金龟(*Oxycetonia jucunda*)、无斑弧丽金龟(*Popillia mutans*)。诱捕到的半翅目害虫1种:茶小绿叶蝉(*Empoasca flavescens*)。具体结果见表2。

表2 糖醋液诱捕油茶害虫结果
Table 2 Trap catches of insect pests of oil-tea by sweet and sour liquid

种名 species	诱集数量(头/诱捕器) numbers of trapped insects		
	红糖 brown sugar	白砂糖 white sugar	对照 CK
相思拟木蠹蛾 *Arbela bailbarana*	1.00±0.71	0.20±0.45	0.00±0.0
茶蓑蛾 *Cryptothelea minuscula*	2.80±2.50	1.60±1.14	0.00±0.0
茶梢蛾 *Parametriates thea*	8.40±3.78	5.00±1.58	0.00±0.0
大蓑蛾 *Cryptothelea vaiegata*	0.40±0.55	0.20±0.45	0.00±0.0
茶蚕 *Andraca bipunctata*	0.40±0.55	0.60±0.89	0.00±0.0
茶枝镰蛾 *Casmara patrona*	7.40±3.36	4.20±1.24	0.00±0.0
茶天牛 *Aeolesthes induta*	1.40±1.14	0.80±0.84	0.00±0.0
油茶金花虫 *Colaspoides dentata*	19.20±6.98	10.80±3.35	0.00±0.0
油茶宽盾蝽 *Poecilocoris latus*	5.20±2.39	4.60±3.05	0.00±0.0
绿鳞象甲 *Hypomeces squamosus*	3.80±3.70	6.40±2.79	2.00±1.22
茶材小蠹 *Xyleborus rornicatus*	0.80±0.84	0.80±1.30	0.00±0.0
小青花金龟 *Oxycetonia jucunda*	9.80±3.70	6.40±1.14	0.00±0.0
无斑弧丽金龟 *Popillia mutans*	4.00±2.12	4.00±1.22	0.00±0.0
茶小绿叶蝉 *Empoasca flavescens*	7.80±5.36	6.00±2.24	1.00±1.0

利用糖醋液诱捕到较多的茶梢蛾、茶枝镰蛾、油茶金花虫、油茶宽盾蝽、小青花金龟、茶小绿叶蝉。但诱捕到的相思拟木蠹蛾、茶蓑蛾、大蓑蛾、茶蚕、茶天牛、绿鳞象甲、茶材小蠹、无斑弧丽金龟数量较少。

3 结果讨论

云南省委省政府提出要全力打造世界一流的"绿色能源""绿色食品"和"健康生活目的地"三张牌。"绿色食品"需要病虫害的"绿色防控"来保障。油茶油是受到市场追捧的高档食用油，因其高高不饱和脂肪酸含量受到消费者的青睐。但油茶也是病虫害频发的一个经济林树种，病虫害防控是茶果生产必不可少的环节。为了保证油茶油的绿色品质，必须采取绿色的病虫害防控技术。油茶害虫的糖醋液诱杀，为我们提供了一个可行的选择。

利用糖醋液诱捕到较多数量的小青花金龟、油茶金花虫、茶小绿叶蝉；也诱捕到明显大于空白对照的相思拟木蠹蛾、茶蓑蛾、大蓑蛾、茶蚕、茶天牛、绿鳞象甲、茶材小蠹、无斑弧丽金龟。其诱捕量与林间踏查和调查中统计的各种害虫发生和危害程度一致。由此可见，糖醋液是油茶害虫较好的食诱剂。

油茶象是广南油茶最严重的害虫，我们的踏查中也发现油茶象在林间活动。曾有报道，糖醋液对油茶象具有较好的诱集作用[8]。但利用糖醋液没有诱捕到油茶象这一危害油茶最为严重害虫。可能的原因是，诱捕装置不合适。油茶象飞翔力弱，其活动时爬行时候多于飞行时候[9]。而以矿泉水塑料瓶作为诱捕器，悬挂于油茶树的侧枝上，不利于以爬行活动为主的油茶象的诱捕。如果以诱杀油茶象为目标，必须改进诱捕装置。

植食性害虫主要取食植物的茎叶、果实、花蜜等，并且常对某些食物表现出明显的偏好性，植物挥发物在害虫食物偏好选择行为中发挥着重要作用[10,11]。利用寄主油茶的挥发物可诱捕油茶害虫[12]。油茶象对其寄主嫩枝、嫩果的挥发物具有明显的趋向反应[13]。如果将糖醋液和油茶嫩枝、嫩果的粗提物联合使用，可能会大大提高油茶害虫的诱捕效果。

有关云南油茶害虫的研究，基本上局限于害虫种类和危害程度的调查。迄今为止，尚无油茶重要害虫生活史的研究。这为油茶害虫的防治带来困扰。如对危害枝干的茶梢蛾、茶枝镰蛾、相思拟木蠹蛾等，往往是发现其幼虫危害时，才采取防控措施，而错过了对成虫的杀灭。所以，进一步提升油茶园的管理水平，对油茶害虫进行精准防治，还需要针对滇西、滇南等不同的油茶种植区主要害虫的年发生动态进行详细的研究。这样才能采取针对性的防控措施。所以，油茶重要害虫的生活史研究，是绿色防控必不可少的基础性已经，也是我们应该立即开展的工作。

参考文献：

[1] 伍建榕,穆丽娇,林梅,等.滇西地区红花油茶主要病虫害种类调查[J].中国森林病虫,2012,31(1):22~26.

[2] 刘凌,闫争亮,周楠,等.云南广南油茶园节肢动物群落组成和多样性特征[J].西部林业科学,2015(5):33~38.

[3] 孙德友.日本卷毛蜡蚧的发生预测和综合防治探讨[J].浙江农林大学学报,1989(1):53~59.

[4] 陈祝安.油茶黑胶粉虱的研究[J].林业科学,1981,17(1):30~36.

[5] 武小恺.糖醋液对苹果园昆虫的诱集作用研究[D].河北农业大学,2016.

[6] 卢辉,卢芙萍,梁晓,等.糖醋液诱杀热带瓜菜害虫轻简化技术研究[C]//中国热带作物学会第九次

全国会员代表大会暨2015年学术年会论文摘要集.2015.

[7] 马惠芬,李勇杰,闫争亮,等.用糖醋酒液引诱陈齿爪鳃金龟试验初报[J].西部林业科学,2010,39(4):92~94.

[8] 周绪佑,肖善斌.茶籽象(虫甲)诱杀试验[J].林业科技通讯,1981(6):25~26.

[9] 赵丹阳,秦长生,徐金柱,等.油茶象甲形态特征及生物学特性研究[J].环境昆虫学报,2015,37(3):681~684.

[10] Bruce T J,Wadhams LJ,Woodcock C M. Insect host location: a volatile situation[J]. Trends in Plant Science,2005,10(6):269~274.

[11] 戴建青,韩诗畴,杜家纬.植物挥发性信息化学物质在昆虫寄主选择行为中的作用[J].环境昆虫学报,2010,32(3):407~414.

[12] 刘俊,周德明,周国英.油茶害虫植物源引诱剂筛选及其诱集的昆虫种群动态[J].植物保护,2017,43(5):174~179.

[13] 赵丹阳,秦长生,徐金柱,等.油茶象甲成虫对油茶寄主选择性研究[J].中国农学通报,2015,31(17):100~104.

(本文发表于《西部林业科学》,2018年)

针叶挥发物及营养物对楚雄腮扁叶蜂产卵选择影响的初步研究

刘凌[1]，闫大琦[2]，毛云玲[1]，祁荣频[1]，胡光辉[1]，闫争亮[1]

(1. 云南省林业科学院，云南 昆明 650204；2. 华中农业大学植物科学技术学院，湖北 武汉 430070)

摘要：摘要：楚雄腮扁叶蜂雄蜂、抱卵雌蜂对云南松针叶粗提物、卵与针叶的混合粗提物、已产卵雌蜂的趋向反应显示，楚雄腮扁叶蜂抱卵雌蜂对这3种气味均表现出明显的趋向；与针叶粗提物相比，抱卵雌蜂对卵与针叶的混合粗提物趋向反应更明显。除针叶粗提物外，雄蜂对卵与针叶混合粗提物、已产卵雌蜂均表现出无趋向反应。趋向实验表明，这些挥发物在楚雄腮扁叶蜂进行产卵寄主选择过程中起重要作用。野外调查表明，楚雄腮扁叶蜂雌蜂在云南松上的产卵数明显高于在云南油杉上的；松针针叶营养物的测定发现，云南松氨基酸含量比云南油杉多，说明寄主氨基酸的丰富度可能是楚雄腮扁叶蜂产卵寄主选择的依据之一。

关键词：楚雄腮扁叶蜂；产卵；寄主选择；针叶挥发物；氨基酸

中图分类号：Q968

Preliminary study on the effects of needle volatiles and nutrients of two conifer species on oviposition host selection of *Cephalica chuxiongnica* Xiao (Hymenoptera：Pamphiliidae)

Liu Ling[1], Yan Da-qi[2], Mao Yun-ling[1], Qi Rong-pin[1], Hu Guang-hui[1], Yan Zheng-liang[1]

(1. Yunnan Academy of Forestry, Kunming, Yunnan 650204, China; 2. College of Plant Science & Technology, Huazhong Agricultural University, Wuhan, Hubei 430070, China)

Abstract：Tests of taxis responses of gravid females and males of *Cephalica chuxiongnica* Xiao to crude extracts of needles of *Pinus yunnanensis*, mixed extract of fresh eggs and needles of the pine, egg-laying females of the pine sawfly were conducted. Gravid females showed significantly taxis responses to the three odors, and preferred to the mixed extract when compared with the needle extract. Males significantly tended to extrats of needles of *P. yunnanensis*, and showed none selection response to the other odors. The results of taxis response tests implied that the volatiles of needles, needles with eggs and females they are ovipositing play important roles in the oviposition host selection of gravid female. Field investigation showed that egg - laying females on *P. yunnanensis* were far greater than on K. evelyniana. In amino acids analysis, there were more acids in needles of *P. yunnanensis* than needles of Keteleeriaevelyniana. This suggested that abundances of amino acids in the needles may also be chemical signals for the gravid females to determine suitable oviposition hosts.

Key words：*Cephalica chuxiongnica*; oviposition; host selection; needle volatile; amino acids

在植食性昆虫的生命周期中，产卵是其繁衍后代、维持种群的重要环节，反映了植食性昆虫与植物之间的相互适应关系,以及昆虫群落构建机制[1]。在一定程度上规定了植食性昆虫对植物的利用策略,从而影响昆虫种群的演化和繁荣[2]。植食性昆虫对寄主植物的选择性是指它对寄主植物的种类、取食和产卵部位等均有一定的偏好性,包括取食选择性和产卵选择性[1]。由于绝大多数种类的初孵幼虫活动能力较弱,其存活有赖于雌成虫对产卵寄主植物种类、植株和特定产卵位置的明智选择。一般认为,亲代雌成虫的产卵选择首先必须有利于后代个体的存活和生长发育,即所谓"优选—效能"理论(Preference-performance)[3]。寄主植物的营养、选择压力、天敌因素、植物气味等均能成为植食性昆虫对寄主植物选择的影响因素[4]。在对产卵基质远距离的搜索定位过程中,产卵基质释放较高浓度的挥发物,雌虫通过嗅觉感受器挥发物来定向产卵基质[5~8]。当接触产卵基质后,雌虫则依靠各种触觉和化学感受器识别基质表面的非挥发性的接触信息化合物(主要是营养物及其衍生物,如氨基酸及其衍生物、水溶性糖等),来决定是否产卵[9~11]。

楚雄腮扁叶蜂(*Cephalica chuxiongnica* Xiao)隶属于膜翅目(Hymenoptera)扁叶蜂科(Pamphiliidae)腮扁叶蜂属(*Cephalica*),仅在中国滇中和滇东北地区有分布,是一类经济重要性较大的昆虫[12]。该虫主要危害松科(Pinaceae)植物,以幼虫取食松树叶片,不仅因取食吐丝形成的灰白色幕巢影响观瞻;还造成树木畸形、生长缓慢;同时,楚雄腮扁叶蜂为害后造成树势衰弱,还易引发小蠹虫等次期性害虫的侵害,危害严重时导致寄主死亡。野外调查发现,寄主植物云南松(*Pinus yunnanensis* Franch)、华山松(*Pinus armandii* Franch)和云南油杉(*Keteleeria evelyniana* Mast)中,云南松被楚雄腮扁叶蜂危害较重,云南油杉则危害较轻,且在云南松上聚集产卵的雌蜂数最多。因此,本文试图从化学生态学角度分析影响楚雄腮扁叶蜂产卵寄主选择的因素。为此,测试了楚雄腮扁叶蜂产卵雌蜂对针叶挥发物、带卵针叶挥发物和正在产卵雌蜂的挥发物的趋向反应,试图探索挥发性物质在其产卵寄主选择中的作用;调查了楚雄腮扁叶蜂在云南松、云南油杉上的产卵雌蜂数;分析了2种针叶中的氨基酸种类和含量。

1 材料与方法

1.1 供试虫源

试虫采于在云南省楚雄州禄丰县和平村。在有虫巢的云南松树冠下方挖10~20cm土层,单头收集刚羽化但未交尾的楚雄腮扁叶蜂成虫,带回室内。将雌、雄虫分别放入养虫笼(40cm×40cm×40cm)内,喂以10%的蔗糖水,备用。养虫室温度为室温,相对湿度(60±10)%。

1.2 林间调查

调查地点:经度102°12′E,纬度25°12′N,海拔1 820m,年平均气温16.2℃,年降水量930mm,寄主植物是云南松、云南油杉。

调查在2012年9月(楚雄腮扁叶蜂的产卵盛期)开展。共选择2块样地,第一块样地随机抽查云南松、云南油杉各15株,在第二块样地随机选择云南松、云南油杉各14株。调查时,统计每株树上的带虫枝数及每个枝梢的产卵雌蜂数。

1.3 寄主树针叶和卵粗提物的提取

采用同时蒸馏法(SDE)提取针叶挥发油[4]。将新鲜的带卵和无卵云南松针叶剪下后,迅速放入磨口玻璃瓶中密封,将玻璃瓶放入加有医用冰袋的保温瓶中。带回实验室后,用医用小刀将针叶剪碎为2mm×2mm的碎屑,各称取350g放入同时蒸馏萃取装置,以二氯甲烷为萃取。同时蒸馏萃取5h后,旋转蒸发仪除去大部分二氯甲烷,将剩余萃取溶液30mL,用无水硫酸钠干燥。将所得提取液放入冰箱中保存,保存温度为-10℃,备用。

1.4 楚雄腮扁叶蜂的趋向试验

利用"Y"形嗅觉仪法[5]测试楚雄腮扁叶蜂的趋向反应。测试时,在一个球形管内放入滴有10μL二氯甲烷的滤纸条(2.0cm×0.5cm)作为对照,另一个球形管内放入滴有10μL的粗提物二氯甲烷溶液的滤纸条作为气味源。试验时,将10头试虫放入"Y"型嗅觉仪主管入口处,用黑色聚乙烯膜覆盖整个嗅觉仪,以免由于趋光性而影响实验结果。启动真空泵,调节"Y"型嗅觉仪的主管内气流流量为$100L \cdot h^{-1}$。5min后检查试虫趋向行为,如果试虫进入作为气味源的支管或梨形干燥管,记为正趋向;试虫进入作为对照的支管或梨形干燥管,记为负趋向,如果试虫停留在"Y"型管适应臂中则记为无选择。用正趋向试虫数占被测试试虫总数的百分数来表示被测试样品的活性(趋向率)。每次测试使用同一批从野外采回的经过饲养的楚雄腮扁叶蜂成虫各10头,5个重复。对每个样品经过一轮测试的试虫不再使用。

测试试虫对针叶和卵的混合粗提物活性时,以针叶粗提物为对照;测试楚雄腮扁叶蜂抱卵雌蜂对已开始产卵的雌蜂的活性时,将5头抱卵雌蜂作为气味源。

1.5 氨基酸测定

氨基酸测定按照GB/T5009.124—2003《食品中氨基酸的测定》标准测定。

2 结果与分析

2.1 楚雄腮扁叶蜂在云南松和云南油杉上产卵雌蜂数的比较

林间调查表明,楚雄腮扁叶蜂产卵雌蜂数在云南松上的显著多于云南油杉上的。在2块云南松、云南油杉的混交林样地中,楚雄腮扁叶蜂在云南松上的平均产卵雌蜂数分别为(12.8±7.7)、(14.7±11.7)头·梢$^{-1}$,而在云南油杉树梢上平均产卵雌蜂数为(2.9±1.7)、(4.9±3.4)头·梢$^{-1}$。云南松上的产卵雌蜂数显著大于云南油杉上的产卵雌蜂数($P<0.01$)。

2.2 楚雄腮扁叶蜂雌蜂对寄主挥发物的趋向反应

楚雄腮扁叶蜂抱卵雌蜂对针叶粗提物、卵及松针的混合粗提物和产卵雌蜂均表现出明显的趋向反应(表1)。松针粗提物与卵和松针混合粗提物相比较,抱卵雌蜂更愿意选择卵和松针混合粗提物,二者趋向率分别为(70±15.8)%和(14.0±20.7)%。抱卵雌蜂对来自待产卵雌蜂的气味也表现出明显的趋向反应,反应率达(60.0±14.1)%,与空白对照差异显著($P<0.05$)。

楚雄腮扁叶蜂雄蜂对以上气味源的趋向反应与抱卵雌蜂不同(表2)。雄蜂对云南松针叶粗提物表现出趋向反应,趋向率(56.0±11.4)%,与空白对照差异显著($P<0.05$)。松针粗提物与卵及松针混合粗提物相比较,楚雄腮扁叶蜂雄蜂没有表现出选择性;楚雄腮扁叶蜂雄蜂对正在产卵的雌蜂没有表现出明显的趋向反应。

表1 楚雄腮扁叶蜂抱卵雌蜂对不同气味源的趋向反应(平均值±标准差)[①]

Table 1 Taxis responses of female *C. chuxiongnica* to different odor source (Mean±SD)

试验组	气味源	趋向反应率(%)
1	针叶粗提物	78.0±8.4**
	CK	0.0±0.0
2	卵及松针粗提物	70.0±15.8*
	松针粗提物	14.0±20.7
3	产卵雌蜂	60.0±14.1*
	CK	10.00±14.1

注:[①]每组处理5个重复,每个重复使用10头试虫。*经T检验,楚雄腮扁叶蜂抱卵雌蜂对本组不同气味源的趋向率差异显著($P<0.05$);**经T检验,楚雄腮扁叶蜂抱卵雌蜂对本组不同气味源的差异极显著($P<0.01$)。

表2 楚雄腮扁叶蜂雄蜂对不同气味源的趋向反应(平均值±标准差)[①]

Table 2 Taxis responses of male *C. chuxiongnica* to different odor source (Mean±SD)

试验组	气味源	趋向反应率(%)
1	针叶粗提物	56.0±11.4*
	CK	24.0±11.4
2	卵及松针粗提物	48.0±25.9
	松针粗提物	46.0±20.7
3	产卵雌蜂	30.0±28.3
	CK	40.00±38.1

注:[①]每组处理5个重复,每个重复使用10头试虫。*经T检验,楚雄腮扁叶蜂抱卵雌蜂对本组不同气味源的趋向率差异显著($P<0.05$)。

2.3 云南松、云南油杉针叶的氨基酸含量

从表3可以看出,2种针叶中均含17种氨基酸。其中,以谷氨酸、天冬氨酸、精氨酸和亮氨酸的含量较高。与云南油杉相比较,不论是各种氨基酸含量还是氨基酸总含量,云南松中的含量都相对较高,就针叶中氨基酸总含量而言,云南松是云南油杉的近2倍。

表3 云南松、云南油杉针叶氨基酸组成和占干重的百分含量

Table 3 The composition and percent contents indry needles of amino acid in needles of *P. yunnanensis* and *K. evelyniana*

氨基酸种类	氨基酸含量(%)	
	云南松	云南油杉
谷氨酸 Glu	1.94	1.01
天冬氨酸 Asp	1.90	0.74

(续)

氨基酸种类	氨基酸含量/%	
	云南松	云南油杉
精氨酸 Arg	1.35	0.86
亮氨酸 Leu	1.13	0.63
赖氨酸 Lys	0.98	0.47
丝氨酸 Ser	0.92	0.56
丙氨酸 Ala	0.91	0.55
脯氨酸 Pro	0.91	0.45
缬氨酸 Val	0.79	0.42
苯丙氨酸 Phe	0.71	0.44
酪氨酸 Tyr	0.67	0.37
异亮氨酸 Ile	0.54	0.27
苏氨酸 Thr	0.46	0.28
甘氨酸 Gly	0.46	0.26
组氨酸 His	0.29	0.15
甲硫氨酸 Met	0.19	0.10
半胱氨酸 Cys	0.12	0.05
氨基酸总量 TAAC	14.28	7.61

3 结论与讨论

楚雄腮扁叶蜂抱卵雌蜂对云南松针叶、云南松针叶和卵、楚雄腮扁叶蜂抱卵雌蜂的挥发性物质均表现出明显的趋向,说明这些物质对楚雄腮扁叶蜂待产卵雌蜂具有一定的引诱作用。云南松针叶及楚雄腮扁叶蜂新卵的混合粗提物对楚雄腮扁叶蜂雌蜂的引诱作用大于云南松针叶的粗提物,说明不仅针叶挥发物对楚雄腮扁叶蜂雌蜂具有引诱作用,且初产卵的挥发物也同样具有引诱作用。寄主挥发物是植食性害虫寻找适宜寄主的主要化学信号,欧洲赤松叶蜂(Diprion pini)抱卵雌蜂可辨别适宜和非适宜的产卵寄主,其抱卵雌蜂偏好选择欧洲赤松(Pinus sylvestris)、欧洲黑松(Pinus nigra)为产卵寄主[13]。对北美短叶松(Pinus banksiana)和北美乔松(Pinus strobus)的针叶挥发性单萜烯进行主成分分析,将不适宜欧洲赤松叶锋产卵的北美短叶松、北美乔松与适宜其产卵的欧洲赤松、欧洲黑分为两类;β-蒎烯、柠檬烯和月桂烯是欧洲赤松叶锋抱卵雌蜂决定适宜和非适宜寄主的主要依据。此外,欧洲新松叶蜂(Neodiprion sertifer)幼虫偏好取食欧洲赤松和欧洲黑松,对意大利五针松(Pinus pinea)则很少为害,其抱卵雌蜂也不在意大利五针松上产卵。意大利五针松针叶挥发物中含较高比例的柠檬烯,而欧洲赤松针叶挥发物则含较高比例的月桂烯。较高浓度的柠檬烯对欧洲新松叶蜂具有驱避作用,但较低浓度时则具有引诱作用[14]。楚雄腮扁叶蜂对云南松针叶挥发物的强烈趋向反应说明,寄主挥发物是其辨别适宜和非适宜寄主化学信号。卵挥发物、初产卵的楚雄腮扁叶蜂雌蜂对待产卵雌蜂的引诱作用说明,楚雄腮扁叶蜂的聚集产卵行为也依靠来自新卵、已产卵雌蜂的信息化学物质为其信号。类似的情况在聚集性危害的植食性昆虫中较为普遍[15]。

楚雄腮扁叶蜂雄蜂只对松针粗提物表现趋向反应,而针叶粗提物中含有初产新卵的粗提物时,雄虫不表现选择性。楚雄腮扁叶蜂雄蜂对已产卵的雌蜂也不表现趋向反应。这可能是因为楚雄腮扁叶蜂雄蜂在寻找交尾对象时,不仅依靠雌蜂释放的性信息素,松树针叶挥发物也在其中起着一定作用。

在对害虫为害与寄主植物氨基酸之间关系的研究发现,杨树木质部氨基酸相对含量越高,越利于青杨脊虎天牛(*Xylotrechus rusticus*)幼虫的生长发育[16],杨树游离氨基酸组成和含量与青杨天牛危害呈一定的相关性[17]。氨基酸是昆虫生长必需的营养物质,马尾松松针内含有的18种水解氨基酸和松针内昆虫10种必需氨基酸的含量在受害后减少[18],野外诱集松墨天牛(*Monodchamus alternatus*)数量最高的黑松松枝中的氨基酸总量明显高于马尾松和湿地松[19]。本文发现,与云南油杉相比较,云南松针叶中氨基酸含量较高;楚雄腮扁叶蜂嗜食云南松而对云南油杉危害较轻,降落于云南松上产卵的楚雄腮扁叶蜂远远多于云南油杉,云南松、云南油杉针叶内的昆虫必需氨基酸总含量分别为6.45%、3.62%;氨基酸含量的高低也可能是楚雄腮扁叶蜂抱卵雌蜂降落于产卵基质后判别适宜寄主和非适宜寄主的依据之一。

参考文献:

[1] 钦俊德.昆虫与植物的关系[J].生物学通报,1985,(10):16~18.

[2] 钦俊德,王琛柱.论昆虫和植物的相互作用和进化的关系[J].昆虫学报,2001,44(3):360~365.

[3] Catta-Preta P D, Zucoloto F S. Oviposition behavior and performanceaspects of *Asciamonuste*(Godart,1919)(Lepidoptera, Pieridae) onkale(*Brassica oleracea var. acephala*)[J]. RevistaBrasileira deEntomologia, 2003, 47: 169~174.

[4] Berdegue M, Reitz S R, Trumble J T. Host plant selection and development in *Spodopteraexigua*: Do mother and offspring know best?[J]. EntomologiaExperimentalis et Applicata, 1998, 89(1): 57~64.

[5] 张庆贺,姬兰柱.植食性昆虫产卵的化学生态学[J].生态学杂志,1994,13(3):39~43.

[6] Gouinguené, S P, Buser H R, Städler E. Host-plant leaf surface compounds influencing oviposition in *Delia antiqua*[J]. Chemoecology, 2005, 15(4): 243~249.

[7] Uechia K, Matsuyamab S, Suzuki T. Oviposition attractants for *Plodia interpunctella*(Hübner)(Lepidoptera: Pyralidae) in the volatiles of whole wheat flour[J]. Journal of Stored Products Research, 2007, 43(2): 193~201.

[8] 樊慧,金幼菊,李继泉,等.引诱植食性昆虫的植物挥发性信息化合物的研究进展[J].北京业大学学报,2004,26(3):76~81.

[9] Nakayama T, Honda K, Omura H, et al. Oviposition stimulants for the tropical swallowtail butterfly, *Papiliopolytes*, feeding on a rutaceous plant, *Toddaliaasiatica*[J]. Journal of Chemical Ecology, 2003, 29(7): 1621~1634.

[10] Honda K, Omura H, Hayashi N, et al. Condufitols as oviposition stimulants for the danaid butterfly, *Paranticasita*, identified froma host plant, *Marsdeniatomentosa*[J]. Journal of Chemical Ecology, 2004, 30(11): 2285~2296.

[11] Bruinsma M, Van Dam N M, Van Loon JJ A, et al. Jasmonicacid-induced changes in Brassica oleracea affect oviposition preference of two specialist herbivores[J]. Journal ofChemical Ecology, 2007, 33(4): 655~668.

[12] 萧刚柔.云南腮扁叶蜂两新种(膜翅目:扁叶蜂科)[J].昆虫分类学报,1984,6(2~3):137~140.

[13] Barre F, Milsant F, Palasse C, et al. Preference and performance of the sawfly *Diprionpinion* host and non-host plants of the genus *Pinus*[J]. EntomologiaExperimentalis et Applicata, 2002, 102(3): 229~237.

[14] Martini A, Botti F, Galletti G, et al. The Influence of pine volatile compounds on the olfactory response by *Neodiprionsertifer* (Geoffroy) females[J]. Journal of Chemical Ecology, 2010, 36(10): 1114~1121.

[15] 闫争亮. 小蠹科害虫化学信息物质及其对侵害寄主等行为的影响[J]. 西部林业科学, 2006, 35(3): 22~33.

[16] 严善春, 李金国, 温爱亭, 等. 青杨脊虎天牛的危害与杨树氨基酸组成和含量的相关性[J]. 昆虫学报, 2006, 49(1): 93~99.

[17] 郭树平, 孙萍. 李海霞. 杨树游离氨基酸组成和含量与青天牛危害的相关分析[J]. 东北林业大学学报, 2008, 36(7): 69~71.

[18] 戈峰, 李典谟, 邱业先, 等. 松树受害后一些化学物质含量的变化及其对马尾松毛虫种群参数的影响[J]. 昆虫学报, 1997, 40(4): 337~342.

[19] 陈向阳, 林雪飞, 汪文俊, 等. 松树内含物与松墨天牛种群数量的关系[J]. 生态学报, 2010, 30(13): 3553~3561.

(本文发表于《福建林学院学报》, 2014 年)

3种菊科入侵植物叶片精油成分的 GC-MS 分析

季梅,泽桑梓[1],孙盟[2],杨斌[2],赵宁[2]

(1. 云南省林业科学院,云南 昆明 650201;2. 西南林业大学,云南 昆明 650224)

摘要:用同时蒸馏萃取法提取薇甘菊、紫茎泽兰和飞机草3种有害入侵植物叶片精油,用气相色谱—质谱进行精油化学组成的分析。结果表明,从3种有害入侵植物叶片精油中共鉴定出78种化合物,主要成分均为单萜类及倍半萜类,其中飞机草鉴定出化学成分50种,紫茎泽兰45种,薇甘菊50种。3种植物精油中都能检测到3-己烯-1-醇、α-蒎烯、β-蒎烯、α-水芹烯、柠檬烯、反-罗勒烯、芳樟醇、α-胡椒烯、β-澄椒烯、β-石竹烯、α-香柠檬烯、α-石竹烯、α-姜烯、双环大香叶烯、荜澄茄油烯醇、橙花叔醇、石竹烯氧化物、库贝醇和马兜铃酮,部分化合物在医药领域开发利用前景广阔。

关键词:菊科;入侵植物;叶片;精油;化合物;GC-MS 分析

GC-MS Analysis of Essential Oils from Leaves of Three Asteraceae Invasive Plants

Ji Mei[1], Ze Sang-zi[1], Sun Meng[2], Yang Bin[2], Zhao Ning[2]

(1. Yunnan Academy of Forestry, Kunming Yunnan 650201, P. R. China; 2. Southwest Forestry University, Kunming Yunnan 650224, P. R. China)

Abstract: This paper reports the result of GC-MS analysis of the essential oils from the leaves of three Asteraceae invasive plants, namely, Mikania micrantha, Eupatorium adenophorum and Eupatorium odoratum. The essential oils from the leaves were extracted by simultaneous distillation and extraction. Totally 78 compounds were identified, monoterpenes and sesquiterpenes were found to be the major components. The kinds of compounds identified from Eupatorium odoratum, Eupatorium adenophorumand Mikania micrantha were 50, 45 and 50 respectively. The common components for three plants were 3-Hexen-1-ol, α-Pinene, β-Pinene, α-Phellandrene, Limonene, trans-β-Ocimene, Linalool, α-Copaene, β-Cubebene, β-Caryophyllene, α-Bergamotene, α-Caryophyllene, α-Zingiberene, Bicyclogermacrene, Cubenol, Nerolidol, Caryophyllene oxide, Cubenol and Aristolane. Some of those compounds had utilization prospect in medicine development.

Key words: Asteraceae; Invasive plant; leaf; essential oil; compound; GC-MS analysis

薇甘菊(*Mikania micrantha*)、紫茎泽兰(*Eupatorium adenophorum*)和飞机草(*Eupatorium odoratum*)是入侵云南热区种群数量较大的3种菊科(Asteraceae)有害植物,薇甘菊为菊科假泽兰属(*Mikania*)植物,紫茎泽兰和飞机草同为菊科泽兰属(*Eupatorium*)植物,3种植物虽然都具有较强的生态入侵性,但入侵能力却存在巨大的差异。与紫茎泽兰、飞机草相比,薇甘菊生长迅速,通过攀援争夺阳光,或分泌化感物质,抑制所攀附植物生长。薇甘菊是目前国家林业局公布的唯一林业重大检疫性植物,2011年在云南省的发生面积为3.3万 hm^2,造成的直接经济损失近6亿元。本项研究对采自相同样点的薇甘菊、紫茎泽兰和飞机草3种植物叶片挥发性成分进行了分析,为研究这3种有害植物的防控及开发利用提供依据。

1 材料与方法

1.1 样品采集

本次实验分析测定的薇甘菊、紫茎泽兰和飞机草3种植物叶片样本均于2012年1月12日采自云南省瑞丽市帕当柠檬地(97°21′20.6″E,24°2′13.8″N)。在野外将3种植物新鲜叶片剪下后,迅速放入磨口玻璃瓶中密封,将玻璃瓶放入加有医用冰袋的保温瓶中带回实验室备用。

1.2 仪器和试剂

仪器为美国Agilent Technologies公司HP6890GC/5973MS气相色谱-质谱联用仪。试剂为正己烷GC级,无水硫酸钠AR级,均采购自百灵威公司。

1.3 叶片精油提取

采用同时蒸馏法(SDE)提取供试植物叶片挥发油[1]。用医用小刀将叶片切成约2mm×2mm的碎屑,各称取1 000g放入同时蒸馏萃取装置的试样烧瓶,添加50ml纯净水,并加入沸石防止暴沸,电热套加热并保持微沸;在溶剂烧瓶中加入20ml色谱纯正己烷,水浴75℃加热蒸馏萃取5h。蒸馏萃取完成后,萃取液倒入分液漏斗中分层,取上层正己烷相,加入10g无水硫酸钠干燥保留24h,过滤后将所得提取液放入冰箱中保存,保存温度为-10℃,备用。

1.4 气相色谱-质谱分析

(1)GC条件:HP-5MS石英毛细管柱(30mm×0.25mm×0.25μm);柱温80~260℃,程序升温3℃·min^{-1};柱流量为1.0mL·min^{-1};进样口温度250℃;柱前压100kPa;进样量0.20μL;分流比10:1;载气为高纯氦气。

(2)MS条件:电离方式EI;电子能量70eV;传输线温度250℃;离子源温度230℃;四极杆温度150℃;质量范围35~500amu;采用wiley7n.1标准谱库计算机检索定性。

2 结果与分析

2.1 3种菊科入侵植物叶片精油的化学成分

经GC-MS分析,谱库检索结合手工检索,3种有害入侵植物叶片精油共鉴定出78种成分,根据保留时间排列,结果见表1。由表1可知:飞机草叶片精油鉴定出化学成分50种,紫茎泽兰45种,薇甘菊50种。鉴定出的3种植物叶片精油成分的含量,分别占各种植物总精油量的81.17%、81.41%、88.15%。

表1 3种菊科入侵植物叶片精油的化学组成
Table 1 Chemical composition of essential oils for leaves 3 Asteraceae invasive plants

序号	化合物(保留时间,min)	精油相对含量(%)		
		飞机草	紫茎泽兰	薇甘菊
1	2-戊烯醛(2.589)	0.12	—	—
2	3-己醇(2.811)	—	—	0.34
3	3-己烯-1-醇(3.360)	0.26	0.06	0.53
4	己醇(3.449)	—	—	0.44
5	2-己烯-1-醇(3.461)	0.20	0.05	—
6	三环烯(4.304)	—	0.08	—
7	α-侧柏烯(4.316)	0.05	—	0.05
8	α-蒎烯(4.525)	8.07	0.67	2.56
9	莰烯(4.789)	0.13	12.52	—
10	香桧烯(5.179)	1.04	—	0.25
11	β-蒎烯(5.325)	4.20	0.83	0.26
12	月桂烯(5.396)	0.87	—	0.61
13	乙酸 3-己烯酯(5.617)	—	—	0.10
14	α-松油烯(5.733)	0.12	—	—
15	2-蒈烯(5.743)	—	8.40	—
16	α-水芹烯(5.816)	0.09	5.40	0.05
17	对聚伞花素(6.259)	0.10	2.89	—
18	柠檬烯(6.380)	0.75	1.77	1.41
19	反-罗勒烯(6.727)	0.97	0.18	11.37
20	γ-松油烯(7.079)	—	—	0.10
21	异松油烯(7.971)	—	0.25	0.38
22	芳樟醇(8.204)	0.09	0.12	0.16
23	4,8-二甲基-1,3,7-壬三烯(8.663)	0.12	0.06	0.58
24	马鞭草烯醇(8.954)	—	—	0.10
25	异龙脑(10.658)	0.14	0.52	—
26	对-薄荷-1,8-二烯-7-醇(11.466)	—	0.49	—
27	α-松油醇(11.533)	0.19	—	—
28	百里香酚甲醚(13.231)	0.05	0.11	—
29	香茅醇(13.995)	—	—	0.10
30	乙酸龙脑酯(15.157)	4.70	11.14	—
31	二甲基环癸三烯(15.519)	3.67	—	—
32	乙酸桃金娘烯醇酯(16.537)	—	0.08	—
33	δ-榄香烯(17.189)	0.12	0.07	—
34	α-澄椒烯(17.684)	—	0.06	0.27

(续)

序号	化合物(保留时间,min)	精油相对含量(%)		
		飞机草	紫茎泽兰	薇甘菊
35	α-胡椒烯(18.862)	0.70	0.25	1.26
36	β-榄香烯(19.187)	0.65	—	—
37	壬醛(19.232)	—	—	0.02
38	β-澄椒烯(19.538)	0.65	2.95	15.66
39	茉莉酮(19.735)	—	—	0.12
40	8,9-脱氢环异长叶烯(19.902)	—	—	0.21
41	β-石竹烯(20.848)	6.49	4.06	10.61
42	顺-β-金合欢烯(21.494)	—	0.44	0.31
43	α-香柠檬烯(21.542)	0.60	0.78	2.06
44	α-石竹烯(22.020)	1.68	1.80	0.32
45	α-杜松烯(22.142)	—	—	2.14
46	反-β-金合欢烯(22.181)	1.03	0.46	—
47	γ-姜黄烯(23.086)	—	1.42	4.31
48	大香叶烯 D(23.441)	10.80	—	6.88
49	表-双环倍半水芹烯(23.534)	—	—	0.79
50	α-姜烯(23.740)	0.45	1.37	7.37
51	双环大香叶烯(23.991)	2.37	1.58	3.20
52	β-红没药烯(24.316)	2.99	2.06	—
53	α-木罗烯(24.323)	—	—	0.69
54	荜澄茄油烯醇(24.738)	0.54	0.31	1.80
55	β-倍半水芹烯(24.825)	—	1.35	—
56	δ-杜松烯(24.958)	2.28	—	1.40
57	脱氢芳萜烯(25.307)	0.40	0.34	—
58	α-红没药烯(25.609)	0.44	0.31	—
59	反-倍半水合桧烯(25.677)	—	—	1.49
60	顺-罗勒烯(25.864)	0.10	—	0.20
61	榄香醇(26.095)	2.42	1.05	—
62	橙花叔醇(26.418)	0.60	0.44	0.77
63	顺-倍半水合桧烯(27.121)	—	—	1.34
64	大香叶烯 D-4-醇(27.161)	0.56	0.44	—
65	α-胡椒烯-8-醇(27.418)	—	—	1.03
66	石竹烯氧化物(27.499)	1.19	0.82	1.03
67	绿花醇(27.884)	0.33	—	—
68	喇叭茶醇(28.265)	—	—	0.27
69	姜醇(28.447)	—	—	0.54

(续)

序号	化合物(保留时间,min)	精油相对含量(%)		
		飞机草	紫茎泽兰	薇甘菊
70	库贝醇(29.680)	8.72	7.23	0.81
71	Tau-依兰油醇(29.726)	—	—	0.62
72	Tau-杜松醇(29.814)	0.91	0.47	—
73	α-杜松醇(30.340)	1.43	—	0.51
74	α-红没药醇(31.491)	3.81	3.22	—
75	马兜铃酮(32.038)	2.66	2.40	0.40
76	松油-4-醇(35.281)	0.10	—	—
77	二十一烷(45.362)	—	—	0.33
78	植醇(45.970)	0.22	0.11	—
	被测定物合计	81.17	81.41	88.15

注:"—"表示未检出。

3种植物叶片的主要精油成分均为单萜类及倍半萜类。飞机草精油主要成分为大香叶烯D(10.80%)、库贝醇(8.72%)、α-蒎烯(8.07%)、β-石竹烯(6.49%)、乙酸龙脑酯(4.70%)和β-蒎烯(4.20%);紫茎泽兰精油主要成分为莰烯(12.52%)、乙酸龙脑酯(11.14%)、2-蒈烯(8.40%)、库贝醇(7.23%)、α-水芹烯(5.40%)和β-石竹烯(4.06%);薇甘菊精油主要成分为β-澄椒烯(15.66%)、反-罗勒烯(11.37%)、β-石竹烯(10.61%)、α-姜烯(7.37%)、大香叶烯D(6.88%)和γ-姜黄烯(4.31%)。

2.2 3种菊科入侵植物叶片精油的相同和特有成分

3种植物叶片精油中含有一些相同成分,都能检测到的成分有3-己烯-1-醇、α-蒎烯、β-蒎烯、α-水芹烯、柠檬烯、反-罗勒烯、芳樟醇、α-胡椒烯、β-澄椒烯、β-石竹烯、α-香柠檬烯、α-石竹烯、α-姜烯、双环大香叶烯、荜澄茄油烯醇、橙花叔醇、石竹烯氧化物、库贝醇和马兜铃酮。

有些植物叶片精油中含有一些特有成分。飞机草精油中检测出的2-戊烯醛、α-松油烯、α-松油醇、二甲基环癸三烯、β-榄香烯、绿花醇和松油-4-醇在另外2种精油中未检出;紫茎泽兰精油中检测出的三环烯、2-蒈烯、对-薄荷-1,8-二烯-7-醇、乙酸桃金娘烯醇酯和β-倍半水芹烯在另外2种精油中未检出;薇甘菊精油中检测出的3-己醇、己醇、乙酸3-己烯酯、γ-松油烯、4,8-二甲基-1,3,7-壬三烯、马鞭草烯醇、香茅醇、壬醛、茉莉酮、8,9-脱氢环异长叶烯、α-杜松烯、表-双环倍半水芹烯、α-木罗烯、反-倍半水合桧烯、顺-倍半水合桧烯、α-胡椒烯-8-醇、喇叭茶醇、姜醇、Tau-依兰油醇和二十一烷在另外2种精油中未检出。

3 结语

袁经权等[2]对飞机草挥发油成分进行分析,其主要化学成分是9-甲基-10-亚甲基-三环[4.2.1.1(2,5)]癸-9-醇(13.66%)、富马酸乙基-2-(2-亚甲基环丙基)丙酯(5.02%)、4-(4-羟基-2,2,6-三甲基-7-氧杂双环[4.1.0]庚-1-基)-3-丁烯-2-酮(3.24%)、4-羟基-2-戊酮(3.16%)等;吴田捷等[3]对紫茎泽兰精油成分进行分析,其主要化学成分是邻正丙基甲苯(13.08%)、内-乙酸龙脑酯(10.81%)、莰烯(5.12%)、β-倍半水芹烯(4.33%)、α-松油烯(4.22%)等。本项研究鉴定的3种有害入侵植物叶片精油成分与上述已报道的特征成分差别较大,这可能与样品的种类、采样季节和采样地点有关。

菊科植物挥发油成分大多具有杀菌消炎的作用,具有多种生理活性,能够在医药领域里得到广泛应用[4~5]。本项研究中,3种植物都含有的一些共同成分,譬如β-石竹烯具有平喘作用,是治疗老年性慢性支气管炎的有效成分之一[6],还具有保护神经细胞和增强记忆能力的作用[7],也是芳香性健胃驱风药[8];α-蒎烯是合成樟脑、冰片、松油醇、香料、树脂等化工产品的重要原料之一,且具有明显的镇咳和祛痰作用,并具有抗真菌作用[9];α-香柠檬烯对淋球菌、葡萄球菌、大肠杆菌和白喉菌均有抑制作用。故对于飞机草、紫茎泽兰和薇甘菊叶片精油中主要化学成分的应用可作进一步的调查和研究,从而提高其开发利用价值。

参考文献:

[1] HuiRuihua, Hou Dongyan, Li Tiechun. Analysis of Volatile Components in PhellodendronChinense Schneid [J]. Chinese Journal of Analytical Chemistry, 2001, 29(3):361~364.

[2] 袁经权,冯洁,杨峻山,等. 飞机草挥发油成分的GC~MS分析[J]. 中国现代应用药学杂志,2008,25(3):202~205.

[3] 吴田捷,杨光忠. 紫茎泽兰精油化学成分的GC/MS研究[J]. 华中师范大学学报(自然科学版),1994,28(1):87~90.

[4] 刘向前,陈素珍,倪娜. 湖南产艾叶挥发油成分的GC-MS研究[J]. 中药材,2005,28(12):1069~1071.

[5] 李丽丽,刘向前,张晓丹. GC-MS研究湖南产奇蒿和猪毛蒿挥发油成分[C]. 湖南省药学会2007年学术年会论文集,2007:137.

[6] 李玉平,龚宁,慕小倩,等. 菊科植物资源及其开发利用研究[J]. 西北农林科技大学学报,2003,31(5):150~160.

[7] 黄如稻,陈化新,张奕斌,等. 菊科植物的利用价值[EB/OL]. 维普资讯,http://www.cqvip.com,2009-08-15.

[8] 郑虎占,董泽宏,余靖,等. 中药现代研究与应用(第六卷)[M]. 北京:学苑出版社,1999.

[9] 夏忠地,徐俊龙. α-蒎烯对白色珠菌生物合成的影响[J]. 中国现代医学杂志,2000,10(1):23~24.

(本文发表于《西部林业科学》,2012年)

高原山区油茶茶苞病的发生与防治研究

贾代顺[1], 卯吉华[1], 卯梅华[2], 陈福[1], 廖永坚[1]

(1. 云南省林业科学院油茶研究所,云南 广南 663300;2. 云南省林业高级技工学校,云南 昆明 650224)

摘要:由细丽外担菌[*Exobasidium gracile*(Shirai)Syd]引起的油茶茶苞病是高原山区油茶主产区分布极为普遍和广泛的病害。林间调查结果表明,该病害在高原山区1年发生2次,症状表现为危害子房形成茶桃、危害叶片形成饼状及危害嫩梢形成泡状3种类型。病害的发生与栽培立地条件、经营管理措施、坡位、坡向、树龄等因素密切相关。针对目前高原山区油茶茶苞病的发生现状及特点,提出相应的防治措施。

关键词:油茶茶苞病;发生;防治;高原山区

Occurrence and Controlof *Camellia oleifera* Gall Disease Induced by *Exobasidium gracile* in Plateau Areas

Jia Dai-shun[1], Mao Ji-hua[1], Mao Mei-hua[2], Chen Fu[1], Liao Yong-jian[1]

(1. Institute of Oil Tea, Yunnan Academy of Forestry, Guangnan Yunnan 663300, P. R. China; 2. ForestryVocational Technical School of Yunnan Province, Kunming Yunnan 650204, P. R. China)

Abstract: Gall disease of *Camellia oleifera* caused by *Exobasidium gracile* was a prevalent disease in oil tea cultivation in plateau areas. The results of field investigation showed that in plateau areas, this disease occurred twice a year, with the symptoms galls at ovary, leaf or shoot. The occurrence of the disease was closely related to the site condition of orchard, management measures, slope position, slope aspect and tree age. Aiming at the present disease condition and characteristics, the corresponding measures for disease control and management were proposed.

Key words: *Camellia oleiferagall*; occurrence; control; plateau area

油茶(*Camellia oleifera* Abel)与油橄榄、油棕、椰子并称世界四大木本油料植物[1],在我国有2 300多年的栽培历史。油茶耐干旱瘠薄、适生范围广、综合利用价值高、生态功能强,是我国特有的生态和经济效益俱佳的优良乡土木本食用油料树种,在我国南方油茶主产区(湖南、江西、广西、浙江、福建、云南、贵州等地)林业产业中具有十分重要的地位和作用[1-3]。充分利用我国丰富的山地资源,大力发展油茶产业,有利于缓解耕地压力、保障粮油安全、改善山区生态环境、增加山区农民收入、提升人民健康水平都具有十分重要的意义[3]。广南县是云南省油茶的最适生区,种植面积达13 300hm^2,居全省第一位,随着油茶栽培面积的不断扩大,油茶茶苞病已成为高原山区油茶种植区极其普遍且危害严重的一种病害。油茶茶苞病又称为油茶饼病、叶肿病、茶桃、茶泡等,是广南地区分布极为普遍和广泛的病害。油茶茶苞病主要危害子房、嫩梢和叶片,导致发病部位肿大变形,造成油茶产量下降,在广南危害株率达23%~45%,发病指数在8~36。在降水量较多,温度相对较低的年份(2012年),油茶茶苞病的危害十分严重,导致油茶减产22%。油茶茶苞病已成为当前油茶生产上亟待需要解决的问题。为此,2010~2012年在广南开展了油茶茶苞病发病情况和防治措施的研究,以期为高山区油茶产业的健康、可持续发展提供理论依据。

1 材料与方法

1.1 试验区概况

试验于2010~2012年在高原山区云南省广南县进行。试验区分设在广南县东风水库坝尾、坝美发利、旧莫里乜三片油茶林内。试验林地属中亚热带气候类型,年平均气温17.5℃,极端最高气温38.2℃,极端最低气温-4.5℃,海拔1 150~1 300m,属高原低缓坡地形,土壤为山地红壤、黄红壤类型,pH值5~6,年平均降水量650~1 100mm。5月下旬至10月下旬为雨季,尤以6~8月降雨最多,11月至翌年5月为干季。

1.2 试验方法

试验采用标准地调查和线路调查相结合,根据地形及调查目的,在每片油茶林内选定3块标准地,共选标准地9块,每块为667m²,有油茶80~120株。油茶树龄为5~65a生幼龄林、壮龄林和老龄林,树高在2.0~4.5m,郁闭度在0.5~0.9,林内平均湿度在65%~85%。调内容主要有发病株率和病情指数两项,病情指数按油茶病害5级分级标准调查统计[4],分级标准见表1。

表1 油茶茶苞病分级标准

病害等级	代表数值	分级标准(上限排外)
Ⅰ	0	健康
Ⅱ	1	25%以下嫩梢、叶或果实感病
Ⅲ	2	25%~50%嫩梢、叶或果实感病
Ⅳ	3	50%~75%嫩梢、叶或果实感病
Ⅴ	4	75%以上嫩梢、叶或果实感病

发病株率和病情指数按照下面的公式计算:
发病株率=发病株数/调查总株数×100%
病情指数=Σ(各级病株数×该病级代表数值)/(调查总株数×发病最高病级代表数值)×100%

2 结果与分析

2.1 油茶茶苞病症状

调查发现,油茶茶苞病侵染源较广,普通油茶的子房、嫩梢、叶片均能危害。广南地区油茶茶苞病的发病形态有三种症状:一是病原菌侵入子房,使感病部位不断膨大呈畸形,肿大如桃,故称为"茶桃"(图1A),茶桃内形成空腔,病组织松软能食,稍有酸、涩、苦、甘味,受害果实的直径可达3~10cm,前期为绿色,以后表皮渐破裂,具有一层白粉,后期逐渐变黑,最后脱落,严重的植株,整株果实都受害;二是病原菌侵入叶片,发病初期叶片上有淡黄色半透明圆斑,后逐渐扩大下陷,叶背渐隆起呈饼状,故称其为"茶苞病"(图1B),表面具有一层白粉,后期表皮裂开,最后变黑枯落;三是嫩梢受病原菌入侵后,感病部位肥肿而短粗,其上的嫩叶肥厚呈泡状,淡红色半透明,多重叠在一起,成丛感病,后期枯死脱落(图1C)。

油茶茶苞病是一种真菌性病害,病原为细丽外担菌[*Exobasidium gracile* Syd][5~7],病菌在受病组织中越冬,担孢子随风及气流传播危害。通常在高原山区(广南)油茶茶苞病一年发病2次,第一次在春末夏初(3~5月)发病,第二次在秋末(9~10月)发病,一般情况第一次发病较第二次发病严重。

图1 油茶茶苞病症状
A 危害子房症状;B 危害叶片症状;C 危害嫩梢症状

2.2 不同栽培点油茶茶苞病危害情况

从表2可知,油茶茶苞病在广南的发病率高,平均发病率达35.7%,平均病情指数达16.5。在3个油茶林分,东风水库坝尾的发病率和病情指数最高,分别为48.8%、23.4,坝美法利的发病率和病情指数居中,而旧莫里乜的发病率和病情指数最低。方差分析结果表明,3个林分油茶茶苞病的发病率和病情指数两两之间的差异达到极显著水平,其中,东风水库坝尾的发病率和病情指数极显著高于坝美法利和旧莫里乜,而坝美法利的发病率和病情指数极显著高于旧莫里乜。东风水库坝尾的油茶发病率和病情指数均是3个林分最严重的,可能是因为这片油茶林处于水库旁,常年雾多,湿度相对较大,因此危害较严重;而旧莫里乜的发病率和病情指数最低,可能与里乜的气候条件有很大关系,里乜的气候条件是最适宜油茶生长发育的,全日光照时数长,空气较干燥,湿度较小,因此油茶茶苞病的危害较轻。

表2 不同栽培点油茶茶苞病发病情况

调查地点	海拔(m)	湿度(%)	调查株数(株)	发病株数(株)	发病率(%)	病情指数
东风水库坝尾	1 220	83	320	156	48.8±3.9aA	23.4±2.8aA
坝美法利	1 100	76	280	111	39.7±2.1bB	18.3±1.9bAB
旧莫里乜	1 250	65	300	56	18.5±1.9cC	7.8±1.8cC
平均					35.7	16.5

注:表中数据为平均值±标准差,每一列后面不同的小写字母(a~b)表示不同栽培点该平均值的多重比较在0.05的水平上有显著差异,不同的大写字母(A~B)表示不同栽培点该平均值的多重比较在0.01水平上有极显著差异。

2.3 不同坡位油茶茶苞病危害情况

从表3可知,油茶茶苞病的发生与坡位有一定的关系。在同一栽培点,山脚的发病率和病情指数最高,分别达54.9%、19.5,其次是山腰,而山顶的发病率和病情指数最低,分别为28.8%、7.5。方差分析结果表明,不同坡位山脚、山腰、山顶两两之间的差异除山脚与山腰的病情指数达显著水平外,其他两两之间的发病率和病情指数均达到极显著水平。主要是因为山脚土壤较山腰和山顶肥厚,植株的生长势旺盛,林分郁闭度大,林内湿度也较大,较有利于病菌的发生和蔓延,因此,导致山脚的发病率和病情指数较大,危害较严重。

表3 不同坡位油茶茶苞病发病情况

调查地点	坡位	海拔(m)	调查株数(株)	发病株数(株)	发病率(%)	病情指数
东风水库坝尾	山脚	1 210	300	165	54.9±2.1aA	19.5±1.9aA
	山腰	1 250	300	131	43.6±1.9bB	14.9±2.3bA
	山顶	1 280	200	58	28.8±1.8cC	7.5±0.7cB

注:表中数据为平均值±标准差,每一列后面不同的小写字母(a~b)表示不同坡位该平均值的多重比较在0.05的水平上有显著差异,不同的大写字母(A~B)表示不同坡位该平均值的多重比较在0.01的水平上有极显著差异。

2.4 不同坡向油茶茶苞病危害情况

从表4可知,油茶茶苞病的发生与坡向有一定的关系。在同一栽培点(东风水库坝尾),北坡(阴坡)的发病率和病情指数最高,分别达53.6%、33.5,其次是东坡(半阳坡),而南坡(阳坡)的发病率和病情指数最低。方差分析结果表明,不同坡向(北坡、东坡、南坡)之间,南坡(阳坡)与东坡(半阳坡)两两之间的发病率和病情指数的差异均未达到显著水平,而北坡(阴坡)与南坡(阳坡)和东坡(半阳坡)两两之间的发病率和病情指数差异均达到极显著水平。主要原因在于,北坡(阴坡)林分较稠密,加之光照较少,导致林分通风透光性差,较阴湿,有利于病菌的发生与传播。因此,北坡(阴坡)的危害较南坡(阳坡)和东坡(半阳坡)严重。

表4 不同坡向油茶茶苞病发病情况

调查地点	坡向	调查株数(株)	发病株数(株)	发病率(%)	病情指数
东风水库坝尾	南坡	300	116	38.8±2.2bB	13.8±1.5bB
	东坡	260	117	44.9±3.6bB	15.6±1.6bB
	北坡	220	118	53.6±2.9aA	33.5±3.3aA

注:表中数据为平均值±标准差,每一列后面不同的小写字母(a~b)表示不同坡向该平均值的多重比较在0.05的水平上有显著差异,不同的大写字母(A~B)表示不同坡向该平均值的多重比较在0.01的水平上有极显著差异。

2.5 不同管理措施油茶茶苞病危害情况

从表5可知,油茶茶苞病的发生与管理措施有一定的关系。在同一栽培点,荒芜(不管理)的发病率和病情指数最高,分别达55.2%、35.5,通过垦复和施肥管理的发病率次之,而通过垦复、施肥并修剪管理的油茶林发病较轻。方差分析结果表明,不同的管理措施(垦复+施肥+修剪、垦复+施肥、荒芜)两两之间的发病率和病情指数的差异均达到极显著水平。主要是因为荒芜(不管理)的油茶林杂草丛生,林内郁闭度大,通风透光不好,植株生长势差,导致发病最严重;而通过垦复、施肥并修剪管理的油茶林植株生长优良,郁闭度小,通风透光性较好,故发病率较低,发病较轻。因此,在生产管理中,加强油茶园的抚育管理,可有效控制病菌的发生与危害。

表5 不同管理措施油茶茶苞病发病情况

调查地点	管理措施	郁闭度	调查株数(株)	发病株数(株)	发病率(%)	病情指数
东风水库坝尾	垦复+施肥+修剪	0.6	300	100	33.3±2.1cC	10.8±1.7cC
	垦复+施肥	0.8	260	114	43.7±3.4bB	21.6±2.4bB
	荒芜	1	200	110	55.2±3.9aA	35.5±3.8aA

注:表中数据为平均值±标准差,每一列后面不同的小写字母(a~b)表示不同管理措施该平均值的多重比较在0.05的水平上有显著差异,不同的大写字母(A~B)表示不同管理措施该平均值的多重比较在0.01的水平上有极显著差异。

2.6 不同树龄油茶茶苞病危害情况

由表6可知,油茶茶苞病的危害与树龄有一定的关系,其中树龄最大(50a以上)的油茶茶苞病发病率最高,达51.7%,发病形态为茶桃,即病原菌主要侵染子房;其次是树龄最小(8a以下)的,发病率达46.5%,发病形态为茶泡,即病原菌主要侵染嫩叶和嫩梢,尤其是砍伐主枝后,重新从主干萌发出来的新梢感病最严重,这与邱建生[5]等在贵州油茶叶肿病的研究结果一致,在广南县东风水库坝尾的油茶高接换优的试验林中,从截干下萌发的枝条感病率达55.8%,病情非常严重;而树龄处于中壮林的(18a和30a)的发病率相对较低。方差分析结果表明,不同树龄之间,除18a和30a之间油茶茶苞病的发病率和病情指数差异不显著外,其他几个树龄段两两之间的发病率和病情指数差异均达到极显著水平。

表6 不同树龄油茶茶苞病发病情况

调查地点	树龄(a)	郁闭度	发病状态	发病率(%)	病情指数
东风水库坝尾	50以上	0.9	茶桃	51.7±3.7aA	21.6±1.9aA
	30	0.8	茶桃、偶有茶泡	37.2±2.1cBC	12.1±1.3cBC
	18	0.7	茶泡、偶有茶桃	34.9±2.3cBC	10.9±1.7cBC
	8以下	0.7	茶泡	46.5±2.7bAB	15.9±1.8bB

注:表中数据为平均值±标准差,每一列后面不同的小写字母(a~b)表示不同树龄该平均值的多重比较在0.05的水平上有显著差异,不同的大写字母(A~B)表示不同树龄该平均值的多重比较在0.01的水平上有极显著差异。

2.7 防治措施

油茶茶苞病在广南地区的发生及危害是很严重的,针对目前的发生现状,提出以下防治措施:

2.7.1 加强苗木质量监督检查及病害的监测预报

在苗木出圃时,应进行严格的抽样调查,若发现苗期就有感病的苗木,禁止山上造林,从源头上控制病害的发生。另外加强病害的监测预报工作,在病菌孢子成熟扩散前,可组织人力集中剪除病果、病叶烧毁及深埋,清除侵染源这样可以减轻下一年的危害,若能连续几年剪除病物,防治效果会更佳。

2.7.2 加强栽培管理措施

油茶茶苞病的发生,其原因除气候这一特殊因素之外,与栽培管理水平有很大关系。建议首先对现有油茶林进行分类经营,对于生长势较差的林分要加强林地土、肥、水管理,提高林木的抗病能力,并进行垦复、整形修剪,保持林内通风透光性好,使其不利于病菌的发生蔓延;对于油茶茶苞病发生严重,大面积发病的林分,在剪除病原物的前提下,应结合化学防治。其次,由于油茶属于阳性树种,在营林规划设计中,尽量设计在光照较好的阳坡及半阳坡山的中上部,这样可以有效减轻病害的发生。另外,在栽植过程中,尽量采用抗病品种并用多品种混栽,株行距适当大一点,保持林内通风透光,以不利于病害的侵染蔓延。

2.7.3 采取"以人工摘除为主,化学药剂防治为辅"的综合防治措施

在担孢子尚未成熟飞散前,连年集中摘除病组织表皮尚未破裂的茶桃、茶苞、茶片,并深埋或烧毁,在病害发病高峰期结合使用1%的波尔多液或500倍敌克松喷雾[7],效果较好。

3 结论与讨论

该试验结果表明,油茶茶苞病在高原山区云南广南发生普遍,危害严重,发病率较高,平均发病率高达35.7%,已对高原山区油茶的健康、可持续发展形成很大的威胁。

病原细丽外担菌在受病组织中越冬,担孢子随风及气流传播危害。该病害侵染源较广,普通油茶的子房、嫩梢、叶片均能危害。通常在高原山区(广南)一年发病2次,第一次在春末夏初(3~5月)发病,第二次在秋末(9~10月)发病,第一次发病较第二次发病严重。春末夏初多雨、多雾(湿度大),阴坡和管理粗放的老龄林油茶茶苞病的发生极其严重。

加强油茶苗木质量的监督检查和病害的监测预报,在营造林时进行合理的规划种植,选择抗病品种,加强经营管理措施,实行集约经营,并采取"以人工摘除为主,化学药剂防治为辅"的综合防治措施,防治病害的发生蔓延。

参考文献:

[1] 庄瑞林. 中国油茶(第2版)[M]. 北京:中国林业出版社,2008.

[2] 韩宁林,赵学民. 油茶高产品种栽培[M]. 北京:中国农业出版社,2008.

[3] 姚小华. 油茶高效实用栽培技术[M]. 北京:科学出版社,2010.

[4] 黄敦元. 油茶病虫害防治[M]. 北京:中国林业出版社,2008.

[5] 邱建生,余金勇,吴跃开,等. 贵州油茶叶肿病研究初报[J]. 贵州林业科技,2011,39(1):19~22.

[6] 阙生全,朱必凤,刘霞. 油茶茶苞病病原菌生物学特性的初步研究[J]. 福建林业科技,2008,35(2):97~99.

[7] 黄飞龙,钟文勇. 油茶茶苞病(*Exobasidium gracile*)的研究[J]. 广西农学院学报,1987,(1):67~74.

(本文发表于《西部林业科学》,2017年)

蒜头果半寄生特性研究

李勇鹏,景跃波,卯吉华,李荣波,李孙玲

(云南省林业和草原科学院,云南 昆明 650201)

摘要:蒜头果为我国特有的单种属濒危保护植物,因其种仁油脂中富含神经酸而具有很高的经济开发价值,但人工种植的成活率和保存率都很低。对蒜头果根系的调查发现蒜头果根部具有特殊的瘤状结构,依形态特征观察判定该结构为吸器,蒜头果是一种根部半寄生植物。通过对天然分布蒜头果寄生情况的调查、寄生结构的切片观察和与寄主植物的盆栽共培养试验,证实了蒜头果的半寄生特性。盆栽试验发现蒜头果幼苗在无寄主植物伴生的条件下也能存活,但能与多种寄主植物形成寄生关系,对寄主选择的专一性不强。蒜头果自身的根系间会互相形成很多的吸器,具有明显的自寄生现象。因寄主植物的不同,蒜头果在寄主植物根系上形成的吸器和自寄生吸器在数量、平均直径、最大直径及自寄生吸器所占比例存在显著差异,且吊兰参试寄主植物显著促进了蒜头果幼苗生物量的积累,表明蒜头果对寄主植物有较明显的偏好性。蒜头果幼苗与吊兰表现出紧密的寄生关系,无论是吸器数量还是苗木生物量指标都显著优于其他各处理,在蒜头果人工种植中具有一定的应用前景。本研究的结果有助于解决目前蒜头果人工种植成活率和保存率低下的问题,促进这一珍稀濒危物种的保护、繁育及利用。

关键词:蒜头果;半寄生植物;寄主植物;吸器

Study on Hemiparasitic Characteristics of *Malania oleifera*

Li Yong-peng, Jing Yue-bo, Mao Ji-hua, Li Rong-bo, Li Sun-ling

(Yunnan Academy of Forestry and Grassland, Kunming Yunnan 650201, P. R. China)

Abstract: *Malania oleifera* is an endangered monotypic species endemic to China, with promising economic value for the high content of nervonic acid in seed. However the planting of this species was seriously limited by the low survival rate in the field. In December 2016, the root system of *Malania oleifera* naturally growing at Guangnan county of Yunnan Province were investigated, warts were prevalently observed on the root, with the same morphological characteristics of haustorium, indicated that *Malania oleifera* was a hemiparasitic plant. Slice observation and pot culture experiment further verified the root hemiparasitism of *Malania oleifera*. In pot culture experiment, *Malania oleifera* could survive without host (the control), in the treatments with host plant, *Malania oleifera* developed haustoria and contacted the roots of varied hosts. It was found that the haustoria contacted not only the host roots, but also the roots of its own. The number and diameter of haustorium, the number percentage of haustorium contacted its own root, seedling biomass varied with tested host plants, showing the preference towards host plant to a certain extend. Among the tested host plants in this study, *Malania oleifera* developed a close parasitic relationship with Chlorophytum comosum, both haustorium index and seedling biomass were significantly higher than other treatments, indicating the prospect of Chlorophytum comosum as a host candidate in cultivation of *Malania oleifera*. The results of this study will contribute to solve the problem of low survival rate of *Malania oleifera* in field planting, and promote the conservation, propagation, cultivation and utilization of this valued tree species.

Key words: *Malania oleifera*; hemiparasitic plant; host plant; haustorium

蒜头果(*Malania oleifera*)为铁青树科(Olacaceae)蒜头果属(*Malania*)常绿乔木,是我国特有的单种属植物[1,2],仅自然分布于云南东南部和广西西部的狭窄区域[3],被列为国家二级保护植物[4]。蒜头果种子富含油脂,其种仁油脂中存在含量高达67%的15-24-碳烯酸(俗称神经酸)[5],是一种具有很高经济价值的树种,具有很好的产业开发前景。但蒜头果的人工种植存活率极低。潘晓芳[6]于1999年进行了蒜头果的育苗和造林试验,发现蒜头果育苗较容易,但是几次的土山营造试验林都不成功。另有一些研究者开展的造林试验也证实,蒜头果在石山造林中的表现极差[7,8]。同时,蒜头果引种也比较困难。肖来云等[9]在气候相似区进行了多次蒜头果的引种都不成功,认为可能是由于蒜头果适应性太弱,生态环境与遗传性不协调所致。

造林成活率与植物根系特征密切相关,因此本课题组于2016年12月对云南省广南县天然分布的蒜头果开展了根系调查,发现蒜头果根系普遍具有白色的瘤状结构,并以该结构吸附或包裹住其他植物的根。对该瘤状结构的形态观察判定这是寄生植物特有的结构~吸器。寄生植物是指从别的活体植物中获取部分或全部养分需求的植物,它们约占被子植物总数的1%,几乎世界各地所有的生物群落中都有存在[10]。吸器是寄生植物特化的寄生器官,是联系寄生植物和寄主植物的桥梁,寄生植物通过吸器穿透寄主植物,连接到寄主植物的输导组织,使寄生植物具有了从寄主植物中吸取水分和营养的能力[11]。具有吸器这一特征表明蒜头果是一种根部半寄生植物,能从其他植物根系上获取水分和养分。因此,本研究在野外调查了解到该树种具有半寄生特性的基础上,进行切片显微观察和寄主植物共培养盆栽试验,对蒜头果的半寄生性进行试验验证,并对其与寄主植物的关系开展初步的研究,以期为该物种的濒危机制研究,解决蒜头果人工种植成活率低的问题等提供强有力的理论依据,进而促进对这一珍稀濒危树种的保护与繁育。

1 研究方法

1.1 野外调查

蒜头果寄生特性的野外调查点位于云南省广南县的旧莫乡、曙光乡和董堡乡,为蒜头果在云南省天然分布较集中的地区。挖掘蒜头果植株的部分侧根,观察根系上吸器的情况及寄主植物种类,要求不能破坏天然资源,尽量少损伤树体。

1.2 吸器的切片观察

采集到的寄生结构吸器通过徒手切片和石蜡切片法进行显微观察,查看吸器与寄主植物根系的关系。

1.3 蒜头果与寄主植物共培养盆栽试验

通过共培养盆栽试验来验证蒜头果的半寄生特性,并筛选优良的寄主植物。盆栽试验在云南省林业科学院昆明树木园温室内开展。选择野外能与蒜头果形成寄生关系或是根系特征近似的共8种植物作为参试寄主植物,分别为:大将军(*Lobelia clavata*)、白花银背藤(*Argyreia seguinii*)、番薯(*Ipomoea batatas*)、降香黄檀(*Dalbergia odorifera*)、华山松(*Pinus armandii*)、洋芋(*Solanum tuberosum*)、吊兰(*Chlorophytum comosum*)和白蒿(*Artemisia argyi*)。

蒜头果苗种植于塑料盆钵中,盆口径190mm,底径140mm,高170mm。用红土和腐殖土按1∶1的体积比均匀混合后作为栽培基质。对照仅种植蒜头果,没有寄主植物伴生。其余每个处理每盆种植1株蒜头果和1株相应的寄主植物,二者相距约3cm。每个处理20盆,一共180盆。经过约8个月的共生培养,调查蒜头果根部的寄生状况和苗木的生长状况。对根部寄生的调查指标包括吸器的数量、平均直径和最大直径。同时调查测定蒜头果苗木的苗高、地径、叶片数、全株干重、地上部分干重、地下部分干重等生长指标。苗高、地径和叶片数直接测定,生物量指标取蒜头果全株,在烘箱烘至恒重后,称量全株干重、地上部分干重和地下部分干重。

共培养盆栽的8种参试寄主植物中,有1种参试寄主植物(大将军)在培养过程中全部死亡,另有3种参试寄主植物(白花银背藤、番薯、降香黄檀)在试验结束时,保存的重复数未达到试验分析的要求,因而没有参与数据分析。因此获得完整数据处理共有5个,处理Ⅰ为无寄主植物(对照),处理Ⅱ为华山松(*Pinus armandii*),处理Ⅲ为洋芋(*Solanum tuberosum*),处理Ⅳ为吊兰(*Chlorophytum comosum*),处理Ⅴ为白蒿(*Artemisia argyi*)。

1.4 数据处理

数据的分析用WPS表格和SPSS 17.0软件进行分析。

2 结果与分析

2.1 蒜头果野外半寄生情况

在云南广南县天然分布区的调查发现,蒜头果的根系能与多种植物形成寄生关系,包括滇油杉(*Keteleeria evelyniana*)、云南松(*Pinus yunnanensis*)、华山松(*Pinus armandit*)、白花油麻藤(*Mucuna birdwoodiana*)、大将军(*Lobelia clavata*)、白花银背藤(*Argyreia seguinii*)、七叶薯蓣(*Dioscorea esquirolii*)、莎草(*Cyperus rotundus*)等植物。这些植物分属不同的科属,有草本、灌木、藤本和木本植物。在各种植物根系上附着的吸器数量和大小不一,直径范围0.2~3.2cm。而在伴生的紫茎泽兰(*Eupatorium adenophora*)、酸模(*Rumex acetosa*)等植物根系上,未观察到蒜头果吸器。

2.2 蒜头果寄生结构吸器的形态特征

蒜头果的吸器为白色肉质组织,圆球形、钟状或扁平,吸附、半包裹或是全包裹寄主植物的根系。经比较,蒜头果的吸器与檀香(*Santalum album*)的吸器[12]形态结构极为相似(图1)。

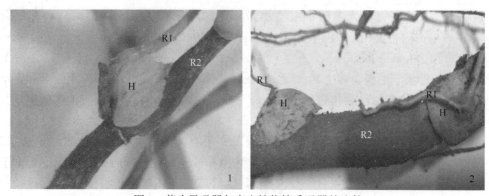

图1 蒜头果吸器与寄生植物檀香吸器的比较

注:1为檀香[12]的寄生结构,2为蒜头果的寄生结构,H为吸器,R1为半寄生植物的根,R2为寄主植物的根

Figure 1 Haustorium morphologicalcharacteristics of *Malania oleifera* and Santalum album

对蒜头果吸器进行了徒手切片(图2)和石蜡切片观察(图3),发现蒜头果的吸器包裹住了寄主植物的根,吸器内有探针状结构穿破了寄主植物根的皮层,伸入到寄主植物根的内部,形成连通的结构,并通过该连通结构吸取寄主植物的水分和养分。

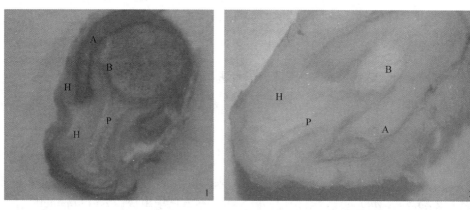

图2 蒜头果吸器徒手切片图

注:1为木质化的寄主植物根,2为细嫩的寄主植物根,
A为寄主植物根的根皮,B为寄主植物根的内部组织,H为吸器,P为吸器的探针状结构

Figure 2 Freehand slice of hemiparasitic structure

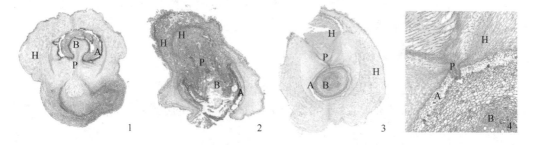

图3 蒜头果寄生结构吸器的石蜡切片图

注:1、2中吸器已伸入寄主植物的内部,3中吸器刚穿破寄主植物根的外皮层,4为3吸器探针状结构放大图,
A为寄主植物的根皮,B为寄主植物根的内部组织,H为蒜头果的吸器,P为吸器的探针状结构

Figure 3 Paraffin slice of hemiparasitic structure

2.3 盆栽试验蒜头果根部的寄生情况

在共培养盆栽试验中,蒜头果能与所有的参试植物形成寄生结构,且能在植物块根上形成较大的吸器。蒜头果的侧根间能形成大量的吸器,通过吸器根系会互相缠绕在一起,自寄生现象明显(图4)。蒜头果根系与其他植物根系形成的典型寄生结构如图5。

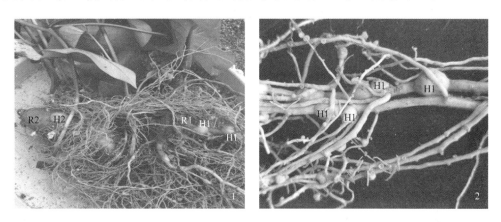

图4 蒜头果在寄主植物上的吸器和自寄生的吸器

注:R1为蒜头果的根,H1是蒜头果根系自寄生的吸器,R2为番薯(Ipomoea batatas)的块根,H2寄生在块根上的吸器

Figure 4 Haustoria of *Malania oleifera* developed on host plant and its own root

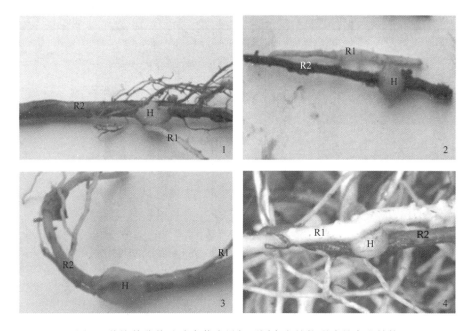

图5 共培养盆栽试验中蒜头果与不同寄主植物形成的寄生结构

注:H为吸器,R1为蒜头果的根,R2为寄主植物的根

Figure 5 Haustoria of *Malania oleifera* with different host plants

2.4 寄主植物共培养盆栽试验中蒜头果的吸器情况

分别对5个处理蒜头果根系间自寄生的吸器以及与寄主植物形成的吸器数量、平均直径、最大直径进行了统计(表1)。结果表明蒜头果能形成非常多的自寄生吸器,每株的自寄生吸器达到12~42个,平均直径约为4mm,最大直径达11mm。不同处理间进行比较,发现处理Ⅱ、Ⅲ和Ⅴ中,蒜头果与寄主植物形成的吸器数量只有3~8个,显著少于自寄生的吸器数量,且吸器直径也明显小于自寄生的吸器。而处理Ⅳ自寄生的吸器数量显著少于其他3个处理,与寄主植物形成的吸器数量及直径均显著大于其他3个处理。在同一处理内,处理Ⅳ蒜头果与寄主植物形成的吸器数量显著多于自寄生的吸器数量,占吸器总数的58.80%;其他处理蒜头果与寄主植物形成的吸器数量占总数的14.80%~19.20%,显著低于处理Ⅳ(图6)。以上结果表明蒜头果具有一定的寄主选择偏好性,在4种植物中,蒜头果与处理Ⅳ的寄主植物形成更紧密的寄生关系。

表1 蒜头果共培养盆栽根系吸器情况
Table.1 number and size of two kinds of haustorium among different host plants

处理号	蒜头果侧根之间形成的自寄生吸器			蒜头果与伴生植物形成的寄生吸器		
	数量(个)	平均直径(mm)	最大直径(mm)	数量(个)	平均直径(mm)	最大直径(mm)
处理Ⅰ	31.00±24.47ab	3.74±0.83a	10.72±3.30a	/	/	/
处理Ⅱ	42.00±15.70a	4.58±0.92a	9.58±2.33ab	7.20±2.86a	2.55±0.34bc	3.68±1.38b
处理Ⅲ	34.20±13.88ab	4.45±0.38a	10.02±3.35ab	8.00±4.85a	3.41±0.87b	9.03±4.11a
处理Ⅳ	12.00±2.87b	4.14±0.57a	6.58±1.51b	17.60±6.18b	5.70±1.04a	9.97±2.50a
处理Ⅴ	20.60±8.88b	4.66±0.86a	11.22±3.00a	3.60±2.07a	2.62±0.34b	3.58±1.32b

注:表中同列相同字母表示不同处理间无显著的差异($P>0.05$),不同字母表示不同处理间有显著的差异($P<0.05$)。

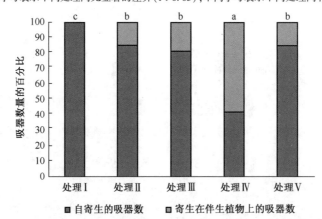

图6 不同处理蒜头果自寄生吸器与在寄主植物上的形成吸器的数量比例

注:图中相同字母表示不同处理间无显著的差异($P>0.05$),
不同字母表示不同处理间有显著的差异($P<0.05$),下同。

Figure 6 Proportion of haustoria on host roots and its own roots

2.5 寄主植物对蒜头果幼苗生长的影响

蒜头果共培养试验5种处理的苗高平均分别为:39.30cm、40.20cm、40.00cm、42.90cm、

40.80cm,处理Ⅳ的苗高略高于其他处理,但差异未达到显著水平(图7)。地径平均值分别为6.95cm、7.07cm、7.01cm、7.20cm、6.83cm,处理Ⅳ的苗木粗度略高于其他处理,但差异未达到显著水平(图8)。处理Ⅳ的叶片数为15.6片,显著高于其他处理,其他4个处理的叶片数为11.2~12.0片,互相之间显著不差异(图9)。

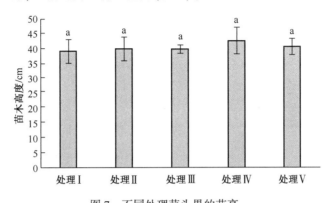

图7 不同处理蒜头果的苗高

Figure 7 Height of *Malania oleifera* seedlings

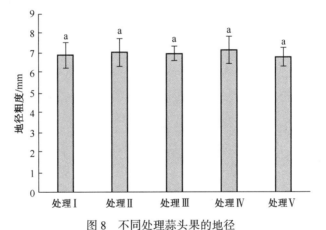

图8 不同处理蒜头果的地径

Figure 8 Diameter of *Malania oleifera* seedlings

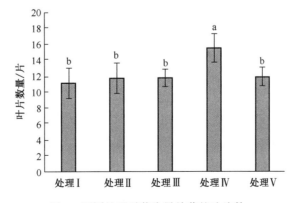

图9 不同处理对蒜头果幼苗的叶片数

Figure 9 Number of leaves of *Malania oleifera* seedlings

各处理生物量的测定结果见图10,其中处理Ⅳ的全株生物量、地上部分干重分别为15.52g、9.61g,显著高于其他各处理。处理Ⅳ的地下部分干重为5.91g,高于处理Ⅲ但差异不显著,显著高于处理Ⅰ、Ⅱ和Ⅴ。以上结果表明处理Ⅳ的寄主植物能显著促进蒜头果的生物量积累,特别是根的生长。其次处理Ⅲ蒜头果的全株干重、地下部分干重显著高于处理Ⅰ、Ⅱ和Ⅴ,表明处理Ⅲ的寄主植物对促进蒜头果的生物量积累也有一定的效果。

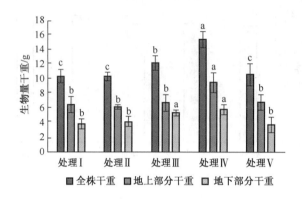

图10 不同处理对蒜头果幼苗的生物量
Figure 10 Biomass of *Malania oleifera*

3 结论与讨论

3.1 结论

在天然分布区的调查发现,蒜头果的根系能与多种植物形成寄生关系,这些植物分属不同的科属,有草本、灌木、藤本和木本植物。在各种植物根系上附着的吸器数量和大小不一,直径范围0.2~3.2cm。蒜头果的吸器为白色肉质组织,圆球形、钟状或扁平,吸附、半包裹或是全包裹寄主植物的根系。蒜头果的侧根间能形成大量的吸器,通过吸器根系会互相缠绕在一起,自寄生现象明显。蒜头果与寄主植物吊兰所形成的吸器数量、平均直径、最大直径均大于其他植物;同时,与吊兰共培养的蒜头果苗木的苗高、地径、叶片数量及其全株生物量、地上生物量、地下生物量均高于其他植物共培养的蒜头果苗木。由此说明,蒜头果与吊兰形成更紧密的寄生关系。

3.2 讨论

目前还未见蒜头果半寄生的文献报道,有一些研究人员发现了蒜头果根系能形成特殊的瘤状结构,但未明确其为寄生植物特有的结构——吸器。赖家业[13]在开展蒜头果的保护生物学研究时,发现了蒜头果根瘤状结构的存在,根瘤状物能像吸盘一样吸附其他植物的根系。切片观察发现根瘤状结构是从根部中柱鞘部位延伸出来的,且根瘤状结构四周均有木质部导管的分布,其中靠近中柱鞘的两侧部位有螺纹导管的存在。从发生来源上看,根瘤状结构应是由蒜头果根木质部的薄壁细胞恢复分裂能力而向外分裂形成,组成根瘤状结构的细胞主要为薄壁细胞。利用乙炔还原法测定了蒜头果根瘤状结构固氮酶活性,发现该结构不是固氮根瘤[13]。吴彦琼[14]也对蒜头果的根瘤状结构作了观察,发现根瘤组织与根紧密相连直到髓心,确定根瘤为植物本身的组织而非附生物。根瘤与根的养分含量无显著差异,确定该种瘤状结构不是植物的养分储存组织。且根瘤状结构也不是线虫和微生物引起[14]。本研究通过野外调查、切片观察和温室共培养盆栽试验,证实了蒜头果根部的瘤状结构为寄生结构——吸器,表明蒜头果是一种根部半寄生植物。

蒜头果在植物分类上属于檀香目(Santalales)铁青树科。檀香目包含桑寄生科(Loranthaceae)、檀香科(Santalaceae)、铁青树科(Olacaceae)、山柚子科(Opiliaceae)4个科,其中桑寄生科、檀香科和山柚子科都有较多种类的寄生或半寄生植物[15,16],铁青树科铁青树属的一种植物 Olax phyllanthi 据报道也是半寄生植物[17]。一般来说,亲缘关系相近的植物种类由于相近的遗传关系,通常具有相似的形态、结构和生理生化特征,如植物胚胎学研究也发现蒜头果成熟胚囊的特征与近缘科檀香科、桑寄生科等相同[18]。基于蒜头果近缘科中有较多的寄生或半寄生植物,蒜头果为半寄生植物是有遗传基础的。

另外,本研究中野外调查和共培养盆栽试验的结果表明,蒜头果能与不同科属的多种植物形成半寄生关系,蒜头果对寄主选择的专一性不强。但盆栽试验发现,蒜头果与不同种类的寄主植物共培养,产生的吸器在数量和大小上存在显著的差异,表明蒜头果对寄主植物的选择存在较明显偏好性;不同寄主植物对蒜头果苗木生物量的积累促进效果也各异。蒜头果幼苗与吊兰表现出更紧密的寄生关系,吊兰作为寄主植物在蒜头果人工种植中具有一定的应用前景。寄生植物的成功栽培离不开适宜的寄主植物,针对蒜头果人工栽培中成活率和保存率低下的问题,今后还将在本研究的基础上进一步筛选优良的寄主植物,开展相应的山地造林试验研究,促进蒜头果这一珍稀濒危物种的保护、繁育与利用。

参考文献:

[1] 李树刚. 油料植物一新属——蒜头果属[J]. 植物研究,1980(1):67~72.

[2] 李树刚. 广西现代植物分类学研究的发展[J]. 广西植物,1995,15(3):256~267.

[3] 国家环境保护局. 珍稀濒危植物保护与研究[M]. 北京:中国环境科学出版社,1991.

[4] 国家环境保护局. 国家重点保护野生植物名录(第一批)[J]. 植物杂志,1999(5):4~11.

[5] 欧乞铖. 一个重要脂肪酸 CIS-TETRACOS-15-ENOIC 的新存在——蒜头果油[J]. 植物分类与资源学报,1981,3(2):59~62.

[6] 潘晓芳. 蒜头果育苗情况初报[J]. 广西农业科学,1999,18(3):236~238.

[7] 陈强,常恩福,李品荣,等. 滇东南岩溶山区造林树种选择试验[J]. 云南林业科技,2001(3):11~16.

[8] 瞿林. 广南县岩溶森林保护及石漠化治理[J]. 林业调查规划,2001,26(4):70~75.

[9] 肖来云,普正和. 珍稀濒危植物的迁地保护研究[J]. 云南林业科技,1996(1):45~53.

[10] H. Heide-Jørgensen. Parasitic flowering plants[M]. Brill,2008.

[11] Irving L J,Cameron D D. You are What You Eat: Interactions Between Root Parasitic Plants and Their Hosts[M]// Advances inBotanical Research. Elsevier Science & Technology,2009:87~138.

[12] 陆俊锟. 印度檀香与寄主植物间寄生关系的研究[D]. 北京:中国林业科学研究院,2011.

[13] 赖家业. 珍稀植物蒜头果保护生物学研究[D]. 成都:四川大学,2006.

[14] 吴彦琼. 蒜头果保护的初步研究[D]. 南宁:广西大学,2002.

[15] 中国科学院中国植物志编辑委员会. 中国植物志:第二十四卷[M]. 北京:科学出版社,1988.

[16] 李松. 寄生被子植物吸器结构研究进展[J]. 内蒙古民族大学学报(自然科学版),2013,28

(1):41~43
[17] Pate J S,Pate S R,Kuo J,et al. 1990. Growth,resource allocation and haustorial biology of the root hemiparasite Olax phyllanthi (Olacaceae). Annals of Botany,65:437~449.
[18] 杨鲁红. 蒜头果属与铁青树科的系统关系[D]. 昆明:云南大学,2004.

(本文发表于《西部林业科学》,2019 年)

Gravid Females of *Cephalcia chuxiongica* (Hymenoptera: Pamphiliidae) are Attracted to Egg-carrying Needles of *Pinus yunnanensis*

YAN Zheng-Liang, MA Hui-Fen, MAO Yun-Ling, LIU Ling

Institute of Forestry Protection, Yunnan Academy of Forestry, Kunming 650201, China

Abstract: *Cephalcia chuxiongica* Xiao is one of the most dangerous defoliators of *Pinus yunnanensis* and other pine species in Yunnan province, resulting in serious losses. Its distinguishing characteristics are the females' aggregation oviposition and larvae's aggregation feeding. In order to explore the mechanism of aggregation oviposition in this sawfly, preliminary olfactory bioassay was conducted in laboratory. In in-cage choice tests, on average vast majority gravid females selected the shoots that had been loaded and oviposited by a 'pioneer' female. In one -choice tests in laboratory by a Y-tube olfactometer, the gravid females were attracted by the odors of eggs-carrying shoots (PE), shoots with one delivering female and her eggs (PGE), needles' extract (NE), and fresh eggs' eluent (EL); the virgin females were attracted by odors of fresh needles (P), PE, PGE, and NE, but repelled by odors of virgin and gravid females. In two-choice tests, the odors were tested in pairs for gravid females. When compared with odors of gravid females (G) or P, gravid females showed significantly more tendency to odors of PE or PGE. When given odors EL vs. NE, gravid females preferred the odors of NE, but they did not make obvious selection between G vs. P, and PE vs. PGE. Based on the results, our conjectures were: (1) Delivery female, as a pioneer, could summon her conspecific gravid females to aggregate in the same pine shoot; (2) Pine needles' odors were attractive for both the virgin and gravid females; (3) Gravid females could be attracted by odors released by the pioneer gravid females; (4) The olfactory sensation of the females may be changed by mating.

Keywords: *Cephalcia chuxiongica*; Aggregation oviposition; Host volatile; Accessory gland volatile; Olfactometer bioassay.

Introduction

Cephalcia chuxiongica Xiao is an important defoliator of pine trees in Yunnan Province of southwest China. The morphology of the insect was described by Xiao Gangrou in 1984 (Xiao, 1984), and its life-history was briefly reported by Liu *et al.* (2014). The full-grown larvae drop down to the ground from trees and drills into the soil below the canopy from August to October, and then spends two winters in the soil before pupation. Adults' emergence occurs from June to September in the third year. The females strongly prefer conifer species to lay eggs, and *Pinus yunnanensis* and *Pinus armandii* are more suitable for the oviposition than *Keteleeria evelyniana* Mast, a native conifer tree species in Yunnan Province (Cheng *et al.* 2014).

Our previous field investigation found that females of *C. chuxiongica* gathered on the shoots of *P. yunnanensis* to lay eggs (Liu *et al.* 2015). In further observation, we noticed an interesting phenomenon: the pioneer gravid female randomly laid her eggs in a pine shoot, but the subsequent gravid females selected the shoot on which the pioneer female had laid. Thus, we speculated that, besides the pine needles' volatiles, other odors may play a role in attracting the subsequent gravid females of *C. chuxiongica*. In order to verify our surmise, laboratory bioassay was conducted.

Materials and Methods

Insects. Tested insects were obtained from a 20-year-old plantation of *P. yunnanensis*, located on a small mountain near Lufeng town in the Yunnan Province, in 2016. Many webbing leftovers were observed on the ground below the pine tree canopies. New adult males and females were directly dug out of the soil below the pine trees where webbing leftovers were scattered. Collected females and males were discriminated (cervical sclerites, pronotum, and scutoprescutum of the female are russet, but piceous in the male), and reared separately by 5% sugar solution in insect cages under natural condition until they were tested the next day.

Before bioassay, the candidate females were divided into two groups. One group was kept as virgin; the other group was mated with males in a transparent plastic cup, in order to get gravid females. After a round of trials was performed, the tested insects were discarded. *Crude extracts*. Crude extracts of needles of *P. yunnanensis* were taken by a simple distilling apparatus. 100 g of needles were cut into small pieces and placed in a 500 mL round-bottom flask with 200 mL water plus 50 mL trichloromethane. The flask was attached to a reflux condenser. The solution was heated and refluxed for 2 h. After cooling down room temperature, the organic phases were isolated, filtrated, and dried by sodium sulfate. The solvent was removed by rotary vacuum evaporator below 40℃.

The crude extract of fresh eggs was obtained by elution of fresh eggs with trichloromethane. The fresh eggs are bright yellow, but lose their luster after a day. Shoots with fresh eggs were cut and collected from the field, then carried back to the laboratory in sealed polyethylene taker-bags. The needles with eggs were plucked; the eggs were drip-washed with trichloromethane, taking care to avoid staining by the metabolites from the needles' wounds. A single bunch of needles (three needles in one bunch, about 85 eggs laid by each female on average) was washed by 10 μL trichloromethane. The eluents were combined and stored at 4℃ until testing.

In-cage choices of oviposition. Choice tests were conducted inside 80 cm × 80 cm × 80 cm screen cages with 0.42 mm mesh openings. The cages were placed in the laboratory and the shoots of *P. yunnanensis* were tested. Fresh shoots were cut with 25 cm length, and the new cross sections were coated with white latex to prevent the shoots from withering. A pine shoot was hung upside down by a thin wire in a corner of the cage, and the other shoot in the opposite corner; both shoots were 20 cm high from the ground. One gravid female was introduced into the cage, imitating a 'pioneer' as in the field condition, and allowed to select her oviposition shoot and lay her eggs randomly. Subsequently, other 10 candidate gravid females were introduced into the cage one-by-one. Once the preceding candidate female had selected her oviposition shoot, it was taken out of the cage; and the next candidate gravid female was introduced. The number of females (except for the 'pioneer') on the shoot with vs. without eggs was counted. In order to eliminate influences of visual cues, six cages were used: in three cages, the shoots that bore the 'pioneer' females were hung in one corner of the cages, and in the other three cages the shoots bearing the 'pioneer' females were hung in the opposite corner.

Behavioral Responses with a Y-tube. Olfactory responses of gravid females were tested in a glass Y-tube olfactometer (50 mm diameter, main tube length 50 cm, arm length 25 cm), with 120° arm angles, which was placed horizontally. Incoming air was filtered through activated charcoal and humidified with doubly distilled water. The filtered air was split between two holding chambers (Blackmer et al. 2004).

Fresh pine shoots were cut at 20 cm length, and the new cross sections were coated with white latex to prevent the shoots from withering. In one-way choices, one chamber of the Y-tube olfactometer held one pine shoot (P), one virgin female (V), one gravid female (G), one pine shoot each with one female and her fresh eggs on the needles (PGE), or one pine shoot with fresh eggs without protecting female (PE), as odor sources. The other chamber, for controls, held no odor source, only clean air passing through it. Air-flow through the system was maintained at 200 mL^{-1} · min by an inline flow meter. A prior smoke test showed laminar airflow in both arms and throughout the olfactometer. To eliminate visual cues, the Y-tube setup was surrounded by a 100 cm × 80 cm black fabric enclosure. Approximately 30 min before trials were initiated, mated C. chuxiongica females were introduced into a separate holding container, so they would not be exposed to test odors before their release. In order to determine whether the olfactory behavior of the females changed after mating, virgin females were also tested simultaneously. The candidate mated or virgin females were tested one-by-one with 25 females as a group (one treatment), and each treatment was duplicated six times. The females were given 10 min to respond to the treatment; the females that went past the Y junction that led to the treated chamber were noted as attracted, those that remained in the main tube were noted as no-response, and those that went into the other arm tube were noted as repelled. The trend rate of females to the odor was calculated as the percentage of attracted females of the total females that were attracted or repelled. When crude extracts were tested, the needles' extract (NE) was diluted with trichloromethane to 1mL/100 mL, the fresh eggs' eluent (EL) was used directly. 10 μL solution of NE or EL was dropped on a slip of filter paper (0.5 cm ×3.0 cm), and left for 10 seconds until solvent evaporation. Then, the filter paper was put into one of the chambers of the Y-tube olfactometer as a treatment; the other chamber was provided a slip of filter paper dropped with the solvent only and then left to volatize as a control. All the females were used only once for each of the treatments.

In two-way choices, the odor sources were given in pairs, and olfactory responses of gravid females were tested. Thus, the odor sources were: G vs. P, G vs. PGE, G vs. PE, P vs. PGE, P vs. PE, PGE vs. PE, and NE vs. EL. The olfactometer was washed by n-hexane; the treatment and control sides were alternated after every trial.

Data analysis. Nonparametric Wilcoxon matched pairs test was used for paired comparison (SPSS), statistical significance was accepted 0.05. Figures were drawn by Microsoft Excel 2007.

Results

Results of in-cage oviposition tests of the gravid females are given in Figure 1. The pioneer gravid females' selection of oviposition pine shoot was random, no obvious preference between the two shoots in the opposite corners was observed. However, the majority of delivery females selected the shoots that bore a 'pioneer' female (and her eggs) as their oviposition sites (9.00±1.26 females/shoot on average); only sporadic females (0.33±1.26) selected the shoots without the 'pioneer' female.

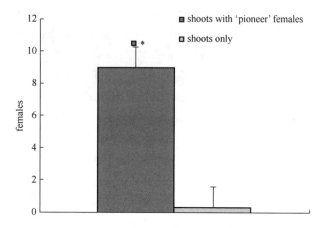

Figure 1 Numbers of gravid females of *C. chuxiongica* in pine shoots with vs. without 'pioneer' females. *Significantly different between the treatment and clean air based on nonparametric Wilcoxon matched pairs test (two-tailed), $P<0.05$.

Trend rates of gravid females to odors from PE, PGE, NE, and EL were significantly larger than those to clean air; there was no significant difference between the trend rates to the odor of P, V, and G and to clean air (Figure 2). The trend rates of virgin females to P, PGE, PE, and NE were obviously larger than that to clean air; the virgin females did not make choice between odor of EL and clean air. Compared with the odors of V and G, the virgin females were more inclined to clean air (Figure 3).

In two-way choice tests, only gravid females were tested (Figure 4). When compared with odors of G or P, gravid females were significantly more attracted to odors of PE or PGE. Gravid females did not make obvious choice between P vs. G, and PE vs. PGE; but preferred to the odors of NE than EL.

Discussion

Delivery female can summon the conspecific gravid females to aggregate in the same pine shoot. In in-cage test of oviposition choices, vast majority of gravid females selected the shoot in which the 'pioneer' females had laid eggs (Figure 5). This affirmed our surmise that the delivery females of C. chuxiongica in pine shoots can attract more other conspecific gravid females to aggregate in the same pine shoot for oviposition.

Pine needles' odors were attractive for both the virgin and gravid females. Host volatiles and

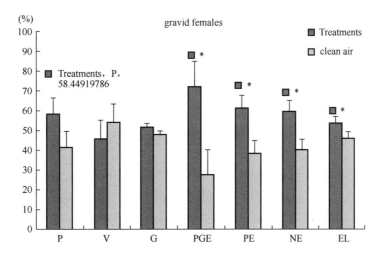

Figure 2 Trend rates of gravid females of *C. chuxiongica* to volatiles in one-way choice tests.
P: one pine shoot, V: one virgin female, G: one gravid female, PGE: one pine shoot with
one female and her fresh eggs on the needles, PE: one pine shoot with fresh eggs but no female;
NE: needles' extract, EL: eggs' eluent.
*Significantly different ($P<0.05$) between the treatment and clean air based on nonparametric Wilcoxon
matched pairs test (two-tailed).

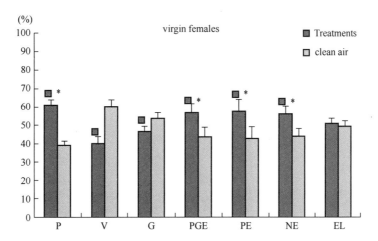

Figure 3 Trend rates of virgin females of *C. chuxiongica* to volatiles in one-way choice tests.
P: one pine shoot, V: one virgin female, G: one gravid female, PGE: one pine sho
ot with one female and her fresh eggs on the needles, PE: one pine shoot with fresh eggs but no female;
NE: needles' extract, EL: eggs' eluent.
*Significantly different ($P<0.05$) between the treatment and clean air based on nonparametric
Wilcoxon matched pairs test (two-tailed).

their chemical profiles play an indispensable role in oviposition preference in phyllophagous sawflies. Females of the sawfly *Diprion pini* can discriminate between pine species for their oviposition

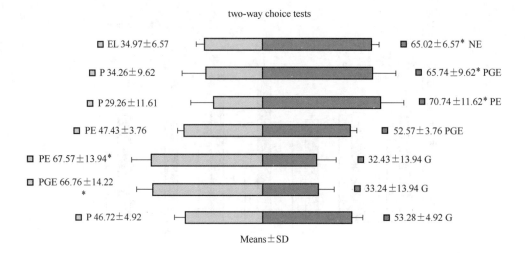

Figure 4　Paired comparison of trend responses of gravid females of *C. chuxiongica* to attractive odors
P: one pine shoot, G: gravid female, PGE: one pine shoot with one female and herfresh eggs
on the needles, PE: ne pine shoot with fresh eggs but no female. NE: needles' extract, EL: eggs' eluent.
*Represents the trend rate to the odor being higher than that to the other odor in this pair
($P<0.05$, nonparametric Wilcoxon matched pairs test, two-tailed).

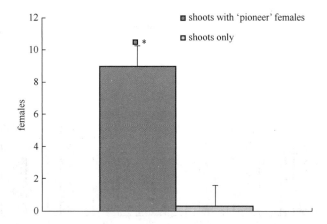

Figure6.　Numbers of gravid females of *C. chuxiongica* in pine shoots with vs. without 'pioneer' females.
*Significantly different between the treatment and clean air based on nonparametric Wilcoxon matched
pairs test (two-tailed), $P<0.05$.

sites (Barre et al., 2002); *Pinus banksiana* and *Pinus strobus*, whose needles release significantly higher relative concentrations of limonene, β-pinene and myrcene than needles of *Pinus sylvestris* and *Pinus nigra*, are not suitable for *D. pini* oviposition (Buda, 2011); the high-carene chemotype of *P. sylvestris* captured more numbers of *Sirex noctilio* females than the low-carene chemotype (Böröczky, 2012); In an olfactometer bioassay, a low amount of limonene was attractive for *Neo-*

diprion sertifer females, while a repellent effect was evident when higher amounts were used (Martini et al., 2010). Trend rates of virgin females (Figure 3) of *C. chuxiongica* to odors from P or NE were significantly larger than those to clean air; gravid females did not show obvious tendency to odors of P, but significantly inclined to odors of NE. This indicates that volatiles of needles are attractive for both the virgin and gravid females. The reason for the inconspicuous tendency of gravid females to odors of P ($P = 0.0789$) may be due to the lower concentration of odors of undamaged needles. However, the crude extracts of needles of *P. yunnanensis*, or the monoterpene β-pinene or myrcene captured a few of females in field tests (data unpublished). There are huge differences between compositions of volatile extracts obtained by steam distillation (Yang, 2009) and by head-space adsorption (Wu et al., 2010); the volatile composition of crude extract of needles did not simulate the chemical profile of needles' odor in the natural state adequately. This maybe a probable reason of the low captures. In fact, the pioneer females randomly selected their oviposition shoots in cages and in the field; but the subsequent gravid females were summoned to the shoot that had born delivery females. Thus, we surmise that, in addition to host volatiles, other factors may also play a role in the aggregation oviposition in this sawfly.

Pine sawfly *D. pini* oviposition can change the volatile composition of host *P. sylvestris* twigs quantitatively or constitutively (Hilker et al., 2002, Mumm et al., 2003; Mumm et al., 2004); and the terpenoid volatile pattern of systemically oviposition-induced pine twigs changes quantitatively after an induction time of 3 d compared to controls (Mumm et al., 2003). The difference of volatiles of *P. yunnanensis* needles before and after *C. chuxiongica* females' oviposition remains to be analyzed.

Gravid females could be attracted by odors released from the fresh eggs. The results showed (Figure 2) that the trend rates of gravid females to odors from EL were significantly larger than those to clean air. Based on the results of Günthardt-Goerg (2010), the volatile of the needles' eluent contains negligible monoterpenes. All these indicate that odors released from fresh eggs are attractive for the gravid females. This can be further corroborated by the results in Figure 4; the trend rate of the gravid females to odors of PE was significantly larger than to odors of P; also, the trend rate to odor PGE was significantly larger than to odor of P. Adult's aggregation oviposition and larvae's aggregation feeding is the obvious feature in *C. chuxiongica* (Fang, 2010); the former is a necessary premise for the latter. These prompted us to further confirm that the volatiles released from fresh eggs contain some attractive components, which are not the monoterpenes themselves. Blümke and Anderbrant (1997) mentioned that the presence of *N. sertifer* eggs and extracts obtained from eggs had no inhibitory effect on the females' egg-laying, but they did not record whether the odor of eggs' extract was attractive for females. However, reproductive accessory gland secretions of some insect species, such as in *Delia radicum* (Gouinguené et al., 2006) and *Schistocerca gregaria* (Kahoro et al, 1997; Torto et al, 2015), which are attached to the surface of the eggs, contain stimulating pheromone that induces the same female adult to aggregate and lay eggs. The constituents of the secretion of *C. chuxiongica* remain to be analyzed and the attractive components remain to be identified.

The gravid females did not show a significant tendency to odors of G (Figure 2). However, this did not mean that the gravid females did not release odors attractive for the conspecific females, since the odors of fresh eggs were attractive; we suspect that females, as odor sources in chamber of the Y-tube olfactometer, were constantly disturbed in the bioassay, consequently the odor release was disturbed.

Theolfactory sensation of the females may be changed by mating. Firstly, there was no difference between the trend rates of virgin females to EL and to clean air (Figure 3), while an obvious large trend rate of gravid females to EL (Figure 2). Secondly, a relatively low trend rate to V was observed in virgin females (Figure 3), but gravid females did not express any selectivity between V and clean air (Figure 2). This means that odor of V was repellent for the virgin females, whereas odor of EL was attractive for the gravid females. Thirdly, the odor of P (Figure 3) was significantly attractive for the virgin females, but its attraction to gravid females was not obvious (Figure 2). Therefore, we infer that the physiological state, such as olfactory receptor of the *C. chuxiongica* females in this case, had changed. The truth remains to be confirmed by further research.

On the whole, we can imagine a scene for *C. chuxiongica*: the females select their suitable host by distinguishing the differences among volatiles of different tree species (Cheng *et al.*, 2014). The pioneer female randomly locates a pine shoot as its oviposition site; after mating, she informs her conspecific gravid females to gather in the same pine shoot by sending some signals (such as attractive chemicals) through her reproductive accessory gland; the attractive chemicals maybe part of the components of secretion attached on the surfaces of fresh eggs. The aggregated females lay their eggs on the same shoot; the later aggregation feeding of the low instar larvae might be beneficial to resist the resistance of the host. This conjecture is consistent with the behaviors of this sawfly observed in the field (Liu *et al.*, 2014, 2015). If this conjecture does indeed reflect the truth about the mechanism by the sawfly's aggregation oviposition, many details in the process of insect damage remain to be studied in depth.

Acknowledgments

This study was funded by the Applied Basic Research Programs of Yunnan Provincial Science and Technology Department (2011FA027).

References

Barre F, Milsant F, Palasse C, Géri C (2002) Preference and performance of the sawfly Diprion pini on host and non-host plants of the genus Pinus. Entomologia Experimentalis et Applicata 102: 229~237.

Blackmer JL, Rodriguezsaona C, Byers JA, Shope KL, Smith JP (2004) Behavioral response of Lygus hesperus to conspecifics and headspace volatiles of alfalfa in a Y-tube olfactometer. Journal of Chemical Ecology 30: 1547~1564.

Blümke A & Anderbrant O (1997) Oviposition pattern and behaviour of the pine sawflyNeodiprion sertifer (Hymenoptera: Diprionidae). Bulletin of Entomological Research 87:231~238.

Böröczky K, Zylstra KE, Mccartney NB, Mastro VC, Tumlinson JH (2012) Volatile profile differences and the associated Sirex noctilio activity in two host tree species in the Northeastern United States. Journal of Chemical Ecology 38: 213~221.

Buda V (2011) Comparative analysis of monoterpene composition in four pine species with regard to suitability for needle consumer Diprion pini L. Ekologija 57:163~172.

Cheng L, Yan DQ, LiuL, Qi RP, Yan ZL (2014) Host-selection preference of Cephalcia chuxiongicaXiao (Hymenoptera: Pamphiliidae) to different tree species. Hubei Agricultural Sciences 53: 4316~4319(in Chinese with English abstract).

Fang S (2010) Spatial distribution pattern and optimal sampling number of Cephalcia chuxiongica larvae. Forest Inventory & Planning 35(6): 88~92(in Chinese with English abstract).

Gouinguené S, Poiger T, Städler E (2006) Eggs of cabbage root fly stimulate conspecific oviposition: Evaluation of theactivity and determination of an egg-associated compound. Chemoecology 16:107~113.

Günthardt-Goerg MS (2010) Epicuticular wax of needles ofPinus cembra, Pinus sylvestris and Picea abies. Forest Pathology 16:400~408.

Hilker M, Kobs C, Varama M, Schrank K (2002) Insect egg deposition induces Pinus sylvestris to attract egg parasitoids. Journal of Experimental Biology 205:455~461.

Kahoro H, Odongo H, Saini RK, Hassanali A, Rai MM (1997) Identification of components of the oviposition aggregation pheromone of the gregarious desert locust, Schistocerca gregaria, (Forskal). Journal of Insect Physiology 43: 83~87.

Liu L, Yan DQ, Qi RP, Liu DY, Yan ZL, Liu HP (2014) Circadian rhythm of emergence and reproduction ofCephalcia chuxiongica Xiao (Hymenoptera: Pamphiliidae). Forest Pest & Disease 33(6): 5~8(in Chinese with English abstract).

Liu L, Yan DQ, Mao YL, Qi RP, Hu GH, Yan ZL (2015) Effects of needle volatiles and nutrients on the host for oviposition selection ofCephalica chuxiongnica Xiao (Hymenoptera: Pamphiliidae). Journal of Fujian Agriculture & Forestry University 44: 14~17(in Chinese with English abstract).

Martini A, Botti F, Galletti G, Bocchini P, Bazzocchi G, Baronio P, Burgio G (2010) The influence of pine volatile compounds on the olfactory response by Neodiprion sertifer (Geoffroy) females. Journal of Chemical Ecology 36: 1114~1121.

Mumm R, Kai S, Wegener R, Schulz S, Hilker M (2003) Chemical analysis of volatiles emitted by Pinus sylvestris, after induction by insect oviposition. Journal of Chemical Ecology 29: 1235~52.

Mumm R, Tiemann T, Schulz S, Hilker M (2004) Analysis of volatiles from black pine (Pinus nigra): significance of wounding and egg deposition by a herbivorous sawfly. Phytochemistry 65: 3221~3230.

Torto B, Assad YO, Njagi PG, Hassanali A (2015) Semiochemical modulation of oviposition behaviour in the gregarious desert locust Schistocerca gregaria. Pest Management Science 55: 570~571.

Wu SR, Zhou PY, Li ZY, Fu RJ, Yuan SY (2010) Analysis the volatile constituents of pine needles from health and debility Pinus yunnanensis. Natural Product Research & Development 22: 1048~1052 (in Chinese with English abstract).

Xiao GR (1984) Two new species of Cephalcinae (Hymenoptera: Pamphiliidae) from China. Entomotaxonomia 6: 137~140 (in both Chinese and English).

Yang Y (2009) Chemical constituents of volatile from pine needles of Pinus yunnanensis. Scientia Silvae Sinicae 45: 173~177(in Chinese with English abstract).

抱卵的楚雄腮扁叶蜂可被着卵云南松针叶所引诱

摘要：楚雄腮扁叶蜂(*Cephalcia chuxiongica* Xiao)是西南地区重要的松树食叶害虫,造成严重的损失。其危害的显著特点是:雌蜂的聚集产卵和幼虫的聚集取食为害。利用性信息素诱杀其雄蜂,并不能降低其在林间的产卵量,因而也无助于减少下一世代的虫口数量。在成虫产卵阶段消灭产卵雌蜂,可大大减少其下一代的虫口数量,进而减少其下一世代的危害。室内单向和双向选择趋向试验表明:楚雄腮扁叶蜂未交尾雌蜂只对松针挥发性气味具有趋向反应;而已交尾雌蜂则不仅对松针挥发物,而且对产卵雌蜂及其产卵器副腺的挥发物具有趋向反应。实验结果为进一步研发楚雄腮扁叶蜂的产卵引诱剂提供了基础。

(**本文发表于** *Journal of Hymenoptera Research* 2018)

Attack Pattern and Reproductive Ecology of the Pine Shoot Beetle *Tomicus brevipilosus* on Yunnan Pine (*Pinus yunnanensis*) in Southwestern China

CHEN Peng[1,2], LU Jun[1], Robert A. Haack[3c], Hui Ye[1d]

[1] Laboratory of Biological Invasion and Transboundary Ecosecurity, Yunnan University, Kunming 650091, China

[2] Yunnan Academy of Forestry, Kunming 650201, China

[3] USDA Forest Service, Northern Research Station, 3101 Technology Blvd., Suite F, Lansing, MI 48910, USA

Abstract: *Tomicus brevipilosus* (Eggers) (Coleoptera: Curculionidae, Scolytinae) was recently discovered as a new pest of Yunnan pine (*Pinus yunnanensis* Franchet) in Yunnan province in southwestern China. However, little was known on its reproductive biology and pattern of trunk attack on Yunnan pine. The objectives of the present study were to better understand the reproductive biology of *T. brevipilosus* by investigating the seasonality of trunk attacks by parent adults for the purpose of reproduction (i. e., breeding attacks) and the within-tree pattern of these attacks. Our results showed that *T. brevipilosus* breeding attacks in *P. yunnanensis* generally started in early March and ended in early June in Anning County, Yunnan. *Tomicus brevipilosus* exhibited two general patterns of infestation. From early March to mid-April, *T. brevipilosus* bred preferentially in the trunks of Yunnan pine trees that were already infested by *Tomicus yunnanensis* Kirkendall and Faccoli and *Tomicus minor* (Hartig), colonizing spaces along the trunk (mostly in the mid- and lower-trunk) that were not already occupied by the other two Tomicus species. Later, from about mid-April to early June, when there were no Yunnan pine trees newly infested by *T. yunnanensis* and *T. minor*, *T. brevipilosus* attacked Yunnan pine by itself, infesting the lower parts of the trunk first and then infesting progressively upward along the trunk into the crown. Infestation by *T. brevipilosus* extends the total period that *P. yunnanensis* trees are under attack by Tomicus beetles in southwestern China, which helps explain why Yunnan pine has suffered high levels of tree mortality in recent decades.

Keywords: *Tomicus brevipilosus*; *Tomicus yunnanensis*; *Tomicus minor*; Pine shoot beetle; Within-tree attack pattern; *Pinus yunnanensis*

Introduction

Bark beetles in the genus Tomicus (Coleoptera: Curculionidae, Scolytinae), with the exception of *T. puellus* (Reitter) that infests primarily Picea trees, are well known pests of pine (Pinus) trees in most Eurasian countries where pine is native (Bakke, 1968; Ye, 1991; Faccoli, 2007; Kirkendall et al., 2008; Li et al., 2010; Lieutier et al., 2015). These univoltine bark beetles are commonly called shoot beetles because the newly emerged adults feed inside living shoots of their host trees during the summer months as they become sexually mature, which often results in death of the infested shoots (Långström, 1983; Ye, 1994a; Kohlmayr et al., 2002). The Palaearctic genus *Tomicus* contains eight recognized species worldwide, of which five occur only in Asia, one only in Europe, and two in Eurasia (Kirkendall et al., 2008; Li et al., 2010; Lieutier et al., 2015). The Eurasian species, *T. piniperda* (Linneaus) is the only Tomicus species to be introduced beyond its native range, being first found in the Great Lakes region of North America in 1992 (Czokajlo et al., 1997; Haack and Poland, 2001; Lieutier et al., 2015).

Tomicus brevipilosus (Eggers) is native to Asia where it has been reported to occur in China, India, Japan, Korea, and Philippines (Kirkendall *et al*., 2008; Lu *et al*., 2014). This bark beetle has been recorded to infest several species of Pinus, including *Pinus yunnanensis* (Franchet), *P. koraiensis* Siebold & Zucc., and *P. kesiya* Royle ex Gordon, but severe damage has seldom been reported (Murayama, 1959; Kirkendall *et al*., 2008; Lu *et al*., 2014). In China, *T. brevipilosus* has been generally regarded as a secondary pine pest, mainly infesting pine trees weakened by drought, defoliation, and other environmental stressors (Kirkendall *et al*., 2008; Lu *et al*., 2014). However, since the late 1990s, *T. brevipilosus* along with *Tomicus yunnanensis* Kirkendall and Faccoli and Tomicus minor (Hartig) have reached outbreak levels, resulting in widespread tree mortality with more than 200,000 ha of infested *P. yunnanensis* forests in southwestern China (Ye, 1991; Duan *et al*., 2004; Liu *et al*., 2010; Lu *et al*., 2014).

In Yunnan, *T. brevipilosus* completes its life cycle in Yunnan pine (*P. yunnanensis*) (Lu *et al*., 2014). After exiting the host material in which the brood developed, newly emerged adults fly to the crowns of nearby pine trees, usually beginning in June, where they feed in shoots for the next 9–10 months and become sexually mature (Lu *et al*., 2014). Each adult usually feeds inside and kills 3–5 shoots (Lu, 2011). Extensive shoot feeding by Tomicus adults can cause growth loss and lower a tree's natural resistance to the point where individual trees are predisposed to infestation (i.e., breeding attacks) during the beetles' reproductive phase (Lieutier *et al*., 2003; Lu *et al*., 2014).

Sexually mature adult beetles eventually depart the shoots to reproduce in the trunks of *P. yunnanensis* trees, primarily in April and May (Lu *et al*., 2014). It is not known what initiates the reproductive phase in *T. brevipilosus*. Adult females initiate attack by constructing individual longitudinal galleries in the phloem tissue, and depositing eggs in niches along the gallery walls. After hatching, larvae feed transversely in individual galleries in the phloem and outer sapwood and then pupate at the ends of the galleries. Larval feeding disrupts nutrient flow within the phloem tissue (Långström, 1983; Fernández *et al*., 1999; Ye and Ding, 1999; Långström *et al*., 2002).

Gallery construction and reproduction by *T. brevipilosus* and at times other Tomicus species are regarded as the direct causes for *P. yunnanensis* tree mortality, particularly when mass attack occurs on individual trees (Ye and Ding, 1999; Lu *et al*., 2014). As one aspect of developing a *T. brevipilosis* management program, the objective of our study was to clarify the timing and duration of the beetle's reproductive period and within-tree attack pattern, as well as other aspects of the beetle's reproductive biology. A further aim of the present study was to improve our understanding of the reproductive ecology of *T. brevipilosus* with respect to other Tomicus species that infest *P. yunnanensis* in southwestern China.

Materials and Methods

Study area

The field study was carried out (primarily by PC, JL, and HY) in a *P. yunnanensis* stand in Anning County (24.97°N, 102.33°E, 1 800 m a.s.l), approximately 80 km west of Kunming, the capital of Yunnan, in southwestern China. The stand was along the lower slope of a hillside and covered about 300 ha. The *P. yunnanensis* trees in this stand were originally planted by aerial seeding in the mid-1970s. Most of the pine trees were 24~28cm in diameter and 7.5~8.5 m tall at the time of this field study (2007~2009). Low numbers of *T. yunnanensis* and *T. minor* were first recorded infesting trees in this stand in the early 1990s, followed by *T. brevipilosus* within a few years (Duan et al., 2004). All of these three Tomicus species have been active in this stand since the late 1990s (Duan et al., 2004; Lu et al., 2014). It is important to note that in Yunnan, *T. yunnanensis* was considered as *T. piniperda* due to their morphological similarities before 2008 (e.g., Ye 1991; 1994a; 1994b; 1995; Lieutier et al., 2003; Långström et al., 2002; Ye and Lieutier, 1997; Ye and Ding, 1999; Duan et al., 2004), when *T. yunnanensis* was first described as a new Tomicus species (Kirkendall et al., 2008).

Experimental procedures

During January to July in each year from 2007 to 2009, 3~4 *Pinus yunnanensis* trees with evidence of recent bark beetle breeding attacks were sampled every 10 days, for a total of 10 trees per month. Most of the sampled trees were cut from the interior portion of the stand and were among the larger trees present. We selected trees that appeared to have been recently infested based on the texture and color of the frass and resin that was present near the entry hole where the Tomicus beetles entered the bark along the trunk surface (Lu, 2011). The sampled trees were felled by cutting the trunk near the ground with a chainsaw, and then we cut the trunk into 50-cm-long logs, starting at the base and stopping near mid-crown. The logs were marked in the field to identify from which tree and which part of the trunk they were cut, and then transported to the laboratory. In the laboratory, usually within 1~2 days after cutting, we measured the surface area of each log and then carefully removed the outer bark to look for bark beetles and their galleries. We recorded several parameters related to beetle reproduction and development, including length and width of the maternal gallery using a vernier caliper (HMCT 6202-01, Harbin, China), number of maternal galleries for each Tomicus species, and the number of eggs, larvae and at times pupae for each maternal gallery. All parent bark beetles were collected from the maternal galleries, placed individually into labelled plastic bags, and identified to species under a stereo-microscope (Nikon-smz 500) based on morphological characteristics of the various Tomicus species (Kirkendall et al., 2008; Li et al., 2010). We recorded our observations for each Tomicus species on a per log basis, using data from all the bark beetles galleries that were present, regardless of their age.

In addition, detailed observations were made on 106 *T. brevipilosus* maternal galleries (in which the parent female was still present) at 10-day intervals from March to July 2009, from which the timing of first occurrence of eggs, larvae and pupae were recorded. These parameters were used to estimate the duration of egg, larval and pupal development in the field. At the same time, several other maternal gallery parameters were measured and recorded including gallery length, gallery width, and numbers of eggs or larvae present (for details see Lu 2011).

Data analysis

Correlation analysis was conducted between the length of the maternal egg gallery and the number of eggs present in the galleries constructed by *T. brevipilosus*. Statistical analyses were performed using SPSS (version 13 for Windows) with an alpha level of 0.05. Temperature data was obtained from Yunnan Meteorological Bureau, which had a weather recording station about 8 km from our field site.

Results

Breeding attacks

Tomicus brevipilosus breeding attacks, including sister broods, on *P. yunnanensis* trees were initiated in early March and ended in early June, spanning a period of about 3 months (Fig. 1). The timing and duration of breeding attacks was similar over the 3-year study period from 2007 to 2009 (Fig. 1).

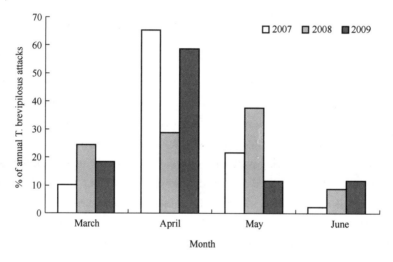

Figure 1 Percent of annual *Tomicus brevipilosus* breeding attacks that were initiated monthly on *Pinus yunnanensis* trees (n = 10 trees per month) that were sampled at the study site in Yunnan from January through July in 2007-2009.

The attacking population of *T. brevipilosus* parent females varied from month to month, based on the total number of maternal galleries found each year for each Tomicus species while debarking the logs, but formed an approximately normal distribution curve each year during the period from March to June (Fig. 1). Typically, breeding attacks in March accounted for about 10-25% of the annual attacks (10.25% in 2007, 24.44% in 2008, and 18.39% in 2009). From March to May, the number of attacks increased rapidly, peaking in April in 2007 and 2009, or in May in 2008 (Fig. 1). The peak month of breeding attacks in each year of study accounted for a relatively high percentage of the total annual attacks: 65.38% in 2007, 37.78% in 2008, and 58.62% in 2009. In June, when new Tomicus infestations ended, the June breeding attacks accounted for only 2.56% of all annual attacks in 2007, 8.89% in 2008, and 11.49% in 2009 (Fig. 1).

The daily minimum and maximum air temperature for the recording station about 8 km from our field site are shown in Fig. 2 for 1 January through 30 June during 2007~2009. Considering areas of the world where Tomicus species experience cold winters, spring flight is usually initiated when air temperatures exceed 10~12℃ (Bakke, 1968; Salonen, 1973; Långström, 1983; Haack et al., 2000; 2001; Haack and Poland, 2001; Ye et al., 2002; Lieutier et al., 2015). Considering a flight threshold of 12℃ at our study site, it is clear from the data presented in Fig. 2 that this temperature was exceeded on almost every day from January through June during 2007-2009. Even air temperatures of 15℃ were exceeded nearly every month (Fig. 2).

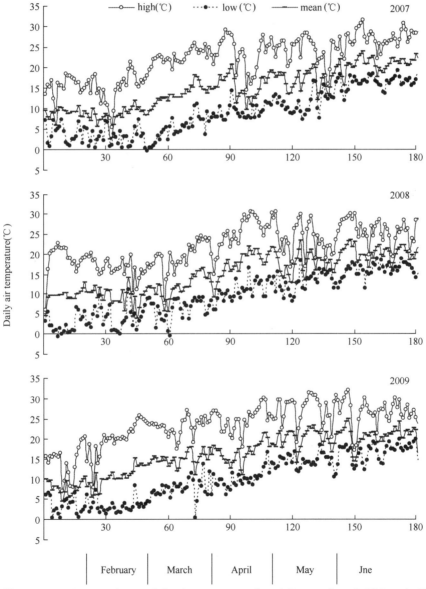

Figure 2 Maximum, minimum, and mean daily air temperatures from 1 January through 30 June in 2007-2009 in Anning County (24.97°N, 102.33°E, 1800 m a.s.l), that were recorded at an official weather station about 8 km from our study site.

T. brevipilosus breeding attacks in relation to other Tomicus species

The log dissections during this 3-year study indicated that breeding attacks by *T. brevipilosus* occurred in *P. yunnanensis* trees both with and without co-infestation by the other two Tomicus species that were present in the area (Figs. 3 and 4). For example, in March, when *T. brevipilosus* commenced breeding attack, there already existed several Yunnan pine trees that were infested by *T. yunnanensis* and *T. minor*, which had initiated breeding earlier in the year than *T. brevipilosus* (Lu et al. 2014). In this situation, *T. brevipilosus* only attacked trees that were already infested by *T. yunnanensis* and *T. minor* (Fig. 3). In April, as the number of trees newly infested by *T. yunnanensis* and *T. minor* decreased, *T. brevipilosus* continued to infest pine trees in which *T. yunnanensis* and *T. minor* already occurred as well as initiated attack by itself on other pine trees (Fig 3). During May and June, when there were no other pine trees with new *T. yunnanensis* or *T. minor* breeding attacks, *T. brevipilosus* attacked only by itself on previously uninfested pine trees (Fig. 3).

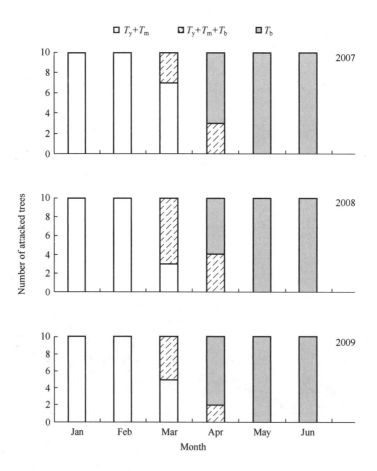

Figure 3 Number of *Pinus yunnanensis* trees with recent attacks by various combinations of three Tomicus species (*T. brevipilosus*=Tb, *T. minor*=Tm, and *T. yunnanensis*=Ty) based on sampling 10 trees per month in Yunnan that were cut from January to July, 2007-2009.

Within-tree distribution on breeding attacks

Tomicus brevipilosus was able to colonize the entire trunk of *P. yunnanensis* trees, usually starting about 50 cm above the ground level and reaching the mid-crown level (Fig. 3). However, the colonization pattern varied dramatically, depending if the other two Tomicus species were already present on the tree trunks or not. In those cases when both *T. yunnanensis* and *T. minor* were already present along the trunk at the time of *T. brevipilosus* infestation, which usually happened in March to April (Fig. 3), *T. brevipilosus* tended to colonize the open spaces of inner bark (phloem) where the other two Tomicus species and their brood had not already occupied. In most cases, *T. yunnanensis* appeared to be the first species to colonize the trees, occupying primarily the upper portions of the trunk, while *T. minor* tended to colonize trees already infested by *T. yunnanensis*, infesting primarily the lower portions of the trunk (Ye and Ding, 1999) (Fig. 4A). There was some overlap of both *T. yunnanensis* and *T. minor* in the mid–trunk region of most trees (Fig. 4A). Given this situation, the early-season *T. brevipilosus* attacks were scattered along the entire trunk between the gallery systems of the two other Tomicus species, but being concentrated mostly in the mid- and lower-trunk samples (Figs. 1, 4A). Therefore, during the early portion of the *T. brevipilosus* flight season, *T. brevipilosus* attacks tended to overlap mostly with *T. minor* galleries in the lower trunk and to a much lesser degree with *T. yunnanensis* galleries in upper trunk, and with both of these Tomicus species in the mid-trunk sections (Fig 4A).

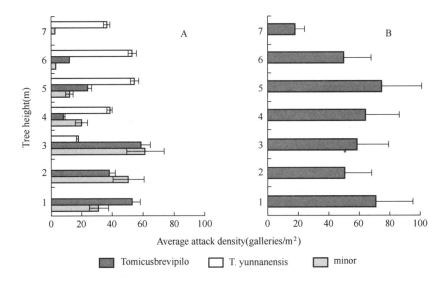

Figure 4 Average attack density of *Tomicus brevipilosus*, *T. minor* and *T. yunnanensis* at various heights along the main trunk of *Pinus yunnanensis* trees in Yunnan when sampled monthly from January through June 2007–2009. Figure 4A shows the vertical distribution for the three Tomicus species individually when all three species were present on the same trees (March and April). 4B shows the vertical distribution for *T. brevipilosus* when it was the only beetle present on the sampled trees (April–June). The data were pooled for all months and years.

In those cases where *T. brevipilosus* was the only *Tomicus* species present to initiate attack, which usually happened from mid-April to early June (Fig. 3), *T. brevipilosus* tended to initiate attack along the lower trunk first and then move progressively upward along the trunk (Fig. 4B). When the tree trunks were fully colonized by *T. brevipilosus*, the attack density was broadly similar over much of the trunk surface area (Fig. 4B).

Reproduction

Tomicus brevipilosus adult females initiated oviposition within a few days of starting gallery construction on the trunks of *P. yunnanensis* trees given that the first eggs were observed in early March. Overall, considering all *T. brevipilosus* breeding attacks observed in 2009, oviposition occurred from early March to early June (Table 1), about 3 to 3.5 months, peaking in April to May. The period of active oviposition closely matched the period of tree colonization (i.e., breeding attacks; Fig. 1).

Table 1 Typical life cycle of *Tomicus brevipilosus* in Anning County (24.97°N, 102.33°E, 1800 m a.s.l), Yunnan, China, based primarily on field work conducted in 2009

Life stage	Jan F M L	Feb F M L	Mar F M L	April F M L	May F M L	June F M L	July F M L	Aug F M L	Sept F M L	Oct F M L	Nov F M L	Dec F M L
Adult	+ + +	+ + +	+ + +	+ + +								
Egg			0 0 0	0 0 0	0 0 0	0						
Larva			− −	− − −	− − −	− − −						
Pupa					¤ ¤ ¤	¤ ¤ ¤	¤ ¤					
Adult						+	+ + +	+ + +	+ + +	+ + +	+ + +	+ + +

F: First 10-day period of month; M: Middle 10-day period; L: Last 10-day period.

In 2009, the first *T. brevipilosus* eggs were found in early March and the first larvae in mid-March, indicating that the egg incubation period was less than 2 weeks in the field during March. The first pupae were recorded in early May, with most pupating by mid-May, when average daily air temperature was about 18℃ (Fig. 2). Therefore, *T. brevipilosus* larvae appeared to require about 40~50 days to complete larval development. The pupation period lasted about 10-20 days at the study site. Callow adults of *T. brevipilosus* were first observed under the bark of the sampled trees in late May, with emergence of the new brood adults starting in early June and peaking during mid- to late June (Table 1).

The typical maternal gallery of *T. brevipilosus* is a single longitudinal tunnel with an entrance hole at the bark surface that leads to a slightly enlarged mating chamber and then a single gallery in which eggs are deposited along the gallery walls. The average length of apparently fully constructed *T. brevipilosus* maternal galleries was 7.2±2.3 (mean±SE, n = 106) cm long (14.0 cm maximum), and varied in width from 4.7 to 5.4 mm. The average number of eggs laid per gallery was 50.9±18.7 (mean±SE, n = 106), and ranged from 22 to 128 eggs. The number of the eggs per gallery was significantly and positively correlated with maternal gallery length (R^2 = 0.486, n=106,P<0.0001)(Fig. 5).

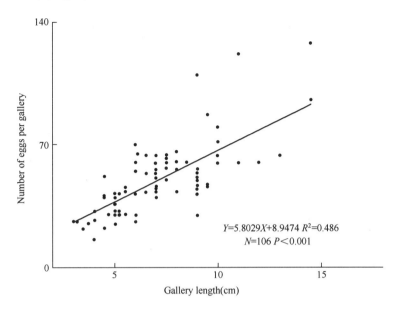

Figure 5 Linear relationship between number of eggs within a single *Tomicus brevipilosus* maternal gallery and the length of the corresponding maternal gallery.

Discussion

The presence of two *Tomicus* species co-occurring in the same pine forest has been commonly reported in many Eurasian countries (Bakke,1968;Långström,1983;Ye and Ding,1999), but the co-occurrence of three Tomicus species infesting the same pine trees has seldom been reported in the world (Lu *et al.*,2014). In Yunnan,in southwestern China,where three Tomicus species can coexist in the same *P. yunnanensis* forest,it was not known how these three Tomicus species would interact. That is,would they compete with each other for the same resource or would they reduce competition by partitioning their breeding resource either spatially or seasonally (Paine *et al.*, 1981;Flamm *et al.*,1987;Wilkinson and Haack,1987;Haack *et al.*,1989;Amezaga and Rodrı′guez,1998;Ayres *et al.*,2001). Obviously,answering this question will help in understanding the damage potential caused by the various Tomicus species in southwestern China.

The present study, along with the results presented by Lu *et al.* (2014), indicated that the three Tomicus species that coexist in *P. yunnanensis* forestsare in large part separated seasonally in their reproductive cycles,and when they do overlap seasonally,they partition themselves spatially along the main trunk. For example, Lu *et al.* (2014) noted that *T. yunnanensis* reproduces from early November to March of the next year; while *T. minor* started to reproduce about 1-2 weeks later,from late November to late March or early April, and usually only infested trees already under attack by *T. yunnanensis* (Ye and Ding,1999;Chen *et al.*,2010). By contrast,breeding duration of *T. brevipilosus* was primarily from early March to early June,thus overlapping with that of the other two Tomicus species by only about one month (Fig. 3). These differences in the peak breeding season for the above three Tomicus species in Yunnan likely reduce interspecific competition among these Tomicus species on the one hand, but also lengthen the total period of time that pine trees can be under attack by Tomicus beetles each year and thereby can result in more serious damage to Yunnan pine stands (Lu et al., 2014).

The initiation of Tomicus breeding attacks in northern latitudes that experience cold winters is closely related to maximum daily air temperatures. For example, in northern Europe and North America, *T. piniperda* typically initiates flight and breeding attacks when daily air temperatures reach 10–12℃, and T. minor when temperatures reach 12–14℃ (Bakke, 1968; Salonen, 1973; Långström, 1983; Schlyter and Löfqvist, 1990; Haack et al., 2000; 2001; Haack and Poland, 2001; Ye et al., 2002). In northeastern Italy, *Tomicus destruens* (Wollaston) initiates spring flight when daily mean temperatures exceed 12℃ (Faccoli et al., 2005), but in central Italy, T. destruens flies to breeding sites in October and November when the maximum daily temperatures are 18–22℃ and daily minimum temperatures are 7–14℃ (Sabbatini Peverieri et al., 2008).

By contrast, in Yunnan, little is known about the circumstances that trigger spring flight in the four local Tomicus species. In the case of T. *yunnanensis*, adults initiate breeding attacks in November apparently by simply departing the shoots and walking along the branches to the main trunk of the same tree on which they shoot fed (Liu et al., 2010). Later, usually beginning in January, *T. yunnanensis* adults will fly to new hosts after first experiencing periods of cold and then having mean maximum air temperatures once again reaching or exceeding 12℃ (Liu et al., 2010). The threshold temperature for spring flight of T. minor has not been studied in Yunnan, but it is likely slightly higher than that of T. *yunnanensis* given that T. minor initiates spring flight soon after T. *yunnanensis* (Lu et al., 2014).

The threshold temperature for spring flight of T. *brevipilosus* is not known, but it is likely considerably higher than that of T. *yunnanensis* and T. *minor* given that T. *brevipilosus* spring flight starts about two months later, in March (Lu et al., 2014; Fig 3). Perhaps initiation of T. *brevipilosus* spring flight is more closely linked to daily mean temperature rather than the daily maximum as reported for T. destruens in Italy (Faccoli et al., 2005). An examination of the temperature data in Fig. 2 shows that in March of each year the daily mean air temperature ranged over 11.8–19.6℃ in 2007, 8.7–18.5℃ in 2008, and 11.9–18.2℃ in 2009. If we consider a mean air temperature of 15℃ as an arbitrary threshold value for T. *brevipilosus* flight, then there were 15 days that could have supported flight in March and 16 in April 2007, 10 and 25 days in 2008, and 18 and 26 in 2009, respectively. Of course, other factors besides simply air temperature could influence initiation of spring flight in T. *brevipilosus*, such as day length, ovarian development, or some physiological change within the host tree that is detected by adults as they overwinter in the shoots.

As shown in Fig. 1, T. *brevipilosus* peak flight was apparently shifted one month later in 2008 as compared with 2007 and 2009. It is not clear if this shift is simply an artefact of the trees that were sampled in the different years, or if this apparent shift reflects some major temperature differences among the three sampling years. If the difference in the flight season is related to temperature, it is difficult to identify a major difference in the 2008 temperature data shown in Fig. 2 as compared with the data from 2007 and 2009. One difference, as mentioned above, is that there were fewer days in March 2008 when the mean daily temperature exceeded 15℃ as compared to March in 2007 or 2009. However, mean temperatures in April 2008 were broadly similar to April 2009 but warmer than April 2007. Such year to year variation in the timing of the initial spring flight of overwintering Tomicus adults is common and has been shown to be related to annual differences in spring temperatures (Bakke, 1968; Salonen, 1973; Långström, 1983; 1986; Haack and Lawrence, 1995; Haack et al. ,2000; Faccoli et al. ,2005).

Among the eight species of Tomicus worldwide, *T. piniperda* usually breeds in wind-blown trees, fire-damaged trees, stumps, and other severely stressed pine trees (Bakke, 1968; Ye, 1991; Haack et al. ,2000; Haack et al. ,2001; Kirkendall et al. ,2008). *Tomicus armandii* Li and Zhang may infest weakened Pinus armandii Franchet trees in Yunnan (Li et al. 2010). *Tomicus yunnanensis* is regarded as one of the most aggressive species of Tomicus, being able to attack and kill live, apparently healthy *P. yunnanensis* trees (Ye and Lieutier, 1997; Lu et al. , 2014). *Tomicus minor* is often considered to be more of an opportunist, usually infesting host trees that are already infested by other Tomicus species such as *T. piniperda* or *T. yunnanensis*, which could facilitate tree death (Bakke 1968; Långström, 1983; Eidmann, 1992; Ye and Ding, 1999; Lieutier et al. ,2003; Chen et al. ,2009). In the present study, we discovered that *T. brevipilosus* preferred to breed in trunks of *P. yunnanensis* trees that were already infested by both *T. yunnanensis* and *T. minor*. However, later in the season, *T. brevipilosus* was able to attack apparently healthy *P. yunnanensis* trees on its own, indicating that *T. brevipilosus* is more similar to *T. yunnanensis* in its aggressiveness as compared with *T. minor* (Chen et al. ,2009,2010).

The spatial distribution of *T. brevipilosus* breeding attacks in the trunks of *P. yunnanensis* trees was greatly influenced by the presence or absence of *T. yunnanensis* and *T. minor* breeding in the same tree (Fig4). In trees that were already infested by the other two Tomicus species, *T. brevipilosus* colonized areas of the trunk that were not already occupied, resulting in a scattered distribution of *T. brevipilosus* attacks along the trunks. The ability of *T. brevipilosus* to adjust its infestation pattern in response to other Tomicus species likely decreases interspecific competition among Tomicus species, as well as better utilizes the limited inner bark resources. In addition, it is possible that the early-season attack pattern of *T. brevipilosus* could enhance the reproductive success of *T. yunnanensis* and *T. minor* by attacking unoccupied areas of the tree trunk and thereby further reducing any residual host tree resistance. However, later in the flight season, when *T. yunnanensis* and *T. minor* brood are already well advanced, the arrival of *T. brevipilosus* would likely not help much in further reducing host vigor.

In those cases when *T. brevipilosus* initiates attack by itself, the first attacks are generally located along the lower trunk with subsequent attacks occurring throughout the entire trunk. The within-tree attack pattern varies among other Chinese Tomicus species as well (Ye and Ding 1999). In Yunnan, the first attacks by *T. yunnanensis* tend to be along the upper trunk, while

T. minor attacks are more concentrated along the lower trunk (Ye, 1995; Ye and Ding, 1999; Chen et al., 2010). By contrast, in northern Europe, Långström (1983) reported that *T. piniperda* tended to initiate attack along the lower trunk of local pines (mostly *Pinus sylvestris* L.), where the bark was thicker, while *T. minor*, which flies later, tended to colonize the upper portions of the trunk and branches where the bark is thinner. In the above two situations, it appears that *T. minor* is simply colonizing trees already infested by other Tomicus species but concentrating its attack on those portions of the trunk that are less occupied and thereby reducing interspecific competition. It is reasonable to assume that the attack pattern in any one world area is a reflection of the Tomicus and host-tree species present, as well as differences in beetle aggressiveness, threshold temperatures for flight, and adult body size (Kirkendall et al., 2008; Li et al., 2010; Lieutier et al., 2015). For the three Tomicus species infesting *P. yunnanensis* in Yunnan, the adult beetles are broadly similar in body size, with *T. yunnanensis* being slightly larger, and *T. brevipilosus* and *T. minor* being very similar (Kirkendall et al., 2008; Lieutier et al., 2015).

It is unclear why the attack pattern differs between *T. yunnanensis* (starting in the upper trunk) and *T. brevipilosus* (starting in the lower trunk) when each is the first bark beetle to infest the same species of pine, although such variation in attack pattern is well-recognized among bark beetles (Rudinsky 1962). Nevertheless, both attack patterns can result in tree death. This topic still requires further exploration.

The average length and number of eggs in *T. brevipilosus* maternal galleries reported in the present study are broadly similar to findings reported for other Tomicus species that make longitudinal galleries (all species except for *T. minor*). For example, in comparison to the 7.2-cm-long average *T. brevipilosus* maternal gallery in the present study, others have reported mean values of 4-8cm for *T. destruens* (Faccoli, 2007), 7.1-11.0 cm for *T. piniperda* (Långström and Hellqvist, 1985), and 9.6~10.6cm for *T. yunnanensis* (Ye and Ding, 1999). However, an inverse relationship has been reported between attack density and average length of maternal galleries in some Tomicus species (Salonen, 1973; Långström, 1984; Faccoli, 2009), but this relationship has not yet been investigated in *T. brevipilosus*. Clearly much more research is needed to fully elucidate the life history of *T. brevipilosus* and its interactions with other Tomicus species in Yunnan.

Acknowledgements

The present study was funded by the Natural Science Foundation of China (31160095), Yunnan Provincial Natural Science Foundation (2013FA055), the opening project of the Life Sciences College of Yunnan University, Yunnan Key Laboratory of International Rivers and Transboundary Eco-Security, and Key Laboratory for Animal Genetic Diversity and Evolution of Higher Education in Yunnan Province. We thank Toby R. Petrice for his valuable comments on the preliminary manuscript.

References

Amezaga I, Rodrı′guez MÁ. 1998. Resource partitioning of four sympatric bark beetles depending on swarming dates and tree species. Forest Ecology and Management 109: 127~135.

Ayres BD, Ayres MP, Abrahamson MD, Teale SA. 2001. Resource partitioning and overlap in three sympatric species of Ips bark beetles (Coleoptera: Scolytidae). Oecologia 128: 443~453.

Bakke A. 1968. Ecological studies on bark beetles (Coleoptera: Scolytidae) associated with Scots pine (Pinus sylvestris L.) in Norway, with particular reference to the influence of temperature. Meddelelser fra det Norske Skogforsøksvesen 21,441~602.

Chen P, Li L, Liu H. 2009. Interspecific competition between Tomicus yunnanensis and T. minor (Col. Scolytidae) during shoot – feeding period in Yunnan of China. Journal of West China Forestry Science 3:52~58

Chen P, Li L, Liu H. 2010. Host preference and competition in bark beetles, Tomicus yunnanensis and T. minor in breeding period. Journal of West China Forestry Science 1: 15~20.

Czokajlo D, Wink RA, Warren JC, Teale SA. 1997. Growth reduction of Scots pine, Pinus sylvestris, caused by the larger pine shoot beetle, Tomicus piniperda (Coleoptera, Scolytidae) in New York State. Canadian Journal of Forest Research 27:1394~1397.

Duan YH, Kerdelhue C, Ye H, Lieutier F. 2004. Genetic study of the forest pest Tomicus piniperda(Col., Scolytinae) in Yunnan province (China) compared to Europe: new insights for the systematics and evolution of the genus Tomicus. Heredity 93:416~422.

Eidmann H. 1992. Impact of bark beetles on forests and forestry in Sweden. Journal of Applied Entomology 114:193~200.

Faccoli M. 2007. Breeding performance and longevity ofTomicus destruens on Mediterranean and continental pine species. Entomologia Experimentalis et Applicata 123:263~269.

Faccoli M. 2009. Breeding performance ofTomicus destruens at different densities: the effect of intraspecific competition. Entomologia Experimentalis et Applicata 132:191~199.

Faccoli M, Battisti A, Masutti L. 2005. Phenology of Tomicus destruens (Wollaston) in northern Italian pine stands. In: Lieutier F, Ghaioule D (eds) Entomological Research in Mediterranean Forest Ecosystems, pp 185~193. Rabat (Morocco), 6~10 May 2002. INRA Editions, France.

Fernández M, Alonso J, Costas J. 1999. Shootfeeding and overwintering in the lesser pine shoot beetle Tomicus minor (Col., Scolytidae) in north west Spain. Journal of Applied Entomology 123:321~327.

Flamm RO, Wagner TL, Cook SP, Pulley PE, Coulson RN, McArdle TM. 1987. Host colonization by cohabitingDendroctonus frontalis, Ips avulsus, and I. calligraphus (Coleoptera: Scolytidae. Environmental Entomology 16: 390~399.

Haack RA and Lawrence RK. 1995. Spring flight of Tomicus piniperda in relation to native Michigan pine bark beetles and their associated predators. In: Hain FP, Payne TL, Raffa KF, Salom SM and Ravlin FW (eds) Proceedings: Behavior, Population Dynamics and Control of Forest Insects, pp 524~535. Maui, Hawaii, USA, 6~11 February 1994, Ohio State University Press, Columbus, Ohio

Haack A, Poland TM. 2001. Evolving management strategies for a recently discovered exotic forest pest: the pineshoot beetle, Tomicus piniperda (Coleoptera). Biological Invasions 3: 307~322.

Haack RA, Billings RF, Richter AM. 1989. Life history parameters of bark beetles (Coleoptera: Scolytidae) attacking West Indian pine in the Dominican Republic. The Florida Entomologist 72:591~603.

Haack RA, Lawrence RK, Heaton GC. 2000. Seasonal shoot-feeding by Tomicus piniperda (Coleoptera: Scolytidae) in Michigan. Great Lakes Entomologist 33:1~8.

Haack RA, Lawrence RK, Heaton GC. 2001. Tomicus piniperda (Coleoptera: Scolytidae) shoot-feeding characteristics and overwintering behavior in Scotch pine Christmas trees. Journal of Economic Entomology 94:

422~429.

Kirkendall LR, Faccoli M, Ye H. 2008. Description of the Yunnan shoot borer, Tomicus yunnanensis Kirkendall & Faccoli sp. n. (Curculionidae, Scolytinae), an unusually aggressive pine shoot beetle from southern China, with a key to the species of Tomicus. Zootaxa 1819:25~39.

Kohlmayr B, Riegler M, Wegensteiner R, Stauffer C. 2002. Morphological and genetic identification of the three pine pests of the genusTomicus (Coleoptera, Scolytidae) in Europe. Agricultural and Forest Entomology 4: 151~157.

Långström B. 1983. Life cycles and shoot feeding of the pine shoot beetles. Studia Forestalia Suecica 163: 1~29.

Långström B. 1984. Windthrown Scots pines as brood material forTomicus piniperda and T. minor. Silva Fennica 18:187~198. Långström B. 1986. Attack density and brood production of Tomicus piniperda in thinned Pinus sylvestris stems as related to felling date and latitude in Sweden, Scandinavian Journal of Forest Research 1: 351~357.

Långström B, Hellqvist C. 1985. Pinus contorta as a potential host for Tomicus piniperda L. and T. minor (Hart.)(Col., Scolytidae) in Sweden. Zeitschrift für Angewandte Entomologie 99:174~181.

Långström B, Li LS, Liu HP, Chen P, Li HR, Hellqvist C, Lieutier F. 2002. Shoot feeding ecology of Tomicus piniperda and T. minor (Col., Scolytidae) in southern China. Journal of Applied Entomology 126:333~342.

Li X, Zhang Z, Wang HB, Wu W, Cao P, Zhang PY. 2010. Tomicus armandii Li & Zhang (Curculionidae, Scolytinae), a new pine shoot borer from China. Zootaxa 2572:57~64.

Lieutier F, Ye H, Yart A. 2003. Shoot damage byTomicus sp. (Coleoptera: Scolytidae) and effect on Pinus yunnanensis resistance to subsequent reproductive attacks in the stem. Agricultural and Forest Entomology 5: 227~233.

Lieutier F, Långström B, Faccoli M. 2015. The genusTomicus. In: Vega FE, Hofstetter RW (eds), Bark Beetles: Biology and Ecology of Native and Invasive Species, in press. Elsevier, Amsterdam, Netherlands.

Liu H, Zhang Z, Ye H, Wang H, Clarke SR, Jun L. 2010. Response of Tomicus yunnanensis (Coleoptera: Scolytinae) to infested and uninfested Pinus yunnanensis bolts. Journal of Economic Entomology 103:95~100.

Lu J. 2011. On the occurrence, distribution and damage mechanisms of fourTomicus species in southwestern China. Doctoral Dissertation, School of Life Sciences, Yunnan University, Kunming, Yunnan, China.

Lu J, Zhao T, Ye H. 2014. The shoot-feeding ecology of three Tomicus species in Yunnan Province, southwestern China. Journal of Insect Science 14:37. Available online: http://www.insectscience.org/14.37

Murayama JJ. 1959. Description ofBlastophagus khasianus, new species (Coleoptera: Scolytidae), Bulletin of the Brooklyn Entomological Society 54: 75~76.

Paine TD, Birch MC, Svihra P. 1981. Niche breath and resource partitioning by four sympatric species of bark beetles (Coleoptera: Scolytidae). Oecologia 48: 1~6.

Rudinsky JA. 1962. Ecology of Scolytidae. Annual Review of Entomology 7:327~348.

Sabbatini Peverieri G, Faggi M, Marziali L, Tiberi R, 2008. Life cycle of Tomicus destruens in a pine forest of central Italy. Bulletin of Insectology 61:337~342.

Salonen K. 1973. On the life cycle, especially on the reproduction biology of Blastophagus piniperda L. (Col., Scolytidae). Acta Forestalia Fennica 127: 1~72.

Schlyter F, Löfqvist J. 1990. Colonization pattern in the pine shoot beetle, Tomicus piniperda: effects of host declination, structure and presence of conspecifics. Entomologia Experimentalis et Applicata 54:163~172.

Wilkinson RC, Haack RA. 1987. Within-tree distribution of pine bark beetles (Coleoptera: Scolytidae) in Honduras. Ceiba 28: 115~133.

Ye H. 1991. On the bionomy of Tomicus piniperda (L.) (Col., Scolytidae) in the Kunming region of China. Journal of Applied Entomology 112:366~369.

Ye H. 1994a. The distribution of Tomicus piniperda (L.) population in the crown of Yunnan pine during the shoot-feeding period. Acta Entomologica Sinica 3:21~26.

Ye H. 1994b. Influence of temperature on the experimental population on the pine shoot beetle, Tomicus piniperda L. (Col., Scolytidae), Journal Applied Entomology (117): 190~194.

Ye H. 1995. Preliminary observations on the trunk attacks by Tomicus piniperda L. on Yunnan pine in Kunming, China, Journal Applied Entomology (119): 1~33.

Ye H, Ding XS. 1999. Impacts of Tomicus minor on distribution and reproduction of Tomicus piniperda (Col., Scolytidae) on the trunk of the living Pinus yunnanensis trees. Journal of Applied Entomology 123:329~333.

Ye H, Lieutier F. 1997. Shoot aggregation by Tomicus piniperda L (Col: Scolytidae) in Yunnan, southwestern China. Annals of Forest Science 54:635~641.

Ye H, Haack RA, Petrice TR. 2002. Tomicus piniperda within and between tree movement when migrating to overwintering sites. The Great Lakes Entomologist 35:183~192.

中国西南地区松芽小蠹蛀害云南松的攻击模式及繁殖生态学

摘要：松芽小蠹 *Tomicus brevipilosus* 是中国西南部新近发现的一种危害云南松 *Pinus yunnanensis* 的害虫。然而，松芽小蠹蛀害云南松的繁殖生物学及蛀干攻击模式鲜有了解。为了探明松芽小蠹的繁殖生物学，分别研究松芽小蠹成虫单独及多种小蠹共存时，其对寄主云南松的繁殖攻击模式。结果表明，在云南安宁，松芽小蠹蛀干繁殖期通常始于3月上旬结束于6月上旬；其表现出两种蛀干繁殖模式：3月上旬至4月中旬，松芽小蠹更多是选择已经被云南切梢小蠹 *T. yunnanensis* 和横坑切梢小蠹 *T. minor* 危害树干进行蛀害，其蛀干繁殖区域是还没有被蛀害占用的空间（多在树干的中下部）；随后的4月中旬至6月上旬，此时林间已没有 *T. yunnanensis* 和 *T. minor* 蛀干危害的林木，松芽小蠹单独蛀干危害云南松，它首先攻击树干的下部，然后由下往上蛀害直至树冠层。松芽小蠹在中国西南部对云南松的蛀干繁殖延长了蠹害蛀干危害期，多种小蠹（3种）的联合持续攻击加重了其对寄主的伤害，这也解释了多年来中国西南部云南松林蠹害猖獗的成因。

[本文发表于《Journal of Insect Science》,2015,15(1)]

第六篇 经济林研究

板栗新品系的生物学特性

陆斌,邵则夏,杨卫明,宁德鲁,杜春花

(云南省林业科学院,云南 昆明 650201)

摘要:通过两轮优株无性系品比试验,筛选出 24 个板栗新品系在全省做区域化栽培试验,并进行物候期观测、授粉试验、发枝结果习性、修剪反应等试验研究,最终选育出云丰等 6 个板栗新品种。本试验结果为生产上选配主栽品种和授粉品种、幼树 1 年生壮枝修剪等提供了可靠的依据。

关键词:板栗新品系;授粉试验;花粉直感

Biological Characters of Chinese Chestnut New Lines

Lu Bin, Shao Zexia, YangWeiming, Ning Delu, Du Chunhua

(Yunnan Academy of Forestry, Kunming Yunnan 650201 China)

Abstract: Studies were carried out with 24 new lines of Chinese chestnut. After comparing the results of their phonological observation, pollination experiment, branching and bearing habits and pruning reaction, Yunfeng and the other 5 strains were finally selected. Trails proved they are suitable to be recommended for commercial growing in different regions of Yunnan province.

Key Words: Chinese chestnut; New chestnut line; Pollination test; Metaxenia

20 世纪 80 年代以来,我国板栗生产发展迅速,面积和产量均远远超过世界各国[1,2]。我国板栗研究多集中于良种选育和早期丰产栽培技术方面[2],对板栗授粉、发枝结果习性、修剪反应等生物学特性的研究很少报道。我们旨在通过对自己选育板栗优株无性系生物学特性的研究,为所选育的板栗新品种制定合理的栽培技术提供科学依据。

1 材料和方法

1.1 材料来源

1991~1992 年,在云南板栗种质资源调查中,根据丰产、优质、抗性及不同成熟期等指标,选择了 122 个优树,经比较筛选,确定了 43 个优树。1993 年春,分别采集接穗,高接于峨山县森警队 25 年生的实生栗园,为第 1 轮优株无性系测定圃。1995 年春又将这些优系嫁接于峨山舍郎 2 年生 2hm² 实生栗园,为第 2 轮优株无性系测定圃。经多年对优株无性系测产考种和其他指标综合评定,选择出 24 个新品系在全省不同地区和不同海拔高度做新品系区域化试验及生物学特性研究。

1.2 试验地基本情况

第1轮和第2轮优株无性系测定圃位于峨山县城附近,约24°11′N,102°24′E,年均温15.9℃,1月份均温8.4℃,7月份均温21.1℃,绝对最高气温32.6℃,绝对最低气温-5℃,年均≥10℃活动积温5 084.1℃。年均降水量986.5℃,降水主要集中于5~9月。年均蒸发量1 708.1。全年平均日照时数2 272.5h。试验地海拔1 650~1 670m,为缓坡台地,土壤为微酸性红壤,质地偏粘,极少部分为沙砾土,肥力偏低。新品系发枝结果习性、修剪反应试验设置在石林县板栗新品系比较试验园,约24°44′N,103°16′E,海拔1 600m,为缓坡耕作红壤,土质为沙壤,较肥沃。为实生栽培7年生栗园,平均树高2.5~3m,冠幅1.5m×1.5m~2m×2m,极少结果,1997年3月上旬嫁接改造为24个板栗新品系品种比较试验园。

1.3 研究方法

1.3.1 物候期观测

观测地点为峨山县森警队板栗第1轮优株无性系测定圃。观测时间1996~1998年。并按记载标准进行记载[3]。

1.3.2 授粉试验

1997~1998年在峨山县森警队第1轮优株无性系测定圃进行。供试材料为云栗1、6、8、9、15、16、17、22、24、26、27、29、30、32、33、38、42、44号共18个优株无性系(1号仅作授粉树)。试验方法:授粉试验前采集花粉,将已出现的雌花簇柔荑花序结果枝进行去雄套袋。雌花套袋前,先将该果枝上的雄花序以及双性花上雄花摘除,待雌花开放时进行授粉,授粉时间上午9:00~11:00。试验内容:17个优株无性系的自花授粉试验、异花授粉和雌花隔离孤雌生殖试验。

1.3.3 发枝结果习性

试验地设置在石林县新品系区试园,树龄9年,嫁接年龄3年。
试验处理:选择生长发育正常植株,对24个新品系结果母枝的发枝数量及结果情况调查统计。

1.3.4 修剪反应观察

试验对24个板栗新品系1年生粗壮结果母枝进行修剪反应试验,采用4种处理:①轻剪,剪掉枝条长度的1/3;②中剪,剪掉枝条长度的1/2;③重剪,剪掉枝条长度的2/3;④对照,不剪。每处理设置3个重复,处理时间为1999年1月28~29日。停止生长后调查处理结果。

1.3.5 测产考种

1994~2000年,对峨山县森警队第1轮优株无性系鉴定圃、舍郎第2轮优株无性系鉴定圃以及各区域化品比试验园进行测产和考种。于果实成熟前7月上中旬,测定每供试植株结实总苞数,测定株数3~10株。待果实成熟时进行考种,以考种资料和单株平均结实总苞数计算产量。

2 结果与分析

2.1 物候期及成熟期划分

萌芽期一般在3月1~15日,萌芽后3~5d开始展叶;3月26~30日,新发幼叶先端开始抽新梢,生长迅速;3月25~4月30日雄花序不断地出现和伸长;5月1~5日,当结果新梢出现第7~11个雄花序时,上部1~4个花序基部出现雌花簇幼体,新梢下部柔荑花序雄花部分花丝直立,花药开裂进入雄花初期;5月6~20日为雄花盛期,此时也是雌花盛花期,雌雄花期一致,自然授粉条件基本相符。5月21~25日,雄花末花期,雌花柱头大部分反卷授粉结束;6月10日后授粉的果实开始膨大,总苞直径3cm左右时,未受精的雌花簇枯萎脱落。8月中下旬大多数新品系总苞由青绿色变为黄褐色,部分总苞开裂,进入果实成熟期,11月底~12月中旬多数新品系落叶,进入休眠期。

根据国内有关板栗新品种记载方法与标准[3]:24个新品系可划分为3个类型:早熟栗,7月底至8月中旬成熟,如云栗17号;中熟栗,8月20~31日成熟,如云栗4、5、6、8、15、16、22、23、25、30、32、33、37、38、41、42、44号;晚熟栗,9月上旬以后成熟,如云栗9、24、26、27、31、34号。

2.2 授粉试验

17个优株无性系自花授粉率平均为37.6%,多数优株无性系在25%~38%之间,自花授粉率最高的为云栗17号,达68.8%,自花授粉率在45%以上有云栗15、32、33号;在隔离授粉试验中,云栗22、24、29、30号4个优株无性系单性结苞率分别为14.3%、11.1%、13.3%和15.6%,其总苞有所发育,壳加厚,坚果成秕籽,不能发育成胚。其余无性系总苞均枯黄脱落;各优株无性系异花授粉的授粉率平均为84.1%,较自花授粉提高了46.5%。各优株无性系间异花授粉率变幅平均为78.1%~90.9%,变异系数平均11.4%,变幅7.5%~20.6%(表1)。

表1 板栗新品系自花传粉结果统计表
Table 1 Self-pollination statistics of Chinese chestnut new vatieties

新品系母株号	授粉枝	雌花株	结果结苞率	授粉结苞率	空苞数	成实率
6	5	11	7	63.6	3	35.4
8	5	13	7	53.8	2	38.5
9	5	13	5	38.5	1	30.7
15	5	15	9	60	2	46.7
16	3	16	7	43.8	3	25
17	3	16	13	81.3	2	68.8
22	4	12	5	41.7	1	33.3
24	4	15	7	46.7	2	33.3
26	3	10	5	50	2	30
27	5	14	9	64.3	4	35.7
29	3	11	5	45.5	2	27.3
30	5	13	7	53.8	3	30.1

(续)

新品系母株号	授粉枝	雌花株	结果结苞率	授粉结苞率	空苞数	成实率
32	4	13	8	61.5	2	46.2
33	3	11	7	63.6	2	45.5
38	3	11	6	54.5	2	36.4
42	5	14	9	64.3	5	28.6
44	3	13	6	46.2	1	38.5
平均	4	13	7.18	55.2	2.29	37.6

花粉直感现象明显。不同新品系间授粉,对坚果平均重有明显的影响。父本坚果大,用其给母本授粉,母本坚果相对变大;反之亦然。如用坚果大的云栗42号作父本,给8号授粉,坚果平均重为16.2g;用坚果小的24号作父本给8号授粉,坚果平均重11.1g,二者相差5.1g。云栗6、24、8、42号自花授粉的平均重分别为10.6、10.7、15.1、15.2g,用16个新品系作父本,分别给其授粉,其坚果平均重分别为11.64、12.13、13.31、13.56g。

2.3 发枝结果习性

在24个新品系中,结果母枝萌发健壮结果枝的数量,除云栗17、24、25、29和42号外,均超过了4个,结果枝占当年萌发芽总数50%以上,且连续结果性能好。最佳主栽品系和授粉品系组合见表2。

表2 优株无性系主栽系号和授粉系号适宜组合
Table 2 Plant and pollinating new lines of Chinese chestnut

主栽系号	授粉系号	雌花株	结果结苞率
6	1	15	33
8	1	15	33
9	1	15	38
15	8	16	38
16	8	22	26
17	1	33	42
22	8	15	42
24	8	9	26
26	1	9	16
27	8	15	22
30	1	8	15
32	15	33	42
33	1	15	42
38	8	15	42
42	15	22	38
44	1	15	42

2.4 结果母株修剪反应

板栗1年生健壮结果母枝不同修剪处理试验结果统计数据表明:1年生健壮结果母枝轻剪有利于发枝和坐果,每结果母枝平均发枝数为2.9,平均结苞总数为8,分别较对照高11.2%和37.9%;较中剪高20.8%和56.9%;较重剪高31.8%和2.81倍。由此可见,轻剪和不剪(缓放)有利于发枝和结果,中剪次之,重剪发枝数和结苞数最少。不同修剪强度的反应同品种(系)有密切关系,重剪结果好的有云栗4、25、30、32、37、42和44号。另外,据观察极重短剪(留基部5~6芽)结果好的有23号和44号。

2.5 新品系生长及结实情况

新品系生长结实情况受各区试点热量状况、土壤条件、管理水平和砧木状况制约。在海拔低、热量资源丰富、土壤肥力高、管理较细、嫁接时砧木粗壮的栗园,新品系生长量大,结苞数量多,反之则生长量小,结苞数量少。如石林等区试点,砧木5年生,嫁接后第2年结果,第3年平均株产4.95kg;而立地条件差,海拔高度在2 200m以上的丽江等试点,植株生长量小,结果差,有的系号不能成熟。

3 小结

(1)板栗新品系自花授粉的结实率低,平均为36.7%;孤雌生殖不能形成种仁;运用不同新品系间异花授粉平均结实率82.9%,较自花授粉高44.5%。

(2)板栗花粉直感现象明显,正确选择主栽品种,合理配置授粉品种,可提高产量和质量。

(3)板栗新品系母本有较强的遗传性。

(4)24个板栗新品系结果母枝萌发结果枝数量平均为5.8,除云栗17号等4个系号外均超过了4个,其中结果枝占当年萌发总枝条的50%以上,且连续结果性能好。

(5)对1年生结果母枝轻剪有利于发枝和结果,中剪次之,重剪最少。但不同品系间有差异,重剪结果好的有云栗4、25、32、37、42和44号,极重短剪结果好的有23号和44号。

(6)云丰6号、云腰9号、云富15号、云早22号、云良33号、云珍44号6个早实丰产的新品种可在生产上大面积推广应用。

参考文献:

[1] 章继华,何永进. 国内外板栗科学研究进展及其发展趋[J]. 世界林业研究,1999(2):8~12
[2] 柳鎏. 世界板栗业与21世纪我国发展板栗的思考[J]. 河北林果研究,1999(1):89~92.
[3] 邵则夏,杨卫明. 板栗良种选育与早实丰产栽培技术[M]. 昆明:云南大学出版社,2000.

(本文发表于《果树学报》,2004年)

青刺果生物学特性观察及人工栽培技术

范志远[1]，习学良[1]，欧阳和[2]，陈武[2]，廖永坚[1]，邹伟烈[1]

(1. 云南省林业科学院，云南 昆明 650204；
2. 丽江青刺果天然营养植物油有限公司，云南 丽江 674100)

摘要：在对野生木本油料植物青刺果(Prinsepiautilis Royle)生态学、植物学、生物学特性调查观察基础上，试验初步形成了青刺果人工栽培采种、育苗、造林、早实丰产管理技术，为建立青刺果人工原料基地提供了技术依据。

关键词：青刺果；生物学特性；人工栽培技术

Botanical Characteristics and Cultivation Technology of Prinsepiautilis

Fang Zhi-yuan[1], Xi Xue-liang[1], Ouyang He[2], Chen Wu[2], Liao Yong-jian[1], Zou Wei-lie[1]

(1. Yunnan Academy of Forestry, Kunming Yunnan 650204, P. R. China; 2. Natural Vegetable Oil Company Ltd. of Lijiang, Lijiang Yunnan 674100, P. R. China)

Abstract: The botani calcharacteristics of *Prinsepia utilis* were described from morphological characteristics, habit of growth, distribution, habitatand seednatural variation. The cultivation technology such as seed collec-tion, seedling raising, planting, and the methods for promoting the early fruiting and high yield of young forest was introduced as well.

Keywords: *Prinsepia utilis*; botanical characteristics; cultivation technology

青刺果(*Prinsepia utilis* Royle)，别名青刺尖、扁核木、打油果等，为蔷薇科扁核木属的一种常绿灌木，自然生长于云南、四川、贵州等省的冷凉山地。云南省丽江、大理的纳西族、白族和摩梭人利用野生青刺果作为食用油、化妆护肤品，历史悠久，保健护肤效果奇特。1998年12月，云南省成立了丽江青刺果天然营养植物油有限公司。该公司开发出青刺果高级食用油、护肤品等十来个产品投放市场，供不应求。针对野生资源远不能满足加工需求的严峻局面，公司就着手原料基地建设。结合公司原料基地建设，我们开展了野生青刺果分布特征调查，青刺果植物学、生物学特性观察，实生变异调查，以及繁育技术、造林技术、早实丰产栽培管理技术试验。现做如下总结，以便及时为青刺果人工栽培提供技术支撑。

1 青刺果分布特征及其对生态环境的基本要求

1.1 分布基本特征

调查表明：从地域来看，青刺果广为分布在云南省丽江、迪庆、怒江、保山、大理、楚雄、昆明、昭通、玉溪、红河、曲靖等地州市，以及与之毗邻的四川省、贵州省的部分区域，但重点分布在丽江、大理、昭通、曲靖、迪庆、楚雄、怒江、昆明等地的冷凉区域。从海拔高度来看，在我省海拔 1 800~3 200m 的区域内均可见分布,但又集中分布在 2 100~2 800m 海拔地段。从伴生树种看,青刺果分布较多地区森林,多生长槲树林、云南松、华山松、旱冬瓜、麻栎、栓皮栎、元江栲、高山栲、滇青冈、滇毛青冈、滇石栎、滇油杉等树种。从小生境看,青刺果喜生长在森林植被较好、山涧开阔土壤深厚潮湿的沟边、溪边。

1.2 对生态环境的基本要求

依据青刺果自然分布特征及生物学特性,参照青刺果主要分布区气候指标,初步提出青刺果对生态环境基本要求如下:气候上,要求年均温 10~15℃,年降水量 800mm 以上,年日照时数 1 900 小时以上。该树种有耐寒、耐旱、耐涝、并抗风等抗性,但作为经济栽培,仍要求土壤深厚肥沃(厚度 1m 以上)潮湿。

2 青刺果植物学、生物学特性

2.1 观察地点

观察点设在丽江县古城区龙山乡罗足村,海拔 2 600m,有青刺果野生及 1~6 年生人工栽培林。

2.2 植物学特性

(1)树冠:自然树冠为多主干丛生状,呈扁圆形,一般冠高 2~3m,冠幅 5~12m^2,少数冠高 4~7m,冠幅 30~50m^2。

(2)树干和枝:多年生主干呈灰褐色,浅纵裂,主干上有多个主枝,各主枝又密生多级侧枝;1~2 年枝呈灰绿色,扁圆,密布腋生针状硬枝刺,枝刺长 1~8cm,长枝刺上长有正常叶片。

(3)叶片:叶片矩圆状卵形、矩圆形或长椭圆形;单叶互生或丛生,叶柄长 0.5~1.0cm,无毛,托叶细小,宿存或早落;叶片长 3.0~6.5cm,宽 1.0~3.0cm,先端渐尖或短尖,基部钝,宽楔形或近圆形,边缘有细锯齿或全缘,两面均光滑无毛;叶草质、亚草质或厚纸质,羽状网脉,上面颜色较深,下面颜色较浅。

(4)花:冬末春初开花;总状花序(无限花序),腋生,有花 3~8 朵,花梗长约 0.6cm,花直径 0.8~1.0cm;萼筒杯状,无毛,顶端 5 裂(花萼 5 片),裂片三角状卵形,全缘或有浅齿,宿萼,花后反折;花瓣 5 枚,白色,阔倒卵形或矩圆状倒卵形。雄蕊多数、多列。雌蕊一心皮(极少数二心皮),子房一室间或二室,子房上位(下位花),雌雄同株,雌雄同花(两性花);花分批开放,下部先开,渐及上部,边缘的花先开,渐及中央。

(5)果:果为核果,椭圆形,初绿色,成熟后暗紫红色,有粉霜,基部有花后膨大的萼片。果实 4~5 月成熟。果实成熟跟海拔和气温有很大关系,一般低海拔和气温偏高的区域成熟早,随海拔和气温的逐渐升高,成熟期逐渐推迟。果实成熟后自然脱落。

(6)根:主根不明显,侧须根发达。土层 4~15cm 范围根系占总根量的 90%,根展范围为树冠幅的 1.5 倍左右。

2.3 生物学特性

(1)生长结果习性:树势强。3~4 年生实生树,在较好肥水条件下,1 年生枝平均长 80cm,离枝基 5cm 处平均直径 0.7~0.8cm,平均节间长 1.57cm;中果枝占 20%,长果枝 30%,徒长性结果枝 50% 左右;实生树 2~3 年生开始结果,6~8 年进入盛果期。

(2)物候期:9 月初陆续显蕾,1 月上旬初花,2 月初盛花,3 月中旬终花,花期长达 70 天。2 月下旬萌芽,3 月中旬展叶,新梢 4 月初开始生长,5~8 月为速生期,9 月后陆续停止生长,12 月 2 年生以上叶落叶。2~3 月陆续坐果,随即进入速生期,4 月初果核变硬,4 月中旬日果实退绿,4 月底至 5 月上旬陆续成熟。果实从终花到成熟,发育天数为 60 天左右。

3 青刺果自然变异

千百年来,在复杂多变的环境下,青刺果生长繁衍,产生多样性变异,形成了今天这样一个稳定而又庞大的变异群体。我们初步考察了经济价值最高的种子在云南省不同区域及丽江市不同区域的变异现象。

表 1 丽江市不同区域青刺果种子变异表明:在环境条件和青刺果自身遗传特性共同作用下,青刺果种子质量(大小、出仁率、饱满程度等)产生了多样性变异,为青刺果优良单株无性系品种选择提供了广阔空间。

表 1 丽江市不同区域青刺果种子变异

种子来源	海拔(m)	千粒重(g)	千粒仁重(g)	粒(kg)	出仁率(%)	饱满情况			
						饱满比例(%)	较饱比例(%)	不饱比例(%)	空粒(%)
玉龙县九河乡	2 430	136	108	7 353	79.79	30	66	3	1
玉龙县巨甸镇	1 900	154	120	6 493	77.92	26	63	10	1
玉龙县白沙乡	2 400	131	111	7 634	84.73	35	60	5	
玉龙县鸣音乡	3 000	162	129	6 173	79.63	20	70	10	
玉龙县大具乡	2 600	168	135	5 952	80.38	25	70	5	
宁蒗县蝉战河乡	2 600	128	103	7 812	80.86	20	70	10	
华坪县通达乡	2 200	156	125	6 410	80.19	20	75	5	
永胜县顺州乡	2 400	144	100	6 944	70.00	15	71	10	4
华坪县通达乡	2 200	136	101	7 353	73.97	15	80	4	1

在收购上来的种子中分区域混合抽样检测种子质量得表 2。从表 2 可知:云南省不同区域种子质量同样存在较大变异,在一定程度上表明青刺果在长期的自然演变过程中,适应不同类型气候环境形成了多样性地理生态型或地理种源。有必要筛选优良种源提供给生产发展。

表 2 云南省不同区域青刺果种子变异

产地名	平均千粒重(g)	平均千粒仁重(g)	平均出仁率(%)	平均含油率(%)
丽江玉龙	154	120	77.9	36
大理云龙	132	110	83.3	34
怒江兰坪	155	128	82.6	32

(续)

产地名	平均千粒重(g)	平均千粒仁重(g)	平均出仁率(%)	平均含油率(%)
楚雄大姚县	144	114	79.2	29
曲靖会泽县	136	102	75.0	32
昭通镇雄县	129	96	74.4	28

4 青刺果人工栽培技术

4.1 采种育苗技术

4.1.1 采种育苗技术

(1)苗圃地选择基本要求:①背风向阳,霜冻轻;②土壤为肥沃的沙质壤土;③排灌水方便,农家肥充足;④交通方便。根据上述要求,我们选择丽江市玉龙县拉市乡海东村作为试验苗圃,面积15亩。

(2)苗圃地、容器袋准备:选好苗圃地后,3~5月份,每亩施腐熟农家肥3 000~4 000kg深耕,经细致整地后作高床,床面宽1~1.2m,床间宽30cm,长度视地块而定,并规划设置了排水沟渠。用于培育裸根苗。4~5月份,配制育苗基质,基质基本组成:3份云南松林肥沃表土+1份腐殖土。用5cm×20cm,厚度1mm的无毒聚乙烯容器装袋,开宽1m低床(低于地面20cm)整齐排列。

(3)采种:采集充分成熟的青刺果肉果,洗净果肉,用于0.5%的高锰酸钾溶液浸泡60min,捞出后用清水洗净,晾干水分,去除瘪子,得种子。

(4)播种:采用开沟点播方法播种:开深5cm沟,沟(行)间距15cm,种子距离3~4cm左右;容器苗每袋播种2粒。播后盖2~3cm细土,浇足水后,再盖一层松针保湿。

(5)播后管理:以保持土壤湿润为度,视需要补浇水,并及时除草。苗出齐后,苗木速生期(出苗后1~5个月)叶面喷施5‰尿素3~4次,速生后期叶面喷施5‰磷酸二氢钾2~3次。雨季(7~9月),每15天叶面喷施300倍多菌灵或甲基托布津,防治叶斑病;每15天叶面喷施600倍敌杀死防治蚜虫。苗出齐后,以株距6~8cm左右及时除过密苗木。注意防止鼠害。

4.1.2 技术效果

表3列出了青刺果不同播种期出苗成苗情况。青刺果种子成熟期随采随播出苗率可达71%~75%,每亩可出苗4万株左右;随着储藏时间后移,出苗率显著下降,干藏或沙藏至来年2月播种,出苗率下降为0。干藏与沙藏对种子出苗率影响差异不明显。

表4青刺果不同播种期苗木生长情况。5月种子成熟时随采随播到当年雨季造林时(9月)苗木嫩小,不宜上山造林,但容器苗此时可造林;到翌年5月,裸根苗、容器苗平均高分别可达67.6cm、71.5cm,平均粗达0.78cm、0.79cm,大部分可用于雨季造林(7~9月)。采用特早熟种子当年3月播种,到当年5月份,容器苗可用于造林;到8月份,裸根苗也可用于造林。

表3 青刺果苗木不同播种期出苗成苗情况

苗木类型	播种时间	播种面积或播种量	出苗率(%) 干藏	出苗率(%) 湿藏	出苗率(%) 随采随播	成苗数
裸根苗	2000年3月	2亩、10kg/亩			71	8.21万株
裸根苗	2000年5月	2亩、10kg/亩			72	8.26万株
裸根苗	2000年8月	5kg	52	56		
裸根苗	2000年10月	5kg	25	26		
裸根苗	2000年12月	5kg	5	6		
裸根苗	2001年2月	5kg	0	0		
容器苗	2000年3月	1万袋、2粒/袋			67	0.91万袋
容器苗	2001年5月	1万袋、2粒/袋			75	0.98万袋

表4 青刺果不同播种期苗木生长情况

苗木类型	播种时间	2000年9月 平均高(cm)	2000年9月 平均粗(cm)	2000年12月 平均高(cm)	2000年12月 平均粗(cm)	2001年5月 平均高(cm)	2001年5月 平均粗(cm)	2001年8月 平均高(cm)	2001年8月 平均粗(cm)
裸根苗	2000年5月	23.4	0.34	47.6	0.52	67.7	0.78	97.2	1.21
裸根苗	2001年3月					20.4	0.31	46.1	0.53
容器苗	2000年5月	24.7	0.37	49.2	0.53	71.5	0.79	98.4	1.23
容器苗	2001年3月					21.2	0.31	47.2	0.53

4.2 造林技术

4.2.1 造林设计

(1)造林地选择:试验地设在永胜县顺州乡西场村青刺果公司基地,面积300亩。海拔2400m,土地较平坦肥沃,光照充足,肥料来源广泛,周围植被良好。

(2)造林设计:①整地:统一为大穴整地,规格60cm×60cm×60cm。②造林密度:设计1m×4m、2m×4m、3m×4m、1m×5m、2m×5m、3m×5m两组株行距。③造林时间:设计雨季裸根苗5、6、7、8、9、10、11月分次造林。④造林苗木:设计同一规格裸根苗、容器苗对比。

4.2.2 造林技术

(1)造林密度选择

表5不同密度造林初步结果显示,青刺果栽植密度可考虑2m×4m、3m×4m、2m×5m、3m×5m。但从长远考虑,以3m×5m最适宜。考虑到早期丰产,前期可先按2m×4m、2m×5m栽植,株间过密后可隔株间除,改造成4m×4m、4m×5m。

表5 不同密度造林试验

栽植密度	2年生冠幅 株向×行向	3年生冠幅 株向×行向	4年生冠幅 株向×行向	初步评价
1m×4m	0.41m×0.42m	0.72m×0.92m	1.11m×1.52m	3年后株间过密。
2m×4m	0.42m×0.41m	0.82m×0.91m	1.32m×1.56m	前期(2~6年生)密度适宜。后期株间过密时可隔株间除,改造成4m×4m。

（续）

栽植密度	2年生冠幅 株向×行向	3年生冠幅 株向×行向	4年生冠幅 株向×行向	初步评价
3m×4m	0.42m×0.42m	0.91m×0.92m	1.32m×1.61m	适宜密度。但后期（7~10年后）可能行间过密不便于施肥及种子采摘。
1m×5m	0.43m×0.42m	0.91m×1.10m	1.31m×1.61m	3年后株间过密。
2m×5m	0.42m×0.43m	0.91m×1.10m	1.32m×1.61m	前期（2~6年生）密度适宜。后期株间过密时可隔株间除，改造成4m×5m。
3m×5m	0.42m×0.43m	0.92m×1.10m	1.32m×1.63m	适宜密度。

注：2000年7月造林，裸根苗，每种密度造林10亩。

(2) 雨季裸根苗造林时间选择

由于云南旱季降雨量较少，加上山区人工灌溉困难，不能较好满足造林后苗木对水的需求，因而选择雨季造林。表6雨季裸根苗不同造林时间试验表明：青刺果雨季造林5月、6月、9月、10月、11月成活率不太理想，而7月、8月造林成活率较高，故青刺果雨季裸根苗造林宜在7~8月间进行。

表6 雨季裸根苗不同造林时间试验

造林时间	造林面积	2002年5月成活率
2001年5月5~10日	8亩	43.5%
2001年6月5~10日	10亩	61.6%
2001年7月5~10日	13亩	87.7%
2001年8月5~10日	15亩	93.4%
2001年9月5~10日	10亩	71.5%
2001年10月5~10日	8亩	54.3%
2001年11月5~10日	8亩	23.3%

注：裸根苗造林，苗木规格：高60~70cm，粗（地径）0.7~0.9cm。

(3) 容器苗造林

表7容器苗造林试验表明：相对于裸根苗，容器苗造林有成活率高、无缓苗期、长势好等突出优点，宜采用容器苗造林。

表7 容器苗造林试验

苗木类型	2001年7月造林成活率	2002年12月新梢生长情况		2003年12月新梢生长情况	
		平均长（cm）	平均粗（cm）	平均长（cm）	平均粗（cm）
容器苗	99.6%	31.4	0.54	54.2	0.67
裸根苗	85.4%	14.1	0.31	31.1	0.42

注：苗木规格：高60~70cm，粗（地径）0.7~0.9cm。

4.3 促进实生幼林早实早丰技术

青刺果实生繁殖条件下，如何达到早实早丰是我们要解决的关键技术问题。参考核桃等油料干果树种促进早实早丰技术,我们开展了以下几方面技术探索与选择。

4.3.1 肥水综合管理技术选择

表8肥水综合管理表明:肥水对促进青刺果早结实早丰产非常关键。花前(12月)、采果前后(4~5月)各施有机肥一次(10kg·株$^{-1}$·次$^{-1}$),并在旱季(1~5月)灌水3~4次,可实现青刺果3年生全部植株挂果投产,3年生平均株产0.7kg,折算亩产56kg;实现4年生平均株产1.2kg,折算亩产96kg。

不施肥灌水,青刺果表现出结果慢、产量低等问题。同时,通过肥水综合管理,青刺果种子质量明显提高。

表8 肥水综合管理试验

肥水管理技术方案	挂果株率			平均株产(kg)		种子质量	
	2年生	3年生	4年生	3年生	4年生	千粒重(g)	千粒仁重(g)
花前(12月)、采果前后(4~5月)各施有机肥1次(10Kg/株次)	20%	60%	100%	0.4	0.7	146	118
花前(12月)、采果前后(4~5月)各施有机肥1次(10Kg/株次)。并在旱季(1~5月)灌水3~4次	40%	100%	100%	0.7	1.2	168	137
对照(不灌水施肥,仅松土除草)	0	15%	50%	0.02	0.2	129	104

注:2000年7月裸根苗造林,株行距2m×5m,试验面积30亩。

4.3.2 间作技术选择

表9间作结果初步显示:雨季间种绿肥(绿肥开沟覆盖)、间种低杆粮食作物,对实现青刺果实生幼林早结果、早丰产和提高种子质量作用明显。青刺果在放任管理条件下,表现出结果晚、产量低、种子质量偏差等问题。

表9 间作试验

间作方案	挂果株率			平均株产(kg)		种子质量	
	2年生	3年生	4年生	3年生	4年生	千粒重(g)	千粒仁重(g)
间作绿肥(6月播种苕子,9月开沟压青覆盖)	20%	80%	100%	0.2	0.6	141	114
间作马铃薯(对马铃薯施肥)	20%	70%	100%	0.2	0.5	143	114
对照1:不间种,但进行松土除草	10	15%	50%	0.01	0.2	134	107
对照2:不间种,不进行松土除草	0	10%	30%		0.1	129	104

注:2000年7月裸根苗造林,株行距2m×5m,试验面积20亩。

4.3.3 树盘旱季覆盖技术选择

从生物学特性看,12月~翌年5月是青刺果开花坐果期,良好的水肥条件是实现其丰产优质的根本保证。而此期又恰值云南旱季,大部分山区灌溉较为困难,如何保持青刺果林地水分显得十分重要。实践表明:旱季覆盖是保持土壤水分的有效途径。表10反映青刺果树盘旱季覆盖试验情况:与不覆盖比,覆盖能显著促进青刺果早实早丰和提高种子质量;在各种覆盖方案中,以地膜覆盖效果最佳,其次为松针、杂草、树木嫩枝叶等混杂覆盖与松土后提土覆盖。

表10 青刺果树盘旱季覆盖试验

覆盖方案	挂果株率(%)			平均株产(kg)			种子质量		
	2年生	3年生	4年生	2年生	3年生	4年生	千粒重(g)	千粒仁重(g)	空瘪率(%)
松针、杂草、树木嫩枝叶等混杂覆盖	15	40	90	0.01	0.11	0.31	139	113	5
地膜覆盖	25	60	100	0.02	0.22	0.42	144	117	2
松土后提土覆盖	10	30	80	0.01	0.12	0.33	141	114	5
不覆盖	0	15	55	0	0.01	0.15	131	106	10

注:2000年7月裸根苗造林,株行距2m×5m,试验面积25亩。

4.3.4 青刺果人工栽培种子与野生种子质量比较

采集野生青刺果种子,在相同生境条件下直播造林,3~4年后得人工栽培种子,其种子质量情况列表11。表11说明,野生青刺果通过人工栽培,由于水肥条件改善,其种子质量(大小、出仁率、含油率)稳步提高。

表11 人工栽培种子与野生种子质量比较

种子来源	千粒重(g)	千粒仁重(g)	出仁率(%)	含油率(%)
野生1号	134	107	79.85	34.3
人工栽培1号	146	118	80.82	34.9
野生2号	132	106	80.30	33.7
人工栽培2号	143	115	80.41	34.1

5 小结

(1)通过调查,弄清了野生青刺果在云南分布的地域、海拔、伴生树种特征,并据此提出青刺果对环境条件的基本要求,为确定青刺果人工栽培适宜区域提供了基本依据。通过定点观察,弄清了青刺果植物学、生物学特性,为制订青刺果人工栽培技术措施提供了基本依据。

(2)青刺果种子成熟期随采随播有较高出苗率(71%~75%),随着储藏时间后移,出苗率显著下降,干藏或沙藏至来年2月播种,出苗率下降为0。故青刺果育苗宜随采随播。

(3)青刺果栽植密度可考虑2m×4m、3m×4m、2m×5m、3m×5m。但从长远考虑,以3m×5m最适宜。为实现早期丰产,前期可先按2m×4m、2m×5m栽植,株间过密后可隔株间除,改造成4m×4m、4m×5m。裸根苗雨季造林宜在7~8月间进行。相对于裸根苗,容器苗造林有成活率高、无缓苗期、长势好等突出优点,宜采用容器苗造林。

(4)肥水、间种、覆盖等综合技术管理对促进青刺果早结实早丰产非常关键。花前(12

月)、采果前后(4~5月)各施有机肥一次(10kg·株$^{-1}$·次$^{-1}$),并在旱季(1~5月)灌水3~4次,可实现青刺果3年生全部植株挂果投产,3年生平均株产0.7kg,折算亩产56kg;实现4年生平均株产1.2kg,折算亩产96kg。不施肥灌水,青刺果表现出结果慢、产量低等问题。同时,通过肥水管理,青刺果种子质量明显提高。雨季间种绿肥(绿肥开沟覆盖)、间种低杆粮食作物,对实现青刺果实生幼林早结果、早丰产和提高种子质量作用明显。青刺果在放任管理条件下,表现出结果晚、产量低、种子质量偏差等问题,不宜提倡。与不覆盖比,覆盖能显著促进青刺果早实早丰和提高种子质量;在各种覆盖方案中,以地膜覆盖效果最佳,其次为松针、杂草、树木嫩枝叶等混杂覆盖与松土后提土覆盖。

(5)千百年来,在复杂多变的环境下,青刺果生长繁衍,产生多样性变异。下一步有必要筛选优良种源及无性系品种提供给生产发展。

参考文献:

郑万钧主编. 中国树木志(第二卷)[M]. 北京:中国林业出版社,1985:1164~1167.

(本文发表于《西部林业科学》,2005年)

云南榛树资源及其开发利用

宁德鲁,陆 斌,邵则夏,杨卫民,杜春花

(云南省林业科学院,云南 昆明 650204)

The Corylus Species in Yunnan and TheirUtilization

Ning De-lu,Liu Bin,Shao Ze-xia,Yang Wei-min,Du Chun-hua

(Yunnan Forestry Academy,Kunming Yunnan 650204,China)

  榛树为桦木科(Betulaceae)榛属(*Corylus* Linn.),全世界约20种榛属植物,分布于亚洲、欧洲及北美洲。我国是榛树的原产地之一,栽培和利用榛子已有悠久的历史。我国约有榛属植物9种,分布于东北、华北、西北及西南。云南产6种,即:藏刺榛(*Corylus thibetica* Batal.)、滇刺榛(*Corylus ferax* Wall.)、维西榛(*Corylus wangii* Hu)、滇榛(*Corylus yunnanensis* A. Camus)、喜马拉雅榛(*Corylus jacquemontii* Decne.)、华榛(*Corylus chinensis* Franch.)其中滇榛分布最为广泛,分布于昆明、楚雄、大理、曲靖、昭通、丽江、迪庆、文山、保山等地州市1 700~3 700m海拔高度的山坡灌丛中。云南省的榛树资源,尚处于野生或半野生状态,人工栽培园很少。榛树资源的开发利用,对于解决云南省高寒山区经济林树种单一、山区群众脱贫致富无疑是一个比较好的项目。

1 云南榛树的一般生物学特性

  榛属,灌木或乔木。芽卵圆形,具多数覆瓦状鳞片。单叶,互生,有锯齿或浅裂,叶脉羽状,第三脉与侧脉常垂直;托叶膜质,分离,早落。花单性;雌雄同株;雄花序为柔荑花序,常2~3年生于上年生枝的顶端或叶痕腋,下垂,苞鳞覆瓦状排列,每苞鳞内具2与苞鳞贴生的苞片及1雄花,雄花无花被,雄蕊4~8,插生于苞片之上,花丝短,分叉,花药2室,药室分离,顶端生簇毛;雌花序聚生呈短穗状,每苞鳞内2对生花,均包藏于总苞之内,仅2红色花柱伸出,每花基部具1枚2裂的苞片(果期发育成果苞),具花被,花被与子房贴生,顶端不规划小齿裂,子房下位,2室,每室1胚珠发育,另1胚珠常败育,花柱2袭,柱头钻形。果苞钟状或管状,顶端裂片有的种类硬化成针刺状。坚果近球形,大部或全部为果苞所包,外果皮木质。种子1,子叶肉质。

  (1)藏刺榛(*Corylus thibetica* Batal.)又称山板栗(昭通)、山榛子(东川)、大树榛子(维西)。小乔木,高约8m。树皮灰褐色,平滑;小枝褐色,无毛,具明显皮孔。叶厚纸质,宽卵形或倒宽卵形,长约15cm,宽约4cm,先端渐尖,基部圆或斜心形,叶缘具不规则的重锯齿,上面幼时疏被长柔毛,后脱落无毛,下面沿脉疏被长柔毛,余无毛,侧脉8~14对;叶柄长1.5~2.5cm,被长毛,近叶片基部有腺体。果3~6聚生,果苞棕色,分枝刺状裂片无毛,或仅基部疏被绒毛,针刺细密,几乎全部遮盖果苞。坚果近球形,长1~1.5cm,上部被短柔毛。

果期10~11月。产昭通、镇雄、大关、彝良、东川、禄劝、瀍、维西、鹤庆、丽江等地,生于海拔1 500~3 000m的山地林中。

(2)滇刺榛(*Corylus ferax* Wall.)乔木或小乔木,高达12m。树皮灰黑色或黑色;幼枝褐色,疏被长柔毛,基部较密,有时具或疏或密的刺状腺体。叶厚纸质,长圆形或倒卵状长圆形,稀宽倒卵形,长5~15cm,宽3~9cm,先端渐尖,基部近心形或圆形,有时两侧不对称,叶缘具刺毛状重锯齿,上面仅幼时疏被长柔毛,后脱落无毛,下面沿脉密被淡黄色长柔毛,脉腋间稀具髯毛,侧脉8~14对;叶柄长1~3.5cm,被或疏或密的长柔毛,稀无毛。雄花序1~5,排列成总状;苞鳞外面密被长柔毛,花药紫红色。果通常3~6聚生,稀单生;果苞钟状,外面密被绒毛,偶有刺状腺毛,上部分枝裂片成针刺状,针刺粗短,总梗较长,果苞明显可见。坚果球形,长1~1.5cm,顶端被短柔毛。果期10月。产维西、德钦、贡山、碧江、中甸、丽江、剑川、片马、镇雄、永善和禄劝等地,生于海拔2 000~3 200m的杂木林中。

(3)维西榛(*Gorylus wangii* Hu)小乔木,高约7m。幼枝褐色,被长柔毛及刺状腺体。叶厚纸质,长圆形或卵状长圆形,稀宽椭圆形,长5~10cm,宽3~6.5cm,先端渐尖或骤尖,基部心形或斜心形,叶缘具不规则的锐锯齿,两面疏被毛或无毛,下面沿脉被长柔毛,侧脉8~10对;叶柄长1~2cm,密被长柔毛或刺状腺体。果4~8聚生,序梗长约1cm,密被长柔毛,果苞钟状,木质,显露,背面具多数条肋和针状腺体,疏被短绒毛或无毛,上部裂片深裂,条形,密被黄色短柔毛及刺状腺体,分枝呈鹿角状的针刺。坚果卵圆形,两侧稍扁,无毛,长约1.3cm。果期11月。产维西等地,生于海拔3 000~3 400m的山间林地。

(4)滇榛(*Corylus yunnanensis* A. Camus)灌木或小乔木,高1~7m。树皮暗灰色。小枝褐色,密被黄色柔毛和被或疏或密的刺状腺体。叶厚纸质,近圆形或宽倒卵形,长5~10cm,宽4~8cm,尖端骤尖或短尾状,基部心形,叶缘具不规划的细锯齿,上面疏被短柔毛,幼时具刺状腺体;下面密被绒毛,幼时沿主脉下部生刺状腺体,侧脉5~7对;叶柄长7~12mm,密被绒毛,幼时密生刺状腺体,雄花序2~3排列成总状,下垂;苞鳞三角形,外面密被短柔毛。果单生或2~3聚生成极短的穗状;果苞厚纸质钟状,通常与果等长或稍长,外面密被黄色绒毛和刺状腺毛,上部浅裂,裂片三角形,边缘疏具数齿。坚果球形,长1.5~2cm,密被绒毛。花期2~3月,果期9月。产昆明、蒿明、安宁、大姚、楚雄、武定、元谋、富民、路南、师宗、文山、禄劝、巧家、彝良、镇雄、丽江、寻甸、中甸、维西、洱源、鹤庆、漾濞、大理、永平、腾冲等地,生于海拔1700~3700m的山坡灌丛中。

(5)喜马拉雅榛(*Corylus jacquemontii* Decne.)小乔木,高约7m。树皮灰黄色,幼枝紫黑色,密被长柔毛及刺状腺体,老枝具浅色皮孔,无毛。叶厚纸质,宽卵形或宽长圆形,长9~15cm,宽7~13cm,先端钝圆或骤尖,基部斜心形,两侧耳常相交或重叠,叶缘具不规则的齿牙状锯齿,幼时两面疏被长柔毛,下面较密,后脱落,仅沿脉被或疏或密的短柔毛,侧脉8~10对,基部的一对常向外具3支脉;叶柄长1~2cm,密被短柔毛或刺状腺体。果3~5聚生,序梗长1.5~2cm,被短柔毛及刺状腺体;果苞厚纸质,在果的顶端缢缩,形成不明显的短管,裂片深裂,条形,全缘,反折,长约2cm,宽约2mm。坚果球形,长约1.5cm。产维西等地,生于海拔2300~2800m的湿润山谷林中。

(6)华榛(*Corylus chinensis* Franch.)乔木,高可达20m。树皮灰褐色,纵裂;幼枝褐色,密被长柔毛及刺状腺体,稀无毛无腺体,通常基部具淡黄色长毛。叶厚纸质,宽椭圆形或宽卵形,长8~15cm,宽6~10cm,先端聚尖至短尾状,基部心形,两侧显著不对称,叶缘具不规划的钝锯齿,上面无毛,下面沿脉疏被淡黄色柔毛,有时具刺状腺体,侧脉 7~10 对;叶柄长 2~2.5cm,密被淡黄色长柔毛及刺状腺体。雄花序 2~8 排列总状,长 2~5cm;苞鳞三角形,钝尖,顶端具 1 易脱落的刺状腺体。果 2~6(10)聚生或短穗状,长 3~5cm,直径约 1.5cm;果苞管状,于果之上部溢缩,较果长 2 倍,外面具纵肋,疏被柔毛及刺状腺毛,稀无毛无腺体,上部裂片 3~5,深裂,呈镰状条形,通常裂片顶端又分叉成数小裂片。坚果球形,长 1~1.5cm,无毛。果期 10 月。产丽江、中甸、德钦、维西、鹤庆、大理、镇雄、嵩明等县,生于海拔 2 000~3 400m 的湿润山坡林中。

2 化学成分及利用价值

2.1 化学成分

据测定,每 100g 榛仁的营养成分及含量平均为:脂肪 58.03g、蛋白质 21.12g、碳水化合物 6.91g、灰分 2.4g、热量 665.11J、维生素 C33.9mg、维生素 A14.2mg、钙 307.0mg、磷 91.1mg、铁 7.85mg、钾 581.4mg,其营养价值相当于牛肉的 9 倍。据吉林农业大学测定,榛叶含水分 10.37%、蛋白质 15.9%、粗脂肪 4%、粗纤维 15.03%、无氮抽出物 46.08%、灰分 8.18%、钙 1.172%、磷 0.27%。据吉林大学和中国科学院林业土壤研究所分析,其种子油比重(20℃)0.912、折光率 1.471、碱化值 203.93~335.1、碘值 46.7~76.6、酸值 2.97~3.84、酯值 206.9~298.3。树皮和叶总苞含鞣质,其中叶含 5.95%~14.58%,并有黄酮反应。树皮、总苞和叶片均含单宁 8.5%~14.5%。

2.2 利用价值

2.2.1 食用

我国榛子的利用历史悠久,陕西半坡村原始人遗址发现大量已碳化的榛子果壳,说明距今 6000 年前我们的祖先就已采集榛子为食。而早在 3000 年前《诗经》中说有文字记载,如"山有榛""尸鸟鸠在桑,其子在榛"等。公元 5 世纪(北魏)的农书《齐民要术》记载了"榛……栽种与栗同"。每 100kg 榛实可出榛仁 30kg,榨油 15kg。榛子含油量约为大豆的 2~3 倍,榛油清亮,橙黄色,叶香,是高级食用油和高级钟表油。榛仁可炒食、加工制成各种巧克力、糖果、糕点等食品及榛子乳、榛子粉等高级营养品。出口 1t 榛仁可换回钢材 8t。

2.2.2 药用

榛仁可入药,据《开宝本草》记载:"榛仁性味甘、平、无毒,有调中、开胃、明目的功用。"临床配方:①治疗病后虚弱、食少无力、疲乏,用榛子 60g、山药 30g、党参 12g、陈皮 6g、水煎服。②取榛子若干,炒熟常食之,可以明目、开胃、健体、美皮肤。榛子花与玉米面、麦芽、神曲合剂对肝硬化腹水、肾炎等有一定疗效。美国俄勒冈州波特兰大学霍夫曼教授领导的研究小组经过长期研究后发现榛树本身能抵抗若干癌症细胞的生存,榛树的嫩枝、果实及树皮中也含有一种 PACLITAXE 的化学成分,是制造著名治癌药物紫杉醇的主要原料,而这种原料以往只能从红豆杉树皮中提炼出来,据报道,榛子中所含 PACLIT AXE 的成分仅为红豆杉的 10%。但由于榛子产量大,种植方便且生长迅速,科学家相信,从榛子可以提炼足够的 PACLIT AXE 原料,今后可望生产更廉价的治疗癌药物,对癌症患者是一大福音。榛树的雄花干粉有止血、消炎的作用。

2.2.3 其他用途

榛叶可养柞蚕,嫩叶采摘晒干贮藏后可做畜禽饲料。树皮、总苞和叶片可提取栲胶和生物碱。榛油还可用于制造化妆品、肥皂、蜡烛等。榛木坚硬致密,可制手杖和伞把,根条可作编织用品。此外,榛树还是优良的水土保持树种和较好的蜜源植物。果壳可作为制造活性炭的原料。英国纽卡斯尔大学的 Murat Dogru 在研究怎样处理榛子壳时,发现榛子壳中可以提取有用的氢气为燃料电池提供动力,而燃料电泡利用氢气可以产生电流驱动汽车。

3 资源的开发利用

3.1 开发利用现状

(1)争相采青,掠夺式经营鉴于野生资源缺乏合理的保护制度和措施,因此多数人都想"先下手为强",常出现争相抢收的局面。据辽宁省经济林研究所调查,每年野生榛林因过早进山掠青而造成产量损失达 28%,全省少收获榛果 70 万 kg。黑龙江省嫩江市对商品榛子抽样调查,由于提上抢采,榛果饱满仁率只占 30%,空壳率占 11%。

(2)生态条件恶劣,生产效率低下由于长时期的实生繁殖和自然变异,野生果树个体间良莠不齐现象非常突出,加之生境条件恶劣、无人管护、杂草杂木丛生,导致榛子的产量和质量低而不稳,商品率和经济效益普遍较低,而这种现状反过来又影响了人们管护和开发利用榛树的积极性。

(3)边远山区的榛树尚无人问津据果树工作者考察,长白山区有大面积的天然榛林,果实成熟期没人去采,大量坚果被鸟兽食用;生长在陕西秦岭山区的川榛,因当地群众没有食用习惯,尽管果实累累,但无人采集;在云南分布最多最广的滇榛,由于知道其坚果可食用的群众很少,每年大量的榛子被浪费。

3.2 开发利用前景

云南省野生榛树资源非常丰富,但目前多处于自生自灭状态,大面积人工栽培尚未实行,出口和内销量几乎是空白。榛子是世界四大干果(扁桃、核桃、榛子、腰果)之一,经济价值较高。榛子的含油率高于油料作物花生、大豆和芝麻等,是高级食用油。榛叶、果壳、树皮、果苞、榛木等都有不同的开发利用价值。可见开发榛树资源,对于食品加工业、制油工业及其他加工业都有促进和发展作用,对致富贫困山区,促进出口创汇等具有重要意义。

3.3 开发利用途径

(1)保护资源,科学开发保护利用野果资源是搞活商品经济的重要组成部分。因此,各级政府主管部门以及广大果树工作者,都应重视这一具有战略意义的工作。应逐步建立健全科技组织和科研队伍,认真开展对榛树野生资源经济价值、开发利用的研究,充分发挥其各有用组织和器官的利用价值,以提高资源的利用率和经济效益。

(2)以深细加工为龙头,带动产业化全方位开发榛子在营养含量、保健功效、天然风味和没有污染等方面占有明显优势,这一特点决定了榛树在"绿色"食品和营养保健食品开发方面的前途。榛子的开发应以精细加工适销对路的高附加值"绿色"保健食品为突破口,以高效益骨干加工企业为龙头,带动科研、生产、加工、供销的全面繁荣,实现规模化和产业开发。

(3) 野生榛林的垦复更新 要恢复和提高野生榛林的产量,就必须进行垦复和科学管理,垦复后一般可增产80%以上。垦复的方法是:于早春和7月上旬清除野生榛林中的柞、桦、胡枝子等非经营树种;选集中成片,分布比较均匀,榛树覆盖度在85%以上的榛林,于4月初沿等高线进行带状平茬,平茬带宽1m,保留带宽1m、1.5m或2m;榛林平茬后,用镐将平茬在带内的老茬全部刨除,并深翻30cm,整成水平沟,沟内土培在其上方的保留带上或留在原地;整地后于4月中旬在保留带上每平方米施尿素0.1kg。在上述各项工作中把榛丛内部的过密株、病虫害株、机械损伤株、生长不良的瘦弱株以及新的无用萌条除掉。水平沟整地和水平沟施肥的3年生榛树最佳密度为每亩9 338株,带状整地的最佳密度为每亩2 001株。

(4) 加快良种化和科学管理步伐,实现高产、优质、高效栽培 野生榛树分布面积大而零散,产果量低,采集时费时费工。应因地制宜选择果大、丰产、适口性强、营养价值高的优良单株或类型,对云南省的野生榛林进行改造,迅速实现规模化和良种化,同时通过清理树盘、合理间伐及修剪施肥等由简入繁,逐步实现管理科学化,大幅度提高榛树的单产、品质和商品率。目前,国内外榛属植物中表现最佳的是欧榛(*Corylus avellana* L.),它具有坚果大、营养丰富的特点,其榛仁含脂肪55%~77%、蛋白质18%~25%、碳水化合物6%,还含有维生素及微量元素。坚果平均直径达1.7~2.1cm,单粒重达2~4g,出仁率可达45%~60%,而且果皮薄(0.7~1mm),外观美,具有较高的商品价值。云南省林业科学院自1997年开始,先后从美国俄勒冈州引入6个品种,从安徽省林科所引入21个品系,通过近几年的试验研究,所引进品种(系)已初步表现出较好的适应性,我们用滇榛作砧木,欧榛作接穗,嫁接成活率平均可达89.5%,并且生长结实基本正常,可见,应用欧榛良种改造现有榛林,繁殖方法简单可行。2001年11月又从辽宁省经济林研究所引进了20个品系,通过"十五"的研究,可望推出适合云南发展的欧榛新品种2~3个。

参考文献:

[1] 西南林学院,云南省林业厅. 云南树木图志(中)[M]. 昆明:云南科学技术出版社,1990:394~401.
[2] 刘阵军. 中国野生果树[M]. 北京:中国农业出版社,1998:110~114.
[3] 李家福,高崇学,肖柏森. 野果开发与利用[M]. 北京:科学技术文献出版社,1989:110~115.
[4] 徐迎春,杨福兰,房义福,等. 榛子及其在山东的发展前景[J]. 山东林业科技,1997(3):16~19.

(本文发表于《经济林研究》,2002年)

10个引种核桃品种嫁接成活率及苗期生长性状研究

熊新武[1],李俊南[1],刘恒鹏[1],张群[2],陈伟[2],孟梦[2],冯弦[2],李江[2]

(1.云南省林业科学院漾濞核桃研究院,云南 漾濞 672500;2.云南省林业科学院,云南 昆明 650201)

摘要:以1a生铁核桃(*Juglans siggillata* L.)苗和1a生新疆核桃(*Juglans regia* L.)实生苗作砧木,试验研究10个国外引种核桃品种和2个本地核桃(对照)品种嫁接成活率的差异性和对1a生嫁接苗苗高和地径的定株、定期观测数据进行综合分析,采用Logistic生长曲线方程对其生长过程进行拟合和样本聚类,分析苗高和地径生长动态规律。结果表明,①采用1a年生铁核桃苗和1年生新疆核桃实生苗作砧木,10个引种品种和对照品种间的嫁接成活率存在显著差异,1a年生铁核桃苗的嫁接成活率高于1a生新疆核桃实生苗作砧木的嫁接成活率;②10个引种品种中,Hansen和Axel 2个品种适合于采用1年生新疆核桃实生苗作砧木,Broadview、Buccaneer、Coenen、Fernelle和Rita适合于采用1年生铁核桃作砧木;③本地核桃品种(对照品种)采用1年生新疆铁核桃实生苗作砧木的嫁接成活率极低,不宜采用。④苗高和地径系统聚类结果很明显的划为不同的2类;⑤苗高和地径生长规律基本一致,符合"S"型曲线,呈现"慢~快~慢"节律,且有二次生长;⑥苗高和地径生长可划为分生长前期、速生期、生长后期3个时期,从嫁接至停止生长期大约为205天,速生期大约为112 d,生长前期大约为33d,生长后期大约为70d;⑦*Juglans regia* L.和*Juglans siggillata* L.两个核桃种间苗高和地径相对于种内变异较大,种内变异不大。

关键词:核桃嫁接苗;*Juglans regia*;*Juglans siggillata*;Logistic方程;嫁接成活率;苗期生长性状

Study on the Grafting Survival Rate and Growth Characteristics of 10 Introduced Walnut Grafting Seedling Varieties

Xiong Xin-wu[1], Li Jun-nan[1], Liu Heng-peng[1],
Zhang Qun[2], Chen Wei[2], Meng Meng[2], Feng Xian[2], Li Jiang[2]

(1. Yangbi Walnut Research Institute Under Yunnan Academy of Forestry Yangbi Yunnan 672500,
2. Yunnan Academy of Forestry, Kunming Yunnan 650201, P. R. China)

Abstract: The 1a seeding of Juglans siggillata L and 1a seeding of Juglans regia L are rootstock. Based on the grafting survival rate difference of 10 introduced walnut varieties of Juglans regia and 2 of Juglans siggillata (local ones), comparison was made on 1a walnut grafting seedlings height and ground diameter, the Logistic growth curve equation was used for the sample clustering for their growth process, analysis of seedling height and ground diameter growth dynamic law as well as accurately differentiate various periods in the process of growth. The results showed that (1) using 1a seeding of Juglans siggillata L and 1a seeding of Juglans regia L as stock, the grafting survival rate issignificant difference in 10 introduction walnut varieties and contrast walnut varieties. 1a seeding of Juglans siggillata Lsurvival rate was higher than that grafted 1a seeding of Juglans regia L as rootstocks survival rate. (2) in 10 introduction varieties, the seedling rootstock of Hansen and Axel two varieties suitable for 1a seeding of Juglans regia L, Broadview, Buccaneer, Coenen, Fernelle and Rita are suitable for 1a seeding of Juglans siggillata L. (3) local Walnut Cultivars (CK) by grafting seedlings as rootstocks 1a seeding of Juglans regia L survival rate is extremely low, should not be used. (4) the seedling height and ground diameter system clustering results is clearly divided into 2 classes; (5) the seedling height and ground diameter growth in line with the "S" type curve, rendering fast to slow and slow rhythm; (6) according to the growth change law and combining with the ordered sample clustering, seedling height and ground diameter growth process almost same, can be divided into three stages, initial growth stage, fast growth period, the later growth period, from grafting to stop growing season about 205 days, seedling height and ground diameter fast-growing period of about 112d, the innitial growth is about 33d and about70d at later growth stage. (7) For Juglans regia and Juglans siggillata, seedling height and ground diameter in Interspecific variation was siginificent while the small Intraspecific variation.

Key words: walnut grafting seedling; *Juglans regia*; *Juglans siggillata*; Logistic equation; grafting survival rate ;growth characteristics.

核桃(*Juglans regia* L.)是我国传统的出口物资之一,具有极高的食用价值和商品价值[1]。研究分析表明,核桃脂肪含量为65.1%~68.4%,其中90%以上由油酸、亚油酸、亚麻酸和花生四烯酸等不饱和脂肪酸组成,高含量的不饱和脂肪酸有利于降低人体胆固醇、防止动脉硬化和保护心脏。核桃蛋白质含量为13.3%~15.6%,其中包括18种氨基酸,除8种人体必需氨基酸含量较高外,还含有较多的精氨酸,此种氨基酸能刺激脑垂体分泌生长素,控制多余脂肪形成。此外,核桃还含有多种矿物质和维生素,具有极高的医疗价值[2]。

核桃属(*Juglans*)植物约有20多种,我国现有12个种,其中分布广且栽培较多的是核桃(*J. regia*)和铁核桃(*J. sigillata*)[2,3]。核桃目前是中国栽培范围最广的种,而铁核桃是我国特有种,分布在云南、贵州、四川等地[4~7]。云南核桃是铁核桃(*J. sigillata*)种,已有3 500多年的栽培历史[8,9],核桃种植面积已达1 536万hm²[9],核桃产业已在云南山区经济发展中占有重要地位[10]。云南核桃主栽品种由于不耐寒、栽培中容易受晚霜危害[11]。近几年,国内核桃新品种的选育和引种工作较多[12],云南省也在选育各地区当地耐寒和避晚霜的良种工作[13~15]。为了丰富云南省核桃品种资源以及在耐寒和避晚霜核桃良种上有所突破,在国家林业局"948"项目的资助下,以抗寒和适宜高海拔地区(2 100m以上)为思路,引进了欧洲一些核桃品种或无性系,以期选育出适应云南寒冷、避晚霜、高海拔区栽培的核桃品种,为云南核桃产业发展奠定基础,同时满足云南核桃产业发展的需要。

目前,分别以 *J. regia* 和 *J. sigillata* 两个种的实生苗作砧木,研究其嫁接成活率和苗期生长性状的研究尚未见报道。孟丙南、张俊佩等[16]研究过两个种的砧木对核桃光合特性的影响,表明两个种的砧木净光合速率存在显著差异,其他树种如伯乐、麻栎、半印度紫檀、土沉香等[17~24]的苗期生长性状、苗期生长表现和苗期生长节律的研究报道较多。

通过对两个核桃种的砧木进行嫁接成活率试验研究,为核桃苗木培育中砧木的选择、良种扩繁提供科学依据。同时,通过对1年生嫁接苗苗期生长特性的研究,能够有效指导核桃苗期的科学管理,提高苗木的培育质量。

1 试验材材料与方法

1.1 试验地概况

试验地位于漾濞县苍山西镇下普村,地处99°36′~100°07′E,25°12′~2554′N,海拔

1 535m，年平均气温 17.2℃，极端最低气温 −3.9℃，极端最高气温 34.6℃，全年无霜期 301d，年平均降水量 1 055.5mm，日照时数 2 300h；试验地为大田农耕地，壤土，土层厚度，土质疏松，土壤肥沃，灌溉方便。

1.2 试验材料及来源

试验材料为2012年、2013年分别从荷兰 Kwekerij Westhof 公司引进 34 个品种，根据品种抗寒性和综合嫁接情况选取 10 个作为试验研究对象。同时，以本地核桃品种漾濞大泡核桃和娘青核桃为对照。2012年的接穗由 wekerij Westhof 公司通过邮寄方式引入，2013年的接穗由项目组成员考察学习引入。10 个引种（品系）基本情况见表1。

表1 10 个引进品种及对照品种基本情况
Table 1 The 10 introduced walnut varieties of *Juglans regia* and 2 of *Juglans siggillata* situation

	序号	品种/无性系	种	引种年份	特点
引种品种	1	Buccaneer	*Juglans regia* L.	2012年 2013年	荷兰北部主栽品种，原产加拿大西海岸，抗寒性强，生长较慢
	2	Coenen	*Juglans regia* L.	2012年、2013年	原产德国，壳极薄
	3	Axel	*Juglans regia* L.	2012年 2013年	原产于荷兰，Buccaneer 是最佳授粉树，植株生长良好，果实巨大，树形大，可作为观赏核桃栽培。
	4	Fernelle	*Juglans regia* L.	2013年	原产法国
	5	Parisieane	*Juglans regia* L.	2012年 2013年	荷兰实生选品种，果光滑、饱满
	6	Lara	*Juglans regia* L.	2012年 2013年	原产荷兰，实生选育品种
	7	*J.regialociniala*	*Juglans regia* L.	2012年 2013年	美国品种，丰产
	8	Broadview	*Juglans regia* L.	2012年 2013年	荷兰北部主栽品种，原产加拿大西海岸，抗耐寒性强，生长较慢
	9	Hansen	*Juglans regia* L.	2012年 2013年	日本品种，早实
	10	Rita	*Juglans regia* L.	2012年 2013年	原产比利时，丰产
本地品种（对照）	11	漾濞大泡 YangBiDaPao	*Juglans siggillata* L.		云南主栽品种、晚实
	12	娘青 niangqin	*Juglans siggillata* L.		漾濞地方栽培品种，较耐寒、晚实

1.3 试验方法

1.3.1 苗木嫁接

试验主要于2013年进行。2月10~14日分别采用1年生铁核桃（*Juglans sigillata* L.）苗和1年生新疆核桃（*Juglans regia* L.）实生苗作砧木，单芽硬枝、移砧嫁接的方法嫁接后定植于圃地内。1年生铁核桃苗作砧木的每品种嫁接40株，3次重复，共计120株；1年生新疆核桃实生苗作砧木的每品种嫁接28株,3次重复,共计84株。对照品种为漾濞大泡核桃和娘青核桃,两种砧木均为每品种嫁接60株,3次重复,共计180株。定植采用高床,床宽1.6m,株行距20cm×30cm,圃地按常规核桃育苗方法进行管理。

1.3.2 嫁接成活率及性状测定

嫁接后每5d观察一次苗木萌芽情况,待苗木基本出齐、展叶后,调查各品种采用两种类型的砧木的嫁接成活率。并以1年生铁核桃作砧木成活的苗木为观测对象,每个品种和对照品种随机选取5株、3次重复共15株定株挂牌,每14d观测记录一次苗高,地径生长值,直至10月初结束。

1.3.3 数据处理

1.3.3.1 方差分析

用SPSS统计软件进行方差分析,最小显著差异数法(LSD法)进行多重比较。

1.3.3.2 Logisitc生长曲线方程的拟合

Logisitc生长曲线是典型的"S"型曲线,它的方程为 $\hat{y}=\dfrac{K}{1+ae^{-bx}}$,式中 x 为苗高和地径生长定期观测时间,K 时间无限延长的终极生长量,a、b 为参数,e 为自然对数的底数。经比较 $|rxy'|>r_{0.05}=0.874$,达到极显著水平,说明 x 与 y' 的直线关系是极显著的,适合Logisitc生长曲线方程的拟合[18,19]。

1.3.3.3 系统聚类和有序样本聚类。

用SPSS统计软件进行系统聚类,用Excel制表制图。

2 结果分析

2.1 嫁接成活率差异性分析

2.1.1 1年生铁核桃砧木嫁接成活率差异性分析

对10个引种品种和对照品种采用1a生铁核桃作砧木嫁接的成活率进行差异性分析,分析结果得出各引种品种以及对照品种间嫁接成活率差异极显著。多重比较结果见表2,从表中可以看出,采用1a生铁核桃苗作砧木,10个引种核桃品种的嫁接成活率与对照品种的嫁接成活率相比,Buccaneer、Coenen、Broadview与娘青核桃间差异不显著,与漾濞大泡核桃间差异极显著;Parisieane和Rita与两个对照品种间差异不显著;Axel、Fernelle、Lara、J.regialociniala和Hansen与两个对照品种间差异极显著;10个引种品种间相比,Broadview与Axel、Fernelle、Parisieane、Lara、J.regialociniala、Hansen、Rita间差异极显著;Rita与Axel、Fernelle、Hansen、Lara、J.regialociniala间差异极显著;Hansen和Axel间差异极显著。

10个引种核桃品种中,嫁接成活率最高的是Broadview,达94.17%,比对照品种娘青核桃高16.94%,比漾濞大泡核桃高23.06%;嫁接成活率最低的是Axel,比对照品种娘青核桃低54.72%,比漾濞大泡核桃低48.61%;对照品种的平均嫁接成活率(74.17%)比引种品种的平均嫁接成活(56.25%)率高。

2.1.2 1年生新疆核桃砧木嫁接成活率差异性分析

对10个引种品种和对照品种采用1年生新疆核桃实生苗作砧木嫁接的成活率进行差异性分析,差异极显著。多重比较结果见表3。从表3中可以看出,采用1a生新疆核桃实生苗

表2 1a生铁核桃砧木嫁接成活率多重比较
Table 2 The L grafting survival rate of 1a seeding of *Juglans siggillata* L multiple comparisons

品种	砧木类型	平均嫁接株数(株)	平均成活株数(%)	平均成活率(%)
Buccaneer		40	32	80.83AB
Coenen		40	31	78.33AB
Axel		40	9	22.50D
Fernelle		40	15	37.50CD
Parisieane		40	25	61.67BC
Lara		40	15	37.50CD
J.regialociniala	1年生铁核桃砧	40	12	30.00CD
Broadview		40	38	94.17A
Hansen		40	18	45.83C
Rita		40	30	74.17B
漾濞大泡 YangBiDaPao		60	43	71.11B
娘青 niangqin		60	46	77.22AB

作砧木,10个引种核桃品种的嫁接成活率与对照品种的嫁接成活率相比,Hansen和Rita与漾濞大泡核桃间差异极显著,其他品种与两个对照品种相对差异均不显著;10个引种核桃品种间,Hansen和Rita与Broadview、Fernelle、Parisieane、Lara、*J.regialociniala*间差异极显著,其他品种间差异均不显著。

10个引种核桃品种中,嫁接成活率最高的是Rita,为72.62%,比对照品种娘青核桃高22.62%,比漾濞大泡核桃高27.62%;嫁接成活率最低的是Fernelle,比对照品种娘青核桃低27.38%,比漾濞大泡核桃低22.38%;对照品种的平均嫁接成活率(47.5%)比引种品种的平均嫁接成活率(45.6%)高。

表3 1年生新疆核桃砧木种嫁接成活率多重比较
Table 3 The grafting survival rate of 1a seeding of *Juglans regia* L multiple comparisons

品种	砧木类型	平均嫁接株数	平均成活株数	平均成活率(%)
Buccaneer		28	17	59.52AB
Coenen		28	14	51.19AB
Axel		28	15	53.57AA
Fernelle		28	6	22.62B
Parisieane		28	9	32.14B
Lara		28	8	27.38B
J.regialociniala	1年生新疆核桃砧	28	7	25.00B
Broadview		28	12	41.67B
Hansen		28	20	70.24A
Rita		28	20	72.62A
漾濞大泡 YangBiDaPao		60	27	45.00B
娘青 niangqin		60	30	50.00AB

2.1.3 不同砧木嫁接成活率比较分析

从表1、表2可以看出,1年生铁核桃苗作砧木的嫁接成活率明显高于1年生新疆核桃实生苗作砧木的嫁接成活率,前者嫁接成活率最高可达94.17%(Broadview),最低为22.5%(Axel),仅有Axel1个品种嫁接成活率低于30%,而后者最高为72.62%(Rita),最低为22.62%(Fernelle)嫁接成活率低于30%的有3个品种,成活率分别是27.38%(Lara),25%(*J.regialociniala*),22.62%(Fernelle)。Axel和Hansen2个品种采用1年生新疆核桃实生苗作砧木的嫁接成活率比采用1年生铁核桃苗作砧木的嫁接成活率高,其他品种均是采用1年生铁核桃苗作砧木的嫁接成活率高于采用1年生新疆核桃实生苗作砧木的嫁接成活率。

2.2 苗木生长性状

2.2.1 10个引种进品种1年生嫁接苗系统聚类分析

以铁核桃苗作砧木的1年生嫁接苗的苗高和地径2个性状为指标,对10个引种品种和对照品种进行系统聚类分析。根据系统聚类图(图1)并结合各品种嫁接苗的生长情况,可以明显地看出,10个引种品种和对照品种分为2类。第1类本地核桃品种即漾濞大泡核桃和娘青核桃,生长状况较好,其平均苗高104.3cm和,平均地径为1.89cm,第2类为引种的10个品种即 Buccaneer、Parisieane、Coenen、Hansen、Axel、Broalview、Rita、*J.regialociniala*、Lara、Parisieane、Fernelle,其平均苗高71.63cm,平均地径1.82cm,10个品种中生长最好的为Parisieane,苗高为84.2cm,地径为1.97cm,生长最差的为*J.regialociniala*,苗高为61cm,地径1.7cm。

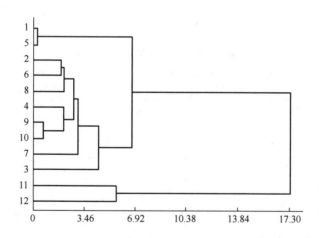

图1 10个引进品种及对照品种系统聚类图

Figure 1 10 introduced walnut varieties of *Juglans regia* and 2 of *Juglans siggillata* systematic clustering

2.2.2 10个引种品种1年生嫁接苗生长特性

2.2.2.1 苗高和地径生长规律

试验根据不同调查日期的苗高增长和地径增长绘制成曲线,结果如图2、图3。从图2

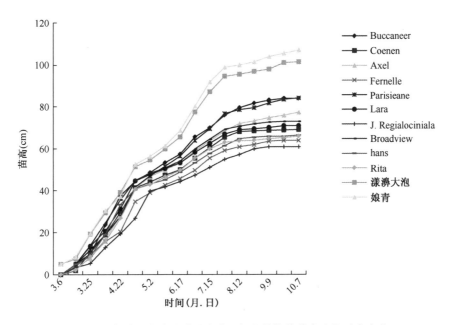

图 2 10个引进品种及对照品种1年生嫁接苗苗高生长动态变化
Figure 2 1a walnut grafting seedlings height of 10 introduced walnut varieties of *Juglans regia* and 2 of *Juglans siggillata* growth dynamic change

和图3可以看出,10个引种品种的苗高和地径增长趋势基本一致,两个对照品种的苗高和地径增长趋势也基本一致。从6月17日以后两个对照品种的苗高增长幅度比10个引种苗高增长幅度大。3月6~25日,各品种处于萌芽期阶段,3月26日~4月8日,苗高和地径的生长较慢,4月9日~6月30日,苗高和地径的增长较快,是苗高和地径生长的第一个高峰期,7月1日~7月29日,各品种的苗高有开始二次生长,出现第二次生长高峰。利用Logistic方程对10个引种品种和对照品种1年生嫁接苗苗高和地径年生长节律进行拟合,其拟合效果达到极显著水平($P<0.01$)(表4)。从拟合方程和图2、图3可知,10个引种品种和对照品种1年生嫁接苗生长节律符合"S"型曲线,各品苗高生长和地径生长均表现为"慢-快-慢"S型生长规律。其苗高和地径的生长时期可划分为生长前期(含萌芽期)、速生期、生长后期3个时期(表5)。从表5可知,10个引种品种和对照品种1年生嫁接苗从嫁接至停止生期大约205d,速生期约占112d,生长前期约占33d,生长后期约占70d。10个引种品种和对照品种从嫁接至基本出齐大约要30~40d,期间要经历愈伤组织产生、愈合、萌芽、展叶(出现复叶)等各个阶段,此阶段受管理状况、气温影响较大。10个引种品种和对照品种中速生期苗高累积生长量最高为对照品种娘青核桃(61.6cm),最低是Hansen(28.1cm),累积百分率最高是Rita(71.82%),最低是Broadview(61.10%);地径累积生长量最高是Rita(1.2cm),最低是对照品种娘青核桃(0.84cm),累积百分率最高是Rita(71.82%),最低是对照品种娘青核桃(50.79%);苗高日平均生长量最高是对照品种娘青核桃和漾濞大泡核桃(0.61cm和0.58cm),地径日平均生长基本相当。

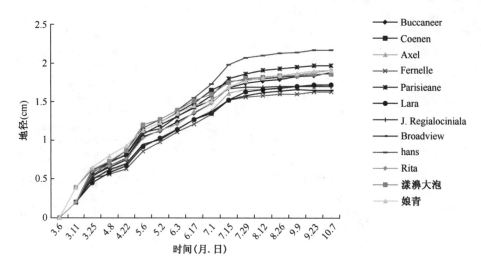

图 3 10 个引进品种及对照品种 1 年生嫁接苗地径生长动态变化

Figure 3 1a walnut grafting seedlings ground diameter of 10 introduced walnut varieties of *Juglans regia* and 2 of *Juglans siggillata* growth dynamic change

表 4 10 个引进品种 1 年生嫁接苗苗高和地径 Logisitic 回归

Table 4 Logisitic regression of 1a walnut grafting seedlings height and ground diameter of 10 introduced walnut varieties of *Juglans regia*

品种		K	a	b	P
Buccaneer	H	84.940	9.008	0.033	<0.01
	D	1.954	5.53E+16	0.377	<0.01
Coenen	H	70.011	1.75E+23	0.523	<0.01
	D	1.875	1.46E+29	0.656	<0.01
Axel	H	78.276	9.800	0.031	<0.01
	D	1.791	7.78E+18	0.425	<0.01
Fernelle	H	65.043	8.402148	0.0314	<0.01
	D	1.6736	3.8E+21	0.485	<0.01
Parisieane	H	85.735	8.720	0.030	<0.01
	D	2.025	3.45E+20	0.462	<0.01
Lara	H	71.874	1.63E+24	0.545	<0.01
	D	1.775	1.65E+19	0.432	<0.01
J.regialociniala	H	61.531	12.475	0.036	<0.01
	D	1.718	7.41E+34	0.784	<0.01
Broadview	H	73.8996	6.224	0.032	<0.01
	D	1.685	9.97E+24	0.56	<0.01
hans	H	67.170	7.014	0.032	<0.01
	D	2.226	2.52E+20	0.460	<0.01
Rita	H	66.358	9.355	0.035	<0.01
	D	2.019	6.65E+14	0.334	<0.01

（续）

品种		K	a	b	P
漾濞大泡 YangBiDaPao	H	104.268	6.581	0.027	<0.01
	D	1.876	3.447	0.029	<0.01
娘青 niangqing	H	110.301	4.39E+14	0.329	<0.01
	D	2.009	1.89E+18	0.411	<0.01

表 5 10 个引进品种苗期生长特征值
Table 5 The value seedling growth character Of 10 introduced walnut varieties of *Juglans regia*

	品种	项目	生长前期 3月6日~4月8日		速生期 4月9日~7月29日			生长后期 7月30日~10月7日	
			累积生长量 (cm)	累积百分率 (%)	累积生长量 (cm)	累积百分率 (%)	日平均生长量 (cm)	累积生长量 (cm)	累积百分率 (%)
引进	Buccaneer	H	37.6	28.21	38.4	62.26	0.47	8.00	9.52
		D	0.75	34.57	0.99	57.98	0.01	0.14	7.45
	Coenen	H	29.7	27.31	35.9	67.49	0.42	3.60	5.20
		D	0.82	38.17	0.97	58.06	0.01	0.07	3.76
	Axel	H	28.00	22.09	40.5	66.41	0.46	8.90	11.50
		D	0.69	34.68	0.97	61.27	0.01	0.07	4.05
	Fernelle	H	20.5	24.53	38.7	67.97	0.39	4.80	7.50
		D	0.63	34.36	0.93	61.35	0.01	0.07	4.29
	Parisieane	H	27.8	22.45	49.00	68.76	0.52	7.40	8.79
		D	0.77	34.01	1.09	60.41	0.01	0.11	5.58
	Lara	H	31.4	29.30	35.6	65.07	0.41	4.00	5.63
		D	0.67	33.72	0.96	61.05	0.01	0.09	5.23
	J.regialociniala	H	19.4	21.15	35.6	69.02	0.38	6.00	9.84
		D	0.81	42.35	0.88	57.06	0.01	0.01	0.59
	Broadview	H	35.6	33.84	33.7	61.10	0.40	3.70	5.07
		D	0.7	37.58	0.88	58.18	0.01	0.07	4.24
	Hansen	H	33.4	28.72	28.1	63.76	0.38	5.00	7.52
		D	0.87	32.72	1.2	62.67	0.01	0.10	4.61
	Rita	H	26.9	23.79	36.2	71.82	0.42	2.90	4.39
		D	0.76	35.08	1.02	58.12	0.01	0.13	6.81
对照	漾濞大泡 YangBiDapao	H	34.2	25.39	55.5	67.56	0.58	6.80	7.05
		D	0.87	39.25	0.93	57.53	0.01	0.06	3.23
	娘青 niangqing	H	32.3	24.98	61.6	66.99	0.61	8.20	8.03
		D	0.93	41.88	0.84	50.79	0.01	0.14	7.33

2.2.2.2 10个引种品种苗高和地径变异分析

10个引种品种和对照品种的1年生嫁接苗苗高和地径变异分析可以看出(表6)，*Juglans regia* L. 和 *Juglans siggillata* L. 两个核桃种间苗高和地径相对于种内变异较大，10个引种品种和对照品种种内1年生嫁接苗苗高和地径变异不大。种间苗高和地径变异系数为19.9%和15.4%，种内10个引种品种苗高和地径变异系数为13.8%和10.6%，对照品种苗高和地径变异系数为18.9%和9.5%。

表6 10个引进品种的苗高和地径变异系数

Table 6 Height and ground diameter coefficient variation of 10 introduced walnut varieties of *Juglans regia*

性状	种间	种内	
		10个引种品种	对照
苗高	19.9%	13.8%	18.9%
地径	15.4%	10.6%	9.5%

3 结论与讨论

试验得出，采用1年生铁核桃苗作砧木和采用1年生新疆核桃实生苗作砧木对各引种品种和对照品种间存在显著差异。采用1年生铁核桃苗作砧木嫁接的成活率明显高于采用1年生新疆核桃实生苗作砧木的嫁接成活率。1年生铁核桃苗作砧木嫁接的成活率最高可达94.17%，而1年生新疆核桃实生苗作砧木的嫁接成活率最高仅达72.6%。

引种的10个核桃品种中，Hansen、Rita适合于采用1a生新疆核桃实生苗作砧木培育苗木，Broadview、Buccaneer、Coenen、Fernelle、Rita适合于采用1a生铁核桃作砧木进行苗木培育。本地核桃品种(对照品种)不宜采用1年生新疆核桃作砧木进行苗木培育。

无论引种的10个核桃品种和本地核桃品种1a生嫁接苗苗期生长表现基本一致。从嫁接至苗木停止生长大约经历205天，生长的整个过程可分为生长前期→速生期→生长后期，苗高和地径生的生长呈现"慢-快-慢"的节律，符合"S"型曲线，且6月下旬有开始二次生长，7月初出现第二次生长高峰。本地核桃品种(对照品种)在速生期的生长量比引种的核桃品种在速生期的生长量大，且速生期与生长前期和生长后期间的差异较大，而引种核桃多数差异不明显。同时，生长整齐度和苗高生长状况没本有本地核桃品种(对照品种)好。

引种的10个核桃品种1a年生嫁接苗苗高和地径种内生长变异小，*Juglans regia* L. 和 *Juglans siggillata* L. 种间变异较大，这对于引种和选育具有较好的意义。

从理论上，引种核桃品种与新疆核桃实生苗同属于 *Juglans regia* L. 种，其亲缘关系更接近，亲和力更强，嫁接成活率应更好。而通过这次试验研究表明，1a生铁核桃苗作砧木的嫁接成活率比1年生新疆核桃实生苗作砧木的嫁接成活率更好，刚好与理论相反，其原因可能是多方面的，相关方面的研究未见报道。作者认为这可能与嫁接地的气温条件、管理有很大的关系。同时与砧木的适宜性也有很大关系，此次1年生新新疆核桃实生苗是从云南剑川引入试验地的，引种地气候寒冷，而试验地气温较高，对嫁接成活率会有一定的影响。

试验表明，引种核桃品种多数适宜采用1年生铁核桃苗作砧木培育嫁接苗，个别品种适宜1年生新疆核桃实生苗作砧木培育嫁接苗，这对于引种品种苗木培育具有重要的指导意义。

引种品种和本地核桃品种1年生嫁接苗苗高和地径生长规律表现为基本一致,且出现第二次生长。这可能与试验地的气候条有很大关系,云南6~11月是雨季,雨水较多,为二次生长的出现创造了条件,但在实际观测中发现有部分苗木二次生长的枝干木质化程度低,在实际生产中,应加强二次生长期的管理,提高苗木的质量。

通过此次引种不同砧木类型嫁接成活率对比研究以及苗期生长性状研究,能有效掌握苗木的培育方法和指导苗木管理。此次研究是从1年生嫁接苗的苗期生长性状出发,反映了苗木的生长过程,但不能完全反映其物候期,而对于引种核桃品种的物候期与本地核桃品种的物候期的差异有待进一步研究。品种的引种能否成功,取决于品种的遗传基础、气候的差异性和具体的管理水平,其过程相当漫长,只有充分了解品种的适应性、抗逆性,经多年驯化并结合当地的气候特点,才能选育出优良品种。

参考文献:

[1] 郗荣庭.中国果树志·核桃卷[M].北京:中国林业出版社,1996:12~58.

[2] 张宏潮,李明亮.我国核桃坚果品质标准化的研究[J].经济林研究.1987,(S1):241~246.

[3] 匡可任,路安民.中国植物志:第21卷[M].北京:科学出版社,1979:30~35.

[4] 刘湘林,付艳华,等.南方核桃栽培技术体系的研究[J].中南林业科技大学学报.2011,8:35~39.

[5] 曲泽洲,孙云蔚.果树种类论[M].北京:农业出版社,1990:183~191.

[6] 王滑,郝俊民,等.中国核桃8个天然居群遗传多样性分析[J].林业科学.2007,43(7):120~124.

[7] Fjellstrom R G,Parfitt D E. Walnut(Juglans spp.) genetic diversity determined by restrict ion fragment length polymorphisms. Geno me,1994,37:690~700.

[8] 杨源.核桃丰产栽培技术[M].昆明:云南科技出版社,2002.

[9] 宁德鲁;马庆国,等.云南省核桃品种遗传多样性的FISH-AFLP分析[J].林业科学研究.2011,24(2):189~193.

[10] 熊新武.陆斌,等.栽培技术措施对山地核桃中幼树的促花促果作用[J].中南林业科技大学学报.2012,32(10):79~82.

[11] 刘妖,范志远,等.云南省鲁甸县核桃选优初报[J].中国南方果树.2010,39(6):46~48.

[12] 赵登超,侯立群,等.我国核桃新品种选育研究进展[J].经济林研究.2010,28(1):118~121.

[13] 肖良俊,马婷,等.云南镇雄核桃优良单株选择初报[J].西部林业科学.2011,40(3):69~72.

[14] 熊新武,陆斌,等.云南省剑川县新疆核桃选优初报[J].北方园艺.2012(18):53~56.

[15] 张雨,毛云林,等.滇东北地区铁核桃种群优良单株的选择[J].经济林研究.2010,28(1):62~68.

[16] 孟丙南,张俊佩,等.不同砧木对核桃光合特性的影响[J].经济林研究.2013,31(2):32~37.

[17] 田华林,刘宗媚,等.3个地理种源的伯乐树苗期生长性状[J].贵州农业科学.2011,39(10):171~174.

[18] 董章凯,邢世岩,等.麻栎半同胞家系苗期特性分析[J].东北林业大学学报.2011,39(10):27~28.

[19] 沙二,朱先成,等.印度紫檀的育苗技术及其在西双版纳的苗期生长表现[J].广西林业科学.2008,37(2):92~94.

[20] 梁国校,钟瑜,等.土沉香实生苗苗期生长节律研究[J].安徽农业科学.2011,39(30):18509~18511.

[21] 李春喜,邵云等.生物统计学(第四版)[M].北京:科学出版社,2008.

[22] 戴宏芬,邱燕萍,等.储良龙眼果实发育的Logistic生长曲线方程[J].广西农业科学.2006,(3):

15~16.

[23] Bossart J L, Prowell D P. Genetic est imates of population structure and gene f low：limitations, lessons and new direct ions[J]. Trends in Ecology& Evolut ion, 1998, 13：202~206

[24] Bossart J L, Prowell D P. Genetic est imates of population structure and gene f low：limitations, lessons and new direct ions[J]. Trends in Ecology& Evolut ion, 1998, 13：202~206

(本文发表于《中南林业科技大学学报》,2015年)

珍稀濒危蒜头果资源保护与产业化发展瓶颈研究

徐德兵,陈福,郭晓春,廖永坚,袁其琼,张林涛,贾代顺,宋顺超

(云南省林业科学院,云南昆明 650201)

摘要:蒜头果是我国特有珍稀濒危单属单种古老植物,种群数量少且消失速度较快。近年来,随着蒜头果在医药、保健品、香料等方面的开发利用不断取得新突破,产品经济价值极高,市场需求量大,从而导致蒜头果市场收购价格逐年攀升,也进一步加剧了野生蒜头果资源的人为破坏性,资源保护与开发利用之间的矛盾日益突出。为此,本文针对目前蒜头果种群数量急剧减少、种子自然传播萌发困难、自然更新能力差、人工栽培尚未取得突破性进展,产业前期基础薄弱,资源匮乏,可持续开发利用困难等产业发展瓶颈。提出如何加大对野生资源的保护力度,建立自然保护小区,开展良种选育工作;如何建立种苗快繁育体系,开展原地回归种植,扩大种群数量;如何突破人工栽培瓶颈,建立标准化生产基地,实现全产业链规划布局等可持续发展对策及建议。

关键词:蒜头果;珍稀濒危;资源保护;开发利用

Research on the bottleneck of resource protection and industrialization development of rare and endangered *Malania oleifera*

Xu De-Bing,Cheng Fu,Guo Xiao-chun,Liao Yong-jian,Yuan Qi-qiong,Zhang Lin-tao,Jia Dai-shun,Song Shun-chao

(Yunnan Academy of Forestry,Kunming Yunnan 650204,P. R. China)

Abstract:Background *Malania oleifera* is a rare and endangered monophyletic single ancient plant in China with a small number of populations and a rapid rate of disappearance. In recent years, new break throughs have been made in the development and utilization of garlic seeds in medicine, health products, and spices. The economic value of products is extremely high, and the market demand is large. As a result, the purchase price of garlic head fruit market has been rising year by year, which has further exacerbated the wild. The man-made fruit is destructive, and the contradiction between resource protection and development and utilization is becoming increasingly prominent. At the same time, due to the weak basis of the development prospects of garlic head fruit, insufficient reserves of improved varieties and technologies, no break through has been made in the construction of artificial standardized bases, and the raw materials for product development and utilization are still mainly dependent on the existing wild garlic head fruit resources. Methods On the basis of past research, investigations and production practices, the relevant literature was consulted, combined with the current development status of garlic head fruit industry, and the important links in the development of garlic head fruit industry were investigated to comprehensively analyze the main problems and constraints in the development of garlic head fruit industry. , and propose countermeasures and development ideas for the existing problems and constraints. Results The investigation and research found that in the process of the conservation and industrialization of the rare and endangered garlic head fruit resources, some difficulties and key constraints are still present,which are mainly manifested in: (1)The number of populations is drastically reduced, and the natural regeneration capacity is poor,and the Guangnan County is wild. The number of garlic head fruit was sharply reduced from 38,706 strains in 2000 to 7,941 strains,of which only 1249 were natural regeneration seedlings. (2)Development and utilization of resources are scarce and man-made destruction phenomenon exists. In recent years, along with the development and utilization value of garlic fruit,the purchase price of garlic head fruit has also increased year by year. Under the temptation of high purchase price,the phenomenon of destructive fruit picking occurs. occur. (3)Seed germination is difficult to spread naturally, and artificial cultivation has not yet achieved a breakthrough. The base construction is in its infancy. (4)The industry has a weak foundation in the early stage, and reserves of scientific research,service,and technology for industrial development have made it difficult for the industry to develop and use it sustainably. In view of these existing problems and difficulties, it is proposed that the protection of wild resources should be strengthened in the future, natural protection communities should be established, selection and breeding of improved varieties should be carried out, and scientific and technological support should be strengthened; a modern plant seedling rapid propagation system should be established to carry out the local return cultivation. , to expand the number of populations; to solve the bottleneck of artificial cultivation of garlic, to establish a standardized production base, to achieve a full industrial chain planning and layout; to establish a strong multi-sector linkage support mechanism to improve the industrial development environment and conditions and other sustainable development strategies and suggest. Conclusion and discussion Garlic fruit industry chain interlocking and mutual restraint, in the process of industrial development, it is necessary to establish an effective incentive mechanism, combined with the current fight against poverty, integrate and use funds for agriculture to support the development of garlic head industry, especially for funds already invested Planting enterprises, cooperatives, large households and other groups. It is necessary to actively encourage and support various leading enterprises and business entities. Through the construction of the modern garlic head fruit industry park, the core radiation will be brought into full play so that forest farmers can truly integrate into the industrial chain to obtain benefits, so as to boost industrial poverty alleviation and consolidate the effect of poverty alleviation.

Keywords: *Malania oleifera*; rare and endangered; resource protection; development and utilization

蒜头果(*Malania oleifera* Chun et Lee)属铁青树科蒜头果属常绿乔木,是我国特有单属单种古老植物,国家二级珍稀濒危保护木本油料树种,云南省极小种群重点拯救保护对象[1~3]。蒜头果生长迅速,但因受其自身生物学特性、气候及特定环境的限制,自然资源分布狭窄,仅零星或小片状间断分布于云南东南部的广南县、富宁县及广西西部的石灰岩山区。目前全国仅存有野生蒜头果资源15 000多株,其中广南县分布有12 945株(人工种植3 755株),广西约5 000株,种群数量极少[4]。蒜头果树体高大,木材优良,种子富含油脂,种仁含油率高达60%以上,油脂可食用,经皂化后含有40%~50%的神经酸,是迄今发现含神经酸物质最高的木本植物[5,6]。神经酸对神经末梢活性的恢复,促进和刺激神经细胞的生长发育功效显著,市场需求量大[7];同时也是合成麝香酮、十五内酯、十五环酮、配制高级香水和化妆品的优质原料,经济价值极高,并已用于医药工业[8~12]。《云南日报》曾经以"价值万金的蒜头果"报道,《科技报》也曾以"21世纪的摇钱树"加以介绍[13,14]。国外无蒜头果资源分布,也无相关研究报道;国内的研究工作主要集中在蒜头果油脂提取和分离、油脂的开发利用[15~17],生物生态学、种苗繁育、栽培等方面的研究较薄弱[18~22],种群濒危方面的研究则更少[14,23,24],资源保护与可持续利用方面的研究尚属空白。对广南县野生蒜头果种群数量的调查发现,近20年来,广南县蒜头果种群数量急剧减少,濒危速度极快。因此,针对目前资源匮乏而开发利用价值较高的瓶颈问题,很有必要深入调查分析蒜头果资源现状,研究蒜头果种群濒危机制和可持续发展方式,为今后蒜头果资源的保护和利用提供一定的参考。

1 资源保护与开发利用现状

我国蒜头果资源主要分布于 $22°23'\sim24°48'N$,$105°30'\sim107°30'E$ 区域内,云南主要分布于海拔 300~1 640m,广西主要分布于海波 300~1 000m。广南县是我国蒜头果资源的原产地、分布中心,广南县蒜头果资源数量占全国总量的85%以上。通过调查发现,蒜头果主要为野生分布为主,人工栽培极少,现仅有广南县、富宁县开展过少量栽培试验,其中广南县栽培面积最大,2012年以来,先后种植过500亩左右,平均每亩种植密度约45株。但由于人工造林保存率低,特别是,造林成活后逐年枯黄死亡现象十分明显,从而导致目前广南县仅保留有人工栽培成活幼树3 755株。现有野生蒜头果资源,大多零星分布于山坡的中下部,富含Ca元素,瘠薄的微酸性土壤上,且以石灰岩石山为主,土质山坡也有一定数量分布。调查中还发现,现有野生蒜头果植株结果大小年(隔年结果)现象十分明显,大部分植株盛产1年后,需要缓存2~3年的能量才能再次盛产,加之现有零星分布的野生蒜头果资源常年缺乏人为管护,土壤瘠薄,结果大小年现象尤为突出。通过对蒜头果收购企业近10年收购数据分析可知,全国蒜头果年产量约25t左右,最多的年份可收购35t,少的年份仅有5吨左右,每年结果产量不稳定,波动性大。郭方斌对广南县野生蒜头果结实量调查发现,广南县野生蒜头果植株间,结实量差异较大,单株结实量从几十个至几千个,变异系数高达136.38%[25]。

以往人们对蒜头果的利用主要是取材制作家具等,也有少量民间榨油食用习俗[13],油脂合成麝香酮、制作高级香水等方面的开发利用较少。直至20世纪90年代,浙江大学首次从蒜头果种仁中成功分离、提取出了高纯度的神经酸后,蒜头果资源的珍贵性引起了人们的关注[7]。目前,国际市场上销售的高纯度神经酸原粉价格在8 000~10 000美元·kg^{-1}。杭州施惠泰食品化学有限公司近年来生产的神经酸,销售价格8 500美元·kg^{-1};南京圣诺生物科技有限公司生产的神经酸胶囊(净含量/规格:48g/240mg×50粒×4瓶),每盒售价1 980元,其中神经酸含量仅为31.25%,其他辅料为淀粉、蔗糖-硬脂酸酯,该公司生产原料来源于广南县的蒜头果;美国市场,高纯神的经酸价格约15万美元/kg,市场价格极高。尤其是,利用神经酸合成的麝香酮、高级香料等产品,其价格与黄金相当。此外,从蒜头果新鲜枝叶、果皮和提油后的残渣中还可提取挥发油、木质素、多糖、蛋白质、总黄铜及抗氧化物质等[26~29]。其中,多糖成分对抑制肿瘤效果明显,并已成功应用于肺肿瘤、子宫肿瘤、鼻咽肿瘤的治疗,且对人体肝细胞无其他毒副作用[26,28]。同时,从蒜头果果皮、果肉及枝叶中提取的挥发性油,80%以上的成分为天然苯甲醛,目前,苯甲醛市场销售价格约90万元·t^{-1}[26,27]。蒜头果收购价格也随着研发应用的不断突破,而一路飙升,近5年来,蒜头果种子的收购价格翻了近10倍,2016年单个蒜头果鲜果收购价格在1.5~2元,单个蒜头果平均鲜果重仅有35~50g,经济价值极高。

2 资源保护与开发利用的瓶颈

2.1 种群数量急剧减少,自然更新能力差

为准确了解蒜头果种群数量变化,本文以广南县为例,详细调查了广南县蒜头果资源分布数量,并与2000年陆树刚的调查数据进行对比。2000年,陆树刚对广南县蒜头果数量调查报道,广南县6个乡(镇),20个村委会,75个自然村分布有上百个蒜头果种群,植株38 706株[23]。本次调查发现,近20年来,广南县蒜头果种群数量急剧减少,现仅存有植株12 945株(见表1),其中人工种植3 755株,野生植株7 941株,自然更新幼苗优树1 249株,许多原分布有蒜头果资源的村寨已不复存在。现有蒜头果种群中,幼树小苗极小,多以大树、老树为主,自然更新能力差,自然更新幼树主要生长在具有一定荫蔽的林下。这说明,广南县蒜头果种群数量结构不合理,属于衰退型种群,加之种群自然更新能力差,消失速度极快,种群已处于极濒危状态,种群自然更新环境条件难以满足。梁月芳对广西蒜头果种群的研究也发现,目前广西蒜头果种群也在下降,种群内植株年龄结构不连续,多为老龄大树或萌芽树,极少幼苗幼树,蒜头果分布区内生境质量日益下降,一些20世纪80年代分布有蒜头果资源的分布点,现已无蒜头果个体存在[24]。我国蒜头果资源的保护已迫在眉睫。

表1 广南县蒜头果资源数量及其分布

乡镇名称	野生株数(株)	人工林株数(株)	自然更新幼树(株)	合计(株)	主要分布位置
董堡乡	1 979		225	2 204	零星分布、少量位于林内。
旧莫乡	4 314	2 100	443	6 857	零星分布、少量位于林内。
莲城镇	857		80	937	零星分布、约60%位于林内。
署光乡	698		467	1 165	主要集中分布于林内
南屏镇	54		34	88	零星分布、少量位于林内。
杨柳井乡	39			39	零星分布、少量位于林内。
那洒镇		55	34	55	位于山坡中部的低洼处,四周为松、栎类树木,林内阴湿,集中种植在一起。
珠琳镇		1 600		1 600	
合计	7 941	3 755	1 249	12 945	

2.2 开发利用资源匮乏,人为破坏现象存在

近年来,随着蒜头果开发利用价值的不断凸显,蒜头果收购价格也逐年攀升,从而导致各地在蒜头果采收、收购过程中比较混乱,加之现有野生蒜头果资源数量极少,且为零星分布,资源保护工作十分困难。特别是,在高收购价的诱惑下,野生蒜头果资源遭到了无序哄采,由于野生蒜头果树体高大,采摘困难,"杀鸡取卵"破坏性、毁灭性采果现象时有发生,保护与开发利用之间的矛盾日益突出,资源保护工作面临着严峻的挑战。同时,还由于野生蒜头果树体高大、树干粗壮、木材优良、且耐腐,深受当地群众喜欢。陈金凤对蒜头果木材的解剖研究表明,蒜头果木材不仅纹理斜、结构细、材身平滑、材色红黄,越近随心色越浓,无气

味,而且木材干燥容易,加工容易,削面光滑,固钉力较弱,油漆、胶合性能良好[2]。正是这一优良特性,也加剧了野生蒜头果树被砍伐取材的速度。调查发现,近20年来,广南县野生蒜头果植株从2000年的38 706株锐减到9 190株,除部分自然死亡外,大部分还是人为破坏。在走访过程中,我们还了解到,在野生蒜头果资源分布最多的董堡乡,当地少数民族有用蒜头果木材打造新婚新床的习俗。目前,随着蒜头果收购价的不断上升,砍树取材的现象有所缓解,但少数民族民间利用习俗仍在延续。

2.3 种子自然传播萌发困难,人工栽培尚未取得突破

蒜头果鲜果大而重,扁球形果实横径4.0~5.0cm、纵径3.5~4.5cm,单果重35~50g。在自然状况下,风力传播困难,且传播范围有限。调查发现,由于蒜头果种子富含油脂,鼠类动物喜食,动物取食成为了一种有效传播途径,部分动物储藏在洞内遗漏的种子,当条件适宜时能萌芽成苗生长。目前,蒜头果自然传播困难、自然更新能力差是导致种群濒危的重要原因。吴彦琼对蒜头果花朵的研究发现,蒜头果不仅花粉萌发率低,而且花粉管生长速度慢,易弯曲,结实率较低。气干种子易失水而失去生活力,在沙藏过程中,种子腐烂情况严重[14]。梁月芳的调查也发现,蒜头果因种实较大而难于进入土层,加之蒜头果颗粒大,种胚小,果实成熟期以后,蒜头果资源分布片区普遍降水较少,正常情况下很难满足种子发芽所需要的水分,通常只有丢落在疏松且含水较多的地方,能幸免萌发成苗[2]。刘钊权1981年、1982年,在广南县董那孟乡开展过直播造林,本次调查时,当年造林的地块现已无蒜头果踪影[19]。为进一步了解蒜头果种子的发芽率、造林成效,本文从2012年开始,陆续采用广南县收集的种子开展沙藏、栽培试验。沙藏试验结果表明,蒜头果种子在沙藏过程中,种子霉坏率高达42.66%,发芽芽苗根腐率为9.67%,正常芽苗仅有47.67%;造林试验表明,蒜头果小苗造林保存率低,造林后第三年保存率仅为24.6%;用沙藏刚冒芽种子点播能显著提高造成保存率,点播造林后第三年保存率为73.82%。这说明,高含油率的蒜头果种子,在储藏过程中容易酸败霉坏,小苗移栽保存率差,加之造林成活后逐年枯黄死亡现象明显,成为蒜头果种群濒危的另一关键因素。通过6年的栽培试验证明,用沙藏冒芽种子点播是目前有效提高造成保存率的有效方式,种苗无性繁育,良种标准化基地建设道路徘徊不前。

2.4 产业前期基础薄弱、可持续开发利用困难

由于每年蒜头果产量有限且不稳定,人工标准化基地栽培也尚未取得突破性进展,目前产业基础还比较薄弱。特别是,蒜头果种植后产果周期比较长,一般种植后需要6~7年时间才能挂果,10年左右才能盛产,加之现有产业开发的科研、技术等储备不足,种苗繁育、良种选育、丰产栽培等工作还处于起步阶段,产业链前端的许多问题还未得到有效解决。以致现有企业、林农等种植群体不愿在种苗、种植方面投入太多,投资仅集中在后端产品的生产和销售上。蒜头果产业链建设是一个集苗木培育、种植基地建设、原料收集、生产加工、宣传销售等环节相接的系统工程,产业链上游的原料生产,中游的生产加工,下游的宣传销售是一个相互制约的产业体系。因此,在蒜头果产业链前端还没有形成一定规模和数量的基础上,仅依靠现有野生资源产量,很难扩大和改变目前蒜头果全产业链的分布格局。尽管产业链后端的迫切需求,导致前端果实的哄采和价格的飙升,但这不能改变现状,反而加大了人

们对野生蒜头果资源的依赖程度,加剧了野生资源的破坏程度。以牺牲原始野生资源消耗来实现产业发展,必将是不可持续的。同时,在调查中还发现,尽管蒜头果后期生产加工比较精细化,但目前企业的榨油方式还比较落后,生产企业在收购果实后,采取就得榨油方式取油。由于收购地现有榨油工艺、流程、技术都比较落后,且生产设备陈旧,家庭小作坊式的榨油方式必然导致部分油脂残留在榨油后的油枯中,造成很多的浪费,提取工艺急需升级改造。

3 资源保护与可持续利用发展对策

3.1 加大野生资源保护力度,建立自然保护小区

针对目前野生蒜头果资源零星、或小片状间断分布的实际情况,在资源保护过程中,应结合实际加强对野生蒜头果资源的保护力度。一方面可在资源分布比较集中的小片区,建立自然保护小区,实施就地保护,以减少人畜破坏,为蒜头果野生种群的自我维持和自然更新创造条件;另一方面可建立种质资源收集圃,收集各地优良、特异种质资源,为下一步良种选育和科研的开展工作奠定基础。只有保护了丰富的蒜头果种质资源,生产中才能容易获得优良、特异的种质资源,种质资源的收集、保存、评价和利用,不仅是保护珍稀濒危物种的根本需要,也是推动蒜头果产业可持续发展的物质基础,丰富的种质资源和遗传变异,能为今后蒜头果育种获得更大的遗传增益。同时,各级部门也要加强法制管理措施,加大执法力度,把野生蒜头果保护工作纳入议事日程,并积极做好宣传教育工作,普及保护蒜头果的科学知识,使当地群众能充分认识蒜头果资源保护意义,自觉珍惜和爱护这一宝贵珍稀资源。

3.2 开展良种选育工作,建立苗木快繁育体系

通过近 10 年蒜头果收购数据可知,全国蒜头果年产量仅 25 吨左右,蒜头果产量极低,平均每株。除去结果大小年影响,其中还存在部分植株不结果或结果极少的状况。对广南县野生蒜头果调查发现,现有野生资源中也有一些高产植株,例如在广南县董堡乡有一株平均年产鲜果量可达 170kg 左右,这些高产植株为下一步良种选育工作的开展创造了条件。生产中,良种是核心关键,为有效推进蒜头果产业的可持续发展,要选育出优质高产的蒜头果良种,以产量、种仁含油率、鲜出籽率、适应性和稳定性等选育指标为依据,严格经济林木良种选育程序,选育出适应性前的特色区域性良种,推动基地建设良种化。同时,还要在现有的技术基础上,进一步解决蒜头果规模化、工厂化育苗关键技术,通过大棚及田间试验掌握无性繁殖(嫁接、扦插、组培)及采穗圃营建的关键技术,形成成熟的现代化、规模化、工厂化育苗技术,形成强有力的科技创新支撑能力,助推和提升产业发展动力。

3.3 开展原地回归种植,扩大种群数量

由于人类干扰、气候环境变化等多种因素的影响,野生蒜头果种群数量消失较快,加之大部分野生蒜头果零星分布在偏远山区或陡峭山坡上,保护工作十分困难。目前,通过人工繁育种苗,在自然保护小区、蒜头果种群原生地进行回归种植,这将会是一种有效扩大种群数量的方式。尽管全国都没有开展蒜头果过大面积回归种植尝试,近年来,仅有云南省林业科学院在广南县旧莫乡,开展过林下回归种植试验,取得了一定成效,但总体回归种植成活

率不高。其中用沙藏刚冒芽种子点播成活率相对较高,能达到75%左右;用小苗种植成活率则相对较低,仅有30%左右。在试验中还发现,蒜头果幼苗、小树喜阴,种植后如果没有一定的郁闭度和湿度,种植成活率极差,且成活植株逐年枯黄死亡十分明显。因此,在回归种植过程中,要根据蒜头果的生长习性,尽量将种苗种植或点播在自然或半自然的生境中,使其尽可能恢复到历史的状态,例如幼苗、小树喜阴,种植时就要尽量为其创造一个阴凉的环境,以后随着树苗的长大,再逐渐增加其光照度。最终通过回归种植个体的成活、种群的建立和扩散,逐渐使回归种群具备自我维持的能力,以增加回归种群的分布和多度,改进基因流。

3.4 突破人工栽培瓶颈,建立标准化生产基地

蒜头果果实大,自然传播困难、自然更新能力差现阶段种群濒危的重要原因。近年来,在广南县通过沙藏、栽培试验发现,蒜头果种子在沙藏过程中霉坏率高,正常芽苗仅有47.67%,小苗造林成活、保存率低,特别是成活后小苗逐年枯黄死亡现象十分明显。目前,移栽成活率低,标准化栽培技术体系不配套,产业前期科研、技术等储备不足,给蒜头果产业发展带来一定的障碍,蒜头果标准化栽培还处于探索、起步阶段。因此,基于上述困难,在生产中,要积极探讨如何通过控制光照强度、密度、肥料、水分等条件,如何通过整形修剪、激素、覆膜等抚育措施提高蒜头果生长发育、适应性及经济性状。同时,立足蒜头果基地规范化种植技术需要,针对高原山地气候条件、山地栽培实际,开展种植模式设计与试验,进行山地垦复、土壤培肥、水土流失控制、配方施肥、山地灌溉与保水、树体结构培育与调控、病虫害综合防治等综合管理技术试验,形成高原山地蒜头果丰产栽培及管理模式研究与高效管理系列技术。力争做到苗木繁育、种植、水肥、抚育管理标准化,采收标准化等,从而形成完善的标准化生产技术体系,全面解决蒜头果人工栽培困难,标准化基地生产建设停滞不前的现状。

3.5 全产业链布局推进,建立多部门协助机制

大力发展蒜头果产业,不仅是生态环境脆弱地区石漠化生态治理、产业扶贫的有效途径,也是产业带动扶贫开发,产业巩固扶贫成效的有效模式。由于蒜头果自然分布区域都是我国石漠化最严重的片区,区域内片区县和扶贫重点县位居全国首位,是全国脱贫攻坚的主战场,这些区域不仅发展起点低、发展压力大,更是我国重要的生态安全屏障区,承担着保护和修复生态环境的重任。因此,对于这一古老珍稀濒危高价值树种,在生产中,应建立全产业链发展规划布局,建立合理的保护开发利用机制。特别是,在产业发展过程中,要建立有效的激励机制,并结合当前脱贫攻坚战,全面统筹整合使用好产业发展中的涉农资金,在科技研发、濒危资源保护、种苗生产、生产基地建设、加工等重要环节给以扶持,以改善蒜头果产业发展环境。切实落实好生产中的林业财政补贴政策、优惠投资政策、税收政策、金融政策等,把蒜头果产业发展与水土保持、石漠化治理等重大生态建设工程紧密结合,形成多部门上下联动机制,从项目、政策、资金等方面给予照顾倾斜[30]。

同时,要通过外引内联等途径,引导企业、经营主体积极参与蒜头果产业,通过示范园区建设,促进专业化分工,形成"龙头企业+专业合作社+科研机构+林农"等产业化经营模式,激发林农积极参与。最终形成全社会合力助推蒜头果产业发展的良好局面。

4 小结

蒜头果是我国特有古老珍稀濒危木本植物,其形态解剖上既有原始特征,也有进化成分,是云南、广西两省特定狭窄区域范围内,特殊气候环境条件下遗留的活化石[2]。近年来,随着蒜头果多部位提取物在医疗、保健方面的特殊功效被不断被挖掘,蒜头果资源的珍贵性越显凸出,这也进一步加剧了野生蒜头果资源的人为破坏性,野生种群数量急剧减少,资源保护已迫在眉睫。因此,面对资源保护与开发利用相互矛盾的瓶颈问题,我们应进一步加大对野生蒜头果资源的保护力度,深入开展研究蒜头果种群濒危机制、生长习性等,从根源上有效解决蒜头果种苗繁育、原生地回归种植、人工标准化栽培、基地建设、加工利用等方面的制约问题,努力促进蒜头果产业化发展,实现"在发展中保护、在保护中发展"的可持续发展模式。综观全球珍稀濒危植物产业化发展历程,只有加强对野生种质资源的收集、保护,科学开展种苗繁育、人工标准化栽培才是最有效的解决途径之一,也只有这样才能保证原材料的产量和质量,才能减少人们对野外资源的依赖程度,才能完成对野生濒危资源实施最有效、最直接的保护。

参考文献:

[1] 李树刚. 油料植物-新属-蒜头果属[J]. 东北林学院植物研究室汇刊,1980,1(6):67~72.

[2] 陈金凤. 广西蒜头果属及青皮木属木材解剖研究[J]. 广西植物,1994,14(4):373~375.

[3] 傅立国. 中国植物红皮书—稀有濒危植物(第1册)[M]. 北京:科学出版社,1992:480.

[4] 广西林业局保护站,广西植物研究所. 广西壮族自治区重点保护野生植物资源调查报告[R]. 2001:61.

[5] 马柏林,梁淑芳,赵德义,等. 含神经酸植物的研究[J]. 西北植物学报,2004,24(12):2362~2365.

[6] 王性炎,樊金栓,王淋清. 中国含神经酸植物开发利用研究[J]. 中国油脂,2006,31(3):69~71.

[7] 侯镜德,陈至善. 神经酸与脑健康[M]. 北京:中国科学技术出版社,2006:2~14.

[8] 李用华,朱亮锋,欧乞碱,等. 蒜头果油合成庸香酮简报[J]. 云南植物研究,1983,5(3):238.

[9] 李伟光,周永红,李统茂. 气象色谱法测定十五内酯[J]. 化学世界,2001,(8):401~402.

[10] 田萍,胡彦. 蒜头果化学成分及应用的研究进展[J]. 农技服务,2017,13(34):11~12.

[11] 黄林华,刘雄民,李伟光,等. 蒜头果油中长链脂肪酸选择性合成大环内酯[J]. 应用化工,2011,40(1):58~61.

[12] 刘雄民,李伟光,李飘英,等. 用蒜头果油脂合成大环内酯的新方法[J]. 高等化学学报,2007,28(5):897~899.

[13] 陆树刚. 蒜头果的民间利用[J]. 植物杂志,1998(1):12~13.

[14] 吴彦琼. 蒜头果保护的初步研究[D]. 广西大学,2002.

[15] 薛冰,邵志凌. 云南蒜头果油的特征指标及脂肪酸组成研究分析[J]. 粮食储藏,2015(1):44~48.

[16] 欧乞鏚. 一个重要脂肪酸CIS-TETRACOS-15-ENOIC的新存在-蒜头果油[J]. 云南植物研究,1981,3(2):181~184.

[17] 赵劲平,欧乞鏚. 蒜头果仁油的应用研究[J]. 中国油脂,2010,35(7):12~16.

[18] 吴彦琼,黎向东,胡玉佳. 蒜头果生殖生物学特性研究[J]中山大学学报(自然科学版),2004,43(2):81~83.

[19] 刘钊权. 蒜头果石山造林初报[J]. 云南林业科技,1984(1):28~30.

[20] 余慧嵘. 珍稀濒危树种蒜头果引种育苗技术及生长节律研究[J]. 黄山学院学报,2011,13(3):

50~52.
- [21] 卯吉华,贾代顺,陈福,等.中国特有珍稀植物蒜头果嫁接繁殖技术[J].林业科技通讯,2016(2):35~37.
- [22] 潘晓芳.蒜头果育苗情况初报[J].广西农业生物科学,1999,18(3):236~238.
- [23] 陆树刚,雷林斌,杨芹生,等.滇东南蒜头果的保护现状和濒危原因[C].生物多样性保护与区域可持续发展.第四届全国生物多样性保护与利用研讨会论文集.北京:中国林业出版社,2002:169~172.
- [24] 梁月芳.蒜头果的濒危原因研究及挽救对策[D].广西大学,2001.
- [25] 郭方斌,王四海,王娟,等.珍稀植物蒜头果野生植株结实量及果实特征研究[J].广西植物,2018,38(1):57~64.
- [26] 唐婷范.蒜头果有效成分及其生物活性研究[D].广西大学,2013.
- [27] 唐婷范,刘雄民,凌敏,等.蒜头果皮果肉抗氧化成分提取及其抗氧化性质研究[J].食品科学,2012,33(2):16~19.
- [28] 戴晓畅,毛宇,黄晓麒,等.用于人类癌症治疗的蒜头果蛋白质:中国,200510010633.1[P].2006-08-09.
- [29] 刘雄民,李伟光,李飘英,等.从蒜头果的果皮果肉提取精油和天然苯甲醛的方法:中国,02139271.4[P].2003-4-16.
- [30] 徐德兵,袁其琼,陈福,等.油茶产业转型升级的瓶颈及对策研究[J].林业经济问题,2017,37(6):78~83.

(本文发表于《林业经济问题》,2018年)

云南早实早熟杂交核桃新品种——云新90303号的选育

赵廷松,方文亮,范志远,习学良,张雨

(云南省林业科学院,云南 昆明 650204)

摘要:核桃新品种云新90303号,亲本为三台核桃×新早13号(新疆早实优株)。经过10余年在不同生态条件地区比较试验,该新品种果实为椭圆形,外观好,平均单果质量13 g,最大果质量16 g,仁色白,壳厚0.9 mm,出仁率59.40%,含油率68.2%,食味香醇。昆明地区9月上旬成熟,早实、丰产、质优、抗寒、树体矮化。

关键词:核桃;新品种;云新9030;早实;杂交

Yunxin 90303, a promising new ear ly walnut selection

Zhao Ting-song, Fang Wen-liang, Fan Zhi-yuan, Xi Xue-liang, Zhang Yu

(Yunnan Academy of Forestry, Kunming Yunnan 650204, China)

Abstract: Yunxin 90303 is a newhybrid walnut selection, which was derived from the cross between *Juglans sigillata* cv. Santai and *Juglans regia* cv. Xinzao No. 13. Trials at different sites with different growing conditions for more than 10 years proved it is a promising selection. The fruit shape is desirable ellipsoid, with the average weight of 13 g, but reaching 16 g. The nut color is whitely, with a shell thickness of 0.9 mm. The kernel is white in color, containing a fat rate of 68.2%. The flavor is rich aromatic, of very good eating quality. The kernel rate is 59.40%. In Kunming area, it matures in early September. The trees are dwarfing, hardy, productive and resistant to diseases.

Key words: Walnut; Newcultivar; Yunxin 90303; Early season; Hybrid

云南是我国核桃起源地之一,分布广泛,种质资源丰富,作为云南两大主栽品种漾濞泡核桃、三台核桃具有个大、壳薄、仁色白、食味香醇、品质优良等特点,但结实晚、成熟晚、效益慢、种壳欠美观、不耐寒。而我国北方新疆早实核桃具有种壳光滑、结实早、耐寒等特点。针对此问题,云南省林业科学院采用南北核桃两大种群进行种间杂交,开展了核桃早实、早熟、丰产、优质、种壳光滑、耐寒新品种的选育研究工作,培育出了符合上述目标的早实、早熟新品种核桃云新90303号。

1 品种选育过程

云新90303号是云南省林业科学院采用种间杂交育种方法,于1990年用三台核桃(*Juglans sigillata* cv. Santai)×新早13号(*Juglans regia* cv. Xinzao No. 13)为亲本杂交,1991年1月育苗,1992年2月定植于云南省林业科学院昆明选择圃内,1993年3月该单株开始结果,1996年评定为优株,定名为云新90303号,1997~2003年进行区域性试验。经过5年对区域试验植株的生长、结果、丰产性、果实经济性状的观察,结果表明,该株系各种性状稳定,且优良性状突出,早结果、丰产;种壳光滑、品质优、抗寒、树体矮化。该品种于2004年12月通过云南省林木良种审定委员会认定。

2 果实经济性状

云新90303号果实为大果型,三径均值3.20cm,果型为椭圆形,种壳光滑,壳厚0.90mm,整仁取出,仁色白,粒质量13.0g,种仁饱满,出仁率59.23%,含油率68.2%,食味香醇(与亲本比较见表1)。

表1 新品种核桃云新90303号与杂交亲本的坚果品质比较
Table 1 Nut quality comparison between Yunxin 90303 and its parents

品种	三径均值	种壳外观	仁色	壳厚	粒质量(g)	仁质量(g)	出仁率(%)	含油率(%)	风味
云新90303	3.20	光滑	白	0.90	13.0	7.7	59.23	68.2	香醇
三台核桃(母本)	3.60	较光滑	黄白	1.00	12.6	6.8	54.00	70.0	香醇
新早13号(父本)	3.10	光滑	黄白Y	0.80	8.7	5.0	57.00	65.0	稍涩

3 植物学特征

云新90303号植株5年生树高2.62m、干径6.1cm、冠幅7.57m,与父本新早13号相差不大;复叶长36cm、小叶数9~11,多为9;叶形披针形,介于两亲本之间;顶芽圆锥形、腋芽圆锥形、有芽距,偏向于父本;树形为自然开心形,树势中等,树干树皮灰色,表面光滑,1年生枝深绿色。

4 生物学特性

4.1 生长结果习性

云新90303号幼树生长旺盛,果枝率90.3%,顶果枝率8.7%,侧果枝率81.6%,以侧枝结果为主,每果枝坐果2个以上,坐果率80%以上;云新90303号丰产性较好,5年生树平均株产3.7kg,是早实类型核桃国家标准(GB7907-87:5年生1.5kg)的2.47倍,继承了母本三台核桃果枝平均坐果多(2.31个)的优良性状,又继承了父本新早13号分枝力强、侧枝结果力高,且以短果枝结果为主的丰产性状。云新90303号嫁接苗定植1~2年开花结果,表现出较好的早实性。

4.2 物候期

根据多年观察,在云南地区物候期基本相同,2月20~25日萌动;3月5~10日发芽;3月24~29日展叶抽梢;4月3~6日雄花盛花;4月8~12日雌花盛花;5月10~15日幼果形成;5~7月为枝、叶、果实速生期,9月6~10日果实成熟;11月20~25日落叶;年生长期274d左右。

4.3 适应性和抗寒性

区域性试验表明,在云南省昆明市、丽江市、云县、漾濞县、鲁甸县等地土层深厚肥沃潮湿的地区栽培,植株表现生长健壮,结果正常,产量高,坚果大,核仁饱满。经历了1999年和2000年冬季的2次低温(最低温−15℃)后,试验植株均未受冻害,而鲁甸县引种的1~2年生漾濞泡核桃80%以上冻死,表明云新90303号同时具有很强的抗寒性。

5 栽培技术要点

5.1 环境条件

年平均气温12.2~17.5℃,年降雨量1 000mm左右,海拔高度1 500~2 400m的温凉地区;微酸微碱中性土壤。依据其生态习性最好选择阳坡或半阳坡、深厚湿润、排水良好的壤土或沙壤土,水源条件较好的地带作为种植园地。

5.2 种植密度

采用果粮间作栽培模式,行距10~12m,株距4~6m;采用园艺栽培模式,株行距4m×4m或4m×5m。

5.3 整形修剪

采用自然开心形,定干高1.0~1.5m,培养成2~3大主枝开心形的圆头树冠。每年疏除病枝、枯枝、弱枝、交叉枝、重叠枝、过密枝、下垂枝和平行枝等。

5.4 施肥浇水

云新90303号结果早,边结果边生长,消耗的养分较多,必须进行有效补充。增施肥料务必采取有机肥和化肥相结合施用。每年秋末冬初施基肥(厩肥较佳),追肥每年2次,在干旱季节(1~6月)有条件的地区浇水2~3次。

5.5 病虫害防治

在云南地区危害核桃的主要害虫有金龟子、核桃木蠹蛾、核桃毛虫等。虫害发生严重时,对核桃的生长、产量及品

质影响很大,应及时采取综合防治措施:对虫害的发生进行预测预报工作;加强园地的综合管理,促进树体健壮;适时清

除虫源虫枝;进行生物及化学防治。病害主要是根腐病,在雨季及时进行挖沟排涝可防治根腐病的发生。

(本文发表于《果树学报》,2007年)

美国山核桃在云南的引种表现及丰产栽培技术

习学良,范志远,张雨,邹伟烈,廖永坚,董润泉

(云南省林业科学院经济林研究所,云南 昆明 650204)

美国山核桃(*Carya illinoensis*),又名薄壳山核桃、长山核桃,为胡桃科,山核桃属落叶乔木,原产美国和墨西哥,是世界上重要的油料干果树种之一。目前全世界年产美国山核桃坚果18万t(仅为核桃105万t的1/6),主产国美国(占世界产量70%左右)坚果售价3~8美元/kg;仅零星栽培,市场上很难看到销售,属珍稀干果。20世纪初首先由西方传教士引入我国栽培,近10年来,国家及云南等南方省份开始立项研究,广泛从国外引进品种和栽培技术。云南省林业科学院1974年开始先后从国内和美国引种54个品种,进行栽培研究。现将早期从国内引入的品种中筛选出的云光(引自浙江金华,原取名金华)、云星(引自浙江绍兴,原取名绍兴)两个优良品种的特性和配套丰产栽培技术介绍如下。

1 引种研究过程及试验地环境

1974从浙江金华引种4个品种嫁接苗栽培在云南省林业科学院下属的漾濞研究站内,经评比筛选出云光、云星两个品质比较优良的品种。1989年用引种树所产种子育砧,1990年采其大树枝条做接穗培育嫁接苗,1991~2003年进行丰产栽培试验。试验地海拔1 540m,年平均气温16.8℃,最冷月(1月)平均温8.8℃,最热月(7月)平均温21.5℃;年降水量1 055mm,年平均相对湿度72%,11月至次年5月为旱季,降水仅151mm;年日照2 238.9h;土壤为冲积土,土层厚度大于2m,pH值6.8。

2 引种表现

2.1 根系生长习性

美国山核桃苗期主根发达,侧须根少,1~3年生苗主根长度可达地上部分的1.5~2倍。嫁接苗定植头两年主根生长快,侧根生长慢,第三年开始侧根生长明显加快。大树根系很发达,具有良好的固土功能。10年生树主根深达5m,根系分布范围达98m^2,吸收根集中分布在地表下25~50cm距树干基部4m半径的范围内。美国山核桃有外生菌根。

2.2 树体生长结果习性

2.2.1 物候期:2003年对两个品种的物候期进行系统观测,两个品种发芽较晚,各自的雄花散粉时间与雌花可接受花粉时期部分重叠,属雌雄同熟,两品种间也可相互授粉(见表1)。

表1　美国山核桃云光和云星的物候期(月.日)

品种	芽			雄花				雌花				果实采收期	开始落叶期
	萌动	绽开	始叶	初花	盛花	末花	落花	显蕾	初花	盛花	末花		
云光	3.20	3.29	4.9	5.4	5.8	5.12	5.16	4.18	4.25	5.3	5.8	10.2~10.22	11.6
云星	3.25	4.26	4.9	5.1	5.4	5.8	5.13	4.10	4.22	4.27	5.4	10.3~10.30	11.4

2.2.2　生长结果习性：在较好的管理下，1年生嫁接苗定植第2年有1/3雄花；第3年有50%植株开雄花，1/3植株开雌花；4年有80%植株结果。云光10年生树树高11.2m，主干直径25.2cm，冠幅8.4m；29年生树树高11.2m，干径46cm，冠幅11m。云星10年生树树高12.2m，干径26cm，冠幅9.2m；29年生树树高15.2m，干径54.1cm，冠幅13.2m。各品种的丰产性见表2。

表2　美国山核桃丰产性相关指标

品种	树龄(年)	发枝率(%)	果枝率(%)	每穗果数(个)	着果率(%)	株产(kg)	产仁(kg/m²)
云光	10	1.45	33.4	2.8	86.4	18.8	0.19
	29	1.43	68.4	2.6	82.4	45.6	0.25
云星	10	1.68	42.2	3.2	78.6	23	0.17
	29	1.57	74.2	3.1	86.0	67.4	0.23

2.3　坚果质量

测定坚果质量，结果见表3。样品经农业部农产品质量监督检验测试中心(昆明)分析结果：美国山核桃每100g种仁含粗脂肪76g(其中棕榈酸5.57%、硬脂酸2.43%、油酸67.29%、亚油酸23.05%、α-亚麻酸1.10%、二十碳烯酸0.36%)、淀粉1.9g、粗纤维5.85g、蛋白质9.7g(其中天门冬氨酸0.62g、苏氨酸0.24g、丝氨酸0.39g、谷氨酸1.37g、甘氨酸0.56g、丙氨酸0.49g、胱氨酸0.03g、缬氨酸0.41g、蛋氨酸0.01g、异亮氨酸0.32g、亮氨酸0.48g、酪氨酸0.15g、苯丙氨酸0.30g、赖氨酸0.24g、组氨酸0.17g、精氨酸0.62g、脯氨酸0.67g)，100g种仁含硫88.4mg，磷255.9mg，钾388.3mg，钙45.8mg，镁103.0mg，铁1.78mg，锌2.77mg，铜1.23mg，锰4.25mg。

表3　美国山核桃坚果质量

品种	粒重(g)	出仁率(%)	壳厚(mm)	取仁难易程度	仁色	种仁饱满度	食味
云光	6.8	52.7	0.9	易取整仁	黄白	特饱满	细腻、香润、爽口
云星	5.8	48.8	0.9	易取半仁	黄白	特饱满	细腻、香润、爽口

2.4　抗性

30年栽培观测结果表明，美国山核桃树体生长快，能很好地适应当地气候土壤环境，从未发生严重影响产量质量的病虫害，其病虫害种类与本地核桃相同。美国山核桃发芽晚，能避开晚霜对新梢的危害，比核桃更耐低温。美国山核桃主根分布深，根系发达，抗旱能力强。

2.5 适宜栽培范围

利用生产的种子培育出嫁接苗 3 万余株，在云南省 54 个县(市)扩大栽培试验看出，在 1 月平均气温 7~11℃，≥10℃年积温 4 500~6 500℃，6~9 月每旬平均气温在 19~25℃，年降水量 700~1 400mm，生长季(4~10 月)降水量大于 600mm，年相对湿度大于 60%，无霜期 280~330 天，4~11 月日照时数大于 800 小时的环境中能正常开花结果，果实品质优良。7 月平均温度 20℃以下的环境下，树体生长旺盛，但坚果偏小、欠饱满。不论是沙壤土或壤土，只要厚度大于 1m 以上，pH 值 5.6~7.5 自然或人为(浇灌)保持土壤湿润，都能生长结实良好。由此得出，中、南亚热带地区(海拔 1 000~1 600m)都能找到适于栽培美国山核桃的区域。另外，在四川兴文，贵州盘县，湖南凤凰，湖北宜昌和重庆江津等地扩大栽培试验初步看出，引种的美国山核桃品种在西南地区凡适于栽培柑橘的区域都可发展种植。

2.6 综合评价

美国山核桃树体生长快，根系发达，具有良好的水土保持功能，比核桃更耐热，适宜在普通核桃适生海拔下限以下的海拔地段发展，可解决过去低海拔(中、南亚热带气候)区域没有适宜发展的油料干果树种栽培的问题。美国山核桃仁营养成分丰富、全面(尤其是脂肪的成分与世界上质量最好的植物油——油橄榄油极相似)，有较高的食疗保健价值，产品市场前景十分看好。干果易贮藏运输，适于西南广大山区发展。云光的坚果质量优于云星，生产中应以发展云光为主，云星则作为培育砧木采种树和授粉树。

3 配套栽培技术

3.1 苗木培育

采用本砧嫁接育苗，采收的鲜种子用 GA3100mg·L^{-1} 浸泡 7 天后，播种在塑料营养袋(12cm×15cm)中，选择光照好的苗床集中催芽，苗床上搭拱棚覆盖薄膜，出苗后霜期加盖 70%遮阳网防冻害，2~3 个月后出苗率可达 90%以上。霜害较轻时(一般在 2 月中下旬)，将培育出的山核桃营养袋苗移栽到年平均气温 16~20℃、肥力好的土地中培育。如有金龟子成虫为害，在傍晚叶面喷洒敌敌畏等药剂，进入雨季后每 5~7 天喷一次药剂防治卷叶蛾，6~7 月土壤中金龟子幼虫取食期在苗木根部浇施辛硫磷等长效杀虫剂。如此，1 年生苗平均地径可达 0.8cm 以上。枝条用 100℃石蜡加蜂蜡液处理，采用单芽枝腹接法嫁接，1 年生砧苗的可嫁接率达 90%以上，大批量嫁接的成活率可达 90%。嫁接苗出圃时在离地 50~80cm 处定干，剪口用薄膜条包扎，可以提高今后的造林成活率。

3.2 定植

选择适宜的气候土壤环境地区建园，以云光为主栽品种，云星为授粉品种，按 10∶1 配置。定植最佳时期为 12 月至翌年 2 月上旬，采取规模种植和"四旁"(地旁、屋旁、沟旁、路旁等)种植两种模式，规模化种植的株行距为 8m×10m，每 667m^2 种 8 株，"四旁"种植株距控制在 6~8m。种植穴规格：80cm×80cm×80cm，定植时每穴施腐熟农家肥 25kg，合肥 0.2kg 作为底肥，土、肥混匀。苗木栽植深度以土埋至嫁接口以下为宜，踩实土壤后浇定根水，覆膜增温保湿，旱季补水 2~3 次。美国薄壳山核桃苗期主根发达(1~3 年生苗一般长达 1~2m)，侧须根较少，根系恢复困难，为提高裸根苗造林成活率和缩短缓苗期，定植前可在断根切口韧皮部与木质部间插入蘸有吲哚丁酸 1 000mg·L^{-1} 的牙签 1~2 根，以诱导新根生长。

3.2　整形修剪

离地面 1.5~2m 处定干，一般选留 3~4 个主枝。通过拉枝加大分枝角度，6 月扭新梢和摘心，可明显提早结果。4 年生后，每年冬季去除顶枝，促进树姿开张。对旺长枝进行环割倒贴皮有抑制树体旺长促进结果的作用。

3.3　肥水管理

进行果粮(烤烟或绿肥)间种，每年秋末冬初穴施基肥，每年每株土施秸秆 20kg 改土，根据树体大小，每厘米主干直径施氮 15g、磷 7g、钾 6g 和适量的锌、硼等微量元素，施肥穴上用薄膜覆盖，3~5 月在穴内浇水 3~5 次。

3.4　虫害防治

美国山核桃树的害虫主要有木蠹蛾两种、毒刺蛾 3 种、天牛两种和卷叶蛾 1 种。7 月中旬幼虫孵出后，用烟雾机以溴氰菊酯加柴油熏烟对毒刺蛾防治效果极佳。天牛在美国山核桃树上产卵时间一般在 6~7 月，孵化前削除树皮上虫卵能大大减少蛀入虫孔幼虫密度；对天牛蛀孔塞入蘸有敌敌畏或乐果等药剂的纸团或棉纱熏杀效果很好。卷叶蛾在雨季危害嫩叶及嫩梢，使其卷缩，严重时停止生长，喷敌敌畏 800 倍液效果良好。5 月用黑光灯诱杀成虫和及时剪除并烧毁虫枝能有效控制木蠹蛾危害。

3.5　采收、干燥、贮藏

10 月上中旬，美国山核桃进入种子采收期，待树上 80% 果实的青皮开裂时，先在树下垫塑料布等，后用竹竿在树上敲打采收。采收的种子在自然通风的室内晾干或 35~38℃ 热风烘干至含水 5% 以下即可。干燥的种子在 10~15℃ 环境中贮藏 10 个月或在 3~5℃ 冷库中贮藏两年种仁仍可食用。

(本文发表于《中国果树学报》，2004 年)

腾冲红花油茶花器官的数量性状变异研究

郭玉红[1]，司马永康[1]，徐德兵[1]，江期川[1]，吴兴波[2]，钱迎新[3]，张立新[1]，周凤林[1]

(1. 云南省林业科学院，云南 昆明 650201；2. 云南省腾冲市林业局林业技术推广站，云南 腾冲 679100；3. 云南省楚雄州林业科学研究所，云南 楚雄 675000)

摘要：以楚雄及腾冲33株野生和半野生的腾冲红花油茶作为观察研究对象，对其花器官数量性状进行观察，分析腾冲红花油茶的花器官数量特征及其变异特点。结果表明：腾冲红花油茶花器官数量均有变异，苞片和萼片5~13片，花瓣5~12片，雄蕊66~174条，花柱2~5条，子房室2~5室，每室胚珠1~10粒，每果胚珠6~23粒，变异系数为12.46%~26.37%。其中，子房室数量的变异系数最小，而每室胚珠数量的变异系数最大。腾冲红花油茶花器官数量差异主要发生在株间，花瓣和花柱的数量差异主要发生在居群间，而且差异均为极显著。

关键词：腾冲红花油茶；花器官；数量性状；变异

Floral Organs Quantitative Characteristics Variation of *Camellia reticulata* f. *simplex*

Guo Yu-hong[1], Sima Yong-kang[1], Xu De-bing[1], Jiang Qi-chuan[1], Wu Xing-bo[2], Qian Ying-xin[3], Zhang Li-xin[1], Zhou Feng-lin[1]

(1. Yunnan Academy of Forestry, Kunming Yunnan 650201, China; 2. Station for Forestry Technology Extension, Forestry Bureau of Tengchong Municipality, Tengchong Yunnan 679100, China; 3. Forestry Institute of Chuxiong Prefecture, Chuxiong Yunnan 675000, China)

Abstract: The wild and semi-wild Camellia reticulata from Chuxiong and Tengchong was used as the observation object, and the quantitative characteristics of flower organs were observed. The quantitative characteristics and variation characteristics of flower organs of C. reticulata were analyzed. The results show that the floral organ numbers of the form are varied. The numbers of bracteoles and sepals are from 5 to 13, those of petals from 5 to 12, those of stamens from 66 to 174, those of stylets from 2 to 5, those of locules from 2 to 5, those of ovules per locule from 1 to 10 and those of ovules per fruit from 6 to 23. Their variation coefficients are 12.46% ~ 26.37%, of which the least one is of the numbers of locules and the most one is of that of ovules per locule. The floral organ number variances between the individuals are more than those in an individual, petal and stylet number variances between the populations more than those in a population and they all are very significant.

Key words: *Camellia reticulata* f. *simplex* Sealy; floral organ; numerical character; variation

腾冲红花油茶(*Camellia reticulata* f. *simplex*)又名红花油茶，是具有巨大经济价值的油用和观赏常绿阔叶树种，花白色至深红色，花瓣5~12枚[1]，具9瓣以下的花，是英国植物学家 J. R. Sealy[2]依据花瓣数量于1958年发表的滇山茶"单"瓣花类型，并被认为在滇山茶中最为原始。腾冲红花油茶分布于云南西部和中部、四川西南部和贵州西部，生长于海拔1 000~3 200m的阔叶林或混交林中[3]，是云南重要的优良木本油料树种和云南八大名花

"山茶花"的原始类型[4]。目前,有关腾冲红花油茶的研究文献多集中于分类与地理分布[5]、细胞学[6]、植物化学与生物活性[7]、资源调查与保护利用[8]、引种栽培繁殖技术[9]、遗传育种[10]、病虫害[1]等方面,但有关花器官数量性状变异的研究却很少。本研究通过对腾冲红花油茶花器官数量性状变异的观察研究,旨在为腾冲红花油茶的分类学、形态学、传粉生物学以及油用或观花优良品种的选育等实际工作提供参考。

1 材料和方法

1.1 实验材料

以云南省楚雄市(24°51′57″~25°00′46″N,101°03′43″~101°25′15″E)和腾冲市(24°56′10″~24°57′37″N,98°35′08″~98°35′40″E)挂牌的33株野生和半野生的腾冲红花油茶作为观察研究对象,其中楚雄市有11株,腾冲市有22株。

1.2 实验方法

在花期和果期,随机采集5朵花或5个果实进行观察,而个别植株由于结果太少,不能采足5个果实,则全部采集观察。每朵花观察记数其苞片、萼片、花瓣、雄蕊和花柱;每个果实观察记数子房室、每室胚珠和每果胚珠。其中,由于苞片和萼片在位置和形态上极难区别,则两者合起来记数。在所观察的花器官结构数量中,由于子房室数、每室胚珠数和和每果胚珠数等3个项目在花期难于观察而计数不准确,故子房室数、每室胚珠数和每果胚珠数等3个项目的观察计数以果期的观察结果为准。

1.3 数据分析

根据以上观察记录的原始数据,采用生物统计学的方法,利用Microsoft Excel和统计大师标准版计算花器官数量性状的最小值、最大值、平均值、标准差和变异系数[11~13];绘制花器官数量性状的变异式样图[11,12];通过方差分析比较株间和居群间花器官数量性状的差异[11,12]。

2 结果和分析

2.1 变异程度

由表1可知,腾冲红花油茶7个数量性状的变异系数为12.46%~26.37%,其从小到大的排序是子房室数<花柱数<苞片和萼片数<雄蕊数<花瓣数<每果胚珠数<每室胚珠数。其中,子房室数的变异系数最小,仅12.46%,而每室胚珠数的最大,达26.37%。根据司马永康等[14]对变异系数的分级标准,子房室、花柱、苞片和萼片、雄蕊、花瓣和每果胚珠等6个数量性状的变异系数很小,均不足25%,数量很稳定;而每室胚珠数量性状的变异系数小,为25%~50%,数量较稳定。可见,腾冲红花油茶花器官数量性状的变异都不是很大。

2.2 变异式样

从腾冲红花油茶花器官各数量性状的变异式样(图1~图6)看,都呈一定的正态分布,但离散程度不一样,从小到大的排序与变异系数是一致的,即子房室数<花柱数<苞片和萼片数<雄蕊数<花瓣数<每果胚珠数<每室胚珠数。其中,苞片和萼片的数量为5~13片,通常为7~10片,出现频率达86.05%;花瓣的数量为5~12片,通常为6~7片,出现频率达78.18%;雄蕊的数量为66~174条,通常为86~135条,出现频率达88.47%;花柱的数量为2~5条,通常为3条,出现频率达83.65%;子房室的数量为2~5室,通常为3室,出现频率达85.29%;每室胚珠的数量为1~10粒,通常为4~6粒,出现频率达81.88%;每果胚珠的数量为6~23粒,通常为11~18粒,出现频率达87.50%。可见,变异式样所显示的通常数绝大部分与前人所研究的记载数吻合。

表1 腾冲红花油茶花器官数量性状变异情况
Table 1 Numerical character variation of floral organs of *C. reticulata* f. *simplex*

观察项	样本数	最小值	最大值	平均值	标准差	变异系数(%)
苞片和萼片(片)	165	5	13	8.79	1.42	16.15
花瓣(片)	165	5	12	6.93	1.21	17.46
雄蕊(条)	165	66	174	109.60	17.72	16.17
花柱(条)	159	2	5	3.12	0.46	14.74
子房室(室)	136	2	5	3.13	0.39	12.46
每室胚珠(粒)	425	1	10	4.74	1.25	26.37
每果胚珠(粒)	136	6	23	14.82	2.80	18.89

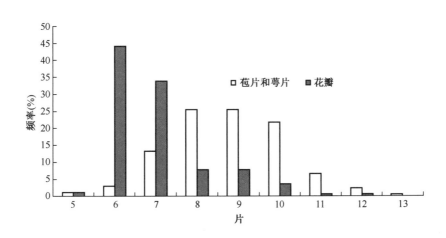

图1 腾冲红花油茶2个花结构数量变异式样

Figure 1 The patterns of number variation in 2 floral structures of *C. reticulata* f. *simplex*

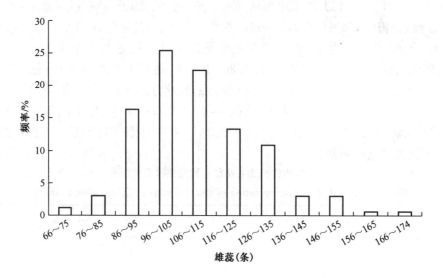

图 2 腾冲红花油茶雄蕊数量变异式样

Figure 2 Thepattern of number variation in stamens of *C. reticulata f. simplex*

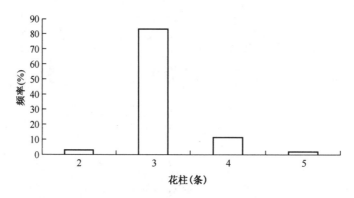

图 3 腾冲红花油茶花柱数量变异式样

Figure 3 Thepattern of number variation in stylets of *C. reticulata f. simplex*

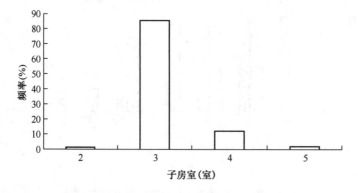

图 4 腾冲红花油茶子房室数量变异式样

Figure 4 Thepattern of number variationinlocules of *C. reticulata f. simplex*

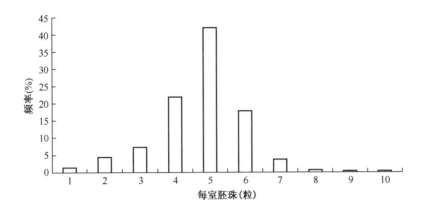

图 5 腾冲红花油茶每室胚珠数量变异式样

Figure 5 The pattern of number variation in ovules per locule of *C. reticulata* f. *simplex*

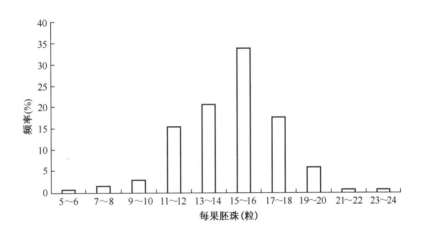

图 6 腾冲红花油茶每果胚珠数量变异式样

Figure 6 The pattern of number variation in ovules per fruit of *C. reticulata* f. *simplex*

2.3 株内和株间差异

由腾冲红花油茶花器官数量方差分析结果可知,苞片和萼片、花瓣、雄蕊、花柱、子房室、每室胚珠和每果胚珠等 7 个花器官观察结构的数量差异主要来自株间,而 7 个花器官观察结构的数量均为差异极显著($P<0.01$)。可见,在所观测的腾冲红花油茶中,单株间在苞片和萼片、花瓣、雄蕊、花柱、子房室、每室胚珠和每果胚珠等 7 个花器官观察结构的数量上差异极显著($P<0.01$)。因此,此 7 个花器官观察结构的数量性状作标准可筛选出符合特定选育目标要求的迥异单株。

2.4 居群内和居群间差异

由腾冲红花油茶居群花器官数量方差分析可知,苞片和萼片、子房室、每室胚珠和每果胚珠等4个数量性状的差异主要来自居群内;而花瓣、雄蕊和花柱等3个数量性状的差异主要来自居群间。其中,只有花瓣和花柱等2个结构的数量差异为极显著($P<0.01$),而其余5个结构的数量差异均为不显著。可见,在所观测的腾冲红花油茶中,只有花瓣和花柱等2个结构的数量在居群间存在显著差异,而且只有依据花瓣和花柱的数量才能筛选出符合特定选育目标要求的迥异居群及单株。

3 结论和讨论

从本研究的结果看,腾冲红花油茶的苞片和萼片5~13片,通常为7~10片;花瓣5~12片,通常为6~7片;雄蕊66~174条,通常为86~135条;花柱2~5条,通常为3条;子房室2~5室,通常为3室;每室胚珠1~10粒,通常为4~6粒;每果胚珠6~23粒,通常为11~18粒。曾有记载苞片和萼片8~10片[3]或9~11片[5],花瓣5~7片[3,5,15],雄蕊142~197条[15],花柱3~5条[3,5]或3~7条[15],子房室3~5室[3,15]。其中,花柱数和子房室数其实与心皮数有关。腾冲红花油茶的心皮连生或合生形成子房,花柱基部至近顶部连生或合生,看似浅裂至深裂。而1个心皮具有1条花柱和1个子房室,因此花柱数与子房室数相同。而前人所记载花柱有6~7条的情况[15]与其所记载的子房室数不相同,而且本研究和其他学者也未观察到,可能是该前人观察记录或撰写时笔误的结果。至于在本研究结果中,存在花柱和子房室的数量相同时而出现频率略有不同的情况,是由于两者在观察时,一个在花期,一个在果期,所记数目不是来自同一朵花的结果。总之,前人所记载的这些数据虽与本研究的结果略有不同,但均在本研究各数量性状的变异范围之内,而前人记载的数据范围普遍偏小,且多限于通常范围,可能是由于观察量没有本研究多的结果。因此,为保证数据的真实可靠性,在观察时应尽可能地增加观察样本量。

在遗传育种的实际工作中,变异尤其是差异显著的变异是选育新品种和优良品种或优良无性系的重要依据。在植株水平,腾冲红花油茶的苞片和萼片、花瓣、雄蕊、花柱、子房室、每室胚珠和每果胚珠等7个花器官观察结构的数量变异主要来自株间,而且差异均为极显著,为特定选育目标要求单株的筛选提供了重要的理论依据和不可或缺的基础数据。在居群水平,腾冲红花油茶仅有花瓣和花柱等2个观察结构的数量在居群间差异为极显著,表明以花瓣和花柱等数量为特定选育目标要求单株可在居群间筛选出来。而在靳高中[16]的研究中,居群间差异极显著的是花萼数和雄蕊数,与本研究的结果为花瓣数和花柱数不同。这可能是由于本研究统计的是苞片和萼片数,而且各自对腾冲红花油茶的分类概念定义不同而造成观察对象不同,因为靳高中采用的腾冲红花油茶的分类概念是广义的而本研究是狭义的。总之,本研究的结果可为腾冲红花油茶新品种和优良品种或优良无性系的选择和培育提供必要的科学指导。

(致谢:在野外工作中,得到了云南省楚雄州林业科学研究所施庭有、周庆宏、陆启华、王之华、毛发恩,云南省楚雄市中山乡林业工作站王云峰等同志的热心支持和积极参与,在此一并致谢!)

参考文献：

[1] 昆明市科协．滇之奇葩：云南山茶花[M]．昆明：云南科技出版社，2010．

[2] Sealy JR. A revisionof thegenusCamellia [M]. London：RoyalHorticultural Society, 1958.

[3] 闵天禄．世界山茶属的研究[M]．昆明：云南科技出版社，2000．

[4] 沈立新，梁洛辉，王庆华，等．腾冲红花油茶自然类型及其品种类群划分[J]．林业资源管理，2009（6）：75~79．

[5] Min T L (Ming TL), Bartholomew B M. Theaceae [A]. In：Wu ZY (Wu CY), Raven PH, HongDY. Flora of China, 12 [M]. Beijing：Science Press；St. Louis：Missouri Botanical Garden Press, 2007：366~478.

[6] 顾志建．云南山茶花四倍体的首次发现及其科学意义[J]．植物分类学报，1997，35（2）：107~116．

[7] 郭磊，王嘉新，皮佳玉，等．腾冲红花油茶多糖体外抗氧化活性研究[J]．中国油脂，2014，39（3）：37~39．

[8] 袁其琼，郭玉红，司马永康，等．云南省腾冲红花油茶资源及其开发利用[J]．陕西林业科技，2017（4）：64~67．

[9] 岳元彦，段成波，李自蕊．腾冲红花油茶不同无性系嫁接成活率研究[J]．林业调查规划，2014，39（3）：149~154．

[10] 袁其琼，郭玉红，司马永康，等．腾冲红花油茶良种选育指标的探讨[J]．林业科技通讯，2017（10）：29~31．

[11] 唐浩君，司马永康，郝佳波，等．极危植物多脉含笑的花部变异研究[J]．西部林业科学，2012，41（3）：14~23．

[12] 朱云凤，郝佳波，司马永康，等．多花含笑的花部数量变异研究[J]．西部林业科学，2012，41（6）：58~62．

[13] 李颖林，董蒙蒙，陈辉，等．锥栗主栽农家品种表型性状变异及选择研究[J]．西南林业大学学报，2018，38（3）：36~43．

[14] 司马永康，余鸿，杨桂英，等．云南省三尖杉属植物的地理分布与环境因子的关系[J]．林业调查规划，2004，29（1）：83~87．

[15] 云南省林业科学研究所．云南主要树种造林技术[M]．昆明：云南人民出版社，1985．

[16] 靳高中．腾冲红花油茶主要性状变异分析[D]．重庆：西南大学，2012．

（本文发表于《西南林业大学学报》，2018年）

油橄榄品种'皮瓜尔'的引种选育

李勇杰,宁德鲁,贺娜,张艳丽,马婷,耿树香

(云南省林业科学院经济林木研究所,云南省木本油料工程技术研究中心,
云南省林业科学院木本油料研发省创新团队,云南 昆明 650201)

摘要:'皮瓜尔'原产西班牙哈恩省,是西班牙主栽的油橄榄品种,我国1979年引入接穗,分别嫁接于湖北武昌、陕西城固、四川三台、西昌及云南昆明等省区。果实卵圆形,平均单果重5.38g,果顶圆,柱头遗存,果基平,果顶具嘴,不对称,鲜果含油率24.69%,油酸含量77.5%,油质佳。在云南省金沙江干热河谷冷冬地区,果实9月下旬成熟。2016年通过云南省林木品种审定委员会良种审定。

关键词:油橄榄;品种;皮瓜尔;引种

油橄榄(*Olea europaea* L.)属木犀科木犀榄属常绿乔木,又名齐墩果,原产小亚细亚,是世界著名的速生、高产、果实含油率高的木本油料树种[1~5]。其鲜果直接冷榨而成的天然食用植物油,营养丰富、抗氧化性强,产品用途广泛,是世界上公认的"植物油皇后",在欧美备受消费者青睐,被誉为"飘香的软黄金"[6~8]。云南省从20世纪60年代初开始油橄榄引种[9],从中选择产量高、含油率高、油质好、抗性强的良种提供生产。

1 引种选育经过

我国1979年3月从西班牙科尔多瓦引进油橄榄良种'皮瓜尔'1~2年生枝条1 033根,分别在湖北武昌、陕西城固、四川三台、西昌及云南昆明地区繁殖。2007年云南省林业科学院又从希腊引入'皮瓜尔'扦插苗,现已正常开花结果。2004~2016年在云南省林业科学院昆明树木园、云南绿原实业发展有限公司永仁油橄榄基地、丽江市三全油橄榄开发有限公司大具油橄榄基地和迪庆藏族自治州德钦县奔子栏乡对'皮瓜尔'进行了区域试验和生产试验。试验结果表明,'皮瓜尔'品种在各地均表现出良好的适应性,生长结实状况良好,早实、丰产、耐寒、大小年现象不严重、抗性好。2016年12月通过云南省林木品种审定委员会良种审定,定名为'皮瓜尔'。

2 主要性状

2.1 植物学特征

'皮瓜尔'树势旺盛,树冠圆锥形。隐芽萌发力强,3~4年生枝甚至老干上都能萌生出新枝条。叶对生,叶面灰绿色,背面银灰色,叶片狭披针形,长5.1cm,宽0.95cm,叶形指数5.4,叶尖渐尖,叶基楔形,全缘,革质,中脉明显。花序中长20~26cm,小花15~20朵,花着生紧密,完全花比率83.4%。

2.2 果实主要经济性状

'皮瓜尔'单果重5.38g,卵圆形,纵径2.45cm,横径1.88cm,果形指数1.3,果顶圆,具嘴,不对称,果基平,果面粗糙,果点明显,大而凹陷。果肉率89.40%,鲜果含油率24.69%,油质佳,不饱和脂肪酸含量84.94%,其中油酸含量77%,维生素E含量高,为380mg·kg^{-1}。果核重0.57g,长椭圆形,纵径1.50cm,横径0.70cm,核形指数2.1,核面粗糙,核纹数量中等。果肉细嫩,可加工成盐渍青果或黑橄榄。

'皮瓜尔'单果重高于对照品种'佛奥',纵径和横径均比'佛奥'大;果核重与'佛奥'一致,纵径和横径均'佛奥'相近;果肉率和含水率均比'佛奥'高;鲜果含油率比'佛奥'低(表1)。

表1 '皮瓜尔'与对照品种'佛奥'的果实主要经济性状

品 种	果径		核径		单果重(g)	果核重(g)	果肉率(%)	含水率(%)	鲜果含油率(%)
	纵径(cm)	横径(cm)	纵径(cm)	横径(cm)					
皮瓜尔	2.45	1.88	1.50	0.70	5.38	0.57	89.40	48.75	24.69
佛奥(对照)	2.23	1.51	1.59	0.74	3.16	0.57	82.02	46.46	25.96

2.3 生长结果特性

'皮瓜尔'扦插苗定植后第3年树高2.2m,东西冠径1.46m,南北冠径1.68m,干径7.54cm。自花结实率低,0.8%。早实、丰产,大小年现象不明显。在云南省楚雄州永仁县油橄榄良种繁育基地,定植后第2年开花结果,第3年株产鲜果6.19kg,平均每667m^{2}产鲜果204.27kg;第5年株产鲜果11.73kg,平均每667m^{2}产鲜果387.09kg;第6年株产鲜果18.38kg,平均每667m^{2}产鲜果606.54kg。

2.4 物候期

在云南省永仁县,'皮瓜尔'芽萌动期2月2~10日,抽梢期2月15~20日,始花期3月5~12日,盛花期3月12~17日,末花期3月17~27日,果实转色期7月18日,9月下旬开始成熟。

2.5 抗性和适应性

'皮瓜尔'是油橄榄栽培品种中适应性最强的,适宜在云南省金沙江干热河谷区冬季冷凉地带及滇中地区种植。'皮瓜尔'对栽培条件要求不严,具有耐寒、耐湿和耐盐碱的特点,在云南海拔1 500~2 200m,年均气温14~21℃,年降水量500~1 000m的地区生长最好,适宜在长日照、夏季雨水偏少、土壤通透性良好的地区栽培。夏季高温、高湿可造成'皮瓜尔'落叶,影响其成花,对油橄榄孔雀斑病敏感。

3 栽培技术要点

3.1 建园

山地或缓坡地建园均可,行株距一般为5m×4m,冬末春初苗木生长相对缓慢期定植为宜。定植前全园深翻 0.8m,按行株距挖 50cm×50cm×50cm 的定植穴,将腐熟农家肥 30~50kg、钙镁磷肥 1kg、生石灰 1kg 与熟土拌匀后施入穴内。苗木定植后浇足定根水,穴面覆盖塑料薄膜。'皮瓜尔'与大多数油橄榄品种一样,自花结实率低,所以定植时需配置授粉品种,授粉品种宜选'莱星',主栽品种与授粉品种的比例可用 4~5∶1 配置。

3.2 整形修剪

树形宜选择自然开心形,定干高度 0.6~0.8m。由下而上选留 3 个生长健壮、分布均匀、与主干夹角 45℃的枝条做主枝,培养成 3 大主枝开心形树冠。每年采果后疏除过密骨干枝,剪去病虫枝、背上枝、交叉枝、平行枝,回缩结果枝组。

3.3 土肥水管理

每年采果后,结合施农家肥扩穴 1 次,穴沟深 30~40cm,宽 30~40cm,每穴施有机肥 30~50kg,追肥 3 次:第 1 次在 1 月油橄榄树萌动后,以氮肥为主;第 2 次在 3 月,第 3 次在 6 月,以磷钾肥为主,同时辅以适量钙肥和硼肥。施肥方法可采用条状沟施肥或放射状施肥。灌排水根据土壤墒情进行。

3.4 病虫害防治

'皮瓜尔'在云南的主要病害为油橄榄孔雀斑病和油橄榄细菌性炭疽病,常见害虫为介壳虫、木蠹蛾和金龟子等。本着"预防为主、科学治理"的方针,在加强肥水基础上,做好油橄榄病虫害的防治。

参考文献:

[1] 徐纬英. 中国油橄榄种质资源与利用[M]. 长春:长春出版社,2001.
[2] 邓明全,俞宁. 油橄榄引种栽培技术[M]. 北京:中国农业出版社,2000.
[3] 贾瑞芬,肖千文,李进峰. 油橄榄种质资源研究进展[J]. 北方园艺,2006(5):64~66.
[4] 徐莉,王若兰,高雪琴. 橄榄油的类型和特性[J]. 中国食物与营养,2006(10):46~48.
[5] Delgado A,Benlloch M,Fernandezescobar R. Mobilization of boron in olive trees during flowering and fruit-development [J]. Hortscience,1994,29(6):616~618.
[6] Blekas G,Vassilakis C,Harizanis C,et al.Biophenols in table olives[J]. Journal of Agricultural and Food Chemistry,2002,50(13):3688~3692.
[7] 谢普军,黄立新,张彩虹,等. 佛奥油橄榄叶营养成分测定与分析[J]. 林产化学与工业,2014,34(4):97~101.
[8] 王成章,高彩霞,叶建中,等. HPLC 研究油橄榄叶中橄榄苦苷的含量变化规律[J]. 林产化学与工业,2008,28(6):39~43.
[9] 宁德鲁,杨卫明. 油橄榄良种选育与栽培[M]. 昆明:云南科技出版社,2013.

(本文发表于《中国果树》,2017 年)

Transcriptomics and Comparativeanalysis of Three Juglans Species, *J. regia*, *J. sigillata* and *J. cathayensis*

WU Tao, XIAO Liang-jun, CHEN Shao-yu, NING De-lu*

Institute of Economic Forest, Yunnan Academy of Forestry, Kunming 650201, China

Abstract: Walnut (Juglans) has been globally cultivated for its valuable nut, which has abundant polyunsaturated fatty acids and proteins. In China, only the Persian or English walnut (*J. regia*) and Yunnan or iron walnut (*J. sigillata*) are commercially cultivated for nut production, and Chinese butternut (*J. cathayensis*) is commonly used as rootstock and potential breeding material. However, few genomic resources are available for these non-model plants, particularly the last two species. Hence we present the sequencing, de novo assembly and annotation of transcriptomes from fresh leaves of the three Juglans species by RNA-seq technology and bioinformatics analysis to discover a collection of SSR and SNP markers, and suggestion of interesting differences for further genetic improvements. In total, 59 035 134 (7.38 G bp), 43 949 544 (5.5 G bp) and 58 609 226 (7.32 G bp) high quality clean reads were generated from cDNA libraries of *J. regia*, *J. sigillata* and *J. cathayensis*, respectively. A total of 192 360 unigenes longer than 200 bp were de novo assembled, 92 858 (48.20%) unigenes were annotated, and 32 110 CDSs (16.70%) were deduced. The potential function of each unigene was classified based on COG and GO database. 5 683 differentially expressed genes (DEGs) were enriched in KEGG pathways. A total of 41 141 SSRs and 206 355 SNPs were identified as potential molecular markers. The raw reads of transcriptome from *J. regia* (accession number SRR1767234), *J. sigillata* (accession number SRR1767236) and *J. cathayensis* (accession number SRR1767237) were deposited in NCBI database. Our transcriptome data enrich the genomic resource of Juglans species and will be essential to accelerate the process of molecular research and breeding.

Key word: Walnut; Juglans; Transcriptome; High-throughput sequencing; Differential expression genes.

Abbreviations: BLAST_Basic Local Alignment Search Tool; CDS_Coding Region Sequence; CO G_Cluster of Orthologous Groups of proteins; GO_Gene Ontology; KEGG_Kyoto Encyclopedia of Gene and Genomes; DEGs_differentially expressed genes; SSR_simple sequence repeat; SNP_single nucleotide polymorphism

Introduction

All species of walnuts (Juglans) produce nuts, but the Persian or English walnut (*J. regia*) is the most widespread species cultivated for nut production. It is globally popular and valued for its nutritional, health and sensory attributes (Martínez et al., 2010). China is considered to be one of the main walnut production countries (Britton et al., 2008). Annual world walnut production totaled 3 418 559 metric tons in 2012, of which, China produced 1 700 000 tons, followed by the Iran (450 000 T), USA (425 820 T), Turkey (194 298 T), Mexico (110 605 T), Ukraine (96 900 T), India (40 000 T), Chile (38 000 T), France (36 425 T), and Romania (30 546 T) (FAOSTAT, updated 24 December 2014, http://faostat.fao.org/faostat/).

In China, walnut is distributed from 21°29′ to 44°54′ north latitude and from 75°15′ to 124°21′ east longitude, and there are six species in the walnut genus (Juglans). They are *J. regia*, *J. sigillata*, *J. cathayensis*, *J. mandshurica*, *J. cordiformis* and *J. hopeiensis*, and all species produce nuts, but only *J. regia* and *J. sigillata* are commercially cultivated for nut production, and no less than 17 authorized or approved cultivars of *J. sigillata* have been popularized, including 'Yangpao', 'Santai', 'Niangqing'; the others are wild. Nevertheless, the Chinese walnut or Chinese butternut, *J. cathayensis*, is a vigorous tree, commonly used as rootstock for *J. regia* to provide tolerance of biotic/abiotic stresses in regions of the Yangtze River in China. Because of climatic diversity, high heterozygosity and sexual propagation, there exist very rich genetic materials among Chinese walnut population (McGranahan and Leslie, 1990, 2009). The considerable variations between *J. regia* and *J. sigillata*, particularly in nut size and shape, led taxonomists to describe other additional species that have not been widely accepted but that illustrate some of the diversity (Dode, 1909). Undoubtedly, the best genotypes selected for nut improvement and the expansion of commercial growth range are probably possible based on germplasm evaluation and breeding.

Although walnut has been cultivated for centuries, walnut breeding starts recently and only a few systemic molecular studies on walnut have been reported (McGranahan and Leslie, 1990, 2009; Britton *et al.*, 2008). Because of its commercial value, far more gene sequences are available for *J. regia* than other members of the same genus, even though, the number of nucleotide sequences of *J. regia* is still smaller than other crops (Dandekar *et al.*, 2005). As of December 2014, GenBank (National Center for Biotechnology Information, NCBI), the public repository for DNA sequence data in the USA, listed 6 287 nucleotide sequences for *J. regia*, including nuclear and chloroplast genes, and expressed sequence tags (ESTs). The derived information is sporadic and limited, and insufficient to use, especially to picture a global transcriptome profile for genetic improvements on the important agronomic and economic traits, such as improved climate adaptation (late budbreak, low chilling requirement or winter hardiness), early fruiting and high productivity (lateral fruitfulness), and disease tolerance (blight and anthracnose). This situation is merely a question of weeks with the increasing availability of low cost, high-throughput sequencing technologies (Metzker, 2010; Grabherr *et al.*, 2011).

In this study, total RNAs, extracted from the leaves of three Juglans species, *J. regia*, *J. sigillata* and *J. cathayensis*, were sequenced using RNA-seq technology, and the assembled sequences were analyzed. In addition, a collection of cDNA-derived SSR and SNP markers were developed and characterized. The data obtained here might be beneficial for cloning genes of interest and genetic improvements of Juglans.

Results and Discussions

Illumina sequencing and de novo assembly

The cDNA libraries of *J. regia* (JRE), *J. sigillata* (JSI) and *J. cathayensis* (JCA) were generated using the mRNA-Seq procedure for transcriptome sequencing on an Illumina Hiseq^{TM} 2000 platform. Using Illumina paired-end sequencing technology, each sequencing feature can yield 2 × (100±25 bp) independent reads from both ends of a 200—300 bp cDNA fragment. After data filtering, a total of 59 035 134, 43 949 544 and 58 609 226 high quality clean reads from cDNA libraries of JRE, JSI and JCA (a total of 7.38×10^9, 5.50×10^9 and 7.32×10^9 base pairs) were obtained, respectively (Tab. 1). A total of 113 043, 115 036 and 123 567 transcripts longer than 200 bp were generated with a mean length of 1 103 bp, 1 066 bp and 1 063 bp and a N50 of 1 886 bp, 1 814 bp and 1 869 bp for JRE, JSI and JCA, respectively. To remove any redundancies, transcripts were clustered using the TGICL (Pehtea et al., 2003) and sequences not extended on either end were defined as unigenes. A total of 60 710, 62 415 and 69 235 unigenes longer than 200 bp were generated by the clustering, with a mean length of 712 bp, 700 bp and 674 bp and a N50 of 1 300 bp, 1 283 bp and 1 204 bp for JRE, JSI and JCA, respectively (Tab. 1). The raw reads of transcriptome from JRE, JSI and JCA were deposited in NCBI database (Accession numbers: SRR1 767 234 for JRE, SRR1 767 236 for JSI, SRR1 767 237 for JCA).

Table 1 Statistics for pyrosequencing of the three Juglans species.

Species	J. regia	J. sigillata	J. cathayensis
Total no. of reads	59 035 134	43 949 544	58 609 226
Total size (bp)	7.38×10^9	5.50×10^9	7.32×10^9
Total no. of transcripts	113 043	115 036	123 567
Total size wihtin transcripts (bp)	124 693 119	122 644 019	131 347 815
Average length of transcripts (bp)	1 103	1 066	1 063
N50 of transcripts	1 886	1 814	1 869
N90 of transcripts	447	426	408
Total no. of unigenes	60 710	62 415	69 235
Total size wihtin unigenes (bp)	43 208 532	43 696 782	46 668 510
Average length of unigenes (bp)	712	700	674
N50 of unigenes	1 300	1 283	1 204
N90 of unigenes	273	269	262
Unigenes with full-length ORF	9 162 (15.09%)	11 266 (18.05%)	11 682 (16.87%)
No. of unigenes annotated	26 704 (43.98%)	32 055 (51.35%)	34 099 (49.25%)

On average, this assembly produced a substantial number of large unigenes: 22 708 unigenes (35.41%) longer than 500 bp, 12 333 unigenes (19.23%) longer than 1 000 bp, and 4 837 unigenes (7.54%) longer than 2 000 bp (Fig. 1). These results suggested that the number of unigenes assembled from JRE, JSI and JCA were close to each other. The genome size of one of three walnut species (J. regia, C = 0.62 pg, approximately 606 Mb) was recently described (Bennett and Leitch, 2012), but the percentages of the transcribed genomes remain unknown. Thus, it is difficult to predict the depth of coverage of the walnut transcriptome by our de novo assembled sequences.

One of the most important questions in this type of study is whether the short reads were correctly assembled. However, the presently available publications on de novo assembly are very limited, especially for non-model organisms. As no generally accepted protocol for evaluating such an assembly exists, it is therefore a great challenge to make a reliable judgement or validation of an optimal assembly without the reference genomic sequence. Fortunately, many ESTs and genomic survey sequences deposited in NCBI database can serve as a reference. An alternative approach to experimental validation is to conduct computational analysis. We selected 49 unigenes from this de novo assembly data, which could be matched withnucleotide sequences of *J. regia* in GenBank (as reference sequence) through megablast search. There was 32 783 bps (96.0%, 32 783/34 188) of 49 unigenes producing identical alignments (Unigene sequences available in Supplementary File 1, Table S1). Moreover, although a total of 3 219 out of 4 837 unigenes longer than 2 000 bp had significant matches that covered more than 90% of their corresponding subjects, they were classified as putative, hypothetical or predicted protein. Overall, these results suggested that our data assembly was of high quality, but also indicated that the major assembled unigenes have not been sequenced previously or far not well been characterized.

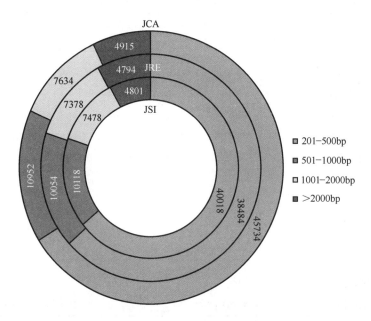

Figure 1 Length distribution of assembled unigenes of JRE, JSI and JCA. The outer, middle and inner cycles represent the distribution of JCA, JRE and JSI unigenes, respectively. The data in each colour region indicates the number of unigenes in this range of nucleotide length. Blue region indicates unigene size ranges from 201 to 500 bp, red 501-1 000 bp, yellow 1 001-2 000 bp, green above 2 000 bp.

Annotation and CDS prediction

For annotation, homologs of the unigenes were searched in seven databases: NT, NR, Swiss-Prot, KOG, KO, GO and PFAM. A total of 26 704 (43.98%) JRE unigenes, 32 055 (51.35%) JSI unigenes and 34 099 (49.25%) JCA unigenes were annotated to at least one of the seven databases (Tab. 2). Among them, most unigenes could be annotated to the NR database, followed by the Swiss-Prot, GO, PFAM, NT, KOG and KO database. Based on NR annotation and E-value distribution, average 70.58% of the mapped unigenes had high homology (E-value<1e^{-30}), and 33.94% showed very strong homology (E-value<1e^{-100}) with known protein sequences (Fig. 2A). The similarity distribution showed 77.34% of the annotated sequences have a similarity greater than 60% (Fig. 2B). Averagely, 80.73% of unigenes could be annotated to the top 5 species in the species distribution (Fig. 2C). After BLASTx analysis, 96.33% of the unigenes over 2 000 bp had BLASTx hits, while only 41.28% of the unigenes shorter than 500 bp had homologs. It indicates that the length of query sequence was important for determining the level of significance of the BLASTx match. Longer unigenes were more likely to have BLASTx hits in protein databases.

Table 2 Summary of the unigene annotations of the three Juglans species.

Annotation database	J. regia	J. sigillata	J. cathayensis
NR	25 970 (42.77%)	29 417 (47.13%)	31 095 (44.91%)
GO	11 398 (18.77%)	22 471 (36%)	23 923 (34.55%)
SwissProt	17 883 (29.45%)	20 179 (32.33%)	21 006 (30.34%)
PFAM	8 816 (14.52%)	19 074 (30.55%)	20 222 (29.2%)
NT	15 100 (24.87%)	16 323 (26.15%)	16 835 (24.31%)
KOG	8 964 (14.76%)	10 374 (16.62%)	10 600 (15.31%)
KO	7 703 (12.68%)	8 847 (14.17%)	9 101 (13.14%)
Annotated in all Databases	3 118 (5.14%)	4 151 (6.65%)	4 319 (6.23%)
Annotated in at least one Database	26 704 (43.98%)	32 055 (51.35%)	34 099 (49.25%)
Total Unigenes	60 710 (100%)	62 415 (100%)	69 235 (100%)

The coding sequence (CDS) of all unigenes was predicted by ESTScan or BLAST with an E-value threshold of 10^{-5} in the NR and Swiss-Prot protein database. The front and back sequences beyond CDS in the unigene were considered as the 5' and 3' UTR sequences. A total of 11 682 JCA unigenes (16.87%) and 11 266 JSI unigenes (18.05%) contained the 5' and 3' UTR sequence, and the full-length open reading frames (ORF), whereas the number of unigenes containing these in JRE transcriptome was only 9 162 (15.09%) (Table 1).

Functional classification

In order to classify the potential function of each unigene, all walnuts unigenes were aligned to the KOG database. The result showed that all the JRE, JSI and JCA unigenes covered all 26 KOG functional categories (Fig 3). The top 10 categories were R, O, T, J, C, K, U, A, G, I, respectively. Another alternative approach, GO assignments were used to classify the functions of all unigenes. Based on sequence similarity, 322 098 unigenes of all walnuts were assigned to one or more ontologies. Totally, 147 316 unigenes were grouped under biological processes, 102 789 unigenes under cellular components, 71 993 unigenes under molecular functions. Binding (32 676 unigenes, 45.39%) and catalytic activity (28 793 unigenes, 40.00%) were the most highly represented classes under the molecular function category. For the biological process class, the

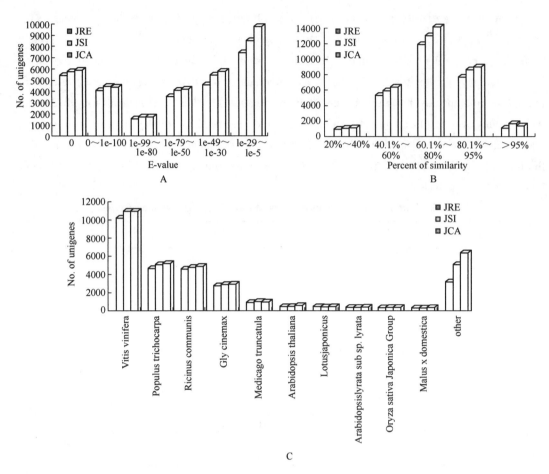

Figure 2 Homology alignments of unigenes of JRE, JSI and JCA against NR database.
E-value distribution (A), similarity distribution (B) and species distribution.

assignments were mainly given to the cellular process (34 592 unigenes, 23.48%) and metabolic process (33 192 unigenes, 22.53%). In the cellular components category, the largest proportion of transcripts was involved in the 'cell' (47.78%), followed by the 'cell part' (33.21%), 'organelle' (5.35%) 'macromolecular complex' (7.48%) and membrane (5.35%) (Fig. 4). The GO results were similar to herbaceous plants Eleusine indica (Shu et al., 2015), Daucus carota var. sativus (Iorizzo et al., 2011) and woody plants Camellia sinensis (Shi et al., 2011), Hevea brasiliensis (Xia et al., 2011), which suggested that our unigenes are broadly representative of the flesh plant leaf transcriptome. The concordance in the overall distributions trend of three species suggests that our library sampled widely across GO categories and provided a good representation of the Juglans species leaf transcriptome. These KOG and GO annotations provide a valuable clue for investigating the specific processes and molecular function of Juglans species.

It can be seen from Figure 4 that the number of classified unigenes of *J. regia* is robustly smaller than other two species. One possible explanation for this is that a unigene or its product can be annotated to more than one term of each ontology, at any level within each ontology. Another possible reason is that many of the assembled sequences may represent distinct non-overlapping regions of the same genomic locus. Finally, it is likely due to the sequence data of *J. regia* far outnumber *J. sigillata* and *J. cathayensis* in GO database, many of short unigenes have been assembled together into full-length unigenes, reducing the total number of unigenes in the transcriptome.

When we collected the samples for sequencing, the walnut trees were actively growing. For example *J. regia* under the fourth level GO terms, vigorous cell growth and metabolic processes were reflected by the assignation of 2 152 transcripts to 'cellular biosynthetic process', 2 270 to 'organic substance biosynthetic process', 2 192 to 'organic cyclic compound metabolic process', 2 133 to 'protein metabolic process', 2 121 to 'cellular nitrogen compound metabolic process', 2 071 to 'cellular aromatic compound metabolic process' and 1 343 to 'organic acid metabolic process'. The plant has a strong self-regulation ability to respond to stimuli, and this was supported by the assignation of 2 685 transcripts to 'response to stimulus', 3 339 to 'biological regulation' and 3 112 to 'regulation of biological process'.

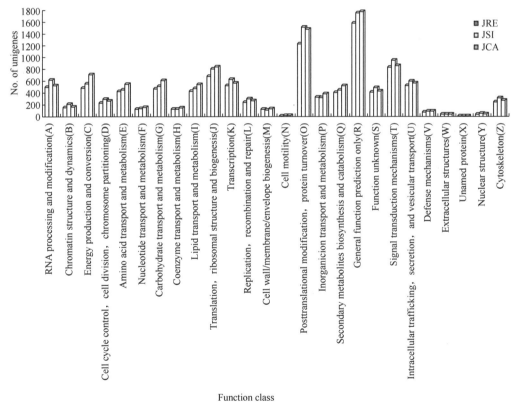

Figure 3 Functional classifications of KOG terms of unigenes from JRE, JSI and JCA.

Comparative analysis of differential expressed genes (DEGs)

The coefficient of determination (R^2) of gene expression profile reflects the similarity between samples. The higher the R^2, the stronger the similarity. The R^2 was higher between JSI

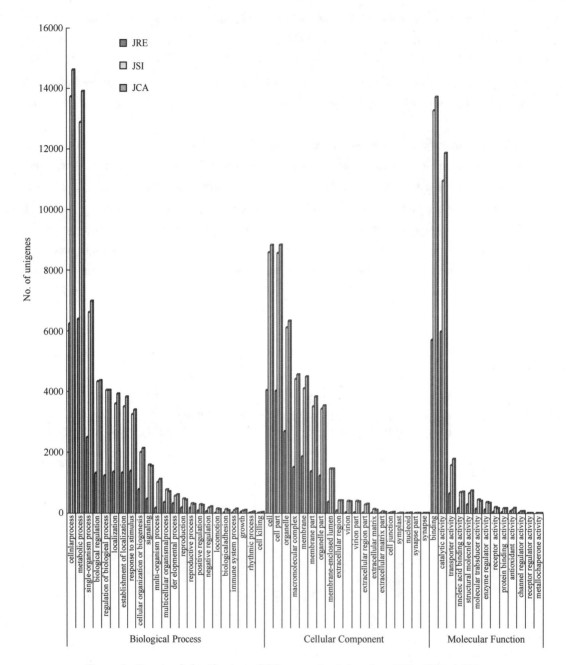

Figure 4 Functional classifications of GO terms of unigenes from JRE, JSI and JCA.

and JRE ($R^2 = 0.708$) than that between JRE and JCA ($R^2 = 0.533$) or JSI and JCA ($R^2 = 0.507$). The result of hierarchical clustering analysis within three species showed also the gene expression pattern was closer between JSI and JRE than between JSI and JCA or JRE and JCA (Fig. 5). The results about gene expression of three species, to sum up, agreed with traditional botany taxonomy, in which, JRE and JSI belong to same section Juglans but JCA belong to another section Cardiocaryon (Manning, 1978; Wang et al., 2015).

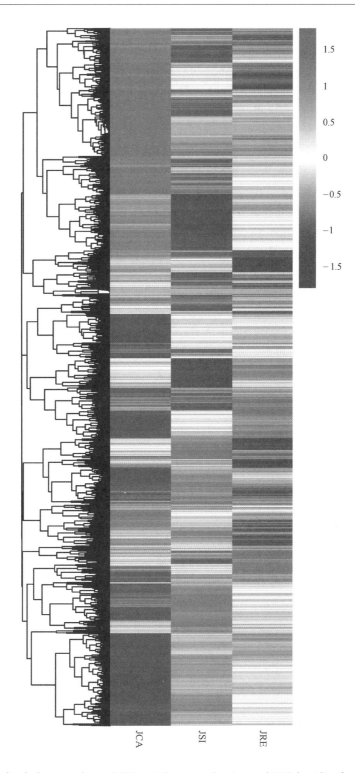

Figure 5 Hierarchical cluster analysis of differential expressed unigenes (DEGs) within three species. Expression differences are shown in different colours. The colour from blue to red means that transcript abundance of unigenes was from relatively low to relatively high. The tags and counts of DEGs were listed in supplementary file.

To study the conservation and divergence of global expression patterns, we performed comparative analysis for all unigenes from three species. As shown in Volcano plot (Fig. 6) and Venn diagram (Fig. 7), a total of 684 genes were differentially expressed between JREVs JSI analysis, of which, 587 and 97 genes showed the common and exclusive differential expression with other analyzed groups respectively. Totally 2 215 DEGs were identified in JCA Vs JSI analysis, of that 1 841 and 374 genes showed the common and exclusive differential expression with rest of the analyzed groups respectively. Highest number of genes, 2 784, found differentially expressed in JRE Vs JCA analysis, of which 1 846 and 938 genes showed the common and exclusive differential expression with rest of the analyzed groups respectively. 211 genes were commonly expressed in between both JRE and JCA samples compared with JSI one. 1 470 genes were commonly expressed in between JCA Vs JSI and JRE Vs JCA samples, while 216 genes showed the common expression in between JRE Vs JCA and JRE Vs JSI samples. Of all, only 160 genes were commonly expressed in all three species, among that, there were 20 ones involved oxidation-reduction process, 11 ones involved regulation of transcription, 19 ones were hypothetical in nature and have not yet been annotated for coding specific protein. The DEGs having least p values and higher fold changes within JRE, JSI and JCA species were furthersorted out based on the functional involvement in fatty acid metabolism, regulation of transcription and response to stress, important genes are listed in Table 3.

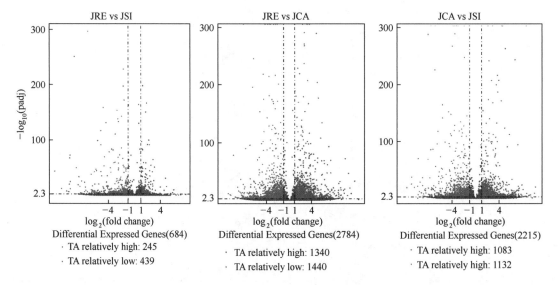

Figure 6 Volcano plot of differential expressed unigenes (DEGs) between two species. X axis represents the fold change (log transformed), and Y axis represents the p value (log transformed). If the intensity ratio of a unigene between two species is more than 2 fold (Transcript Abundance of unigenes was relatively high, red dot) or less than 2 fold (TA of unigenes was relatively low, green dot) having p value less than 0.05, considered as a DEG. Those dots shown in blue are unigenes that did not show obviouschanges and did not identified as DEGs.

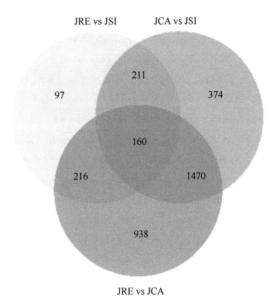

Figure 7 Venn analysis of differentially expressed unigenes (DEGs) within JRE, JSI and JCA.

Table 3 The differential expressed genes (DEGs) identified from transcriptome data within JRE, JSI and JCA

Gene ID	Gene Name	Gene description	Size (bp)	Fold change / log2	
				JRE Vs JSI	JRE Vs JCA
comp115501_c0	FAD8	omega-3 fatty acid desaturase (delta-15 desaturase)	1636	-2.9833	-4.2523
comp125017_c2	FAD2	omega-6 fatty acid desaturase (delta-12 desaturase)	1859	-1.0242	-2.3015
comp118997_c0	KCS	3-ketoacyl-CoA synthase 6	913	-2.1354	-3.7428
comp116221_c0	KCS	3-ketoacyl-CoA synthase 12	1749	-3.2335	-4.9476
comp120968_c0	KCS	3-ketoacyl-CoA synthase 21	2265	-1.6482	-3.563
comp114842_c0	CHS	naringenin-chalcone synthase	1656	1.0658	-3.3973
comp115430_c0	CHS	Chalcone synthase	1590	-1.2052	-2.7808
comp125700_c0	WSD1	Wax ester synthase-like Acyl-CoA acyltransferase	1971	-2.2993	-3.7916
comp117542_c0	GDE1	Glycerophosphoryl diester phosphodiesterase	1822	-1.7483	-3.0505
comp116141_c0	CICLE	GDSL-like Lipase/Acylhydrolase	1631	1.3438	3.7606
comp114457_c0	AUX28	Auxin-induced protein	1281	-1.1681	-2.2551
comp119129_c0	TEM1	AP2/ERF and B3 domain-containing transcription represso	1563	-1.0104	1.2174
comp121653_c0	HSFF	Heat stress transcription factor A-2b	2841	-2.5351	-1.1395
comp123626_c0	HSF30	Heat shock factor protein	2684	-1.6521	1.3662
comp95810_c0	NAC29	NAC transcription factor 29	1318	1.6982	4.2963
comp123093_c0	TAG	WRKY transcription factor 40	4696	1.4162	2.7957
comp114659_c0	HSP20	Heat shock protein	1052	-1.7824	5.7675
comp101719_c0	ARG2	Indole-3-acetic acid-induced protein	1866	1.1289	3.1685

KEGG pathway enrichment analysis of differential expressed genes (DEGs)

KEGG is a database resource for understanding high-level functions from molecular-level information, especially large-scale molecular datasets generated by high-throughput experimental technologies (Kanehisa et al., 2008). DEGs were subjected to KEGG pathway enrichment analysis, and 18.99% (1 079/5 683) of the DEGs could be annotated, which were associated with 417 KEGG pathways (Supplementary File1, Table S2, Table S3, Table S4,). The 20 top KEGG pathways with the highest representation of the DEGs are shown in Table 4. The Biosynthesis of amino acids (ko01230), Starch and sucrose metabolism (ko00500), Phenylpropanoid biosynthesis (ko00940), Cysteine and methionine metabolism (ko00270), Phenylalanine metabolism (ko00360), Photosynthesis (ko00195), Phenylalanine, tyrosine and tryptophan biosynthesis (ko00400), Flavonoid biosynthesis (ko00941) and Fatty acid elongation (ko00062) pathways are significantly enriched both JRE Vs JCA and JSI Vs JCA. Nevertheless, Protein processing in endoplasmic reticulum (ko04141), Phagosome (ko04145), Gap junction (ko04540) and Phenylpropanoid biosynthesis (ko00940) pathways are significantly enriched in JRE Vs JSI, and the number of DEGs was less than other comparisons.

SSR and SNP discovery

Transcriptomes are an important resource for the rapid and cost-effective development of genetic markers (Liao et al., 2014). The molecular markers derived from the transcribed regions are more conservative, providing a greatest potential for identifying functional genes. Among the various molecular markers, simple sequence repeats (SSRs) and single nucleotide polymorphisms (SNPs) are highly polymorphic, easier to develop, and serve as a rich resource of diversity (Parchman et al., 2010; Ruperao & Edwards, 2015). To detect new molecular makers, all of the unigenes were used to mine potential SSRs motifs using the MISA software. In total, 34 011 unigenes from 185 912 sequences examined (116 625 163 bp) contained 41 141 SSRs, from which 5 835 unigenes had more than one SSR marker signature (Detailed statistics of SSRs available in Supplementary File 1, Table S5). The mono-nucleotide (19 356, 47.05%), di-nucleotide (16 103, 39.14%) and tri-nucleotide (5 120, 12.45%) repeat motifs had the highest frequencies. All SSRs were further counted based on the number of repeat units (Fig 8). After designing and filtering primers, 19 784 SSR markers were found to have at less one primer (6 674 for JRE, 6 077 for JSI and 7 033 for JCA) (Supplementary File 1, Table S6, Table S7, Table S8). This data could lay a platform for better understanding the polymorphisms of Juglance species.

GATK2 software was used to perform SNP calling (McKenna et al., 2010). In total, 61 422, 91 101 and 53 832 candidate SNPs were identified in *J. regia*, *J. sigillata* and *J. cathayensis*, respectively (Table 5). The average SNP frequency was one SNP per 266 base pairs. The majority of SNPs (76.56%) were detected in transcripts ranging from 100 bp to 1 100 bp. Howerer, the number of SNPs per transcripts increased with transcripts size, indicating that larger datasets with greater transcripts size could be use to identify more SNPs. The distribution of substitution types was shown in Table 5. A greater number of transitions (122 839) than transversions (83 516) were identified, and the ratio between transitions and transversions was 1.47.

Table 4 10 top KEGG pathways with high representation of the DEGs.

Rank	JRE vs JCA			JSI vs JCA			JRE vs JSI		
	pathway	Pathway ID	No. of DEGs	pathway	Pathway ID	No. of DEGs	pathway	Pathway ID	No. of DEGs
1	Biosynthesis of amino acids	ko01230	44 (8.66%)	Biosynthesis of amino acids	ko01230	39 (9.18%)	Protein processing in endoplasmic reticulum	ko04141	22 (15.17%)
2	Phenylpropanoid biosynthesis	ko00940	25 (4.91%)	Starch and sucrose metabolism	ko00500	21 (4.94%)	Phagosome	ko04145	9 (6.21%)
3	Plant hormone signal transduction	ko04075	25 (4.91%)	Cysteine and methionine metabolism	ko00270	18 (4.24%)	Gap junction	ko04540	7 (4.83%)
4	Starch and sucrose metabolism	ko00500	25 (4.91%)	Phenylpropanoid biosynthesis	ko00940	18 (4.24%)	Phenylpropanoid biosynthesis	ko00940	7 (4.83%)
5	Cysteine and methionine metabolism	ko00270	19 (3.73%)	Phenylalanine metabolism	ko00360	14 (3.29%)	MAPK signaling pathway	ko04010	6 (4.14%)
6	Phenylalanine metabolism	ko00360	15 (2.95%)	Flavonoid biosynthesis	ko00941	12 (2.82%)	alpha-Linolenic acid metabolism	ko00592	6 (4.14%)
7	Photosynthesis	ko00195	14 (2.75%)	Phenylalanine, tyrosine and tryptophan biosynthesis	ko00400	12 (2.82%)	Endocytosis	ko04144	6 (4.14%)
8	Flavonoid biosynthesis	ko00941	13 (2.55%)	Photosynthesis	ko00195	11 (2.59%)	Peroxisome	ko04146	6 (4.14%)
9	Phenylalanine, tyrosine and tryptophan biosynthesis	ko00400	13 (2.55%)	Sulfur metabolism	ko00920	8 (1.88%)	Flavonoid biosynthesis	ko00941	5 (3.45%)
10	Fatty acid elongation	ko00062	12 (2.36%)	Fatty acid elongation	ko00062	8 (1.88%)	Cysteine and methionine metabolism	ko00270	5 (3.45%)

Among the transition, the number of C/T transitions was a little greater than that of G/A in JRE and JSI, however, the opposite happened in JCA. Taken as a whole, A/T transversions were more infrequent than other three types among all species. Similar results were found in *J. regia* (Liao *et al.*, 2014) and Citrus clementina (Terol *et al.*, 2008). The availability of large numbers of SNPs should facilitate population genetics and gene-based association studies in Juglans species.

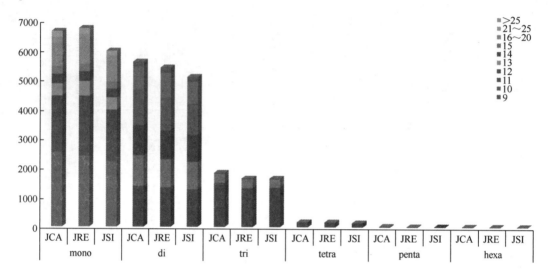

Figure 8 SSRs counts based on the number of motif repeat units.

Table 5 Statistics of SNPs generated among three Juglans species

Types	JRE	JSI	JCA
Transitions	36408 (59.28%)	54458 (59.78%)	31973 (59.39%)
A-G	18167 (29.58%)	27476 (30.16%)	16197 (30.09%)
C-T	18241 (29.70)	26982 (29.62%)	15776 (29.31%)
Transversions	25014 (40.72%)	36643 (40.22%)	21859 (40.61%)
A-C	6104 (10.00%)	9131 (10.02%)	5248 (9.75%)
A-T	7688 (12.52%)	10996 (12.07%)	6923 (12.86%)
C-G	4783 (7.79%)	7036 (7.72%)	4229 (7.86%)
T-G	6439 (10.48%)	9480 (10.41%)	5459 (10.14%)
non coding SNP	44459 (72.38%)	65190 (71.56%)	36683 (68.14%)
coding SNP	16963 (27.62%)	25911 (28.44%)	17149 (31.86%)
synonymous	16883 (27.49%)	25806 (28.33%)	17060 (31.69%)
nonsynonymous	80 (0.13%)	105 (0.12%)	89 (0.17%)
Total	61422 (100%)	91101 (100%)	53832 (100%)

qPCR validation

The experimental validation has been done on 9 randomly selected transcripts (Supplementary File 1, Table S9) in order to confirm DEGs analysis of RNA-seq data by qPCR. Analysis results in leaves of three Juglans species confirmed the relative amounts of these transcripts observed with RNA-seq, with a high correlation ($R^2 = 0.89$) of fold change between RNA-seq and qPCR data (Figure 9).

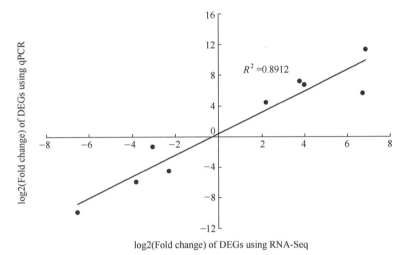

Figure 9 qPCR validation. EGs detcted through RNA-Seq in our study were validated using qPCR. Nine DEGs were randomly selected for validation. The corralation between RNA-Seq and qPCR was shown.

Materials and Methods

Plant materials and RNA extraction

In July 2014, fresh leaves of *J. regia*, *J. sigillata* and *J. cathayensis* were collected from the walnut germplasm resources garden of Yunnan Academy of Forestry. Tissue samples were frozen in liquid nitrogen and stored at −80℃ until RNA extraction.

Total RNA of plant materials was extracted using a RNeasy Mini kit (QIAGEN, Shanghai, China). The concentration and integrity of the RNA was measured and assessed using the Qubit 2.0 Flurometer (Life Technologies, CA, USA) and the Agilent Bioanaylzer 2 100 system (Agilent Technologies, CA, USA). Finally, one of the best RNA samples (OD260/280 ⩾ 1.8, concentration ⩾ 100ng/μl, RIN (RNA Integrity Number) ⩾ 8.0) for each material was used for the subsequent experiments.

Library construction and sequencing

A total amount of 3 μg total RNA per sample was used to construct the cDNA library using NEBNext Ultra™ RNA Library Prep Kit for Illumina (NEB, USA) following manufacturer's recommendations and index codes were added to attribute sequences. Three cDNA libraries were constructed and sequenced on an Illumina Hiseq™ 2 000 platform and paired-end reads were generated at Novogene Bioinformatics Technology Co. Ltd, Beijing, China (www.novogene.cn).

De novo assembly and gene annotation

Raw data (raw reads) of fastq format were firstly processed through in-house perl scripts. After filtering the raw reads, de novo assembly of the transcriptome was carried out with a short reads assembling program - Trinity (Grabherr et al., 2011). Trinity connected the transcripts and obtains sequences for as long as possible. Sequences not extended on either end were defined as unigenes.

The generated unigenes of *J. regia* (JRE), *J. sigillata* (JSI) and *J. cathayensis* (JCA) were each searched against the public databases for annotation, including the NCBI non-redundant protein sequences (NR) database, NCBI nucleotide sequences (NT) database, eukaryotic ortholog groups (KOG) database, KEGG ortholog (KO) database, Swiss-Prot protein database, Gene Ontology (GO) database, and protein family (PFAM) database. The CDS of unigenes was predicted according to the Wang's method (Wang et al., 2011).

Identify and enrichment analysis of differentially expressed genes (DEGs)

Prior to differential gene expression analysis, for each sequenced library, the read counts were adjusted by edgeR program package through one scaling normalized factor. Differential expression analysis of two samples was performed using the DEGseq R package (Wang et al., 2010). Those with q value < 0.005 and |log2(fold change)| > 1 were set as the threshold for significantly differential expression. The statistical enrichment of DEGs in KEGG pathways was performed using the KOBAS software (Mao et al., 2005).

SSRs and SNPs marker identification

Using MISA software (http://pgrc.ipk-gatersleben.de/misa/) (Thiel et al., 2003), the potential SSR markers with motifs ranging from mono- to hexa-nucleotides in size were detected among the unigenes. The minimum of repeat units were set as follows: ten for mononucleotide, six for di and five for tri-, tetra-, penta- and hexa-nucelotides. Primer pairs flanking each SSR loci were designed using the Primer3 program (http://primer3.ut.ee/). Sequencing data between samples were compared with the unigene database, using GATK2 software (McKenna et al., 2010), which builds consensus sequence, and then analyzes samples to get SNP loci.

qPCR validation

A set of nine genes, five be of relatively high abundance and four be of relatively low abundance in RNA-seq analysis, was selected for real-time reverse transcription-PCR (qPCR) in order to validate the DEGs analysis results (Supplementary File 1, Table S9). Total RNA was extracted from the leaves using Trizol (Invitrogen), and then be subjected to RQ1 RNase-free RDNase (Promega) digestion in order to remove any residual genomic DNA contamination. The first strand cDNA was synthesized by RevertAidTM First Strand cDNA Synthesis Kit (Fermentas) using the Oligo(dT)18 Primer performed as protocols. The primers used in qRT-PCR (see Supplementary File 1, Table S9) were designed using Primer Express 3.0 (Applied Biosystems) with melting temperatures of 58-60 ℃ and amplicon sizes of 100-200 bp. qPCR was performed with an ABI 7 500 instrument (Applied Biosystems). *J. mandshurica* ribosomal protein L32 (JmaRPL32, accession number: HM466693) was selected as the reference gene (Bai et al.,

2010). Data were collected, and threshold cycle (Ct) values were analyzed using $2^{-\Delta\Delta Ct}$ method.

Conclusion

This study reports the transcriptomes and their comparative analysis for two economically important nut crops and one potential breeding material: *J. regia* (JRE), *J. sigillata* (JSI) and *J. cathayensis* (JCA). We de novo assembled 192 360 unigenes (60 710 for JRE, 62 415 for JSI and 69 235 for JCA) with a mean size of 695 bp (712 for JRE, 700 for JSI and 674 for JCA), and on average 48.2% (30 952) of unigenes showed significant similarities to known sequences in protein databases. A total of 5 683 unigenes were found to be differentially expressed genes (DEGs) within three Juglans species and the highest representation KEGG pathways of the DEGs was biosynthesis of amino acids (ko01230). Welocated and predicted 41 141 SSRs and 206 355 SNPs as potential molecular markers in our assembled and annotated sequences. Overall, these unigenes assembled and markers identified in this study will serve as usefulgenomic resource for mining and cloning interested genes and understanding the polymorphisms of Juglance species.

Acknowledgments

This work was supported by the National Science & Technology Pillar Program (Grant No. 2011BAD46B01), Construction Plan for Science & Technology Innovation Platform of Yunnan Province (2013DH007) and Program for Science & Technology Innovation Talents (2012HC009).

References

Bai WN, Liao WJ, Zhang DY. 2010. Nuclear and chloroplast DNA phylogeography reveal two refuge areas with asymmetrical gene flow in a temperate walnut tree from East Asia. New Phytol. 188 (3), 892~901.

Bennett MD, Leitch IJ (2012) Angiosperm DNA C-values database (release 8.0, Dec. 2012). http://www.kew.org/cvalues/

Britton MT, Leslie CA, Caboni E, Dandekar AM, McGranahan GH (2008) Persian Walnut. In: Chittaranjan K and Timothy CH (ed) Compendium of transgenic crop plants: transgenic temperate fruits and nuts. Wiley-Blackwell, Massachusetts.

Dandekar A, Leslie C, McGranahan G (2005) Juglans regia walnut. In: Litz RE (ed). Biotechnology of fruit and nut crops (Biotechnology in Agriculture Series, No. 29). Cromwell, Trowbridge.

Dode LA (1909) Contribution to the study of the genusJuglans. Bull Soc Dendrologique de France. 11:22~90.

Grabherr MG, Haas BJ, Yassour M, Levin JZ, Thompson DA, Amit I, Adiconis X, Fan L, Raychowdhury R, Zeng Q, Chen Z, Mauceli E, Hacohen N, Gnirke A, Rhind N, di Palma F, Birren BW, Nusbaum C, Lindblad-Toh K, Friedman N, Regev A (2011) Full-length transcriptome assembly from RNA-Seq data without a reference genome. Nat Biotechnol. 29: 644-652.

Iorizzo M, Senalik D, Grzebelus D, Bowman M, Cavagnaro P, Matvienko M, Ashrafi H, Van Deynze A, Simon PW (2011) De novo assembly and characterization of the carrot transcriptome reveals novel genes, new markers, and genetic diversity. BMC Genomics. 12(1): 389.

Kanehisa M, Araki M, Goto S, Hattori M, Hirakawa M, Itoh M, Katayama T, Kawashima S, Okuda S, Tokimatsu T, Yamanishi Y (2008) KEGG for linking genomes to life and the environment. Nucleic Acids Res. 36: D480~484.

Liao Z, Chen Y, Dai X, Li S, Yin T (2014) Genome-wide discovery and analysis of single nucleotide polymorphisms and insertions/ deletions inJuglans regia by high - throughput pyrosequencing. Plant Omics. 7: 445~449.

Manning WE (1978) The classification within the Juglandaceae. Ann Missouri Bot Garden. 65:1058~1087.

Mao X, Cai T, Olyarchuk JG, Wei L (2005) Automated genome annotation and pathway identification using the KEGG Orthology (KO) as a controlled vocabulary. Bioinformatics. 21: 3787~3793.

Martínez ML, Labuckas DO, Lamarque AL, Maestri DM (2010) Walnut (Juglans regia L.): genetic resources, chemistry, by-products. J Sci Food Agr. 90:1959~1967.

McGranahan GH, Leslie CA (1990) Walnut (Juglans L.). In: Moore JN, Ballington JR (ed) Genetic resources of temperate fruit and nut crops, vol 2. International Society for Horticultural Science, Wageningen.

McGranahan GH, Leslie CA (2009) Breeding walnuts (Juglans regia). In: Jain SM, Priyadarshan PM (ed) Breeding plantation tree crops: temperate species. Springer, New York.

McKenna A, Hanna M, Banks E, Sivachenko A, Cibulskis K, Kernytsky A, Garimella K, Altshuler D, Gabriel S, Daly M, DePristo MA (2010) The genome analysis toolkit: a MapReduce framework for analyzing next-generation DNA sequencing data. Genome Res. 20: 1297~1303.

Metzker ML (2010) Sequencing technologies-the next generation. Nat Rev Genet. 11: 31~46.

Parchman TL, Geist KS, Grahnen JA Benkman CW, Buerkle CA (2010) Transcriptome sequencing in an ecologically important tree species: assembly, annotation, and marker discovery. BMC Genomics. 11:180.

Pehtea G, Huang X, Liang F, Antonescu V, Sultana R, Karamycheva S, Lee Y, White J, Cheung F, Parvizi B, Tsai J, Quackenbush J (2003) TIGR gene indices clustering tools(TGICL): a software system for fast clustering of large EST datasets. Bioinformatics. 19: 651~652.

Ruperao P, Edwards D (2015) Bioinformatics: Identification of Markers from Next-Generation Sequence Data. In: Batley J (ed) Plant Genotyping: Methods and Protocols - Methods in Molecular Biology (Vol. 1245). Springer, New York.

Shi CY, Yang H, Wei CL, Yu O, Zhang ZZ, Jiang CJ, Sun J, Li YY, Chen Q, Xia T, Wan XC (2011) Deep sequencing of theCamellia sinensis transcriptome revealed candidate genes for major metabolic pathways of tea-specific compounds. BMC Genomics. 12: 131.

Shu C, McElroy JS, Dane F, Peatman E (2015) Optimizing transcriptome assemblies forEleusine indica leaf and seedling by combining multiple assemblies from three de novo assemblers. Plant Genome. 8(1): 1~10.

Terol J, Naranjo MA, Ollitrault P, Talon M (2008) Development of genomic resources forCitrus clementina: characterization of three deep-coverage BAC libraries and analysis of 46000 BAC end sequences. BMC Genomics. 9: 423.

Thiel T, Michalek W, Varshney RK, Graner A (2003) Exploiting EST database for the development and characterization of gen - derived SSR - markers in barley (Hordeum vulgare L.). Theor Appl Genet. 106: 411~422.

Wang H, Pan G, Ma Q, Zhang J, Pei D (2015) The genetic diversity and introgression ofJuglans regia and Juglans sigillata in Tibet as revealed by SSR markers. Tree Genet Genomes. 11: 804.

Wang L, Feng Z, Wang X, Wang X, Zhang X (2010) DEGseq: an R package for identifying differentially expressed genes from RNA-seq data. Bioinformatics. 26: 136~138.

Wang XW, Luan JB, Li JM, Su YL, Xia J, Liu SS (2011) Transcriptome analysis and comparison reveal divergence between two invasive whitefly cryptic species. BMC Genomics. 12: 458.

Xia Z, Xu H, Zhai J, Li D, Luo H, He C, Huang X (2011) RNA-Seq analysis and de novo transcriptome assembly ofHevea brasiliensis. Plant Mol Biol. 77(3): 299~308.

普通核桃、深纹核桃和野核桃的转录组比较分析

摘要：核桃坚果含有大量不饱和脂肪酸及蛋白质，食用价值高，在全球都广泛栽培。在中国，只有普通核桃（*Juglans regia*）和深纹核桃（*J. sigillata*）作为经济林采收坚果而进行栽培，野核桃（*J. cathayensis*）通常仅作为砧木和潜在的育种材料。它们都是非模式植物，已知的基因组信息非常少，深纹核桃和野核桃尤其少。本项研究中，我们通过转录组测序技术和生物信息学分析方法，对上述3种核桃的叶片进行转录组从头测序和数据组装、基因功能注释、比较分析等工作。普通核桃、深纹核桃和野核桃的转录组原始数据已提交至NCBI数据库，收录号分别是SRR1767234、SRR1767236和SRR1767237，总的原始测序数据超过20 G bp，经数据过滤后，分别获得59 035 134 bp、43 949 544 bp和58 609 226 bp的"清洁"数据（clean data）。经过从头组装后，获得192 360条读长超过200 bp的单一序列（unigene），其中92 858条单一序列（48.20%）被注释，获得32110条（16.70%）编码序列（coding region sequence，CDS）。所有的单一序列经过COG、GO和KEGG分类。KEGG pathways分析结果表明，3种核桃间有5683个差异表达基因（DEGs）。共获得41 141个和206 355个潜在的SSR和SNP分子标记。本项研究所获得的转录组数据，极大地丰富了核桃属植物的基因组资源，为核桃功能基因发掘和分子技术辅助育种奠定了坚实基础。

[本文发表于 *Plant Omics Journal*, 2015, 8(4)]